Structural Analysis Fundamentals

Structural Analysis Fundamentals

Ramez Gayed
Amin Ghali

CRC Press is an imprint of the
Taylor & Francis Group, an **informa** business

First edition published 2022
by CRC Press
6000 Broken Sound Parkway NW, Suite 300, Boca Raton, FL 33487-2742
and by CRC Press
2 Park Square, Milton Park, Abingdon, Oxon, OX14 4RN

CRC Press is an imprint of Taylor & Francis Group, LLC

Library of Congress Cataloging-in-Publication Data

Library of Congress Cataloging-in-Publication Data
Names: Gayed, Ramez B., author. | Ghali, A. (Amin), author.
Title: Structural analysis fundamentals / Ramez B. Gayed, Amin Ghali.
Description: First edition. | Boca Raton: CRC Press, 2022. | Includes bibliographical references and index.
Identifiers: LCCN 2021011034 (print) | ISBN 9780367252625 (hbk) | ISBN 9780367252618 (pbk) | ISBN 9780429286858 (ebk)
Subjects: LCSH: Structural analysis (Engineering)
Classification: LCC TA645 .G37 2022 (print) | LCC TA645 (ebook) | DDC 624.1/71--dc23
LC record available at https://lccn.loc.gov/2021011034
LC ebook record available at https://lccn.loc.gov/2021011035

ISBN: 978-0-367-25262-5 (hbk)
ISBN: 978-0-367-25261-8 (pbk)
ISBN: 978-0-429-28685-8 (ebk)

DOI: 10.1201/9780429286858

Typeset in Sabon
by Deanta Global Publishing Services, Chennai, India

Access the support material: www.routledge.com/9780367252625

Contents

Introduction to Structural Analysis Fundamentals

This book contains fundamental procedures of structural analysis that are needed in the design of structures and in courses teaching modeling structures of all types: beams, plane and space trusses and frames, grids and cable nets, and assemblages of finite-elements. Elastic, plastic, and time-dependent responses of structures to static loading are treated. The fundamentals of dynamic analysis of structures and their response to earthquakes are covered. Solved examples demonstrate all procedures. The text includes basic and advanced material for use in undergraduate and higher courses. A companion set of computer programs, with their source codes and executable files, assist in thorough understanding and applying the analysis procedures.

Students, lecturers, and engineers internationally employ, in six languages, the seventh edition of Ghali and Neville, *Structural Analysis: A unified classical and matrix approach*, 933 pages. The closely related shorter text, ***Structural Analysis Fundamentals*** contains the procedures that are most used in modern teaching and practice and covers more material.

Many solved examples and a number of problems at the ends of chapters, with answers given at the end of the book, illustrate the introduced techniques of analysis. No examples or problems are governed by any code or system of units. However, where it is advantageous, a small number of examples and problems use SI units as well as Imperial or British units.

Preface to Structural Analysis Fundamentals

This book contains fundamental procedures of structural analysis that are needed in the design of structures and in courses teaching modeling structures of all types: beams, plane and space trusses and frames, grids and cable nets, and assemblages of finite-elements. Elastic, plastic, and time-dependent responses of structures to static loading are treated. The fundamentals of dynamic analysis of structures and their response to earthquakes are covered. Solved examples demonstrate all procedures. The text includes basic and advanced material for use in undergraduate and higher courses. A companion set of computer programs, with their source codes and executable files, assist in thorough understanding and application of the analysis procedures.

Students, lecturers and engineers, internationally employ, in six languages, the seventh edition of Ghali and Neville, *Structural Analysis: A unified classical and matrix approach*, 933 pages. The closely related shorter text, ***Structural Analysis Fundamentals*** contains the procedures that are most used in modern teaching and practice and covers more material.

Many solved examples and a number of problems at the ends of chapters, with answers given at the end of the book, illustrate the introduced techniques of analysis. No examples or problems are governed by any code or system of units. However, where it is advantageous, a small number of examples and problems use SI units as well as Imperial or British units.

Chapter 1 discusses structural analysis modeling by idealizing a structure as a beam, a plane or a space frame, a plane or a space truss, a plane or eccentric grid, or as an assemblage of finite-elements. The strut-and-tie models are for the analysis of reinforced concrete structures after cracking. We consider internal forces, deformations, sketching deflected shapes, and corresponding bending moment diagrams. We compare internal forces and deflections in beams, arches, and trusses.

The chapter on modeling is followed by a chapter on the analysis of statically determinate structures, intended to help students better prepare. In Chapters 3, 4, and 5, we introduce two distinct general approaches of analysis: the force method and the displacement method. Both methods involve the solution of linear simultaneous equations relating forces to displacements. The emphasis in these three chapters is on the basic ideas in the two methods without obscuring the procedure by the details of derivation of the coefficients needed to form the equations. Instead, use is made of Appendices A, C, and B, which give, respectively, displacements due to applied unit forces, forces corresponding to unit displacements, and fixed-end forces in straight members due to various loadings.

In Chapter 6, we calculate long-term displacements in reinforced concrete structures. We consider effects of cracking, creep, shrinkage, and relaxation of prestressed reinforcement. The computer programs TDA and EGRID, combined, give long-term deflection of concrete floors (Chapter 20). We delay the methods of calculation of displacement or force by energy and virtual work theories to Chapters 7 and 8, by which time the need for this material is clear. This sequence of presentation of material is particularly suitable when the reader is already acquainted with some of the methods of calculating the deflection of beams. If, however, it is preferable to deal with methods of calculation of displacements first, Chapters 7 and 8 should be studied before Chapters 3 through 6; this will not disturb the continuity.

The classical method of moment distribution, suitable for hand calculations, continue to be useful for preliminary calculation. In a short Chapter 9, we present the moment distribution analysis of plane frames without joint translations. In Chapter 10, we present the effects of axial forces on the stiffness characteristics of members of framed structures and apply them in the determination of the critical buckling loads of continuous frames.

Chapter 11 deals with the analysis of shear walls, commonly used in modem buildings. The provision of outriggers is an effective means of reducing the drift and the bending moments due to lateral loads in high-rise buildings. A solved example of a 50-story building demonstrates the benefits of outriggers at two levels.

Chapter 12 uses finite differences in analysis of beams on elastic foundations and extends the procedure in computer analysis for concrete walls of cylindrical tanks (CTW in Chapter 20). We idealize axi-symmetrical shell of revolution as assembly of conical finite-element shells and apply the five steps of the displacement method. Chapter 12 derives the stiffness matrix of a shell element and employs it in a computer program in Chapter 20.

Chapters 13 presents two- and three-dimensional finite-elements employed in practice in the analysis of walls, slabs, plates, shells, and mass structures. We should use commercial software only after mastering the basis of the finite-element technique.

Plastic analysis, in Chapter 14, gives information about the behavior of structures after the development of flexural plastic hinges. Using computer program PLASTICF (Chapter 20), we determine the load levels that successively develop plastic hinges and ultimately cause collapse of plane frames. Chapter 15 deals with plastic analysis of reinforced concrete slabs. The yield-line theory gives an upper bound of the ultimate load capacity.

Chapter 16 studies response of structures to dynamic loading produced by machinery, gusts of wind, blasts, or earthquakes. We discuss first free and forced vibrations of a system with one degree-of-freedom and extend the treatment to multi degree-of-freedom systems, including finite-element models. Chapters 16 and 17 continue to discuss the dynamic response of structures to earthquakes. Nonlinear static (pushover) analysis considers the inelastic behavior of ductile structure during an earthquake event.

Cable nets and fabrics, trusses, and frames with slender members having large deformations require analysis that considers equilibrium in the real deformed configurations. Chapter 18 employs the Newton–Raphson iterative technique in the required geometric nonlinearity analysis. The same chapter also introduces the material-nonlinearity analysis, in which the stress–strain relation of the material is nonlinear.

The companion of *Structural Analysis Fundamentals* includes linear and nonlinear computer programs. The significance of the companion programs in teaching and learning structural analysis justifies their use in solved examples and in problem solving. Chapter 20 presents the name and short description of each computer program. All programs are available through the publisher or Gayed Engineering Consultants.

Frequently used data are presented in the appendices. Dynamic analysis of structures involves solution of the eigenvalue problem. Computer program EIGEN1 or EIGEN2 is adequate for examples and problems for Chapters 16 and 17.

Chapters 1 through 11, 14, and 15 contain basic material covered in first courses. The remainder of the book is suitable for more advanced courses. The book is suitable, not only for the student but also for the practicing engineer who wishes to obtain guidance on the most convenient methods of analysis for a variety of types of structures.

Ramez Gayed
Amin Ghali
Calgary, Alberta, Canada
January 2021

Notations

The following is a list of symbols which are common in the various chapters of the text; other symbols are used in individual chapters. All symbols are defined in the text when they first appear.

A	Any action, which may be a reaction or a stress resultant. A stress resultant at a section of a framed structure is an internal force – bending moment, shearing force, or axial force.
a	Cross-sectional area.
D_i or D_{ij}	Displacement (rotational or translational) at coordinate i. When a second subscript j is provided, it indicates the coordinate at which the force causing the displacement acts.
E	Modulus of elasticity.
EI	Flexural rigidity.
F	A generalized force: a couple or a concentrated load.
FEM	Fixed-end moment.
f_{ij}	Element of flexibility matrix.
G	Modulus of elasticity in shear.
g	Gravitational acceleration.
I	Moment of inertia or second moment of area.
i, j, k, m, n, p, r	Integers.
J	Torsion constant (length^4), equal to the polar moment of inertia for a circular cross section.
l	Length.
M	Bending moment at a section, e.g. M_n = bending moment at sections. In beams and grids, a bending moment is positive when it causes tension in bottom fibers.
$[m]$	Mass matrix.
M_{AB}	Moment at end A of member AB. In plane structures, an end-moment is positive when clockwise. In general, an end-moment is positive when it can be represented by a vector in the positive direction of the axes x, y, or z.
N	Axial force at a section or in a member of a truss.
OM	Overturning moment in an earthquake.
P, Q	Concentrated loads.
q	Load intensity.
R	Reaction.
S_a, S_v or S_d	Spectral acceleration, velocity, or displacement in structure's response to earthquakes.
s_{ij}	Element of stiffness matrix.

s	Used as a subscript, indicates a statically determinate action.
T	Twisting moment at a section.
u	Used as a subscript, indicates the effect of unit forces or unit displacements.
V	Shearing force at a section.
W	Work of the external applied forces.
ε	Strain.
ζ	Damping ratio in vibration analysis.
η	Influence ordinate.
ν	Poisson's ratio.
σ	Stress.
τ	Shearing stress.
$[\phi]$	Mode shapes of vibration.
$[\Phi]$	Normalized mode shapes of vibration.
ω	Natural frequency of vibration (radian/second).
$\{\}$	Braces indicate a vector, i.e. a matrix of one column. To save space, the elements of a vector are sometimes listed in a row between two braces.
$[\,]$	Brackets indicate a rectangular or square matrix.
$[\square]^T_{n \times m}$	Superscript T indicates matrix transpose. $n \times m$ indicates the order of the matrix which is to be transposed resulting in an $m \times n$ matrix.
$\twoheadrightarrow$	Double-headed arrow indicates a couple or a rotation; its direction is that of the rotation of a right-hand screw progressing in the direction of the arrow.
$\rightarrow$	Single-headed arrow indicates a load or a translational displacement.
z, x, y (axes diagram)	Axes: the positive direction of the z-axis points away from the reader.

The SI System of Units of Measurements

Length	meter	m
	millimeter = 10^{-3}m	mm
Area	square meter	m^2
	square millimeter = $10^{-6}m^2$	mm^2
Volume	cubic meter	m^3
Frequency	hertz = 1 cycle per second	Hz
Mass	kilogram	kg
Density	kilogram per cubic meter	kg/m^3
Force	newton	N
	= a force which applied to a mass of one kilogram gives it an acceleration of one meter per second, i.e. 1 N = 1 $kg\text{-}m/s^2$	
Stress	newton per square meter	N/m^2
	newton per square millimeter	N/mm^2
Temperature interval	degree Celsius	deg C;°C

Nomenclature for multiplication factors

10^9	giga	G
10^6	mega	M
10^3	kilo	k
10^{-3}	milli	m
10^{-6}	micro	μ
10^{-9}	nano	n

IMPERIAL EQUIVALENTS OF SI UNITS

Length

meter (m) — 1 m = 39.37 in
1 m = 3.281 ft

Area

square meter (m^2) — 1 m^2 = 1550 in^2
1 m^2 = 10.76 ft^2

Volume

cubic meter (m^3) — 1 m^3 = 35.32 ft^3

Moment of inertia	
meter to the power four (m^4)	1 m^4 = 2403 × 10^3 in^4
Force	
Newton (N)	1 N = 0.2248 lb
Load intensity	
Newton per meter (N/m)	1 N/m = 0.06852 lb./ft
Newton per square meter (N/m^2)	1 N/m^2 = 20.88 × 10^{-3} lb/ft^2
Moment	
Newton meter (N-m)	1 N-m = 8.851 lb-in
	1 N-m = 0.7376 × 10^{-3} kip-ft
	1 kN-m = 8.851 kip-in
Stress	
Newton per square meter (pascal)	1 Pa = 145.0 × 10^{-6} lb/in^2
	1 MPa = 10^6 N/m^2 = 145.0 lb/in^2 = 0.1450 ksi
Curvature	
$(Meter)^{-1}$	1 m^{-1} = 0.0254 in^{-1}
Temperature change	
Degree Celsius (°C)	1° C = (5/9)° Fahrenheit
Energy and power	
Joule (J) = 1 N-m	1 J = 0.7376 lb-ft
Watt (W) = 1 J/s	1 W = 0.7376 lb-ft/s
	1 W = 3.416 Btu/h
Mass	
Kilogram (kg)	1 kg = 2.205 lb/gravitational acceleration
	1 kg = 5.710 × 10^{-3} lb/(in./s^2)

Authors

Ramez B. Gayed is a Civil Engineering Consultant and Adjunct Professor at the University of Calgary. He is expert on analysis and design of concrete and steel structures.

Amin Ghali is Emeritus Professor at the University of Calgary. He is consultant on major international structures. He is inventor of several reinforcing systems for concrete. He has authored over 300 papers and eight patents. His books include *Concrete Structures* (2012), *Circular Storage Tanks and Silos* (CRC Press, 2014), and *Structural Analysis* (CRC Press, 2017).

Chapter 1

Structural analysis modeling

1.1 INTRODUCTION

This book may be used by readers familiar with basic structural analysis and also by those with no previous knowledge beyond elementary mechanics. It is mainly for the benefit of people in the second category that Chapter 1 is included. It will present a general picture of the analysis but, inevitably, it will use some concepts that are fully explained only in later chapters. Readers may therefore find it useful, after studying Chapter 2 and possibly even Chapter 3, to reread Chapter 1.

The purpose of structures, other than aircraft, ships, and floating structures, is to transfer applied loads to the ground. The structures themselves may be constructed specifically to carry loads (for example, floors or bridges) or their main purpose may be to give protection from the weather (for instance, walls or roofs). Even in this case, there are loads (such as self-weight of the roofs and also wind forces acting on them) that need to be transferred to the ground.

Before a structure can be designed in a rational manner, it is essential to establish the loads on various parts of the structure. These loads will determine the stresses and their resultants (internal forces) at a given section of a structural element. These stresses or internal forces have to be within desired limits in order to ensure safety and to avoid excessive deformations. To determine the stresses (forces/unit area), the geometrical and material properties must be known. These properties influence the self-weight of the structure, which may be more or less than originally assumed. Hence, iteration in analysis may be required during the design process. However, consideration of this is a matter for a book on design.

The usual procedure is to idealize the structure by one-, two-, or three-dimensional elements. The lower the number of dimensions considered, the simpler the analysis. Thus, beams and columns, as well as members of trusses and frames, are considered as one-dimensional; in other words, they are represented by straight lines. The same applies to strips of plates and slabs. One-dimensional analysis can also be used for some curvilinear structures, such as arches or cables, and also certain shells. Idealization of structures by an assemblage of finite-elements, considered in Chapter 13, is sometimes necessary.

Idealization is applied not only to members and elements but also to their connections to supports. We assume the structural connection to the supports to be free to rotate(i.e. treat the supports as hinges), or to be fully restrained, that is, built-in or encastré. In reality, perfect hinges rarely exist, because of friction and also because nonstructural members such as partitions restrain free rotation. At the other extreme, a fully built-in condition does not recognize imperfections in construction or loosening owing to temperature cycling.

Once the analysis has been completed, members and their connections are designed: the designer must be fully conscious of the difference between the idealized structure and the actual outcome of construction.

DOI: 10.1201/9780429286858-1

The structural idealization transforms the structural analysis problem into a mathematical problem that can be solved by computer or by hand, using a calculator. The model is analyzed for the effects of loads and applied deformations, including the self-weight of the structure, superimposed stationary loads or machinery, live loads such as rain or snow, moving loads, dynamic forces caused by wind or earthquake, and the effects of temperature as well as volumetric change of the material (e.g. shrinkage of concrete). This chapter explains the type of results that can be obtained by different types of models.

Other topics discussed in this introductory chapter are: transmission (load path) of forces to the supports and the resulting stresses and deformations; axial forces in truss members; bending moments and shear forces in beams; axial and shear forces, and bending moments in frames; arches; the role of ties in arches; sketching of deflected shapes and bending moment diagrams; and hand checks on computer results.

1.2 TYPES OF STRUCTURES

Structures come in all shapes and sizes, but their primary function is to carry loads. The form of the structure, and the shape and size of its members are usually selected to suit this load-carrying function, but the structural forces can also be dictated by the function of the system of which the structure is part. In some cases, the form of the structure is dictated by architectural considerations.

The simplest structural form, the beam, is used to bridge a gap. The function of the bridge in Figure 1.1 is to allow traffic and people to cross the river: the load-carrying function is accomplished by transferring the weight applied to the bridge deck to its supports.

A similar function is provided by the arch, one of the oldest structural forms. Roman arches (Figure 1.2a) have existed for some 2000 years and are still in use today. In addition to bridges, the arch is also used in buildings to support roofs. Arches were developed because of confidence in the compressive strength of the material being used, and this material, stone, is plentiful. An

Figure 1.1 Highway bridge.

Figure 1.2 Arch bridges. (a) Stone blocks. (b) Concrete.

arch made of stone remains standing, despite there be no cementing material between the arch blocks because the main internal forces are compressive. However, this has some serious implications, as we shall see later. The arch allows longer spans than beams with less material: today, some very elegant arch bridges are built of concrete (Figure 1.2b) or steel.

The third, simple form of structural type, is the cable. The cable relies on the tensile capacity of the material (as opposed to the arch, which uses the compressive capacity of the material) and hence its early use was in areas where natural rope-making materials are plentiful. Some of the earliest uses of cables are in South America where local people used cables to bridge gorges.

Subsequent developments of chain, wire, and the strand, permit bridges to span great lengths with elegant structures; today, the world's longest bridges are cable supported (Figures 1.3a, b, and c).

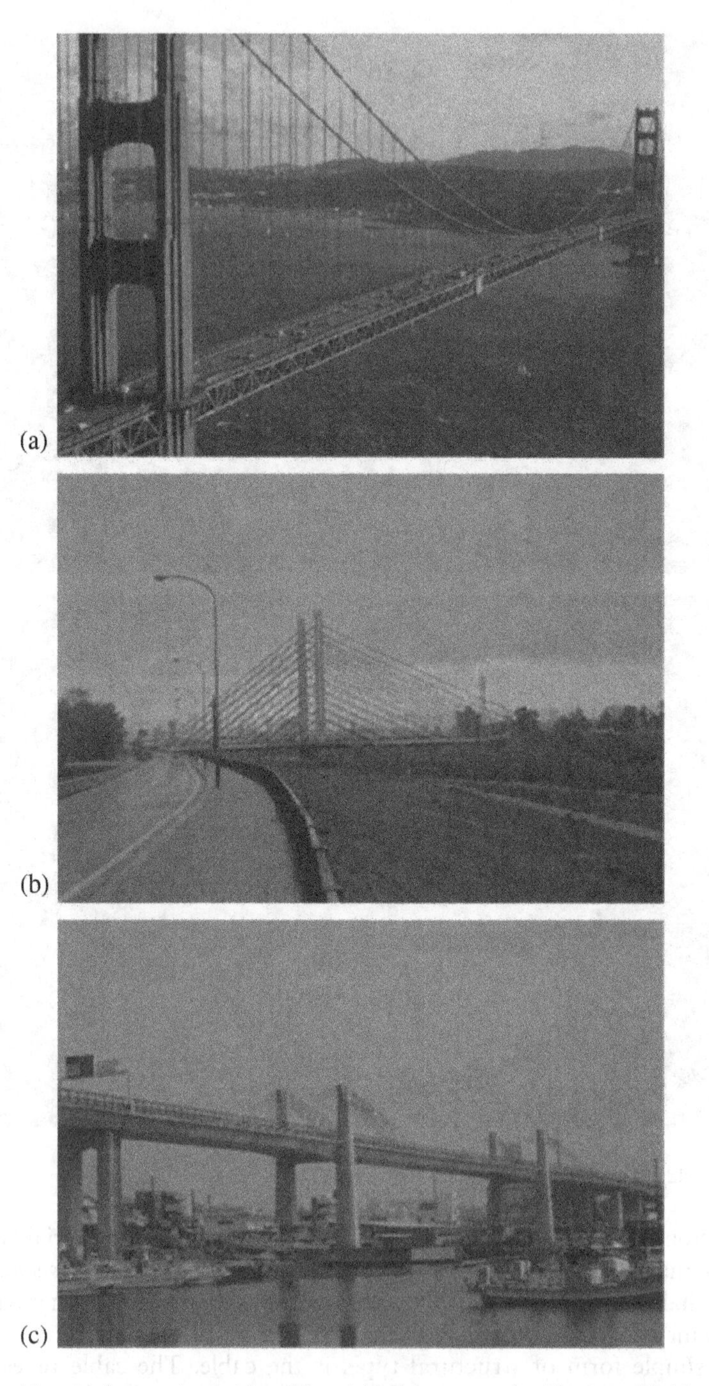

Figure 1.3 Use of cables. (a) Suspension bridge. (b) and (c) Two types of cable-stayed bridges.

The shapes of the arch and suspended cable structures show some significant similarities, the one being the mirror image of the other. Cable systems are also used to support roofs, particularly long-span roofs where the self-weight is the most significant load. In both arches and cables, gravity loads induce inclined reactions at the supports (Figures 1.4a and b). The arch reactions produce compressive forces in the direction (or close to the direction) of the arch axis at the ends. The foundation of the arch receives inclined outward forces equal and

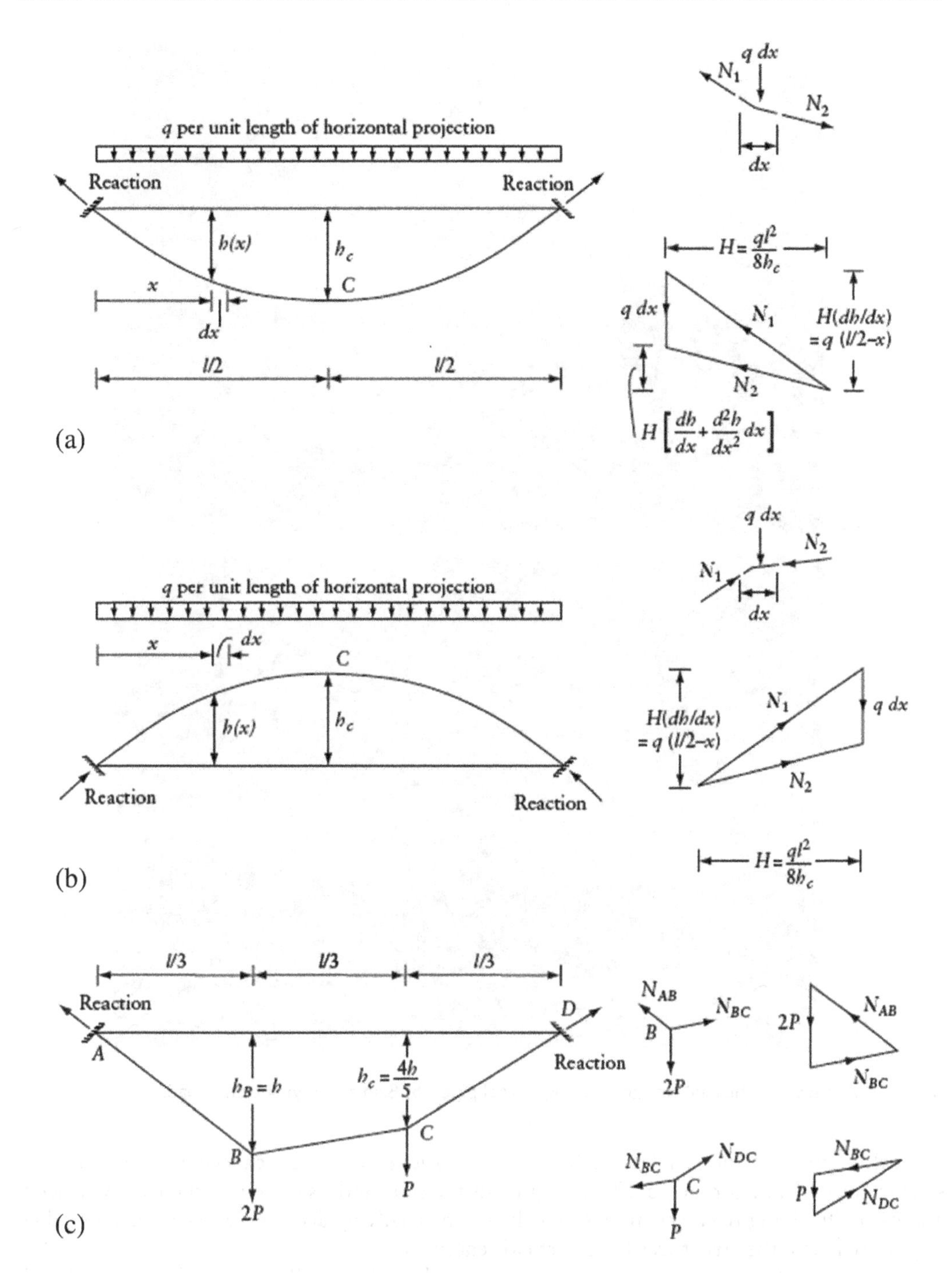

Figure 1.4 Structures carrying gravity loads. (a) Cable – tensile internal force. (b) Arch – compressive internal force. (c) Funicular shape of a cable carrying two concentrated downward loads.

opposite to the reactions. Thus, the foundations at the two ends are pushed outwards and downwards; the subgrade must be capable of resisting the horizontal thrust and therefore has to be rock or concrete block. A tie connecting the two ends of an arch can eliminate the horizontal forces on the supports (Figure 1.25d; e.g. the pedestrian bridge shown on the front cover of this book). Thus, an arch with a tie subjected to gravity load has only

Figure 1.5 Arches with ties. (a) Bridge carrying a steel pipe. (b) Bridge carrying traffic load.

vertical reactions. Figures 1.5a and b show arch bridges with ties, carrying, respectively, a steel water pipe and a roadway. The weights of the pipe and its contents, the roadway deck and its traffic, hang from the arches by cables. The roadway deck serves as a tie. Thus, due to gravity loads, the structures have vertical reactions.

The cable in Figure 1.4a, carrying a downward load, is pulled upward in an inclined direction of the tangent at the two ends. Again, the arrows in the figure show the directions of the reactions necessary for equilibrium of the structure. Further discussion on cables and arches is presented in Section 1.2.1.

The introduction of the railroad, with its associated heavy loads, resulted in the need for a different type of structure to span long distances – hence, the development of the truss (Figure 1.6a). In addition to carrying heavy loads, trusses can also be used to span long

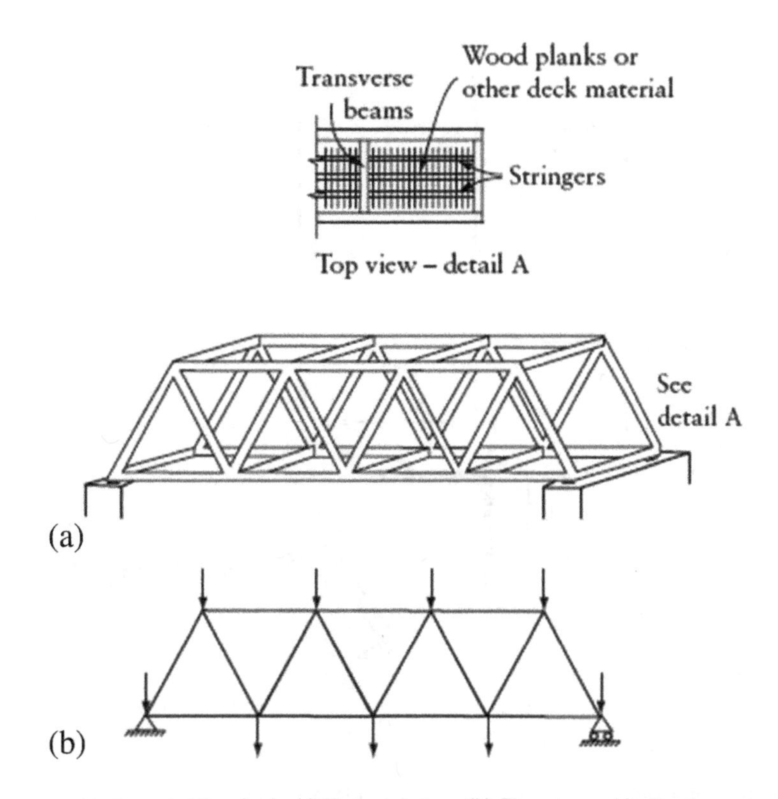

Figure 1.6 Truss supporting a bridge deck. (a) Pictorial view. (b) Plane truss idealization.

distances effectively and are, therefore, also used to support long-span roofs; wood roof trusses are extensively used in housing. Trusses consist of straight members. Ideally, the members are pin-connected and subjected to external forces only at the connections; the internal forces in the members are axial tension or compression. In modern trusses, the members are usually nearly rigidly connected (although assumed pinned in analysis).

If the members are connected by rigid joints, then the structure is a frame – another common form of structural system frequently used in multistory buildings. Similar to trusses, frames come in many different configurations, in two or three dimensions. Figures 1.7a and b show a typical rigid joint connecting two members before and after loading. Because of the rigid connection, the angle between the two members remains unchanged when the joint translates and rotates and the structure deforms under load.

Another category of structures is plates and shells whose one common attribute is that their thickness is small in comparison to their other dimensions. An example of a plate is a concrete slab, monolithic with supporting columns, which is widely used in office and apartment buildings and parking structures. Cylindrical shells are plates curved in one direction. Examples are storage tanks, silos, and pipes. As with arches, the main internal forces in a shell are in the plane of the shell, as opposed to shear force or bending moment.

Axisymmetrical domes, generated by the rotation of a circular or parabolic arc, carrying uniform gravity loads, are mainly subjected to membrane compressive forces in the middle surface of the shell. This is the case when the rim is continuously supported such that the vertical and the radial displacements are prevented; however, the rotation can be free or restrained. Again, similar to an arch or a cable, the reaction at the outer rim of a dome is in the direction or close to the direction of the tangent to the middle surface (Figure 1.8a).

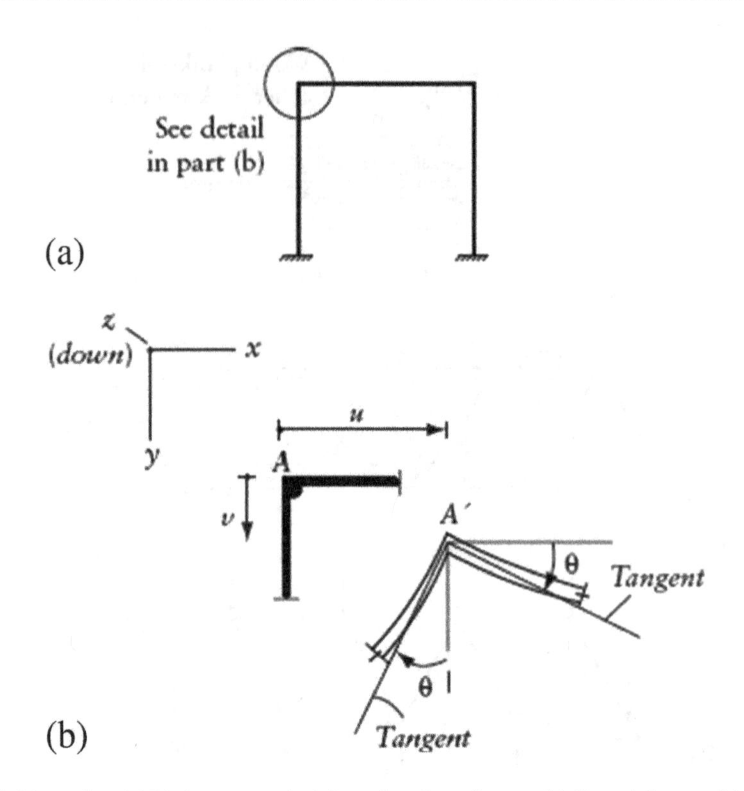

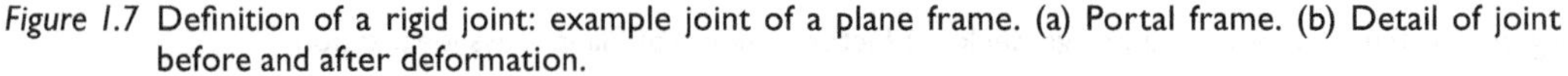
Figure 1.7 Definition of a rigid joint: example joint of a plane frame. (a) Portal frame. (b) Detail of joint before and after deformation.

A means must be provided to resist the radial horizontal component of the reaction at the lower rim of the dome; most commonly this is done by means of a circular ring subjected to hoop tension (Figure 1.8b).

In Figure 1.8c, the dome is isolated from the ring beam to show the internal force at the connection of the dome to the rim. A membrane force N per unit length is shown in the meridian direction at the rim of the dome (Figure 1.8a). The horizontal component per unit length, ($N \cos \theta$), acts in the radial direction on the ring beam (Figure 1.8c) and produces a tensile hoop force = $rN \cos \theta$, where r is the radius of the ring and θ is the angle between the meridional tangent at the edge of the dome and the horizontal. The reaction of the structure (the dome with its ring beam) is vertical and of magnitude ($N \sin \theta = qr/2$) per unit length of the periphery. In addition to the membrane force N, the dome is commonly subjected to shear forces and bending moments in the vicinity of the rim (much smaller than would exist in a plate covering the same area). The shear and moment are not shown for simplicity of presentation.

Arches, cables, and shells represent a more effective use of construction materials than beams. This is because the main internal forces in arches, cables, and shells are axial or membrane forces as opposed to shear force and bending moment. For this reason, cables, arches, or shells are commonly used when it is necessary to enclose large areas without intermediate columns, as in stadiums and sports arenas. Examples are shown in Figures 1.9a and b. The comparisons in Section 1.13 show that to cover the same span and carry the same load, a beam needs to have a much larger cross section than arches of the same material.

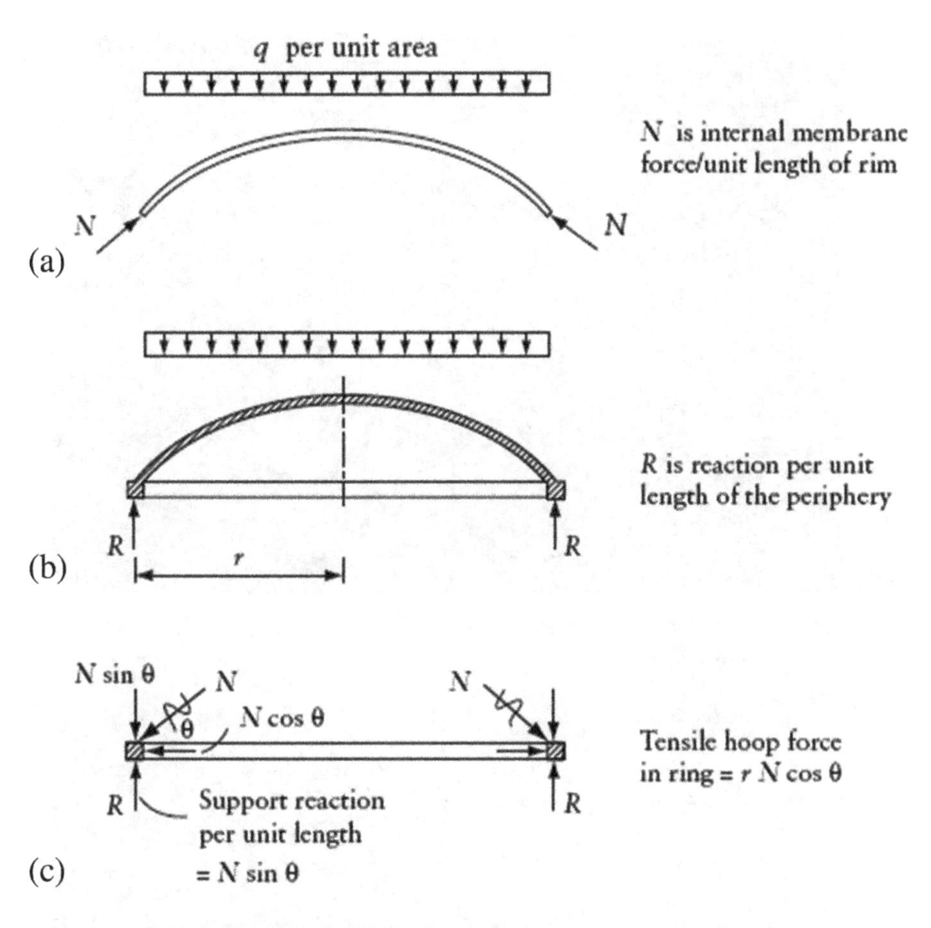

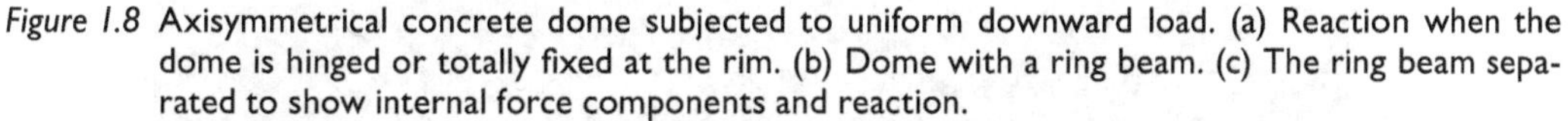

Figure 1.8 Axisymmetrical concrete dome subjected to uniform downward load. (a) Reaction when the dome is hinged or totally fixed at the rim. (b) Dome with a ring beam. (c) The ring beam separated to show internal force components and reaction.

1.2.1 Cables and arches

A cable can carry loads, with the internal force axial tension, by taking the shape of a curve or a polygon (funicular shape), whose geometry depends upon the load distribution. A cable carrying a uniform gravity load of intensity q/unit length of horizontal projection (Figure 1.4a) takes the shape of a second degree parabola, whose equation is:

$$h(x) = h_C \left[\frac{4x(l-x)}{l^2} \right] \tag{1.1}$$

where h is the absolute value of the distance between the horizontal and the cable (or arch), h_C is the value of h at mid-span.

This can be shown by considering the equilibrium of an elemental segment dx as shown. The sum of the horizontal components or the vertical components of the forces on the segment is zero. Thus, the absolute value, H of the horizontal component of the tensile force at

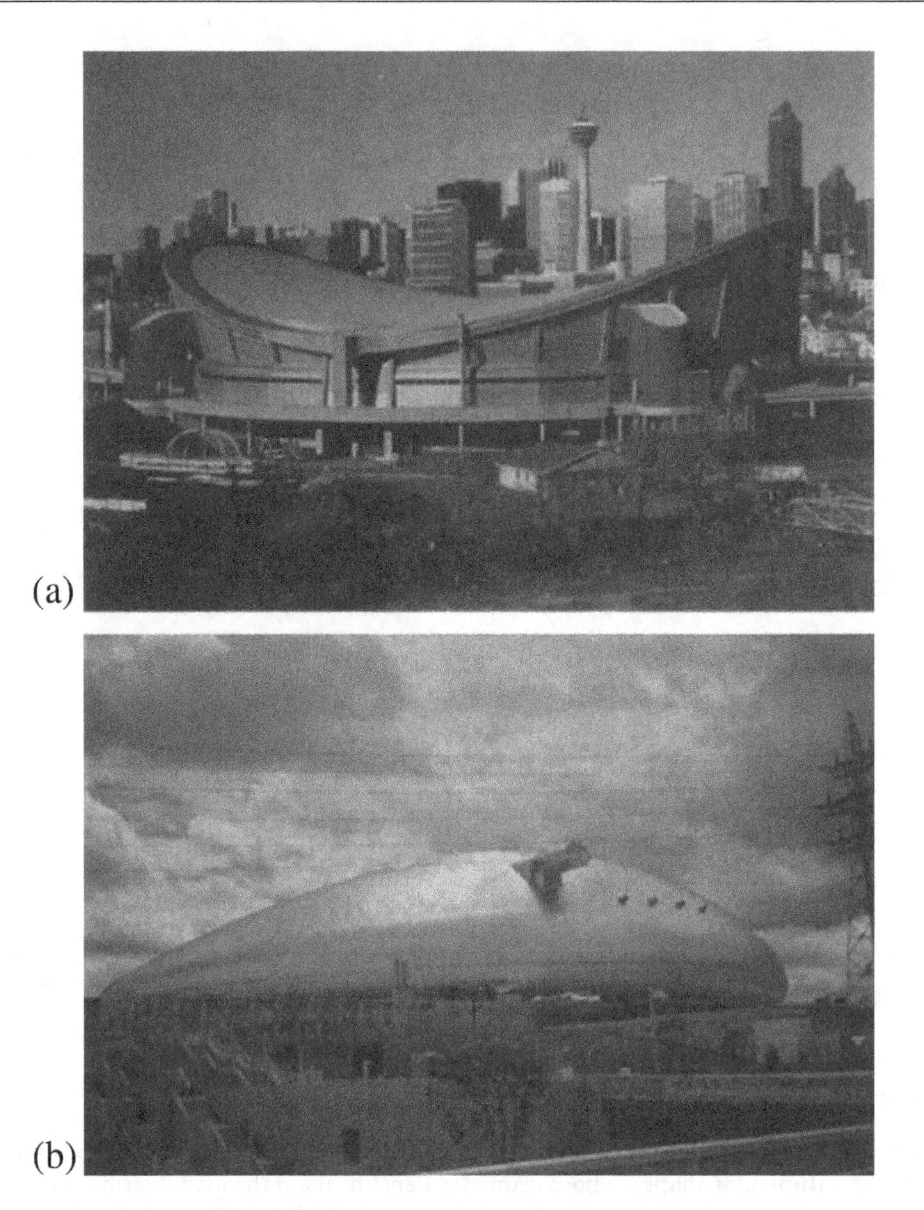

Figure 1.9 Shell structures. (a) The "Saddle dome," Olympic Ice Stadium, Calgary, Canada. Hyperbolic paraboloid shell consisting of precast concrete elements carried on a cable network. (b) Sapporo Dome, Japan. Steel roof with stainless steel covering 53000 m², housing a natural turf soccer field that can be hovered and wheeled in and out of the stadium.

any section of the cable is constant; for the vertical components of the tensile forces N_1 and N_2 on either side of the segment we can write:

$$q\,\mathrm{d}x + H\left[\frac{\mathrm{d}h}{\mathrm{d}x} + \frac{\mathrm{d}^2 h}{\mathrm{d}x^2}\mathrm{d}x\right] = H\left[\frac{\mathrm{d}h}{\mathrm{d}x}\right] \tag{1.2}$$

$$\frac{\mathrm{d}^2 h}{\mathrm{d}x^2} = -\frac{q}{H} \tag{1.3}$$

Double integration and setting $h = 0$ at $x = 0$ and $x = l$; and setting $h = h_C$ at $x = l/2$ gives Equation 1.1 and the horizontal component of the tension at any section:

$$H = \frac{ql^2}{8h_C} \tag{1.4}$$

When the load intensity is $\bar{q}$ per unit length of the cable (e.g. the self-weight), the funicular follows the equation of a catenary (differing slightly from a parabola). With concentrated gravity loads, the cable has a polygonal shape. Figure 1.4c shows the funicular polygon of two concentrated loads $2P$ and P at third points. Two force triangles at B and C are drawn to graphically give the tensions N_{AB}, N_{BC}, and N_{CD}. From the geometry of these figures, we show below that the ratio of $(h_C/h_B) = 4/5$, and that this ratio depends upon the magnitudes of the applied forces. We can also show that when the cable is subjected to equally spaced concentrated loads, each equal to P, the funicular polygon is composed of straight segments connecting points on a parabola (see Problem 1.6).

In the simplified analysis presented here, we assume that the total length of the cable is the same, before and after the loading, and is equal to the sum of the lengths of AB, BC, and CD; and after loading, the applied forces $2P$ and P are situated at third points of the span. Thus, h_B and h_C are the unknowns that define the funicular shape in Figure 1.4c.

By considering that for equilibrium the sum of the horizontal components of the forces at B is equal to zero and doing the same at C, we conclude that the absolute value, H of the horizontal component is the same in all segments of the cable; thus, the vertical components are equal to H multiplied by the slope of the segments. The sum of the vertical components of the forces at B or at C is equal to zero. This gives:

$$\begin{aligned} 2P - H\left(\frac{3h_B}{l}\right) - H\left[\frac{3(h_B - h_C)}{l}\right] &= 0 \\ P + H\left[\frac{3(h_B - h_C)}{l}\right] - H\left(\frac{3h_C}{l}\right) &= 0 \end{aligned} \tag{1.5}$$

$$h_C = \frac{4}{5}h_B \quad \text{and} \quad H = \frac{Pl}{1.8h_B} \tag{1.6}$$

The value of h_B can be determined by equating the sum of the lengths of the cable segments to the total initial length.

Figure 1.4b shows a parabolic arch (mirror image of the cable in Figure 1.4a) subjected to uniform load q/unit length of horizontal projection. The triangle of the three forces on a segment dx is shown, from which we see that the arch is subjected to axial compression. The horizontal component of the axial compression at any section has a constant absolute value H, given by Equation 1.4.

Unlike the cable, the arch does not change shape when the distribution of load is varied. Any load other than that shown in Figure 1.4b produces axial force, shear force, and bending moment. The most efficient use of material is achieved by avoiding the shear force and the bending moment. Thus, in designing an arch, we select its profile such that the major load (usually the dead load) does not produce shear or bending.

It should be mentioned that in the above discussion, we have not considered the shortening of the arch due to axial compression. This shortening has the effect of producing shear and bending of small magnitudes in the arch in Figure 1.4b.

Cables intended to carry gravity loads are frequently prestressed (stretched between two fixed points). Due to the initial tension combined with the self-weight, a cable will have a sag depending upon the magnitudes of the initial tension and the weight. Subsequent application of a downward concentrated load at any position produces displacements u and v at the point of load application, where u and v are translations in the horizontal and vertical directions, respectively. Calculations of these displacements and the associated changes in tension, and in cable lengths, are discussed in Chapter 18.

1.3 LOAD PATH

As indicated in the previous section, the primary function of any structure is the transfer of loads to the final support being provided by the ground. Loads are generally categorized as dead or live. Dead loads are fixed or permanent loads – loads that do not vary throughout the life of the structure. Very often, the majority of the dead load derives from the self-weight of the structure. In long-span bridges, the dead load constitutes the larger portion of the total gravity load on the structure. Live loads and other transient loads represent the effects of occupancy or use, traffic weight, environment, earthquakes, and support settlement.

One of the objectives of the analysis of a structure is to determine the internal forces in its various elements. These internal forces result from the transfer of the loads from their points of application to the foundations. Understanding load paths is important for two reasons: providing a basis for the idealization of the structure for analysis purposes, as well as understanding the results of the analysis. Virtually all civil engineering structures involve the transfer of loads, applied to the structure, to the foundations, or some form of support.

A given structure may have different load paths for different applied loads. As an example, consider the water tank supported by a tower that consists of four vertical trusses (Figure 1.10). The weight of the tank and its contents is carried by the columns on the four corners (Figure 1.10b), the arrows illustrating the transfer of the weight by compression in the four columns. However, the wind load on the tank (Figure 1.10b) causes a horizontal force that must be transferred to the supports (equally) by the two vertical trusses that are parallel to the wind direction. This produces forces in the diagonal members and also increases or decreases the axial forces in the columns. The other two trusses (at right angles) will not contribute simultaneously to this load path but would become load paths if the wind direction changed 90°.

Figure 1.11a represents a plane-frame idealization of a cable-stayed bridge. The frame is composed of beam AB, rigidly connected to towers CD and EF. The beam is stayed by cables that make it possible to span a greater distance. Figure 1.11c shows the deflected shape of the frame due to traffic load on the main span. A typical cable is isolated in Figure 1.11c as a free body; the arrows shown represent the forces exerted by the beam or the tower on the cable. The cables are commonly pretensioned in construction; the arrows represent an increase in the tension force due to the traffic load. Figure 1.11c shows the deflected shape due to traffic load on the left exterior span.

Understanding load paths allows a considerable simplification in the subsequent analyses. It is also an important component in the determination of the forces to be applied to the structure being analyzed.

Consider the example of a bridge deck supported by two parallel trusses (Figures 1.6a and b). The deck is supported on a series of transverse beams and longitudinal beams (stringers) that eventually transfer the weight of the traffic to the trusses. Here, we want to examine how that transfer occurs and what it means for the various components of the bridge deck. These components can then be analyzed and designed accordingly.

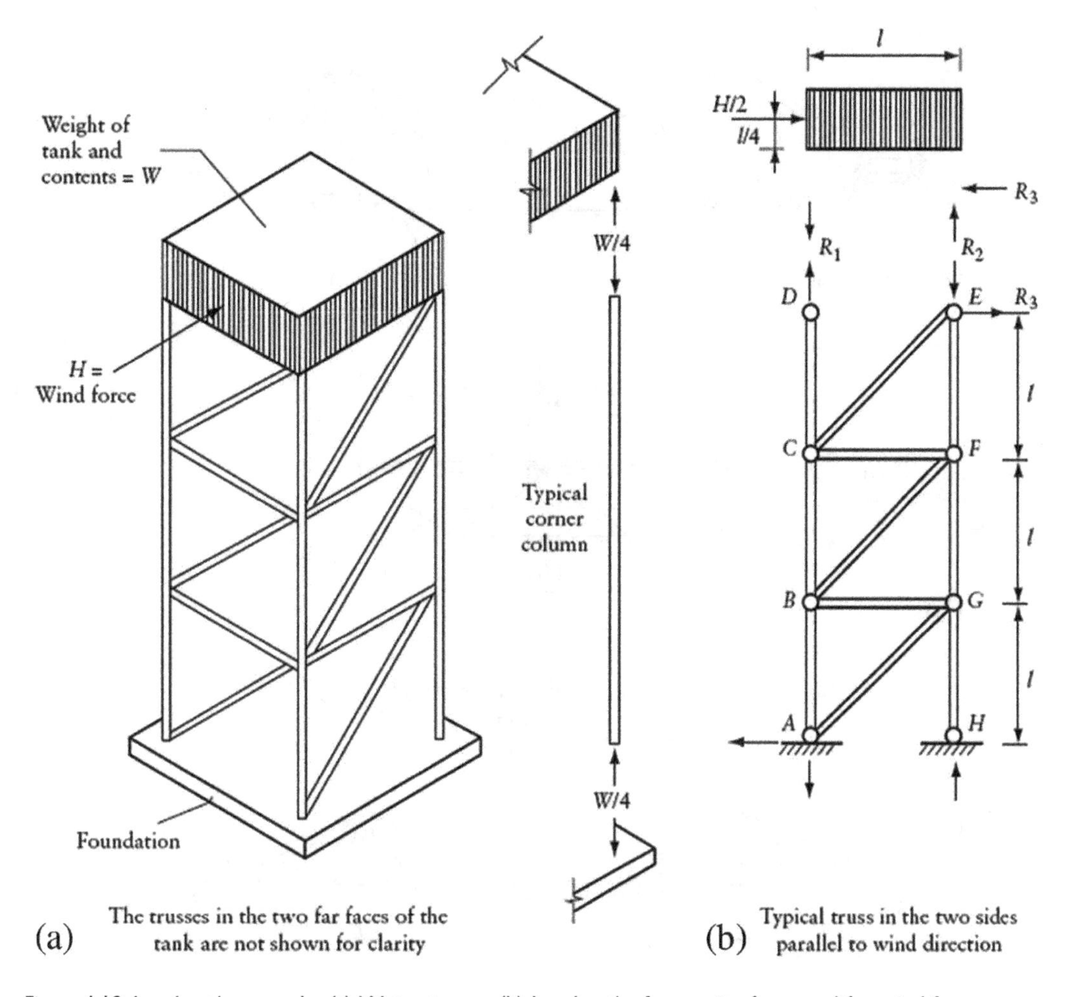

Figure 1.10 Load path example. (a) Water tower. (b) Load paths for gravity force and for wind force.

In a timber bridge deck, the deck beams are only capable of transferring the wheel loads along their lengths to the supports provided by the stringers. The loading on the stringers will come from their self-weight, the weight of the deck, and the wheel loads transferred to it. Each stringer is only supported by the two adjacent floor beams and can, therefore, only transfer the load to these floor beams. These are then supported at the joints on the lower chord of the trusses. We have selected here a system with a clear load path. The transfer of the load of the deck to the truss will be less obvious if the wooden deck and stringers are replaced by a concrete slab. The self-weight of the slab and the wheel loads are transferred in two directions to the trusses and the transverse beams (two-way slab action). However, this difference in load path is often ignored in an analysis of the truss.

Understanding the load path is essential in design. As we have seen in the preceding section, the downward load on the arch in Figure 1.4b is transferred as two inclined forces to the supports. The foundation must be capable of resisting the gravity load on the structure and the outward thrust (horizontal component of the reactions in reversed directions). Similarly, the path of the gravity load on the dome in Figure 1.8b induces a horizontal outward thrust and a downward force in the ring beam. The horizontal thrust

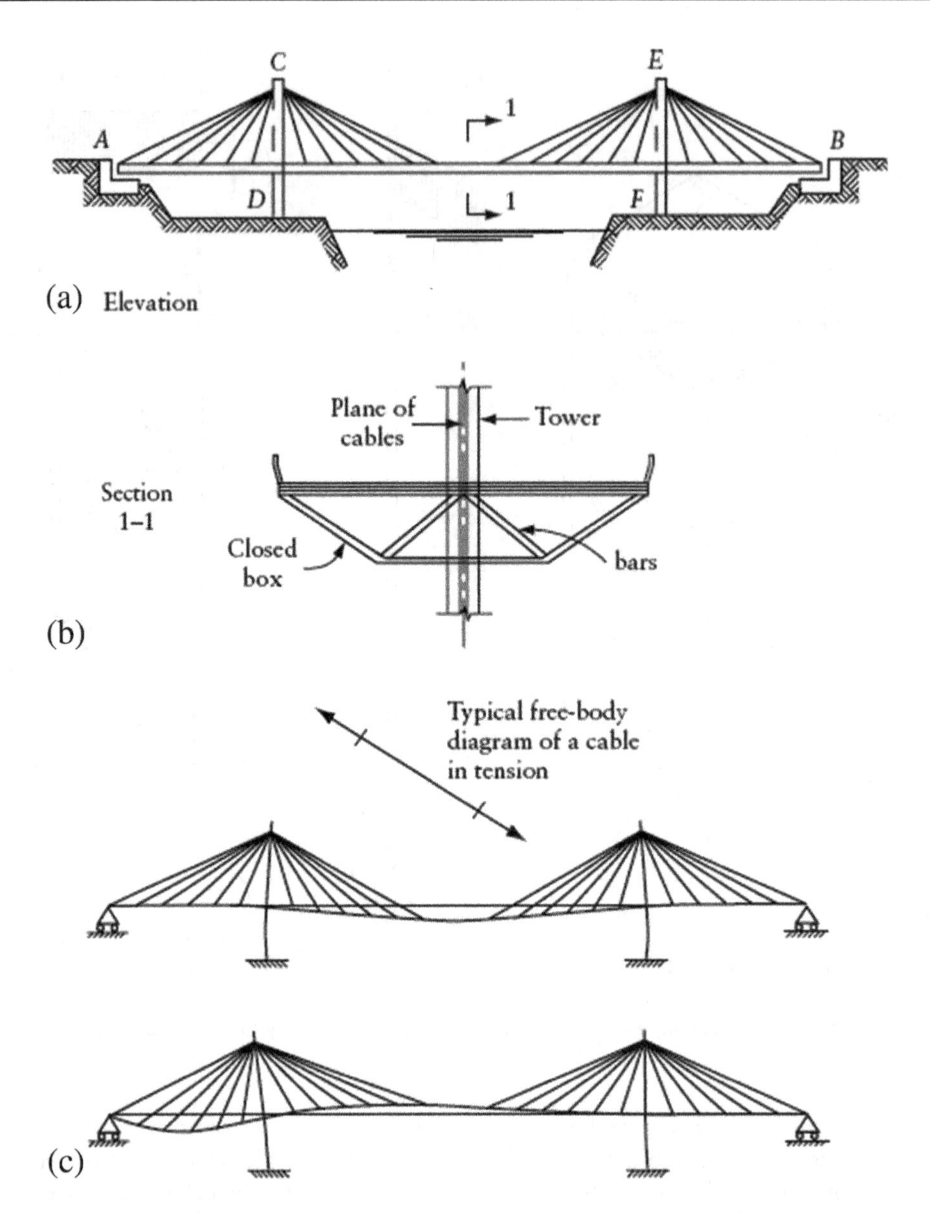

Figure 1.11 Behavior of a cable-stayed bridge. (a) and (b) Elevation of plane frame idealization and bridge cross section. (c) Deflected shape and forces in cables due to traffic load on interior or exterior span.

is resisted by the hoop force in the ring beam, and the structure needs only a vertical reaction component.

Some assessment of the load path is possible without a detailed analysis. One example of this is how lateral loads are resisted in multistory buildings (Figure 1.12). These buildings can be erected as rigid frames, typically with columns and beams, in steel or concrete, connected in a regular array represented by the skeleton shown in Figure 1.12a. The walls, which may be connected to the columns and/or the floors, are treated as nonstructural elements. The beams carry a concrete slab or steel decking covered with concrete; with both systems, the floor can be considered rigid when subjected to loads in its plane (in-plane forces). Thus, the horizontal wind load on the building will cause the floors to move

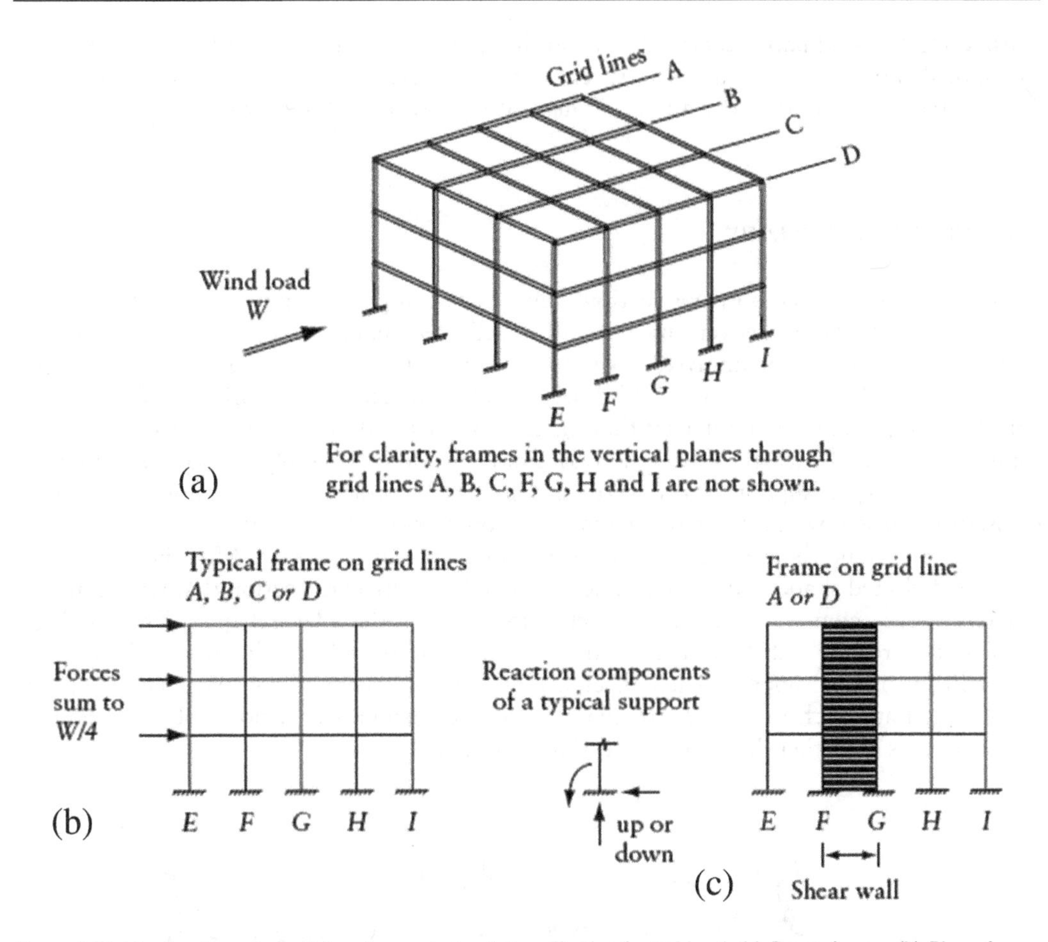

Figure 1.12 Idealization of a building for analysis of the effects of wind load. (a) Space frame. (b) Plane frame idealization. (c) Plane frame containing shear wall.

horizontally as rigid bodies, causing all the columns at each floor level to deflect equally. If the columns and beams in each two-dimensional frame on grid lines A to D have the same dimensions, then all the frames will have the same stiffness and any lateral wind load will be distributed equally (one-quarter) to each frame. The plane frame idealization of one frame shown in Figure 1.12b is all that is required for analysis.

If, however, the frames are different, then the wind load resisted by each frame will depend on its stiffness. For example, consider the same structure with the two outer frames A and D, containing shear walls (Figure 1.12c). Because plane frames A and D are much stiffer than plane frames B and C, the stiffer frames will tend to "attract" or resist the major part of the wind load on the structure (see Chapter 11).

Analysis of the plane frame shown in Figure 1.12b will give the internal forces in all the members and the support reactions. These are the forces exerted by the column bases on the foundations. Typically, at built-in (encastré) supports to the foundation, there will be vertical, horizontal, and moment reactions (Figure 1.12b). The horizontal reactions must add up to $W/4$, the wind load applied to this frame in the case considered. Also, the sum of the moments of the applied forces and the reactions about any point must be zero.

However, the individual vertical reactions will not be zero: the vertical reactions at E and I will be downward and upward, respectively. We conclude from this discussion that considering the load path helps to decide on the structural analysis model that can be used as an idealization of the actual structure, the type of results that the analysis should give, and the requirements that the answers must satisfy. Structural idealization is further discussed in Section 1.5.

1.4 DEFLECTED SHAPE

As we have noted in the previous section, deflections can play an important part in understanding load paths. Moreover, understanding deflected shapes also aids in the interpretation of the results of our analyses. In the very simplest case, if we walk across a simply supported wooden plank (Figure 1.13a), our weight will cause the plank to deflect noticeably. What is perhaps less obvious is that the ends of the plank rotate – because of the simple supports at the ends, which provide no restraint to rotation. A simply supported beam usually has one *hinged* support and one *roller* support. The roller support allows horizontal translation caused by the thermal expansion or contraction of the beam.

If we now extend the beam to become a two-span continuous beam, ABC (Figure 1.13b), then the deflected shape will change. The downward (positive) deflection that we feel under our feet is accompanied by a negative deflection (rise) in the adjacent span. There will be a downward reaction at A; so there must be some way of holding down the plank at A, or it will lift off the support. The deflected shape has concave and convex parts; the point of inflection at which the curvature changes sign corresponds to a point of zero bending moment. This is further discussed in Sections 1.10 and 1.11.

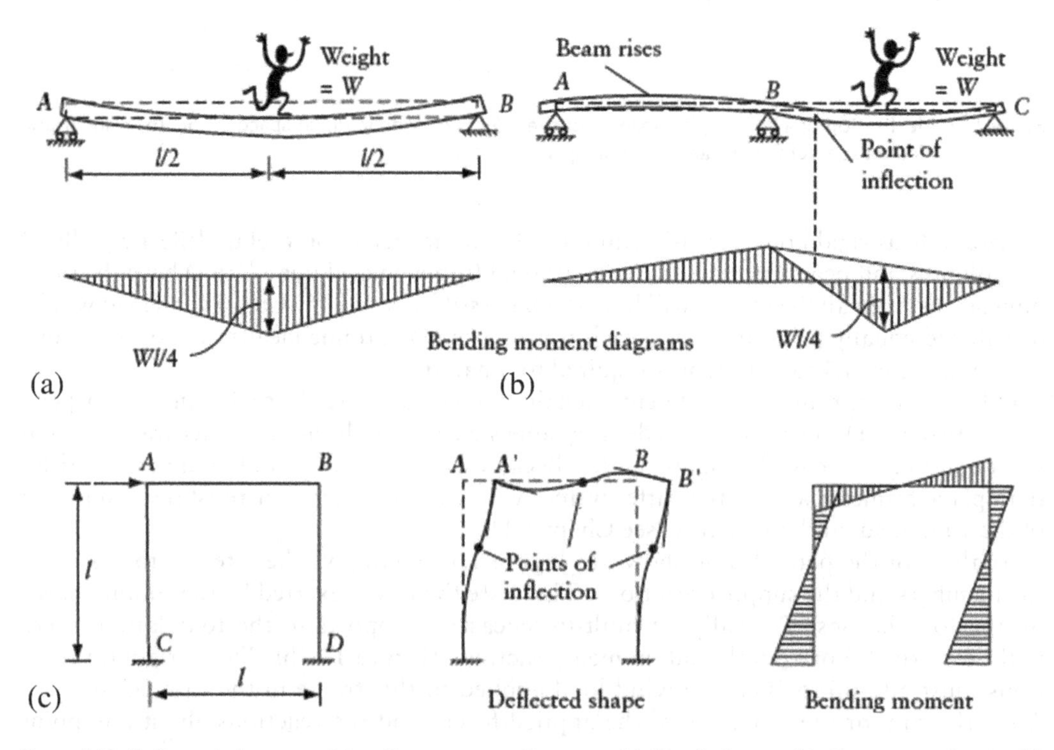

Figure 1.13 Deflected shapes and bending moment diagrams. (a) Simple beam. (b) Continuous beam. (c) Plane frame.

Under the action of a horizontal load, the frame in Figure 1.13c will obviously deflect to the right. The joints at A and B will also rotate as they translate sideways. Because the joint is rigid, the angle between the column and beam remains a right angle and, therefore, the entire joint must rotate. The beam and the columns must deflect with double curvature – the point between the two curves in each member is a point of inflection. This point corresponds to a point of zero bending moment. Deflected shapes are drawn throughout this book because they aid in the understanding of how a given structure behaves.

It is important to recognize that the displacements experienced by structures are small when compared to the dimensions of the structure, and we therefore exaggerate them when drawing them. The deflected shape and the original shape of the structure are drawn on the same figure, with the deflections drawn to a larger scale than the structure. Design guidelines suggest that the maximum deflection of a simple beam should not exceed (span/300). The public would not thank us for beams that deflect more than this. More discussion and examples of deflected shapes are presented in Section 1.11.

1.5 STRUCTURAL IDEALIZATION

Structural idealization is the process in which an actual structure is represented by a simpler model that can be analyzed. The model consists of *elements*, whose force/displacement relationship is known (or can be generated), that are connected at joints (nodes). For the analysis of framed structures, the elements are one-dimensional bars, oriented in one-, two-, or three-dimensional space. The joints can be rigid or hinged. For plates, membranes, shells, and massive structures (such as gravity dams), the elements are termed *finite-elements* and may be one-, two-, or three-dimensional, connected by nodes at the corners and/or sides (Figure 12.1). In any structural idealization, the loads must also be modeled – as must be the supports.

For analysis purposes, we draw the model as a free body and show by arrows a system of forces in equilibrium. The system consists of the externally applied loads and the reactions. Internal forces at a section can be shown and represented by pairs of opposite arrows; each pair consists of forces of equal magnitude and opposite direction (see Figure 1.10b).

Loads on the structural model produce translations and rotations of the nodes; these are referred to as *displacements*. From these displacements, the internal forces and the support reactions can be determined; this information can then be used in the design of the structure.

1.6 IDEALIZATION EXAMPLES: GRID ANALOGY

Horizontal floors consisting of solid slabs with or without beams and drop panels can be idealized as grids consisting of short prismatic beam elements and having a horizontal centroidal principal axis. Floors with uniform thickness, subjected to gravity loads, can be analyzed elastically using three degrees-of-freedom per node (plane grid, Chapter 19). To analyze prestressed concrete floors or slabs with thickness change (with drop panels or beams) and to consider the effects of cracking, creep, shrinkage of concrete, and relaxation of prestressed reinforcement, five degrees-of-freedom are used (ecentric grid, Chapter 19); the nodal displacements are two translations and two rotations about Cartesian axes and a downward deflection. The internal forces at a cross section of an eccentric grid member are a vertical shearing force, a bending moment about horizontal principal centroidal axis, a normal force at centroid, and a twisting moment. Cracking significantly reduces the torsional stiffness. Thus, ignoring the torsional stiffness, while maintaining equilibrium:

- Enables the use of the equations of beams in analyzing effects of cracking and time-dependent deformations; and
- Overestimates deflection; the overestimation is small in flat plates supported on columns, compared to slab panels supported on edges (e.g. spandrel beams).

Computer programs PLANEG and EGRID (Chapter 20) apply to analysis of the two grid types.

Analysis using non-zero twisting moment is suitable for deflection prediction. However, for ultimate strength design of shear and flexural reinforcements, it is conservative to use the results of a torsionless grid.

1.7 FRAMED STRUCTURES

All structures are three-dimensional. However, for analysis purposes, we model many types of structures as one-, two-, or three-dimensional skeletons composed of bars. Thus, an idealized *framed* structure can be one-dimensional (a beam), two-dimensional (a plane frame or truss), or three-dimensional (a space frame, truss, or a grid). The skeleton usually represents the *centroidal axes* of the members; the reason for use of centroidal axis is given in Section 1.12. The seven types of structures are defined below.

Beam: The idealized structure is a straight line (the centroidal axis). A simple beam covers a single span and has a hinged and a roller support. Commonly, a continuous beam covers more than one span and has one support hinged or one end support built-in (encastré) and the remaining supports are rollers. A straight line can be used to model beams, slabs, and composite beam/slabs. Bridge superstructures are often modeled as straight lines, regardless of the shape of their cross sections. Figure 1.14a illustrates beam AB that can be an idealization of structures having, as examples, any of the cross sections in Figures 1.14b to h. For each cross section, the centroidal principal axes are shown. All the applied loads, the reactions, and the deflected axis of the beam are in one plane through a centroidal principal axis. A structure for which the loads are not applied through the centroidal principal axes must be idealized as a spatial structure.

Plane frame: A plane frame consists of members, connected by rigid joints. Again, the cross section of any member has a centroidal principal axis lying in one plane; all applied loads and reactions are in the same plane. Simple and continuous beams are special cases of plane frames. Thus, the computer program PLANEF applies to plane frames and beams. Plane frames are widely used as idealizations of structures such as industrial buildings, multistory buildings, and bridges. Figures 1.11a, b, and c show a cable-stayed bridge and its idealization as a plane frame. Figures 1.12a, b, and c show a multistory building and its idealization as a plane frame for analysis of the effects of wind loads. Figure 1.15 shows a concrete bridge and its plane frame idealization.

Plane truss: A plane truss is similar to a plane frame but all joints are assumed pin-connected, and the loads are usually applied at the joints (still in the plane of the truss). Figures 1.6b and 1.25e are examples of plane truss idealizations of bridges; Figure 1.10b of a tower; and Figure 2.1c of a roof truss. Often, we do not draw circles at the joints (e.g. Figure 1.25e), but the assumption that the members are pin-connected is implied. Note the triangulated nature of trusses – a requirement for their stability. The two assumptions of pin connections and forces applied at joints mean that truss members are subjected only to axial forces, without shear force or bending moment.

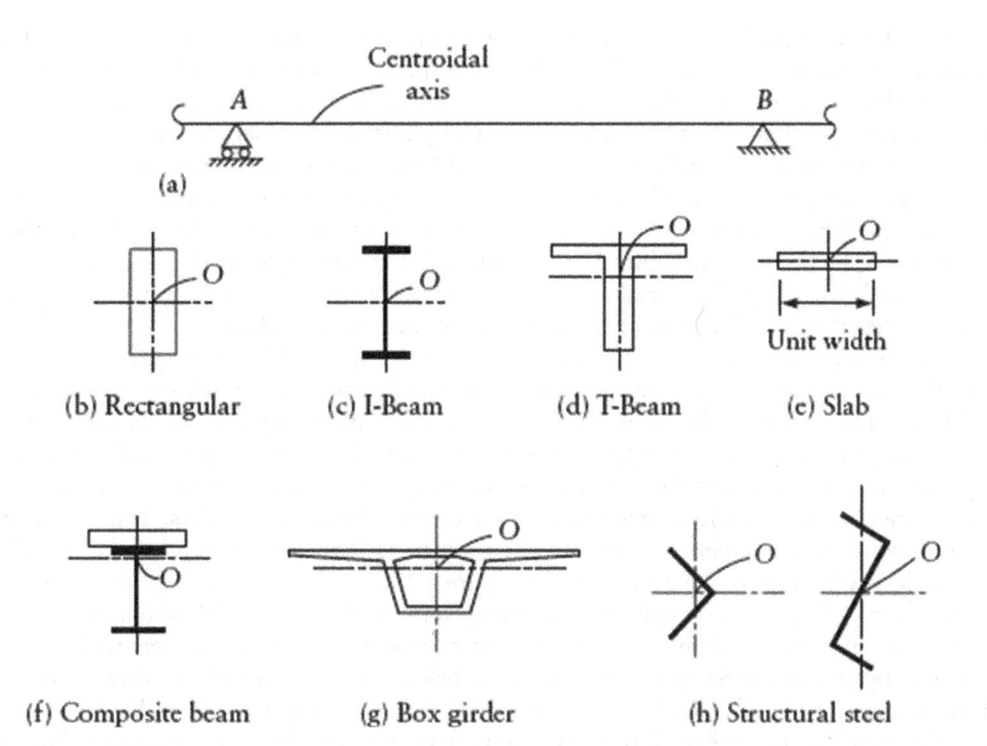

Figure 1.14 Beam idealization. (a) Interior span of a continuous beam. (b) to (h) Possible cross sections.

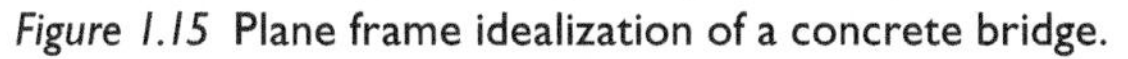

Figure 1.15 Plane frame idealization of a concrete bridge.

Space frame: Members and loads are now spatial (in three dimensions) with members connected by rigid joints. Most tall buildings are space frames – certainly if constructed in reinforced concrete or steel. However, much simpler structures often have to be modeled as space frames, particularly if the loads are out-of-plane, or the members or the structures lack symmetry. As an example, the box-girder of Figure 1.14g must be idealized as a space frame if the bridge is curved. The same box-girder bridge, even when straight, must be idealized as a space frame when being analyzed for horizontal wind loads or for load of traffic on a side lane. The building in Figure 1.16, which is of the same type as that shown in Figure 1.12a, should be modeled as a space frame because of the lack of symmetry, when the analysis is for the effect of wind load in the direction of the shown grid lines.

Space truss: This is like a plane truss, but with members and loads in three dimensions. All joints are pin-connected. Space trusses are most commonly used in long-span roofs.

Plane grid: A plane grid is a special case of a space frame except that all the members are in one plane and the loads are applied perpendicular to that plane. Figure 1.17a shows the top view and sectional elevation of a concrete bridge deck having three simply supported main girders monolithically connected to three intermediate cross-girders. Traffic load will be transmitted to the supports of all three main girders. The internal forces in all members can be determined using the grid model of Figure 1.17b.

Eccentric grid: Eccentric grid members are connected at the nodes, situated in a horizontal reference plane. All beams have a T- or rectangular-section. The centroidal axis of a typical beam runs horizontally at a distance below the reference plane. Five degrees-of-freedom are considered at each node. The nodal displacements are related by geometry to the displacement at member ends by assuming fictitious rigid vertical elements connecting nodes to beam ends. Because the beam elements are short (e.g. one-tenth of span), it is sufficiently accurate to represent both gravity and prestress loadings by nodal forces in a direction normal to the plane of the grid.

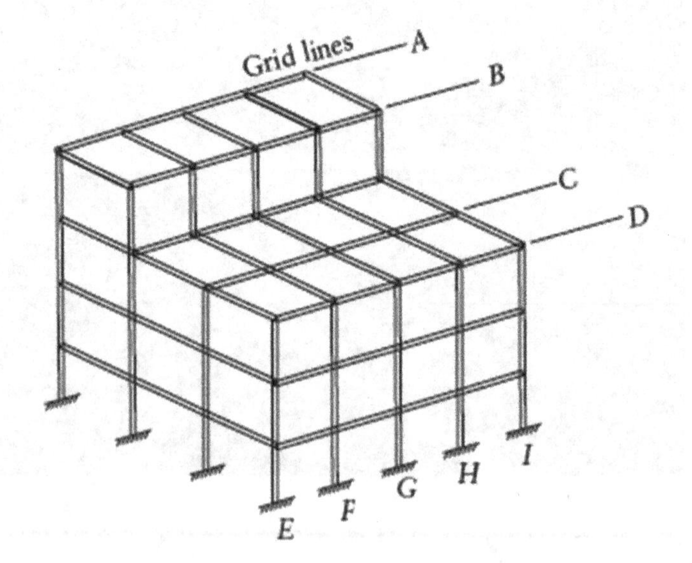

Figure 1.16 Space frame lacking symmetry, when analyzed for the effects of wind load in directions of the shown grid lines.

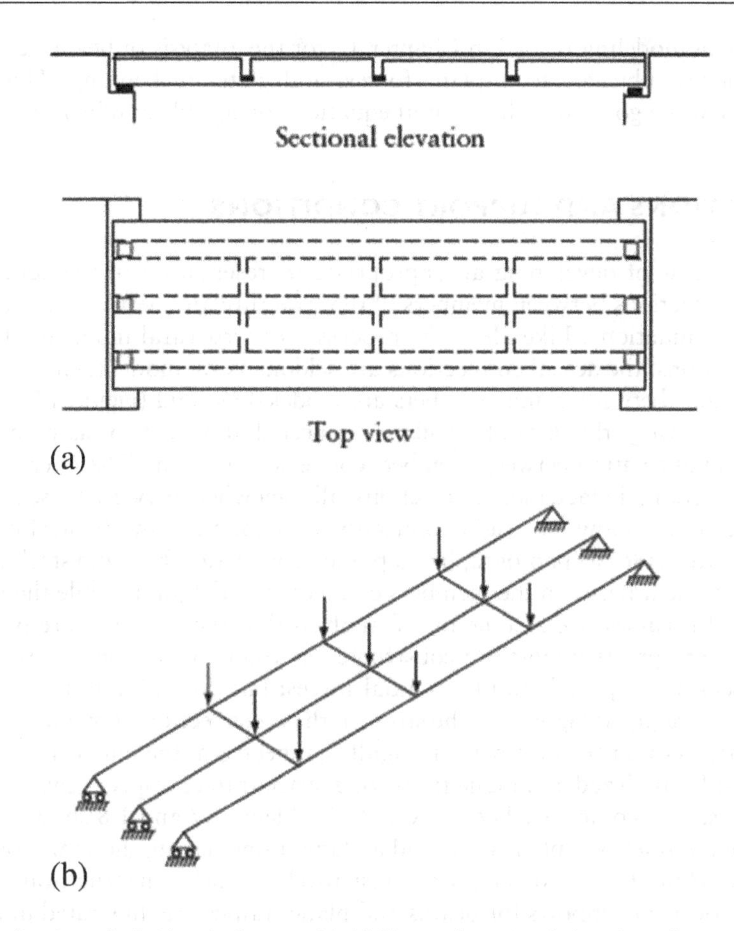

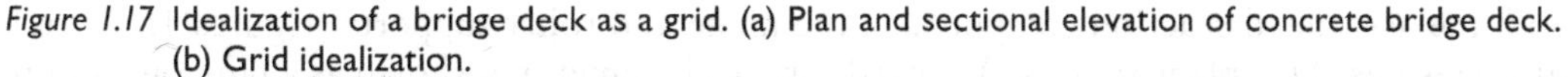

Figure 1.17 Idealization of a bridge deck as a grid. (a) Plan and sectional elevation of concrete bridge deck. (b) Grid idealization.

1.7.1 Computer programs

Chapter 19 presents computer programs that can be used as companions to this book. These include programs PLANEF, PLANET, SPACEF, SPACET, PLANEG, and EGRID for analysis of plane frames, plane trusses, space frames, space trusses, plane grids, and eccentric grids, respectively. Use of these programs is encouraged at an early stage of study. No computer program can eliminate the need for designers of structures to learn structural analysis. Thus, it is recommended to use the programs of this book to execute procedures that will be learned.

1.8 NON-FRAMED OR CONTINUOUS STRUCTURES

Continuous systems, such as walls, slabs, shells, and massive structures can be modeled using *finite-elements* (Figure 12.1). It is also possible to model the shear walls of Figure 1.12c as part of a plane frame model (Chapter 11). We have seen how a strip or unit width of a one-way slab can be modeled as a beam (Figure 1.14e). Bending moments and shear forces in two-way slabs can also be obtained using a plane or eccentric grid model. Solid slabs and their supporting columns are sometimes modeled as plane frames.

Another type of modeling is used in Chapter 12 for the analysis of beams, circular cylindrical shells, plates subjected to in-plane forces, and plates in bending. This is through approximations to the governing differential equations of equilibrium by *finite differences*.

1.9 CONNECTIONS AND SUPPORT CONDITIONS

Much of the process of developing an appropriate representation of the actual structure relates to the connections between members of the structure and between the structure and its supports or foundations. Like all of the processes of structural modeling, this involves some approximations: the actual connections are seldom as we model them.

Most connections between frame members are modeled as rigid (Figure 1.7), implying no relative rotation between the ends of members connected at a joint, or as pinned, implying complete freedom of rotation between members connected at a joint. These represent the two extremes of connections; in fact, most connections fall somewhere between these two extremes.

Figures 1.18a and b show two rigid connections, in concrete and in steel construction, respectively. Figure 1.18c is a photograph of a pinned connection in an old steel truss. A rigid joint transfers moment between the members connected to the joint, while the pinned joint does not do so. In trusses, we assume in the analysis that the members are pin-connected although in modern practice most are constructed as rigid connections. With rigid joints, the members will be subjected mainly to axial forces; but, in addition, there are bending moments that are frequently ignored. The stress in the cross section of members is no longer uniform. To consider that the members are rigidly connected at the joints, a plane truss or a space truss must be analyzed as a plane frame or a space frame, respectively. This will commonly require use of a computer. The answers to Problems 1.7 and 1.8 show a comparison of the results of analyses of a plane truss and a plane frame having the same configuration.

Connections to foundations, or supports, must also be modeled in structural analysis. The three common forms of supports for beams and plane frames are illustrated in Figure 1.19, with the associated reaction components; the hinge and roller supports (Figures 1.19c and d) are also used for plane trusses. A hinged support is sometimes shown as a pin-connection to a rigid surface (Figure 1.19c). Figure 1.19a shows the displacement components at a typical joint, and Figures 1.19b to d indicate the displacement components that are prevented, and those that are free, at the support. The roller support (Figure 1.19d) is used when the

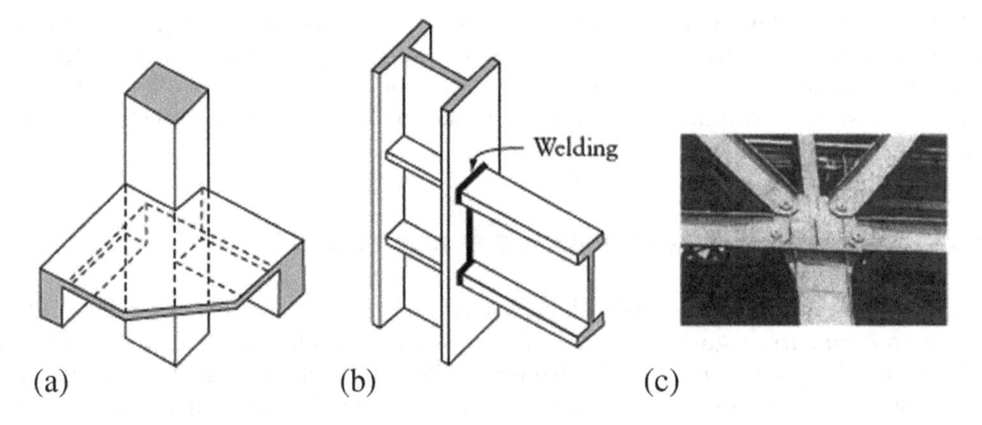

Figure 1.18 Connections. (a) Rigid connection of a corner column with slab and edge beams in a concrete building. (b) Rigid connection of steel column to beam. (c) Pin-connected truss members.

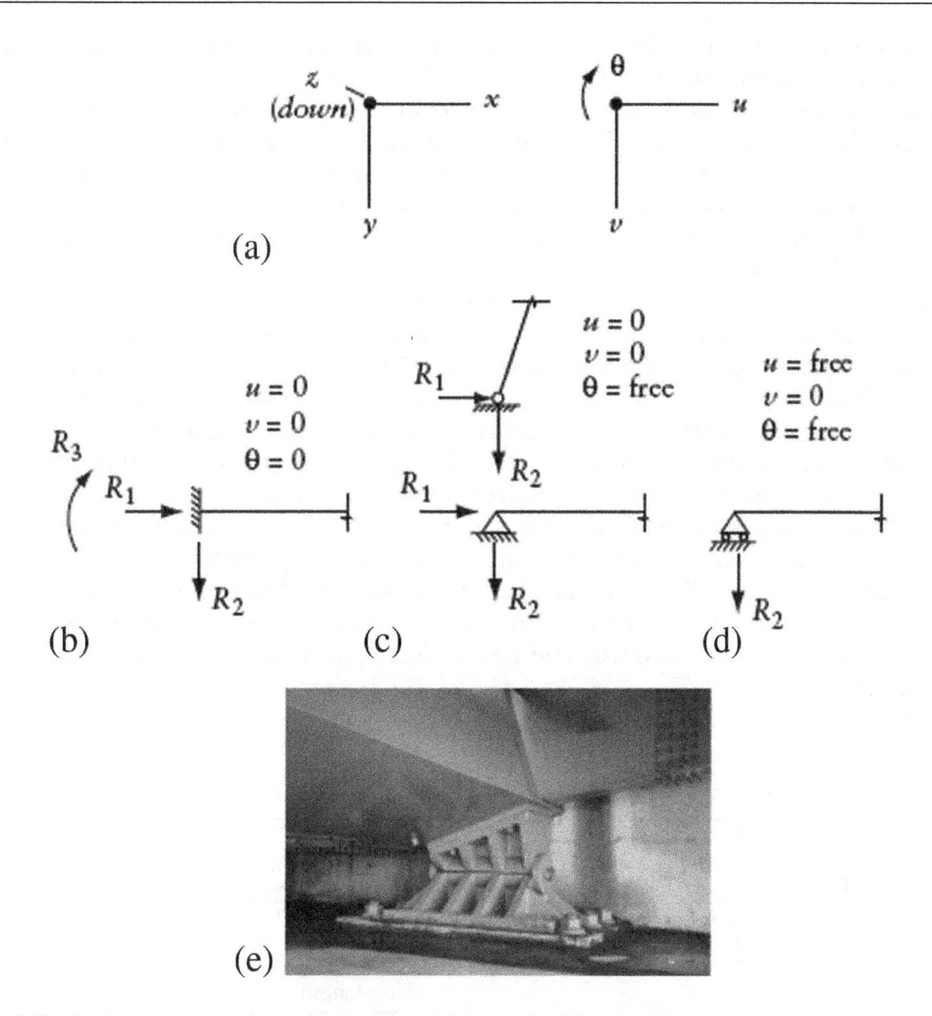

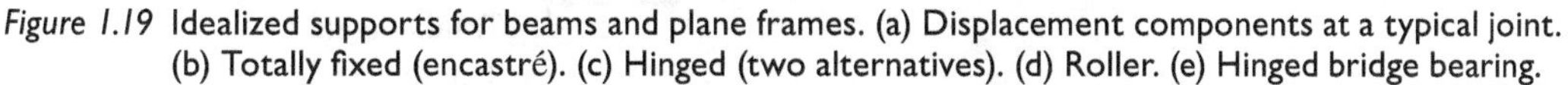

Figure 1.19 Idealized supports for beams and plane frames. (a) Displacement components at a typical joint. (b) Totally fixed (encastré). (c) Hinged (two alternatives). (d) Roller. (e) Hinged bridge bearing.

structure must accommodate axial expansion or contraction, usually due to thermal effects, but also due to creep, shrinkage, and prestressing effects.

In many cases, the actual form of support is different from the idealizations shown in Figure 1.19. For example, short-span beams and trusses in buildings are often provided with no specific support system at their ends. The absence of specific means of restraining end rotations or axial displacements justifies their analysis as simply supported ends.

Special supports (bearings) to ensure that translation or rotation can freely occur as assumed in the analysis are commonly provided for longer spans (Figure 1.19e). A wide variety of bridge bearings, which vary in complexity depending on the magnitude of the reactions and the allowed displacements, is commercially available.

1.10 LOADS AND LOAD IDEALIZATION

Several causes of stresses and deformations need to be considered in structural analysis. Gravity load is the main cause. Examples of gravity loads are the weight of the structure, of nonstructural elements such as floor covering and partitions, of occupants, furniture, equipment, as well as traffic and snow. Pressure of liquids, earth, granular materials, and wind is another type of load. Prestressing produces forces on concrete structures (see Appendix H).

We can idealize loads as a set of concentrated loads. The arrows shown on the truss nodes in Figure 1.6b can represent the self-weight of the members and the weight of the deck and the traffic load it carries. Alternatively, the analysis may consider load distribution over length or over an area or a volume, defined by the load intensity (force/unit length or force/unit area or force/unit volume).

In many cases the deformations due to temperature variation cannot occur freely, causing stresses, internal forces, and reactions. This is briefly discussed in Section 1.10.1.

If one of the supports of a simple beam (Figure 1.20a) settles downward, the beam rotates as a rigid body and no stress is developed. But if the beam is continuous over two or more spans, a support settlement produces internal forces and reactions (Appendix D). Figure D.1 shows the deflected shape, the reactions, and bending moments over the supports due to settlement of an interior support of a beam continuous over three spans.

For some materials, such as concrete, a stress increment introduced and sustained thereafter produces not only an immediate strain but also an additional strain developing gradually with time. This additional strain is creep. Similar to thermal expansion or contraction, when creep is restrained, stresses and internal forces develop. Analysis of the effects of creep, shrinkage, and temperature – including the effect of cracking – is discussed elsewhere.[1]

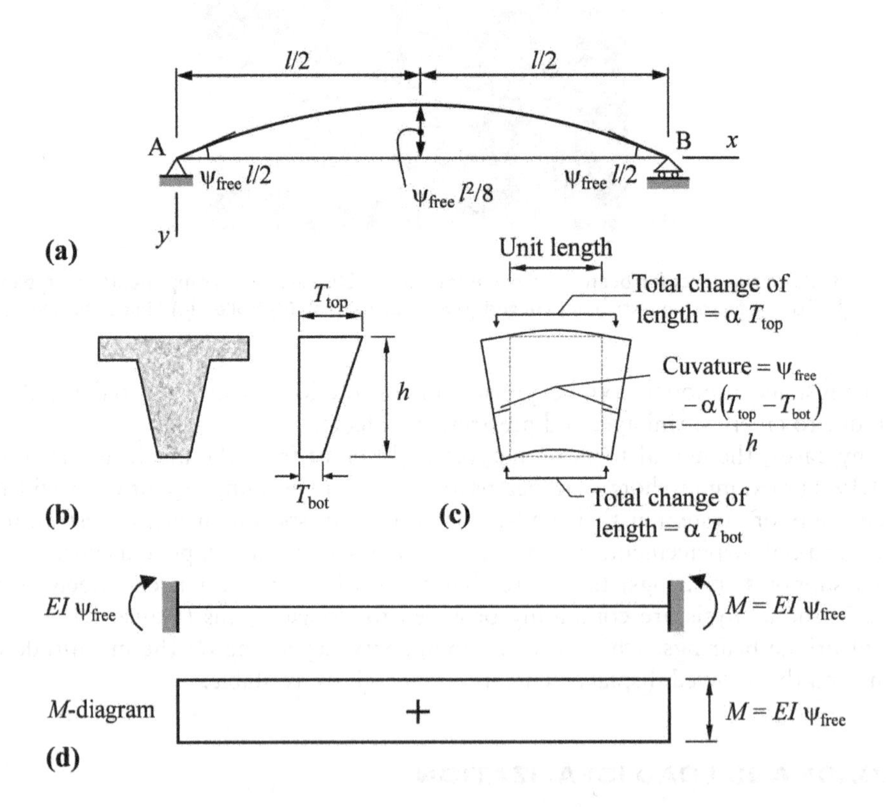

Figure 1.20 Deformation of a simple beam due to temperature change. (a) Deflected shape. (b) Cross section and variation in a rise of temperature over its height. (c) Free change in shape of a segment of unit length. (d) Effect of the temperature rise in part (b) on the beam with ends encastré.

Ground motion in an earthquake produces dynamic motion of the structure causing significant inertial forces. Analysis of the internal forces can be done by a dynamic computer analysis of the idealized structure, subjected to a recorded ground motion. In lieu of the dynamic analysis, codes allow a static analysis using equivalent lateral forces, commonly applied at floor levels. Because these forces represent inertial effect, their values are proportional to the mass of the structure commonly lumped at each floor level.

1.10.1 Thermal effects

Thermal expansion or contraction produces deformation, but no stress, when it can occur freely, without restraint. This is the case of the simple beam in Figure 1.20a, subjected to temperature rise varying linearly between T_{top} at top fiber and T_{bot} at bottom fiber (Figure 1.20b). When $T_{top} = T_{bot} = T$, the beam elongates and the roller at B moves outward a distance equal to αTl; where α is coefficient of thermal expansion and l is beam length. When $T_{top} \neq T_{bot}$ the beam will have constant free curvature:

$$\psi_{free} = \frac{\alpha\left(T_{bot} - T_{top}\right)}{h} \tag{1.7}$$

Figure 1.20c shows the change in shape of a segment of unit length (with $T_{top} > T_{bot}$). The two sections at the limits of this length rotate relative to each other, and the angle ψ_{free} is equal to the curvature:

$$\psi_{free} = -\frac{d^2y}{dx^2} \tag{1.8}$$

where y is the deflection (positive when downward); x is the distance between the left-hand end of the beam (support A) and any section. In the case considered, the beam deflects upwards as shown in Figure 1.20a; the end rotations and the deflection at mid-span given in the figure can be checked by integration of Equation 1.8. The negative sign in Equation 1.8 results from the directions chosen for the axes and the common convention that sagging is associated with positive curvature.

When a beam, continuous over two spans, is subjected to the same rise of temperature as presented in Figure 1.20b, the upward deflection cannot occur freely; a downward reaction develops at the intermediate support and the deflected shape of the beam is as shown in Figure 4.2c.

Restraint of thermal expansion can develop relatively high stresses. A uniform change in temperature of T degrees in a beam with ends encastré develops a stress:

$$\sigma = -\alpha ET \tag{1.9}$$

where E is modulus of elasticity. For a concrete beam with $E = 40$ GPa (5.8×10^6 psi), $\alpha = 10 \times 10^{-6}$ per degree centigrade ($10\ (5/9) \times 10^{-6}$ per degree Fahrenheit), $T = -15°$ C ($-27°$ F), with the minus sign indicating temperature drop, the stress is $\sigma = 6$ MPa (870 psi). This tensile stress can cause the concrete to crack; the restraint is then partially or fully removed and the thermal stress drops. Analysis of the effect of temperature accounting for cracking requires nonlinear analysis. Linear analysis of the effect of temperature is treated in Chapters 4, 5, and 8. Shrinkage or swelling can be treated in the same way as thermal contraction or expansion, respectively.

We conclude the discussion on the effects of temperature and shrinkage (similar to settlement of supports) by stating that these produce stresses and internal forces only in statically indeterminate structures, such as continuous beams; no stresses or internal forces develop in statically determinate structures, such as simple beams.

The topic of statical indeterminacy is discussed in more details in Section 3.2. We should also mention that a temperature rise that varies nonlinearly over the depth of section of the simple beam in Figure 1.20a produces self-equilibrating stress; this means that the resultant of this stress (the internal force) is nil (Section 5.9).

1.11 STRESSES AND DEFORMATIONS

Structures deform under the action of forces. The column of Figure 1.21a is acted on by forces P and Q as shown. Figure 1.21b shows the deflected shape of the column. The applied forces cause internal forces that are stress resultants. The resultants of stresses at any section are bending moment, M, shear force, V, and axial force, N. To show these internal forces, the structure must be cut (Figure 1.21c). Now we have two free body diagrams, with the internal forces shown at the location of the cut. Note that we must show the internal forces on both sides of the cut as pairs of arrows in opposite directions. Applying equilibrium equations to the upper free body diagram (part AC), we get:

$$N = -Q; \quad V = P; \quad M = Px \tag{1.10}$$

The same answers can be obtained by considering equilibrium of part DB. The sign convention for the internal forces and their calculation is discussed in detail in Section 2.3. The above equations tell us that N and V are constant throughout the height of the column, but M varies linearly from top to bottom (Figure 1.21e). These internal forces are resultants of stresses that produce strains. Each strain results in a form of deformation – in this case, axial, shear, and bending (flexural). Generally, the first two are small in comparison to the flexural deformation. For example, if the column is made of timber, 0.1 m (4 in.) square, and

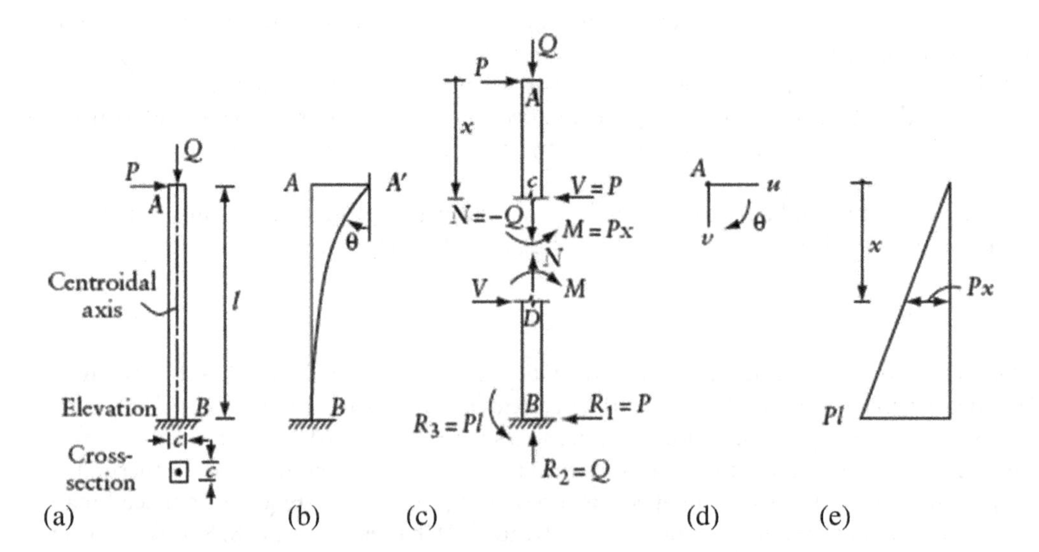

Figure 1.21 Deflected shape and internal forces in a column. (a) Elevation and cross section. (b) Deflected shape. (c) Cut to show internal force at a section. (d) Displacement components at the tip. (e) Bending moment diagram.

3 m (10 ft) long, and the two forces, P and Q, are each 1.00 kN (225 lb), then the displacement components at the top of the column are (Figure 1.21d):

$$u = 0.11 \text{ m (4.3 in.)}; \; v = 30 \times 10^{-6} \text{ m (0.0012 in.)}; \; \theta = 0.054 \text{ rad}$$

These values are calculated by virtual work (Chapters 7 and 8), assuming the modulus of elasticity, $E = 10$ GPa (1.45×10^6 psi), and the shear modulus, $G = 4$ GPa (0.6×10^6 psi). The displacement component v is the axial shortening of the column due to the axial force, N; u is the sidesway at the top due to the shear force, V, and to the moment, M (the latter is generally much larger than the former; in this case, by a ratio of 1200 to 1); and θ is the angular rotation due to the moment. A knowledge of deflections and the deflected shape aids us to understand the behavior of the structure.

When we sketch deflected shapes, we generally consider only the bending deformation and, as mentioned in Section 1.4, we show the deflections on a larger scale than the structure itself. Thus, we show the deflected position A′ of the tip of the column at its original height A, indicating that the change in length of the column is ignored (Figure 1.21b). For the same reason, we show in the deflected shape in Figure 1.13c A and A′ at the same level, and likewise for B and B′. Also, because we ignore the change of length of member AB, the distance $\overline{AA'}$ is shown the same as $\overline{BB'}$. For the same reason, the length of member AB in Figure 1.21 does not change and A moves on an arc of a circle to A′. But, because the deflections are small compared to the member lengths, the displacement $\overline{AA'}$ is shown perpendicular to AB; in other words, the arc of circle is shown as a straight line $\overline{AA'}$ perpendicular to the original direction of AB. When drawing the deflected shape of the frame in Problem 1.1c, the movement of node B should be shown perpendicular to AB (downward) and the movement of C perpendicular to CD (upward).

We often draw diagrams showing the variations of M and V over the length. These are termed *bending moment* and *shear force diagrams.* The sign convention for the bending moment diagram is important. Bending or flexure produces tension on one face of the cross section, and compression on the other. Examining the deflected shape of Figure 1.21b, it is not difficult to imagine the tension on the left-hand side of the column, and, therefore, the bending moment diagram is drawn on that (tension) side (Figure 1.21e). The axial force N and the bending moment M produce a normal stress – tensile or compressive. Calculation of normal stresses is discussed in the following section.

1.12 NORMAL STRESS

Figures 1.22a and b represent the elevation and cross section of a column subjected at the top section to a distributed load of intensity p/unit area. The resultant of the load has the components:

$$N = \int p \, da; \quad M_x = \int p y \, da; \quad M_y = \int p x \, da \tag{1.11}$$

where N is normal force at the centroid of the cross-sectional area; M_x and M_y are moments about the *centroidal axes x* and *y*, respectively. The positive-sign convention for N, M_x, and M_y is shown in Figure 1.22; the reference point O is chosen at the centroid of the cross section. In the deformed configuration, each cross section, originally plane, remains plane and normal to the longitudinal centroidal axis of the member. This assumption, attributed to

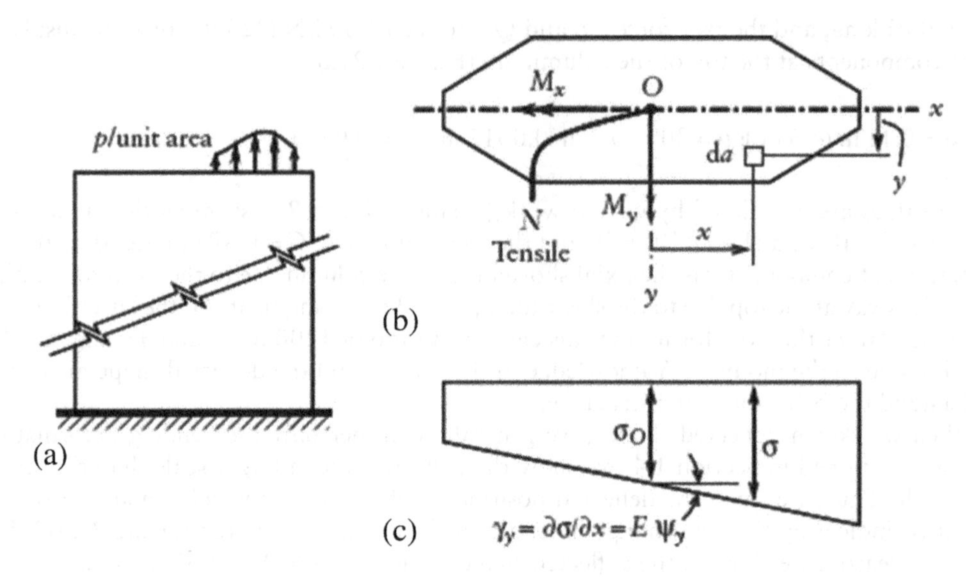

Figure 1.22 Normal stress distribution in a column cross section. Positive sign convention. (a) Elevation. (b) Cross section. (c) Stress variation along the x axis.

Bernoulli (17th–18th century), is confirmed by experimental measurements. The stress and strain are considered positive for tension and elongation, respectively.

It can be assumed that any section away from the top remains plane after deformation. Thus, considering a linearly elastic material with a modulus of elasticity E, the strain ε and the stress σ at any point can be expressed as:

$$\varepsilon = \varepsilon_O + \psi_x y + \psi_y x; \quad \sigma = \sigma_O + \gamma_x y + \gamma_y x \tag{1.12}$$

where ε_O is the strain at the centroid O; $\psi_x = d\varepsilon/dy$ is the curvature about the x axis; $\psi_y = d\varepsilon/dx$ is the curvature about the y axis; $\sigma_O = E\varepsilon_O$ is the stress at O; $\gamma_x = E\psi_x = d\sigma/dy$; $\gamma_y = E\psi_y = d\sigma/dx$ (Figure 1.22c). Tensile stress is considered positive.

The stress resultants are:

$$N = \int \sigma \, da; \quad M_x = \int \sigma y \, da; \quad M_y = \int \sigma x \, da \tag{1.13}$$

Substitution of Equation 1.12 in Equation 1.13 and the solution of the resulting equations gives the parameters σ_O, γ_x, and γ_y:

$$\sigma_O = \frac{N}{a}; \quad \gamma_x = \frac{M_x I_y - M_y I_{xy}}{I_x I_y - I_{xy}^2}; \quad \gamma_y = \frac{M_y I_x - M_x I_{xy}}{I_x I_y - I_{xy}^2} \tag{1.14}$$

Thus, the stress at any point (x, y) is:

$$\sigma = \frac{N}{a} + \left(\frac{M_x I_y - M_y I_{xy}}{I_x I_y - I_{xy}^2} \right) y + \left(\frac{M_y I_x - M_x I_{xy}}{I_x I_y - I_{xy}^2} \right) x \tag{1.15}$$

where $a = \int da$ is the area of the cross section; $I_x = \int y^2 \, da$ is the second moment of area about the x axis; $I_y = \int x^2 \, da$ is the second moment of area about the y axis; and $I_{xy} = \int xy \, da$ is the product of inertia. When x and y are *centroidal principal axes*, $I_{xy}=0$, and Equation 1.15 simply becomes:

$$\sigma = \frac{N}{a} + \frac{M_x}{I_x}y + \frac{M_y}{I_y}x \qquad (1.16)$$

The derivation of Equation 1.14 will involve the first moments of area about the x and y axes, $B_x = \int y \, da$ and $B_y = \int x \, da$; however, both terms vanish because the axes x and y are chosen to pass through the centroid O.

1.12.1 Normal stresses in plane frames and beams

A plane frame is a structure whose members have their centroidal axes in one plane; the cross sections of all members have a centroidal principal axis in this plane, in which all the forces and reactions lie. Figures 1.23a and b show a segment of unit length and the cross section

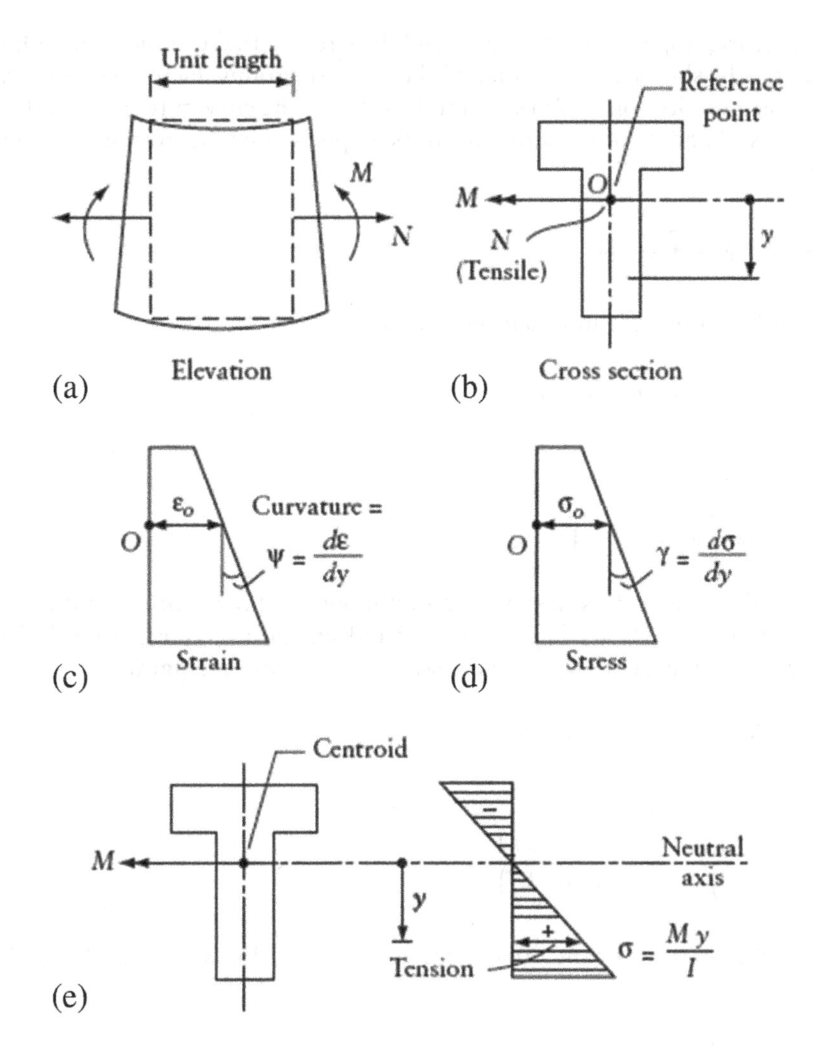

Figure 1.23 Normal stress and strain in a cross section of a beam. (a) Elevation of a segment of unit length. Positive sign convention of N and M. (b) Cross section. (c) Strain diagram. (d) Stress diagram. (e) Stress in a section subjected to M without N.

of a plane frame. A reference point O is chosen arbitrarily on a centroidal principal axis of the cross section (in this case, a vertical axis of symmetry). The general case in Figure 1.22b becomes the same as the case in Figure 1.23c, when M_y is absent and M is used to mean M_x; the equations in Section 1.12 apply when O is chosen at the centroid of the section and M_y, ψ_y, γ_y, and I_{xy} are set equal to zero. Because the case in Figure 1.23b is of frequent occurrence, we derive below equations for the distributions of strain ε, stress σ, and the curvature corresponding to the stress resultants N and M (the internal axial force and bending moment), with the reference point O not necessarily the centroid of the section. The positive sign conventions for N, M, and the coordinate y of any fiber are indicated in Figure 1.23b.

The two sections of the segment in Figure 1.23a are represented by vertical dashed lines before deformation; after deformation, the sections are represented by the rotated solid straight lines. The variations of the strain ε and the stress σ over the depth of the section shown in Figures 1.23c and d are expressed as:

$$\varepsilon = \varepsilon_O + \psi y; \quad \sigma = \sigma_O + \gamma y \tag{1.17}$$

where ε_O and σ_O are, respectively, the strain and the stress at the reference point O; $\psi = d\varepsilon/dy$ and $\gamma = d\sigma/dy$ are the slopes of the strain and the stress diagrams. Here, the symbols ψ and γ stand for ψ_x and γ_x in Section 1.12; ψ, of unit length^{-1}, is the curvature in the vertical plane.

The resultants N and M of the stress σ can be expressed by integration over the area of the section:

$$N = \int \sigma \, da; \quad M = \int \sigma y \, da \tag{1.18}$$

Substitution of Equation 1.17 in Equation 1.18 gives:

$$N = \sigma_O \, a + \gamma B; \quad M = \sigma_O \, B + \gamma I \tag{1.19}$$

where

$$a = \int da; \quad B = \int y \, da; \quad I = \int y^2 \, da \tag{1.20}$$

a, B, and I are the area and its first and second moment about a horizontal axis through the reference point O. Solution of Equation 1.19 and substitution of Equation 1.17 gives the parameters $\{\sigma_O, \gamma\}$ and $\{\varepsilon_O, \psi\}$ that define the stress and the strain diagrams:

$$\sigma_O = \frac{IN - BM}{aI - B^2}; \quad \gamma = \frac{-BN + aM}{aI - B^2} \tag{1.21}$$

$$\varepsilon_O = \frac{IN - BM}{E\left(aI - B^2\right)}; \quad \psi = \frac{-BN + aM}{E\left(aI - B^2\right)} \tag{1.22}$$

When O is at the centroid of the cross section, $B = 0$ and Equations 1.21 and 1.22 simplify to:

$$\sigma_O = \frac{N}{a}; \quad \gamma = \frac{M}{I} \tag{1.23}$$

$$\varepsilon_O = \frac{N}{Ea}; \quad \psi = \frac{M}{EI} \tag{1.24}$$

Substitution of Equation 1.23 in Equation 1.17 gives the stress at any fiber whose coordinate is y with respect to an axis through the centroid:

$$\sigma = \frac{N}{a} + \frac{My}{I} \tag{1.25}$$

When a section is subjected to bending moment, without a normal force (Figure 1.23e), the normal stress at any fiber is:

$$\sigma = \frac{My}{I} \tag{1.26}$$

with y measured downward from the centroidal axis and I is the second moment of area about the same axis. The normal stress is tensile and compressive at the fibers below and above the centroid respectively. Thus, the neutral axis is a centroidal axis, only when N is zero. The bottom fiber is subjected to tension when M is positive; but when M is negative the tension side is at the top fiber.

We note that Equations 1.23 to 1.26 apply only with the reference axis through O being at the centroid. We recall that framed structures are analyzed as a skeleton representing the centroidal axes of the members (Section 1.7); this is to make possible use of the simpler Equations 1.23 and 1.24 in lieu of Equations 1.21 to 1.22, which must be used when O is not at the centroid.

1.12.2 Examples of deflected shapes and bending moment diagrams

The curvature ψ is the change in slope of the deflected shape per unit length. From Equation 1.24, it is seen that the curvature ψ is proportional to the bending moment M. Figure 1.24a shows segments of unit length subjected to positive and negative bending moments. The corresponding stress distributions are shown in the same figures, from which it is seen that positive M produces a concave curvature and tension at bottom fiber; negative M produces a convex curvature and tension at top fiber.

Figure 1.24b shows the deflected shape and bending moment diagram for a beam. Without calculation, we can sketch the deflected shape by intuition. The bending moment diagram can then be drawn by following simple rules: at a free end, at an end supported by a hinged support or roller, and at an intermediate hinge, the bending moment is zero. The bending moment is also zero at points of inflection, where the curvature changes from concave to convex. The bending moment diagram is a straight line for any straight member not subjected to distributed load; see parts AC, CB, and BD in Figure 1.24b and parts BC and CD in Figure 1.24e. The bending moment graph is a second degree parabola for a part of a member carrying a uniform transverse load; see part AB in Figure 1.24e.

Throughout this book, we plot the ordinates of the bending moment diagram in a direction perpendicular to the axis of the member and on the tension side. The hatching of the M-diagram indicates the direction in which the ordinates are plotted. We can now verify that the M-diagram in Figures 1.13a to c and 1.24b to e follow these rules.

We recall that in drawing the deflected shapes, we show only the bending deformation. Thus, the members do not change in length. Because neither the columns nor the beam in Figure 1.24d BC change length, joints B and C must remain at the same height and at the

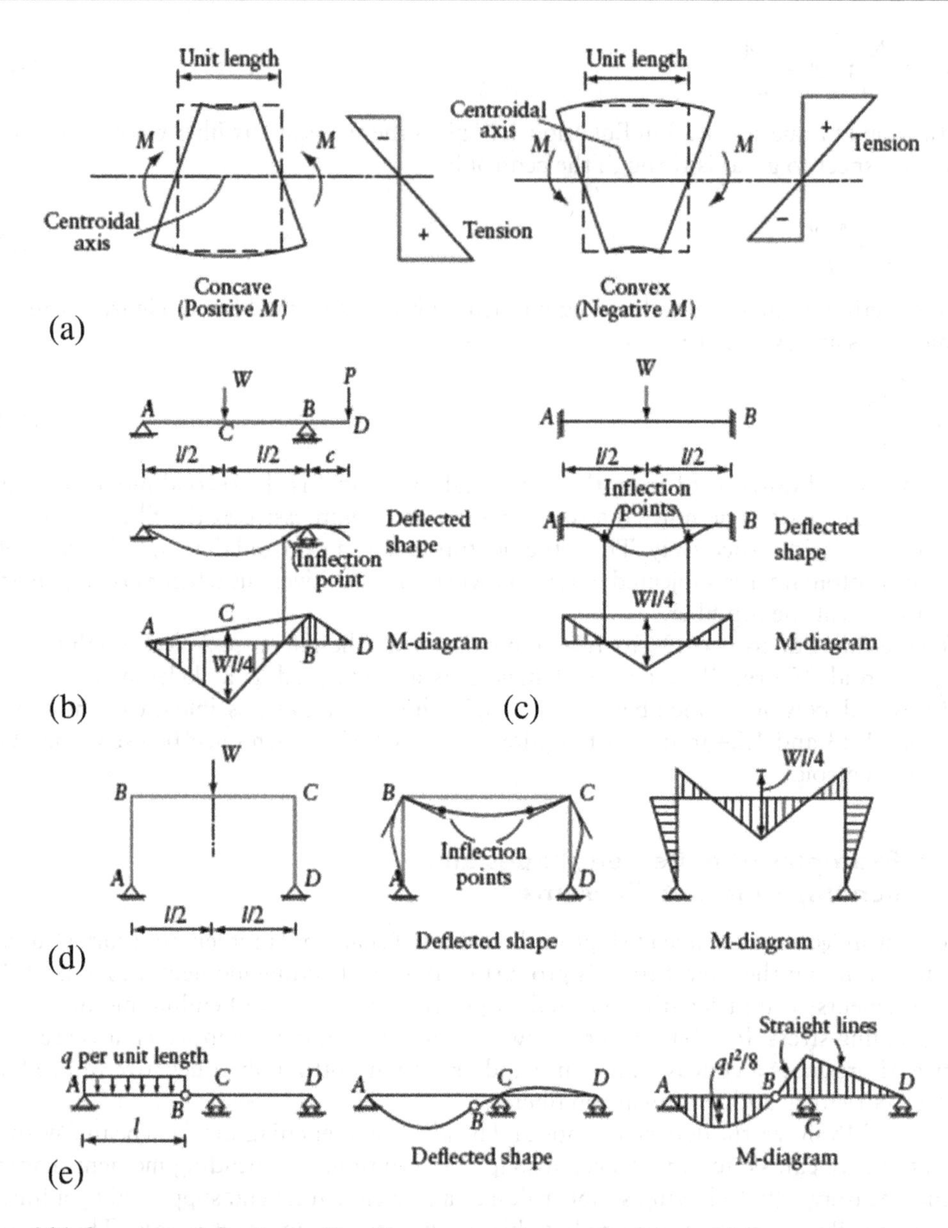

Figure 1.24 Curvature, deflection and bending moment diagrams for beams. (a) Curvature and stress due to bending moment. (b) to (e) Examples of deflected shapes and bending moment diagrams.

same horizontal location (because of symmetry) and therefore only rotate. The deflected shape in Figure 1.24e is discontinuous at the intermediate hinge B; thus, the slopes just to the left and just to the right of the hinge are different.

Some of the ordinates that define the M-diagrams in Figures 1.13 and 1.24 are statically indeterminate and thus require calculations. These are not given in the figures, only the statically determinate values, such as $Wl/4$ and $ql^2/8$ being shown; these represent the bending moment of a simply supported member of length l, carrying a transverse concentrated load W at its middle or a uniformly distributed load of intensity q/unit length.

1.12.3 Deflected shapes and bending moment diagrams due to temperature variation

The simple beam in Figure 1.20a subjected to temperature rise that varies linearly over its depth deflects freely as shown in the figure. A constant curvature (concave), ψ_{free} given by Equation 1.7 occurs, with the bending moment equal to zero at all sections. When $T_{top} > T_{bot}$, Equation 1.7 gives negative curvature (convex, Figure 1.20a). The simple beam is a statically determinate structure in which the thermal expansion or contraction is unrestrained. The same temperature rise in a beam with ends encastré produces reactions and a constant bending moment $M = -EI\psi_{free}$ (Figure 1.20d). This bending moment is positive, producing tension at bottom, and compression at top fibers. The corresponding curvature (Equation 1.24) is: $\psi = M/(EI) = -\psi_{free}$. Thus, the net curvature is zero, and the beam does not deflect. The beam with totally fixed ends is a statically indeterminate structure in which the thermal expansion or contraction is restrained.

We conclude from this discussion that the deflected shape due to temperature variation in a statically indeterminate structure is the sum of the free deflection, in which the thermal expansion or contraction is not restrained and the deflection is due to restraining forces. The deflected shape of the continuous beam in Figure 4.2c is the sum of the free deflected shape in Figure 4.2b and the deflection of simple beam AC, subjected to downward force at B. The net deflected shape due to temperature cannot be used to draw the bending moment diagram; thus, the points of inflection in the deflected shape in Figure 4.2c are not points of zero bending moment.

1.13 COMPARISONS: BEAMS, ARCHES, AND TRUSSES

The examples presented below compare the behavior of different types of plane structures that carry the same load and cover the same span (Figure 1.25). The structures considered are a simple beam, arches with and without ties and with different support conditions, and a simply supported truss. The discussion will show that we can use a lighter structure by avoiding or reducing bending moments.

EXAMPLE 1.1: LOAD PATH COMPARISONS: BEAM, ARCH, AND TRUSS

Each of the five structures shown in Figures 1.25a to e carries a uniformly distributed load of a total value 480 kN (108×10^3 lb). This load is idealized as a set of equally spaced concentrated forces as shown. The five structures transmit the same loads to the supports, but produce different internal forces and different reactions. The objective of this example is to study these differences.

The simple beam in Figure 1.25a, the simply supported arch in Figure 1.25b, and the truss in Figure 1.25e are statically determinate structures. But, the two arches in Figures 1.25c and d are statically indeterminate to the first degree. Chapters 2 and 3 discuss the topic of statical indeterminacy and explain the analysis of determinate and indeterminate structures. Here, we give and compare some significant results of the analysis, with the objective of understanding the differences in the load paths, the internal forces, and the deformations in the structural systems. Some details of the analyses are presented in Examples 2.11 and 4.5.

The analyses can be performed by the use of computer programs PLANEF (for structures in Figures 1.25a to d) and PLANET (for the structure in Figure 1.25e); see Chapter 20. Use of the programs early in the study of this book is encouraged. Simple data

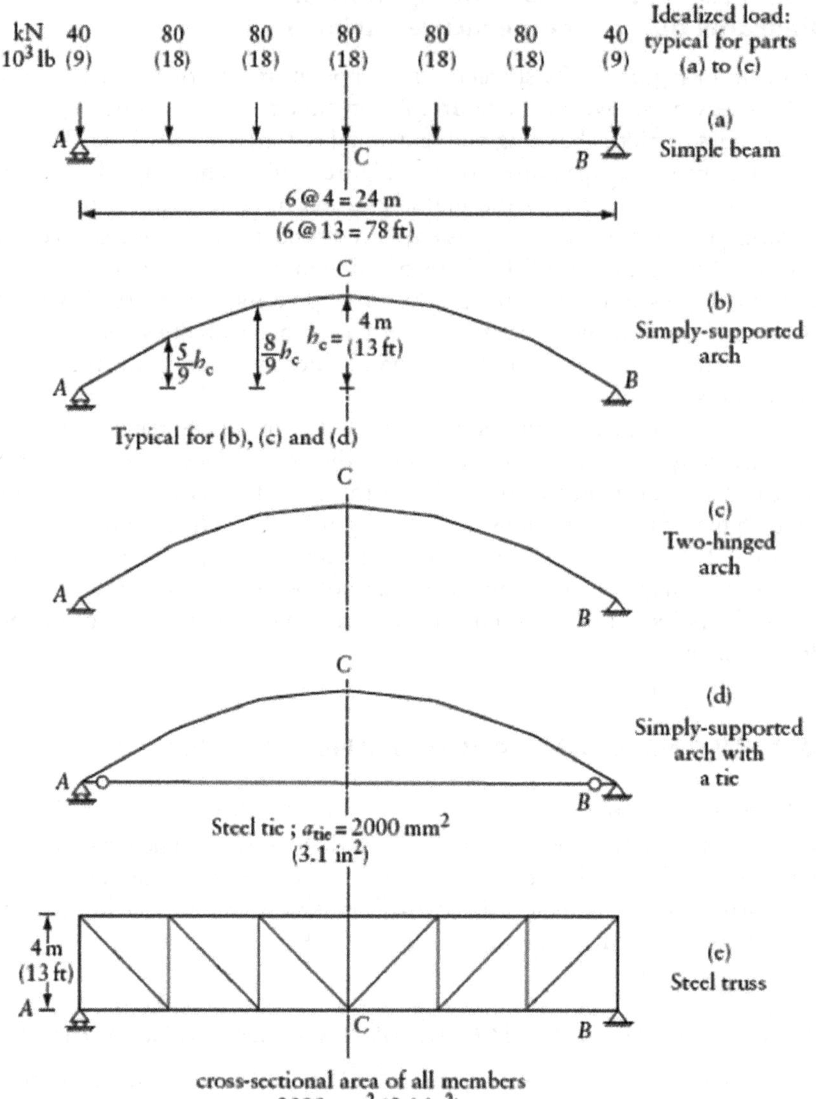

Figure 1.25 Structures covering same span and carrying same load (Example 1.1). (a) Simple beam. (b) Simply supported arch. (c) Two-hinged arch. (d) Simply supported arch with a tie. (e) Truss.

preparation (with explanation included in the appendix) is required for the use of the computer programs.

The simple beam (Figure 1.25a) has an uncracked concrete rectangular section of height 1.2 m and width 0.4m (47×16 in.^{2}). The arches in Figures 1.25b, c, and d have a constant square cross section of side 0.4 m (16 in.). The truss (Figure 1.25e) has steel members of cross-sectional area 2×10^{-3} m^2 (3.1 in.^{2}). The moduli of elasticity of concrete and steel are 40 and 200 GPa respectively (5.8×10^6 and 29×10^6 psi). The analysis results discussed below are for:

1. The mid-span deflections at node C in the five structures.
2. The horizontal displacement of the roller support A in the simply supported arch in Figure 1.25b.
3. The bending moment diagrams for the left-hand half of the simple beam and for the arches (Figures 1.26a to d).
4. The forces in the members of the truss (Figure 1.26e; Figures 1.26a to e also show reactions).

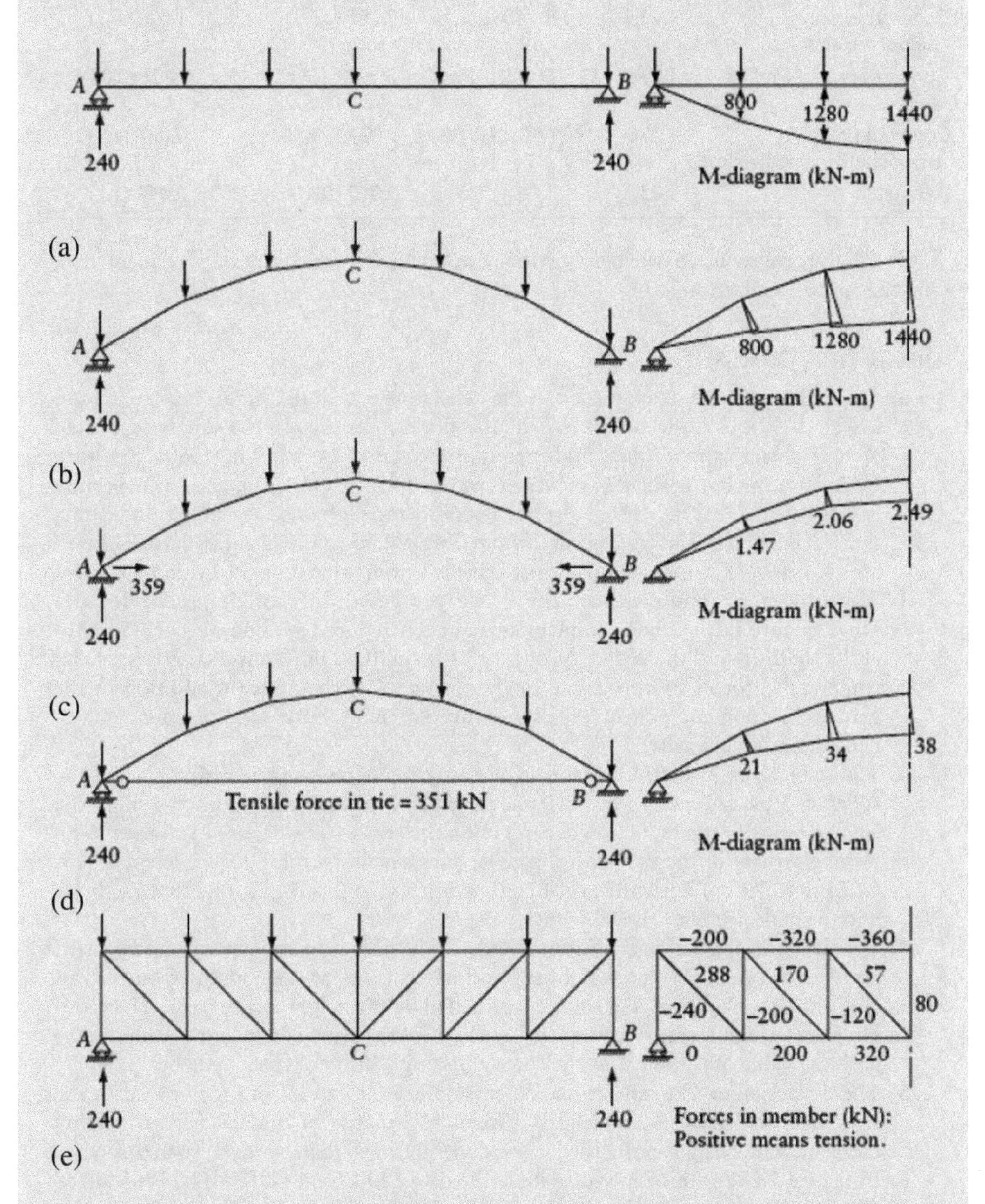

Figure 1.26 Reactions and internal forces in the structures of Figure 1.24. Total load = 480 kN; span = 24 m. (a) Simple beam. (b) Simply supported arch. (c) Two-hinged arch. (d) Simply supported arch with a tie. (e) Truss.

Table 1.1 Bending moments, horizontal displacements at A, and vertical deflections at C in the structures in Figure 1.25.

System	*Figure*	*Bending moment at C kN-m (lb-ft)*	*Horizontal displacement at A m (in.)*	*Vertical deflection at C m (in.)*
Concrete simple beam	1.25a	1440 (1050×10^3)	0	0.037 (1.5)
Concrete simply supported arch	1.25b	1440 (1050×10^3)	0.851 (33.5)	1.02 (40.2)
Concrete two-hinged arch	1.25c	2.5 (1.5×10^3)	0	0.002 (0.08)
Concrete simply supported arch with a tie	1.25d	38 (28×10^3)	0.021 (0.83)	0.027 (1.1)
Steel truss	1.25e	–	0.010 (0.41)	0.043 (1.7)

Table 1.1 gives the values of the bending moments at C, the horizontal displacement at *A*, and the vertical deflection at C.

DISCUSSION OF RESULTS

1. The mid-span bending moment and the deflection at C in the simple beam are 1440 kN-m (1050×10^3 lb-ft) and 0.037 m (1.5 in.), respectively. In the simply supported arch, the bending moment at mid-span is also equal to 1440 kN-m (1050×10^3 lb-ft). In the same arch, the roller A moves outwards 0.851 m (33.5 in.), and the deflection at C is 1.02 m (40.2 in.). These displacements are too high to be acceptable, indicating that a reduction of the cross section, from 0.4×1.2 m^2 in the beam to 0.4×0.4 m^2 in the arch (16×47 to 16×16 in.^{2}), is not feasible by only changing the beam to an arch.
2. When the horizontal displacement at A is prevented, by changing the roller to a hinge (Figure 1.26c), the bending moment at C is reduced to almost zero, 2.5 kN-m (1.9×10^3 lb-ft) and the deflection at C to 0.002 m (0.08 in.). At the same time, axial compressive forces are produced in the members of the arch. In addition to the vertical reaction components, the two-hinged arch has horizontal reaction components pointing inwards.
3. The reductions in deflection and in bending moments achieved by replacing the roller at A by a hinge can be partly achieved by a tie connecting the two supports. The amounts of the reduction depend upon the cross-sectional area and the modulus of elasticity of the tie. With the given data, the horizontal displacement at A is 0.021 m (0.83 in.) outwards; the bending moment at C is 38 kN-m (28×10^3 lb-ft), and the deflection at C is 0.027m (1.1 in.).
4. The forces in the top and bottom chords of the truss are compressive and tensile, respectively, with absolute values approximately equal to the bending moment values in the simple beam (Figure 1.26a) divided by the height of the truss. The absolute value of the force in any vertical member of the truss is almost the same as the absolute value of the shear at the corresponding section of the simple beam.
5. The deflection at C in the truss is also small, 0.043m (1.7 in.), compared to the structures in Figures 1.25a and b. This is so because of the absence of bending moment, which is generally the largest contributor to deflections in structures.
6. In Figure 1.25, a uniformly distributed load $q=20$ kN-m (1385 lb-ft) is idealized as a system of concentrated loads. To show that this idealization is satisfactory, we compare the deflection and the bending moment at C to the shearing force at a section just to the right of A in the simple beam (Figure 1.25a), when subjected to the actual distributed load or to the idealized concentrated loads:

For the uniform load:

$D_C = 0.037$ m (1.5 in.); $M_C = 1440$ kN-m (1050×10^3lb-ft); $V_{A_r} = 240$ kN (54×10^3 lb)

For the concentrated load:

$D_C = 0.037$ m (1.5 in.); $M_C = 1440$ kN-m (1050×10^3 lb-ft); $V_{A_r} = 200$ kN(45×10^3 lb)

We can see that the differences in results between the actual and the idealized loads are small and would vanish, as the spacing between the concentrated forces is reduced to zero. In the values given above, there is no difference in M_C, and the difference in D_C does not appear with two significant figures. At sections mid-way between the concentrated loads, the bending moment values due to uniform load, q exceeds the value due to concentrated loads by $qc^2/8$, with c being the distance between adjacent concentrated loads; this relatively small bending moment slightly increases D_C.

EXAMPLE 1.2: THREE-HINGED, TWO-HINGED, AND TOTALLY FIXED ARCHES

Figure 1.24c and Example 1.1 contain data for a two-hinged arch. The reaction components for this arch are given in Figure 1.26c. Figure 1.27a shows the corresponding results for the same arch but with an intermediate hinge inserted at C; similarly, Figure 1.27c represents the case for the same arch but with the hinged ends becoming encastré (totally fixed, i.e. translation and rotations are prevented). The values given in Figure 1.27 are in terms of load intensity q, span length l, and the height of the arch at mid-span, $h_C = l/6$.

The three-hinged arch in Figure 1.27a is statically determinate; this is discussed in Chapter 2, Example 2.11. Here, we discuss the results of the analysis shown in the figure. The bending moment at any section is equal to the sum of the moments of the forces situated to the left-hand side of the section about that section. Thus, the bending moments at A, D, E, and C are:

$$M_A = 0$$

$$M_D = \left(\frac{ql}{2} - \frac{ql}{12}\right)\frac{1}{6} - \frac{ql^2}{8h_c}\left(\frac{5}{9}h_C\right) = 0$$

$$M_E = \left(\frac{ql}{2} - \frac{ql}{12}\right)\frac{l}{3} - \frac{ql}{6}\left(\frac{1}{6}\right) - \frac{ql^2}{8h_C}\left(\frac{8}{9}h_C\right) = 0$$

$$M_C = \left(\frac{ql}{2} - \frac{ql}{12}\right)\frac{l}{2} - \frac{ql}{6}\left(\frac{1}{3}\right) - \frac{ql}{6}\left(\frac{1}{6}\right) - \frac{ql^2}{8h_C}h_c = 0$$

All the calculated M-values are zero, and the only internal force in the arch members is an axial force. This can be verified by the graphical construction shown on the right-hand side of Figure 1.27a. This is a force polygon in which $\overline{Oa}$ [$= ql^2/(8h_C)$] represents the horizontal reaction; $\overline{ad}$ is the resultant of the vertical forces to the left of D; $\overline{ac}$ those to the left of E; and $\overline{ab}$ those to the left of C. The vectors $\overline{Od}$, $\overline{Oc}$, and $\overline{Ob}$ represent the resultant forces in segments AD, DE, and EC, respectively. It can be verified that the slopes of the three force resultants are equal to the slopes of the corresponding arch segments. Thus, we conclude that this arch is subjected only to axial compression, with the shear force V, and the bending moment M equal to zero at all sections.

In fact, in setting this example, the geometry of the arch axis is chosen such that V and M are zero. This can be achieved by having the nodes of the arch situated on a second-degree parabola, whose equation is:

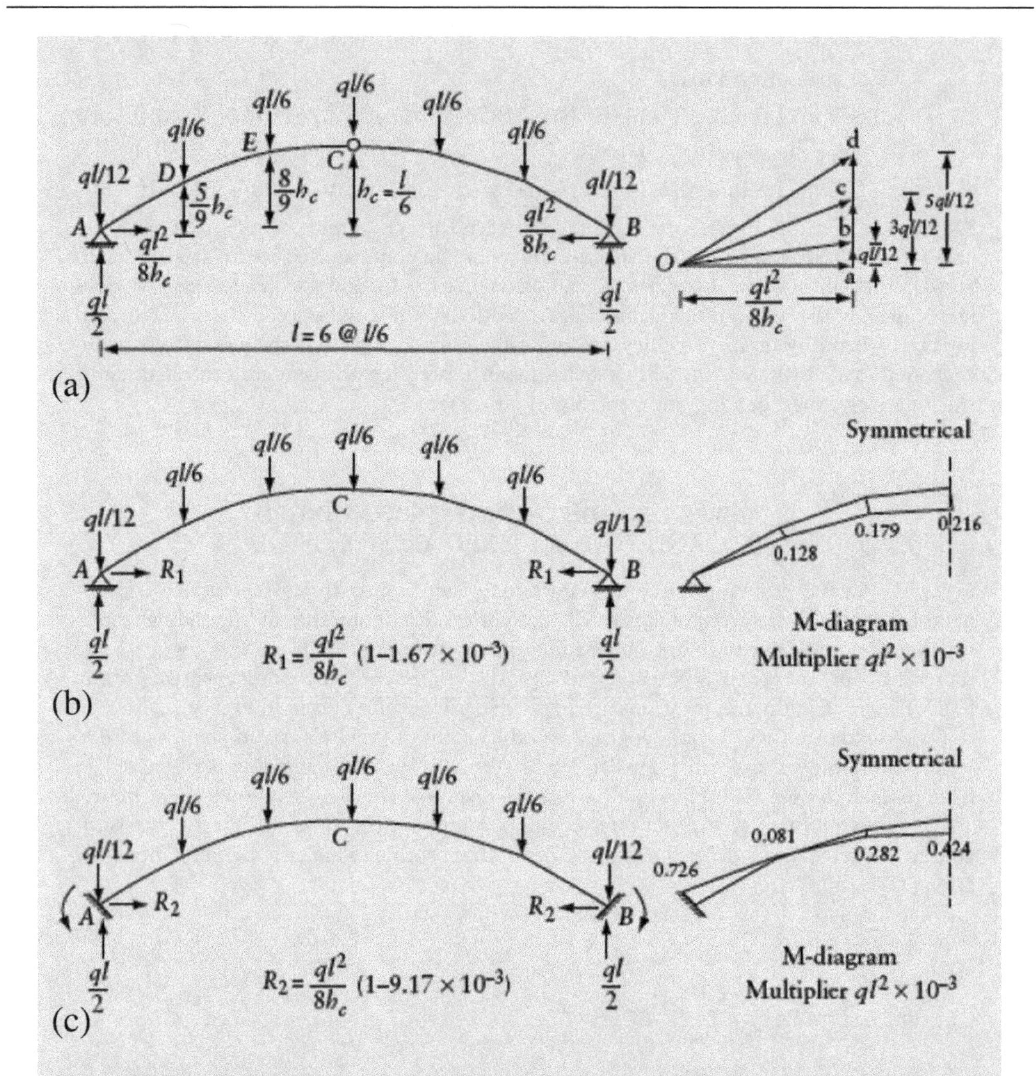

Figure 1.27 Polygonal arches with and without an intermediate hinge. Total load = ql (a) The arch in Figure 1.25c with an intermediate hinge introduced at C. (b) Two-hinged arch. (c) Arch with ends encastré.

$$h(x) = \frac{4x(l-x)}{l^2} h_C \tag{1.27}$$

where $h(x)$ is the height at any section; x is the horizontal distance between A and any node; h_C is the height of the parabola at mid-span. We can verify that each of the three-hinged arches in Figure 1.28, carrying a total load = ql, has a geometry that satisfies Equation 1.27 and has V and M equal to zero at all sections. For all the arches, the horizontal components of the reactions are inward and equal to $ql^2/(8\,h_C)$.

Elimination of the intermediate hinge at C (Figure 1.25c) results in a small change in the horizontal components of the reactions at the ends, and V and M become non-zero. However, the values of V and M in a two-hinged arch are small compared to the values in a simply supported beam carrying the same load (compare the ordinates of M-diagrams in Figure 1.27b with $ql^2/8$).

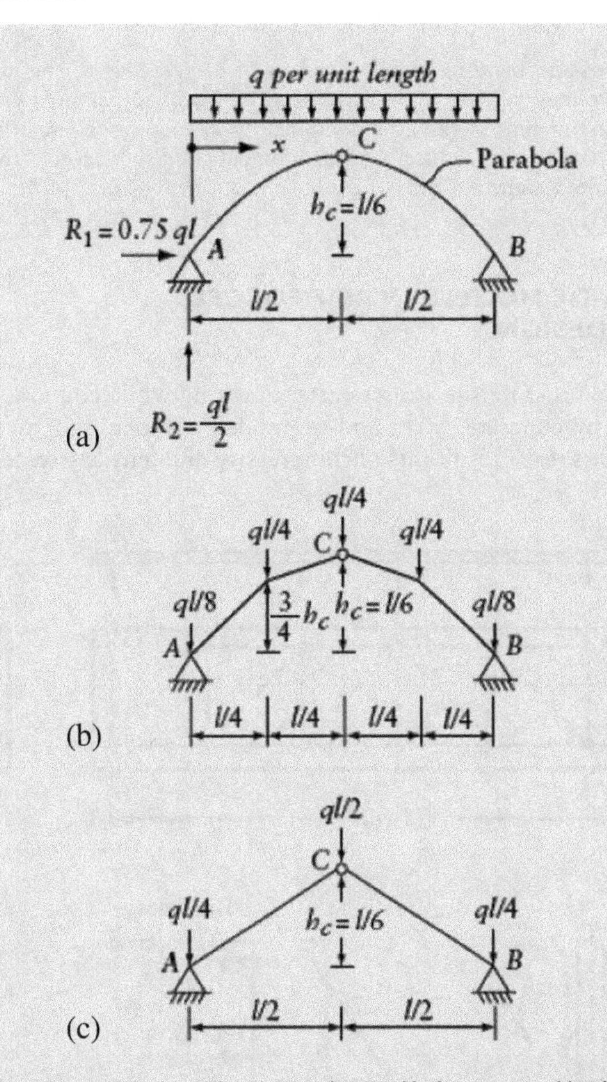

Figure 1.28 Alternative three-hinged arches with zero *M*, due to a total load *ql*.

Similarly, elimination of the hinge at C, combined with making the ends at A and B encastré, changes slightly the horizontal reaction component and produces moment components of the support reactions (Figure 1.27c). Again, we can see that the ordinates of the *M*-diagram in Figure 1.27c are small compared to the ordinates for the simply supported beam (Figure 1.26a). The values given in Figures 1.27b and c are calculated using the same values of E, l, h_C, q, and cross-sectional area as in Example 1.1.

From the above discussion, we can see that arches in which the horizontal reactions at the supports are prevented (or restrained) can transfer loads to the supports, developing small or zero bending moments. For this reason, arches are used to cover large spans, requiring smaller cross sections than beams. The geometry of the axis of a three-hinged arch can be selected such that V and M are zero. With the same geometry of the axis, the intermediate hinge can be eliminated (to simplify the construction), resulting in small bending and shear (much smaller than in a beam of the same span and load).

The arches considered in this example are subjected to a uniform gravity load. We have seen that the geometry of the arch axis can be selected such that the values of the shearing force and bending moment are small or zero due to the uniform load, which can be the major load on the structure. However, other load cases, such as wind pressure on one half of the arch combined with suction on the other half, and non-uniform gravity loads produce bending moments and shear forces that may have to be considered in design.

1.14 STRUT-AND-TIE MODELS IN REINFORCED CONCRETE DESIGN

Steel bars in concrete resist tensile stresses after cracking while compressive stresses continue to be resisted by concrete. Strut-and-tie models are plane (or occasionally space) trusses whose members resist resultants of compressive and tensile stresses. Figure 1.29a is

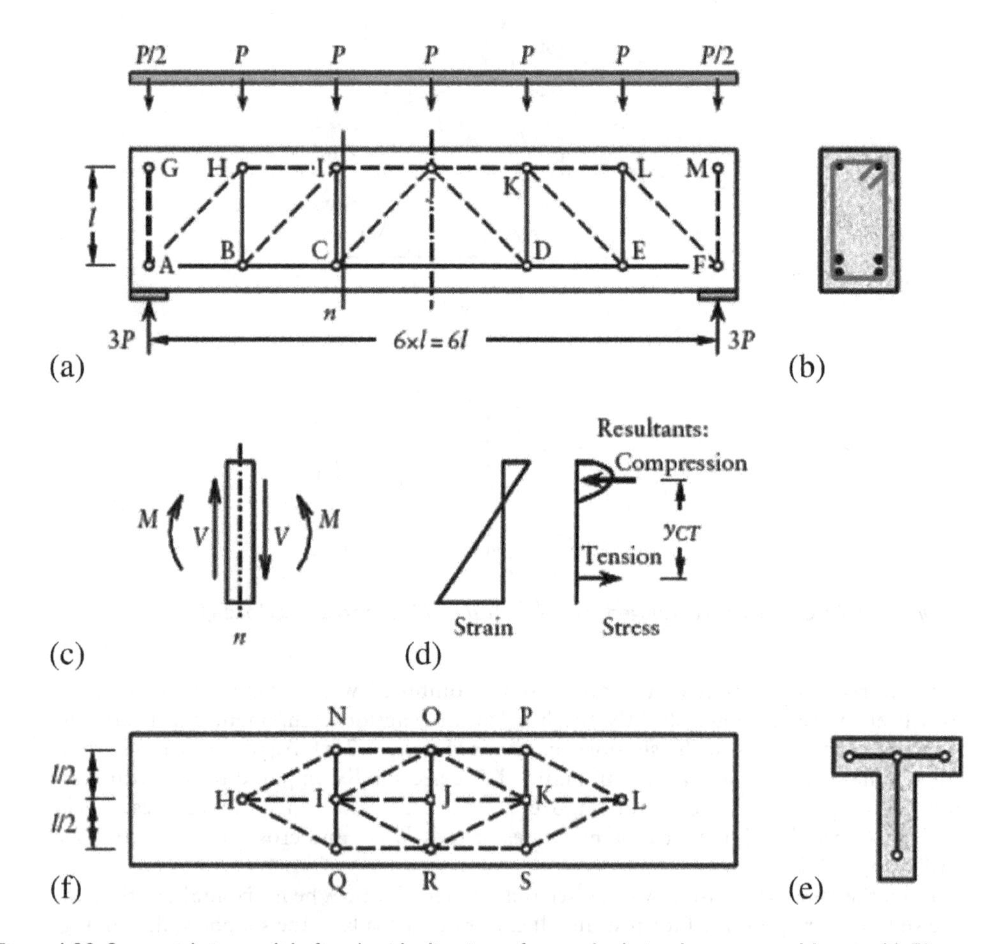

Figure 1.29 Strut-and-tie models for the idealization of a cracked simply supported beam. (a) Plane truss idealization of the beam with rectangular cross section. Also, elevation of a space truss idealization of the beam with T-cross section. (b) Rectangular cross section of beam. (c) Internal forces at section *n*. (d) Strain and stress distribution at *n*. (e) Cross section of spatial truss idealization of a T-beam. (f) Top view of spatial truss.

a strut-and-tie model (a plane truss) idealizing a cracked simple beam, carrying uniformly distributed load, and having rectangular cross section reinforced with longitudinal bars and stirrups (Figure 1.29b). At a typical section *n*, the stress resultants in the beam are a shearing force *V* and a bending moment *M* (Figure 1.29c). The strain and the stress distributions normal to the section are shown in Figure 1.29d. The top horizontal member IJ of the truss carries a compressive force whose absolute value = M/y_{CT}, where y_{CT} is the distance between the resultants of compressive and tensile stresses; *M* is the bending moment at C or I. The vertical member IC carries a tensile force equal to the shearing force *V*. The layout of a strut-and-tie model applies only for a specified load. A member in which the force is known to be zero is often not drawn (e.g. no member is shown connecting nodes G and H in Figure 1.29a); the struts and the ties are shown as dashed and continuous lines, respectively.

If the simple beam considered above has a T-section (Figure 1.29e) and is subjected to the same loading, it can be idealized as a space truss, the elevation and top views of which are shown in Figures 1.29e and f, respectively. Again, this space truss can become unstable with different loading; e.g. a vertical load at any of nodes N, O, P, Q, R, or S would cause partial collapse.

1.14.1 B- and D-regions

The design of a reinforced concrete member whose length is large compared to its cross-sectional dimensions is commonly based on Bernoulli's assumption: a plane section before deformation remains plane. The assumption permits the linear strain distribution shown in Figures 1.23c and 1.29d; also, for materials obeying Hooke's law (Equation 1.11), it permits the linear stress distribution (Equation 1.12) and the derivation of the remaining equations of Section 1.12. The validity of the assumption for members of any material is proven experimentally at all load levels for their major parts, referred to as B-regions, where B stands for "Bernoulli." The assumption is not valid at D-regions, where D stands for "disturbance" or "discontinuity" regions, e.g. at concentrated loads or reactions, sudden changes of cross sections, and openings. The major use of the strut-and-tie models is in the design and the detailing of the reinforcement in the D-regions. The design of the B-regions is mainly done by code equations (some of which are based on truss idealization), without the need to analyze strut-and-tie models. Figure 1.30 shows examples of strut-and-tie models for D-regions representing: (a) and (b), a wall subjected, respectively, to a uniform load and to a concentrated load; (c) a member end with anchorage of a prestressing tendon; (d) a connection of a column and a beam; (e) a tall wall supporting an eccentric load; (f) a beam with a sudden change in depth adjacent to a support; (g) a corbel; and (h) a pile cap (spatial model).

In each application, the strut-and-tie model is an isolated part of a structure subjected, at its boundaries, to a self-equilibrated set of forces of known magnitudes and distributions. The models in Figure 1.30 are statically determinate; the forces in the members can be determined without the need of knowing the *Ea* values (the axial rigidities) of the members. The analysis gives the forces in the ties and hence determines the cross-sectional area of the steel bars for a specified stress. The struts and ties meet at nodes, whose sizes must satisfy empirical rules to ensure that the forces of the struts and the ties do not produce crushing of concrete or failure of the anchorage of the ties.

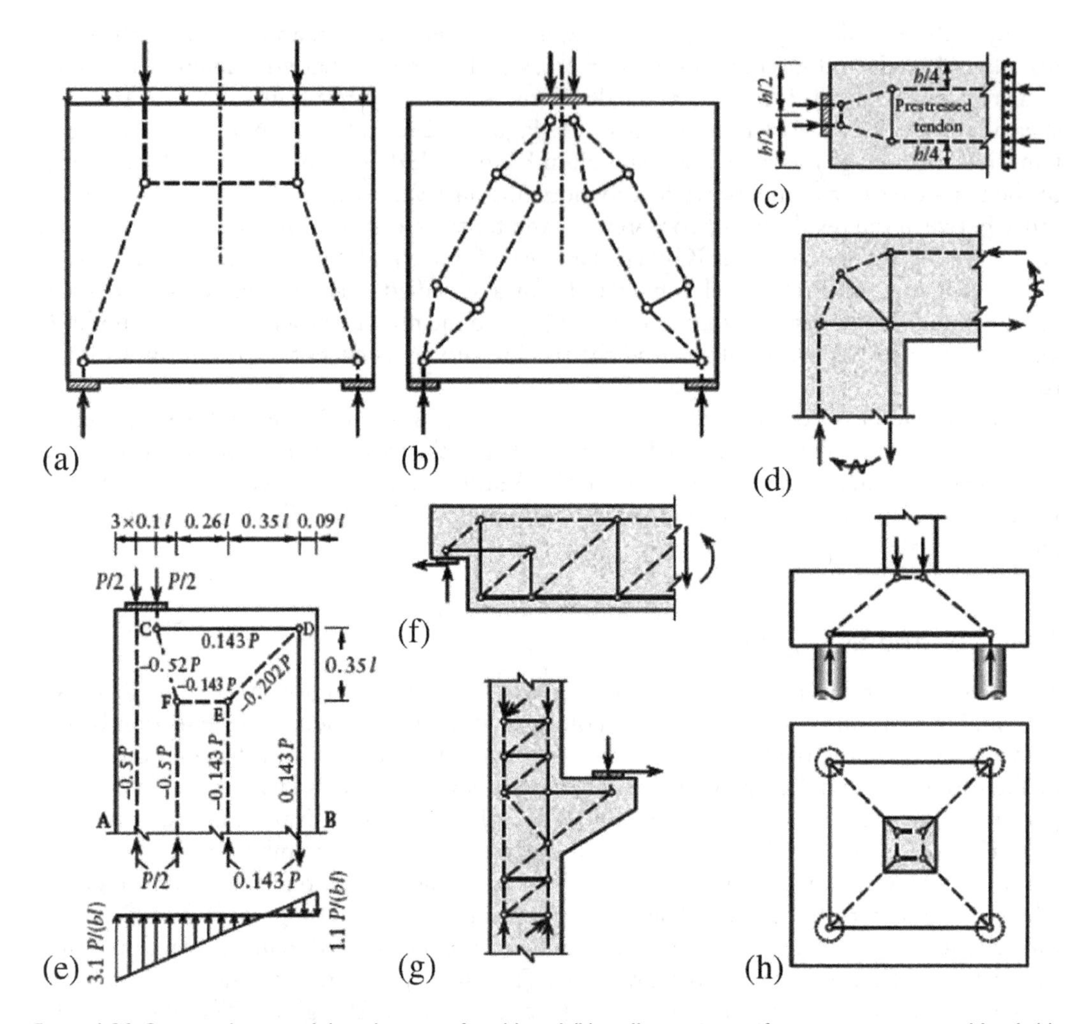

Figure 1.30 Strut-and-tie model applications for: (a) and (b) wall carrying uniform or concentrated load; (c) anchorage of a prestressing tendon; (d) connection of a column with a beam; (e) tall wall supporting a column; (f) sudden change in depth of a beam adjacent to its support; (g) corbel; (h) spatial strut-and-tie model for a pile cap.

EXAMPLE 1.3: STRUT-AND-TIE MODEL FOR A WALL SUPPORTING AN ECCENTRIC LOAD

The wall in Figure 1.30e, of thickness b, is subjected to the eccentric load P at its top. Find the forces in the members of the strut-and-tie model shown.

The strain and stress distributions at section AB can be assumed linear, because the section is far from the D-region. Equation 1.25 gives the stress distribution shown in Figure 1.30e, which is statically equivalent to the four vertical forces shown at AB. It can be verified that the forces shown on the figure maintain equilibrium of each node.

It is noted that the truss is stable only with the set of forces considered. The trapezoidal part CDEF is a mechanism that can be made stable by adding a member EC (or DF); the force in the added member would be nil and the pattern of member forces would be unchanged. The distribution of stress at section AB helps in the layout of the members of the strut-and-tie model. Determination of the stress distribution is not needed in this example because the layout of the strut-and-tie model is given.

1.14.2 Statically indeterminate strut-and-tie models

Analysis of a strut-and-tie model gives a system of forces in equilibrium, without consideration of compatibility. The forces are used in a plastic design that predicts the ultimate strength of the structure and verifies that it is not exceeded by the factored load. With statically determinate models (e.g. the models in Figure 1.30), the members are drawn as centerlines; the axial rigidity, Ea of the members is not needed in the analyses of the forces that they resist. The value of Ea for each member has to be assumed when the analysis is for a statically indeterminate strut-and-tie model. It is also possible to analyze and superimpose the member forces of two or more statically determinate models to represent an indeterminate model. Each determinate model carries a part of the applied loads.

The indeterminate truss in Figure 1.31a can be treated as a combination of two determinate trusses, one composed of members A and B (Figure 1.31b) and the other composed of members A and C (Figure 1.31c). If we assume that the first and the second trusses carry $0.4P$ and $0.6P$, respectively, the forces in the members in the first truss will be $N_A = 0.8P$; $N_B = -0.89P$; in the second truss, $N_A = 0.6P$; $N_C = -0.85P$. Superposition of forces gives: $N_A = 1.4P$; $N_B = -0.89P$; $N_C = -0.85P$ (Figure 1.31a). The solution in Figure 1.31a satisfies equilibrium but does not satisfy the compatibility required in elastic analysis of structures; the loaded node in the trusses in Figures 1.31b and c exhibits non-compatible translations.

The statically indeterminate spatial strut-and-tie model shown in elevation and top view in Figures 1.29e and f, respectively, can be analyzed by summing up the statically determinate forces in the spatial model in absence of members HI, IJ, JK, and KL to the forces in a plane model composed only of the members shown in Figure 1.29a; the applied load has to be partitioned on the two models.

Plastic strength analysis of elastic, perfectly plastic materials is permissible without considering compatibility. However, because plastic deformation of concrete is limited, strut-and-tie models that excessively deviate from elastic stress distribution may overestimate the strength.

Elastic finite-element analysis (Chapter 12) of a D-region can indicate the directions of principal compressive and principal tensile stresses. These should be the approximate directions of the struts and the ties, respectively, in the strut-and-tie models. However, an elastic analysis may be needed only in exceptional cases. The layout of the struts and ties for D-regions of frequent occurrence (e.g. the models in Figure 1.30) and guides on the choice of the models and reinforcement details are available.[2–4]

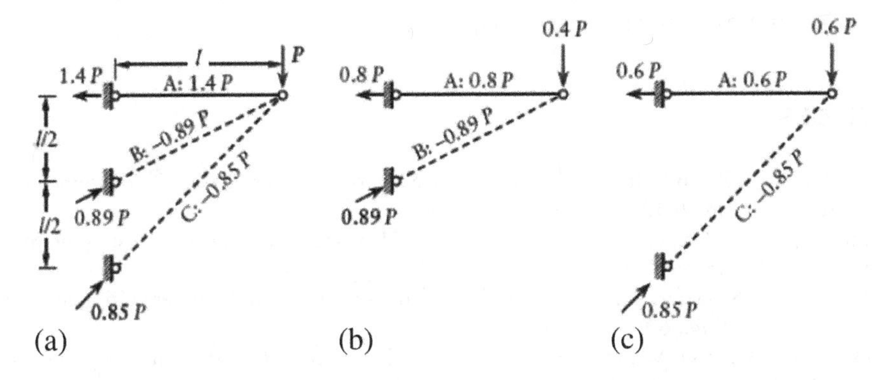

Figure 1.31 Statically indeterminate strut-and-tie model, (a) treated as a combination of two determinate trusses (b) and (c).

1.15 STRUCTURAL DESIGN

Structural analysis gives the deformations, the internal forces, and the reactions that are needed in structural design. The loads that we consider are the probable loads specified by codes or assumed by the designer. The members, their connections, and their supports are then designed following code requirements specifying minimum strength or maximum allowable deformations; these components must have a resistance (strength) that exceeds the load effect. Because of the variability of material strength or the differences between the actual dimensions from those specified by the designer, *strength-reduction factors* (also called "resistance factors") less than 1.0 apply to the computed strength. The loads that the structure must carry are also variables. Thus, *load factors* greater than 1.0 are applied to the expected loads. Codes specify values of the strength-reduction and the load factors to account for some of the uncertainties; the difference between a factor and 1.0 depends upon the probable variance of the actual values from the design values. The calculations that we use for member resistance and for structural analysis are based on simplifying assumptions that are uncertain.

Thus, because of the uncertainties, it is not possible to achieve zero probability of failure. In design codes, the resistance and the load factors are based on statistical models that assume that the chance of a combination of understrength and overload that produces collapse, or failure that the structure performs as required, is at the predicted target level.

1.16 GENERAL

In presenting some of the concepts, it has been necessary to discuss above, without much detail, some of the subjects covered in the following chapters. These include static indeterminacy of structures and calculation of the reactions and internal forces in statically determinate structures. Chapter 2 deals with the analysis of statically determinate structures. After studying Chapter 2, it may be beneficial to review parts of Chapter 1 and attempt some of its problems, if not done so earlier.

Cover Picture: St. Patrick's Bridge, Calgary, AB, Canada runs continuous over three spans. Similar structural system can be a single-span steel arch, with post-tensioned concrete tie serving as a pedestrian bridge. Elastically restrained rollers support the arches. At a higher level, the tie extends a short distance beyond the span. Two steel struts connect tie end to two bearings. The prestressing force limits the long-term movement of the roller to a small value. The choice of the prestressing force considers creep and shrinkage of concrete and relaxation of prestressed steel (see Chapter 6).

REFERENCES

1. Ghali, A., Favre, R. and Elbadry, M., *Concrete Structures: Stresses and Deformations*, 4th ed., CRC Press, London, 2012, 637 pp.
2. Schlaich, J., Schäefer, K. and Jennewein, M., "Toward a Consistent Design of Structural Concrete," *J. Prestressed Concr. Inst.*, 32(3) (1987), pp. 74–150.
3. Schlaich, J. and Schäefer, K., "Design and Detailing of Structural Concrete Using Strut-and-Tie Models," *Struct. Eng.*, 69(6) (1991), 13 pp.
4. MacGregor, J.G. and Wight, J.K., *Reinforced Concrete: Mechanics and Design*, 4th ed., Prentice Hall, Upper Saddle River, NJ, 2005, 1132 pp.

PROBLEMS

1.1 Sketch the deflected shape and the bending moment diagram for the structures shown. In the answers at the back of the book, the axial and shear deformations are ignored and the second moment of area, I, is considered constant for each structure; the rotations and member end moments are considered positive when clockwise; u and v are translations horizontal to the right and downward, respectively.

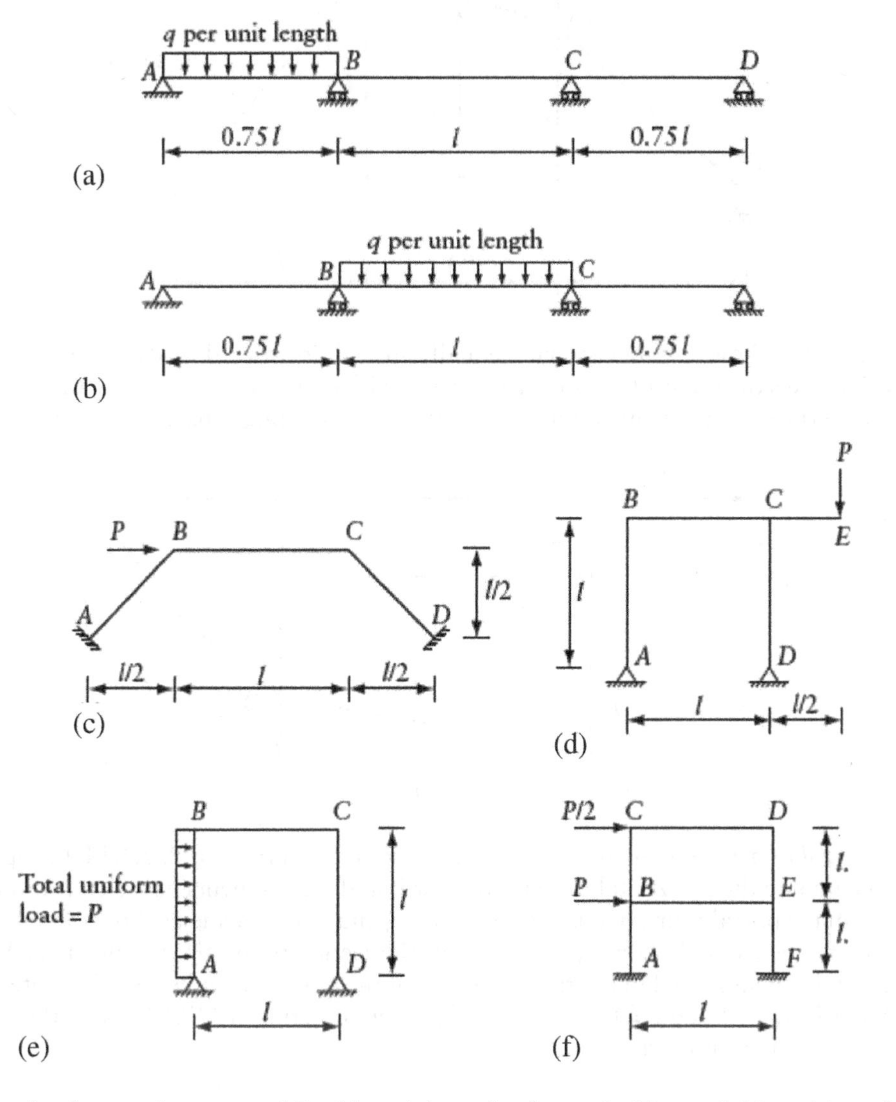

1.2 Apply the requirements of Problem 1.1 to the frame in Figure 1.13c subjected to a downward settlement δ of support D, with no load applied.

1.3 A rectangular section of width b and height h is subjected to a normal tensile force, N, at the middle of the top side. Find the stress distribution by the use of Equation 1.14, with the reference point at the top fiber. Verify the answer by changing the reference point to the centroid.

1.4 For the beam in Figure 1.20a, verify that the deflection at mid-span is $y = \psi_{free} l^2/8$ and the rotations at A and B are equal to $(dy/dx)_A = \psi_{free} l/2 = -(dy/dx)_B$; where ψ_{free} is given by Equation 1.7.

1.5 Select the dimensions h_1 and h_2 such that the three-hinged arch shown has no bending moment or shear force at all sections.

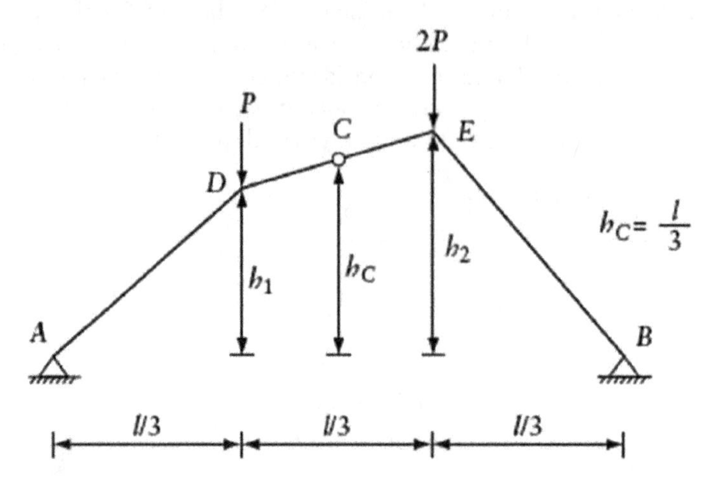

1.6 The figure shows a cable carrying equally spaced downward loads, *P*. Verify that the cable takes the shape of a funicular polygon with $h_D/h_C = 0.75$. What are the horizontal and vertical components of the tensile force in each straight part of the cable?

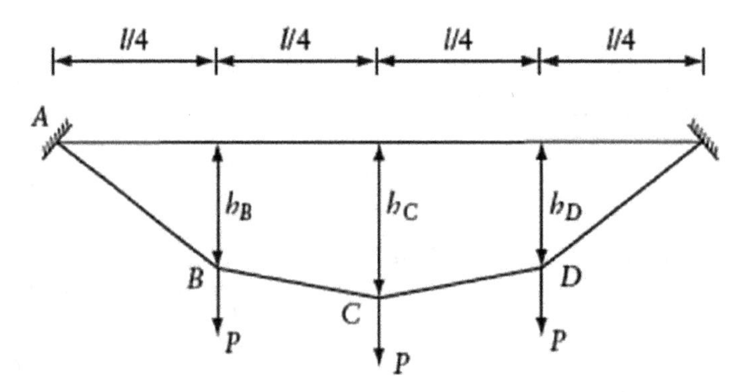

1.7 *Control of drift of a portal frame:* Use the computer program PLANEF (Chapter 20) to compare the horizontal displacement at B and the moments at the ends of member AB in the portal frame in (a), without bracing and with bracing as shown in (b) and (c). All members have hollow square sections; the section properties for members AB, BC, and CD are indicated in (d); the properties of the bracing members are indicated in (e). Enter $l = 1$, $P = 1$ and $E = 1$; give the displacement in terms of $Pl^3/(EI)_{AB}$ and the bending moment in terms of Pl.

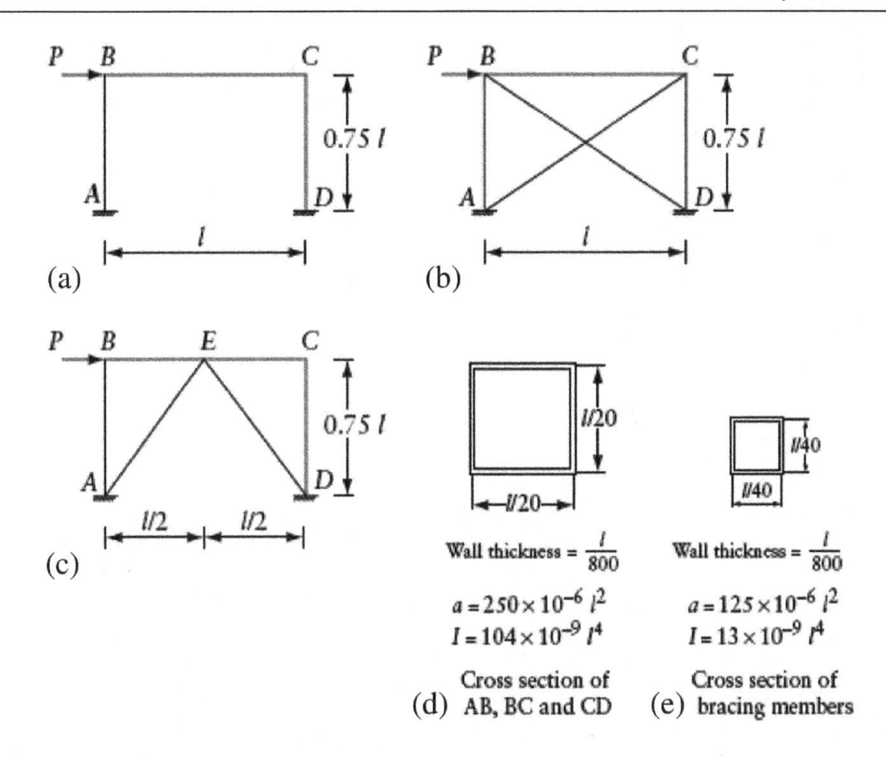

1.8 Use the computer program PLANET to find the horizontal displacement at B and the axial forces in the members treating the structure of Problem 1.7 part (b) as a plane truss. Compare with the following results obtained by treating the structure as a plane frame: Horizontal deflection at B $= 0.968 \times 10^{-3}\ Pl^3/(EI)_{AB}$; $\{N_{AB}, N_{BC}, N_{CD}, N_{AC}, N_{BD}\} = P\{0.411, -0.447, -0.334, 0.542, -0.671\}$.

1.9 Find the forces in the spatial strut-and-tie model shown in elevation in Figure 1.29a and in top view in Figure 1.29f. Omit or assume to be zero forces in the members HI, IJ, JK, and KL.

Chapter 2

Statically determinate structures

2.1 INTRODUCTION

A large part of this book is devoted to the modern methods of analysis of framed structures, that is, structures consisting of members, which are long in comparison to their cross section. Typical framed structures are beams, grids, plane, and space frames or trusses (see Figure 2.1). Other structures, such as walls and slabs, are considered in Chapters 11 and 15.

In all cases, we deal with structures in which displacements – translation or rotation of any section – vary linearly with the applied forces. In other words, any increment in displacement is proportional to the force causing it. All deformations are assumed to be *small*, so that the resulting displacements do not significantly affect the geometry of the structure and, hence, do not alter the forces in the members. Under such conditions, stresses, strains, and displacements due to different actions can be added using the principle of superposition; this topic is dealt with in Section 3.6. The majority of actual structures are designed so as to undergo only small deformations, and they deform linearly. This is the case with metal structures; the material obeys Hooke's law; concrete structures are also usually assumed to deform linearly. We are referring, of course, to behavior under working loads, that is, to elastic analysis; plastic analysis is considered in Chapters 14 and 15.

It is, however, possible for a straight structural member made of a material obeying Hooke's law to deform nonlinearly when the member is subjected to a lateral load and to a large axial force. This topic is dealt with in Chapters 10 and 18. Chapter 18 also discusses nonlinear analysis when the material does not obey Hooke's law.

Although statical indeterminacy will be dealt with extensively in the succeeding chapters, it is important at this stage to recognize the fundamental difference between statically determinate and indeterminate (hyperstatic) structures, in that the forces in the latter cannot be found from the equations of static equilibrium alone: a knowledge of some geometric conditions under load is also required.

The analysis of statically indeterminate structures generally requires the solution of linear simultaneous equations. The number of equations depends on the method of analysis. Some methods avoid simultaneous equations by using iterative or successive correction techniques in order to reduce the amount of computation and are suitable when the calculations are made by hand or by calculator.

For large and complicated structures, hand computation is often impracticable and a computer has to be used. The advent of the computer has shifted the emphasis from easy problem solution to efficient problem formulation: using matrices and matrix algebra,

DOI: 10.1201/9780429286858-2

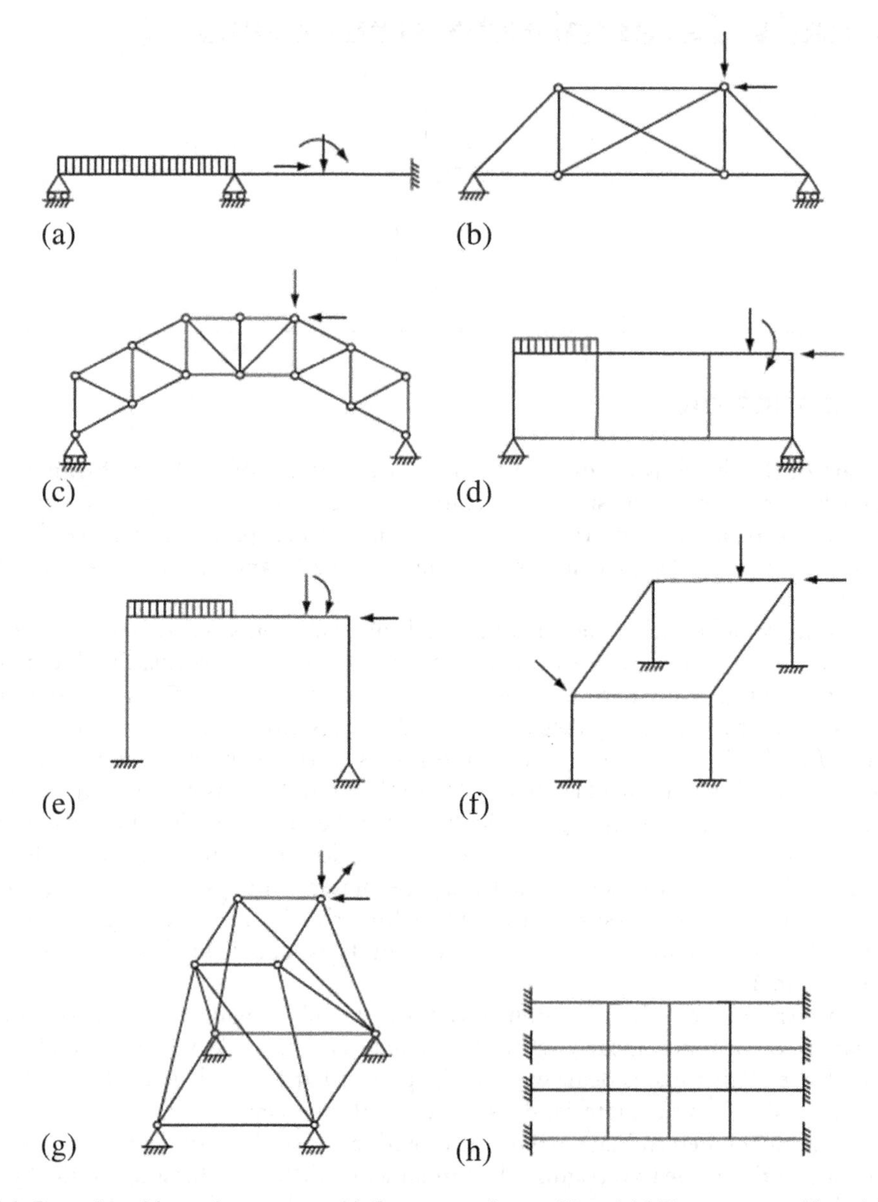

Figure 2.1 Examples of framed structures. (a) Continuous beam. (b) and (c) Plane trusses. (d) and (e) Plane frames. (f) Space frame. (g) Space truss. (h) Horizontal grid subjected to vertical loads.

a large quantity of information can be organized and manipulated in a compact form. For this reason, in many cases, equations in this book are written in matrix form. In the text, the basic computer methods are discussed, but details of programming are not given.

We should emphasize that the hand methods of solution must not be neglected. They are of value not only when a computer is not available but also for preliminary calculations and for checking of computer results. The remaining sections of this chapter are concerned with the analysis of statically determinate structures.

2.2 EQUILIBRIUM OF A BODY

Figure 2.2a represents a body subjected to forces $F_1, F_2, \dots, F_n$ in space. The word *force* in this context means either an action of a concentrated load or a couple (a moment); in the latter case the moment is represented by a double-headed arrow.[1] A typical force F_i acting at point (x_i, y_i, z_i) is shown in Figure 2.2b using a right-handed system[2] of orthogonal axes x, y, and z. Components of F_i in the direction of the force axes are

$$F_{ix} = F_i \lambda_{ix}; \quad F_{iy} = F_i \lambda_{iy}; \quad F_{iz} = F_i \lambda_{iz} \tag{2.1}$$

where F_i is force magnitude; λ_{ix}, λ_{iy}, and λ_{iz} are called *direction cosines* of the force F_i; they are equal to the cosine of the angles α, β, and γ between the force and the positive x, y, and z directions, respectively. Note that the components F_{ix}, F_{iy}, and F_{iz} do not depend upon the position of the point of application of F_i. If the length of $\overline{ID}$ is equal to the magnitude of F_i, projections of $\overline{ID}$ on the x, y, and z directions represent the magnitudes of the components F_{ix}, F_{iy}, and F_{iz}.

The moment of a concentrated load F_i about axes x, y, and z (Figure 2.2b) is equal to the sum of moments of the components F_{ix}, F_{iy}, and F_{iz}; thus:

$$M_{ix} = F_{iz} y_i - F_{iy} z_i; \quad M_{iy} = F_{ix} z_i - F_{iz} x_i; \quad M_{iz} = F_{iy} x_i - F_{ix} y_i \tag{2.2}$$

Note that a component in the direction of one of the axes, say x, produces no moment about that axis (x axis). The positive sign convention for moments used in Equation 2.2 is shown in Figure 2.2b. Naturally, Equation 2.2 applies only when F_i represents a concentrated load, not a moment. When F_i represents a moment (e.g. F_2 in Figure 2.2a), Equation 2.1 can be used to determine three moment components to be represented by double-headed arrows in the x, y, and z directions (not shown in figure).

The resultant of a system of forces in space has six components, determined by summing the components of individual forces, e.g.:

$$F_{x \text{ resultant}} = \sum_{i=1}^{n} F_{ix}; \quad M_{x \text{ resultant}} = \sum_{i=1}^{n} M_{ix} \tag{2.3}$$

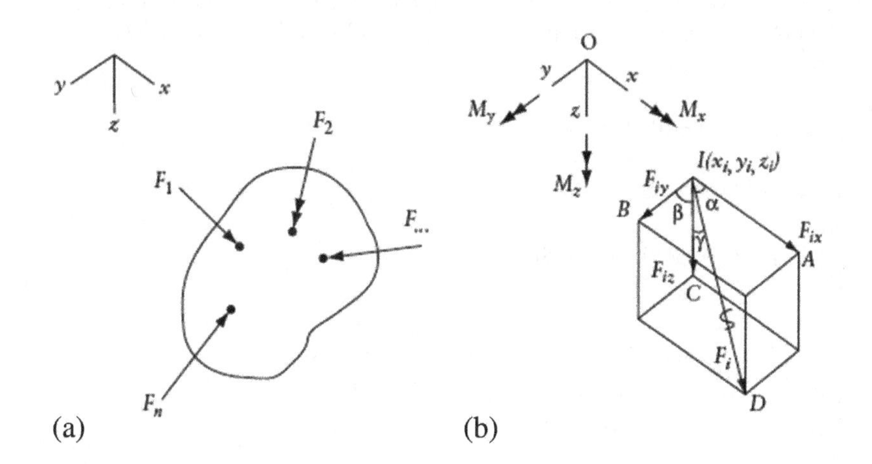

Figure 2.2 Force system and force components. (a) Body subjected to forces in space. (b) Components of a typical force and positive sign convention for M_x, M_y, and M_z.

For a body in equilibrium, components of the resultant in the x, y, and z directions must vanish so that the following equations apply:

$$\left.\begin{aligned} &\sum F_x = 0; \ \sum F_y = 0; \ \sum F_z = 0 \\ &\sum M_x = 0; \ \sum M_y = 0; \ \sum M_z = 0 \end{aligned}\right\} \quad (2.4)$$

The summation in these equations is for all the components of the forces and of the moments about each of the three axes. Thus, for a body subjected to forces in three dimensions, six equations of static equilibrium can be written. When all the forces acting on the free body are in one plane, only three of the six equations of statics are meaningful. For instance, when the forces act in the xy plane, these equations are:

$$\Sigma F_x = 0; \ \Sigma F_y = 0; \ \Sigma M_z = 0 \quad (2.5)$$

When a structure in equilibrium is composed of several members, the equations of statics must be satisfied when applied to the structure as a whole. Each member, joint, or portion of the structure is also in equilibrium, and the equations of statics must also be satisfied.

The equilibrium Equations 2.4 and 2.5 can be used to determine reaction components or internal forces, provided that the number of unknowns does not exceed the number of equations. In trusses with pin-connected members and forces applied only at the joints, the members are subjected to axial forces only; thus, for a truss joint, equations expressing equilibrium of moments included in Equations 2.4 and 2.5 are trivial, but they can be applied to a truss part to determine member forces (see Example 2.5).

EXAMPLE 2.1: REACTIONS FOR A SPATIAL BODY: A CANTILEVER

The prismatic cantilever in Figure 2.3a is subjected, in the plane of cross section at the end, to forces $F_1 = P$, $F_2 = 2Pb$, as shown. Determine components at O of the resultant reaction at the fixed end; point O is at the center of the cross section.

Assume positive directions of reaction components the same as those of the x, y, and z axes. Coordinates of point of application of F_1 are $(3b, 0.5b, -0.75b)$. Direction cosines of F_1 are:

$$\{\lambda_{1x}, \ \lambda_{1y}, \ \lambda_{1z}\} = \{0, \ 0.5, \ 0.866\}$$

Application of Equations 2.1 and 2.2 gives:

$$\{F_{1x}, \ F_{1y}, \ F_{1z}\} = P\{0, \ 0.5, \ 0.866\}$$

$$\begin{Bmatrix} M_{1x} \\ M_{1y} \\ M_{1z} \end{Bmatrix} = Pb \begin{Bmatrix} 0.866 \times 0.5 - 0.5 \times (-0.75) \\ -0.866 \times 3 \\ 0.5 \times 3 \end{Bmatrix} = Pb \begin{Bmatrix} 0.808 \\ -2.596 \\ 1.500 \end{Bmatrix}$$

The applied moment, F_2, has only one component: $M_{2y} = -2Pb$. The equilibrium Equation 2.4 give components of the reaction at O:

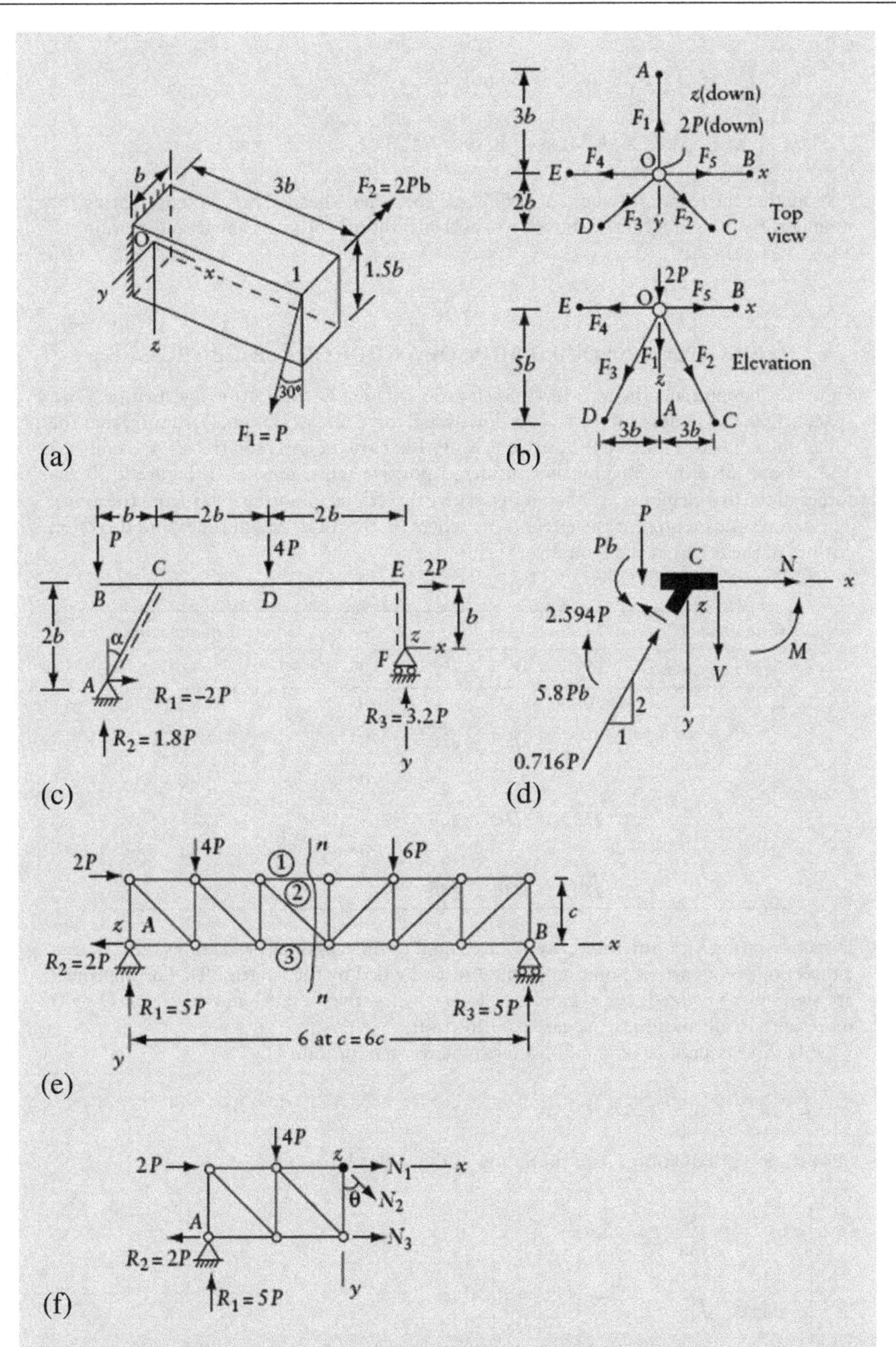

Figure 2.3 Examples 2.1 to 2.5 of uses of equilibrium equations. (a) Space cantilever. (b) Joint of space truss. (c) Plane frame. (d) Joint of a plane frame. (e) Plane truss. (f) Part of plane truss.

$$\{F_{Ox},\ F_{Oy},\ F_{Oz}\} = P\{0,\ -0.5,\ -0.866\}$$

$$\{M_{Ox},\ M_{Oy},\ M_{Oz}\} = Pb\{-0.808,\ 4.598,\ -1.5\}$$

Note that the reactions would not change if the double-headed arrow, representing the moment, F_2 in Figure 2.3a, is moved to another position without change in direction.

EXAMPLE 2.2: EQUILIBRIUM OF A NODE OF A SPACE TRUSS

Figure 2.3b represents the top view and elevation of joint O connecting five members in a space truss. The joint is subjected to a downward force, $2P$, as shown. Assuming that the forces in members 4 and 5 are $F_4 = 2P$, $F_5 = -P$, what are the forces in the other members?

An axial force in a member is considered positive when tensile. In Figure 2.3b the forces in the five members are shown in positive directions (pointing away from the joint). The arrows thus represent the effect of members on the joint. Magnitudes and direction cosines of the forces at the joint are:

Force number	*1*	*2*	*3*	*4*	*5*	*External force*
Force magnitude	F_1	F_2	F_3	$2P$	$-P$	$2P$
λ_x	0	$\frac{3}{\sqrt{38}}$	$\frac{-3}{\sqrt{38}}$	–1	1	0
λ_y	$\frac{-3}{\sqrt{38}}$	$\frac{2}{\sqrt{38}}$	$\frac{2}{\sqrt{38}}$	0	0	0
λ_z	$\frac{5}{\sqrt{38}}$	$\frac{5}{\sqrt{38}}$	$\frac{5}{\sqrt{38}}$	0	0	1

Direction cosines for any force, say F_3, are equal to the length of projections on x, y, and z directions of an arrow pointing from O to D divided by the length OD. The appropriate sign must be noted; for example, λ_x for F_3 is negative because moving from O to D represents an advance in the negative x direction.

Only three equations of equilibrium can be written for joint O:

$$\Sigma F_x = 0;\ \Sigma F_y = 0;\ \Sigma F_z = 0$$

Substitution in Equation 2.1 of the values in the table above gives:

$$\begin{bmatrix} 0 & \frac{3}{\sqrt{38}} & -\frac{3}{\sqrt{38}} \\ -\frac{3}{\sqrt{34}} & \frac{2}{\sqrt{38}} & \frac{2}{\sqrt{38}} \\ \frac{5}{\sqrt{34}} & \frac{5}{\sqrt{38}} & \frac{5}{\sqrt{38}} \end{bmatrix} \begin{Bmatrix} F_1 \\ F_2 \\ F_3 \end{Bmatrix} = P \begin{bmatrix} 3 \\ 0 \\ -2 \end{bmatrix}$$

The right-hand side of this equation is equal to minus the sum of the components of the externally applied force at O and the components of the known member forces F_4 and F_5. The solution is:

$$\{F_1,\ F_2,\ F_3\} = P\{-0.9330,\ 2.3425,\ -3.8219\}$$

The member forces or the reactions obtained from any source can be verified by considering joint equilibrium, as done above. As an example, we verify below the equilibrium of node D of the space truss in Problem 2.2. Given that the forces in members DA and DB are: $\{N_{DA}, N_{DB}\} = \{-4.70, -3.14\}P$; reaction components at D are: $\{R_x, R_y, R_z\}_D = \{-1.78, 0.61, -7.39\}P$.

$$\begin{bmatrix} \begin{Bmatrix} \lambda_x \\ \lambda_y \\ \lambda_z \end{Bmatrix}_{DA} & \begin{Bmatrix} \lambda_x \\ \lambda_y \\ \lambda_z \end{Bmatrix}_{DB} \end{bmatrix} \begin{Bmatrix} N_{DA} \\ N_{DB} \end{Bmatrix} = -\begin{Bmatrix} R_x \\ R_y \\ R_z \end{Bmatrix}_D ;$$

$$\frac{1}{2.121}\begin{bmatrix} -0.683 & -0.183 \\ -0.183 & 0.683 \\ -2.000 & -2.000 \end{bmatrix} \begin{Bmatrix} -4.70 \\ -3.14 \end{Bmatrix} = -\begin{Bmatrix} -1.78 \\ 0.61 \\ -7.39 \end{Bmatrix} P$$

EXAMPLE 2.3: REACTIONS FOR A PLANE FRAME

Determine the reaction components for the plane frame shown in Figure 2.3c.

Select x, y, and z axes as shown and apply Equation 2.5:

$$\Sigma F_x = 0;\ R_1 + 2P = 0$$

$$\Sigma M_z = 0;\ -R_1 b + R_2(5b) - P(5b) - 4P(2b) + 2P(b) = 0$$

$$\Sigma F_y = 0\ ; -R_2 - R_3 + P + 4P = 0$$

The first of the above three equations gives the value of R_1, which, when substituted in the second equation, allows R_2 to be determined. Substitution of R_2 in the third equation gives R_3. The answers are: $R_1 = -2P$; $R_2 = 1.8P$; $R_3 = 3.2P$.

It is good practice to check the answers before using them in design or in further analysis. In this problem, we can verify that $\Sigma M_z = 0$ with the z axis at a different point, e.g. point A. Note that this does not give a fourth equation which could be used to determine a fourth unknown; this is so because the fourth equation can be derived from the above three.

EXAMPLE 2.4: EQUILIBRIUM OF A JOINT OF A PLANE FRAME

Figure 2.3d represents a free body diagram of a joint of a plane frame (e.g. joint C of the frame in Figure 2.3c). The arrows represent the forces exerted by the members on the joint. The forces at sections just to the left and just below the joint C are given. Use equilibrium equations to determine unknown internal forces N, V and M representing normal force, shearing force and bending moment at the section just to the right of C. (Diagrams plotting variation of N, V and M over the length of members of plane frames are discussed in Section 2.3, but here we determine internal forces in only one section.)

Equilibrium Equation 2.5 applies:

$$\sum F_x = 0; \quad N + 0.716P\left(\frac{1}{\sqrt{5}}\right) + 2.594P\left(\frac{-2}{\sqrt{5}}\right) = 0$$

$$\sum F_y = 0; \quad V + P + 2.594P\left(\frac{-1}{\sqrt{5}}\right) + 0.716P\left(\frac{-2}{\sqrt{5}}\right) = 0$$

$$\sum M_z = 0; \quad -M + 5.8Pb - Pb = 0$$

This gives $N = 2P$; $V = 0.8P$; $M = 4.8Pb$.

EXAMPLE 2.5: FORCES IN MEMBERS OF A PLANE TRUSS

Find the forces in members 1, 2, and 3 of the plane truss shown in Figure 2.3e by application of equilibrium Equation 2.5 to one part of the structure to the right or to the left of section $n - n$.

First, we determine the reactions by equilibrium Equation 2.5 applied to all external forces including the reactions: $\Sigma F_x = 0$ gives $R_2 = 2P$; $\Sigma M_z = 0$ gives $R_3 = 5P$; $\Sigma F_y = 0$ gives $R_1 = 5P$.

The part of the structure situated to the left of section $n - n$ is represented as a free body in equilibrium in Figure 2.3f. The forces N_1, N_2, and N_3 in the cut members, shown in their positive directions, are in equilibrium with the remaining forces in the figure. The unknown member forces N_1, N_2, and N_3 are determined by the equilibrium Equation 2.5:

$$\Sigma M_z = 0 \quad ; -N_3c + 5P(2c) + 2Pc - 4Pc = 0; \quad \text{thus} \quad N_3 = 8P$$

$$\Sigma F_y = 0 \quad ; \quad N_2 \cos q - 5P + 4P = 0; \quad \text{thus} \quad N_2 = \surd 2P$$

$$\Sigma F_x = 0 \quad ; \quad N_1 + N_2 \sin q + N_3 - 2P + 2P = 0; \quad \text{thus} \quad N_1 = -9P$$

We may wish to verify that the same results will be reached by considering equilibrium of the part of the structure situated to the right of section $n - n$.

2.3 INTERNAL FORCES: SIGN CONVENTION AND DIAGRAMS

As mentioned earlier, the purpose of structural analysis is to determine the reactions at the supports and the internal forces (the stress resultants) at any section. In beams and plane frames, in which all the forces on the structure lie in one plane, the resultant of stresses at any section has generally three components: an axial force N, a shearing force V, and a bending moment M. The positive directions of N, V, and M are shown in Figure 2.4c, which represents an element (DE) between two closely spaced sections of the horizontal beam in Figure 2.4a. A positive axial force N produces tension; a positive shearing force tends to push the left face of the element upwards and the right face downwards; a positive bending

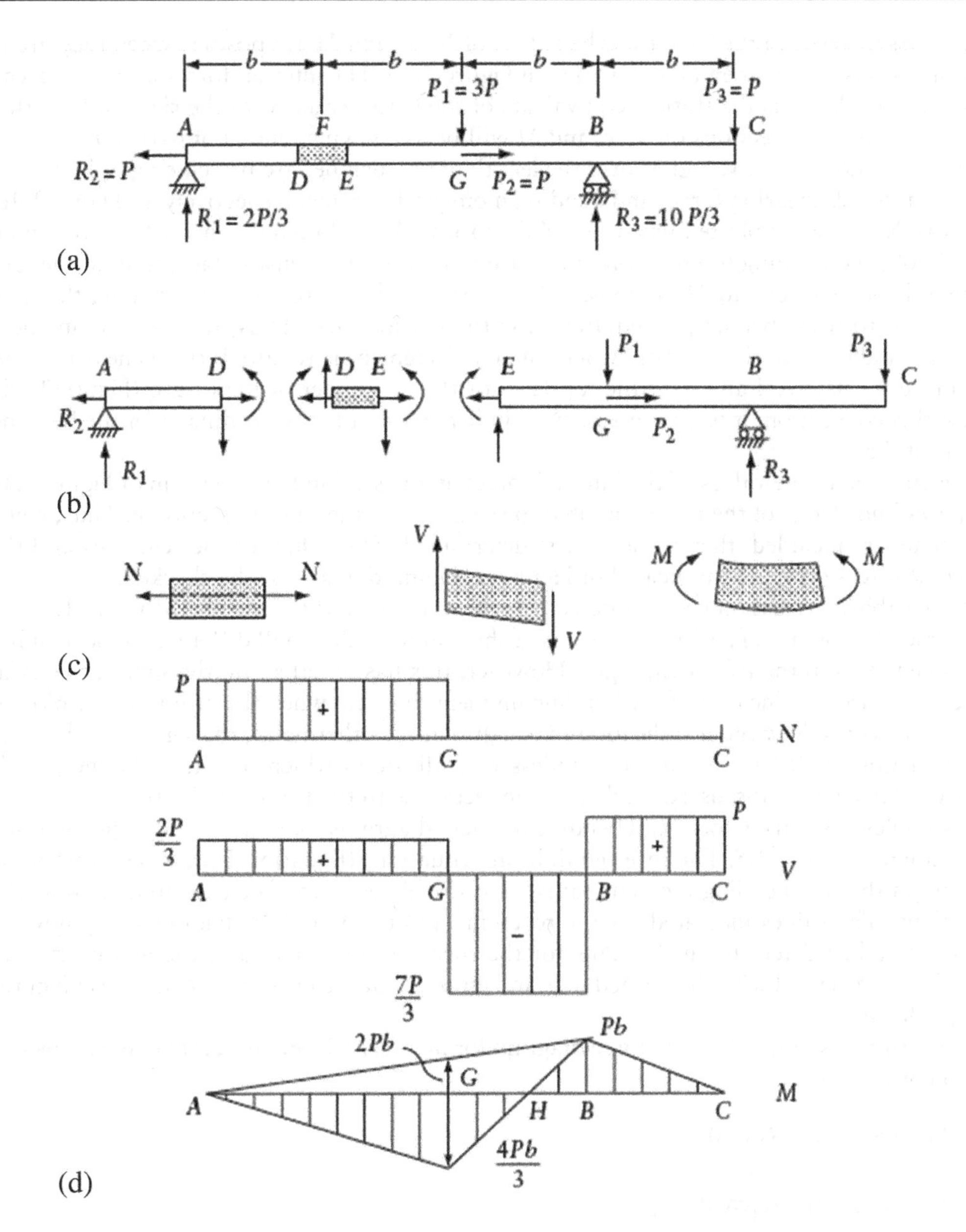

Figure 2.4 Sign convention for internal forces in plane frames and beams. (a) Beam. (b) Free body diagrams. (c) Positive *N*, *V*, and *M*. (d) Axial force, shearing force, and bending moment diagrams.

moment produces tensile stresses at the bottom face and bends the element in a concave shape.

In Figure 2.4b, each of the three parts AD, DE, and EC is shown as a free body subjected to a set of forces in equilibrium. To determine the internal forces at any section F (Figure 2.4a), it is sufficient to consider only the equilibrium of the forces on AD; thus N, V, and M at F are the three forces in equilibrium with R_1 and R_2. The same internal forces are the statical equivalents of P_1, P_2, P_3, and R_3. Thus, at any section F, the values of N, V, and M are, respectively, equal to the sums of horizontal and vertical components and moments of the forces situated to the left of F. The values of N, V, and M are positive when they are in the directions shown at end E of part EC in Figure 2.4b. The internal forces at section F can also be considered as the statical equivalents of the forces situated to the right of F; in that case, the positive directions of N, V, and M will be as shown at end D of part AD.

The variations in N, V, and M over the length of the member are presented graphically by the axial force, shearing force, and bending moment diagrams, respectively, in Figure 2.4d. Positive N and V are plotted upwards, while positive M is plotted downwards. Throughout this book, the bending moment ordinates are plotted on the tension face, that is, the face where the stresses due to M are tensile. If the structure is of reinforced concrete, the reinforcement to resist bending is required near the tension face. Thus, the bending moment diagram indicates to the designer where the reinforcement is required; this is near the bottom face for part AH and near the top face for the remainder of the length (Figure 2.4d). With this convention, it is not necessary to indicate a sign for the ordinates of the bending moment diagram.

Calculation of the values of the internal forces at any section F in the beam of Figure 2.4a requires knowledge of the forces situated to the left or to the right of section F. Thus, when reactions are included, they must be first determined. The values of the reactions and the internal-forces ordinates, indicated in Figures 2.4a and d, may now be checked.

In the above discussion we considered a horizontal beam. If the member is vertical as, for example, the column of a frame, the signs of shear and bending will differ when the member is looked at from the left or the right. However, this has no effect on the sign of the axial force or on the significance of the bending moment diagram when the ordinates are plotted on the tension side without indication of a sign. On the other hand, the signs of a shearing force diagram will have no meaning unless we indicate in which direction the member is viewed. This will be discussed further in connection with the frame in Figure 2.5.

Examples of shearing force and bending moment diagrams for a three-hinged plane frame are shown in Figure 2.5. The three equilibrium equations (Equation 2.5), together with the condition that the bending moment vanishes at the hinge C, may be used to determine the reactions. The values indicated for the reactions and the V and M diagrams may now be checked. When determining the signs for the shearing force diagram, the non-horizontal members are viewed with the dashed lines in Figure 2.5 at the bottom face. (See also Figures 2.8, 2.3c, and 2.9.)

The ordinates of the shearing force diagram for member BC (Figure 2.5b) may be checked as follows:

$$V_{Br} = R_1 \cos\theta - R_2 \sin\theta$$

$$V_{Cl} = \left[R_1 - q(2b)\right]\cos\theta - R_2 \sin\theta$$

where the subscripts r and l refer, respectively, to sections just to the right of B and just to the left of C; θ is the angle defined in Figure 2.5a.

To draw diagrams of internal forces in frames with straight members, it is necessary only to plot the ordinates at member ends and at the sections, where external forces are applied, and then to join these ordinates by straight lines. When a part (or the whole) of the length of a member is covered by a uniform load, the ordinates of the bending moment diagram at the two ends of the part are to be joined by a second-degree parabola (see Figure 2.5c). The ordinate of the parabola at the mid-point is $(qc^2/8)$, measured from the straight line joining the ordinates at the ends; q is the load intensity and c is the length of the part considered. A graphical procedure for plotting a second-degree parabola is included in Appendix E.

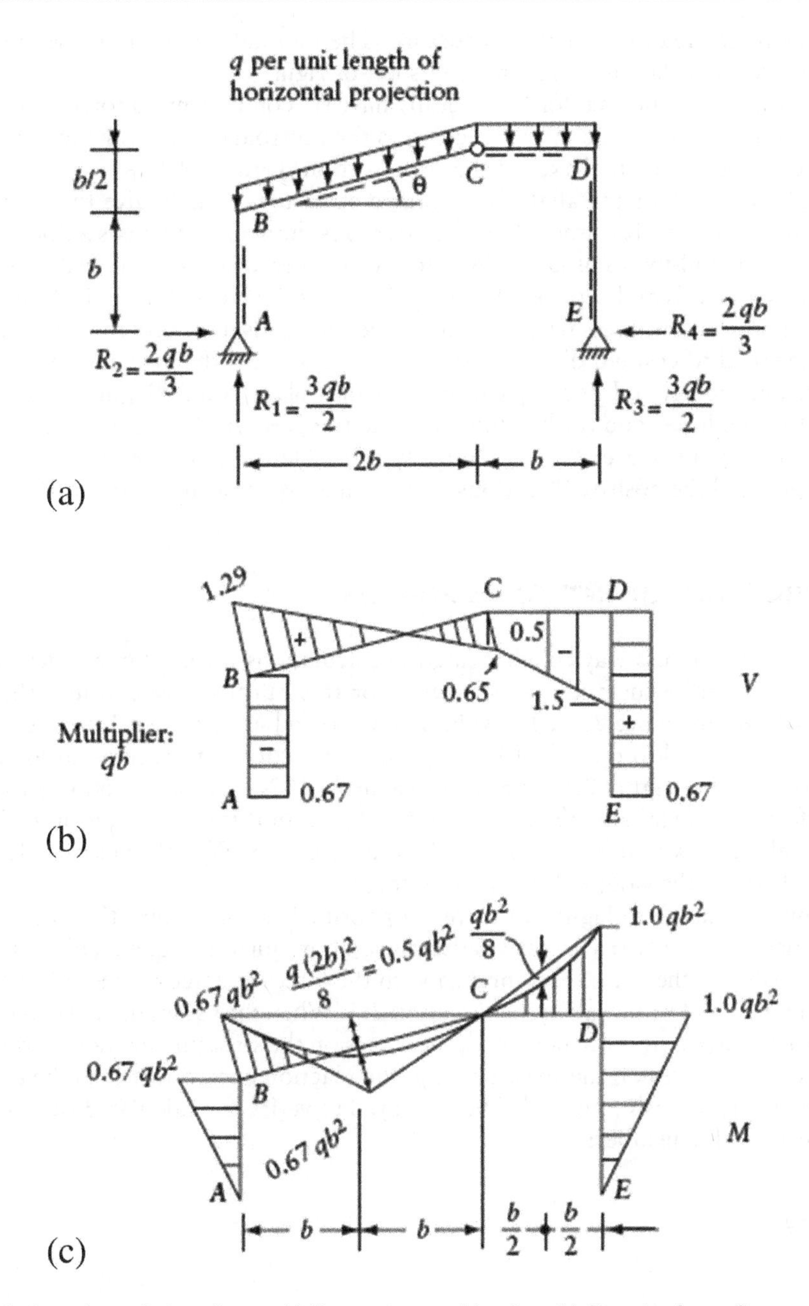

Figure 2.5 A three-hinged plane frame. (a) Dimensions and loading. (b) Shearing force diagram. (c) Bending moment diagram.

It is good practice to plot the ordinates perpendicular to the members and to indicate the values calculated and used to plot the diagram.

The internal forces at any section of a member of a framed structure can be easily determined if the end-forces are known. In Figures 3.8a and b, typical members of plane and space frames are shown. The forces shown acting on each member, being the external applied force(s) and the member end-forces, represent a system in equilibrium. Thus, the member may be treated as a separate structure. The internal forces at any section are the statical equivalents of the forces situated to its left or right.

In a space frame, the internal forces generally have six components: a force and a moment in the direction of x^*, y^*, and z^* axes, where x^* is the centroidal axis of the member, and y^* and z^* are centroidal principal axes of the cross section (Figure 19.2).

Computer programs for the analysis of framed structures usually give the member end-forces (Figures 3.8a and b) rather than the stress resultants at various sections. The sign convention for the end-forces usually relates to the member local axes, in the direction of the member centroidal axis and centroidal principal axes of the cross section. It is important at this stage to note that the stress resultants at the member ends may have the same magnitude as the member end-forces, but different signs, because of the difference in sign conventions. For example, at the left end of a typical member of a plane frame (Figure 3.8a), the axial force, the shearing force, and the bending moment are: $N = -F_1$; $V = -F_2$; $M = F_3$. The stress resultants at the right end are: $N = F_4$; $V = F_5$; $M = -F_6$. Here the member is viewed in a horizontal position and the positive directions of N, V, and M are as indicated in Figure 2.4c.

2.4 VERIFICATION OF INTERNAL FORCES

In this section, we discuss ways of checking the calculations involved in the determination of internal forces in beams and plane frames. For this purpose, we consider the internal forces of the beam in Figure 2.4a. The values of V, N, and M at section F can be calculated from the forces situated to the left of F; the positive directions of the internal forces in this case are shown at E in Figure 2.4b. The same values of V, N, and M at section F can also be calculated form the forces situated to the right of F; but in this case, the positive directions of the internal forces would be as shown at D in Figure 2.4b. This alternative calculation is a means of checking the values of the internal forces.

The arrows at D and E in Figure 2.4b are in opposite directions; thus, the values of N, M, and V calculated by the two alternatives described above must be equal. This is so because the forces situated to the left of F, combined with the forces situated to the right of F, represent a system in equilibrium, satisfying Equation 2.5. When the values of N, M, and V calculated by the two alternatives are not equal, the forces on the structure are not in equilibrium; one or more of the forces on the structure (e.g. the reaction components) can be erroneous.

The following relationships can also be employed to verify the calculated values of V and M at any section of a member:

$$\frac{\mathrm{d}V}{\mathrm{d}x} = -q \tag{2.6}$$

$$\frac{\mathrm{d}M}{\mathrm{d}x} = V \tag{2.7}$$

where x is the distance measured along the axis of the member (in direction AB); q is the intensity of distributed transverse load acting in the direction of the y axis. For a plane frame, the z axis is perpendicular to the plane of the frame, pointing away from the reader. The three axes, x, y, and z, form a right-handed system (see footnote 2 of this chapter). Thus, for the beam in Figure 2.4a, if the x axis is considered in the direction AB, then the y axis will be vertical downward. For this beam, the slope of the shearing force diagram ($\mathrm{d}V/\mathrm{d}x$) is zero at all sections because the beam is subjected to concentrated loads only ($q = 0$). For the same beam, the bending moment diagram is composed of three straight lines, whose slope ($\mathrm{d}M/\mathrm{d}x$) can be verified to be equal to the three shearing force values shown in Figure 2.4d.

As a second example, we will use Equations 2.6 and 2.7 to verify the V and M- diagrams for member BC of the frame in Figure 2.5. We consider the x axis in the direction BC; thus, the y axis is in the perpendicular direction to BC, pointing toward the bottom of the page. The component of the distributed load in the y direction has an intensity equal to:

$$q_{\text{transverse}} = q\cos\theta \tag{2.8}$$

This equation can be verified by considering that the resultant of the load on BC is a downward force equal to $2bq$, and the component of the resultant in the y direction is $2bq\cos\theta$. Division of this value by the length of BC ($= 2b/\cos\theta$) gives Equation 2.8. Thus, for member BC, $q_{\text{transverse}} = 0.94q$, and we can verify that this is equal to minus the slope of the V- diagram:

$$q_{\text{transverse}} = 0.94q = -\left[\frac{(-0.65)-1.29}{2b/\cos\theta}\right]qb$$

Figure 2.5c shows tangents at the ends of the parabolic bending moment diagram of member BC. The slopes of the two tangents are:

$$\left(\frac{\mathrm{d}M}{\mathrm{d}x}\right)_{\mathrm{B}} = \left[\frac{0.67-(-0.67)}{b/\cos\theta}\right]qb^2 = 1.29\,qb;$$

$$\left(\frac{\mathrm{d}M}{\mathrm{d}x}\right)_{\mathrm{C}} = \left[\frac{0-0.67}{b/\cos\theta}\right]qb^2 = -0.65\,qb$$

As expected (by Equation 2.7), the two slope values are equal to the shearing force values at B and C (Figure 2.5b). Equation 2.7 also indicates that, where $V = 0$, the bending moment diagram has zero slope and the value of M is either a minimum or a maximum (e.g. the point of zero shear on member BC in Figure 2.5). Under a concentrated force, the shearing force diagram has a sudden change in value and Equation 2.7 indicates that the M-diagram has a sudden change in slope.

EXAMPLE 2.6: MEMBER OF A PLANE FRAME: V AND M-DIAGRAMS

Figure 2.6a represents a free-body diagram of a member of a plane frame. Three of the member end forces are given: $F_1 = 0.2\,ql$; $F_3 = -0.06\,ql^2$; $F_6 = 0.12\,ql^2$. What is the magnitude of the remaining three end forces necessary for equilibrium? Draw the V and M-diagrams.

The three equilibriums in Equation 2.5 are applied by considering the sum of the forces in the horizontal direction, the sum of the moments of forces about B, and the sum of the forces in the vertical direction:

$$F_1 + F_4 = 0$$

$$-F_2 l + F_3 + F_6 - \frac{ql^2}{2} = 0$$

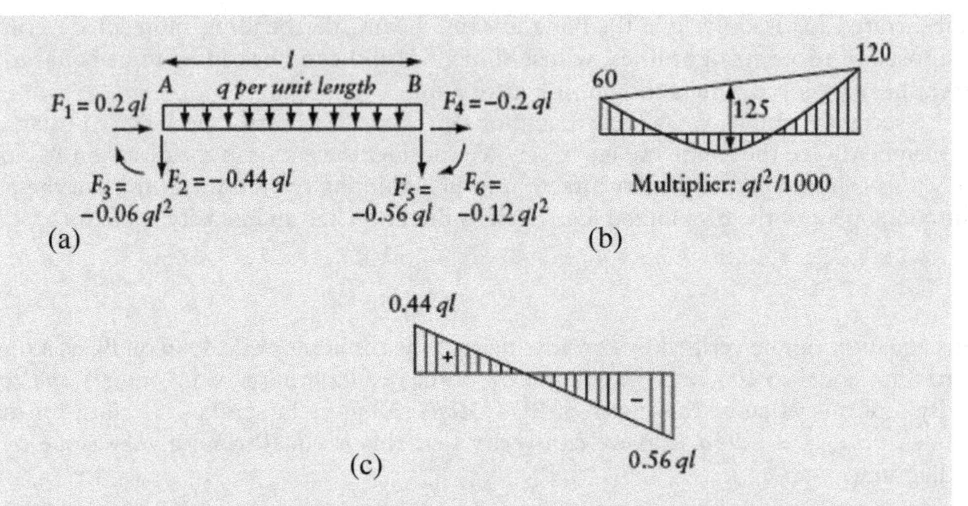

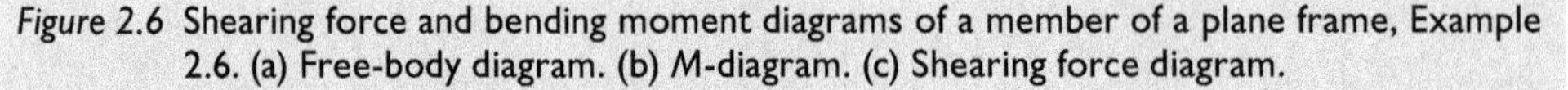

Figure 2.6 Shearing force and bending moment diagrams of a member of a plane frame, Example 2.6. (a) Free-body diagram. (b) *M*-diagram. (c) Shearing force diagram.

$$F_2 + F_5 + ql = 0$$

Substitution of the known values and solution gives:

$$F_2 = -0.44\,ql; \; F_4 = -0.2\,ql; \; F_5 = -0.56\,ql$$

The values of F_3 and F_6 give the bending moment ordinates at A and B, and their signs indicate that the tension face of the member is at the top fiber at both ends. The *M*-diagram in Figure 2.6b is drawn by joining the end ordinates by a straight line and *"hanging down"* the bending moment of a simple beam carrying the transverse load (a parabola).

The forces F_2 and F_5 give the ordinates of the *V*-diagram at the two ends. Joining the two ordinates by a straight line gives the *V*-diagram shown in Figure 2.6c.

EXAMPLE 2.7: SIMPLE BEAMS: VERIFICATION OF V AND M-DIAGRAMS

The simple beams in Figures 2.7a and b are subjected to a transverse distributed load and to end moments, respectively. The simple beam in Figure 2.7c is subjected to a combination of the two loadings. Draw the *V* and *M*-diagrams for the three beams.

The values of the ordinates may be calculated by the reader and compared with the values given in Figure 2.7. Also, it may be verified that the ordinates satisfy the relationships between *q*, *V*, and *M* in Equations 2.6 and 2.7. It can also be noted that the ordinates of the *V* and *M*-diagrams in part (c) of the figure are equal to the sum of the corresponding ordinates in parts (a) and (b). Thus, the diagrams in part (c) can be derived alternatively by the superposition of the diagrams in parts (a) and (b).

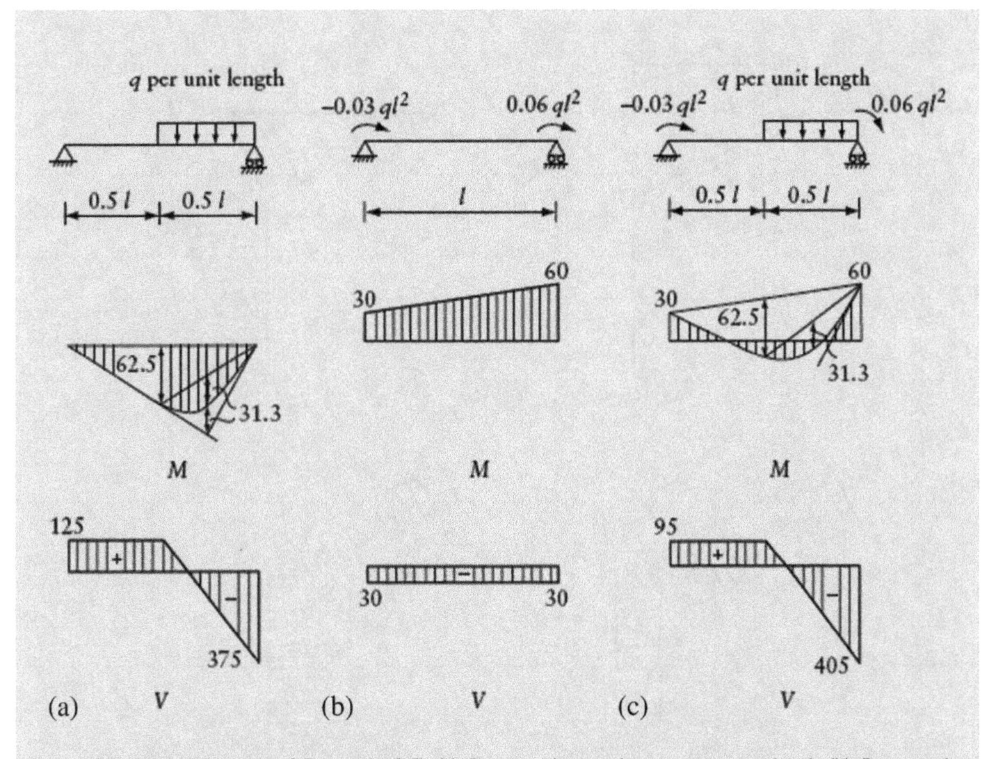

Figure 2.7 Simple beams of Example 2.7. (a) Beam subjected to transverse load. (b) Beam subjected to end moments. (c) Beam subjected to a combination of the loadings in (a) and (b). Multiplier for the *M*-diagrams: $ql^2/1000$. Multiplier for the *V*-diagrams: $ql/1000$.

EXAMPLE 2.8: A CANTILEVER PLANE FRAME

Determine the bending moment, the shearing force, and the axial force diagrams for the cantilever frame in Figure 2.8.

To find the internal forces at any section of a cantilever, consider the forces situated between that section and the free end. In this way, the reactions are not involved. The ordinates of the *M*, *V*, and *N*-diagrams in Figure 2.8 may be verified by the reader. We show below the calculation of the shearing force and the axial force only for a section anywhere in member AB.

$$V = (P/2)\sin\alpha - P\cos\alpha = -0.671P$$

$$N = -(P/2)\cos\alpha - P\sin\alpha = -0.894P$$

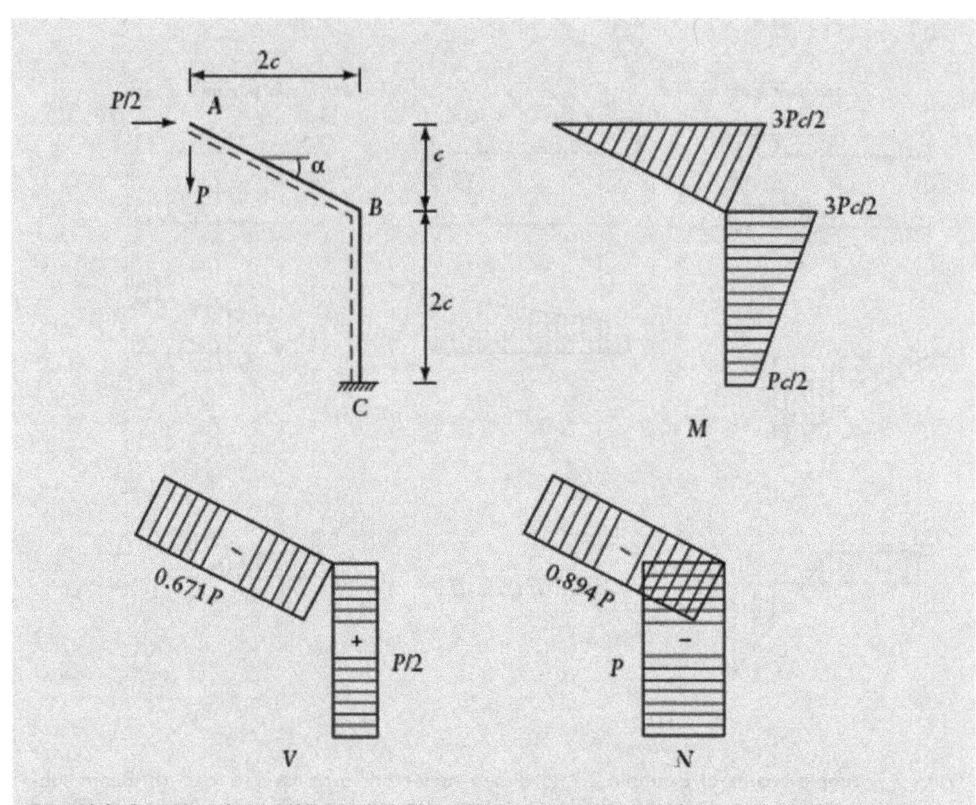

Figure 2.8 Cantilever frame of Example 2.8: bending moment, shearing force, and axial force diagrams.

EXAMPLE 2.9: A SIMPLY-SUPPORTED PLANE FRAME

Determine the bending moment and the shearing force diagrams for the frame in Figure 2.3c. What are the axial forces in the members?

The M and V-diagrams are shown in Figure 2.9. The reader may wish to verify the ordinates given in this figure. See also Example 2.4. The axial forces in the members are:

$$N_{AC} = -R_1 \sin\alpha = -(-2P)\frac{1}{\sqrt{5}} - 1.8P\left(\frac{2}{\sqrt{5}}\right) = -0.716P$$

$$N_{CE} = -R_1 = -(-2P) = 2P$$

$$N_{EF} = -R_3 = -3.2P; \; N_{BC} = 0$$

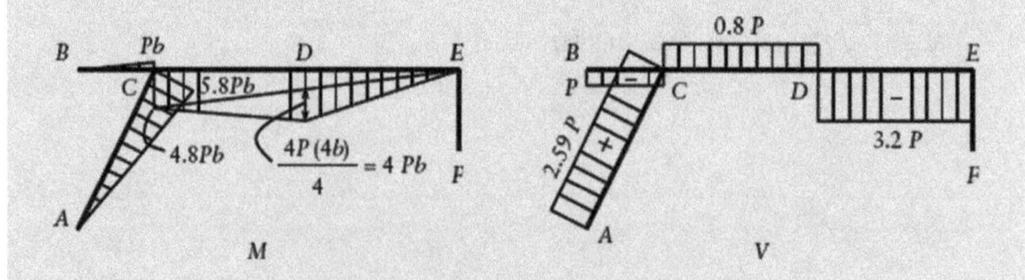

Figure 2.9 Bending moment and shearing force diagrams for the frame in Figure 2.3c.

EXAMPLE 2.10: M-DIAGRAMS DETERMINED WITHOUT CALCULATION OF REACTIONS

Draw the M-diagrams and sketch the deflected shapes for the beams shown in Figure 2.10.

The ordinates of the bending moment diagrams shown in Figure 2.10 require simple calculations as indicated. Note that the calculation of the reactions is not needed to determine the M-diagrams. The sketched deflected shapes show concave and convex parts of the beam. The points of inflection at which the curvature changes sign correspond to points of zero bending moment.

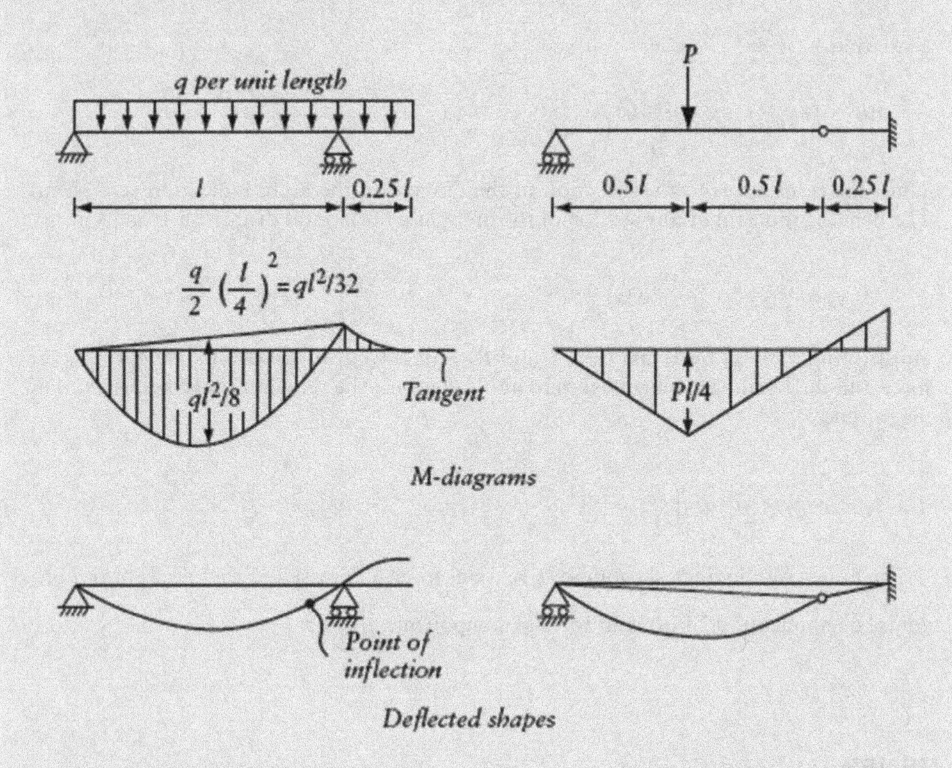

Figure 2.10 Bending moments and deflected shapes for beams of Example 2.10.

EXAMPLE 2.11: THREE-HINGED ARCHES

Figure 1.27a shows a polygonal three-hinged arch subjected to concentrated loads, whose total = ql. Verify that a parabolic three-hinged arch carrying uniform load q per unit length of horizontal projection has the same reactions as those of the polygonal arch. In Example 1.2, it is shown that the shear force V and the bending moment M are zero at all sections of the polygonal arch. Verify that this applies also for the parabolic arch.

Reactions for parabolic arch:

From symmetry, the vertical reaction at A = $ql/2$. At the hinge C, the bending moment = 0. Thus, the sum of moments about C of forces situated to the left of C = 0. This gives:

$$q\frac{l}{2}\left(\frac{l}{2}\right)-\frac{ql}{2}\left(\frac{l}{4}\right)-R_1\,h_C=0;\quad R_1=q\,l^2/(8\,h_C)=0.75\,q\,l$$

Symmetry gives the reactions at B. The reactions are the same for the polygonal arch.

Internal forces in parabolic arch:

The resultant of the forces situated to the left of any section at a horizontal distance x from A has a vertical upward component $= q\left(\frac{l}{2} - x\right)$ and a horizontal component $= R_1 = 0.75\ ql$. The angle between the resultant and the x axis is $\tan^{-1}\left\{ q\left[(l/2) - x\right] \big/ 0.75ql \right\} = \tan^{-1}\left(\frac{2}{3} - \frac{4x}{3l}\right)$. The slope of the arch at the same section is (Equation 1.1):

$$\frac{dh}{dx} = \frac{4h_C}{l^2}(l - 2x) = \frac{4(l/6)}{l^2}(l - 2x) = \frac{2}{3} - \frac{4x}{3l}$$

Thus, the resultant is in the direction of the tangent to the arch, indicating zero shear. The bending moment at any section of the arch, at a horizontal distance x from A is:

$$M(x) = \frac{ql}{2}x - qx\frac{x}{2} - R_1 h(x)$$

Substituting for $h(x)$ by Equation 1.1 and $R_1 = ql^2/8h_C$ gives $M(x) = 0$. Thus, the shear force and the bending moment are zero at all sections; the only internal force is axial of magnitude:

$$N = -\left[(0.75\ ql)^2 + q\left(\frac{l}{2} - x\right)^2\right]^{1/2}$$

The minus sign indicates a compressive axial force, expressed as the resultant of horizontal component, $(0.75\ ql)$, and vertical component, $q\left(\frac{l}{2} - x\right)$

2.5 GENERAL

For statically determinate structures, the equilibrium equations are sufficient to determine the reactions and the internal forces. Indeed, this is why the structures are called statically determinate. Additional equations, considering deformations, are necessary in the analysis of statically indeterminate structures. For all structures, the equilibrium equations apply to the structure as a whole, to an isolated part, and to individual joints. The application of these equations will give unknown components of reactions and internal forces when the number of unknowns does not exceed the number of equilibrium equations.

PROBLEMS

2.1 Find member forces and reaction components in x and y or in x, y, and z directions for the statically determinate trusses shown.

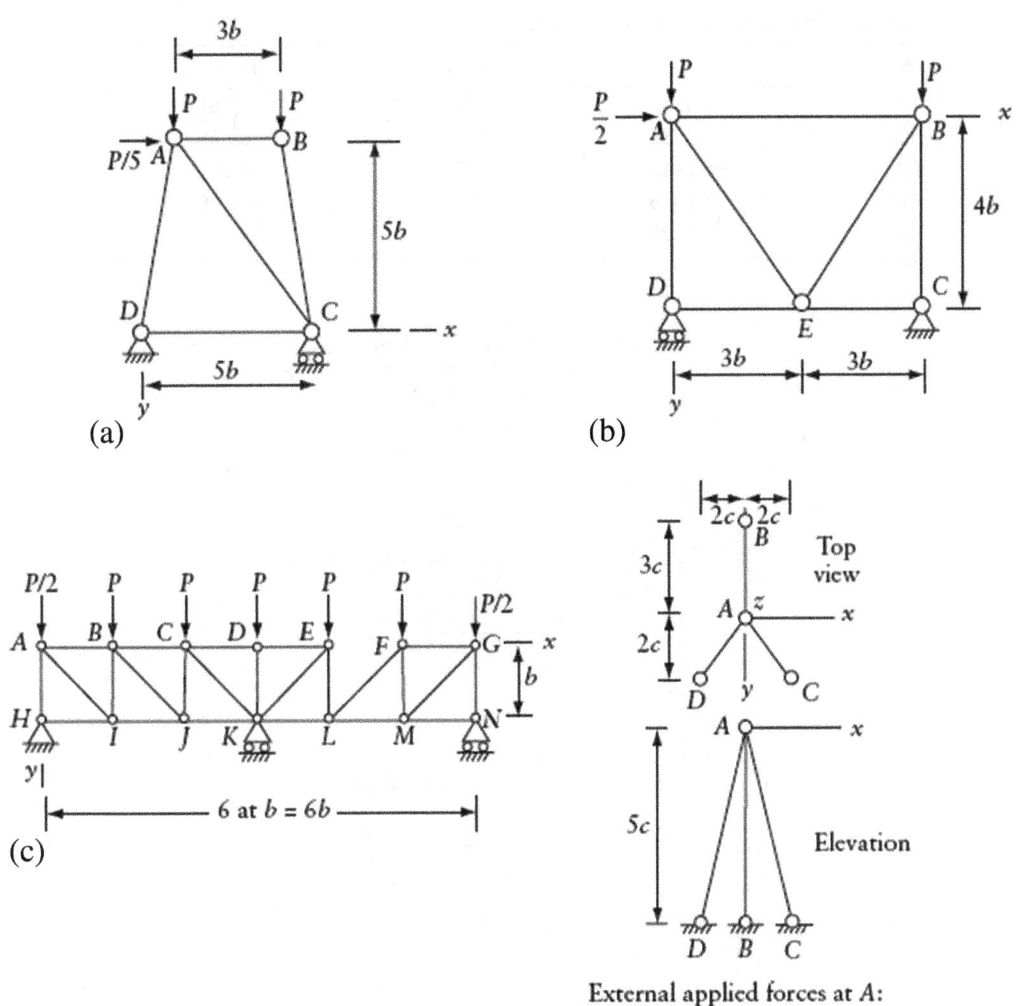

2.2 The space truss shown is composed of three horizontal members AB, BC, and CA and six inclined members connecting A, B, and C to the support nodes E, F, and D. The structure is subjected to forces $\{F_x, F_y, F_z\} = P\{1, 0, 3\}$ at each of joints A, B, and C. Write down nine equilibrium equations from which the reaction components in x, y, and z directions can be determined. Verify the equations by substituting the answers given below. Also verify equilibrium of one of joints A, B, or C using the member forces given in the answers (Tables 2.1 and 2.2).

Table 2.1 Reactions in terms of P

	R_x	R_y	R_z
At D	−1.7835	0.6071	−7.3923
At E	−0.7321	−0.7500	−3.0000
At F	−0.4845	0.1430	1.3923

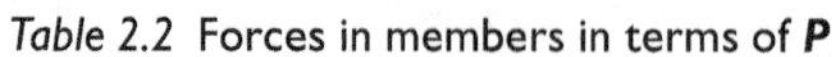

Table 2.2 Forces in members in terms of **P**

Member	*Force*	*Member*	*Force*
AB	1.1585	BE	−0.0381
BC	0.1585	CE	−3.1439
CA	−0.8415	CF	−0.0381
AD	−4.6968	AF	1.5148
BD	−3.1439		

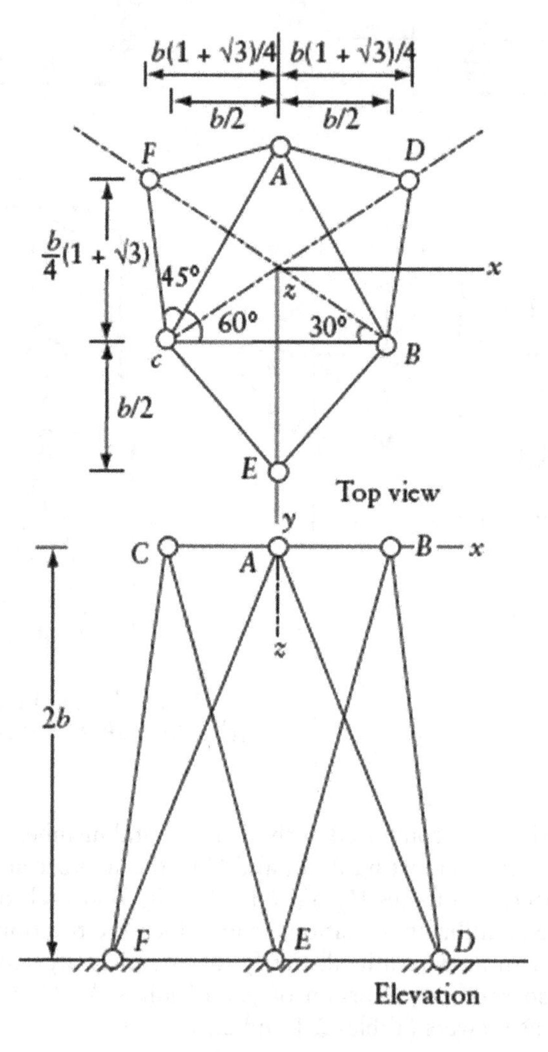

2.3 A bridge deck is schematically presented as a solid prism supported on bearings at A, B, C, and D. The bearings can provide only seven reaction components in the x, y, and z directions as shown. Given $R_2 = -2.5P$, determine the remaining reactions due to the external applied forces shown at E.

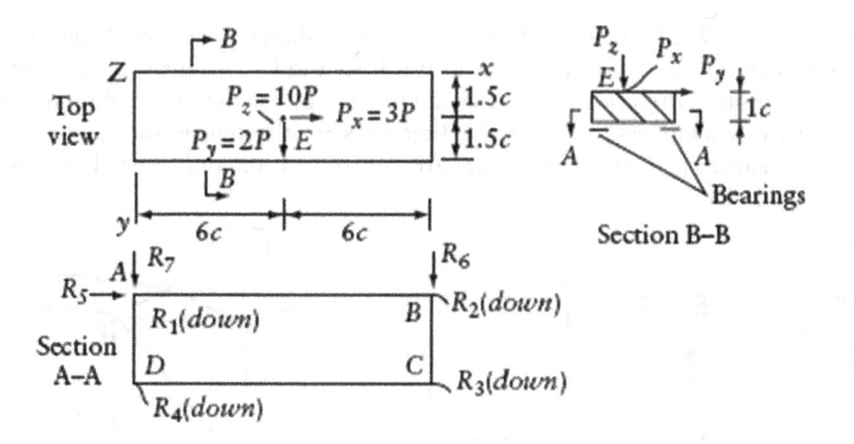

2.4 Obtain the shearing force and bending moment diagrams for the statically determinate beams and frames shown.

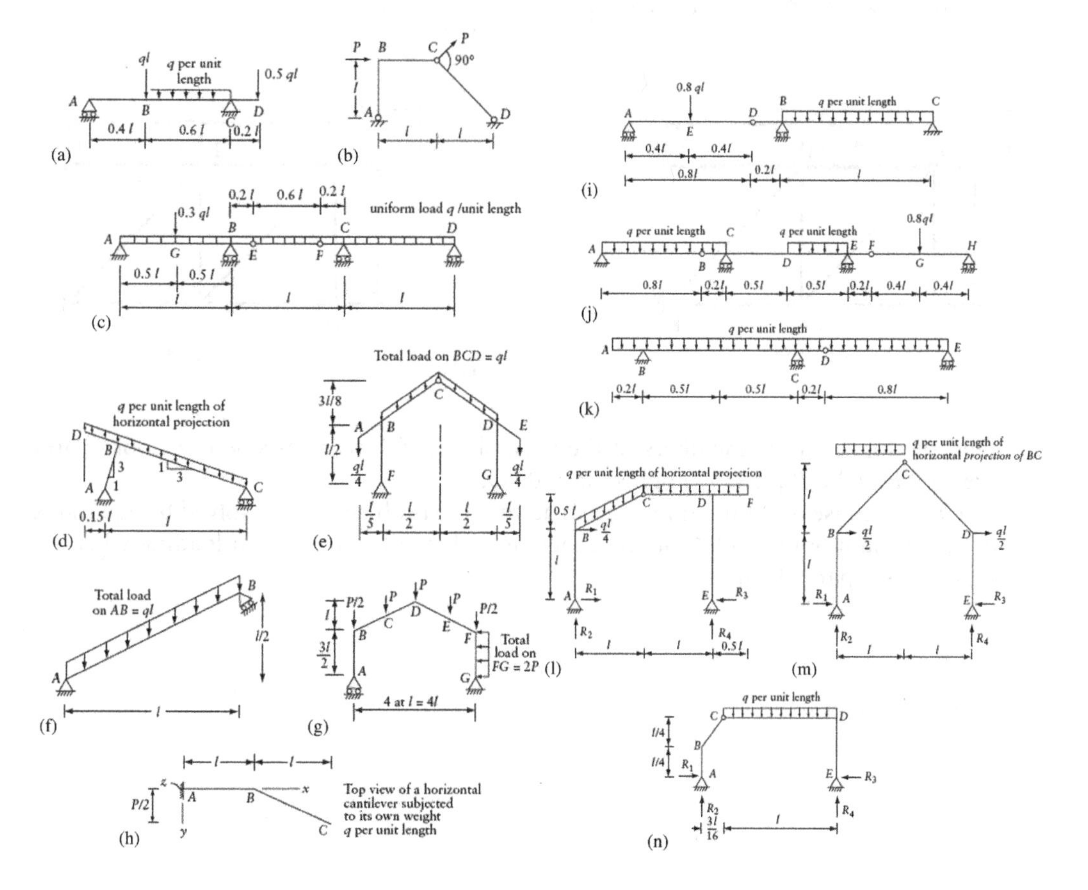

2.5 A prismatic reinforced concrete cantilever subjected to an externally applied twisting couple T is shown in cross section in (a). The structure is idealized as a space truss shown in pictorial view in (b) (strut-and-tie model, see Section 1.14). For clarity, the members in the bottom horizontal plane of the truss, not shown in (b), are shown in top view in (c); similarly, the members in the vertical far side of the truss, not shown in (b), are shown in elevation in (d). The applied couple T is represented by component forces, each = $T/(2l)$ at nodes A, B, C, and D as shown in (a). Find the forces in the members and the reaction components at I, J, K, and L. The answer will show that the members in the x direction are tensile and resisted by the longitudinal reinforcing bars; the transverse members in the y and z directions are also tensile and resisted by the legs of the stirrups; the diagonal members are in compression and resisted by concrete.

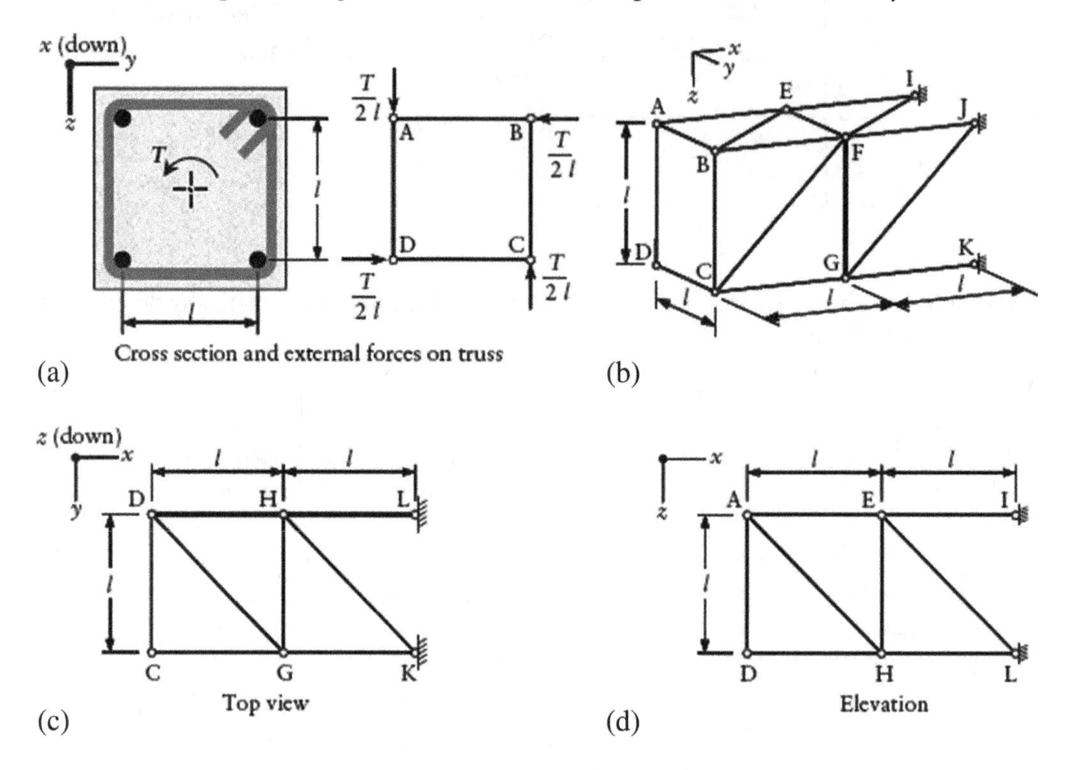

Cross section and external forces on truss

Top view

Elevation

2.6 Find the forces in the members of the truss shown due to a force whose coordinates are: $\{F_x, F_y, F_z\} = P\{1, 1, 7\}$ applied at node O.

Hint: Because of symmetry of the structure, the problem can be solved by equations of equilibrium of node O. Note that there is a plane of symmetry of loading when F_x, F_y, or F_z is separately applied.

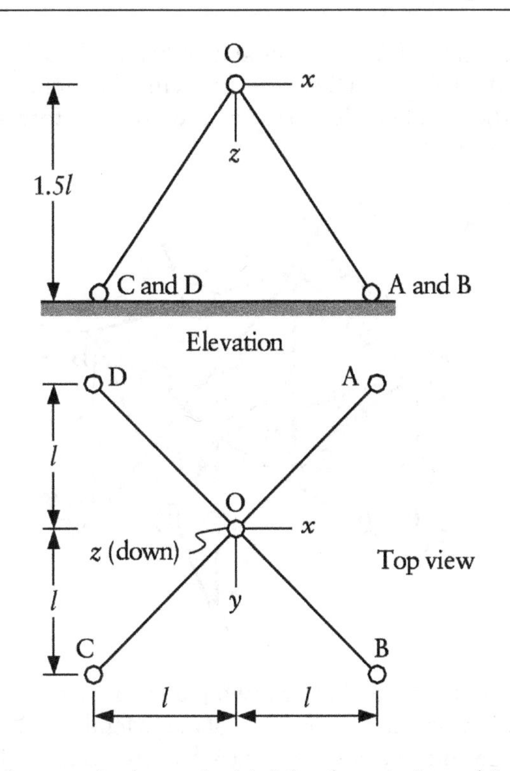

2.7 A pictorial view of a truss is shown in (a). Members 1, 2, and 3 are situated in the horizontal xy plane shown in (b). Members 5, 6, 9, and 10 are in a vertical plane shown in (c). A horizontal force P is applied along the y-axis at node 1. This truss is statically indeterminate; use an equilibrium equation to find the forces in the members in a released structure in which member 3 is cut (making $N_3 = 0$).

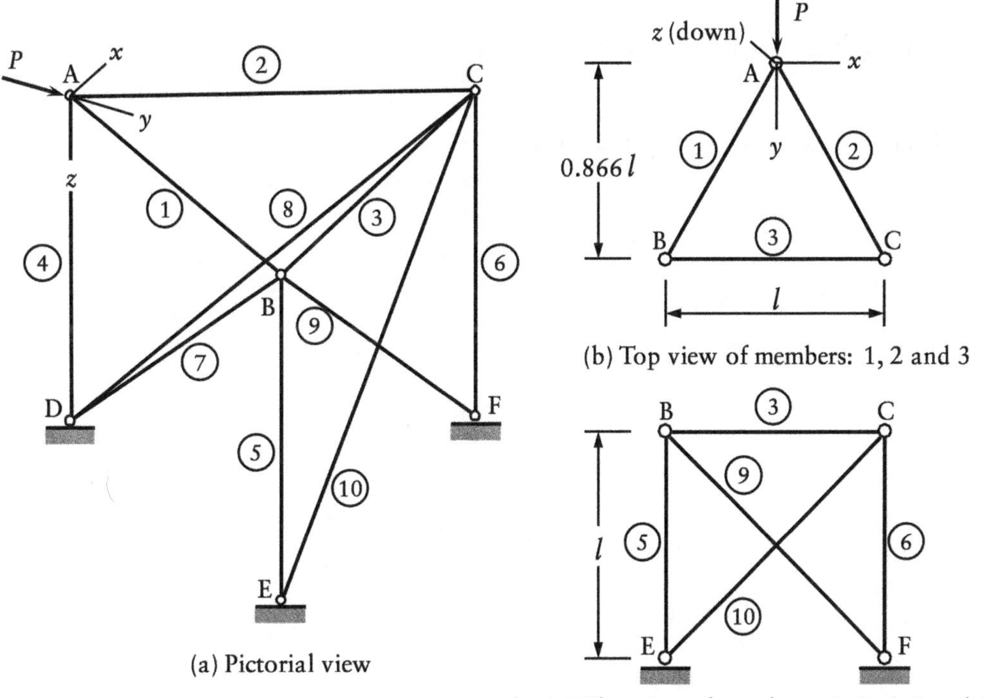

(a) Pictorial view

(b) Top view of members: 1, 2 and 3

(c) Elevation of members: 3, 5, 6, 9 and 10

2.8 The space truss shown has four horizontal members: AB, BC, CD, and DA, four vertical members: AE, BF, CG, and DH and four inclined members: AF, BG, CH, and DE. Find the forces in the members due to equal and opposite forces P applied in the directions of AC and CA, as shown.

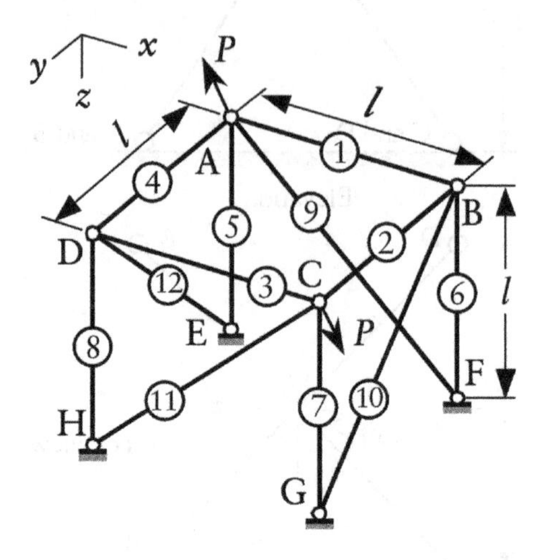

2.9 The figure shows a space truss that has two planes of symmetry: xz and yz. Members AB, BC, CD, and DA are in a horizontal plane. Members AE, BF, CG, and DH are vertical. Each of the four sides of the truss has two diagonal members. For clarity, the diagonal members AH, DE, AF, and BE are not shown in the pictorial view. The truss has pin supports at E, F, G, and H. Each of nodes A, B, C, and D is subjected to a downward force P. In addition, node A is subjected to equal forces P in the x and y directions. (a) Find the components $\{F_x, F_y, F_z\}$ of the reaction components at F and G. Find the forces in the three members meeting at E (members EA, EB, and ED). Given: The reaction components at E and H, in terms of P are:

$$\text{At E}: F_x = -0.203;\ F_y = -0.203;\ F_z = 1.051$$

$$\text{At H}: F_x = -0.309;\ F_y = 0;\ F_z = -1.051$$

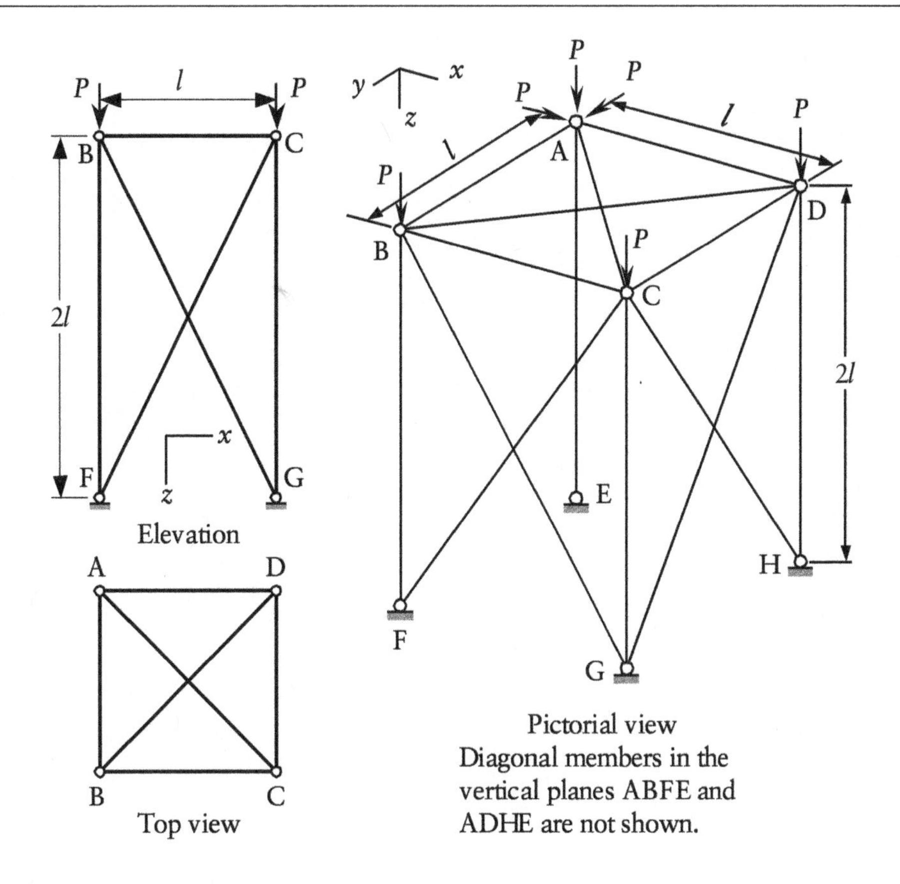

NOTES

1. All through this text, a couple (or rotation) is indicated in planar structures by an arrow in the form of an arc of a circle (see, for example, Figures 3.1 and 3.3). In three-dimensional structures, a couple (or rotation) is indicated by a double-headed arrow. The direction of the couple is that of the rotation of a right-hand screw progressing in the direction of the arrow. This convention should be well understood at this stage.
2. In a right-handed system, directions of orthogonal axes x and y can be chosen arbitrarily, but the direction of the z axis will be that of the advance of a right-handed screw turned through the angle from x axis to y axis. A right-handed system of axes is used throughout this text, for example, see Figures 2.2a and 2.3a.

Chapter 3

Introduction to the analysis of statically indeterminate structures

3.1 INTRODUCTION

This chapter develops concepts which are necessary for the two general methods of analysis of structures: the force method and the displacement method (considered in Chapters 4 and 5, respectively).

3.2 STATICAL INDETERMINACY

The analysis of a structure is usually carried out to determine the reactions at the supports and the internal stress resultants. As mentioned earlier, if these can be determined entirely from the equations of statics alone, then the structure is statically determinate. This book deals mainly with statically indeterminate structures, in which there are more unknown forces than equations. The majority of structures in practice are statically indeterminate.

The indeterminacy of a structure may either be *external*, *internal*, or both. A structure is said to be externally indeterminate if the number of reaction components exceeds the number of equations of equilibrium. Thus, a space structure is in general externally statically indeterminate when the number of reaction components is more than six. The corresponding number in a plane structure is three. The structures in Figures 2.1a, c, e, f, g, and h are examples of external indeterminacy. Each of the beams of Figures 3.1a and b has four reaction components. Since there are only three equations of static equilibrium, there is one unknown force in excess of those that can be found by statics, and the beams are externally statically indeterminate. We define the degree of indeterminacy as the number of unknown forces in excess of the equations of statics. Thus, the beams of Figures 3.1a and b are indeterminate to the first degree.

Some structures are built so that the stress resultant at a certain section is known to be zero. This provides an additional equation of static equilibrium and allows the determination of an additional reaction component. For instance, the three-hinged frame of Figure 3.1c has four reaction components, but the bending moment at the central hinge must vanish. This condition, together with the three equations of equilibrium applied to the structure as a free body, is sufficient to determine the four reaction components. Thus, the frame is statically determinate. The continuous beam of Figure 3.1d has five reaction components and one internal hinge. Four equilibrium equations can therefore be written so that the beam is externally indeterminate to the first degree.

Let us now consider structures which are externally statically determinate but internally indeterminate. For instance, in the truss of Figure 3.2a, the forces in the members cannot be determined by the equations of statics alone. If one of the two diagonal members is removed (or cut) the forces in the members can be calculated from equations of statics. Hence, the

DOI: 10.1201/9780429286858-3

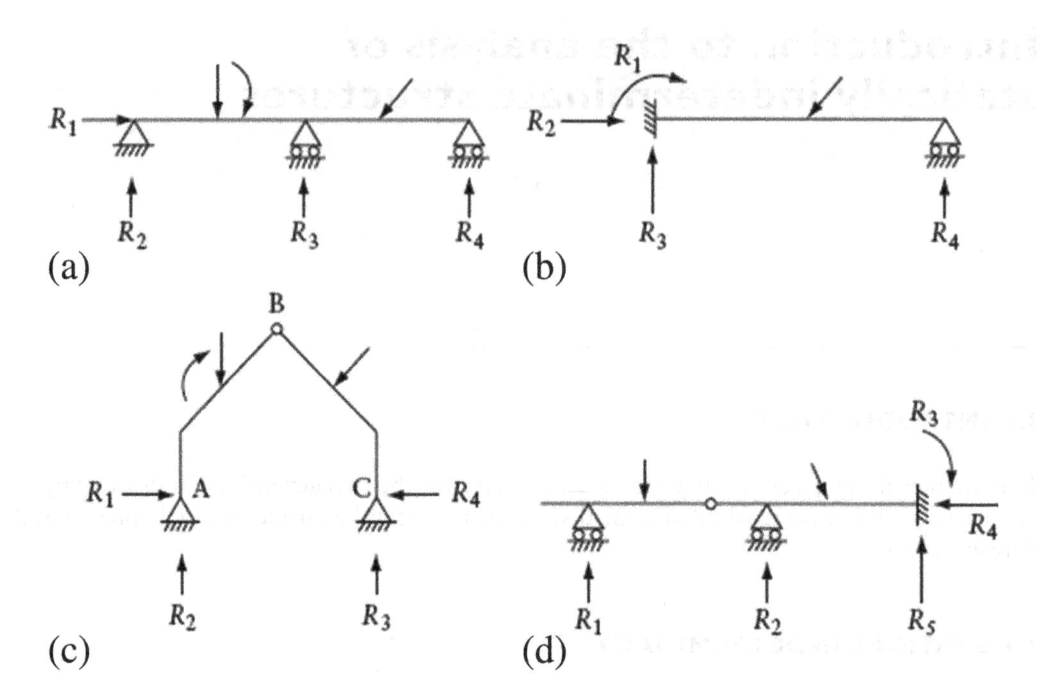

Figure 3.1 (a), (b), and (d) Externally statically indeterminate structures. (c) Statically determinate three-hinged frame.

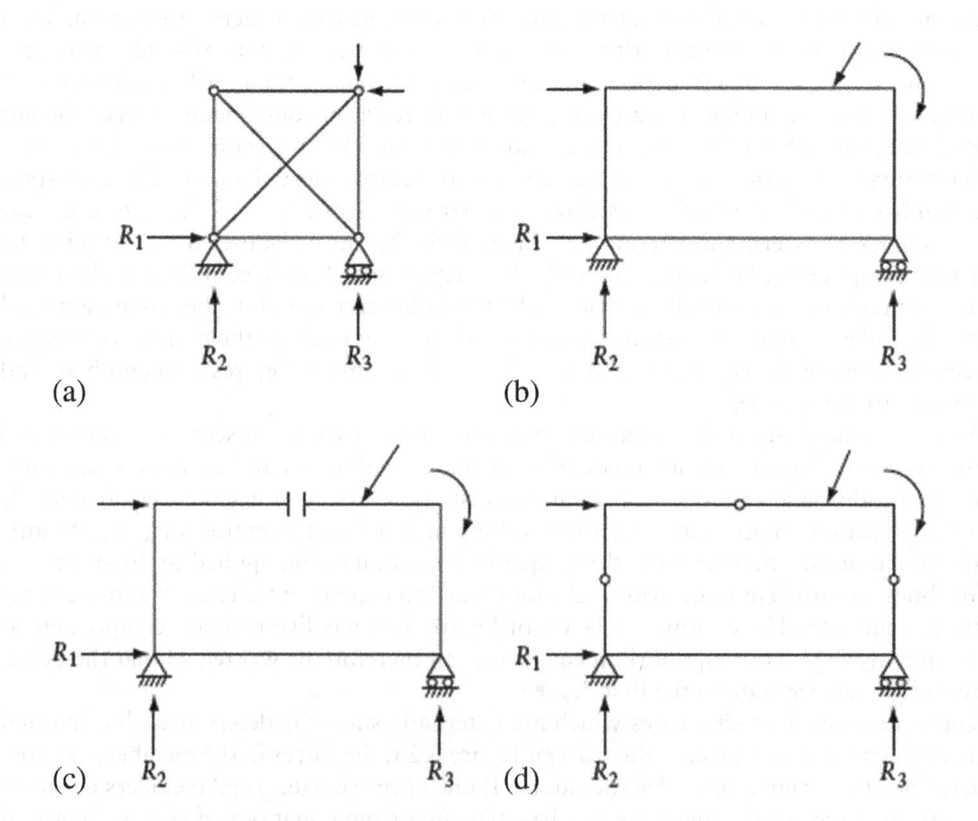

Figure 3.2 Internally statically indeterminate structures.

truss is internally indeterminate to the first degree, although it is externally determinate. The frame in Figure 3.2b is internally indeterminate to the third degree; it becomes determinate if a cut is made in one of the members (Figure 3.2c). The cut represents the removal or *release* of three stress resultants: axial force, shearing force, and bending moment. The number of releases necessary to make a structure statically determinate represents the degree of indeterminacy. The same frame becomes determinate if the releases are made by introducing three hinges as in Figure 3.2d, thus removing the bending moment at three sections.

Structures can be statically indeterminate both internally and externally. The frame of Figure 3.3a is externally indeterminate to the first degree, but the stress resultants cannot be determined by statics even if the reactions are assumed to have been found previously. They can, however, be determined by statics if the frame is cut at two sections, as shown in Figure 3.3b, thus providing six releases. It follows that the frame is internally indeterminate to the sixth degree, and the total degree of indeterminacy is seven.

The space frame of Figure 3.4 has six reaction components at each support: three components X, Y, and Z and three couples M_x, M_y, and M_z. To avoid crowding the figure, the six components are shown at one of the four supports only. The moment vectors are indicated by double-headed arrows.[1] Thus, the number of reaction components of the structure is

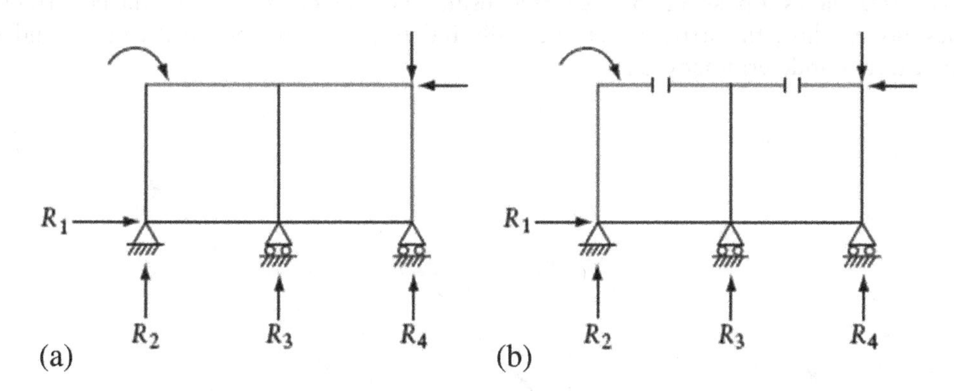

Figure 3.3 Frame that is statically indeterminate both externally and internally.

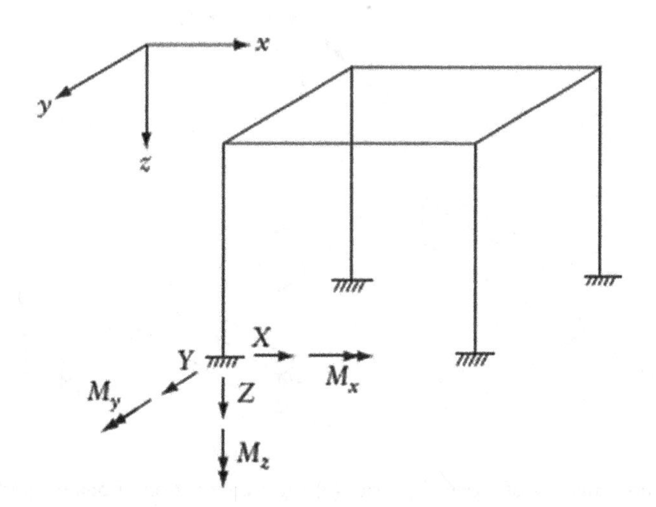

Figure 3.4 Rigid-jointed space frame.

24, while the equations of equilibrium that can be written are six in number (cf. Equation 2.4). The frame is therefore externally indeterminate to the 18th degree. If the reactions are known, the stress resultants in the four columns can be determined by statics, but the beams forming a closed frame cannot be analyzed by statics alone. Cutting one of the beams at one section makes it possible to determine the stress resultants in all the beams. The number of releases in this case is six: axial force, shear in two orthogonal directions, bending moment about two axes, and twisting moment. The structure is thus internally indeterminate to the sixth degree, and the total degree of indeterminacy is 24.

The members of the horizontal grid of Figure 3.5a are assumed to be rigidly connected (as shown in Figure 3.5b) and to be subjected to vertical loads only. Thus, both the reaction components X, Z, and M_y and the stress resultants X, Z, and M_y vanish for all members of the grid. Hence, the number of equilibrium equations which can be used is three only. The reaction components at each support are Y, M_x, and M_z, so that the number of reaction components for the whole structure is $8 \times 3 = 24$. Thus, it is externally statically indeterminate to the 21st degree.

If the reactions are known, the stress resultants in the beams of the grid can be determined by statics alone except for the central part ABCD, which is internally statically indeterminate. Cutting any of the four beams of this part (ABCD) in one location produces three releases and makes it possible for the stress resultants to be determined by the equations of statics alone. Thus, the structure is internally indeterminate to the third degree, and the total degree of indeterminacy is 24.

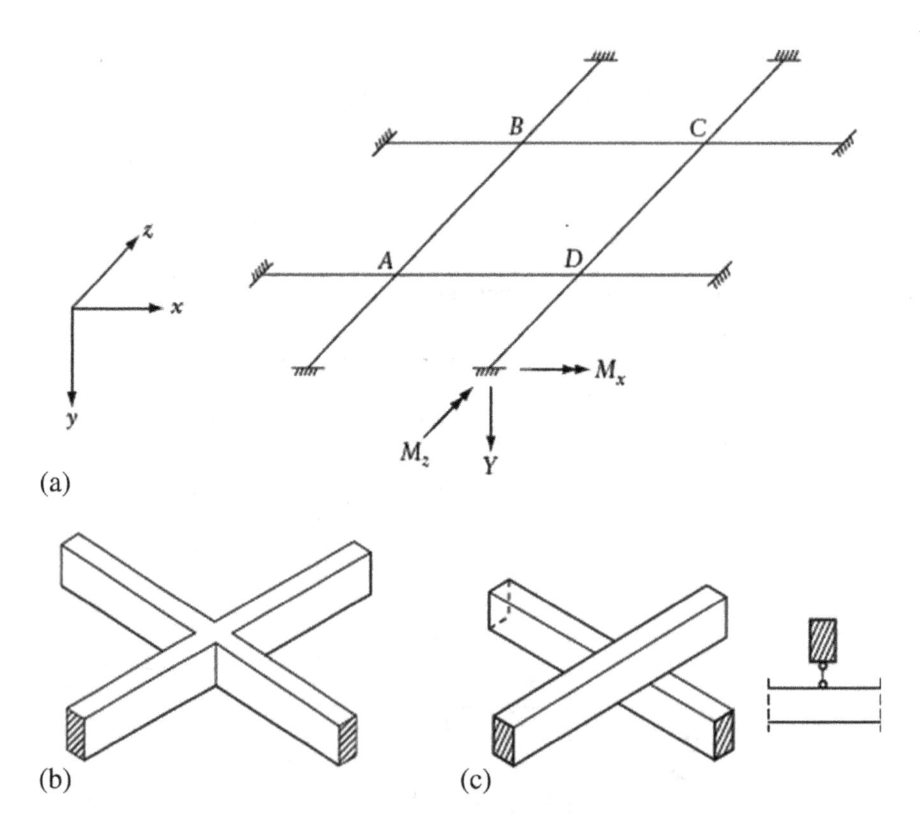

Figure 3.5 Statical indeterminacy of a grid. (a) Grid. (b) Rigid connection of beams. (c) Hinged connection of beams.

If the members forming the grid are not subjected to torsion – which is the case if the beams of the grid in one direction cross over the beams in the other direction with hinged connections (Figure 3.5c) or when the torsional rigidity of the section is negligible compared with its bending stiffness – the twisting moment component (M_z in Figure 3.5a) vanishes and the structure becomes indeterminate to the 12th degree. The grid needs at least four simple supports for stability and becomes statically determinate if the fixed supports are removed and only four hinged supports are provided. Each hinged support has one reaction component in the y direction. Since the number of reaction components in the original grid is 16, it is externally indeterminate to the 12th degree. This number is equal to the number of reaction components minus the four reaction components in the y direction mentioned above. There is no internal indeterminacy, and, once the reactions have been determined, the internal forces in all the beams of the grid can be found by simple statics.

3.3 EXPRESSIONS FOR DEGREE OF INDETERMINACY

In Section 3.2, we found the degree of indeterminacy of various structures by inspection or from the number of releases necessary to render the structure statically determinate. For certain structures, especially those with a great many members, such an approach is difficult, and the use of a formal procedure is preferable.

We assume that pin-connected trusses are subjected to forces concentrated at the connections. Let us consider a *plane* truss with three reaction components, m members and j hinged (pinned) joints (including the supports, which are also hinged). The unknown forces are the three reaction components and the force in each member – that is, $3+m$. Now, at each joint, two equations of equilibrium can be written: the summation being for the components of all the external and internal forces meeting at the joint. Thus, the total number of equations is $2j$.

$$\Sigma F_x = 0; \ \ \Sigma F_y = 0 \tag{3.1}$$

For statical determinacy, the number of equations of statics is the same as the number of unknowns, that is:

$$2j = m + 3 \tag{3.2}$$

Providing the structure is stable, some interchange between the number of members and the number of reaction components r is possible, so that for overall determinacy the condition

$$2j = m + r \tag{3.3}$$

has to be satisfied. The degree of indeterminacy is then:

$$i = (m + r) - 2j \tag{3.4}$$

For the truss shown in Figure 3.6, $r=4$, $m=18$, and $j=10$. Hence, $i=2$.

In the case of a pin-jointed *space* frame, three equations of equilibrium can be written, viz:

$$\Sigma F_x = 0; \ \ \Sigma F_y = 0; \ \ \Sigma F_z = 0 \tag{3.5}$$

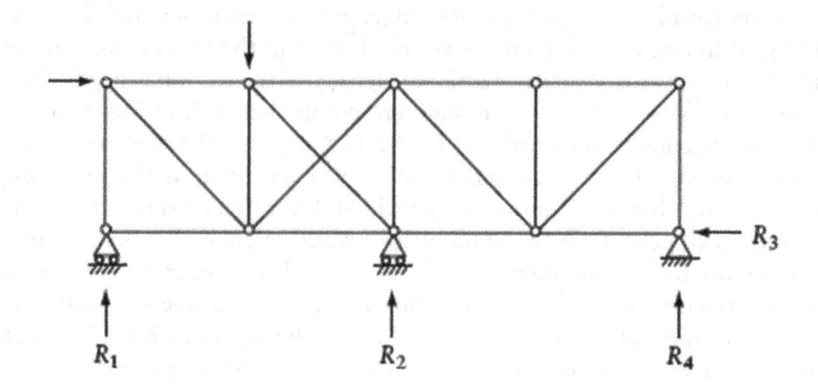

Figure 3.6 Statically indeterminate plane truss.

The summation again is for all the internal and external forces meeting at the joint. The total number of equations is $3j$ and the condition of determinacy is:

$$3j = m + r \tag{3.6}$$

The degree of indeterminacy is:

$$i = (m + r) - 3j \tag{3.7}$$

The use of these expressions can be illustrated with reference to Figure 3.7. For the truss of Figure 3.7a, $j=4$, $m=3$, and $r=9$, there being three reaction components at each support. Thus, Equation 3.6 is satisfied and the truss is statically determinate.

In the truss of Figure 3.7b, $j=10$, $m=15$, and $r=15$. Hence, again Equation 3.6 is satisfied. However, for the truss of Figure 3.7c, $j=8$, $m=13$, and $r=12$, so that from Equation 3.7, $i=1$. Removal of member 5-7 would render the truss determinate.

At a node of a space truss, the three equations of equilibrium (Equation 3.5) are not sufficient to determine the member forces when more than two unknown forces are situated in the same plane; for such a truss, Equation 3.7 underestimates the degree of indeterminacy. For the truss of Problem 2.9, the degree of indeterminacy = 8, while Equation 3.7 gives $i=6$. When the reaction components are known, Equation 3.5 can be used to find the forces in the three members meeting at each of nodes E, F, G, and H. But, the three equilibrium equations are not sufficient to find the forces in the three horizontal members connected to node A, B, C, or D. The truss is externally indeterminate to the sixth degree; when six reaction components are given, the six equilibrium Equation 2.4 can give the remaining six components. The truss is internally indeterminate to the second degree; the indeterminacy can be removed by cutting members AC and BD.

Expressions similar to those of Equations 3.4 and 3.7 can be established for frames with rigid joints. At a rigid joint of a *plane* frame, two resolution equations and one moment equation can be written. The stress resultants in any member of a plane frame (Figure 3.8a) can be determined if any three of the six end-forces F_1, F_2, ..., F_6 are known, so that each member represents three unknown internal forces. The total number of unknowns is equal to the sum of the number of unknown reaction components r and of the unknown internal forces. Thus, a rigid-jointed plane frame is statically determinate if:

$$3j = 3m + r \tag{3.8}$$

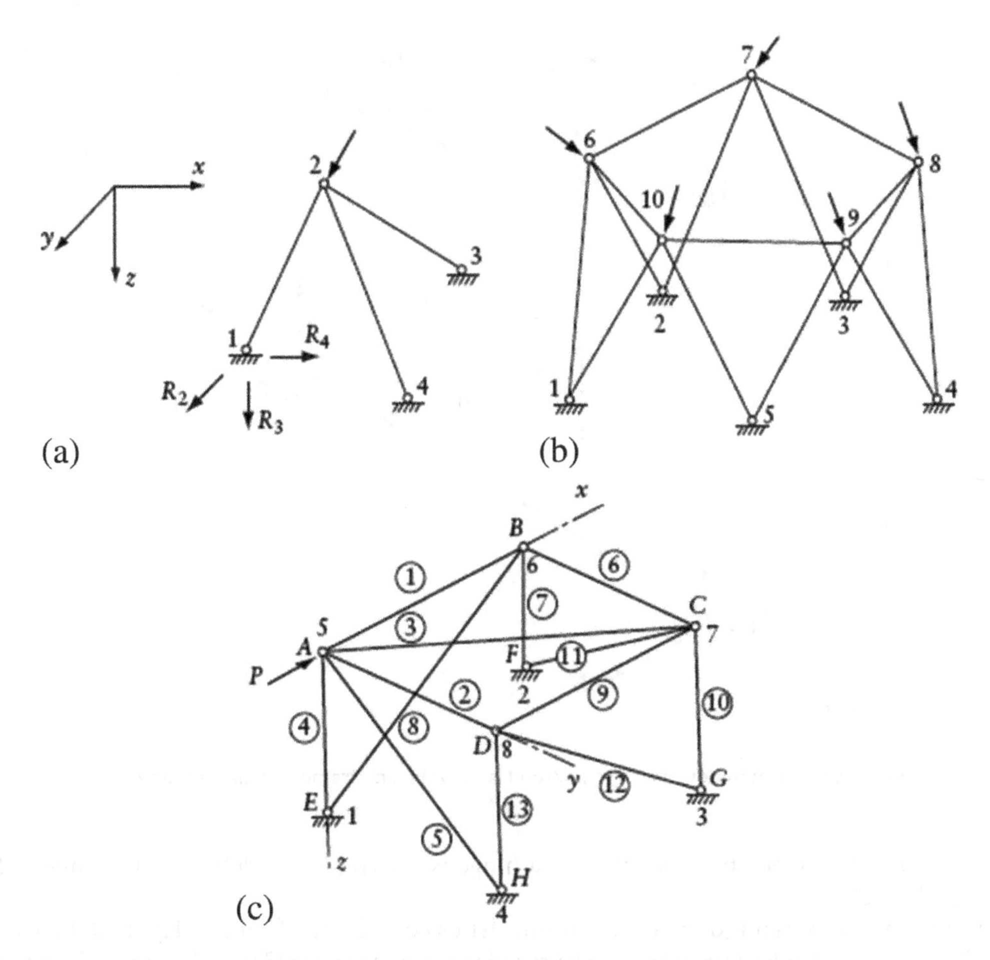

Figure 3.7 Space trusses. (a) and (b) Statically determinate. (c) Statically indeterminate.

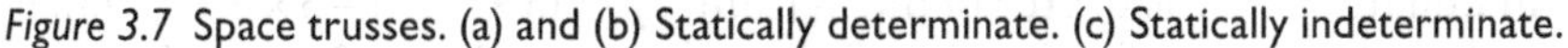

and the degree of indeterminacy is:

$$i = (3m + r) - 3j \tag{3.9}$$

In these equations, j is the total number of rigid joints including the supports, and m is the number of members.

If a rigid joint within the frame is replaced by a hinge, the number of equilibrium equations is reduced by one, but the bending moments at the ends of the members meeting at the joint vanish, so that the number of unknowns is reduced by the number of members meeting at the hinge. We can verify that the frame in Figure 3.1c is statically determinate using Equation 3.9; because of the hinge at joint B, the number of equations and the number of unknowns have to be adjusted as indicated. An alternative method of calculating the degree of indeterminacy of plane frames having pin connections is given in Section 3.3.1. This modification has to be observed when applying Equations 3.8 and 3.9 to plane frames with mixed type joints. We should note that at a rigid joint where more than two members meet and one of the members is connected to the joint by a hinge, the number of unknowns is reduced by one, without a reduction in the number of equilibrium equations. For example, we can verify that the frame of Problem 3.5 is four times statically indeterminate, and the

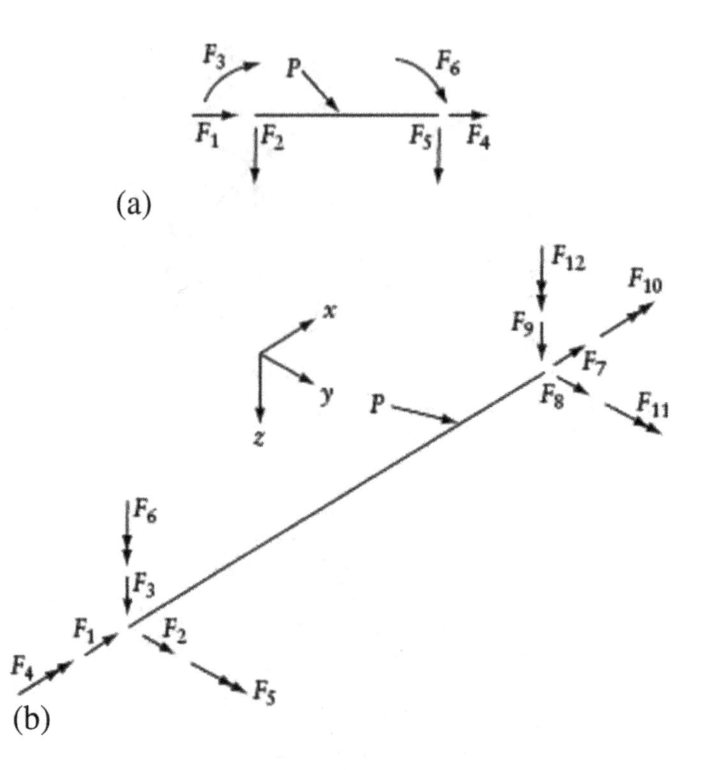

Figure 3.8 End-forces in a member of a rigid-jointed frame. (a) Plane frame. (b) Space frame.

degree of indeterminacy becomes three if a hinge is inserted at the left end of member CF (just to the right of C).

As an example of rigid-jointed plane frame, let us consider the frame of Figure 3.3a: $j=6$, $m=7$, and $r=4$. From Equation 3.9 the degree of indeterminacy $i=(3\times 7+4)-3\times 6=7$, which is the same result as that obtained in Section 3.2. For the frame of Figure 3.1c, $j=5$, $m=4$, and $r=4$. However, one of the internal joints is a hinge, so that the number of unknowns is $(3m+r-2)=14$ and the number of equilibrium equations is $(3j-1)=14$. The frame is therefore statically determinate, as found before.

At a rigid joint of a *space* frame, three resolution and three moment equations can be written. The stress resultants in any members can be determined if any six of the twelve end-forces shown in Figure 3.8b are known, so that each member represents six unknown forces. A space frame is statically determinate if:

$$6j = 6m + r \tag{3.10}$$

and the degree of indeterminacy is:

$$i = (6m + r) - 6j \tag{3.11}$$

Applying Equation 3.11 to the frame of Figure 3.4, we have $m=8$, $r=24$, and $j=8$. From Equation 3.11, $i=24$, which is, of course, the same as the result obtained in Section 3.2.

Consider a grid with rigid connections (Figures 3.5a and b). Three equations of equilibrium can be written at any joint ($\Sigma F_y=0$; $\Sigma M_x=0$; $\Sigma M_z=0$). Member end-forces are three, representing a shearing force in the y direction, a twisting moment, and a bending moment.

The internal forces at any section can be determined when three of the six member end-forces are known; thus, each member represents three unknown forces. A grid with rigid joints is statically determinate when:

$$3j = 3m + r \tag{3.12}$$

when this condition is not satisfied, the degree of indeterminacy is:

$$i = (3m + r) - 3j \tag{3.13}$$

When the connections of grid members are hinged as shown in Figure 3.5c, the members are not subjected to torsion, and the degree of indeterminacy is:

$$i = (2m + r) - (3j + 2\bar{j}) \tag{3.14}$$

Although torsion is absent, three equations of equilibrium can be applied at a joint connecting two or more members running in different directions, e.g. joints A, B, C, and D in Figure 3.5a when the connections are of the type shown in Figure 3.5c. The three equations are: sum of the end-forces in the y direction equals zero and sum of the components of the end moments in x and z directions equals zero. The symbol j in Equation 3.14 represents the number of joints for which three equations can be written. But only two equilibrium equations (one reaction and one moment equation) can be written for a joint connected to one member (e.g. at the supports in Figure 3.5a) or to two members running in the same direction. The symbol $\bar{j}$ in Equation 3.14 represents the number of joints of this type.

We can apply Equation 3.13 or 3.14 to verify that for the grid in Figure 3.5a, $i = 24$ or 12, respectively, when the connections are rigid or hinged (Figure 3.5b or c).

For each type of framed structure, the relation between the numbers of joints, members, and reaction components must apply when the structure is statically determinate (e.g. Equation 3.3 or 3.6). However, this does not imply that when the appropriate equation is satisfied the structure is stable. For example, in Figure 3.7b, the removal of the member connecting nodes 5 and 10, and the addition of a member connecting nodes 5 and 4, will result in an unstable structure even though it satisfies Equation 3.6.

EXAMPLE 3.1: COMPUTER ANALYSIS OF A SPACE TRUSS

The space truss in Figure 3.7c is subjected to a horizontal force P at node 5. For all members, Ea = constant. The computer program SPACET (Chapter 20) can be used for analysis to give the components $\{R_x, R_y,$ and $R_z\}$ of the reactions at supported nodes 1, 2, 3, and 4. The answers are:

$$R_1 = P\{-0.760,\ 0,\ 0.521\};\quad R_2 = P\{0,\ -0.240,\ -0.521\}$$

$$R_3 = P\{-0.240,\ 0, -0.479\};\quad R_4 = P\{0,\ 0.240,\ 0.479\}$$

The forces in members 1, 2, 3, 4, and 5 meeting at node A are: $\{N\} = P\,\{0.760, 0, -0.339, -0.240, 0.339\}$. Use equilibrium equations to verify the answers.

We use Equations 2.4 to check the reactions. Here we verify that $\sum F_x = 0$ and $\sum M_x = 0$. The reader may wish to verify that $\sum F_y = 0$ and $\sum M_y = 0$ or $\sum F_z = 0$ and $\sum M_z = 0$.

$$\sum F_x = P + \sum_{i=1}^{4} F_{ix} = P\left[1.0 + (-0.760) + 0 + 0 + (-0.240)\right] = 0$$

We apply Equation 2.2:

$$\sum M_x = P(0) + \sum_{i=1}^{4} \left(F_{iz}\, y_i - F_{iy}\, z_i\right)$$

$$= 0 + Pl\Big\{\left[0.521(0) - 0(1.0)\right] + \left[-0.521(0) - (-0.240)(1.0)\right]$$

$$+ \left[(-0.479)(1.0) - 0(1.0)\right] + \left[0.479(1.0) - 0.240(1.0)\right]\Big\} = 0$$

The equilibrium equations of a node connected to m members are (Equations 2.1 and 2.4):

$$\sum_{i=1}^{m} (N\lambda_x)_i + F_x = 0; \quad \sum_{i=1}^{m} (N\lambda_y)_i + F_y = 0; \quad \sum_{i=1}^{m} (N\lambda_z)_i + F_z = 0 \tag{3.15}$$

where N_i = axial force in the ith member; $\{\lambda_x, \lambda_y, \lambda_z\}_i$ = direction cosines of a vector along the ith member pointing away from the node. At a node of a space truss, the equilibrium Equations 3.15 can be written as:

$$[\lambda]^{\text{T}} \{N\} = -\{F\} \tag{3.16}$$

where

$$[\lambda]^{\text{T}}_{3\times m} = \begin{bmatrix} \{\lambda_x\}^{\text{T}} \\ \{\lambda_y\}^{\text{T}} \\ \{\lambda_z\}^{\text{T}} \end{bmatrix} \tag{3.17}$$

$\{F\}$ = components of the resultant external force applied at the node.
$\{F\} = \{F_x, F_y, F_z\}$

At node 5 of the space truss considered in this example,

$$\{F\} = P\{1,\ 0,\ 0\}$$

$$[\lambda]^{\text{T}} = \begin{bmatrix} 1.0 & 0 & 1/\sqrt{2} & 0 & 0 \\ 0 & 1.0 & 1/\sqrt{2} & 0 & 1/\sqrt{2} \\ 0 & 0 & 0 & 1.0 & 1/\sqrt{2} \end{bmatrix}$$

Substitution for $\{N\}$ and $\{F\}$ in Equation 3.16 gives:

$$[\lambda]^{\text{T}}\, 10^{-3} P\{-760, 0, -339, -240, 339\} = -P\{1, 0, 0\}$$

Thus, equilibrium equations at node 5 are satisfied. The reader may wish to verify equilibrium of another node (Table 4.1 gives $\{N\}$ for all members).

3.3.1 Plane frames having pin connections

Figure 3.9 shows plane frames having member ends pin-connected to the joints. The external load can be applied anywhere on the frame. The degree of static indeterminacy for planes of this type can be determined in two steps.

Step 1: Introduce artificial restraints to the actual structure to prevent supported joints from rotation and translation. Also, artificially prevent the change in angles between member ends at the joints. The total number of restraints $= n_r$; the resulting frame is referred to as the restrained structure. At a hinged or a roller support, the number of restraints is one or two, respectively. At a joint, the number of restraints equals the number of angles between member ends that can change in the actual structure.

Step 2: Introduce n_s releases to make the restrained structure statically determinate. The degree of indeterminacy of the actual structure:

$$i = n_s - n_r \tag{3.18}$$

Consider the frame in Figure 3.9a. In Step 1, we make the supports A and B encastré by one restrain at each support. We prevent the change in the four marked angles between members; thus, $n_r = 2$ (at supports) $+ 4$ (at angles) $= 6$. In Step 2, we introduce cuts in members BC and in each of the two diagonal members; each cut releases three internal forces; thus, $n_s = 3 \times 3 = 9$. The degree of indeterminacy: $i = n_s - n_r = 9 - 6 = 3$.

Consider the frame in Figure 3.9b. In Step 1, we make the support D encastré by one restrain. We prevent the change in eight marked angles. Thus, $n_r = 1 + 8 = 9$. In Step 2, cutting

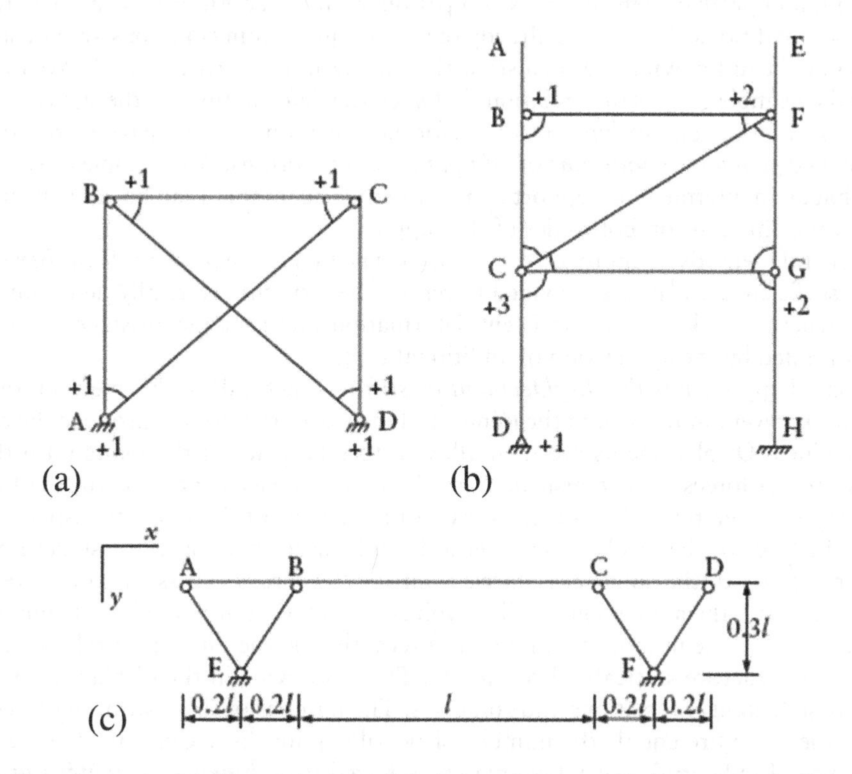

Figure 3.9 Plane frames having pin-connections. Calculation of degree of indeterminacy by Equation 3.15. The numbers with plus signs indicate added restraints in Step 1. (a) Frame having $i = 3$. (b) Statically determinate frame. (c) Frame having $n_r = 8$, $n_s = 9$, and $i = 1$.

the two horizontal members and the diagonal member introduces three releases at each cut; thus, $n_s = 3 \times 3 = 9$. The cuts result in two tree-like statically determinate parts. Thus, the structure in Figure 3.9b is statically determined because $n_s - n_r = 9 - 9 = 0$. A tree is commonly statically determinate; the branches cantilever out from a trunk having one end free and a second end encastré. At any section of a branch or a trunk, the internal forces are statically equivalent to known external forces situated between free end or ends and the considered section.

At joint F in Figure 3.9b, angles BFC and CFG can change independently; but angle EFG is dependent on the other two angles; thus, two restrains are added in Step 1. At joint G of the same structure, angles FGC and CGH can change independently; but the angle between GF and GH is equal to 2π minus the other two angles. In a similar way, we can justify the number of restrains or releases added at the joints in Figures 3.9a and b.

In Step 2, we release the restrained structure whose degree of indeterminacy is higher than the actual structure. Cutting members is a simple way to determine n_s. We may verify that: the frame in Figure 3.9c has $n_r = 8$, $n_s = 9$, and $i = 1$; the frame in Figure 3.1c has $n_r = 3$, $n_s = 3$, and $i = 0$.

3.4 GENERAL METHODS OF ANALYSIS OF STATICALLY INDETERMINATE STRUCTURES

The objective of the analysis of structures is to determine the external forces (reaction components) and the internal forces (stress resultants). The forces must satisfy the conditions of equilibrium and produce deformations compatible with the continuity of the structure and the support conditions. As we have already seen, the equilibrium equations are not sufficient to determine the unknown forces in a statically indeterminate structure and have to be supplemented by simple *geometrical* relations between the deformations of the structure. These relations ensure the *compatibility* of the deformations with the geometry of the structure and are called *geometry conditions* or *compatibility conditions*. An example of such conditions is that at an intermediate support of a continuous beam there can be no deflection, and the rotation is the same on both sides of the support.

Two general methods of approach can be used. The first is the *force or flexibility* method, in which sufficient *releases* are provided to render the structure statically determinate. The released structure undergoes inconsistent deformations, and the inconsistency in geometry is then corrected by the application of additional forces.

The second approach is the *displacement* or *stiffness* method. In this method, restraints are added to prevent movement of the joints, and the forces required to produce the restraint are determined. Displacements are then allowed to take place at the joints until the fictitious restraining forces have vanished. With the joint displacements known, the forces on the structure are determined by superposition of the effects of the separate displacements.

Either the force or the displacement method can be used to analyze any structure. Since, in the force method, the solution is carried out for the forces necessary to restore consistency in geometry, the analysis generally involves the solution of a number of simultaneous equations equal to the number of unknown forces, that is, the number of releases required to render the structure statically determinate. The unknowns in the displacement method are the possible joint translations and rotations. The number of the restraining forces to be added to the structure equals the number of possible joint displacements. This represents another type of indeterminacy, which may be referred to as *kinematic indeterminacy*, and is discussed in the next section. The force and displacement methods themselves are considered in more detail in Chapters 4 and 5.

3.5 KINEMATIC INDETERMINACY

When a structure composed of several members is subjected to loads, the joints undergo displacements in the form of rotation and translation. In the displacement method of analysis, it is the rotation and translation of the joints that are the unknown quantities.

At a support, one or more of the displacement components are known. For instance, the continuous beam in Figure 3.10 is fixed at C and has roller supports at A and B. The fixity at C prevents any displacement at this end, while the roller supports at A and B prevent translation in the vertical direction but allow rotation. We should note that roller supports are assumed to be capable of resisting both downward and upward forces.

If we assume that the axial stiffness of the beam is so large that the change in its length due to axial forces can be ignored, there will be no horizontal displacements at A or at B. Therefore, the only unknown displacements at the joints are the rotations D_1 and D_2 at A and B, respectively (Figure 3.10). The displacements D_1 and D_2 are independent of one another, as either can be given an arbitrary value by the introduction of appropriate forces.

A system of joint displacements is called *independent* if each displacement can be varied arbitrarily and independently of all the others. In the structure in Figure 3.10, for example, the rotations D_1 and D_2 are independent because any of the two, say D_1, can be varied while maintaining D_2 unchanged. This can be achieved by applying a couple of appropriate magnitude at A, while preventing the rotation at B by another couple. The number of the independent joint displacements in a structure is called the *degree of kinematic indeterminacy* or the *number of degrees of freedom*. This number is a sum of the degrees of freedom in rotation and in translation. The latter is sometimes called *freedom in sidesway*.

As an example of the determination of the number of degrees of freedom, let us consider the plane frame ABCD of Figure 3.11a. The joints A and D are fixed, and the joints B and C each have three components of displacement $D_1, D_2, \ldots, D_6$, as indicated in the figure. However, if the change in length of the members due to axial forces is ignored, the six displacements are not independent, as the translation of the joints B and C is in a direction perpendicular to the original direction of the member. If an arbitrary value is assigned to any one of the translation displacements D_1, D_2, and D_4 (or D_5), the value of the other three is determined from geometrical relations.

For example, if D_1 is assigned a certain value, D_2 is equal to $(D_1 \cot \theta_1)$ to satisfy the condition that the resultant translation at B is perpendicular to AB. Once the position B′ (joint B after displacement) is defined, the displaced position C′ is defined, since it can have only one location if the lengths BC and CD are to remain unchanged. The joint displacement diagram is shown in Figure 3.11b, in which the displacements D_2, D_4, and D_5 are determined graphically from a given value of D_1. In this figure, BB′ is perpendicular to AB, BCEB′ is a parallelogram, and EC and CC′ are perpendicular to BC and CD, respectively.

From the above discussion, it can be seen that the translation of the joints in the frame considered represents one unknown, or one degree of freedom.

We should note that the translations of the joints are very small compared with the length of the members. For this reason, the translation of the joints is assumed to be along a

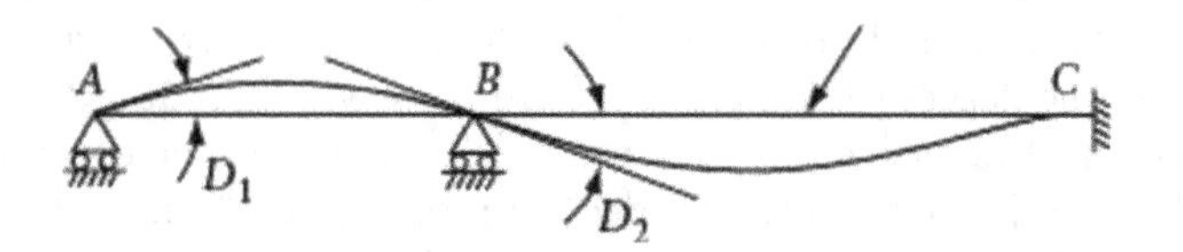

Figure 3.10 Kinematic indeterminacy of a continuous beam.

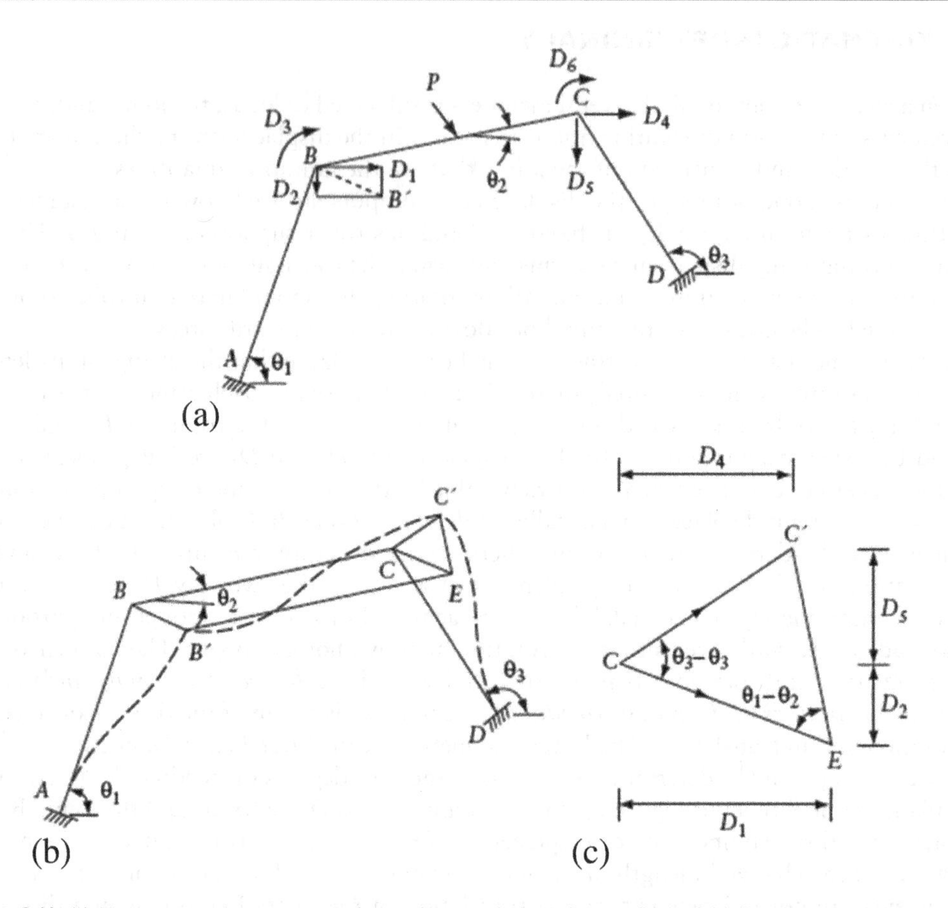

Figure 3.11 (a) Kinematic indeterminacy of a rigid-jointed plane frame. (b) and (c) Displacement diagrams.

straight line perpendicular to the original direction of the member rather than along an arc of a circle. The triangle CC′E is drawn to a larger scale in Figure 3.11c, from which the displacements D_2, D_4, and D_5 can be determined. The same displacements can be expressed in terms of D_1 by simple geometrical relations.

Now, the rotations of the joints at B and C are independent of one another. Thus, the frame of Figure 3.11a has one degree of freedom in sidesway and two degrees of freedom in rotation, so that the degree of kinematic indeterminacy for the frame is three. If the axial deformations are not neglected, the four translational displacements are independent and the total degree of kinematic indeterminacy is six.

The plane frame of Figure 3.12 is another example of a kinematically indeterminate structure. If the axial deformation is neglected, the degree of kinematic indeterminacy is two – the unknown joint displacements being rotations at A and B.

We must emphasize that the kinematic indeterminacy and the statical indeterminacy must not be confused with one another. For instance, the frame of Figure 3.12 has seven reaction components and is statically indeterminate to the fourth degree. If the fixed support at D is replaced by a hinge, the degree of statical indeterminacy will be reduced by one – but at the same time rotation at D becomes possible, thus increasing the kinematic indeterminacy by one. In general, the introduction of a release decreases the degree of statical indeterminacy and increases the degree of kinematic indeterminacy. For this reason, the higher the degree of statical indeterminacy, the more suitable the displacement method for analysis of the structure.

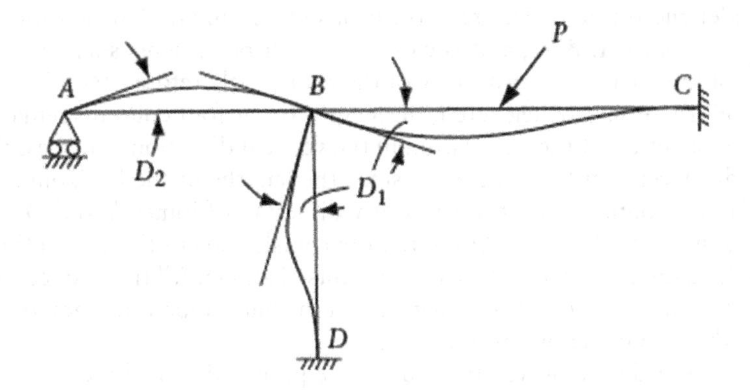

Figure 3.12 Kinematic indeterminacy of a rigid-jointed plane frame.

In a pin-jointed truss with all the forces acting at the joints, the members are subjected to an axial load only (without bending moment or shear) and therefore remain straight. The deformed shape of a plane truss is completely defined if the components of the translation in two orthogonal directions are determined for each joint, and each joint – other than a support – has two degrees of freedom.

Thus, the plane truss of Figure 3.13 is kinematically indeterminate to the second degree, as only joint A can have a displacement that can be defined by components in two orthogonal directions. From Equation 3.4, the degree of statical indeterminacy is three. The addition to the system of an extra bar pinned at A at one end and at a support at the other would not change the degree of kinematic indeterminacy, but would increase the degree of statical indeterminacy by one.

In a pin-jointed space truss loaded at the joints only, the translation of the joints can take place in any direction and can therefore be defined by components in three orthogonal directions, so that each joint – other than a support – has three degrees of freedom. It can be easily shown that the degree of kinematic indeterminacy of the truss of Figure 3.7a is 3, that of the truss of Figure 3.7b is 15, and that of the truss in Figure 3.7c is 12.

Each joint of a rigid-jointed space frame can, in general, have six displacement components: three translations in three orthogonal directions, and three rotations, which can be represented by vectors in each of the three orthogonal directions (double-headed arrows).

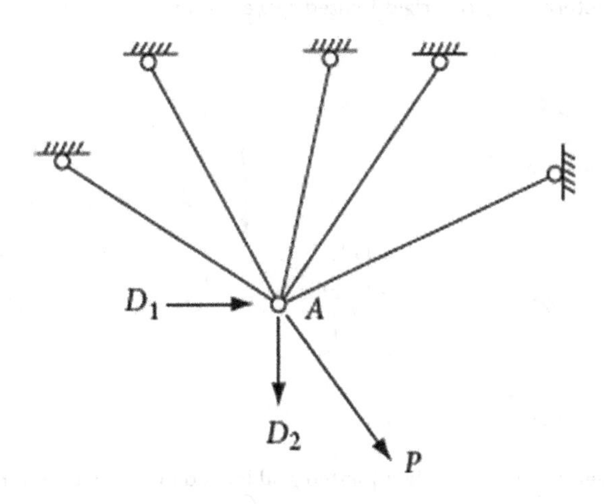

Figure 3.13 Kinematic indeterminacy of a plane truss.

Let us consider the frame of Figure 3.14. It has eight joints, of which four are fixed in space. Each of the joints A, B, C, and D can have six displacements such as those shown at A. The degree of kinematic indeterminacy of the frame is therefore $4 \times 6 = 24$.

If the axial deformations are neglected, the lengths of the four columns remain unchanged so that the component D_3 of the translation in the vertical direction vanishes, thus reducing the unknown displacements by four. Also, since the lengths of the horizontal members do not change, the horizontal translations in the x direction of joints A and D are equal; the same applies to the joints B and C. Similarly, the translations in the y direction of joints A and B are equal; again, the same is the case for joints C and D. All this reduces the unknown displacements by four. Therefore, the degree of kinematic indeterminacy of the frame of Figure 3.14, without axial deformation, is 16.

If a rigid-jointed grid is subjected to loads in the perpendicular direction to the plane of the grid only, each joint can have three displacement components: translation perpendicular to the plane of the grid and rotation about two orthogonal axes in the plane of the grid. Thus, the grid of Figure 3.14 is kinematically indeterminate to the sixth degree. The degree of statical indeterminacy of this grid is 15.

If the beams of a grid in one direction are hinged to the beams in the perpendicular direction in the manner shown in Figure 3.5c, the beams will not be subjected to torsion. Hence, the degree of statical indeterminacy of the grid of Figure 3.15 with hinged connections is eight. On the other hand, the degree of kinematic indeterminacy remains unchanged.

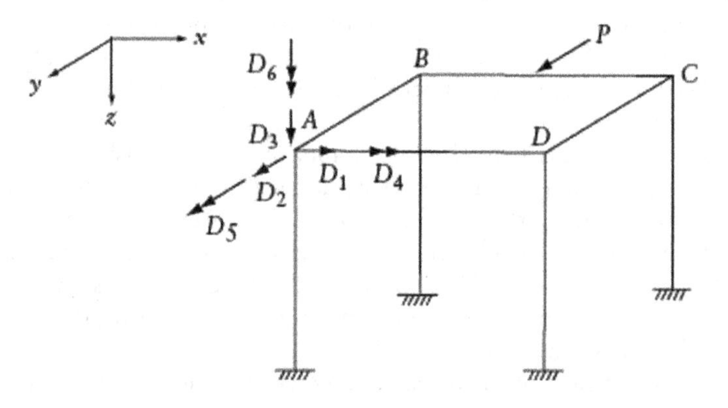

Figure 3.14 Kinematic indeterminacy of a rigid-jointed space frame.

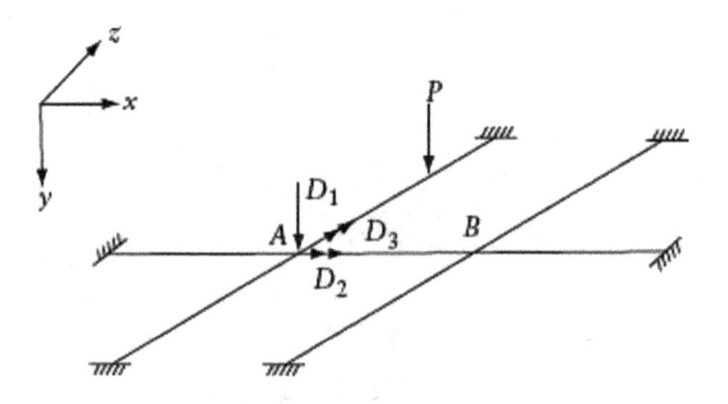

Figure 3.15 Kinematic indeterminacy of a rigid-jointed grid loaded in a direction normal to its plane.

3.6 PRINCIPLE OF SUPERPOSITION

In Section 2.1, we mentioned that when deformations in a structure are proportional to the applied loads the principle of superposition holds. This principle states that the displacement due to a number of forces acting simultaneously is equal to the sum of the displacements due to each force acting separately.

In the analysis of structures it is convenient to use a notation in which a force F_j causes at a point i a displacement D_{ij}. Thus, the first subscript of a displacement describes the position and direction of the displacement, and the second subscript describes the position and direction of the force causing the displacement. Each subscript refers to a *coordinate* which represents the location and direction of a force or of a displacement. The coordinates are usually indicated by arrows on the diagram of the structure.

This approach is illustrated in Figure 3.16a. If the relation between the force applied and the resultant displacement is linear, we can write:

$$D_{i1} = f_{i1}F_1 \tag{3.19}$$

where f_{i1} is the displacement at coordinate i due to a unit force at the location and direction of F_1 (coordinate 1).

If a second force F_2 is applied causing a displacement D_{i2} at i (Figure 3.16b):

$$D_{i2} = f_{i2}F_2 \tag{3.20}$$

where f_{i2} is the displacement at i due to a unit force at coordinate 2.

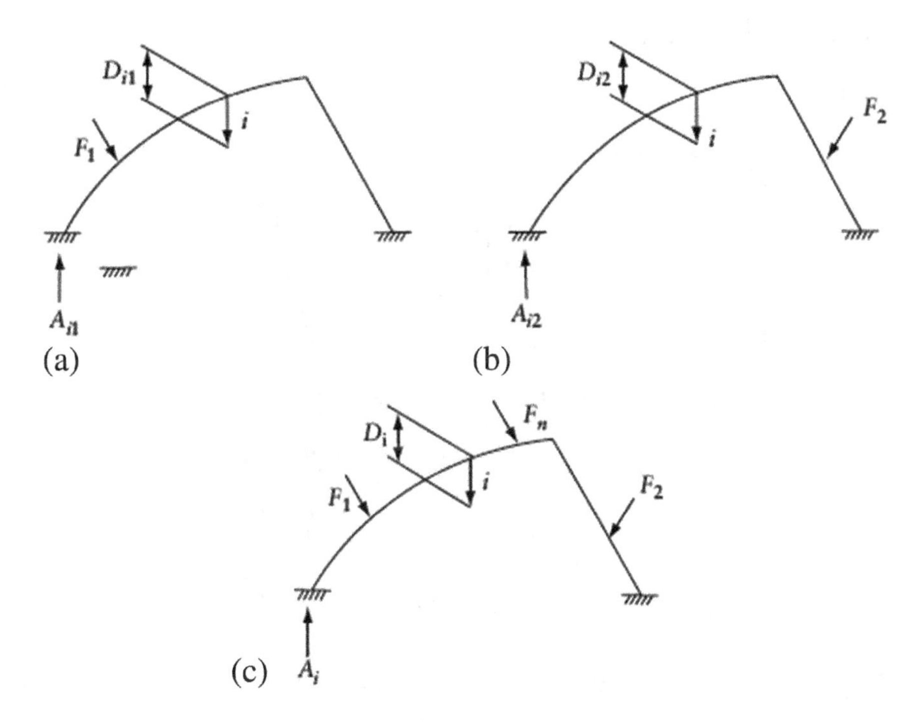

Figure 3.16 Superposition of displacements and forces.

If several forces $F_1, F_2, ..., F_n$ act simultaneously (Figure 3.16c), the total displacement at i is:

$$D_i = f_{i1}F_1 + f_{i2}F_2 + \cdots + f_{in}F_n \tag{3.21}$$

Clearly, the total displacement does not depend on the order of the application of the loads. This, of course, does not hold if the stress–strain relation of the material is nonlinear.

A structure made of material obeying Hooke's law may behave nonlinearly if large changes in the geometry are caused by the applied loads (See Chapter 18). Consider the slender strut in Figure 3.16a subjected to an axial force F_1, not large enough to cause buckling. The strut will, therefore, remain straight, and the lateral displacement at any point A is $D_A = 0$. Now, if the strut is subjected to a lateral load F_2 acting alone, there will be a lateral deflection D_A at point A (Figure 3.17b). If both F_1 and F_2 act (Figure 3.17c), the strut will be subjected to an additional bending moment equal to F_1 multiplied by the deflection at the given section. This additional bending causes additional deflections and the deflection D'_A at A will, in this case, be greater than D_A.

No such bending moment exists, of course, when the loads F_1 and F_2 act separately, so that the combined effect of F_1 and F_2 is not equal to the sum of their separate effects, and the principle of superposition does not hold.

When a structure behaves linearly, the principle of superposition holds for forces as well as for displacements. Thus, the internal stress resultants at any section or the reaction components of the structure in Figure 3.15c can be determined by adding the effects of the forces $F_1, F_2, ..., F_n$ when each acts separately.

Let the symbol A_i indicate a general *action* which may be a reaction, bending moment, shear, or thrust at any section due to the combined effect of all the forces. A general superposition equation of forces can then be written:

$$A_i = A_{ui1}F_1 + A_{ui2}F_2 + \cdots + A_{uin}F_n \tag{3.22}$$

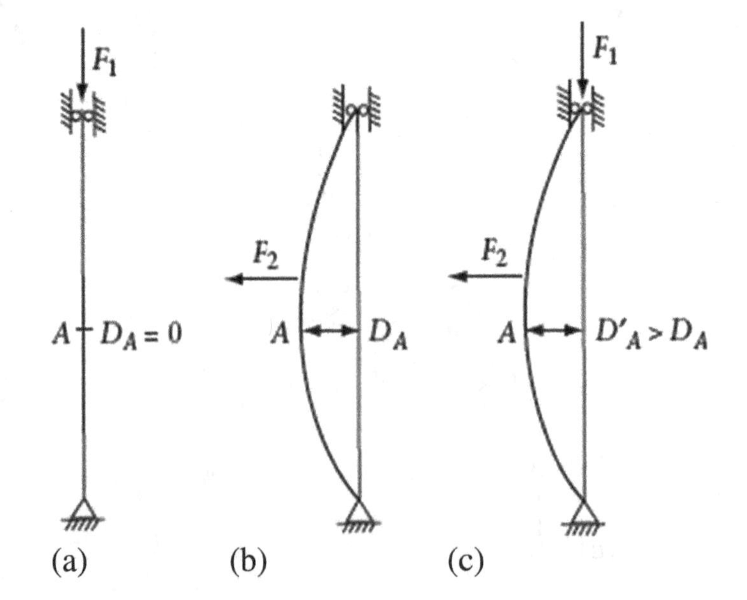

Figure 3.17 Structure to which superposition does not apply.

where A_{ui1} is the magnitude of the action A_i when a unit force is applied alone at coordinate 1. Similarly, A_{ui2}, ..., A_{uin} are the values of the action A_i when a unit force acts separately at each of the coordinates 2, ..., n.

Equation 3.22 can be written in matrix form:

$$A_i = [A_{ui}]_{1\times n} \{F\}_{n\times 1} \tag{3.23}$$

We should note that the superposition of forces of Equation 3.22 holds good for statically determinate structures regardless of the shape of the stress–strain relation of the material, provided only that the loads do not cause a distortion large enough to change appreciably the geometry of the structure. In such structures, any action can be determined by equations of statics alone without considering displacements. On the other hand, in statically indeterminate structures the superposition of forces is valid only if Hooke's law is obeyed because the internal forces depend on the deformation of the members.

3.7 GENERAL

The majority of modern structures are statically indeterminate, and with the flexibility method, it is necessary to establish for a given structure the degree of indeterminacy, which may be external, internal, or both. In simple cases, the degree of indeterminacy can be found by simple inspection, but in more complex or multi-span and multi-bay structures, it is preferable to establish the degree of indeterminacy with the aid of expressions involving the number of joints, members, and reaction components. These expressions are available for plane and space trusses (pin-jointed) and frames (rigid-jointed).

Two general methods of analysis of structures are available. One is the force (or flexibility) method, in which releases are introduced to render the structure statically determinate; the resulting displacements are computed, and the inconsistencies in displacements are corrected by the application of additional forces in the direction of the releases. Hence, a set of compatibility equations is obtained: its solution gives the unknown forces.

In the other method – the displacement (or stiffness) method – restraints at joints are introduced. The restraining forces required to prevent joint displacements are calculated. Displacements are then allowed to take place in the direction of the restraints until the restraints have vanished; hence, a set of equilibrium equations is obtained: its solution gives the unknown displacements. The internal forces on the structure are then determined by superposition of the effects of these displacements and those of the applied loading with the displacements restrained.

The number of restraints in the stiffness method is equal to the number of possible independent joint displacements, which therefore has to be determined prior to the analysis. The number of independent displacements is the degree of kinematic indeterminacy, which has to be distinguished from the degree of statical indeterminacy. The displacements can be in the form of rotation or translation.

The analysis of structures by the force or the displacement method involves the use of the principle of superposition, which allows a simple addition of displacements (or actions) due to the individual loads (or displacements). This principle can, however, be applied only if Hooke's law is obeyed by the material of which a statically indeterminate structure is made. In all cases, the displacements must be small compared with the dimensions of the members so that no gross distortion of geometry of the structure takes place.

PROBLEMS

3.1 to 3.6 What is the degree of statical indeterminacy of the structure shown below? Introduce sufficient releases to render each structure statically determinate.

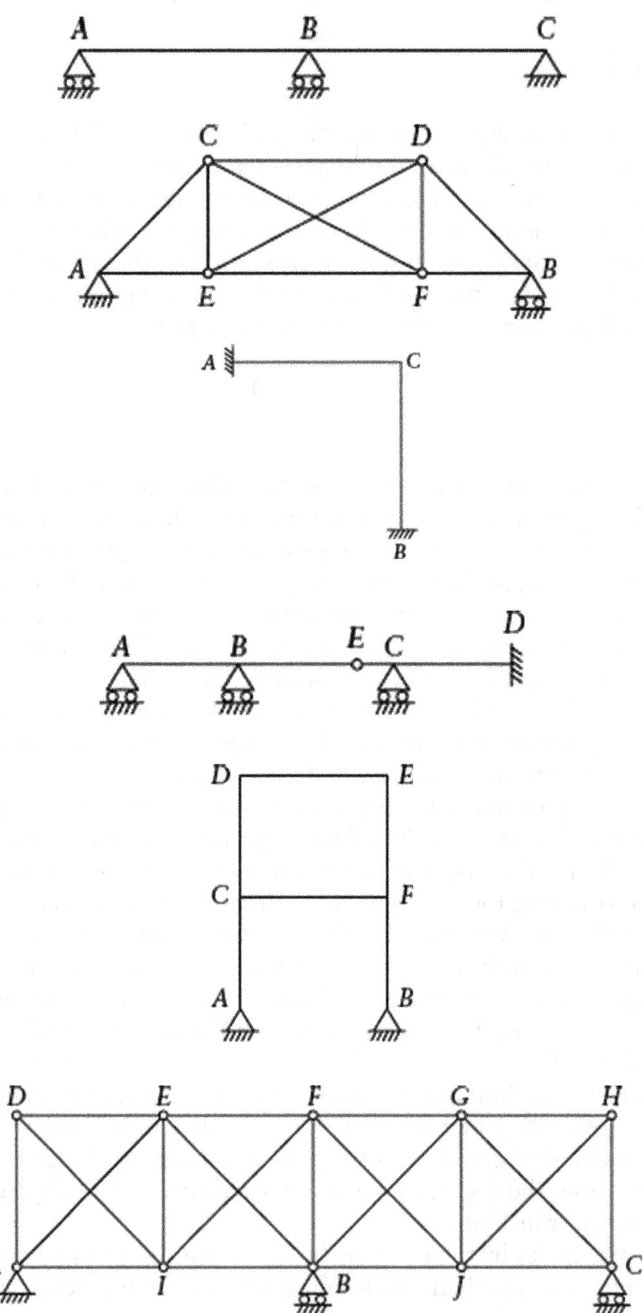

3.7 The horizontal grid shown in the figure is subjected to vertical loads only. What is the degree of statical indeterminacy?

(a) assuming rigid connections at the joints.
(b) assuming connections of the type shown in Figure 3.5c, i.e. a torsionless grid. Introduce sufficient releases in each case to render the structure statically determinate.

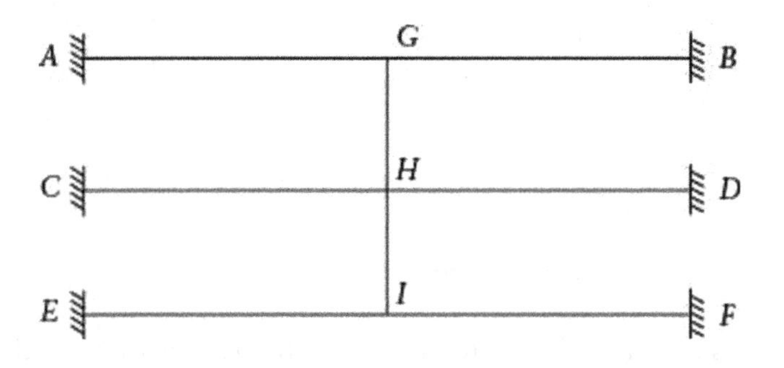

3.8 The figure shows a pictorial view of a space truss pin-jointed to a vertical wall at A, B, C, and D. Determine the degree of statical indeterminacy. Introduce sufficient releases to make the structure statically determinate.

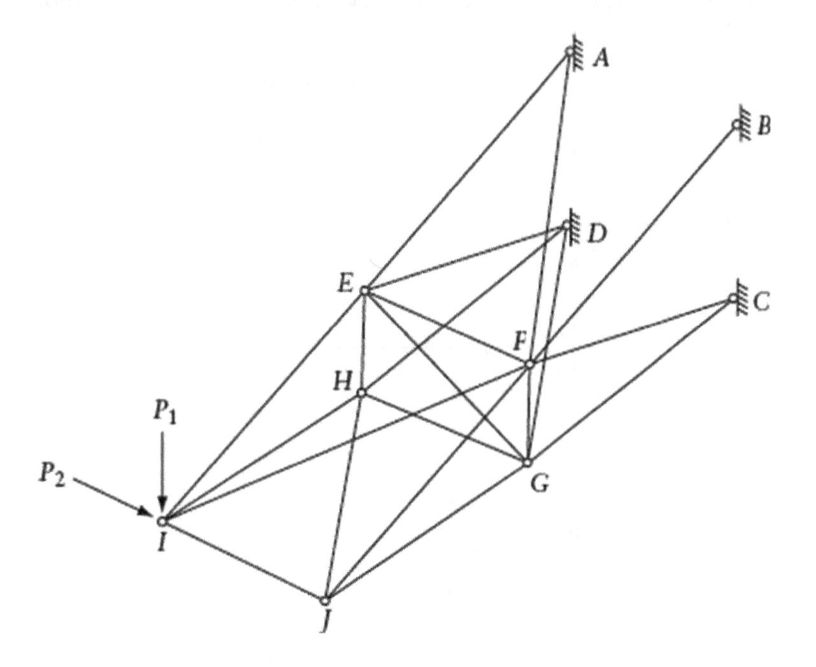

3.9 Determine the degree of kinematic indeterminacy of the beam of Problem 3.1 and indicate a coordinate system of joint displacements. What is the degree of kinematic indeterminacy if the axial deformation is ignored?

3.10 Apply the questions of Problem 3.9 to the frame of Problem 3.5.

3.11 Introduce sufficient releases to render the structure in the figure statically determinate and draw the corresponding bending moment diagram. Draw the bending moment diagram for another alternative released structure.

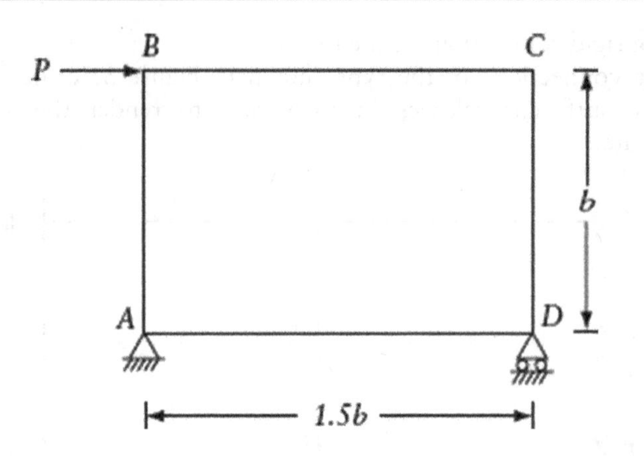

3.12 (a) Introduce sufficient releases to make the frame shown statically determinate. Indicate the releases by a set of coordinates.

(b) Introduce a hinge at the middle of each member and draw the bending moment diagram for the frame due to two horizontal forces, each equal to P, at E and C. Show by a sketch the magnitude and direction of the reaction components at A.

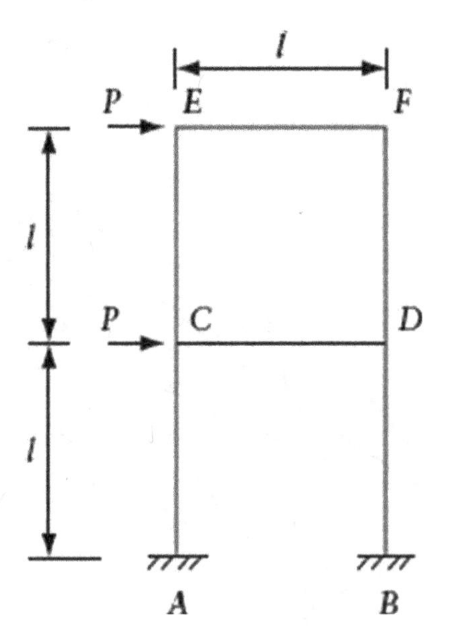

NOTE

1. All through this text, a moment or a rotation is indicated either by an arrow in the form of an arc of a circle (planar structures) or by a double-headed arrow (space structures): see note 1 in Chapter 2.

Chapter 4

Force method of analysis

4.1 INTRODUCTION

As mentioned in Section 3.4, the force method is one of the basic methods of analysis of structures. This chapter outlines the procedure; Chapter 5 will then compare the force and the displacement methods.

4.2 DESCRIPTION OF METHOD

The force method involves five steps. They are briefly mentioned here; but they are explained further in the examples and sections below.

1. First, the degree of statical indeterminacy is determined. A number of releases equal to the degree of indeterminacy is now introduced, each release being made by the removal of an external or an internal force. The releases must be chosen so that the remaining structure is stable and statically determinate. However, we will learn that in some cases the number of releases can be less than the degree of indeterminacy, provided the remaining statically indeterminate structure is so simple that it can be readily analyzed. In all cases, the released forces, which are also called *redundant forces*, should be carefully chosen so that the released structure is easy to analyze.
2. Application of the given loads on the released structure will produce displacements that are inconsistent with the actual structure, such as a rotation or a translation at a support where this displacement must be zero. In the second step, these inconsistencies or "errors" in the released structure are determined. In other words, we calculate the magnitude of the "errors" in the displacements corresponding to the redundant forces. These displacements may be due to external applied loads, settlement of supports, or temperature variation.
3. The third step consists of a determination of the displacements in the released structure due to unit values of the redundants (see Figures 4.1d and e). These displacements are required at the same location and in the same direction as the error in displacements determined in step 2.
4. The values of the redundant forces necessary to eliminate the errors in the displacements are now determined. This requires the writing of superposition equations in which the effects of the separate redundants are added to the displacements of the released structure.
5. Hence, we find the forces on the original indeterminate structure: they are the sum of the correction forces (redundants) and forces on the released structure.

This brief description of the application of the force method will now be illustrated by examples.

DOI: 10.1201/9780429286858-4

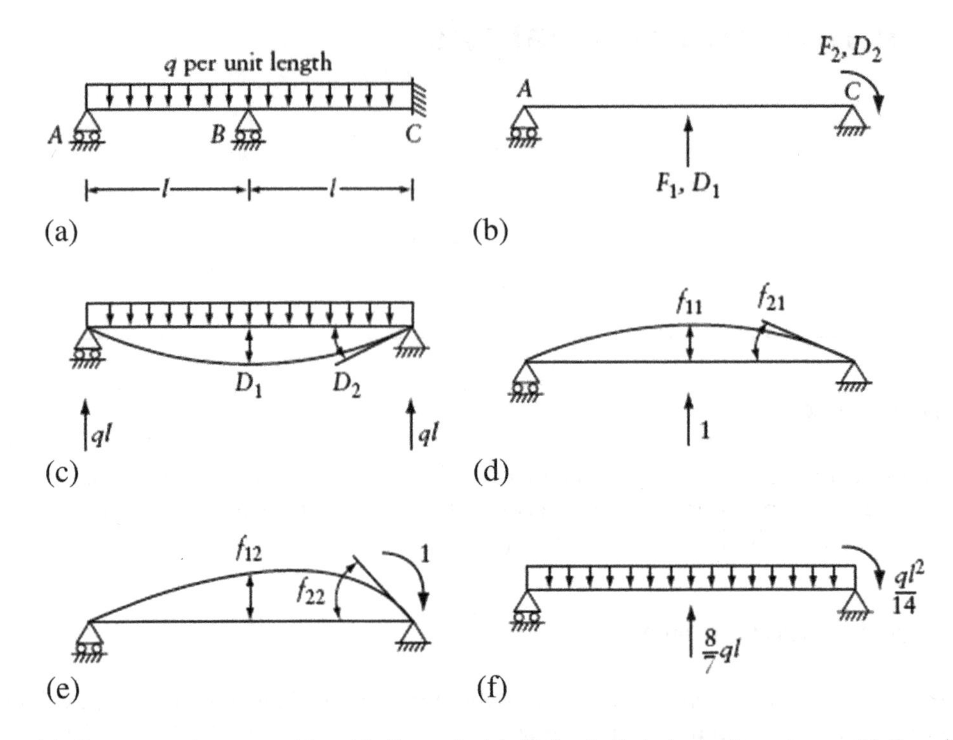

Figure 4.1 Continuous beam considered in Example 4.1. (a) Statically indeterminate beam. (b) Coordinate system. (c) External load on released structure. (d) $F_1 = 1$. (e) $F_2 = 1$. (f) Redundants.

EXAMPLE 4.1: STRUCTURE WITH DEGREE OF INDETERMINACY=2

Figure 4.1a shows a beam ABC fixed at C, resting on roller supports at A and B, and carrying a uniform load of q per unit length. The beam has a constant flexural rigidity EI. Find the reactions of the beam.

COORDINATE SYSTEM

Step 1 The structure is statically indeterminate to the second degree, so that two redundant forces have to be removed. Several choices are possible, e.g. the moment and the vertical reaction at C or the vertical reactions at A and B. For the purposes of this example, we shall remove the vertical reaction at B and the moment at C. The released structure is then a simple beam AC with redundant forces and displacements as shown in Figure 4.1b. The location and direction of the various redundants and displacements are referred to as a *coordinate system.*

The positive directions of the redundants F_1 and F_2 are chosen arbitrarily, but the positive directions of the displacements at the same location must always accord with those of the redundants. The arrows in Figure 4.1b indicate the chosen positive directions in the present case, and, since the arrows indicate forces as well as displacements, it is convenient in a general case to label the coordinates using numerals 1, 2, ..., n.

Step 2 Following this system, Figure 4.1c shows the displacements at B and C as D_1 and D_2, respectively. In fact, as shown in Figure 4.1a, the actual displacements at those points are zero, so that D_1 and D_2 represent the inconsistencies in deformation.

The magnitude of D_1 and D_2 can be calculated from the behavior of the simply supported beam of Figure 4.1c. For the present purposes, we can use Equations A.1 and A.3, Appendix A. Thus,

$$D_1 = -\frac{5ql^4}{24EI} \text{ and } D_2 = -\frac{q^3}{3EI}$$

The negative signs show that the displacements are in directions opposite to the positive directions chosen in Figure 4.1b.

It is good practice to show the selected coordinate system in a separate figure, such as Figure 4.1b, rather than adding arrows to Figure 4.1a. The arbitrary directions selected for the arrows establish the force and displacement sign convention which must be adhered to throughout the analysis. Note that when the release is for an internal force, it must be represented in the coordinate system by a pair of arrows in opposite directions (Figure 4.6b) and the corresponding displacement will be the relative translation or relative rotation of the two sections on either side of the coordinate.

Step 3 The displacements due to the unit values of the redundants are shown in Figure 4.1d and e. These displacements are as follows (Equations A.6, A.7, A.9, and A.12, Appendix A):

$$f_{11} = \frac{l^3}{6EI}; \; f_{12} = f_{21} = \frac{l^2}{4EI}; \; f_{22} = \frac{2l}{3EI}$$

The general coefficient f_{ij} represents the displacement at the coordinate i due to a unit redundant at the coordinate j.

GEOMETRY RELATIONS (COMPATIBILITY EQUATIONS)

Step 4 The *geometry relations* express the fact that the final vertical translation at B and the rotation at C vanish. The final displacements are the result of the superposition of the effect of the external loading and of the redundants on the released structure. Thus, the geometry relations can be expressed as:

$$\left.\begin{aligned} D_1 + f_{11}F_1 + f_{12}F_2 &= 0 \\ D_2 + f_{21}F_1 + f_{22}F_2 &= 0 \end{aligned}\right\} \tag{4.1}$$

A more general form of Equation 4.1 is:

$$\left.\begin{aligned} D_1 + f_{11}F_1 + f_{12}F_2 &= \Delta_1 \\ D_2 + f_{21}F_1 + f_{22}F_2 &= \Delta_2 \end{aligned}\right\} \tag{4.2}$$

where Δ_1 and Δ_2 are prescribed displacements at coordinates 1 and 2 in the actual structure. If, in the example considered, the analysis is required for the combined effects of the given load q and a downward settlement δ_B of support B (Figure 4.1a), we must substitute $\Delta_1 = -\delta_B$, $\Delta_2 = 0$; see Example 4.3, case (2).

FLEXIBILITY MATRIX

The relations of Equation 4.2 can be written in matrix form:

$$[f]\{F\} = \{\Delta - D\} \tag{4.3}$$

where

$$\{D\} = \begin{Bmatrix} D_1 \\ D_2 \end{Bmatrix}; \; [f] = \begin{bmatrix} f_{11} & f_{12} \\ f_{21} & f_{22} \end{bmatrix} \text{ and } \{F\} = \begin{Bmatrix} F_1 \\ F_2 \end{Bmatrix}$$

The column vector $\{\Delta - D\}$ depends on the external loading. The elements of the matrix $[f]$ are displacements due to the unit values of the redundants. Therefore, $[f]$ depends on the properties of the structure and represents the *flexibility* of the released structure. For this reason, $[f]$ is called the *flexibility matrix* and its elements are called *flexibility coefficients*.

We should note that the elements of a flexibility matrix are not necessarily dimensionally homogeneous, as they represent either a translation or a rotation due to a unit load or to a couple. In the above example, f_{11} is a translation due to a unit concentrated load; thus, f_{11} has units (length/force) (e.g. m/N or in./kip). The coefficient f_{22} is a rotation in radians due to a unit couple; thus, its units are (force length)$^{-1}$. Both f_{12} and f_{21} are in (force)$^{-1}$ because f_{12} is a translation due to a unit couple, and f_{21} is a rotation due to a unit load.

The elements of the vector $\{F\}$ are the redundants which can be obtained by solving Equation 4.3; thus,

$$\{F\} = [f]^{-1}\{\Delta - D\} \tag{4.4}$$

In the example considered, the order of the matrices $\{F\}$, $[f]$, and $\{D\}$ is 2×1, 2×2, and 2×1. In general, if the number of releases is n, the order will be $n\times1$, $n\times n$, $n\times1$, respectively. We should note that $[f]$ is a square symmetrical matrix. The generality of this property of the flexibility matrix will be proved in Section 7.7.

In the example considered, the flexibility matrix and its inverse are:

$$[f] = \begin{bmatrix} \dfrac{l^3}{6EI} & \dfrac{l^2}{4EI} \\ \dfrac{l^2}{4EI} & \dfrac{2l}{3EI} \end{bmatrix} \tag{4.5}$$

and

$$[f]^{-1} = \frac{12EI}{7l^3}\begin{bmatrix} 8 & -3l \\ -3l & 2l^2 \end{bmatrix} \tag{4.6}$$

The displacement vector is:

$$\{\Delta - D\} = \frac{ql^3}{24EI}\begin{Bmatrix} 5l \\ 8 \end{Bmatrix}$$

Substituting in Equation 4.4, or solving Equation 4.3, we obtain:

$$\{F\} = \frac{ql}{14}\begin{Bmatrix} 16 \\ l \end{Bmatrix}$$

Therefore, the redundants are:

$$F_1 = \frac{8}{7}ql \quad \text{and} \quad F_2 = \frac{ql^2}{14}$$

The positive sign indicates that the redundants act in the positive directions chosen in Figure 4.1b.

It is important to note that the flexibility matrix is dependent on the choice of redundants: with different redundants, the same structure would result in a different flexibility matrix.

Step 5 The final forces acting on the structure are shown in Figure 4.1f, and any stress resultants in the structure can be determined by the ordinary methods of statics.

The reactions and the internal forces can also be determined by the superposition of the effect of the external loads on the released structure and the effect of the redundants. This can be expressed by the superposition equation:

$$A_i = A_{si} + (A_{ui1} F_1 + A_{ui2} F_2 + \cdots + A_{uin} F_n) \tag{4.7}$$

where:

A_i = any action i, that is, reaction at a support, shearing force, axial force, twisting moment, or bending moment at a section in the actual structure

A_{si} = same action as A_i, but in the released structure subjected to the external loads

$A_{ui1}, A_{ui2}, \ldots, A_{uin}$ = corresponding action due to a unit force acting alone on the released structure at the coordinate 1, 2, ..., n, respectively

$F_1, F_2, \ldots, F_n$ = redundants acting on the released structure.

From Equation 3.21, the term in parentheses in Equation 4.7 represents the action of all the redundants applied simultaneously to the released structure.

Generally, several reactions and internal forces are required. These can be obtained by equations similar to Equation 4.7. If the number of required actions is m, the system of equations needed can be put in the matrix form:

$$\{A\}_{m\times 1} = \{A_s\}_{m\times 1} + [A_u]_{m\times n} \{F\}_{n\times 1} \tag{4.8}$$

The order of each matrix is indicated in Equation 4.8, but it may be helpful to write, on this occasion, the matrices in full. Thus,

$$\{A\} = \begin{Bmatrix} A_1 \\ A_2 \\ \cdots \\ A_m \end{Bmatrix}; \quad \{A_s\} = \begin{Bmatrix} A_{s1} \\ A_{s2} \\ \cdots \\ A_{sm} \end{Bmatrix}; \quad [A_u] = \begin{bmatrix} A_{u11} & A_{u12} & \cdots & A_{u1n} \\ A_{u21} & A_{u22} & \cdots & A_{u2n} \\ \cdots & \cdots & \cdots & \cdots \\ A_{um1} & A_{um2} & \cdots & A_{umn} \end{bmatrix}$$

4.3 RELEASED STRUCTURE AND COORDINATE SYSTEM

In the first step of the force method, it is necessary to draw a figure showing the released structure and a system of numbered arrows. The arrows indicate the locations and the positive directions of the statically indeterminate forces removed from the structure. The removed forces (the releases) can be external, such as the reaction components. As an example, see Figure 4.1b, which indicates the removal of reaction components represented by arrows (coordinates) 1 and 2. The release can also be achieved by the removal of internal forces, e.g. by cutting a member or by introducing a hinge.

When the released force is external, it is to be represented by a single arrow. But, when the released force is internal, e.g. an axial force, a shearing force, or a bending moment, it must be represented by a pair of arrows pointing in opposite directions. Each pair represents a coordinate and thus bears one number (e.g. Figure 4.4b).

The remaining steps of the force method involve the calculation and the use of forces and displacements at the coordinates defined in the first step. Thus, it is impossible to follow or check the calculated forces or displacements (particularly their signs) when the coordinate system is not defined. For this reason, we recommend that the released structure and the coordinate system be represented by a figure that does not show the external applied forces. It should only show a released structure and a set of numbered single arrows, when the released forces are external, or pairs of arrows, when the released forces are internal.

4.3.1 Use of a coordinate represented by a single arrow or a pair of arrows

A coordinate indicates the location and the direction of a force or a displacement. A single arrow represents an external force (e.g. a reaction component) or a displacement in the direction of the arrow. The force can be a concentrated load or a couple; the displacement will then be a translation or a rotation, respectively. Coordinates 1 and 2 in Figure 4.1b represent a vertical reaction at B and a couple (a moment) component of the reaction at C; the same coordinates also represent a vertical translation at B and a rotation at C. The directions of the arrows are arbitrarily chosen; but the choice sets the sign convention that must be followed throughout the analysis.

In Problem 4.1a, the continuous beam in Figure 4.1a is released by the insertion of a hinge at B and the replacement of the totally fixed support at C by a hinged support. The hinge at B releases an internal force (a bending moment), that must be represented by a pair of opposite arrows, jointly denoted coordinate 1; the release of the reaction component at C (a couple) is represented by a single arrow, coordinate 2. The pair of arrows, coordinate 1, represents also the relative rotation (the angular discontinuity) of the two member ends connected by the hinge inserted at B. Coordinate 2 represents also the rotation of the beam end C.

4.4 ANALYSIS FOR ENVIRONMENTAL EFFECTS

The force method can be used to analyze a statically indeterminate structure subjected to effects other than applied loads. An example of such an effect which causes internal stresses is the movement of a support. This may be due to the settlement of foundations or to a differential temperature movement of supporting piers.

Internal forces are also developed in any structure if the free movement of a joint is prevented. For example, the temperature change of a beam with two fixed ends develops an axial force. Stresses in a structure may also be caused by a differential change in temperature.

As an example, let us consider the continuous beam ABC of Figure 4.2a when subjected to a rise in temperature varying linearly between the top and bottom faces. If support B is removed, the beam becomes statically determinate, and the rise in temperature causes it to deflect upward (Figure 4.2b). If the beam is to remain attached to the support at B, a downward force F_1 will develop so as to correct for the error in displacement, D_1. The deflected shape of the beam axis is then as shown in Figure 4.2c.

If a member of a truss is manufactured shorter or longer than its theoretical length and then forced to fit during erection, stresses will develop in the truss. This lack of fit has a similar effect to a change in temperature of the member in question. The effect of shrinkage of concrete members on drying is also similar to the effect of a drop in temperature.

Another cause of internal forces in statically indeterminate structures is the prestrain induced in prestressed concrete members. This may be illustrated by reference to Figure 4.3a, which shows a statically determinate concrete beam of a rectangular cross section. A cable is inserted through a duct in the lower part of the cross section. The cable is then tensioned and anchored at the ends. This produces compression in the lower part of the cross section and causes the beam to deflect upward (Figure 4.3b). If the beam is statically indeterminate,

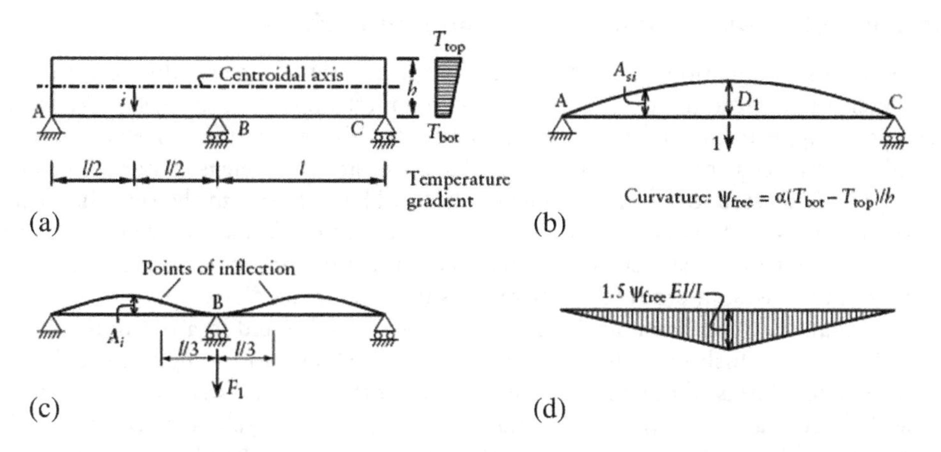

Figure 4.2 Effect of a differential rise in temperature across a continuous beam. (a) Continuous beam subjected to temperature rise. (b) Deflected shape of statically determinate structure. (c) Deflected shape of the statically indeterminate structure in (a). (d) Bending moment diagram.

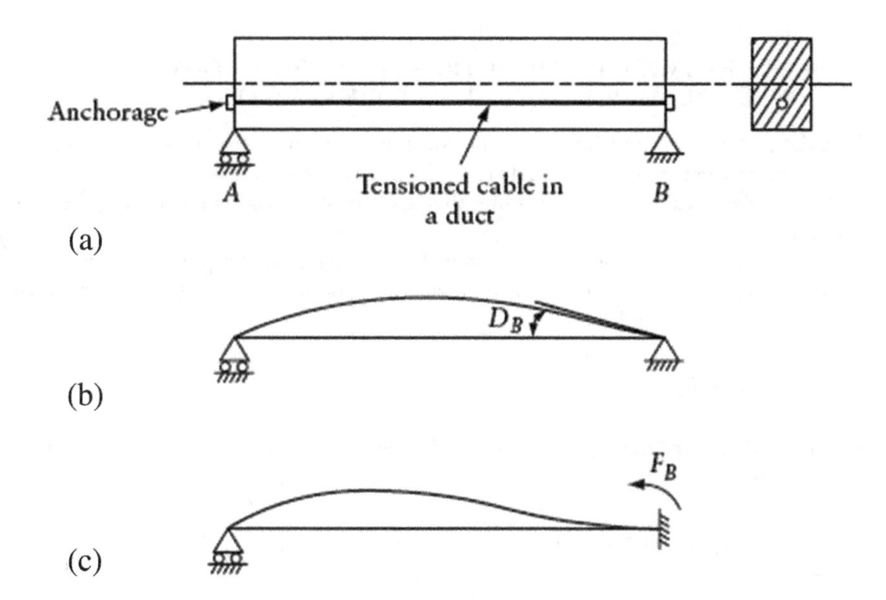

Figure 4.3 Effect of prestrain in a beam.

e.g. if the end B is fixed, the rotation at this end cannot take place freely and a couple will develop at B so as to cause the rotation D_B to vanish (Figure 4.3c).

In all these cases, Equation 4.4 can be applied to calculate the redundant forces, the elements of the matrix $\{D\}$ being the errors in the displacement of the released structure due to the given effect, or to the various effects combined. The effects of temperature, shrinkage, creep, and prestressing are discussed in more detail in Sections 5.9 to 5.11.

We should note that the matrix $\{\Delta\}$ includes the prescribed displacement of the support if this displacement corresponds to one of the coordinates. Otherwise, the effect of the support movement on the displacement of the released structure at a coordinate should be included in the calculation of the displacement $\{D\}$. This is further explained in Example 4.4.

4.4.1 Deflected shapes due to environmental effects

The superposition Equation 4.7 can give the displacement A_i at coordinate i on a statically indeterminate structure due to an environmental effect: temperature, shrinkage, creep, movement of supports, lack of fit, or prestressing. For this application of Equation 4.7, A_i is the required displacement, and A_{si} is the displacement at i of a released statically determinate structure subjected to the environmental effect. The last term in the right-hand side of Equation 4.7 sums up the displacements at i due to the redundants $\{F\}$. We recognize that, in a statically determinate structure, an environmental effect produces strains and displacements without stresses, internal forces (stress resultants), or reactions.

We refer again to the continuous beam in Figure 4.2a, subjected to a rise in temperature, the distribution of which over the depth h is linear as shown in the figure; the deflection A_i of the centroidal axis of the beam is equal to A_{si} of the released structure in Figure 4.2b plus the deflection at i due to the redundant F_1. Note that the released structure deflects while its internal forces and reactions are nil; at any section, the bending moment M is not directly related to the curvature, ψ of the deflected shape in Figure 4.2c, by the relationship: $\psi = M/EI$ (Equation 1.24); e.g. the points of inflection (where $\psi = 0$) on both sides of support B do not correspond to sections of zero bending moment.

EXAMPLE 4.2 DEFLECTION OF A CONTINUOUS BEAM DUE TO TEMPERATURE VARIATION

Find the deflection A_i at coordinate i of the continuous beam in Figure 4.2a subjected to a rise in temperature that varies linearly over the depth h from T_{top} to T_{bot} at top and bottom, respectively, with $T_{top} > T_{bot}$. Assume that EI=constant and consider only bending deformation.

A released structure obtained by the removal of the reaction component at coordinate 1 is shown in Figure 4.2b. The rise of temperature produces constant curvature in the released structure given by Equation 1.7:

$$\psi_{free} = \alpha\left(T_{bot} - T_{top}\right)/h$$

where α is coefficient of thermal expansion. The deflections at B and at i are (Equations A.39 and A.40):

$$D_1 = \psi_{free}\,(2l)^2/8 = 0.5\,\psi_{free}\,l^2$$

$$A_{si} = \psi_{free}\,(0.5l)(1.5l)/2 = 0.375\,\psi_{free}\,l^2$$

Displacements due to $F_1 = 1$ are (Equations A.8 and A.5):

$$f_{11} = (2l)^3/(48EI) = l^3/(6EI)$$

$$A_{u1i} = \frac{l(0.5l)}{6(2l)EI}\left[2l(2l) - l^2 - (0.5l)^2\right] = 0.1146\,l^3/(EI)$$

Equation 4.4 gives the redundant:

$$F_1 = -f_{11}^{-1}\,D_1 = -\frac{6EI}{l^3}\left(0.5\,\psi_{free}\,l^2\right) = -\frac{3EI}{l}\,\psi_{free}$$

The deflection at i is (Equation 4.7):

$$A_i = A_{si} + A_{u1i} F_1$$

$$A_i = 0.375\,\psi_{free}\, l^2 + 0.1146\frac{l^3}{EI}\left(-\frac{3EI}{l}\psi_{free}\right)$$

$$= 31.2\times10^{-3}\,\psi_{free}\, l^2$$

Because $T_{top} > T_{bot}$, ψ_{free} is negative, and the deflection at i is upward.

4.5 ANALYSIS FOR DIFFERENT LOADINGS

When using Equation 4.3 to find the redundants in a given structure under a number of different loadings, the calculation of the flexibility matrix (its inverse) and $[A_u]$ need not be repeated. When the number of loadings is p, the solution can be combined into one matrix equation:

$$[F]_{n\times p} = [f]^{-1}_{n\times n}\,[\Delta - D]_{n\times p} \tag{4.9}$$

where each column of $[F]$ and $[\Delta - D]$ corresponds to one loading.

The reactions or the stress resultants in the original structure can be determined from equations similar to Equation 4.8, viz:

$$[A]_{m\times p} = [A_s]_{m\times p} + [A_u]_{m\times n}\,[F]_{n\times p} \tag{4.10}$$

Each column of $[A]$ corresponds to a loading case.

4.6 FIVE STEPS OF THE FORCE METHOD

The analysis by the force method involves five steps which are summarized as follows:

Step 1 Introduce releases and define a system of coordinates. Also define $[A]_{m\times p}$, the required actions, and define their sign convention (if necessary).

Step 2 Due to the loadings on the released structure, determine $[D]_{n\times p}$, and $[A_s]_{m\times p}$. Also, fill in the prescribed displacements $[\Delta]_{n\times p}$.

Step 3 Apply unit values of the redundants one by one on the released structure and generate $[f]_{n\times n}$ and $[A_u]_{m\times n}$.

Step 4 Solve the geometry equations:

$$[f]_{n\times n}\,[F]_{n\times p} = [\Delta - D]_{n\times p} \tag{4.11}$$

This gives the redundants $[F]_{n\times p}$.

Step 5 Calculate the required actions by superposition:

$$[A]_{m\times p} = [A_s]_{m\times p} + [A_u]_{m\times n}\,[F]_{n\times p} \tag{4.12}$$

At the completion of step 3, all the matrices necessary for the analysis have been generated. The last two steps involve merely matrix algebra. Step 5 may be eliminated when no action besides the redundants is required, or when the superposition can be done by inspection after determination of the redundants (see Example 4.5). When this is the case, the matrices $[A]$, $[A_s]$, and $[A_u]$ are not required.

For quick reference, the symbols used in this section are defined again as follows:

n, p, m = number of redundants, number of loading cases, and number of actions required.
$[A]$ = the required actions (the answers to the problem).
$[A_s]$ = values of the actions due to the loadings on the released structure.
$[A_u]$ = values of the actions in the released structure due to unit forces applied separately at each coordinate.
$[D]$ = displacements of the released structure at the coordinates due to the loadings; these displacements represent incompatibilities to be eliminated by the redundants.
$[\Delta]$ = prescribed displacements at the coordinates in the actual structure; these represent imposed displacements to be maintained.
$[f]$ = flexibility matrix.

EXAMPLE 4.3 A STAYED CANTILEVER

Figure 4.4a shows a cantilever stayed by a link member AC. Determine the reaction components $\{R_1, R_2\}$ at B due to the combined effect of the uniform load shown and a drop of temperature T degrees in AC only. Assume that:

$$a_{AC} = 30 I_{AB}/l^2; \quad \alpha T = \left(ql^3/EI_{AB}\right)/40$$

where a_{AC} and I_{AB} are the cross-sectional area and second moment of area of AC and AB respectively; α is coefficient of thermal expansion (degree^{-1}) for AC; E is modulus of elasticity of the structure.

Step 1 Figure 4.4b shows the structure released by cutting AC. The arbitrarily chosen directions of the arrows representing coordinate 1 indicate that F_1 is positive when the force in AC is tensile. The same arrows also indicate that D_1 is positive when the gap at the cut section closes.

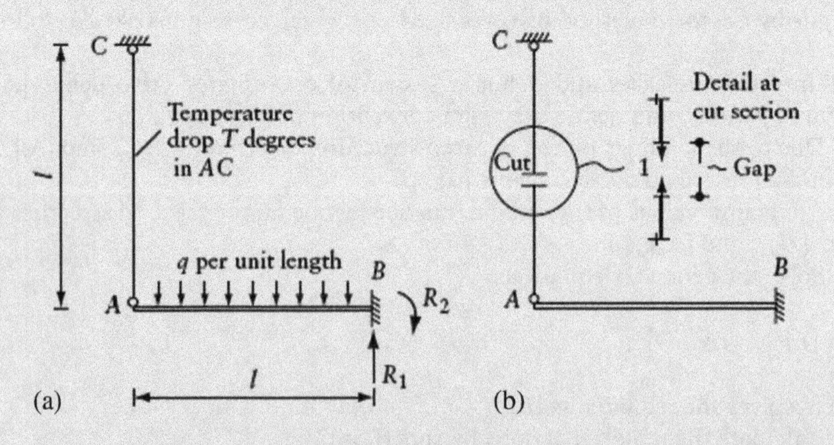

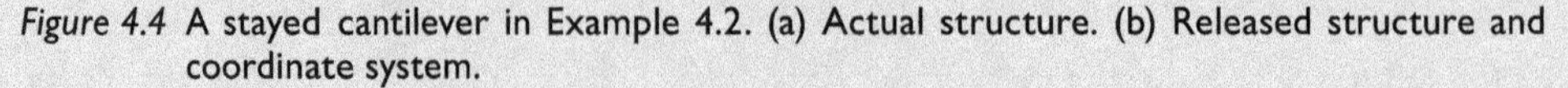
Figure 4.4 A stayed cantilever in Example 4.2. (a) Actual structure. (b) Released structure and coordinate system.

The required actions are:

$$\{A\} = \begin{Bmatrix} R_1 \\ R_2 \end{Bmatrix}$$

Step 2 Application of the load and the temperature change to the released structure produces the following displacement and reactions:

$$D_1 = -\left(\frac{ql^4}{8EI}\right)_{AB} - (\alpha T l)_{AC} = -0.15\frac{ql^4}{EI_{AB}}$$

$$\{A_s\} = \begin{Bmatrix} ql \\ ql^2/2 \end{Bmatrix}$$

The load q produces a downward deflection at A (given by Equation A.27, Appendix A) and opens the gap at the cut section by the same amount. Also, the drop in temperature in AC opens the gap at the cut section. Thus, the two terms in the above equation for D_1 are negative.

Step 3 Application of two equal and opposite unit forces F_1 causes the following displacement and actions in the released structure:

$$f_{11} = \left(\frac{l^3}{3EI}\right)_{AB} + \left(\frac{l}{Ea}\right)_{AC} = \frac{11}{30}\frac{l^3}{EI_{AB}}$$

$$\{A_u\} = \begin{Bmatrix} -1 \\ -l \end{Bmatrix}$$

The first term in the above equation for f_{11} is given by Equations A.19 and A.21.

Step 4 In this example, the compatibility Equation 4.11 expresses the fact that the actual structure has no gap; thus:

$$f_{11}F_1 = \Delta_1 - D_1$$

with $\Delta_1 = 0$.

The solution gives the force necessary to eliminate the gap:

$$F_1 = f_{11}^{-1}(\Delta_1 - D_1)$$

$$F_1 = \left(\frac{11l^3}{30EI_{AB}}\right)^{-1}\left[0 - \left(-0.15\frac{ql^4}{EI_{AB}}\right)\right] = 0.409ql$$

Step 5 The superposition Equation 4.12 gives the required reaction components:

$$\{A\} = \{A_s\} + [A_u]F_1$$

$$\{A\} = \begin{Bmatrix} ql \\ ql^2/2 \end{Bmatrix} + \begin{bmatrix} -1 \\ -l \end{bmatrix} 0.409ql = \begin{Bmatrix} 0.591ql \\ 0.091ql^2 \end{Bmatrix}$$

EXAMPLE 4.4 A BEAM WITH A SPRING SUPPORT

Solve Example 4.3, replacing the link member AC in Figure 4.4a by a spring support below A, of stiffness $K=30(EI)_{AB}/l^3$, where K is the magnitude of the force per unit shortening of the spring. Assume that the spring can produce only a vertical force at the tip A of the cantilever. The loading is q/unit length on AB, as in Example 4.2, combined with a rise in temperature of the spring that would cause it to expand, if it were free, by $[ql^4/(EI)_{AB}]/40$.

Step 1 Release the structure by separating the spring from the cantilever, creating a gap and coordinate 1, represented by a pair of arrows pointing upward on the cantilever and downward on the spring.

Step 2 The displacement D_1, due to the load on AB and the temperature rise of the spring, is:

$$D_1 = -\frac{ql^4}{8(EI)_{AB}} - \frac{ql^4}{40(EI)_{AB}} = -0.15\frac{ql^4}{(EI)_{AB}}$$

The first term on the right-hand side is given by Equation A.27; the second term is the given expansion of the spring. Both terms are negative because they are translations in the opposite direction of coordinate 1.

Step 3 A pair of forces $F_1=1$ produces relative translations (widening the gap):

$$f_{11} = \frac{l^3}{3(EI)_{AB}} + \frac{1}{k} = \frac{l^3}{3(EI)_{AB}} + \frac{l^3}{30(EI)_{AB}} = \frac{11l^3}{30(EI)_{AB}}$$

Steps 4 and 5 The same as in Example 4.3.

EXAMPLE 4.5 SIMPLY SUPPORTED ARCH WITH A TIE

Figure 4.5a represents a concrete arch with a steel tie. Determine the bending moments at C, D, and E due to: (1) the loads shown; (2) the loads shown combined with a rise of

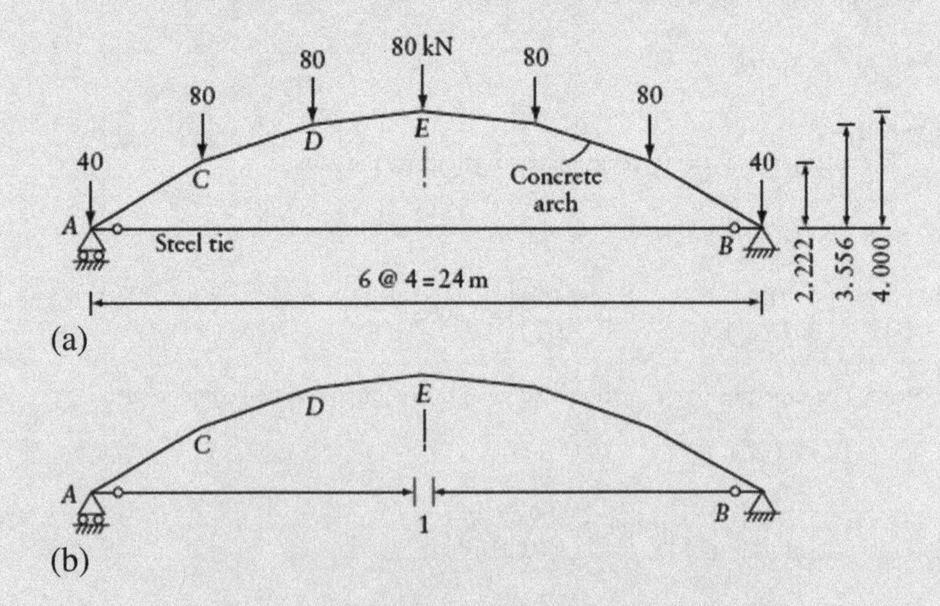

Figure 4.5 Arch with a tie in Example 4.5. (a) Actual structure. (b) Released structure and coordinate system.

temperature, $T = 30°C$ only in the tie AB. Consider only the bending deformation of the arch and the axial deformation of the tie. Given data:

For arch AEB: $I = 2.133 \times 10^{-3}$ m^4; $E = 40$ GPa
For tie AB: $a = 2000$ mm^2; $E = 200$ GPa; $\alpha = 10 \times 10^{-6}$ per degree Celsius

where a and I are area and second moment of area of a cross section; E is modulus of elasticity and α is coefficient of thermal expansion.

We follow below the five steps of the force method. The displacements required in Steps 2 and 3 are presented without calculation. Their values can be verified using virtual work, as discussed in Section 7.7. We concentrate on understanding the steps of the force method, leaving out the details of displacement calculation. The units used below are N and m.

Step 1 The structure is released by cutting the tie. Since the release is an internal force, it must be represented by a pair of arrows, constituting coordinate 1 (Figure 4.5b). With the directions of the arrows arbitrarily chosen in Figure 4.5b, F_1 and D_1 are considered positive when the force in the tie is tensile and when the gap at the cut section closes. The required actions are:

$$[A] = \left[\begin{Bmatrix} M_C \\ M_D \\ M_E \end{Bmatrix}_{\text{Case 1}} \quad \begin{Bmatrix} M_C \\ M_D \\ M_E \end{Bmatrix}_{\text{Case 2}} \right] \tag{a}$$

We consider that positive M produces tension at the inner face of the arch.

Step 2 The following displacements and actions (bending moments) occur when the released structure is subjected to the loading cases (1) and (2):

$$\left[D_{1\,\text{Case 1}} \quad D_{1\,\text{Case 2}} \right] = \left[-0.85109 \quad -0.84389 \right] \text{m} \tag{b}$$

$$[A_s] = \left[\{A_s\}_{\text{Case 1}} \quad \{A_s\}_{\text{Case 2}} \right] = 10^3 \begin{bmatrix} 800 & 800 \\ 1280 & 1280 \\ 1440 & 1440 \end{bmatrix} \text{N-m} \tag{c}$$

We note that with the tie cut, the bending moment values, with or without temperature change, are the same as the values for a simple beam having the same span and carrying the same vertical loads. The downward forces on the arch cause the roller at A to move outwards and the gap at the cut section to open a distance $= 0.85109$ m. The thermal expansion of the tie reduces the opening of the gap by the value $(\alpha T l)_{\text{tie}} = 10 \times 10^{-6}$ (30) (24) $= 7.2 \times 10^{-3}$ m, where l_{tie} is the length of the tie.

Step 3 Opposite unit forces (1 N) at coordinate 1 produce bending moments at C, D, and E equal to {−2.222, −3.556, −4.000} N-m; these values are the elements of $[A_u]$. The same unit forces close the gap at the cut section by a distance equal to:

$$[f_{11}] = [2.4242 \times 10^{-6}] \text{ m/N} \tag{d}$$

This is equal to the sum of an inward movement of the roller at A equal to 2.3642×10^{-6} and an elongation of the tie $= (l/aE)_{\text{tie}} = 24/[2000 \times 10^{-6}(200 \times 10^9)] = 0.0600 \times 10^{-6}$.

$$[A_u] = \begin{bmatrix} -2.222 \\ -3.556 \\ -4.000 \end{bmatrix} \text{N-m/N} \tag{e}$$

In calculating the displacements in Steps 2 and 3, the deformations due to shear and axial forces in the arch are ignored because their effect in this type of structure is small. Calculation of the contribution of any internal force in a framed structure can be done by virtual work (Section 7.7).

Step 4 The compatibility Equation 4.11 and its solution in this example are:

$$[f_{11}]\left[(F_1)_{\text{Case 1}} \quad (F_1)_{\text{Case 2}}\right]=\left[(\Delta-D_1)_{\text{Case 1}} \quad (\Delta-D_1)_{\text{Case 2}}\right]$$

$$\left[(F_1)_{\text{Case 1}} \quad (F_1)_{\text{Case 2}}\right]=[f_{11}]^{-1}\left[(\Delta-D_1)_{\text{Case 1}} \quad (\Delta-D_1)_{\text{Case 2}}\right]$$

$$\begin{aligned}[F]&=\left[2.4242\times10^{-6}\right]^{-1}\left[\{0-(-0.85109)\} \quad \{0-(-0.84389)\}\right]\\ &=\left[351.08 \quad 348.11\right]10^3\text{ N}\end{aligned} \tag{f}$$

In the two loading cases $\Delta_1=0$ because no displacement is prescribed at coordinate 1.

Step 5 The superposition Equation 4.12 gives the required values of bending moments in the two loading cases:

$$[A]=[A_s]+[A_u][F]$$

Substitution of Equations (a), (c), (e) and (f) gives:

$$[A]=\left[\begin{Bmatrix}M_C\\M_D\\M_E\end{Bmatrix}_{\text{Case 1}} \quad \begin{Bmatrix}M_C\\M_D\\M_E\end{Bmatrix}_{\text{Case 2}}\right]=\begin{bmatrix}19.9 & 26.5\\31.6 & 42.1\\35.7 & 47.6\end{bmatrix}10^3\text{ N-m}$$

These bending moments are much smaller than the moments in a simple beam of the same span carrying the same downward forces. This example shows an advantage of an arch with a tie, that is, covering large spans without developing large bending moments.

EXAMPLE 4.6 CONTINUOUS BEAM: SUPPORT SETTLEMENT AND TEMPERATURE CHANGE

Find the bending moments M_B and M_C and the reaction R_A for the continuous beam of Example 4.1 (Figure 4.1), due to the separate effect of: (1) a downward settlement δ_A of support A; (2) a downward settlement δ_B of support B; (3) a rise of temperature varying linearly over the depth h, from T_t to T_b in top and bottom fibers, respectively.

Step 1 We choose the releases and the coordinate system as in Example 4.1, Figure 4.1b. The required actions are:

$$[A]=\left[\begin{Bmatrix}M_B\\R_A\end{Bmatrix}_1 \quad \begin{Bmatrix}M_B\\R_A\end{Bmatrix}_2 \quad \begin{Bmatrix}M_B\\R_A\end{Bmatrix}_3\right]$$

Bending moment is considered positive when it produces tensile stress in the bottom fiber. Upward reaction R_A is positive. The required action M_C does not need to be included in [A] because $M_C=-F_2$ and the values of the redundants {F} will be calculated in Step 4. The subscripts 1, 2, and 3 in the above equation refer to the three load cases.

Step 2 The released structure is depicted in Figures 4.6a and b for cases (1) and (3), respectively. The displacement vectors {Δ} and {D} in the three cases are:

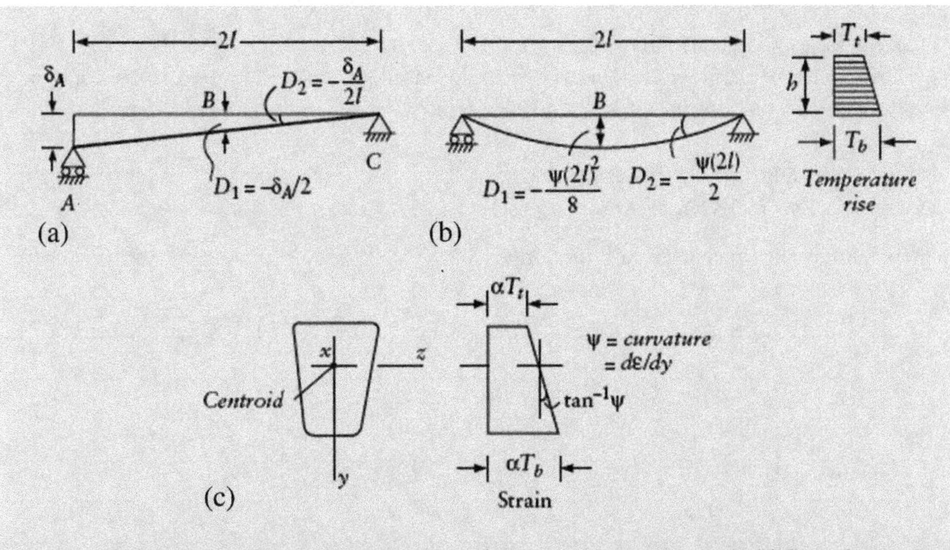

Figure 4.6 Released structure in Example 4.4. (a) Settlement δ_A at support A. (b) Rise of temperature varying linearly over the depth *h*. (c) Strain due to temperature rise.

$$[\Delta]=\begin{bmatrix}0 & -\delta_B & 0\\ 0 & 0 & 0\end{bmatrix};\quad [D]=\begin{bmatrix}-\delta_B/2 & 0 & -\psi(2l)^2/8\\ -\delta_B/(2l) & 0 & -\psi(2l)/2\end{bmatrix}$$

Here ψ is thermal curvature in the released structure (slope of the strain diagram, Figure 4.6c):

$$\psi=\alpha\left(T_b-T_t\right)/h \tag{4.13}$$

where α is the thermal expansion coefficient (degree^{-1}); see Equations A.39 and A.41 for values of $\{D\}$ in case (3).

Note that in case (1), $\{\Delta\}=\{0\}$ because the actual structure has zero displacements at coordinates 1 and 2; however, the released structure has displacements to be eliminated at the coordinates: $\{D\} = \{-\delta_A/2, -\delta_A/2l\}$.

In case (2), the imposed displacements to be maintained are a downward settlement δ_b at coordinate 1 without rotation at coordinate 2; thus, $\{\Delta\}=\{-\delta_b, 0\}$. No load is applied on the released structure in this case; thus, $\{D\}=\{0, 0\}$.

The values of the actions in the released structure are zero in all three cases:

$$[A_s]=[0]_{2\times 3}$$

Step 3 Unit forces applied at the coordinates are represented in Figures 4.1d and e. The flexibility matrix $[f]$ and its inverse, determined in Example 4.1, apply (Equations 4.5 and 4.6). Values of the actions due to $F_1=1$ or $F_2=1$ are

$$[A_u]=\begin{bmatrix}-0.5l & -0.5\\ -0.5 & -1/(2l)\end{bmatrix}$$

Step 4 Solve the geometry for Equation 4.11: $[f]\,[F]=[\Delta - D]$

$$\frac{1}{EI}\begin{bmatrix}l^3/6 & l^2/4\\ l^2/4 & 2l/3\end{bmatrix}[F]=\begin{bmatrix}\delta_A/2 & -\delta_B & \psi l^2/2\\ \delta_A/(2l) & 0 & \psi l\end{bmatrix}$$

The solution for $[F]$ is: $[F]=[f]^{-1}[\Delta - D]$

The inverse of the flexibility matrix of the released structure (Figure 4.1b) is done in Equation 4.6. Substitution for $[f]^{-1}$ and $[\Delta - D]$ gives:

$$[F]=\frac{12EI}{7l^3}\begin{bmatrix} 2.5\delta_A & -8\delta_B & \psi l^2 \\ -0.5l\delta_A & 3l\delta_B & 0.5\psi l^3 \end{bmatrix}$$

Step 5 Apply the superposition Equation 4.12:

$$[A]=[A_s]+[A_u][F]$$

$$[A]=[0]+\begin{bmatrix} -0.5l & -0.5 \\ -0.5 & -0.5l \end{bmatrix}\frac{12EI}{7}\begin{bmatrix} \dfrac{2.5\delta_A}{l^3} & \dfrac{-8\delta_B}{l^3} & \dfrac{\psi}{l} \\ \dfrac{-0.5\delta_A}{l^2} & \dfrac{3\delta_B}{l^2} & 0.5\psi \end{bmatrix}$$

$$=\frac{12EI}{7}\begin{bmatrix} -\delta_A/l^2 & 2.5\delta_B/l^2 & -0.75\psi \\ -\delta_A/l^3 & 2.5\delta_B/l^3 & -0.75\psi/l \end{bmatrix}$$

The elements on the three columns of $[A]$ are the required values of M_B and R_A in the three cases. Because $M_C=-F_2$, reversal of the sign of the elements on the second row of $[F]$ gives the required values of M_C in the three cases:

$$[M_C]=\frac{12EI}{7l^3}\begin{bmatrix} 0.5l\delta_A & -3l\delta_B & -0.5\psi l^3 \end{bmatrix}$$

We should note that R_A, M_B, and M_C are proportional to the value of the product EI. In general, the reactions and the internal forces caused by support settlements or temperature variation of statically indeterminate structures are proportional to the value of EI employed in linear analysis.

Design of concrete structures commonly allows cracking to take place, resulting in a substantial reduction in the moment of inertia, I (the second moment of area) at the cracked sections. Ignoring cracking can greatly overestimate the effects of support movements and of temperature.[1] Also, using a modulus of elasticity E, based on the relation between stress and instantaneous strain, will overestimate the effects of settlement and temperature. This is because such an approach ignores creep, that is, the strain which develops gradually with time (months or years) under sustained stress.

4.7 MOVING LOADS ON CONTINUOUS BEAMS AND FRAMES

The live load on continuous beams and frames is often represented in design by a uniformly distributed load which may occupy any part of the structure so as to produce the maximum value of an internal force at a section or the maximum reaction at a support. We shall now discuss which parts of a continuous beam should be covered by the live load to produce these maxima.

Figure 4.7a and b show the deflected shapes, reactions, and bending moment and shearing force diagrams for a continuous beam due to a uniform live load covering one span only. It can be seen that the deflection is largest in the loaded span and reverses sign, with much smaller values, in adjacent spans. The two reactions at either end of the loaded span are

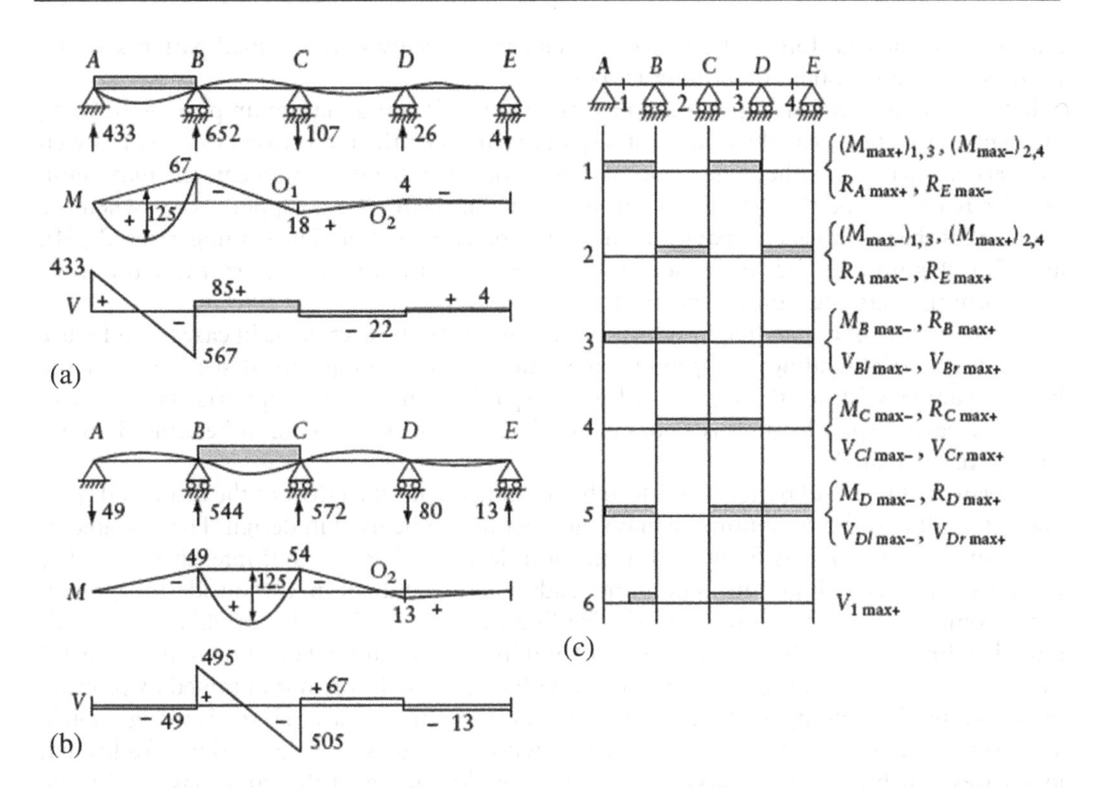

Figure 4.7 Effect of uniform live load on continuous beams. (a) and (b) Deflection, bending moment and shearing force diagrams due to load p per unit length covering one span. The values given are for equal spans l; the multipliers are $10^{-3}\ pl$ for reactions and shears and $10^{-3}\ pl^2$ for moments. (c) Load patterns to produce maximum actions.

upward; the reactions on either side of the unloaded spans are reversed in direction and have much smaller magnitude. The values given in the figures are for the case of equal spans l and load p per unit length; EI is constant.

In a loaded span, the bending moment is positive in the central part and negative at the supports. In the adjacent spans, the bending moment is negative over the major part of the length. The points of inflection on the deflected shapes correspond to the points O_1 and O_2 where the bending moment is zero. These points are closer to the supports C and D in case (a) and (b), respectively, than one-third of their respective spans. This is so because a couple applied at a supported end of a beam produces a straight-line bending moment diagram which reverses sign at one-third of the span from the far end when that end is totally fixed. When the far end is hinged (or simply supported) the bending moment has the same sign over the whole span. The behavior of the unloaded spans BC and CD in Figure 4.7a and of span CD in Figure 4.7b lies between these two extremes; the two extreme deflected shapes are indicated in Appendix A; see Equations A.14–A.17 and Equations A.9–A.13 and the corresponding figures.

The maximum absolute value of shear occurs at a section near the supports of the loaded span. Smaller values of shear of constant magnitude occur in the unloaded spans, with sign reversals as indicated.

In Figures 4.7a and b, we have loaded spans AB and BC. From these two figures, and perhaps two more figures corresponding to the uniform load on each of the remaining two spans, we can decide which load patterns produce the maximum values of the deflections, reactions, or internal forces at any section. Figure 4.7c shows typical load patterns which produce maximum values of various actions.

It can be seen from this figure that maximum deflection and maximum positive bending moment at a section near the middle of a span occur when the load covers this span as well as alternate spans on either side. The maximum negative bending moment, the maximum positive reaction, and the maximum absolute value of shear near a support occur when the load covers the two adjacent spans and alternate spans thereafter. The loading cases (3), (4), and (5) in Figure 4.7c refer to sections just to the left or right of a support, the subscripts l and r denoting left and right, respectively.

Partial loading of a span may produce maximum shear at a section as in case (6) in Figure 4.7c. Also, partial loading may produce maximum bending moments at sections near the interior supports (closer than one-third of the span) but not at the supports. However, in practice, partial span loading is seldom considered when the maximum bending moment values are calculated.

The effect of live load patterns needs to be combined with the effect of the dead load (permanent load) in order to obtain the maximum actions to be used in design. For example, if the beam in Figure 4.7c is designed for uniform dead and live loads of intensities q and p, respectively, we can obtain the maximum bending moment diagram by considering q on all spans combined with p according to the loading cases (1) to (5) in Figure 4.7c. The maximum bending moment diagram due to dead and live loads combined is shown in Figure 4.8 for a continuous beam of four equal spans l, with $q = p$. The diagram is obtained by plotting on one graph the bending moment due to q combined with p in cases (1) to (5) in Figure 4.7c and using for any part of the beam the curve with the highest absolute values. Additional load cases which produce a maximum positive bending moment due to p may need to be considered when p is large compared with q. For example, $M_{B\,\max+}$ occurs when p covers CD only, and $M_{C\,\max+}$ occurs when AB and DE are covered.

In practice, the live load may be ignored on spans far away from the section at which the maximum action is required. For example, the maximum negative moment and the maximum positive reaction at support B may be assumed to occur when the live load covers

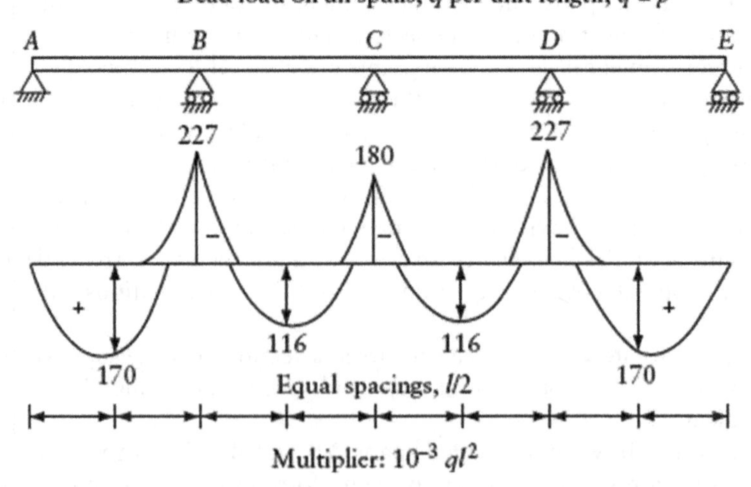

Figure 4.8 Maximum bending moment diagram for a continuous beam due to dead and live loads.

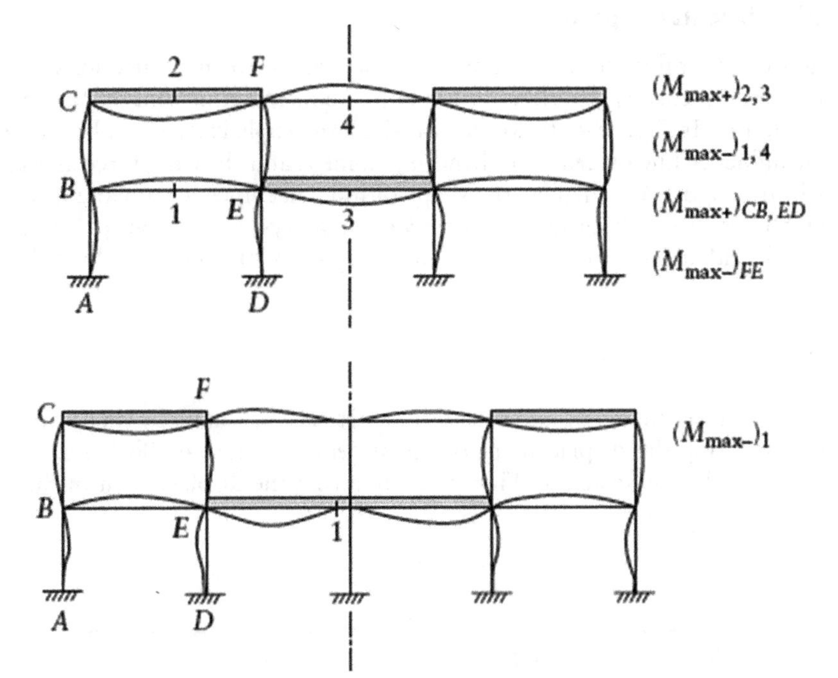

Figure 4.9 Examples of live load patterns to produce maximum bending moments in beams or endmoments in columns of plane frames.

only the adjacent spans AB and BC, without loading on DE (see case (3), Figure 4.7c). This approach may be acceptable because of the small effect of the ignored load, or on the grounds of a low probability of occurrence of the alternate load pattern with the full value of live load.

The alternate load patterns discussed above are typical for continuous beams and are frequently used in structural design. Similar patterns for continuous frames are shown in Figure 4.9. The two loading cases represented produce maximum positive and negative values of the bending moments in the horizontal beams or the end moments in the columns. As followed throughout this book, the bending moment in beams is considered positive when it produces tension at the bottom face; a clockwise member end moment is positive.

A sketch of the deflected shape may help to determine whether or not a span should be considered loaded so as to produce a maximum effect. A span should be loaded if this results in accentuating the deflected shape in all members. However, in some cases the deflected shape is not simple to predict in all parts of the frame, particularly when sidesway occurs. Use of influence lines (Section 4.8) helps in determining the load position for maximum effect, particularly when the live load is composed of concentrated loads.

4.8 INFLUENCE LINES

If the values of any action A are plotted as ordinates at all points of application of a unit load, we obtain the influence line of the action A. The use of Müller-Breslau's principle gives the influence line.

4.8.1 Müller-Breslau's principle

The ordinates of the influence line of any action are equal to the deflections obtained by releasing the restraint corresponding to this action and introducing unit displacement in the remaining structure. In Figure 4.10, Müller-Breslau's principle is used to obtain the general shapes of the influence line of: reaction, bending moment and shearing force at sections of a continuous beam. Figure 4.11 depicts the shape of influence lines for a plane frame.

To prove Müller-Breslau's principle, consider a real system in Figure 4.10a. Apply a unit downward load at B on the virtual system in Figure 4.10b. The unit load theory gives (Section 7.7):

$$\eta_1 P_1 + \eta_2 P_2 + \cdots + \eta_n P_n - 1 \times R_B = F \times 0$$

This equation expresses the fact that the external virtual work done by the system of forces in Figure 4.10a during the displacement of the system in Figure 4.10b is the same as the external work done by the system in Figure 4.10b during the displacement of the system in

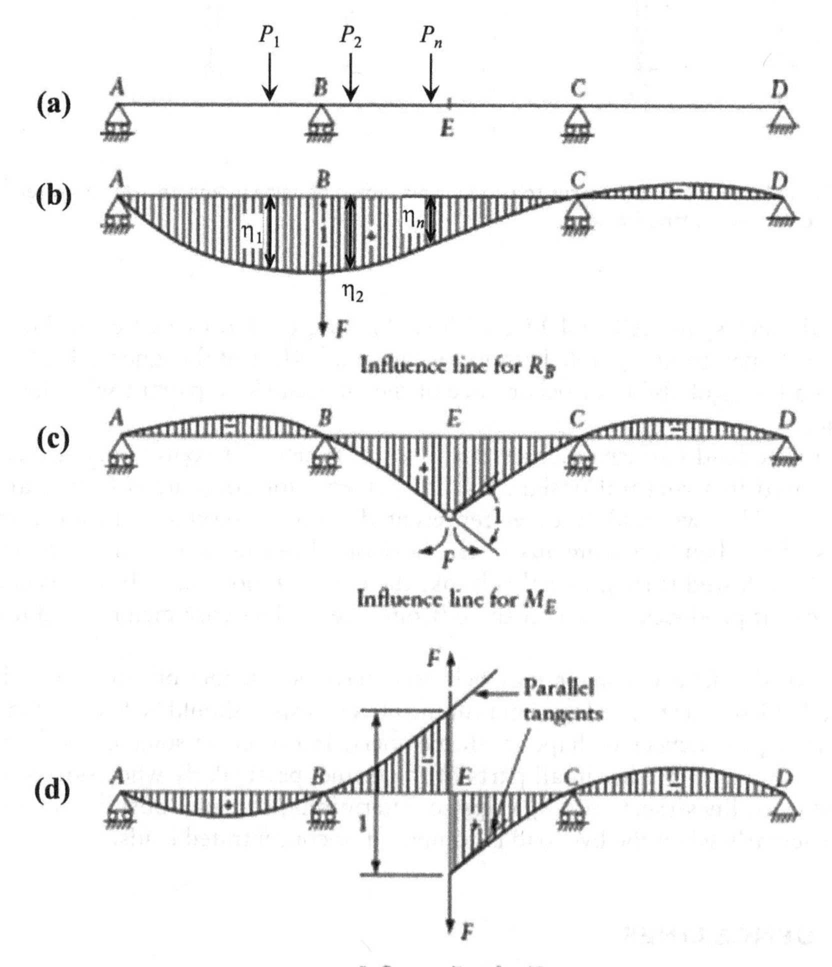

Figure 4.10 Influence lines for a continuous beam. (a) Beam and truck loads. (b) Influence line for R_B. (c) Influence line for M_E. (d) Influence line for V_E.

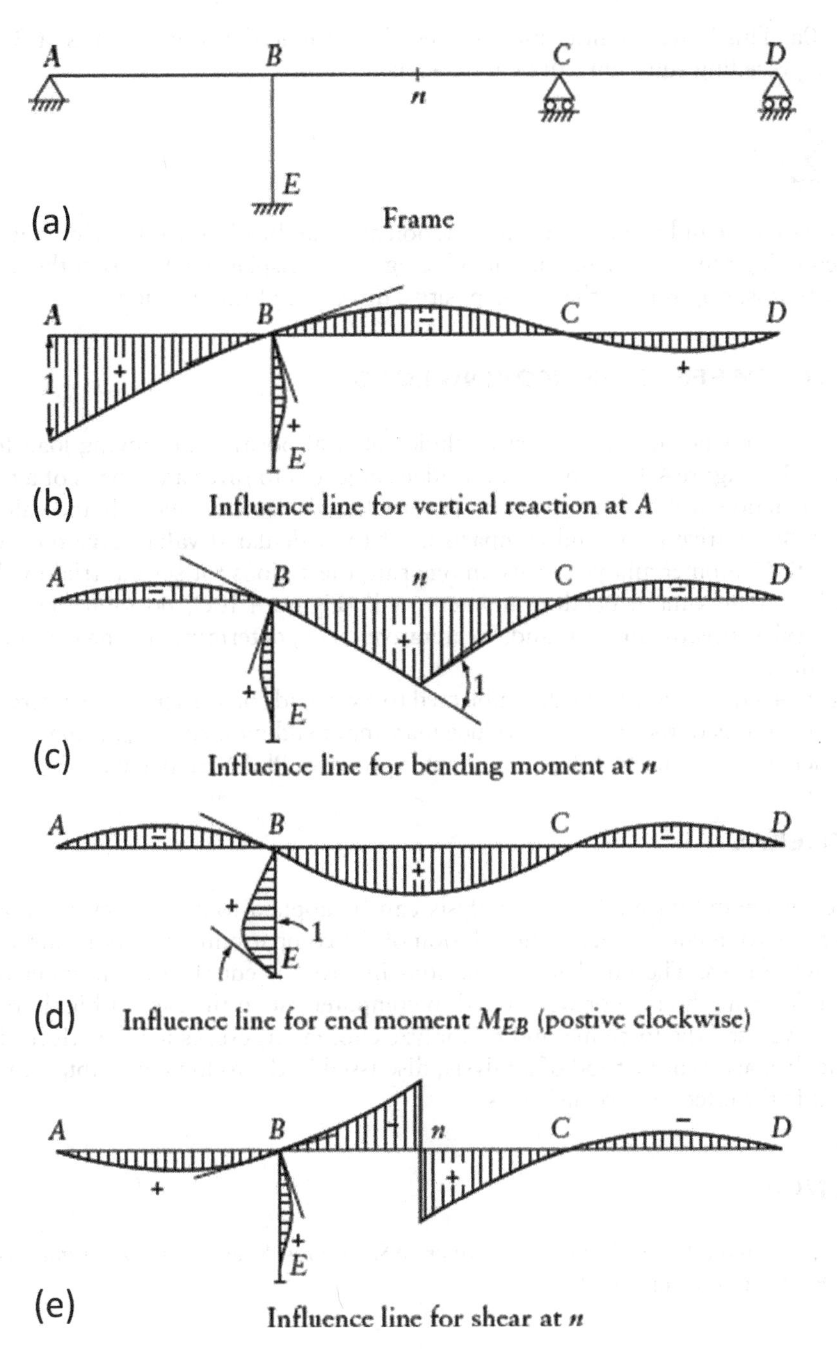

Figure 4.11 Shape of influence lines for a plane frame using Müller-Breslau's principle.

Figure 4.10a. This latter quantity must be zero because no deflection occurs at B in Figure 4.10a. The preceding equation can be written as:

$$R_B = \sum_{i=1}^{n} \eta_i P_i$$

This shows that the influence line of the reaction at B can be obtained by releasing its effect, that is, removing the support B, and introducing a unit displacement at B in the downward direction, that is, opposite to the to the positive direction of the reaction.

4.9 MAXIMUM EFFECT OF MOVING LOAD

The shape of influence line can assist in choice of trial position of moving load for maximum effect. Let Figure 4.10a represent a bridge subjected to three axle loads of a truck. For maximum reaction at B, as trial, position the truck with one axle above B and calculate the reaction at B; iterative trials and comparison of the calculated values give the maximum reaction at B. Computer analysis gives, in one run, the actions for several trial load cases.

Similarly, for maximum bending moment at E (Figure 4.10c), position the truck with one of the axles directly above E and, by iterative trials, determine the maximum bending moment value.

Let Figure 4.11a represent a bridge subjected to axle loads of a truck. As an exercise, guess three trial positions of the truck to produce maximum end moment M_{EB}, using the shape of the influence line in Figure 4.11d and computer program PLANEF (Chapter 20).

4.10 GENERAL

The force (or flexibility) method of analysis can be applied to any structure subjected to loading or environmental effects. The solution of the compatibility equations directly yields the unknown forces. The number of equations involved is equal to the number of redundants. The force method is not well suited to computer use in the case of highly redundant structures. We used the force method to analyze chosen structures for the effect of moving loads. The displacement method of analysis, discussed in the following chapter, can also be used to study the effect of moving loads.

REFERENCE

1. Ghali, A., Favre, R. and Elbadry, M., *Concrete Structures: Stresses and Deformations*, 4th ed., E & FN Spon, London, 2012.

PROBLEMS

The following are problems on the application of the force method of analysis. At this stage, use can be made of Appendix A to determine the displacements required in the analysis, with attention directed to the procedure of the force method rather than to the methods of computation of displacements. These will be treated in subsequent chapters. Additional problems on the application of the force method can be found at the end of Chapter 8, which require calculation of displacements by method of virtual work.

4.1 Write the flexibility matrix corresponding to coordinates 1 and 2 for the structures shown below.

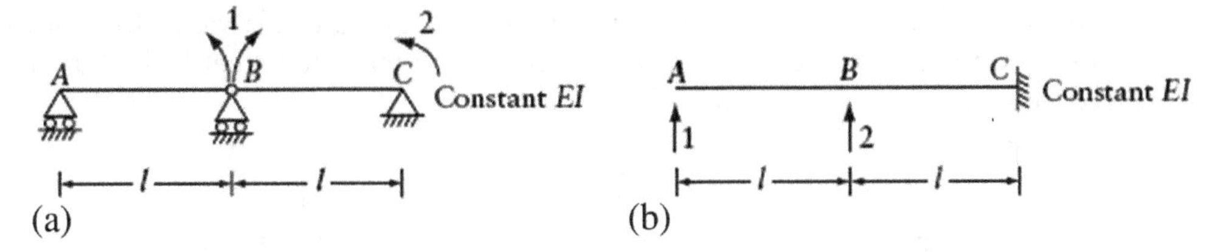

4.2 Use the flexibility matrices derived in Problem 4.1 to find two sets of redundant forces in two alternative solutions for the continuous beam of Example 4.1.

4.3 Use the force method to find the bending moment at the intermediate supports of the continuous beam shown in the figure.

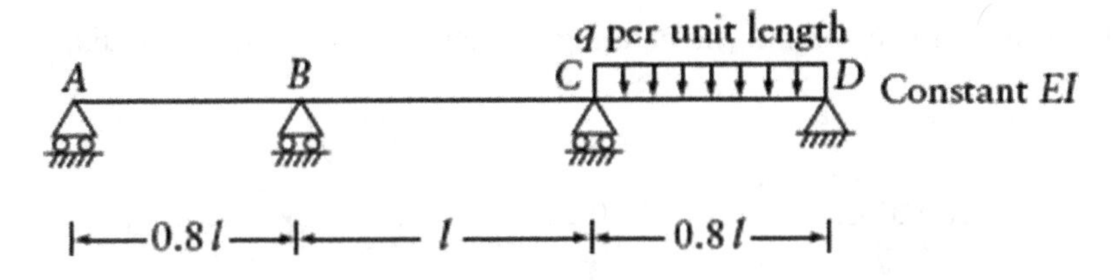

4.4 Obtain the bending moment diagram for the beam of Problem 4.3 on the assumption that support B settles vertically a distance $l/1200$. (No distributed load acts in this case.)

4.5 Use the force method to find the bending moments at the supports of a continuous beam on elastic (spring) supports. The beam has a constant flexural rigidity EI, and the stiffness of the elastic supports is $K = 20EI/l^3$.

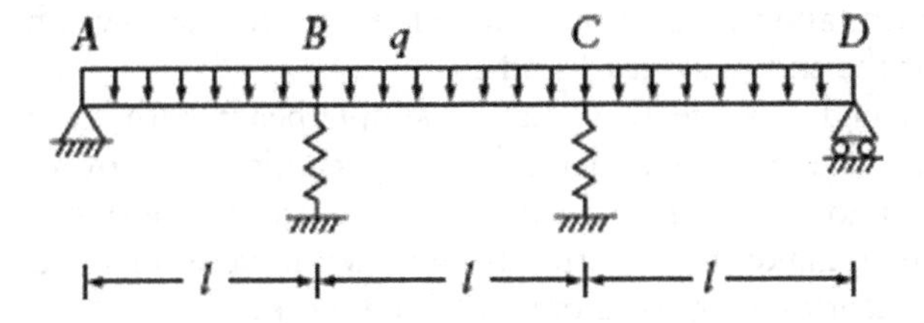

4.6 Use the force method to find the forces in the three springs A, B, and C in the system shown. The beams DE and FG have a constant flexural rigidity EI, and the springs have the same stiffness, $100EI/l^3$.

(a) What are the values of the vertical reactions at D, E, F, and G?

(b) If the springs A, B, and C are replaced by rigid link members, what will be the forces in the links and the reactions?

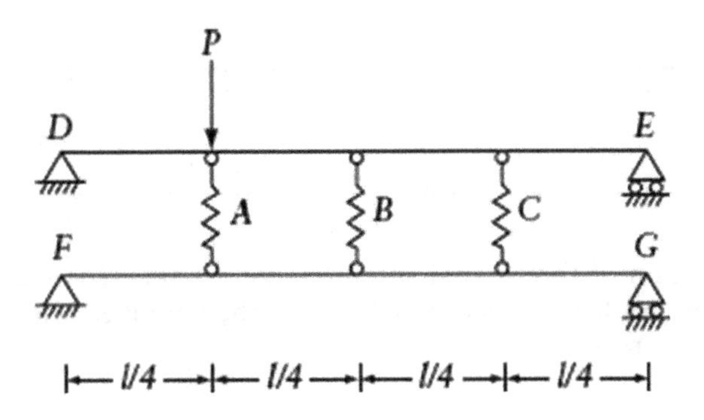

4.7 A steel beam AB is supported by two steel cables at C and D. Using the force method, find the tension in the cables and the bending moment at D due to a load $P = 25$ kN and a drop of temperature of 20 degrees Celsius in the two cables. For the beam $I = 16 \times 10^6$ mm^4, for the cables $a = 100$ mm^2; the modulus of elasticity for both is $E = 200$ GN/m^2, and the coefficient of thermal expansion for steel is 1×10^{-5} per degree Celsius.

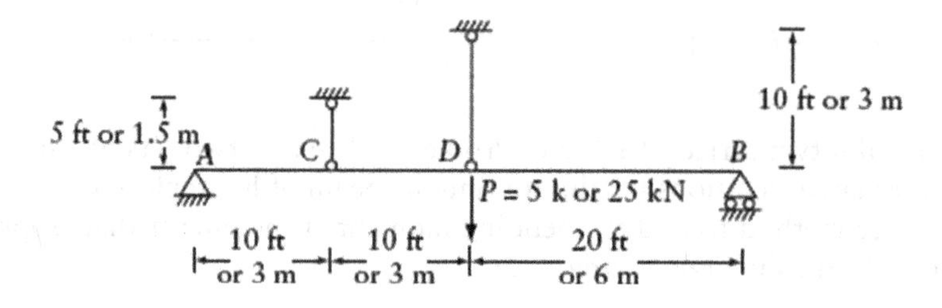

4.8 For the beam shown, obtain the bending moment and shearing force diagrams.

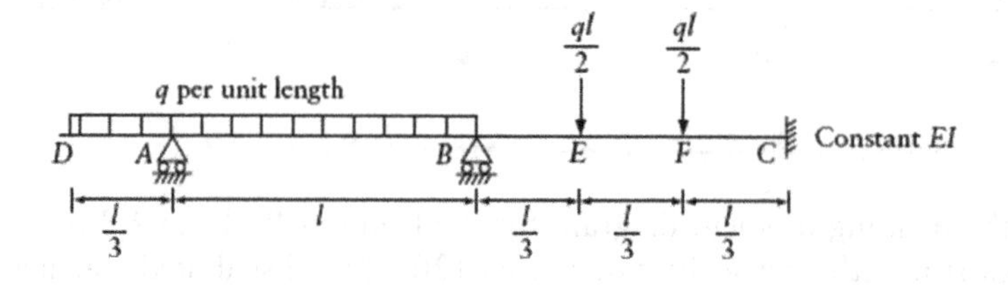

4.9 The reinforced concrete bridge ABC shown in the figure is constructed in two stages. In stage 1, part AD is cast and its forms are removed. In stage 2, part DC is cast and its forms are removed; a monolithic continuous beam is obtained. Obtain the bending moment diagrams and the reactions due to the structure self-weight, q per unit length, immediately at the end of stages 1 and 2.

Hint: At the end of stage 1, we have a simple beam with an overhang, carrying a uniform load. In stage 2, we added a load q per unit length over DC in a continuous beam. Superposition gives the desired answers for the end of stage 2. Creep of concrete tends gradually to make the structure behave as if it were constructed in one stage (see the references mentioned in footnote 1 of Chapter 4).

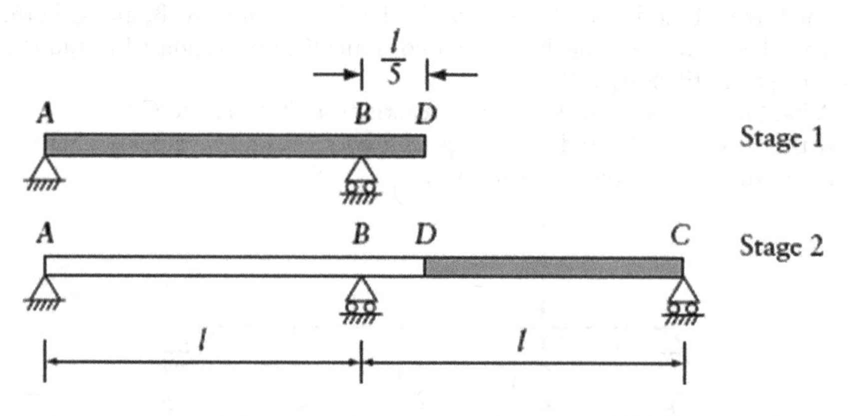

4.10 A continuous beam of three equal spans l and constant EI is subjected to a uniform dead load q per unit length over the whole length, combined with a uniform live load of intensity $p = q$. Determine:

(a) The maximum bending moments at the interior supports and mid-spans.
(b) Diagram of maximum bending moment.
(c) Maximum reaction at an interior support.
(d) Absolute maximum shearing force and its location.

4.11 For the beam of Problem 4.8, find the reactions, and the bending moment and shearing force diagrams due to a unit downward settlement of support B. (The main answers for this problem are included in Table D.3, Appendix D. Note that the presence of the overhang DA has no effect.)

4.12 For the continuous beam shown, determine: (a) the bending moment diagram, (b) the reaction at B, and (c) the deflection at the center of BC.

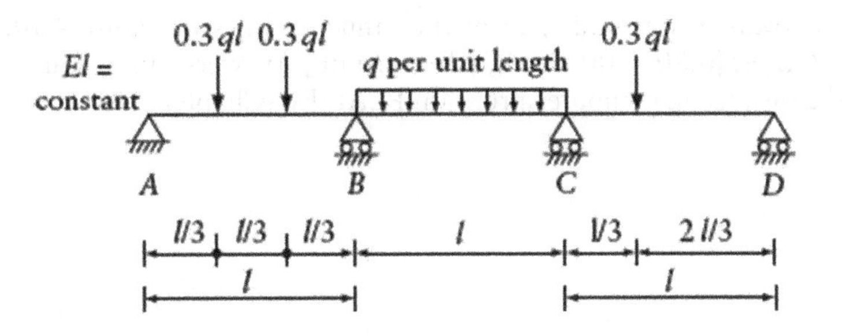

4.13 Find the bending moment diagram for the frame shown. Assume $(Ea)_{\text{Tie BD}} = 130EI/l^2$, where EI is the flexural rigidity of ABCDE. Use the force method cutting the tie BD to release the structure.

Hint: Calculation of the displacements in steps 2 and 3 of the force method has not been covered in earlier chapters. The problem can be solved using given information: the load applied on the released structure would move B and D away from each other a distance $= 0.1258ql^4/(EI)$. Two-unit horizontal forces each $= F_1 = 1$ applied inwards at each of B and D of the released structure would move the joints closer to each other by a distance $= 0.1569l^3/(EI)$.

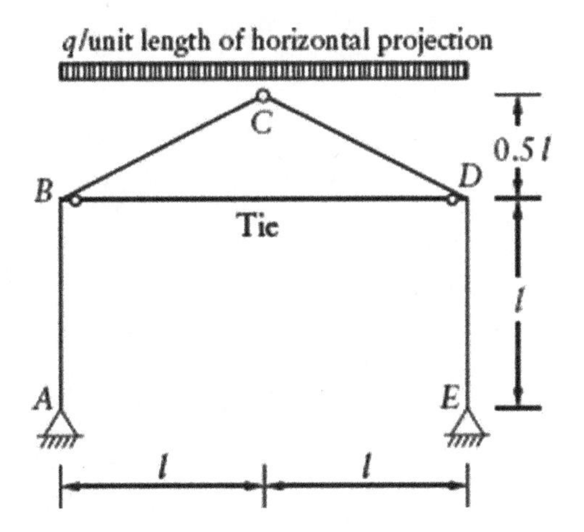

4.14 For the beam in the figure, determine the influence lines for R_B and M_E; distances between supports $= l$. By trial, calculate the maximum values of R_B and M_E due to the truck load in the figure.

Hint: Use Müller-Breslau's method. For statically determinate beams, the influence lines consist of straight lines. The ordinates can be determined by intuition.

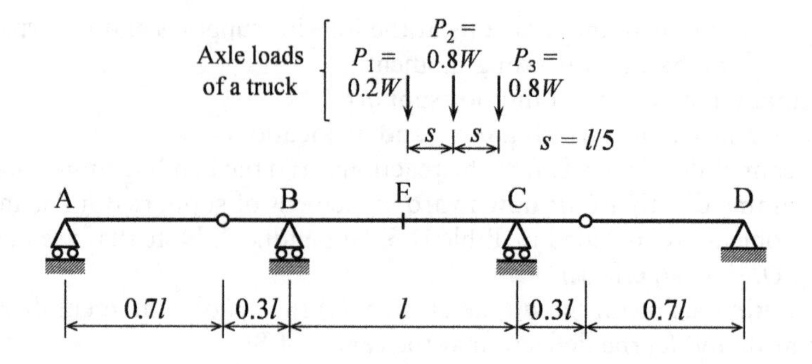

4.15 Find the coordinates at mid-span of the influence lines in Figure 4.10. Span lengths {AB, BC, CD} = l {0.80, 1.00, 0.80}, EI = constant. Answers can be found in one run for two load cases, using computer program PLANEF (Chapter 20).

Chapter 5

Displacement method of analysis

5.1 INTRODUCTION

The mathematical formulation of the displacement method and force method is similar, but from the economy of effort perspective, one or the other method may be preferable. The displacement method can be applied to statically determinate or indeterminate structures, but it is more useful in the latter, particularly when the degree of statical indeterminacy is high.

5.2 DESCRIPTION OF METHOD

The displacement method involves five steps:

1. First, the degree of kinematic indeterminacy has to be found. A coordinate system is then established to identify the location and direction of the joint displacements. Restraining forces equal in number to the degree of kinematic indeterminacy are introduced at the coordinates to prevent the displacement of the joints. In some cases, the number of restraints introduced may be smaller than the degree of kinematic indeterminacy, provided that the analysis of the resulting structure is a standard one and is therefore known. (See remarks following Example 5.2.)

   We should note that, unlike the force method, the above procedure requires no choice to be made with respect to the restraining forces. This fact favors the use of the displacement method in general computer programs for the analysis of a structure.
2. The restraining forces are now determined as a sum of the fixed-end forces for the members meeting at a joint. For most practical cases, the fixed-end forces can be calculated with the aid of standard tables (Appendices B and C). An external force at a coordinate is restrained simply by an equal and opposite force that must be added to the sum of the fixed-end forces.

   We should remember that the restraining forces are those required to prevent the displacement at the coordinates due to all effects, such as external loads, temperature variation, or prestrain. These effects may be considered separately or may be combined.

   If the analysis is to be performed for the effect of movement of one of the joints in the structure, for example, the settlement of a support, the forces at the coordinates required to hold the joint in the displaced position are included in the restraining forces.

   The internal forces in the members are also determined at the required locations with the joints in the restrained position.

DOI: 10.1201/9780429286858-5

3. The structure is now assumed to be deformed in such a way that a displacement at one of the coordinates equals unity and all the other displacements are zero and the forces required to hold the structure in this configuration are determined. These forces are applied at the coordinates representing the degrees of freedom. The internal forces at the required locations corresponding to this configuration are determined.
   The process is repeated for a unit value of displacement at each of the coordinates separately.
4. The values of the displacements necessary to eliminate the restraining forces introduced in (2) are determined. This requires superposition equations in which the effects of separate displacements on the restraining forces are added.
5. Finally, the forces on the original structure are obtained by adding the forces on the restrained structure to the forces caused by the joint displacements determined in (4).

The use of the above procedure is best explained with reference to some specific cases.

EXAMPLE 5.1: PLANE TRUSS

The plane truss in Figure 5.1a consists of m pin-jointed members meeting at joint A. Find the forces in the members due to the combined effect of: (1) an external load P applied at A; (2) a rise of temperature T degrees of the kth bar alone.

The degree of kinematic indeterminacy of the structure is two ($n=2$), because displacement can occur only at joint A, which can undergo a translation with components D_1 and D_2 in the x and y directions. The positive directions for the displacement components, as well as for the restraining forces, are arbitrarily chosen, as indicated in Figure 5.1b. Thus, a coordinate system is represented by arrows defining the locations and the positive directions of displacements $\{D\}_{n\times 1}$ and forces $\{F\}_{n\times 1}$. In this first step of the displacement method, we also define the required actions, $\{A\}_{m\times 1}$; in this example, the elements of $\{A\}$ are the forces in the truss members due to the specified loading.

The joint displacements are artificially prevented by introducing at A, a force equal and opposite to P plus a force $(Ea\alpha T)_k$ in the direction of the kth member. Here, E, a, and α are, respectively, modulus of elasticity, cross-sectional area, and coefficient of thermal expansion of the kth bar. Components of the restraining forces in the directions of the coordinates are:

$$F_1 = -P\cos\beta + \left(Ea\alpha T\cos\theta\right)_k$$

$$F_2 = -P\sin\beta + \left(Ea\alpha T\sin\theta\right)_k$$

where θ_k is the angle between the positive x direction and the kth bar. With the displacements at A prevented, the change in length of any member is zero; thus, the axial force is zero, with the exception of the kth bar. Because the thermal expansion is prevented, an axial force is developed in the kth bar; this is equal to $-(Ea\alpha T)_k$, with the minus sign indicating compression. Denoting by $\{A_r\}$ the axial forces in the bars in the restrained condition, we have:

$$\{A_r\} = \{0, 0, \ldots, -\left(Ea\alpha T\right)_k, \ldots, 0\}$$

All elements of the vector $\{A_r\}$ are zero except the kth. In this second step, we have determined the forces $\{F\}$ necessary to prevent the displacements at the coordinates when the loading is applied; we have also determined the actions $\{A_r\}$ due to the loading while the displacements are artificially prevented.

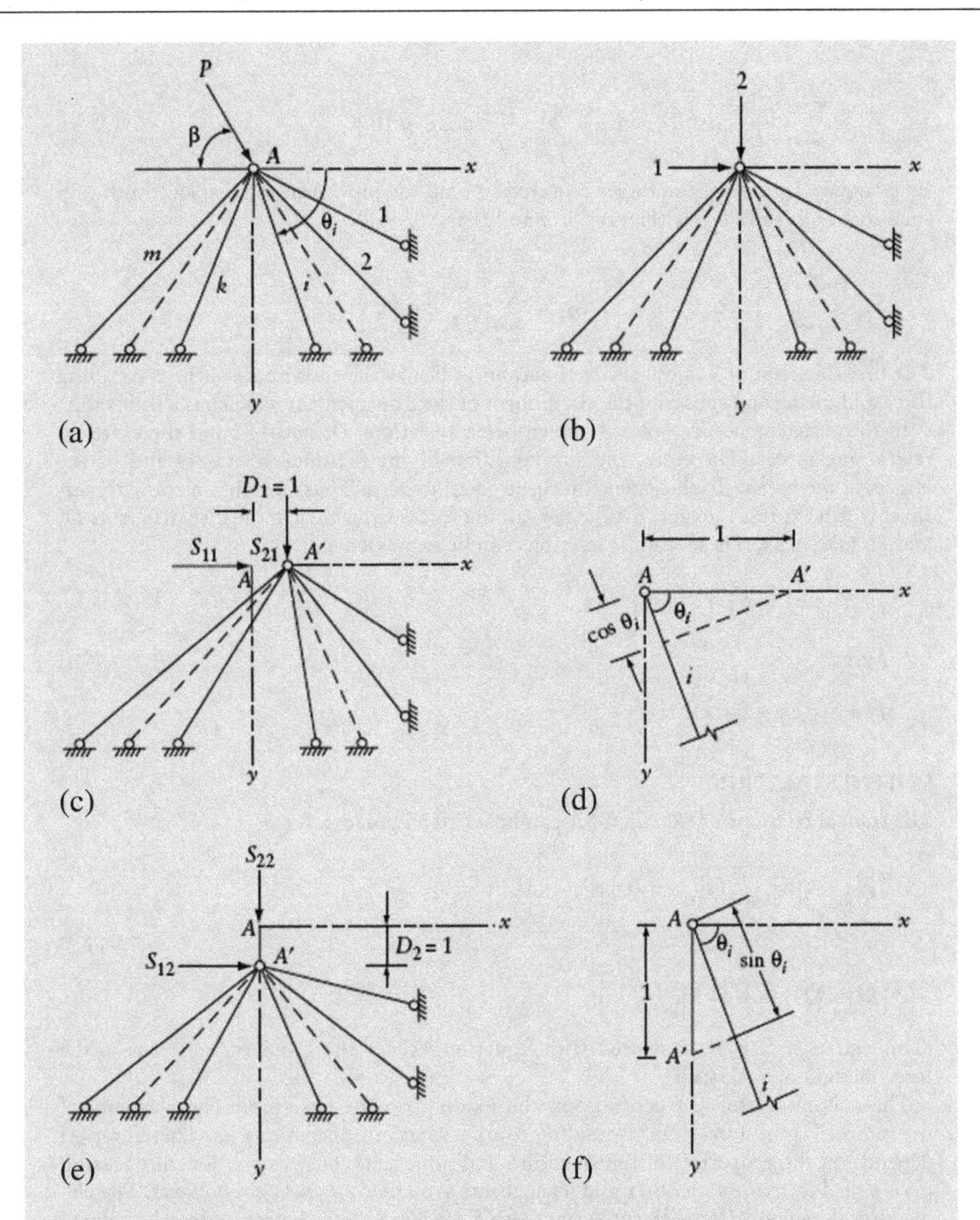

Figure 5.1 Analysis of a plane truss by the displacement method – Example 5.1. (a) Plane truss. (b) Coordinate system. (c) $D_1 = 1$ and $D_2 = 0$. (d) Change in length in the ith member due to $D_1 = 1$. (e) $D_1 = 0$ and $D_2 = 1$. (f) Change in length in the ith member due to $D_2 = 1$.

Figure 5.1c shows the forces required to hold the structure in a deformed position such that $D_1 = 1$ and $D_2 = 0$. Now, from Figure 5.1d, a unit horizontal displacement of A causes a shortening of any bar i by a distance $\cos \theta_i$ and produces a compressive force of $(a_i E_i / l_i) \cos \theta_i$.

Therefore, to hold joint A in the displaced position, forces $(a_i E_i / l_i) \cos^2 \theta_i$ and $(a_i E_i / l_i) \cos \theta_i \sin \theta_i$ have to be applied in directions 1 and 2, respectively. The forces required to hold all the bars in the displaced position are:

$$S_{11}=\sum_{i=1}^{m}\left(\frac{Ea}{l}\cos^2\theta\right)_i;\; S_{21}=\sum_{i=1}^{m}\left(\frac{Ea}{l}\cos\theta\sin\theta\right)_i$$

By a similar argument, the forces required to hold the joint A in the displaced position such that $D_1=0$ and $D_2=1$ (Figures 5.1e and f) are:

$$S_{21}=\sum_{i=1}^{m}\left(\frac{Ea}{l}\cos\theta\sin\theta\right)_i;\; S_{22}=\sum_{i=1}^{m}\left(\frac{Ea}{l}\sin^2\theta\right)_i$$

The first subscript of S in the above equations indicates the coordinate of the restraining force and the second subscript the component of the displacement which has a unit value.

In the actual structure, joint A undergoes translations D_1 and D_2, and there are no restraining forces. Therefore, the superposition of the fictitious restraints and of the effects of the actual displacements must be equal to zero. Thus, we obtain *statical relations* which express the fact that the restraining forces vanish when the displacements D_1 and D_2 take place. These statical relations can be expressed as:

$$\left.\begin{aligned} &F_1+S_{11}D_1+S_{12}D_2=0\\ &\text{and}\\ &F_2+S_{21}D_1+S_{22}D_2=0\end{aligned}\right\}\qquad(5.1)$$

STIFFNESS MATRIX

The statical relations of Equation 5.1 can be written in matrix form:

$$\left.\begin{aligned} &\{F\}_{n\times1}+[S]_{n\times n}\{D\}_{n\times1}=\{0\}\\ &\text{or}\\ &[S]_{n\times n}\{D\}_{n\times1}=\{-F\}_{n\times1}\end{aligned}\right\}\qquad(5.2)$$

(This equation may be compared with Equation 4.3 for the geometry relations in the force method of analysis.)

The column vector $\{F\}$ depends on the loading on the structure. The elements of the matrix $[S]$ are forces corresponding to unit values of displacements. Therefore, $[S]$ depends on the properties of the structure and represents its stiffness. For this reason, $[S]$ is called the *stiffness matrix* and its elements are called *stiffness coefficients*. The elements of the vector $\{D\}$ are the unknown displacements and can be determined by solving Equation 5.2, that is:

$$\{D\}=[S]^{-1}\{-F\}\qquad(5.3)$$

In a general case, if the number of restraints introduced in the structure is n, the order of the matrices $\{D\}$, $[S]$, and $\{F\}$ is $n\times1$, $n\times n$, and $n\times1$, respectively. The stiffness matrix $[S]$ is thus a square symmetrical matrix. This can be seen in the above example by comparing the equations for S_{21} and S_{12}, but a formal proof will be given in Section 7.7.

In the third step of the displacement method, we determine $[S]_{n\times n}$ and $[A_u]_{m\times n}$. To generate any column, j, of the two matrices, we introduce a unit displacement $D_j=1$ at coordinate j, while the displacements are prevented at the remaining coordinates. The forces necessary at the n coordinates to hold the structure in this deformed configuration form the jth column of $[S]$; the corresponding m actions form the jth column of $[A_u]$. In the fourth step of the displacement method, we solve the statical Equation 5.2 (also called *equilibrium equation*). This gives the actual displacements $\{D\}_{n\times1}$ at the coordinates (Equation 5.3).

The final force in any member i can be determined by superposition of the restrained condition and of the effect of the joint displacements.

$$A_i = A_{ri} + \left(A_{ui1}\, D_1 + A_{ui2}\, D_2 + \cdots + A_{uin}\, D_n\right) \tag{5.4}$$

The *superposition* equation for all the members in matrix form is:

$$\{A\}_{m\times1} = \{A_r\}_{m\times1} + [A_u]_{m\times n}\, \{D\}_{n\times1}$$

where the elements of $\{A\}$ are the final forces in the bars, the elements of $\{A_r\}$ are the bar forces in the restrained condition, and the elements of $[A_u]$ are the bar forces corresponding to unit displacements. Specifically, the elements of column j of $[A_u]$ are the forces in the members corresponding to a displacement $D_j = 1$, while all the other displacements are zero.

Since the above equation will be used in the analysis of a variety of structures, it is useful to write it in a general form:

$$\{A\} = \{A_r\} + [A_u]\{D\} \tag{5.5}$$

The fifth step of the displacement method gives the required action $\{A\}$ by substituting in Equation 5.5 matrices $\{A_r\}$, $[A_u]$, and $\{D\}$ determined in steps 2, 3, and 4.

In the truss of Example 5.1, with axial tension in a member considered positive, it can be seen that:

$$[A_u] = \begin{bmatrix} -(Ea\cos\theta/l)_1 & -(Ea\sin\theta/l)_1 \\ -(Ea\cos\theta/l)_2 & -(Ea\sin\theta/l)_2 \\ \cdots & \cdots \\ -(Ea\cos\theta/l)_m & -(Ea\sin\theta/l)_m \end{bmatrix}$$

In a frame with rigid joints, we may want to find the stress resultants in any section or the reactions at the supports. For this reason, we consider the notation A in the general Equation 5.5 to represent any action, which may be shearing force, bending moment, twisting moment, or axial force at a section or a reaction at a support.

5.3 DEGREES OF FREEDOM AND COORDINATE SYSTEM

In the first step of the displacement method, the number n of the independent joint displacements (the number of degrees of freedom, Section 3.5) is determined. A system of n coordinates is defined. A coordinate is an arrow representing the location and the positive direction of a displacement D or a force F. The coordinate system is indicated in a figure showing the actual structure with n arrows at the joint.

The remaining steps of the displacement method involve generating and use of matrices: {*F*}, {*D*}, and [*S*]. The elements of these matrices are either forces or displacements at the coordinates. Thus, it is impossible to follow or check the calculations, particularly the signs of the parameters, when the coordinate system is not clearly defined. For this reason, it is recommended that the coordinate system be shown in a figure depicting only *n* numbered arrows; the forces applied on the structure should not be shown in this figure.

EXAMPLE 5.2: PLANE FRAME

The plane frame in Figure 5.2a consists of rigidly connected members of constant flexural rigidity *EI*. Obtain the bending moment diagram for the frame due to concentrated

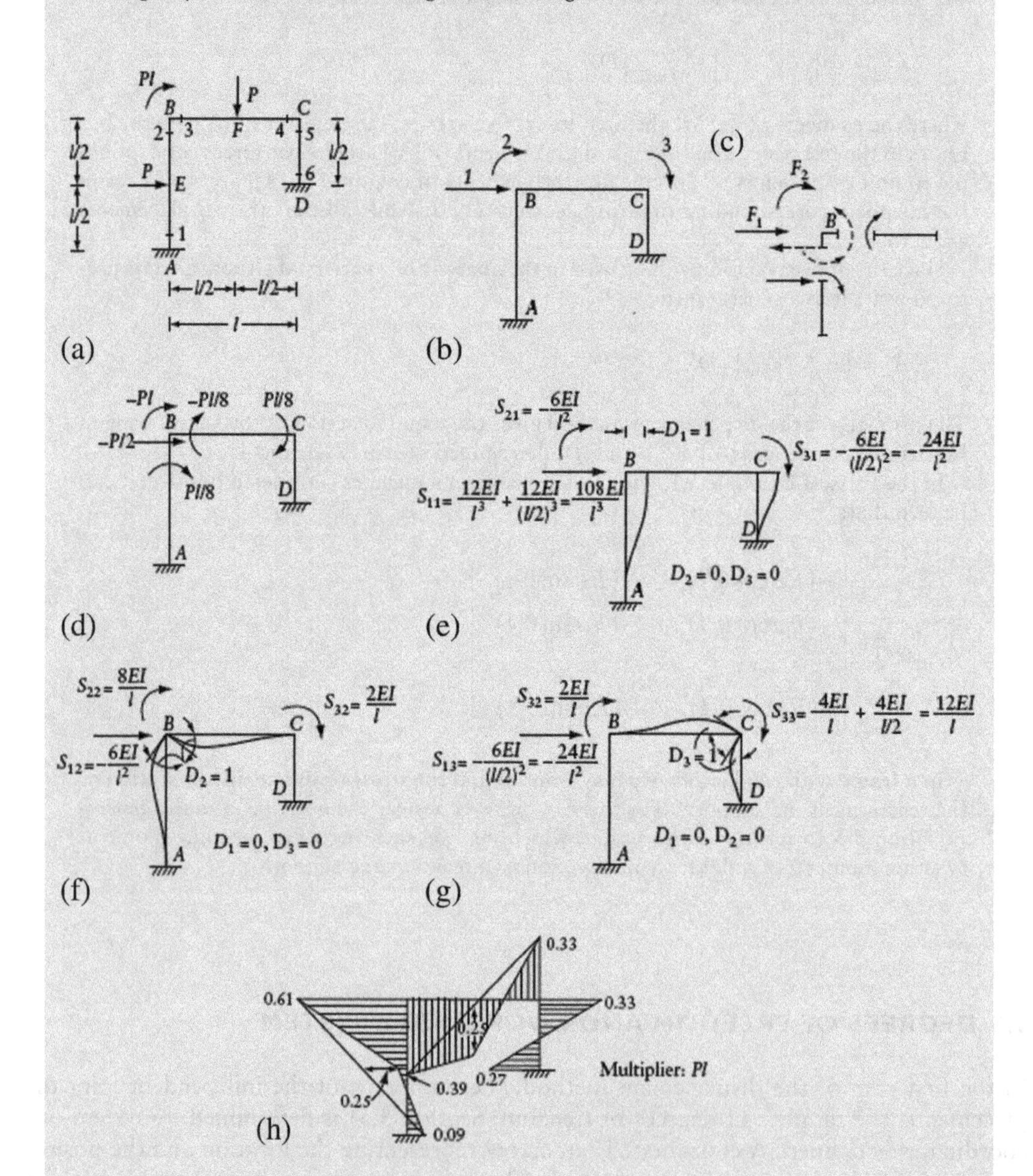

Figure 5.2 Plane analyzed in Example 5.2.

loads P at E and F, and a couple Pl at joint B. The change in length of the members can be neglected.

The degree of kinematic indeterminacy is three because there are three possible joint displacements, as shown in Figure 5.2b, which shows also the chosen coordinate system. The restraining forces, which are equal to the sum of the end-forces at the joints, are calculated with the aid of Appendix B (Equations B.1 and B.2). As always, they are considered positive when their direction accords with that of the coordinates.

To illustrate the relation between the end-forces and the restraining forces, joint B is separated from the members connected to it in Figure 5.2c. The forces acting on the end of the members in the direction of the coordinate system are indicated by full-line arrows. Equal and opposite forces act on the joint and these are shown by dotted-line arrows. For equilibrium of the joint, forces F_1 and F_2 should be applied in a direction opposite to the dotted arrows. Therefore, to obtain the restraining forces, it is sufficient to add the end-forces at each joint as indicated in Figure 5.2d, and it is not necessary to consider the forces as in Figure 5.2c.

The external applied couple acting at B requires an equal and opposite restraining force. Therefore:

$$\{F\} = \begin{Bmatrix} -\dfrac{P}{2} \\ \left(\dfrac{Pl}{8} - \dfrac{Pl}{8} - Pl\right) \\ \dfrac{Pl}{8} \end{Bmatrix} = P\begin{Bmatrix} -0.5 \\ -l \\ 0.125l \end{Bmatrix} \tag{a}$$

To draw the bending moment diagram, the values of the moments at the ends of all the members are required; assume that an end moment is positive if it acts in a clockwise direction. We define the required actions as the member end moments:

$$\{A\} = \{M_{AB}, M_{BA}, M_{BC}, M_{CB}, M_{CD}, M_{DC}\}$$

Throughout this book, a clockwise end moment for a member of a plane frame is considered positive. The two clockwise end moments shown in Figure 3.8a are positive. At the left-hand end of the member, the clockwise moment produces tension at the bottom face; but at the right-hand end, the clockwise moment produces tension at the top face. The member ends 1, 2, ..., 6 are identified in Figure 5.2a. Thus, the values of the six end moments corresponding to the restrained condition are:

$$\{A_r\} = \frac{Pl}{8}\{-1, 1, -1, 1, 0, 0\}$$

Now, the elements of the stiffness matrix are the forces necessary at the location in the direction of the coordinates to hold the structure in the deformed shape illustrated in Figures 5.2e, f, and g. These forces are equal to the sum of the end-forces, which are taken from Appendix C (Equations C.1 to C.5). We should note that the translation of joint B must be accompanied by an equal translation of joint C in order that the length BC remains unchanged. The stiffness matrix is:

$$[S] = \frac{EI}{l}\begin{bmatrix} \dfrac{108}{l^2} & -\dfrac{6}{l} & -\dfrac{24}{l} \\ -\dfrac{6}{l} & 8 & 2 \\ -\dfrac{24}{l} & 2 & 12 \end{bmatrix} \tag{b}$$

To write the matrix of end moments due to the unit displacements, we put the values at the member ends 1, 2, ..., 6 in the first, second, and third column, respectively, for the displacements shown in Figures 5.2e, f, and g (Equations C.1 to C.5, Appendix C).

$$[A_u]=\frac{EI}{l}\begin{bmatrix} -\frac{6}{l} & 2 & 0 \\ -\frac{6}{l} & 4 & 0 \\ 0 & 4 & 2 \\ 0 & 2 & 4 \\ -\frac{24}{l} & 0 & 8 \\ -\frac{24}{l} & 0 & 4 \end{bmatrix} \tag{c}$$

The deflected shapes of the members in Figures 5.2e, f, and g and the corresponding end moments are presented in Appendix C, which is used to determine the elements of $[A_u]$. For example, with $D_3 = 1$, member BC in Figure 5.2g has the same deflected shape as in the second figure of Appendix C. Thus, the end moments $2EI/l$ and $4EI/l$ taken from this figure are equal to elements A_{u33} and A_{u43}, respectively.

Substituting Equations (a) and (b) into Equation 5.3 and solving for $\{D\}$, we obtain:

$$\{D\}=\frac{Pl^2}{EI}\begin{Bmatrix} 0.0087\,l \\ 0.1355 \\ -0.0156 \end{Bmatrix} \tag{d}$$

The final end moments are calculated by Equation 5.5:

$$\{A\}=Pl\begin{Bmatrix} -0.125 \\ 0.125 \\ -0.125 \\ 0.125 \\ 0 \\ 0 \end{Bmatrix}+\frac{EI}{l}\begin{bmatrix} -\frac{6}{l} & 2 & 0 \\ -\frac{6}{l} & 4 & 0 \\ 0 & 4 & 2 \\ 0 & 2 & 4 \\ -\frac{24}{l} & 0 & 8 \\ -\frac{24}{l} & 0 & 4 \end{bmatrix}\frac{Pl^2}{EI}\begin{Bmatrix} 0.0087\,l \\ 0.1355 \\ -0.0156 \end{Bmatrix}=Pl\begin{Bmatrix} 0.09 \\ 0.61 \\ 0.39 \\ 0.33 \\ -0.33 \\ -0.27 \end{Bmatrix} \tag{e}$$

The bending moment diagram is plotted in Figure 5.2h; the ordinates appear on the side of the tensile fiber.

The application of the preceding procedure to a frame with inclined members is illustrated in Example 5.3.

Remarks

1. If, in the above example, the fixed-end A is replaced by a hinge, the kinematic indeterminacy is increased by the rotation at A. Nevertheless, the structure can be analyzed using only the three coordinates in Figure 5.2 because the end-forces for a member hinged at one end and fully fixed at the other are readily available (Appendices B and C).

As an exercise, we can verify the following matrices for analyzing the frame in Figure 5.2, with support A changed to a hinge:

$$\{F\} = (P/16)\{-11, -15l, 2l\}; \quad \{A_r\} = (Pl/16)\{0, 3, -2, 2, 0, 0\}$$

$$[S] = \frac{EI}{l}\begin{bmatrix} 99/l^2 & & \text{Sym.} \\ -3/l & 7 & \\ -24/l & 2 & 12 \end{bmatrix}$$

$$[A_u]^{\mathrm{T}} = \frac{EI}{l}\begin{bmatrix} 0 & -3/l & 0 & 0 & -24/l & -24/l \\ 0 & 3 & 4 & 2 & 0 & 0 \\ 0 & 0 & 2 & 4 & 8 & 4 \end{bmatrix}$$

$$\{D\} = \frac{Pl^2}{EI}\{0.0058l, 0.1429, -0.0227\}$$

$$\{A\} = Pl\{0, 0.60, 0.40, 0.32, -0.32, -0.23\}$$

2. When a computer is used for the analysis of a plane frame, axial deformations are commonly not ignored and the unknown displacements are two translations and a rotation at a general joint. Three forces are usually determined at each member end (Figure 19.2). These can be used to give the axial force, the shearing force, and the bending moment at any section. The superposition Equation 5.5 is applied separately to give six end-forces for each member, using the six displacements at its ends. This is discussed in detail in Chapter 19 for all types of framed structures. We may use the computer program PLANEF (Chapter 20) to analyze the frame in Figure 5.2a with support D totally fixed or hinged. The computer will give the same answers as given here by entering a value = 1.0 for each of P, l, E, and I. Entering a large value for the cross-sectional area of the members (e.g. 1.0E6) will result in negligible change in length of the members.

5.4 FIVE STEPS OF DISPLACEMENT METHOD

The analysis by the displacement method involves five steps which are summarized as follows:

Step 1 Define a system of coordinates representing the joint displacements to be found. Also, define $[A]_{m\times p}$, the required actions as well as their sign convention (if necessary).

Step 2 With the loadings applied, calculate the restraining forces $[F]_{n\times p}$ and $[A_r]_{m\times p}$.

Step 3 Introduce unit displacements at the coordinates, one by one, and generate $[S]_{n\times n}$ and $[A_u]_{m\times n}$.

Step 4 Solve the equilibrium equations to calculate $[D]$:

$$[S]_{n\times n}[D]_{n\times p} = -[F]_{n\times p} \tag{5.6}$$

Step 5 Calculate the required actions by superposition:

$$[A]_{m\times p}=[A_r]_{m\times p}+[A_u]_{m\times n}[D]_{n\times p} \tag{5.7}$$

In a manner similar to the force method (Section 4.6), when Step 3 above is completed, all the matrices necessary for the analysis have been generated. The last two steps involve merely matrix algebra.

For quick reference, the symbols used in this section are defined again as follows:

n, p, m	= number of degrees of freedom, number of loading cases, and number of actions required
$[A]$	= required actions (the answers to the problem)
$[A_r]$	= values of the actions due to loadings on the structure while the displacements are prevented
$[A_u]$	= values of the actions due to unit displacements introduced separately at each coordinate
$[F]$	= forces at the coordinates necessary to prevent the displacements due to the loadings
$[S]$	= stiffness matrix

EXAMPLE 5.3: PLANE FRAME WITH INCLINED MEMBER

Obtain the bending moment diagrams for the plane frame in Figure 5.3a due to the separate effects of: (1) the loads shown; (2) a downward settlement δ_D at support D. Consider EI = constant and neglect the change in length of members.

Step 1 Figure 5.3b defines a coordinate system corresponding to three independent joint displacements (the kinematic indeterminacy; see Section 3.5). The required bending moment diagrams can be drawn from the member-end moments (considered positive when clockwise):

$$[A]=\begin{bmatrix}\begin{Bmatrix}M_{AB}\\M_{BC}\\M_{CD}\\M_{DC}\end{Bmatrix}_1 & \begin{Bmatrix}M_{AB}\\M_{BC}\\M_{CD}\\M_{DC}\end{Bmatrix}_2\end{bmatrix} \tag{a}$$

$M_{BA}=-M_{BC}$ and $M_{CB}=-M_{CD}$; thus, M_{BA} and M_{CB} need not be included in $[A]$.

Step 2 The fixed-end forces for the members are found from Appendices B and C (Equations B.1, B.2, B.7, B.8, C.1, and C.2) and are shown in Figures 5.3c and e. The restraining couples F_1 and F_2 are obtained directly by adding the fixed-end moments. For the calculation of F_3, the shearing forces meeting at joints B and C are resolved into components along the member axes and F_3 is obtained by adding the components in the direction of coordinate 3 (Figure 5.3d). An alternative method for calculating F_3 using a work equation will be explained below. Now, we write the restraining forces:

$$[F]=\begin{bmatrix}-0.417\,Pl & -6\left(EI/l^2\right)\delta_D\\ 0.6\,Pl & -6\left(EI/l^2\right)\delta_D\\ -2.625\,P & -9\left(EI/l^3\right)\delta_D\end{bmatrix} \tag{b}$$

The value of F_3 in case (1) is equal to minus the sum of the horizontal components 0.625P, 1.5P, and 0.5P shown at B and C in Figure 5.3c. If a similar figure is drawn for case (2), the shearing force at end B of member BC will be an upward force equal to ($12EI\delta_D/l^3$). This

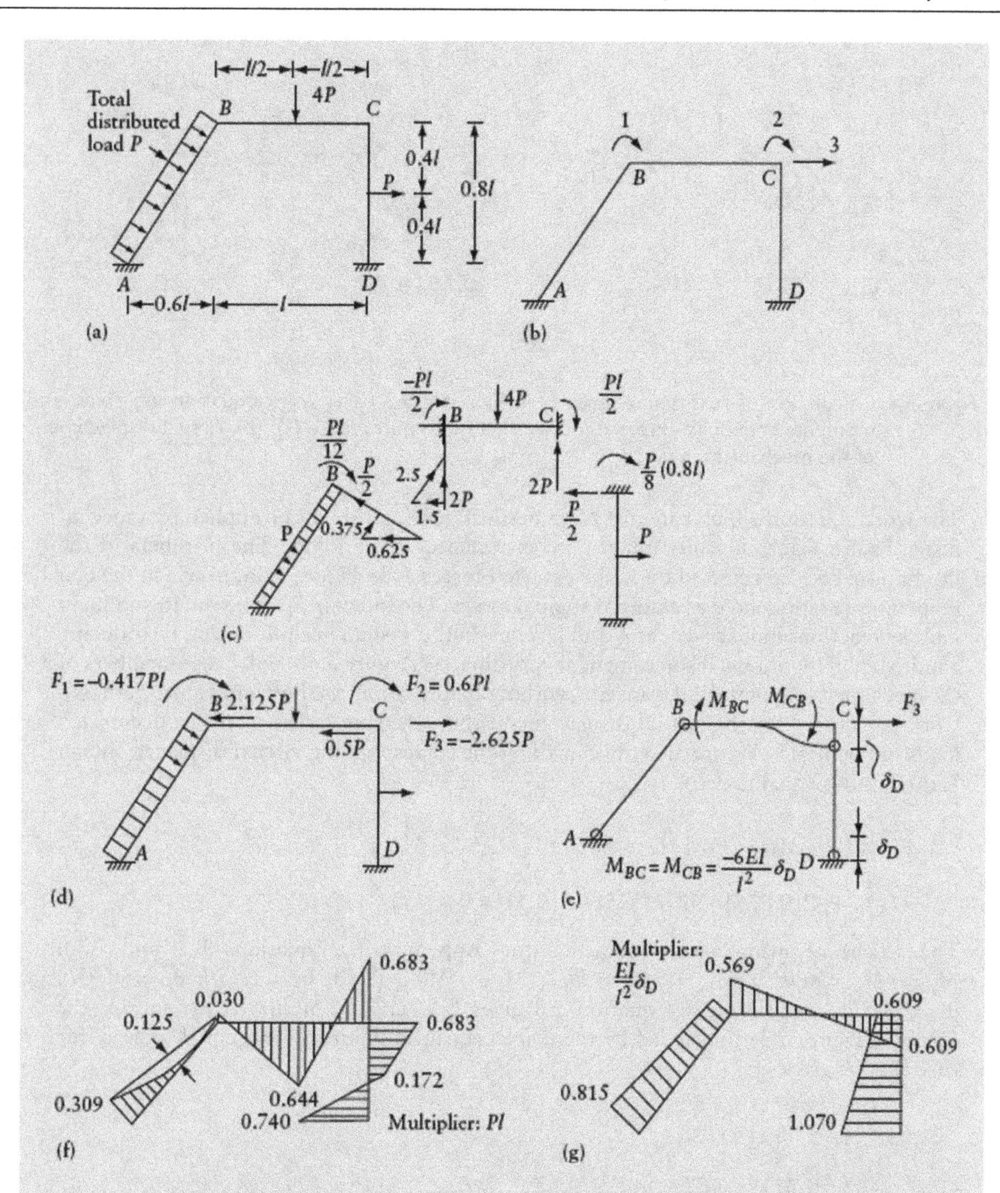

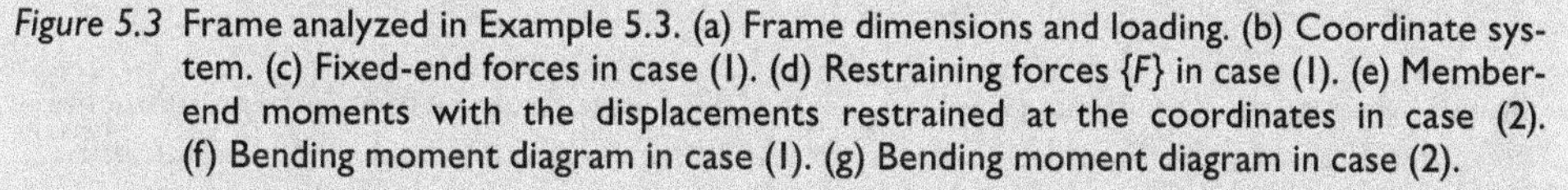

Figure 5.3 Frame analyzed in Example 5.3. (a) Frame dimensions and loading. (b) Coordinate system. (c) Fixed-end forces in case (1). (d) Restraining forces $\{F\}$ in case (1). (e) Member-end moments with the displacements restrained at the coordinates in case (2). (f) Bending moment diagram in case (1). (g) Bending moment diagram in case (2).

force can be substituted by a component equal to $(15EI\delta_D/l^3)$ in AB and a component equal to $(-9EI\delta_D/l^3)$ in direction of coordinate 3. The latter component is equal to F_3 in case (2).

The member-end moments when the displacements at the coordinates are prevented (Figure 5.4a and Figure 5.3e) are:

$$[A_r] = \begin{bmatrix} -0.0833Pl & 0 \\ -0.5Pl & -6\left(EI/l^2\right)\delta_D \\ 0.1Pl & 0 \\ -0.1Pl & 0 \end{bmatrix} \qquad \text{(c)}$$

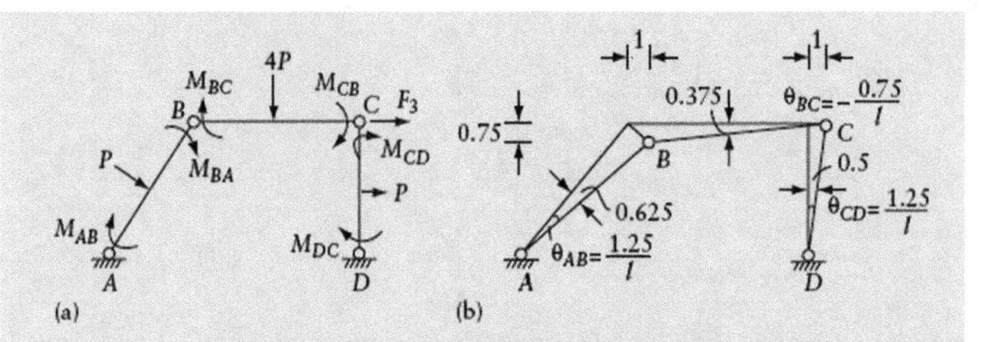

Figure 5.4 Calculation of restraining force F_3 in Example 5.3 using work equation. (a) Fixed-end moments shown as external forces on a mechanism, case (1). (b) Virtual displacement of the mechanism in (a).

The work equation which can give F_3-values in Equation (b) will be applied to a mechanism (Figure 5.4b), in which hinges are introduced at the joints. The members of the mechanism are subjected to the same external forces as in Figure 5.4a, in which the end moments are represented as external applied forces. The forces in Figure 5.4a, including F_3 and the reactions (not shown) at A and D, constitute a system in equilibrium. Introducing a unit virtual (fictitious) displacement at coordinate 3 (Figure 5.4b) will cause members of the mechanism to rotate and translate, without deformation, as shown in Figure 5.4b. No work is required for the virtual displacement; thus, the sum of the forces (concentrated loads or couples) in Figure 5.4a multiplied by the corresponding virtual displacements in Figure 5.4b is equal to zero:

$$\begin{aligned}&\theta_{AB}\left(M_{AB}+M_{BA}\right)+\theta_{BC}\left(M_{BC}+M_{CB}\right)+\theta_{CD}\left(M_{CD}+M_{DC}\right)\\&+F_3\times 1+\left[P(0.625)+4P(0.375)+P(0.5)\right]=0\end{aligned} \tag{5.8}$$

The values of member-end moments are (Appendix B, Equations B.1 and B.7): $M_{AB}=-M_{BA}=-Pl/12$; $M_{BC}=-M_{CB}=-Pl/2$; $M_{CD}=-M_{DC}=Pl/10$; $\theta_{AB}=1.25/l$; $\theta_{BC}=-0.75/l$; $\theta_{CD}=1.25/l$; substitution in Equation 5.8 gives $F_3=-2.625P$. Similarly, the sum of the forces in Figure 5.3e multiplied by the displacement in Figure 5.4b is equal to zero; for case (2) this gives:

$$\theta_{BC}\left(M_{BC}+M_{CB}\right)+F_3\times 1=0$$

$$F_3=-\left(-\frac{0.75}{l}\right)\left(-\frac{6EI}{l^2}\delta_D-\frac{6EI}{l^2}\delta_D\right)=-\frac{9EI}{l^3}\delta_D$$

Step 3 Separate displacements $D_1=1$, $D_2=1$, and $D_3=1$ produce the member-end moments shown in Figures 5.5a, b, and c, respectively (Appendix C, Equations C.1–C.5). Summing up the member-end moments at joints B and C gives the elements of the first two rows of the stiffness matrix $[S]$, Equation (d). Work equations summing the product of the forces in each of Figures 5.5a, b, and c multiplied by the displacements in Figure 5.4b give S_{31}, S_{32}, and S_{33}. The work equation will be the same as Equation 5.8, but without the last term in square brackets; the values of the member-end moments are given in Figures 5.5a, b, and c. The same figures are also used to generate $[A_u]$:

$$[S]=EI\begin{bmatrix}8/l & 2/l & -3/l^2\\ 2/l & 9/l & -4.875/l^2\\ -3/l^2 & -4.875/l^2 & 48.938/l^3\end{bmatrix} \tag{d}$$

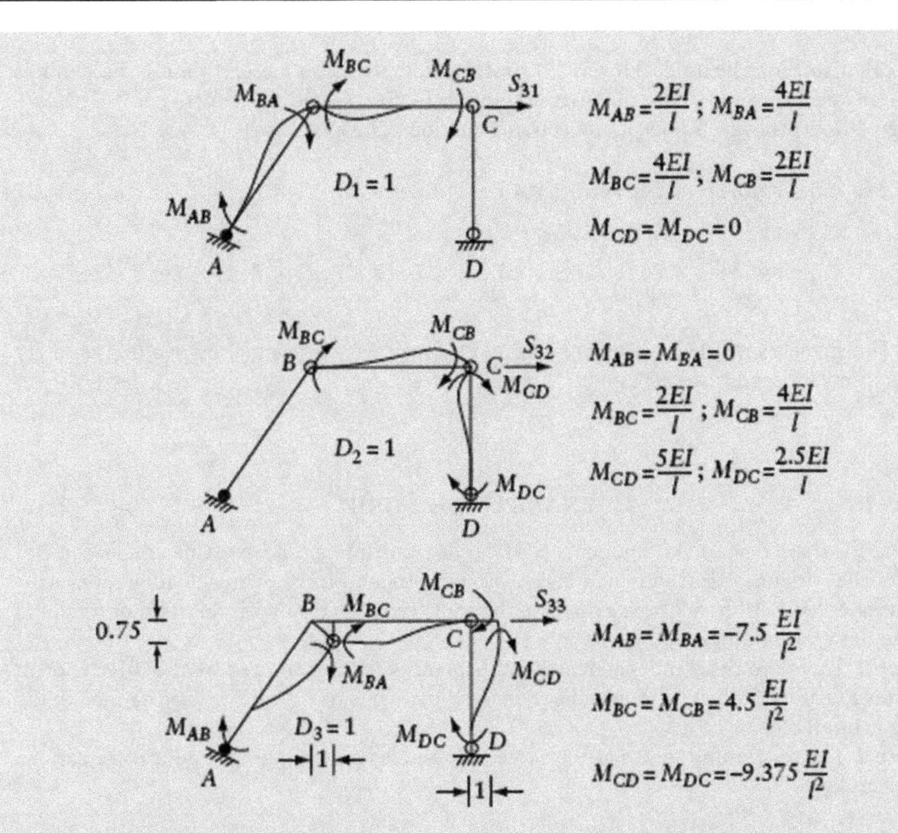

Figure 5.5 Generation of stiffness matrix for the frame of Example 5.3 (Figure 5.3b). (a), (b) and (c) Member-end moments represented as external forces on a mechanism when $D_1 = 1$, $D_2 = 1$ and $D_3 = 1$, respectively.

$$[A_u] = \begin{bmatrix} 2EI/l & 0 & -7.5EI/l^2 \\ 4EI/l & 2EI/l & 4.5EI/l^2 \\ 0 & 5EI/l & -9.375EI/l^2 \\ 0 & 2.5EI/l & -9.375EI/l^2 \end{bmatrix} \qquad \text{(e)}$$

As expected, [S] is symmetrical.

Step 4 Substitution of [S] and [F] in Equation 5.6 and solving for [D] give:

$$[D] = \frac{10^{-3}}{EI} \begin{bmatrix} 133.75l & -26.75l & 5.54l^2 \\ -26.72l & 122.76l & 10.59l^2 \\ 5.54l^2 & 10.59l^2 & 21.83l^3 \end{bmatrix} \begin{bmatrix} 0.417\,Pl & \dfrac{6EI}{l^2}\delta_D \\ -0.6\,Pl & \dfrac{6EI}{l^2}\delta_D \\ 2.65\,P & \dfrac{9EI}{l^3}\delta_D \end{bmatrix} \qquad \text{(f)}$$

$$= \begin{bmatrix} 0.0863\,Pl^2/EI & 0.6920\,\delta_D/l \\ -0.0570\,Pl^2/EI & 0.6717\,\delta_D/l \\ 0.0532\,Pl^3/EI & 0.2932\,\delta_D \end{bmatrix}$$

A calculator may be used to invert $[S]$ considering only the numerical values; the symbols for any element of $[S]^{-1}$ are the inverse symbols in corresponding elements of $[S]$.
Step 5 Substituting $[A_r]$, $[A_u]$, and $[D]$ in Equation 5.7 gives:

$$[A] = \begin{bmatrix} -0.309\,Pl & -0.815\,EI\,\delta_D/l^2 \\ -0.030\,Pl & -0.569\,EI\,\delta_D/l^2 \\ -0.683\,Pl & 0.609\,EI\,\delta_D/l^2 \\ -0.740\,Pl & -1.070\,EI\,\delta_D/l^2 \end{bmatrix}$$

The elements of $[A]$ are the member-end moments used to plot the bending moment diagrams for load cases (1) and (2), Figures 5.3f and g.

EXAMPLE 5.4: A GRID

Find the three reaction components (vertical force, bending, and twisting couples) at end A of the horizontal grid shown in Figure 5.6a due to a uniform vertical load of intensity q acting on AC. All bars of the grid have the same cross section with the ratio of torsional and flexural rigidities $GJ/EI = 0.5$.
Step 1 There are three independent joint displacements represented by the three coordinates in Figure 5.6b. The required actions and their positive directions are defined in the same figure.
Step 2 The restraining forces at the three coordinates are (using Appendix B, Equations B.7 and B.8):

$$\{F\} = \begin{Bmatrix} (-ql/2)_{AE} - (ql/2)_{EC} \\ 0 \\ (ql^2/12)_{AE} - (ql^2/12)_{EC} \end{Bmatrix} = q \begin{Bmatrix} -l/2 \\ 0 \\ l^2/36 \end{Bmatrix}$$

The reaction components at A, while the displacements are prevented at the coordinates, are:

$$\{A_r\} = \left\{(-ql/2)_{AE},\, 0,\, (-ql^2/12)_{AE}\right\} = q\left\{-l/3,\, 0,\, -l^2/27\right\}$$

Step 3 The stiffness matrix is (using Appendix C, Equations C.1 to C.5, C.10, and C.11)

$$[S] = EI \begin{bmatrix} 729/l^3 & & \text{Symm.} \\ -40.5/l^2 & 20.25/l & \\ 40.5/l^2 & 0 & 20.25/l \end{bmatrix}$$

The non-zero elements of this matrix are determined as follows: $S_{11} = \Sigma(12EI/l^3)$, with the summation performed for the four members meeting at E (Figure 5.6b). $S_{21} = (6EI/l^2)_{ED} - (6EI/l^2)_{EB}$.
Similarly, $S_{31} = (6EI/l^2)_{EC} - (6EI/l^2)_{EA}$. The angular displacement $D_2 = 1$ bends and twists beams BD and AC, respectively;

$$S_{22} = (4EI/l)_{EB} + (4EI/l)_{ED} + (GJ/l)_{EC} + (GJ/l)_{EA}$$

Because of symmetry of the structure, S_{33} is the same as S_{22}.
Values of the reactions at A due to separate unit displacements are:

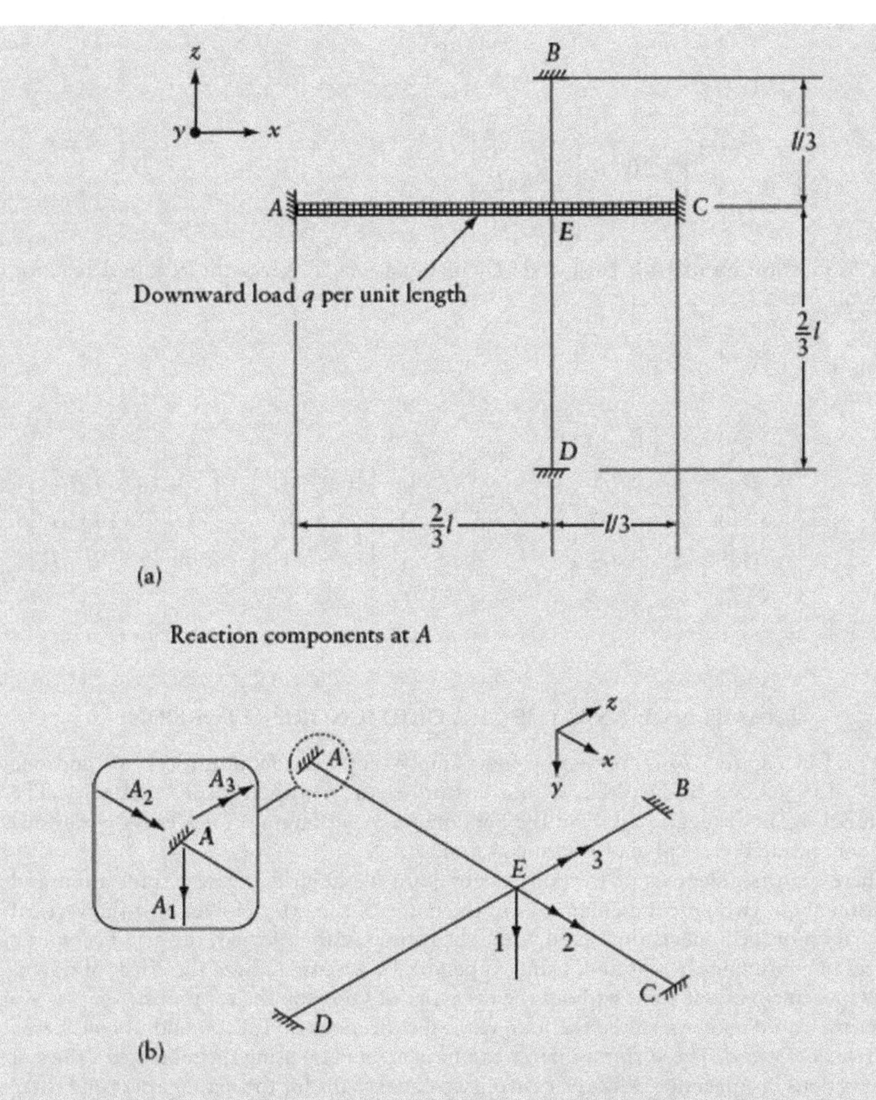

Figure 5.6 Grid analyzed in Example 5.4. (a) Grid plan. (b) Pictorial view showing chosen coordinates 1, 2 and 3 and positive directions of the reaction components at A.

$$[A_u] = \begin{bmatrix} \left(-12EI/l^3\right)_{AE} & 0 & \left(6EI/l^2\right)_{AE} \\ 0 & \left(-GJ/l\right)_{AE} & 0 \\ \left(-6EI/l^2\right)_{AE} & 0 & \left(2EI/l\right)_{AE} \end{bmatrix} = EI \begin{bmatrix} -40.5/l^3 & 0 & 13.5/l^2 \\ 0 & -0.75/l & 0 \\ -13.5/l^2 & 0 & 3/l \end{bmatrix}$$

Equations C.2 and C.1 of Appendix C give A_{u11} and A_{u31}, respectively; Equation C.10 gives A_{u22}; Equations C.5 and C.4 give A_{u13} and A_{u33}, respectively.

Step 4 Substitution of [*S*] and {*F*} in Equation 5.3 and solution gives:

$$[S]^{-1} = \frac{10^{-3}}{EI} \begin{bmatrix} 1.7637l^3 & & \text{Symm.} \\ 3.5273l^2 & 56.437l & \\ -3.5273l^2 & -7.0547 & 56.437l \end{bmatrix}$$

$$\{D\} = -[S]^{-1}\{F\}$$

$$= -[S]^{-1} q \begin{Bmatrix} -l/2 \\ 0 \\ l^2/36 \end{Bmatrix} = \frac{10^{-3} q}{EI} \begin{Bmatrix} 0.9798 l^4 \\ 1.960 l^3 \\ -3.331 l^3 \end{Bmatrix}$$

Step 5 Substitution of $[A_r]$, $[A_u]$, and $\{D\}$ in Equation 5.7 gives the required reaction components:

$$\{A\} = \{A_r\} + [A_u]\{D\}$$

$$\{A\} = \begin{Bmatrix} -\dfrac{ql}{3} \\ 0 \\ -\dfrac{ql^2}{27} \end{Bmatrix} + 10^{-3} q \begin{bmatrix} -\dfrac{40.5}{l^3} & 0 & \dfrac{13.5}{l^2} \\ 0 & -\dfrac{0.75}{l} & 0 \\ -\dfrac{13.5}{l^2} & 0 & \dfrac{3}{l} \end{bmatrix} \begin{Bmatrix} 0.9798 l^4 \\ 1.960 l^3 \\ -3.331 l^3 \end{Bmatrix} = \begin{Bmatrix} -0.4180 ql \\ -1.470 \times 10^{-3} ql^2 \\ -60.26 \times 10^{-3} ql^2 \end{Bmatrix}$$

EXAMPLE 5.5: ANALYSIS OF A GRID IGNORING TORSION

The grid in Figure 5.7a is formed by four simply-supported main girders (m) and one cross-girder (c) of a bridge deck, with a flexural rigidity in the ratio $EI_m{:}EI_c = 3{:}1$. The torsional rigidity is neglected. Find the bending moment diagram for a cross-girder due to a concentrated vertical load P acting at joint 1.

There are three degrees of freedom at any joint of the grid: vertical translation and rotation about two perpendicular axes in the plane of the grid. However, if the vertical translation of the joints is prevented while the rotations are allowed, the grid becomes a system of continuous beams and, using Appendix D, we can analyze the effect of downward movement of one joint without the necessity of knowing the rotations.

Let the coordinate system be the four vertical deflections at 1, 2, 3, and 4, considered positive downward. The stiffness matrix can be generated by using the tabulated values of the reactions in Appendix D for the case of two equal spans for the main girders and three equal spans for the cross-girder. The elements of the first row of the stiffness matrix of the grid are calculated as follows. At ends 1 and 4 of the cross-girder, the bending moment $M_1 = M_4 = 0$. The required actions are the bending moments in the cross girders at 2 and 3.

$$\{A\} = \{M_2, M_3\}$$

The displacements $D_1 = 1$ and $D_2 = D_3 = D_4 = 0$ deform the main girder A and the cross-girder, but girders B, C, and D are not deflected. The rotations at the joints are allowed to take place freely. Then, adding the vertical forces required to hold the girder A and the cross-girder in this deflected shape, we obtain (Table D.1, Appendix D):

$$S_{11} = 6.0 \frac{EI_m}{(3b)^3} + 1.6 \frac{EI_c}{b^3} = 2.267 \frac{EI_c}{b^3}$$

$$S_{21} = -3.6 \frac{EI_c}{b^3}; \; S_{31} = 2.4 \frac{EI_c}{b^3}; \; S_{41} = -0.4 \frac{EI_c}{b^3}$$

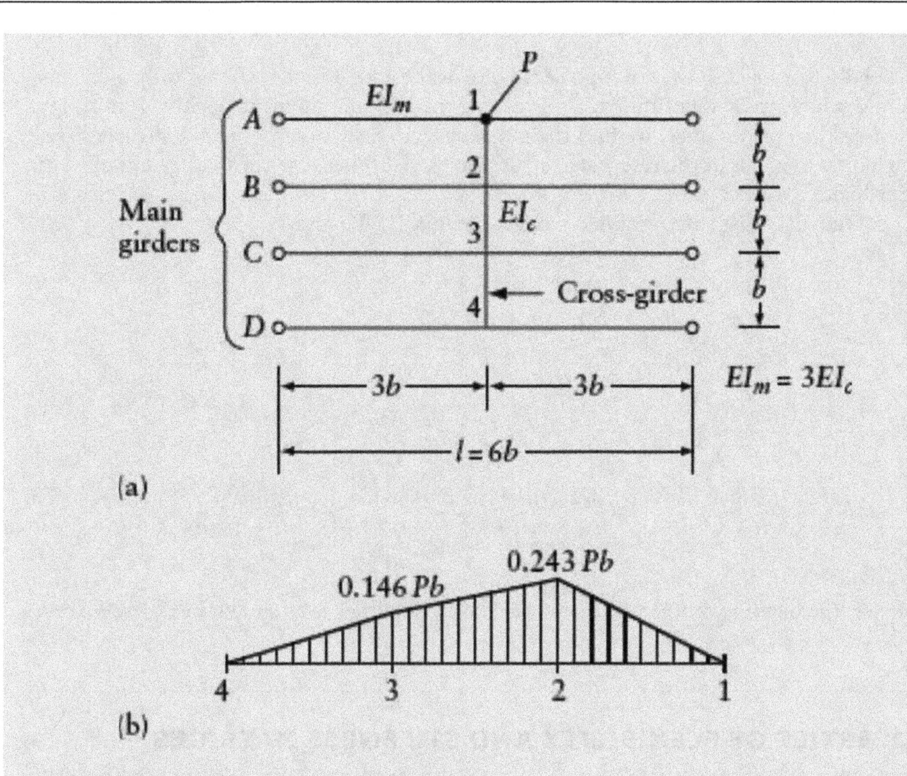

Figure 5.7 Grid considered in Example 5.5. (a) Top view of a torsionless grid. (b) Bending moment diaphragm for the cross-girder.

The elements of other columns of the stiffness matrix are determined in a similar way. Alternatively, because of symmetry of the structure, the elements on the fourth column of [S] are the same as the elements on the first column but reversed in order. From Appendix D, Table D.1, for number of spans = 2 and 3, we determine S_{22}, S_{32}, and S_{42}:

$$S_{22} = 6.0\frac{EI_m}{(3b)^3} + 9.6\frac{EI_c}{b^3} = 10.267\frac{EI_c}{b^3}$$

$$S_{32} = -8.4\frac{EI_c}{b^3};\ S_{42} = 2.4\frac{EI_c}{b^3}$$

We may use the symmetry of [S] to fill up the remaining elements of [S], so that finally:

$$[S] = \frac{EI_c}{b^3}\begin{bmatrix} 2.267 & -3.600 & 2.400 & -0.400 \\ -3.600 & 10.267 & -8.400 & 2.400 \\ 2.400 & -8.400 & 10.267 & -3.600 \\ -0.400 & 2.400 & -3.600 & 2.267 \end{bmatrix}$$

With the load P at coordinate 1, we need only an equal and opposite force at this coordinate to prevent the joint displacements. Thus, $\{F\} = \{-P, 0, 0, 0\}$. Substituting in Equation 5.3 gives:

$$\{D\} = \frac{Pb^3}{EI_c}\begin{Bmatrix} 1.133 \\ 0.511 \\ 0.076 \\ -0.221 \end{Bmatrix}$$

The bending moment in the cross-girder is zero at the ends 1 and 4, so that only moments at 2 and 3 have to be determined. Considering the bending moment positive if it causes tension in the bottom fiber, we find the moment from Equation 5.5. In the present case, the bending moment in the restrained structure $\{A_r\} = \{0\}$ because the load P is applied at a coordinate.

The elements of $[A_u]$ are obtained from Appendix D. We find:

$$[A_u] = \frac{EI_c}{b^2}\begin{bmatrix} -1.6 & 3.6 & -2.4 & 0.4 \\ 0.4 & -2.4 & 3.6 & -1.6 \end{bmatrix}$$

$$\{A\} = \{A_r\} + [A_u]\{D\}$$

$$= \frac{EI_c}{b^2}\begin{bmatrix} -1.6 & 3.6 & -2.4 & 0.4 \\ 0.4 & -2.4 & 3.6 & -1.6 \end{bmatrix}\frac{Pb^3}{EI_c}\begin{Bmatrix} 1.133 \\ 0.511 \\ 0.076 \\ -0.221 \end{Bmatrix} = Pb\begin{Bmatrix} -0.243 \\ -0.146 \end{Bmatrix}$$

Hence, the bending moment diagram for the cross-girder is as shown in Figure 5.7b.

5.5 PROPERTIES OF FLEXIBILITY AND STIFFNESS MATRICES

Consider a force F_i applied gradually to a structure, so that the kinetic energy of the mass of the structure is zero. Let the resulting displacement at the location and in the direction of F_i be D_i. If the structure is elastic, the force-displacement curve follows the same path on loading and unloading, as shown in Figure 5.8a.

Assume now that at some stage of loading, the force F_i is increased by ΔF_i and the corresponding increase in the displacement D_i is ΔD_i. The work done by this load increment is:

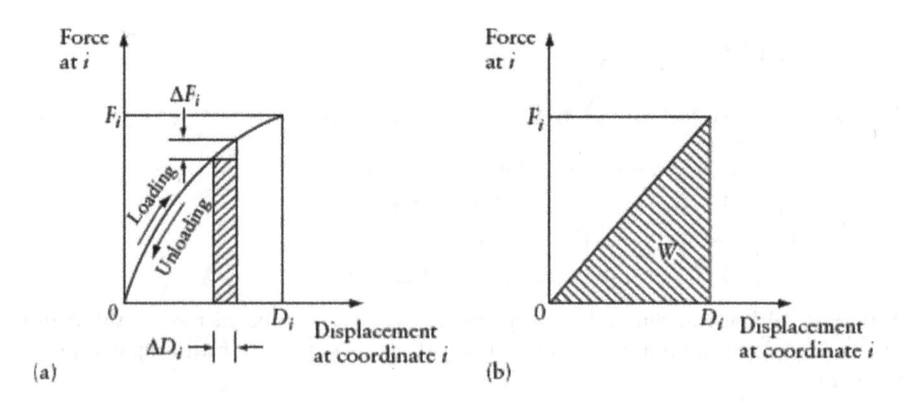

Figure 5.8 Force-displacement relations.

$$\Delta W \cong F_i \, \Delta D_i$$

This is shown as the hatched rectangle in Figure 5.8a. If the increments are sufficiently small, it can be seen that the total external work done by F_i during the displacement D_i is the area below the curve between 0 and D_i.

When the material in the structure obeys Hooke's law, the curve in Figure 5.8a is replaced by a straight line (Figure 5.8b) and the work done by the force F_i becomes:

$$W = \frac{1}{2} F_i \, D_i$$

If the structure is subjected to a system of forces $F_1, F_2, \ldots, F_n$, increased gradually from zero to their final value, causing displacements $D_1, D_2, \ldots, D_n$ at the location and in the direction of the forces, then the total external work is:

$$W = \frac{1}{2}\left(F_1 \, D_1 + F_2 \, D_2 + \cdots + F_n \, D_n\right) = \frac{1}{2}\sum_{i=1}^{n} F_i \, D_i \tag{5.9}$$

This equation can be written in the form:

$$[W]_{1\times1} = \frac{1}{2}\{F\}^{\mathrm{T}}_{n\times1}\{D\}_{n\times1} \tag{5.10}$$

where $\{F\}^{\mathrm{T}}$ is the transpose of the column vector $\{F\}$ representing the forces. Work done is a scalar quantity whose dimensions are (force × length).

The displacements and the forces are related by Equation 4.3. Substituting in Equation 5.10:

$$[W]_{1\times1} = \frac{1}{2}\{F\}^{\mathrm{T}}_{n\times1}[f]_{n\times n}\{F\}_{n\times1} \tag{5.11}$$

Taking the transpose of both sides does not change the left-hand side of the equation. The right-hand side becomes the product of the transpose of the matrices on this side but in reverse order. Therefore:

$$[W]_{1\times1} = \frac{1}{2}\{F\}^{\mathrm{T}}_{n\times1}[f]^{\mathrm{T}}_{n\times n}\{F\}_{n\times1} \tag{5.12}$$

The flexibility matrix and its transpose are equal, that is (Section 7.7.1):

$$[f]^{\mathrm{T}} = [f] \tag{5.13}$$

This means that for a general element of the flexibility matrix:

$$f_{ij} = f_{ji} \tag{5.14}$$

and is known as *Maxwell's reciprocal relation*. In other words, the flexibility matrix is a symmetrical matrix. This property is useful in forming the flexibility matrix because some of the coefficients need not be calculated or, if they are, a check is obtained. The property of symmetry can also be used to save a part of the computational effort required for matrix inversion or for a solution of equations.

The stiffness matrix $[S]$ is a symmetrical matrix; thus, for a general stiffness coefficient

$$S_{ij} = S_{ji} \tag{5.15}$$

This property can be used in the same way as in the case of the flexibility matrix.

Another important property of the flexibility and stiffness matrices is that the elements on the main diagonal, f_{ii} or S_{ii}, must be positive as demonstrated below. The element f_{ii} is the deflection at coordinate i due to a unit force at i. Obviously, the force and the displacement must be in the same direction: f_{ii} is therefore positive. The element S_{ii} is the force required at coordinate i to cause a unit displacement at i. Here again, the force and the displacement must be in the same direction so that the stiffness coefficient S_{ii} is positive.

We should note, however, that in unstable structures – for example, a strut subjected to an axial force reaching the buckling load – the stiffness coefficient S_{ii} can be negative. This is discussed further in Chapter 10.

Let us now revert to Equation 5.11, which expresses the external work in terms of the force vector and the flexibility matrix. We recall that a stiffness matrix relates displacements $\{D\}$ at a number of coordinates to the forces $\{F\}$ applied at the same coordinates by the equation: $[S]\ \{D\} = \{F\}$; substituting the force vector in Equation 5.10, the work can also be expressed in terms of the displacement vector and the stiffness matrix, thus:

$$W = \frac{1}{2}\{F\}^{\mathrm{T}}[f]\{F\} \tag{5.16}$$

or

$$W = \frac{1}{2}\{D\}^{\mathrm{T}}[S]\{D\} \tag{5.17}$$

The quantity on the right-hand side of these equations is referred to as the *quadratic form* in variable F or D. A quadratic form is said to be *positive definite* if it assumes positive values for any non-zero vector of the variable and, moreover, is zero only when the vector of the variables is zero ($\{F\}$ or $\{D\} = \{0\}$). It can also be proven that the determinant of a positive definite symmetrical matrix is greater than zero.

From the above discussion, we can see that the quadratic forms in Equations 5.16 and 5.17 represent the external work of a system of forces producing a system of displacements and this quantity must be positive in a stable structure. Physically, this means that work is required to produce any set of displacements $\{D\}$ by the application of a set of forces $\{F\}$. Thus, the quadratic forms $(1/2)\{F\}^{\mathrm{T}}[f]\{F\}$ and $(1/2)\{D\}^{\mathrm{T}}[S]\{D\}$ are positive definite and the matrices $[f]$ and $[S]$ are said to be *positive definite matrices*. It follows therefore that, for a stable structure, the stiffness and flexibility matrices must be positive definite and the systems of linear equations

$$[S]\{D\} = \{F\}$$

and

$$[f]\{F\} = \{D\}$$

are positive definite. Further, since the determinants $|S|$ or $|f|$ must be greater than zero, for any non-zero vector on the right-hand side of the equations, each system has a single unique solution, i.e. there is only one set of D_i or F_i values which satisfies the first and second sets of equations, respectively.

5.6 STIFFNESS MATRIX FOR A PRISMATIC MEMBER OF SPACE AND PLANE FRAMES

In the examples of this chapter, we see that the elements of the stiffness matrix of a structure are obtained by adding the forces at the ends of the members which meet at a joint. These end-forces are elements of the stiffness matrix for individual members and are derived by the use of Appendix C. In this section, the stiffness matrix for a prismatic member is generated because it is often needed in the analysis of framed structures. We consider 12 coordinates at the ends, representing translations and rotations about three rectangular axes x, y, and z (Figure 5.9a), with the y and z axes chosen to coincide with the principal axes of the cross section. The beam is assumed to be of length l and cross-sectional area a, and to have second moments of area I_z and I_y about the z and y axes, respectively; the modulus of elasticity of the material is E and the torsional rigidity GJ.

If we neglect shear deformations and warping caused by twisting, all the elements of the stiffness matrix can be taken from Appendix C. The elements in any column j are equal to the forces at the coordinates produced by a displacement $D_j = 1$ at coordinate j only. The resulting stiffness matrix is given below:

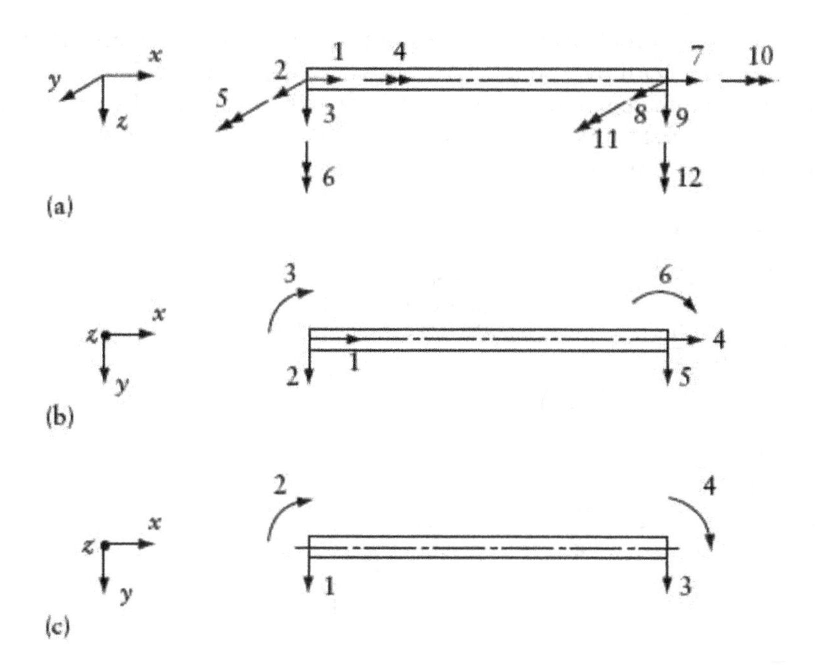

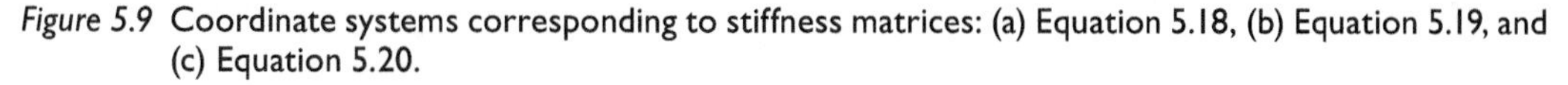
Figure 5.9 Coordinate systems corresponding to stiffness matrices: (a) Equation 5.18, (b) Equation 5.19, and (c) Equation 5.20.

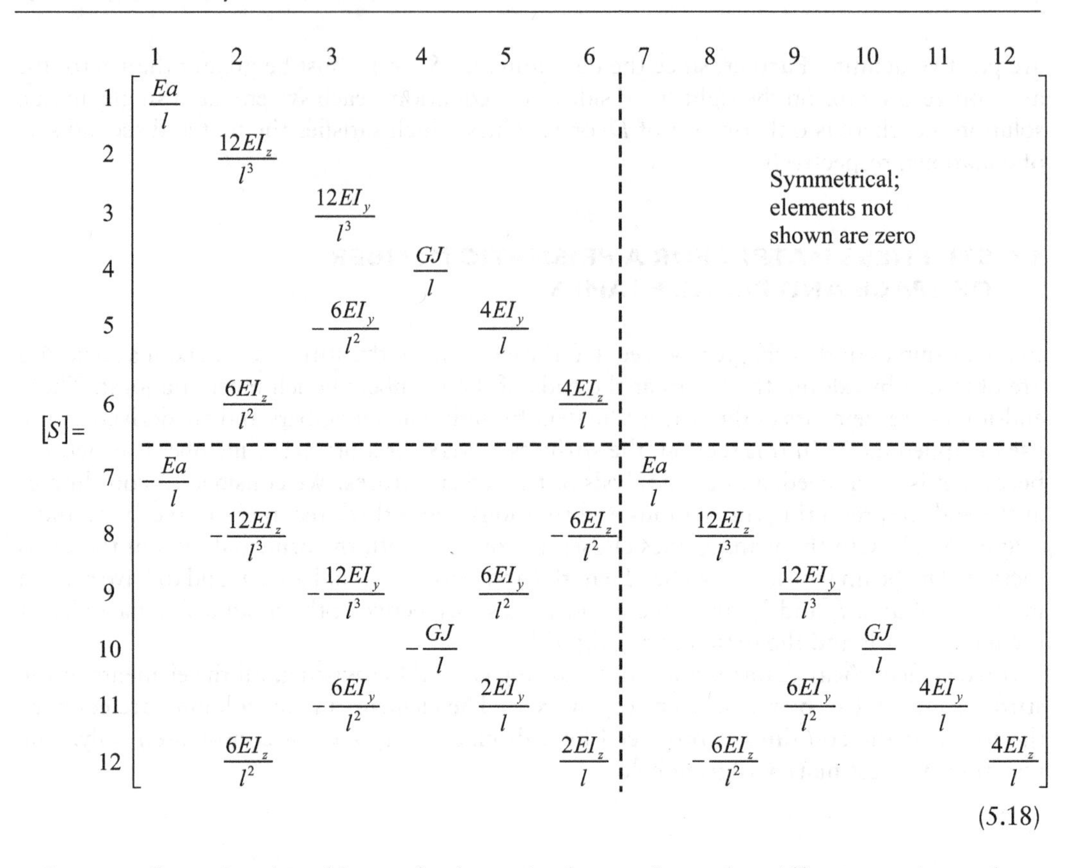

(5.18)

For two-dimensional problems of a frame in the xy plane, the stiffness matrix needs to be considered for six coordinates only: 1, 2, 6, 7, 8, and 12 (Figure 5.9a). Deletion of the columns and rows numbered 3, 4, 5, 9, 10, and 11 from the matrix in Equation 5.18 results in the following stiffness matrix of a prismatic member corresponding to the six coordinates in Figure 5.9b to be used in the analysis of plane frames:

$$[S] = \begin{bmatrix} \frac{Ea}{l} & & & & & \\ & \frac{12EI}{l^3} & & & & \\ & \frac{6EI}{l^2} & \frac{4EI}{l} & & & \\ -\frac{Ea}{l} & & & \frac{Ea}{l} & & \\ & -\frac{12EI}{l^3} & -\frac{6EI}{l^2} & & \frac{12EI}{l^3} & \\ & \frac{6EI}{l^2} & \frac{2EI}{l} & & -\frac{6EI}{l^2} & \frac{4EI}{l} \end{bmatrix} \quad \text{Symmetrical; elements not shown are zero} \tag{5.19}$$

where

$I = I_z$.

If, in a plane frame, the axial deformations are ignored, the coordinates 1 and 4 in Figure 5.9b need not be considered and the stiffness matrix of a prismatic member corresponding to the four coordinates in Figure 5.9c becomes

$$[S] = \begin{array}{c} \\ 1 \\ 2 \\ 3 \\ 4 \end{array} \begin{array}{c} \begin{array}{cccc} 1 & 2 & 3 & 4 \end{array} \\ \begin{bmatrix} \frac{12EI}{l^3} & & \text{Symmetrical} & \\ \frac{6EI}{l^2} & \frac{4EI}{l} & & \\ -\frac{12EI}{l^3} & -\frac{6EI}{l^2} & \frac{12EI}{l^3} & \\ \frac{6EI}{l^2} & \frac{2EI}{l} & -\frac{6EI}{l^2} & \frac{4EI}{l} \end{bmatrix} \end{array} \qquad (5.20)$$

The stiffness matrix of a free (unsupported) structure can be readily generated, as, for example, Equations 5.18 to 5.20 for the beams in Figure 5.9. However, such a matrix is singular and cannot be inverted. Thus, no flexibility matrix can be found unless sufficient restraining forces are introduced for equilibrium.

The criterion of non-singularity of stiffness matrices for stable structures can be used for the determination of buckling loads, as will be discussed in Chapter 10. The presence of a high axial force in a deflected member causes an additional bending moment and if this effect is to be taken into account, the above stiffness matrices must be modified. We shall see that the frame is stable only if its stiffness matrix is positive definite and thus its determinant is greater than zero. If the determinant is put equal to zero, a condition is obtained from which the buckling load can be calculated.

If a symmetrical matrix is positive definite, the determinants of all its minors are also positive. This has a physical significance relevant to the stiffness matrix of a stable structure. If the displacement D_i is prevented at a coordinate i, for example, by introduction of a support, the stiffness matrix of the resulting structure can be obtained simply by deletion of ith row and column from the stiffness matrix of the original structure. Addition of a support to a stable structure results in a structure which is also stable and thus the determinant of its stiffness matrix is positive. This determinant is that of a minor of the original stiffness matrix.

The stiffness matrices derived above correspond to coordinates which coincide with the beam axis or with the principal axes of its cross section. However, in the analysis of structures composed of a number of members running in arbitrary directions, the coordinates may be taken parallel to a set of *global axes* and thus the coordinates may not coincide with the principal axes of a given member. In such a case, the stiffness matrices given above (corresponding to the coordinates coinciding with the principal axes of the members) will have to be transformed to stiffness matrices corresponding to another set of coordinates by the use of transformation matrices formed by geometrical relations between the two sets of coordinates. This will be further discussed in Section 8.8.

5.7 RELATION BETWEEN FLEXIBILITY AND STIFFNESS MATRICES

Consider a system of n coordinates defined on a structure to which the principle of superposition applies (see Section 3.6). In Chapter 4, we have defined the flexibility matrix, $[f]$

and its elements, the flexibility coefficients. We have also defined in this chapter the stiffness matrix, [S] and its elements, the stiffness coefficients. A typical flexibility coefficient, f_{ij} is the displacement at coordinate i due to a force $F_j = 1$ at coordinate j, with no forces applied at the remaining coordinates. A typical stiffness coefficient, S_{ij} is the force at coordinate i when the structure is deformed such that $D_j = 1$, while the displacements are zero at the remaining coordinates. We will show below that the flexibility and the stiffness matrices, *corresponding to the same coordinate system*, are related:

$$[S] = [f]^{-1} \tag{5.21}$$

$$[f] = [S]^{-1} \tag{5.22}$$

Equations 5.21 to 5.25 apply with any number of coordinates. For simplicity of presentation, we consider a system of two coordinates ($n = 2$) shown in Figure 5.10a. The displacements $\{D_1, D_2\}$ can be expressed as the sum of the displacements due to each of forces F_1 and F_2 acting separately (see the superposition Equation 3.21):

$$D_1 = f_{11}\, F_1 + f_{12}\, F_2$$

$$D_2 = f_{21}\, F_1 + f_{22}\, F_2$$

These equations can be put in matrix form:

$$\{D\} = [f]\{F\} \tag{5.23}$$

The forces $\{F\}$ can be expressed in terms of the displacements by solving Equation 5.23:

$$\{F\} = [f]^{-1}\{D\} \tag{5.24}$$

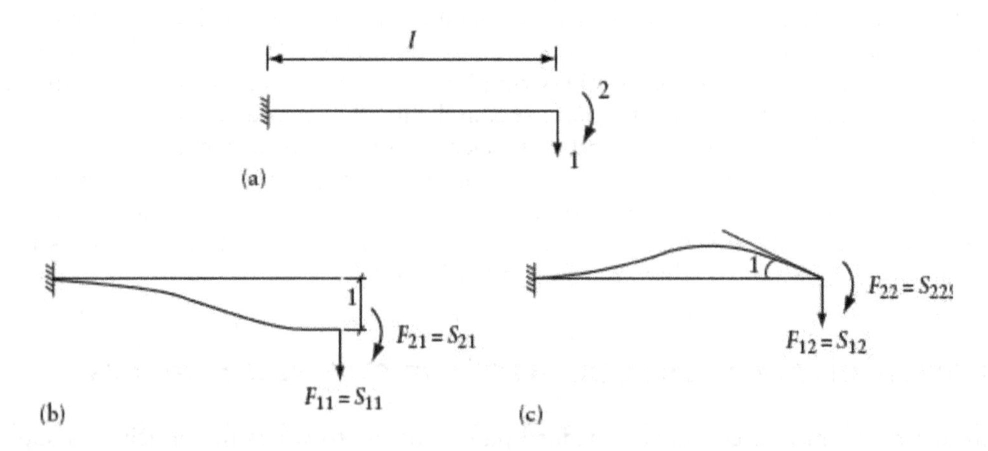

Figure 5.10 A structure with two coordinates. (a) Coordinate system. (b) Forces holding the structure in a deformed configuration with $D_1 = 1$, while $D_2 = 0$. (c) Forces holding the structure in a deformed configuration with $D_1 = 0$, while $D_2 = 1$.

Consider the structure in two deformed configurations: $D_1 = 1$ while $D_2 = 0$ (Figure 5.10b) and $D_1 = 0$ while $D_2 = 1$ (Figure 5.10c).

Application of Equation 5.24 gives:

$$\begin{bmatrix} F_{11} & F_{12} \\ F_{21} & F_{22} \end{bmatrix} = [f]^{-1} \begin{bmatrix} 1 & 0 \\ 0 & 1 \end{bmatrix}$$

The elements of the matrix on the left-hand side of this equation are in fact the stiffness coefficients, as defined above. The last matrix on the right-hand side of the equation is a unit matrix. Thus, the equation can be written in the same form as Equation 5.21. Inversion of both sides of Equation 5.21 gives Equation 5.22. Substitution of Equation 5.21 in 5.24 gives:

$$\{F\} = [S]\{D\} \tag{5.25}$$

Equations 5.21 and 5.22 show that the stiffness matrix is the inverse of the flexibility matrix and vice versa, provided that the same coordinate system of forces and displacements is used in the formation of the two matrices.

We may recall that in the force method of analysis, releases are introduced to render the structure statically determinate. The coordinate system represents the location and direction of these released forces. Now, in the displacement method of analysis, restraining forces are added to prevent joint displacements. The coordinate system in this case represents the location and direction of the unknown displacements. It follows that the two coordinate systems cannot be the same for the same structure. Therefore, the inverse of the flexibility matrix used in the force method is a matrix, whose elements are stiffness coefficients, but not those used in the displacement method of analysis. Furthermore, we note that the flexibility matrix determined in the force method is that of a released structure, whereas in the displacement method, we use the stiffness matrix of the actual structure.

Equation 5.23 must not be confused with the equation: $[f]\{F\} = \{-D\}$ used in the force method of analysis, where we apply unknown redundants $\{F\}$ of such a magnitude as will produce displacements $\{-D\}$ to correct for inconsistencies $\{D\}$ of the released structure.

Similarly, Equation 5.25 must not be confused with the equation: $[S]\{D\} = \{-F\}$ used in the displacement method of analysis, where we seek unknown displacements $\{D\}$ of such a magnitude as will eliminate the artificial restraining forces $\{F\}$.

5.8 ANALYSIS FOR DIFFERENT LOADINGS

We have already made it clear that the stiffness matrix and its inverse are properties of a structure and do not depend on the system of the load applied. Therefore, if a number of different loadings are to be considered, Equation 5.2 can be used for all of them. If the number of cases of loading is p and the number of degrees of freedom is n, the solution can be combined into one matrix equation:

$$[D]_{n \times p} = [S]^{-1}_{n \times n} [-F]_{n \times p} \tag{5.26}$$

with each column of $[D]$ and $[-F]$ corresponding to one loading.

The third step of the displacement method involves generation of $[A_u]_{m \times n}$ in addition to $[S]_{n \times n}$. The elements of any column of $[A_u]$ are values of the required m actions when a unit displacement is introduced at one coordinate, while the displacements are prevented at the remaining coordinates. Thus, similar to $[S]$, the elements of $[A_u]$ do not depend on the applied load. Therefore, the same matrix $[A_u]$ can be used for all loading cases. With p loading cases, the superposition Equation 5.5 applies with the sizes of matrices: $[A]_{m \times p}$, $[A_r]_{m \times p}$, $[A_u]_{m \times n}$, and $[D]_{n \times p}$.

5.9 EFFECT OF NONLINEAR TEMPERATURE VARIATION

Analysis of changes in stresses and internal forces in structures due to a variation in temperature or due to shrinkage or creep can be done in the same way. The distribution of temperature over the cross section of members is generally nonlinear, as shown in Figure 5.11a and b for a bridge girder exposed to the radiation of the sun. In a cross section composed of different materials, such as concrete and steel, the components tend to contract or expand differently because of shrinkage and creep. However, contraction and expansion cannot occur freely and changes in stresses occur. In the following, we consider the effect of temperature rise varying nonlinearly over the cross section of members of a framed structure. The temperature rise is assumed constant over the length of individual members.

In a statically determinate frame, no stresses are produced when the temperature variation is linear; in this case, the thermal expansion occurs freely without restraint. This results in changes in length or in curvature of the members but produces no changes in the reactions

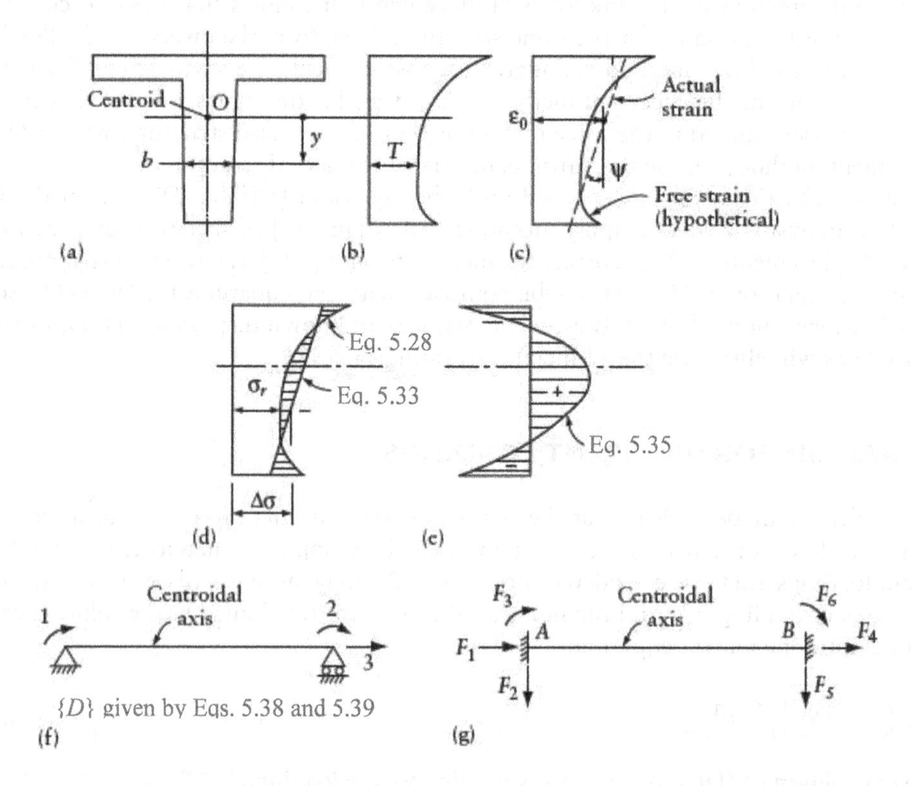

Figure 5.11 Analysis of the effects of nonlinear temperature variation. (a) Cross section of a member. (b) Distribution of temperature rise. (c) Strain distribution. (d) Stresses σ_r and $\Delta\sigma$. (e) Self-equilibrating stresses. (f) Displacements due to a temperature rise in a simple beam. (g) Fixed-end forces due to a temperature rise.

or in the internal forces. However, when the temperature variation is nonlinear, each fiber, being attached to adjacent fibers, is not free to undergo the full expansion and this induces stresses. These stresses must be self-equilibrating in an individual cross section as long as the structure is statically determinate. The *self-equilibrating stresses* caused by nonlinear temperature (or shrinkage) variation over the cross section of a statically determinate frame are sometimes referred to as the *eigenstresses.*

If the structure is statically indeterminate, the elongations and the rotations at the member ends may be restrained or prevented. This results in changes in the reactions and in the internal forces that can be determined by an analysis using the force or the displacement method. We should note that the reactions produced by temperature must represent a set of forces in equilibrium.

Let us now analyze the self-equilibrating stresses in a statically determinate member, e.g. a simple beam of homogeneous material, subjected to a nonlinear rise in temperature (Figure 5.11a). The hypothetical strain which would occur in each fiber, if it were free to expand, is:

$$\varepsilon_f = \alpha\, T \tag{5.27}$$

where α is the coefficient of thermal expansion and $T = T(y)$ is the temperature rise in any fiber at a distance y below the centroid O. If the expansion is artificially prevented, the stress in the restrained condition will be:

$$\sigma_r = -E\, \varepsilon_f \tag{5.28}$$

where E is the modulus of elasticity. Tensile stress and the corresponding strain are considered positive.

The resultant of σ_r may be represented by a normal force N at O and a moment M about the horizontal axis at O, given by:

$$N = \int \sigma_r \, da \tag{5.29}$$

$$M = \int \sigma_r \, y \, da \tag{5.30}$$

N is considered positive when tensile and M is positive when it produces tension in the bottom fiber; the corresponding curvature ψ is positive.

To eliminate the artificial restraint, apply N and M in opposite directions, resulting in the following changes in strain at O and in curvature:

$$\varepsilon_O = -\frac{N}{Ea} \tag{5.31}$$

$$\psi = -\frac{M}{EI} \tag{5.32}$$

where a and I are the area of the cross section and its second moment about a horizontal axis through O, respectively. The corresponding strain and stress at any fiber are:

$$\varepsilon = \varepsilon_O + y\psi \tag{5.33}$$

$$\Delta\sigma = E\left(\varepsilon_O + y\psi\right) \tag{5.34}$$

The addition of σ_r to $\Delta\sigma$ gives the self-equilibrating stress due to temperature:

$$\sigma_s = E\left(-\alpha T + \varepsilon_O + y\psi\right) \tag{5.35}$$

The stress σ_s must have a zero resultant because its components σ_r and $\Delta\sigma$ have equal and opposite resultants. The distribution of the self-equilibrating stress is shown in Figure 5.11e; the ordinates of this graph are equal to the ordinates between the curve σ_r and the straight line $\Delta\sigma$ in Figure 5.11d.

The changes in axial strain and curvature due to temperature are derived from Equations 5.27 to 5.32:

$$\varepsilon_O = \frac{\alpha}{a}\int T b \, dy \tag{5.36}$$

$$\psi = \frac{\alpha}{I}\int T b y \, dy \tag{5.37}$$

where $b = b(y)$ is the width of the section. The actual strain distribution over the depth of the section is presented in Figure 5.11c by a dashed line defined by the values ε_O and ψ. The two values may be used to calculate the displacements at the coordinates in Figure 5.11f (see Appendix A):

$$D_1 = -D_2 = \psi \frac{l}{2} \tag{5.38}$$

$$D_3 = \varepsilon_O \, l \tag{5.39}$$

When using Equation 5.38 we should note that, according to the adopted sign convention, ψ in Figure 5.11c is negative.

If the structure is statically indeterminate, the displacements $\{D\}$, such as those given above, may be used in the force method for the analysis of statically indeterminate reactions and internal forces.

When the analysis is by the displacement method, the values ε_O and ψ (Equations 5.36 and 5.37) can be used to determine the internal forces in a member in the restrained condition and the corresponding member end-forces (Figure 5.11g):

$$N = -E a \varepsilon_O \tag{5.40}$$

$$M = -E I \psi \tag{5.41}$$

$$\{F\} = E\{a\varepsilon_O, 0, -I\psi, -a\varepsilon_O, 0, I\psi\} \tag{5.42}$$

In the special case when the rise in temperature varies linearly from T_{top} to T_{bot} at top and bottom fibers in a member of constant cross section, the fixed-end forces (Figure 5.11g) may be calculated by Equation 5.42, with $\varepsilon_O = \alpha T_O$ and $\psi = \alpha(T_{bot} - T_{top})/h$; where T_O is the temperature at the cross-section centroid and h is the section depth.

The forces F_1 and F_3 are along the centroidal axis, and the other forces are along centroidal principal axes of the member cross section. The six forces are self-equilibrating. The restraining forces at the ends of individual members meeting at a joint should be transformed in the directions of the global axes and summed up to give the external restraining forces which will artificially prevent the joint displacements of the structure. (The assemblage of end-forces is discussed further in Section 16.9.) In the restrained condition, the stress in any fiber may be calculated by Equation 5.28.

When the temperature rise varies from section to section or when the member has a variable cross section, ε_O and ψ will vary over the length of the member. Equations 5.40 and 5.41 may be applied at any section to give the variables N and M in the restrained condition; the member forces are given by:

$$\{F\} = E\left\{\{a\varepsilon_O, 0, -I\psi\}_A, \{-a\varepsilon_O, 0, I\psi\}_B\right\} \tag{5.43}$$

The subscripts A and B refer to the member ends (Figure 5.11g). For equilibrium, a distributed axial load p and a transverse load q must exist. The load intensities (force per length) are given by:

$$p = E\frac{d(a\varepsilon_O)}{dx} \tag{5.44}$$

$$q = E\frac{d^2(I\psi)}{dx^2} \tag{5.45}$$

Positive p is in the direction A to B and positive q is downwards; x is the distance from A to any section. Equations 5.44 and 5.45 can be derived by considering the equilibrium of a small length of the beam separated by two sections dx apart.

The restraining forces given by Equations 5.43 to 5.45 represent a system in equilibrium. The displacements due to temperature can be analyzed by considering the effect of these restraining forces applied in reversed directions. In the restrained condition, the displacement at *all* sections is zero and the internal forces are given by Equations 5.40 and 5.41. These internal forces must be superimposed on the internal forces resulting from the application of the reversed self-equilibrating restraining forces in order to give the total internal forces due to temperature.

The self-equilibrating stresses and the statically indeterminate forces caused by temperature change are proportional to the modulus of elasticity, E. Some materials, such as concrete, exhibit creep (increase in strain), when subjected to sustained stress. The thermal effects will be overestimated when the value of E used in the analysis is based on the relation of stress to instantaneous strain, ignoring creep (see Section 5.10).

Inevitable cracking of reinforced concrete members reduces the effective values of a and I. If cracking is ignored in the values of a and I used in analysis of thermal effects, the results can be greatly overestimated.[1]

EXAMPLE 5.6: THERMAL STRESSES IN A CONTINUOUS BEAM

The continuous concrete beam in Figure 5.12a is subjected to a rise of temperature which is constant over the beam length but varies over the depth as follows:

$$T = T_O + 4.21\,T_{top}\left(\frac{7}{16} - \frac{y}{h}\right)^5 \quad \text{for} \quad -\frac{5h}{16} \le y \le \frac{7h}{16}$$

$$T = T_O \quad \text{for} \quad \frac{7h}{16} \le y \le \frac{11h}{16}$$

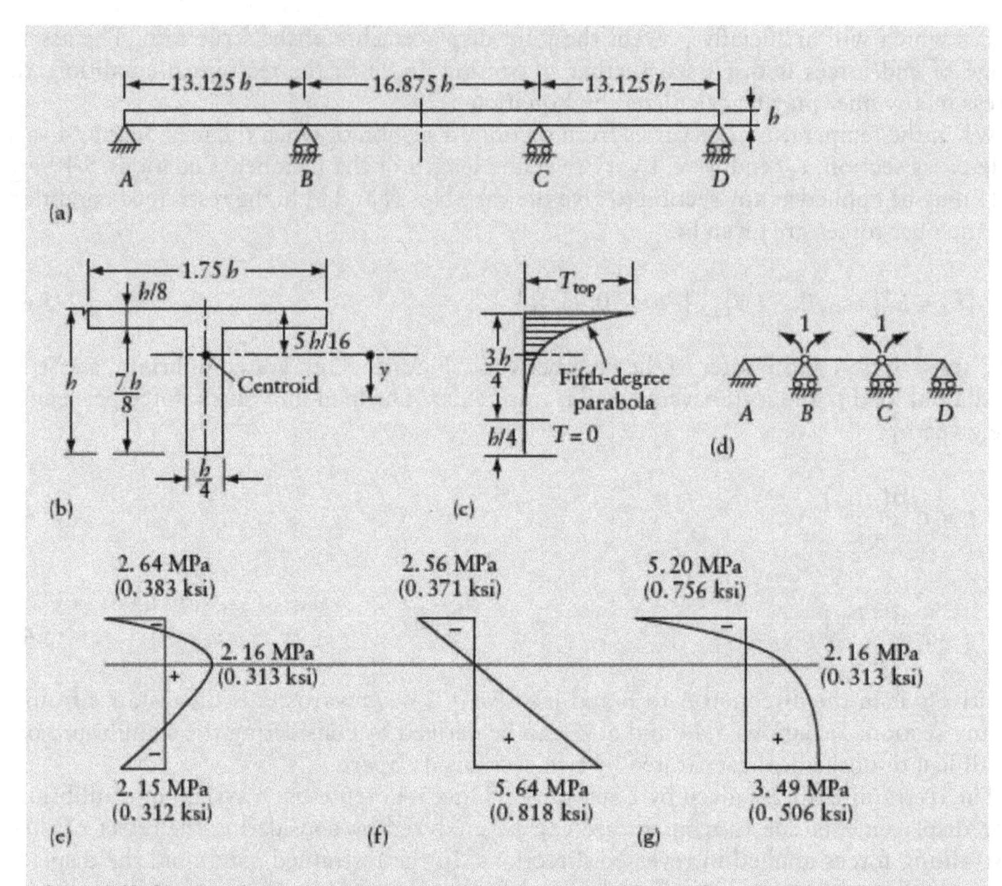

Figure 5.12 Stresses due to a temperature rise in a continuous beam, Example 5.6. (a) Beam elevation. (b) Beam cross section. (c) Temperature rise. (d) Released structure and coordinate system. (e) Self-equilibrating stresses. (f) Continuity stresses. (g) Total stresses.

where T is the temperature rise in degrees, T_{top} is the temperature rise in the top fiber and T_O = constant. The beam has a cross section as shown in Figure 5.12b, with an area $a = 0.4375b^2$ and the second moment of area about the centroidal axis is $I = 0.0416b^4$. Find the stress distribution due to the temperature rise in the section at support B.

Consider $b = 1.6$ m (63 in.), $E = 30$ GPa (4350 ksi), $\alpha = 10^{-5}$ per degree Celsius ((5/9) × 10^{-5} per degree Fahrenheit) and $T_{top} = 25°$ Celsius (45° Fahrenheit).

The above equations represent the temperature distribution which can occur in a bridge girder on a hot summer day. When the temperature rise is constant, the length of the beam will increase freely, without inducing any stress or deflection. Hence, to solve the problem, we may put $T_O = 0$; the temperature rise will then vary as shown in Figure 5.12c.

Application of Equations 5.27 and 5.28 gives the artificial stress which would prevent thermal expansion:

$$\sigma_r = -4.21\, E\, \alpha\, T_{top} \left(\frac{7}{16} - \frac{y}{b} \right)^5 \quad \text{for} \quad -\frac{5}{16} \leq \frac{y}{b} \leq \frac{7}{16}$$

This equation applies for the upper three-quarters of the beam depth, while $\sigma_r = 0$ for the remainder.

If the structure were statically determinate, the axial strain and the curvature at any section would be (Equations 5.36 and 5.37):

$$\varepsilon_O = \frac{\alpha}{0.437h^2}\left[1.75h\int_{-5h/16}^{-3h/16} 4.21\,T_{top}\left(\frac{7}{16}-\frac{y}{h}\right)^5 dy + 0.25h\int_{-3h/16}^{7h/16} 4.21\,T_{top}\left(\frac{7}{16}-\frac{y}{h}\right)^5 dy\right]$$

$$= 0.356\,\alpha\,T_{top}$$

$$\psi = \frac{\alpha}{0.0416h^4}\left[1.75h\int_{-5h/16}^{-3h/16} 4.21\,T_{top}\left(\frac{7}{16}-\frac{y}{h}\right)^5 y\,dy + 0.25h\int_{-3h/16}^{7h/16} 4.21\,T_{top}\left(\frac{7}{16}-\frac{y}{h}\right)^5 y\,dy\right]$$

$$= -0.931\,\alpha\,T_{top}\,h^{-1}$$

The self-equilibrating stresses (Equation 5.35) are:

$$\sigma_s = E\,\alpha\,T_{top}\left[-4.21\left(\frac{7}{16}-\frac{y}{h}\right)^5 + 0.356 - 0.931\frac{y}{h}\right] \quad \text{for} \quad -\frac{5}{16} \le \frac{y}{h} \le \frac{7}{16}$$

$$\sigma_s = E\,\alpha\,T_{top}\left(0.356 - 0.931\frac{y}{h}\right) \quad \text{for} \quad \frac{7}{16} \le \frac{y}{h} \le \frac{11}{16}$$

Substitution of E, α, T, and the y values at: the top fiber, the centroid, and the bottom fiber gives the values of the self-equilibrating stresses shown in Figure 5.12e. This stress diagram is valid for all sections.

Because the structure is statically indeterminate, the temperature rise produces reactions, internal forces, and stresses. The stresses due to the indeterminate forces are referred to as *continuity stresses*. We apply the five steps of the force method (see Section 4.6) to determine the continuity stresses:

Step 1 A released structure and a coordinate system are selected in Figure 5.12d, taking advantage of symmetry (see Section 5.13). The required action is $A=\sigma$, the stress at any fiber of the cross section at B.

Step 2 The displacement of the released structure is the relative rotation of the beam ends at B (or at C). Using Equation 5.38 or Appendix A:

$$D_1 = \frac{\psi\,l_{AB}}{2} + \frac{\psi\,l_{BC}}{2} = -\frac{0.931}{2}\alpha\,T_{top}(13.125 + 16.875)$$

$$= -13.97\,\alpha\,T_{top}$$

The value of the required action in the released structure is the self-equilibrating stress in Figure 5.12e, hence:

$A_s = \sigma_s$

Step 3 Applying $F_1 = 1$ at the coordinates in Figure 5.12d gives the flexibility (see Appendix A):

$$f_{11} = \left(\frac{l}{3EI}\right)_{AB} + \left(\frac{l}{2EI}\right)_{BC} = \frac{1}{E(0.0416h^4)}\left(\frac{13.125h}{3} + \frac{16.875h}{2}\right) = \frac{308}{Eh^3}$$

In this problem, A_u is the stress at any fiber of the cross section at B due to $F_1 = 1$, that is, the stress due to a unit bending moment:

$$A_u = \frac{y}{I} = \frac{y}{0.0416h^4} = 24\frac{y}{h^4}$$

Step 4 The redundant (the connecting moment at B or C) is:

$$F_1 = -f_{11}^{-1} D_1 = -\left(\frac{308}{Eh^3}\right)^{-1} (-13.97\,\alpha\, T_{top}) = 0.0454\,\alpha\, T_{top}\, E h^3$$

Step 5 The stress at any fiber is obtained by the superposition equation:

$$A = A_s + A_u\, F_1$$

Substitution gives:

$$\sigma = \sigma_s + 1.09\,\alpha\, T_{top}\, E y / h$$

The second term in this equation represents the continuity stress plotted in Figure 5.12f. The total stress at section B is plotted in Figure 5.12g, which is a superposition of Figures 5.12e and f.

EXAMPLE 5.7: THERMAL STRESSES IN A PORTAL FRAME

Member BC of the frame in Figure 5.13a is subjected to a rise of temperature which is constant over the beam length but varies over its depth following a fifth-degree parabola, as shown in Figure 5.12c. The cross section of member BC of the frame is the same as in

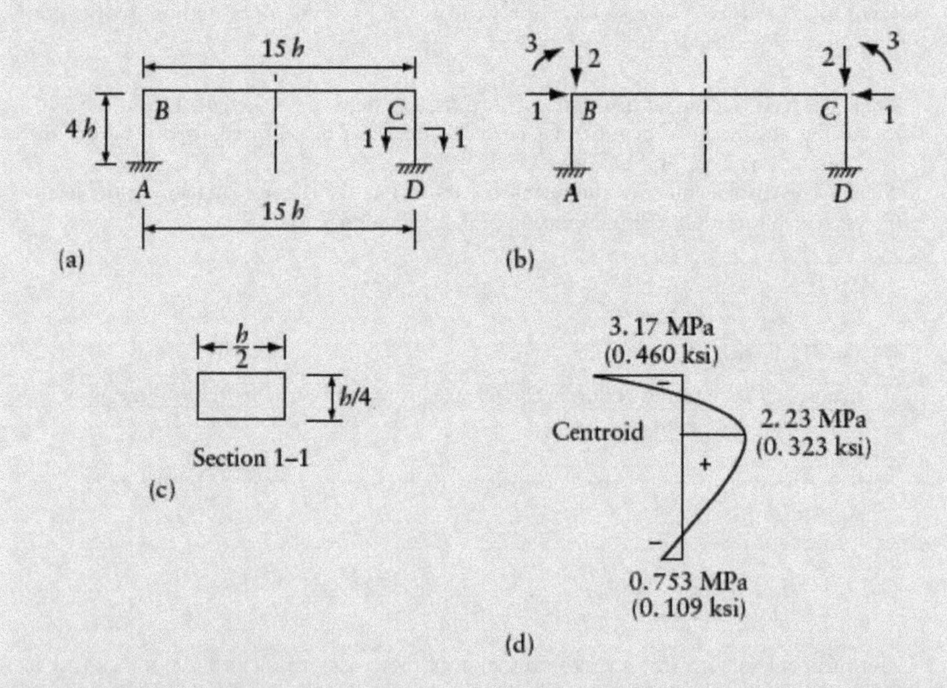

Figure 5.13 Analysis of stresses due to a temperature rise by the displacement method, Example 5.7. (a) Plane frame with a cross section of member BC and a temperature rise as in Figures 5.7b and c. (b) Cross section of columns AB and CD. (c) Coordinate system. (d) Stress distribution at any cross section of BC.

Figure 5.12b. Assuming that the columns AB and CD have a rectangular cross section (Figure 5.13c) and that the rise of temperature is limited to member BC, find the corresponding stress distribution at any cross section of this member. Consider bending and axial deformations. Other data needed for the solution are the same as in Example 5.6.

The five steps of the displacement method (Section 5.2) are:

Step 1 A coordinate system is defined in Figure 5.13c. Taking advantage of symmetry, the number of unknown joint displacements is three. The stress distribution is the same at any section of BC, and the stress at any fiber is the action required; thus, $A = \sigma$.

Step 2 The values $\varepsilon_O = 0.356\alpha T_{top}$ and $\psi = -0.931\alpha T_{top} h^{-1}$ determined in Example 5.6 apply here to member BC. The first three elements of the vector in Equation 5.42 are the restraining forces at the left-hand end of BC, whose cross-sectional area $= 0.4375h^2$ and moment of inertia $= 0.0416h^4$. Because AB and CD are not subjected to temperature change, no restraining forces need to be determined for these two members. The restraining forces at the three coordinates at B or C are (substitute values for α, I, E, ε_O, and ψ in the three elements of the vector in Equation 5.42):

$$\{F\} = E\begin{Bmatrix} 0.4375h^2(0.356\alpha T_{top}) \\ 0 \\ -0.0416h^4(-0.931\alpha T_{top} h^{-1}) \end{Bmatrix} = E\alpha T_{top} h^2 \begin{Bmatrix} 0.156 \\ 0 \\ 0.0387h \end{Bmatrix}$$

The stress at any fiber of member BC with the joint displacement prevented is the same as determined in Example 5.6. Thus:

$$A_r = \sigma_r = -4.21\, E\,\alpha\, T_{top}\left(\frac{7}{16} - \frac{y}{h}\right)^5 \quad \text{for} \quad -\frac{5h}{16} \le y \le \frac{7h}{16}$$

$$= 0 \qquad \text{for remainder of depth}$$

Step 3 The stiffness matrix of the structure (using Appendix C) is:

$$[S] = \begin{bmatrix} \left(\frac{12EI}{l^3}\right)_{AB} + \left(\frac{2Ea}{l}\right)_{BC} & & \text{Symmetrical} \\ 0 & \left(\frac{Ea}{l}\right)_{AB} & \\ -\left(\frac{6EI}{l^2}\right)_{AB} & 0 & \left(\frac{4EI}{l}\right)_{AB} + \left(\frac{2EI}{l}\right)_{BC} \end{bmatrix}$$

Substitute: for AB, $l = 4h$, $a = 0.125h^2$ and $I = 2.60 \times 10^{-3}h^4$; for BC, $l = 15h$, $a = 0.4375h^2$ and $I = 0.0416h^4$.

Hence,

$$[S] = E\begin{bmatrix} 58.82\times10^{-3}h & & \text{Symmetrical} \\ 0 & 31.25\times10^{-3}h & \\ -975.0\times10^{-6}h^2 & 0 & 8.147\times10^{-3}h^3 \end{bmatrix}$$

Displacements $D_1 = 1$ at the two coordinates 1, Figure 5.13b, correspond to strain $= -2/l_{BC}$ and stress $= E(-2/l_{BC})$. The rotations $D_3 = 1$ at the two coordinates 3 in Figure 5.13b produce a constant bending moment $= 2E(I/l)_{BC}$ and a stress at distance y below the centroid (Figure 5.12b) $= [2E(I/l)_{BC}]\, y/I_{BC}$. Thus, the stress at any fiber due to unit displacements at the coordinates is:

$$[A_u] = E\left[-\left(\frac{2}{l}\right)_{BC} \quad 0 \quad \left(\frac{2I}{l}\right)_{BC} \cdot \frac{y}{I_{BC}}\right] = E\left[-\frac{2}{15h} \quad 0 \quad \frac{2y}{15h}\right]$$

Step 4 Substitution for $[S]$ and $\{F\}$ and solution of the equilibrium equation $[S]\{D\} = -\{F\}$ gives:

$$\{D\} = \frac{1}{Eh^2}\begin{bmatrix} 58.82\times10^{-3}/h & & \text{Symmetrical} \\ 0 & 31.25\times10^{-3}/h & \\ -975.0\times10^{-6} & 0 & 8.147\times10^{-3}\,h \end{bmatrix}^{-1} E\alpha T_{top} h^2 \begin{Bmatrix} -0.156 \\ 0 \\ -0.0387\,h \end{Bmatrix}$$

$$= \alpha T_{top}\begin{Bmatrix} -2.736h \\ 0 \\ -5.078 \end{Bmatrix}$$

Step 5 By superposition, the stress at any fiber is:

$$A = A_r + [A_u]\{D\}$$

Substitution for A_r, $[A_u]$ and $\{D\}$ gives the stress at any fiber:

$$A = \sigma = \sigma_r + \frac{E}{h}\left[\frac{-2}{15} \quad 0 \quad \frac{2y}{15}\right]\alpha T_{top} h\begin{Bmatrix} -2.736 \\ 0 \\ -5.078/h \end{Bmatrix}$$

$$= \sigma_r + E\alpha T_{top}\left(0.365 - 0.677\frac{y}{h}\right)$$

$$\sigma = E\alpha T_{top}\left[-4.21\left(\frac{7}{16} - \frac{y}{h}\right)^5 + 0.365 - 0.677\frac{y}{h}\right] \quad \text{for} \quad -\frac{5h}{16} \le y \le \frac{7h}{16}$$

$$= E\alpha T_{top}\left(0.365 - 0.677\frac{y}{h}\right) \quad \text{for} \quad \frac{7h}{16} \le y \le \frac{11h}{16}$$

Substituting for y the values $-5h/16$, 0 and $11h/16$ and using the values of E, α, and T_{top} from the data for Example 5.6 gives the stress values indicated in Figure 5.13d ($E\alpha T_{top} = 7.5$ MPa $= 1.09$ ksi).

5.10 EFFECT OF SHRINKAGE AND CREEP

The phenomena of shrinkage and creep occur in various materials, but in the following discussion we shall refer mainly to concrete because it is widely used in structures.

Shrinkage of concrete is a reduction in volume associated with drying in air. As with a temperature drop, if the change in volume is restrained by the difference in shrinkage of various parts of the structure or by the supports or by the reinforcing steel, stresses develop.

If we imagine a material which shrinks without creep, the analysis for the effect of shrinkage can be performed using the equations of Section 5.9, but replacing the term αT with ε_f, where ε_f is the free (unrestrained) shrinkage. The effect of swelling can be treated in the same manner as shrinkage but with a reversed sign. Swelling occurs in concrete under water.

The strain which occurs during the application of stress, or within a few seconds thereafter, may be referred to as the *instantaneous strain*. For some materials, the strain continues to increase gradually when the stress is sustained without a change in magnitude. The increase in strain with time, under a sustained stress, is referred to as *creep*. For concrete, creep is two to four times larger than the instantaneous strain, depending upon the composition of concrete, the ambient humidity and temperature, the size of the element considered, the age of concrete when the stress is applied, and the length of the period during which the stress is sustained.

If creep is assumed to be equal to the instantaneous strain multiplied by a constant coefficient, creep will have no effect on the internal forces or stresses in a structure made of a homogeneous material. Creep will cause larger displacements, which can be accounted for by the use of a reduced (effective) E, but this has no effect on the reactions even when the structure is statically indeterminate.

When a concrete structure is constructed and loaded in stages, or when member cross sections contain reinforcement, or when the section is composed of a concrete part connected to structural steel, the creep which is different in various components cannot occur freely. Similarly to temperature expansion, restrained creep induces stresses. In statically determinate structures, creep changes the distribution of stresses within a section without changing the reactions or the stress resultants. This is not so in statically indeterminate structures, where creep influences also the reactions and the internal forces.

In concrete structures, shrinkage and creep occur simultaneously. The stress changes caused by these two phenomena develop gradually over long periods, and with these changes there is associated additional creep. Hence, the analysis must account also for the creep effect of the stress which is gradually introduced. Analysis of the time-dependent stresses and deformations in reinforced and prestressed concrete structures is treated in more detail in books devoted to this subject.[1]

When a change in temperature in a concrete structure develops gradually over a period of time (hours or days), the resulting stresses are reduced by creep which occurs during the same period. Ignoring creep overestimates the effects of temperature variations.

5.11 EFFECT OF PRESTRESSING

In Section 4.4, we discussed the effect of prestressing a concrete beam by a cable inserted through a duct and then anchored at the ends. This method is referred to as post-tensioning. In this section, we shall discuss the effects of a post-tensioned tendon which has a nonlinear profile. For simplicity of presentation, we ignore the friction which commonly exists between the tendon and the inner wall of the duct; thus, we assume that the tensile force in the tendon is constant over its length. Let P represent the absolute value of the force in the tendon.

A straight tendon as in Figure 4.3a produces two inward horizontal forces on the end sections, each equal to P. The two forces represent a system in equilibrium, and the reactions in the statically determinate beam are zero. The internal forces at any section are an axial force $-P$ and a bending moment $-Pe$. The used sign convention is shown in Figure 2.4c. The eccentricity e is measured downward from the centroidal axis. The ordinates between the centroidal axis and the tendon profile represent the bending moment diagram with a multiplier $-P$.

Usually, the tendon profile is selected so that the prestressing partly counteracts the effects of forces which the structure has to carry. Whenever a tendon changes direction, a transverse force is exerted by the tendon on the member. The tendon shown in Figure 5.14a

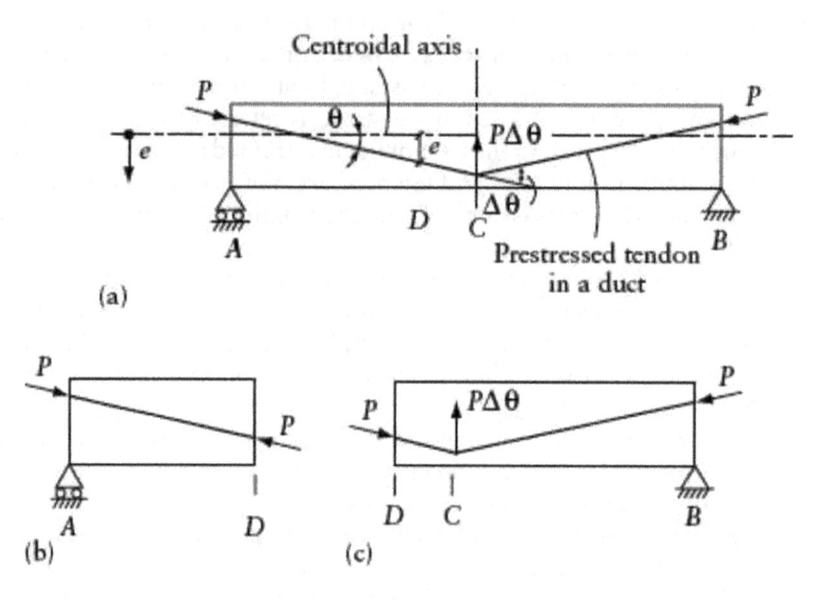

Figure 5.14 Forces on concrete due to prestressing in a statically determinate beam. (a) Representation of prestressing by a system of forces in equilibrium. (b) and (c) Free-body diagrams showing the stress resultant at section D.

produces, at the end anchorages, two inward forces of magnitude P along the tangents to the tendon profile. In addition, an upward force is produced at point C. Thus, the profile in Figure 5.14a may be used in a simple beam carrying a large downward concentrated load.

The three forces shown in Figure 5.14a represent a set of forces in equilibrium. In most practical cases, the angle θ between the centroidal axis and the tangent to the tendon is small, such that $\sin\theta \simeq \theta$ and $\cos\theta \simeq 1$, thus we need to consider only an axial component of the prestressing force, P and a perpendicular component, $P\theta$. It follows that the force at C is equal to $P\Delta\theta$, where $\Delta\theta$ is the absolute value of the change in slope.

A parabolic tendon produces a uniform transverse load and is thus suitable to counteract the effect of the self-weight and other distributed gravity loads. Appendix H gives the magnitude and direction of the forces produced by tendons having profiles commonly used in practice. The forces shown represent the effect of the prestressing tendon on the other components of the member. The forces produced by a prestressed tendon must constitute a system in equilibrium.

Prestressing of a statically determinate structure produces no reactions. It can be shown that the resultant of the internal forces at any section is a force P along the tangent of the tendon profile. This can be seen in Figure 5.14b, where the beam is separated into two parts in order to show the internal forces at an arbitrary section D. The ordinate e between the centroidal axis and the tendon profile, Figure 5.14a, represents the bending moment ordinate with a multiplier $-P$. The axial force is $-P$, and the shearing force is $-P\theta$.

The internal forces, determined as outlined in the preceding paragraph, are referred to as *primary forces*. Prestressing of statically indeterminate structures produces reactions and hence induces additional internal forces referred to as *secondary forces*. The reactions also represent a system of forces in equilibrium. When the structure is statically determinate, the force in the prestressed steel at any section is equal and opposite to the resultant of stresses in other components of the section, and, thus, the total stress resultant on the whole section is zero. This is not so in a statically indeterminate structure.

In the analysis of the effects of prestressing, it is not necessary to separate the primary and secondary effects. The total effect of prestressing may be directly determined by representing the prestressing by a system of external applied self-equilibrating forces (see Appendix H). The analysis is then performed in the usual way by either the force or the displacement method.

EXAMPLE 5.8: POST-TENSIONING OF A CONTINUOUS BEAM

Find the reactions and the bending moment diagram due to prestressing for the continuous beam shown in Figure 5.15a. The prestressing tendon profile for each span of the beam is composed of two second-degree parabolas, ACD and DB, with a common tangent at D (Figure 5.15b). The parabolas have horizontal tangents at B and C. Assume a constant prestressing force P.

The profile shown in Figure 5.15b is often used in practice for the end span of a continuous beam. The condition that the two parabolas have a common tangent at D is required to avoid a sudden change in slope, which would produce an undesired concentrated

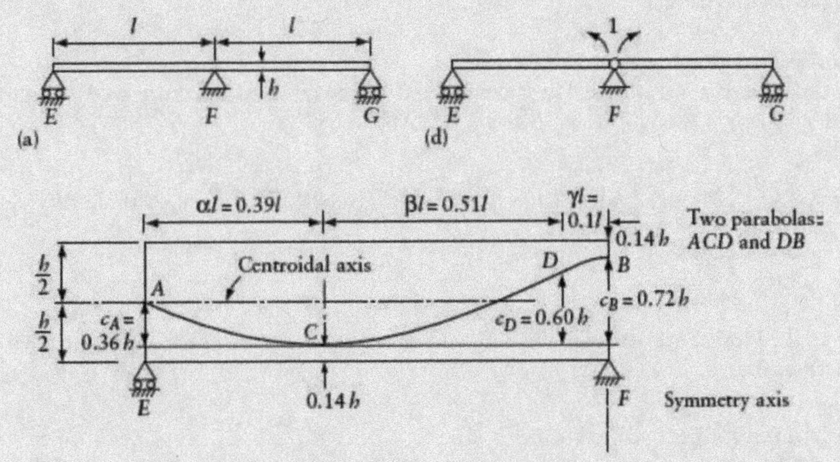

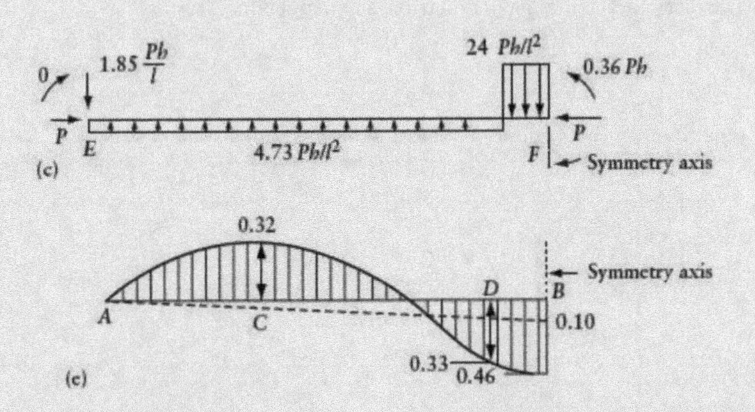

Figure 5.15 Effect of prestressing on a continuous beam, Example 5.8. (a) Beam elevation. (b) Prestressing tendon profile in one-half of the structure. (c) Self-equilibrating forces produced by prestressing. (d) Released structure and coordinate system. (e) Bending moment diagram.

transverse force at D. In design, the geometry of the profile can be obtained by choosing α, c_A, and c_B arbitrarily and determining β and c_D so that:

$$\beta = \gamma \frac{c_D}{c_B - c_D} \quad \text{and} \quad c_D = c_A \frac{\beta^2}{\alpha^2}$$

These two geometrical relations ensure that the slope of the tangents to the two parabolas at D is the same and that ACD is one parabola with a horizontal tangent at C. Solution of the two equations for the unknowns β and c_D may be obtained by trial and error, noting that $\alpha+\beta+\gamma=1$ (Newton-Raphson's technique).

The forces produced by the tendon are calculated by the equations of Appendix H and are shown in Figure 5.15c for the left-hand half of the beam.

The five steps of the force method (Section 4.6) are applied to determine the statically indeterminate reactions:

Step 1 The released structure and the coordinate system are shown in Figure 5.15d. The actions required are:

$$\{A\} = \{R_E, R_F, R_G\}$$

A positive reaction is upwards.

Step 2 The forces in Figure 5.15c when applied on the released structure give the following displacement (Equations A.36 and A.35, Appendix A):

$$D_1 = 2\frac{Phl}{24EI}\left\{24(0.1)^2\left[4-4(0.1)+(0.1)^2\right]-4.73(0.9)^2\left[2-(0.9)^2\right]\right\}+2(0.36Ph)\frac{l}{3EI}$$

$$= -0.068\frac{Phl}{24EI}$$

The applied forces are self-equilibrating and hence produce zero reactions in the released structure; thus:

$$\{A_s\} = \{0\}$$

Step 3 The flexibility coefficient (Equation A.9, Appendix A) is:

$$f_{11} = \frac{2l}{3EI}$$

A unit redundant, $F_1 = 1$, produces the following reactions:

$$[A_u] = \frac{1}{l}\begin{bmatrix} 1 \\ -2 \\ 1 \end{bmatrix}$$

Step 4

$$f_{11}F_1 = -D_1$$

$$F_1 = -\left(\frac{2l}{3EI}\right)^{-1}\left(-0.068\frac{Phl}{EI}\right) = 0.102Ph$$

Step 5 Superposition gives the required reactions:

$$\{A\} = \{A_s\} + [A_u]\{F\}$$

$$\begin{Bmatrix} R_E \\ R_F \\ R_G \end{Bmatrix} = \{0\} + \frac{0.102Ph}{l}\begin{Bmatrix} 1 \\ -2 \\ 1 \end{Bmatrix} = \frac{Ph}{l}\begin{Bmatrix} 0.102 \\ -0.203 \\ 0.102 \end{Bmatrix}$$

The bending moment diagram is plotted in Figure 5.15e. Its ordinate at any section of span EF may be expressed as:

$$M = -Pe + 0.102\frac{Phx}{l}$$

where x is the horizontal distance from E to the section and e is the vertical distance from the centroidal axis to the tendon profile; e is positive where the tendon is below the centroid. Following the convention used throughout this book, the bending moment ordinates are plotted on the tension side of the beam. The dashed line in Figure 5.15e is the statically indeterminate bending moment due to prestressing, called the *secondary bending moment.*

5.12 CONDENSATION OF STIFFNESS MATRICES

We recall that a stiffness matrix relates displacements $\{D\}$ at a number of coordinates to the forces $\{F\}$ applied at the same coordinates by the equation:

$$[S]\{D\} = \{F\} \tag{5.46}$$

If the displacement at a number of coordinates is prevented by the introduction of supports and the matrices in the above equations are arranged in such a way that the equations corresponding to these coordinates appear at the end, we can write Equation 5.46 in the partitioned form:

$$\left[\begin{array}{c:c} [S_{11}] & [S_{12}] \\ \hdashline [S_{21}] & [S_{22}] \end{array}\right]\left\{\begin{array}{c} \{D_1\} \\ \hdashline \{D_2\} \end{array}\right\} = \left\{\begin{array}{c} \{F_1\} \\ \hdashline \{F_2\} \end{array}\right\} \tag{5.47}$$

where $\{D_2\} = \{0\}$ represents the prevented displacements. From this equation, we write:

$$[S_{11}]\{D_1\} = \{F_1\} \tag{5.48}$$

and

$$[S_{21}]\{D_1\} = \{F_2\} \tag{5.49}$$

It is apparent from Equation 5.48 that, if a support is introduced at a number of coordinates, the stiffness matrix of the resulting structure can be obtained simply by deleting the columns and the rows corresponding to these coordinates, resulting in a matrix of a lower order. If the displacements $\{D_1\}$ are known, Equation 5.49 can be used to calculate the reactions at the supports preventing the displacements $\{D_2\}$.

As a simple example, consider the beam in Figure 5.9c and assume that the vertical displacements at coordinates 1 and 3 are prevented, as in the case of a simple beam; the stiffness matrix corresponding to the remaining two coordinates (2 and 4) is obtained by deletion of columns and rows numbered 1 and 3 in the matrix Equation 5.20:

$$[S^*] = \begin{bmatrix} \frac{4EI}{l} & \frac{2EI}{l} \\ \frac{2EI}{l} & \frac{4EI}{l} \end{bmatrix} \tag{5.50}$$

The vertical reactions $\{F_1, F_3\}$ at coordinates 1 and 3 can be calculated from Equation 5.49 by rearrangement of the elements in Equation 5.20 as described for Equation 5.47. Thus:

$$\begin{bmatrix} \frac{6EI}{l^2} & \frac{6EI}{l^2} \\ -\frac{6EI}{l^2} & -\frac{6EI}{l^2} \end{bmatrix} \begin{Bmatrix} D_2 \\ D_4 \end{Bmatrix} = \begin{Bmatrix} F_1 \\ F_3 \end{Bmatrix} \tag{5.51}$$

where the subscripts of D and F refer to the coordinates in Figure 5.9c.

If the forces are known to be zero at some of the coordinates (i.e. the displacements at these coordinates can take place freely), the stiffness matrix corresponding to the remaining coordinates can be derived from the partitioned matrix Equation 5.47. In this case, we consider that the equations below the horizontal dashed line relate forces $\{F_2\}$, assumed to be zero, to the displacements $\{D_2\} \neq \{0\}$ at the corresponding coordinates. Substituting $\{F_2\} = \{0\}$ in Equation 5.47, we write:

$$\left.\begin{aligned} &[S_{11}]\{D_1\} + [S_{12}]\{D_2\} = \{F_1\} \\ &\text{and} \\ &[S_{21}]\{D_1\} + [S_{22}]\{D_2\} = \{0\} \end{aligned}\right\} \tag{5.52}$$

Using the second equation to eliminate $\{D_2\}$ from the first, we obtain:

$$\left[[S_{11}] - [S_{12}][S_{22}]^{-1}[S_{21}]\right]\{D_1\} = \{F_1\} \tag{5.53}$$

Equation 5.53 can be written in the form:

$$[S^*]\{D_1\} = \{F_1\} \tag{5.54}$$

where $[S^*]$ is a condensed stiffness matrix relating forces $\{F_1\}$ to displacements $\{D_1\}$ and is given by:

$$[S^*] = [S_{11}] - [S_{12}][S_{22}]^{-1}[S_{21}] \tag{5.55}$$

The condensed stiffness matrix $[S^*]$ corresponds to a reduced system of coordinates 1^*, 2^*, ..., eliminating the coordinates corresponding to the second row of the partitioned matrix

Equation 5.47. Using the symbols $\{F^*\}=\{F_1\}$ and $\{D^*\}=\{D_1\}$, Equation 5.54 becomes $[S^*]$ $\{D^*\}=\{F^*\}$.

The stiffness matrix of a structure composed of a number of members is usually derived from the stiffness matrix of the individual members and relates the forces at all the degrees of freedom to the corresponding displacements. In many cases, however, the external forces on the actual structure are limited to a small number of coordinates, and it may therefore be useful to derive a matrix of a lower order $[S^*]$ corresponding to these coordinates only, using Equation 5.55. Examples 5.9 and 5.10 demonstrate applications of Equation 5.55.

EXAMPLE 5.9: DEFLECTION AT TIP OF A CANTILEVER

Consider the prismatic member in Figure 5.9c as a cantilever encastré at the right-hand end, such that $\{D_3, D_4\}=\{0, 0\}$. Apply Equation 5.55 to find $[S^*]$ and its inverse $[f^*]$ corresponding to a single coordinate, 1.

Deletion of columns and rows numbered 3 and 4 from the matrix of Equation 5.20 gives the stiffness matrix of the cantilever corresponding to coordinates 1 and 2:

$$[S]=\begin{bmatrix} \frac{12EI}{l^3} & \frac{6EI}{l^2} \\ \frac{6EI}{l^2} & \frac{4EI}{l} \end{bmatrix} \tag{a}$$

Partition the matrix on the right-hand side of Equation (a) as indicated; then apply Equation 5.55:

$$[S^*]=[S_{11}]-[S_{12}][S_{22}]^{-1}[S_{21}] \tag{b}$$

$$[S^*]=\left[\frac{12EI}{l^3}\right]-\left[\frac{6EI}{l^2}\right]\left[\frac{4EI}{l}\right]^{-1}\left[\frac{6EI}{l^2}\right]=\left[\frac{3EI}{l^3}\right]$$

$[S^*]=S_{11}^*=$ the force at the tip of the cantilever that produces a unit deflection, while the rotation at the tip occurs freely. The inverse $[S^*]^{-1}=[f^*]=[l^3/(3EI)]=$ the deflection at the tip of the cantilever due to a unit downward force at the tip (the results are, of course, the same as Equations A.19 and C.7 of Appendices A and C).

EXAMPLE 5.10: END-ROTATIONAL STIFFNESS OF A SIMPLE BEAM

With the displacements at coordinates 1 and 3 prevented for the beam in Figure 5.9c, as in a simple beam, what is the moment necessary to produce a unit rotation at coordinate 2, while the rotation is free to occur at coordinate 4? See the fourth figure in Appendix C (Equation C.8).

Consider the beam in Figure 5.9c with simple supports at the ends. Define new coordinates $\bar{1}$ and $\bar{2}$ (not shown) in the same direction and locations as coordinates 2 and 4, respectively. The stiffness matrix for the new system is given in Equation 5.50. We now condense this stiffness matrix (using Equation 5.55) to obtain a 1×1 stiffness matrix corresponding to a single coordinate:

$$S_{\bar{1}\bar{1}}=\frac{4EI}{l}-\frac{2EI}{l}\left(\frac{4EI}{l}\right)^{-1}\left(\frac{2EI}{l}\right)=\frac{3EI}{l}$$

$S_{\bar{1}\bar{1}}$ is the end-rotational stiffness of a simple beam; it is equal to the moment necessary to introduce a unit rotation at one end, while the rotation at the other end is free to occur.

5.13 ANALYSIS OF SYMMETRICAL STRUCTURES BY DISPLACEMENT METHOD

Advantage can be taken of symmetry to reduce the number of unknown displacement components when the analysis is done by the displacement method. Because of symmetry, the displacement magnitude at a coordinate is zero or is equal to the value at one (or more) coordinate(s). For a zero displacement, the coordinate may be omitted; any two coordinates where the displacement magnitudes are equal may be given the same number. Figure 5.16 shows examples of coordinate systems which may be used in the analysis by the displacement method of symmetrical structures subjected to symmetrical loading.

Figures 5.16a to d represent plane frames in which axial deformations are ignored. In Figure 5.16a the rotations at B and C are equal and are therefore given the same coordinate number, 1. Because sidesway cannot occur under a symmetrical load, the corresponding coordinate is omitted. In Figure 5.16b the rotations at B and F are equal, while the rotation at D and the sidesway are zero; hence, this frame has only one unknown displacement component. No coordinate systems are shown for the structures in Figures 5.16c and d because the displacement components are zero at all nodes. Thus, no analysis is needed; the member end-forces are readily available from Appendix B.

The beam over spring supports in Figure 5.16e has only two unknown displacement components representing the vertical translation at the top of the springs. Because of symmetry, no rotation occurs at B. Also, no coordinates are shown for the rotations at A and C because the end-forces are readily available for a member with one end hinged and the other fixed (Appendix C, Equations C.7, C.8, and C.9). The stiffness matrix for the structure (assuming equal spans l and EI = constant) is:

$$[S]=\begin{bmatrix} \dfrac{3EI}{l^3}+K & -\dfrac{3EI}{l^3} \\ -2\left(\dfrac{3EI}{l^3}\right) & 2\left(\dfrac{3EI}{l^3}\right)+K \end{bmatrix}$$

where K is the spring stiffness (assumed the same for the three springs). The element S_{11} represents the force at coordinate 1 (at A and C) when unit downward displacement is introduced simultaneously at A and at C; S_{21} is the corresponding force at 2. We should note that $S_{21}=2S_{12}$ so that the stiffness matrix is not symmetrical. This is so because the system has two coordinates numbered 1 but only one coordinate numbered 2. The symmetry of the equilibrium equations, $[S]\{D\}=-\{F\}$, can be restored by division of the second row by 2.

If axial deformations are considered, additional coordinates must be used. In Figure 5.16a, a, horizontal and a vertical arrow will have to be added at B and C. Because of symmetry, each of the two corresponding arrows takes the same number, bringing the number of unknown displacements to three. Each of the frames in Figures 5.16c and d will have one unknown displacement: a vertical translation at B.

The horizontal grids shown in Figures 5.16f and g have one or more vertical planes of symmetry. The coordinate systems shown may be used for the analysis of the effects of symmetrical loads.

Each of the structures in Figures 5.16a to g may be analyzed by considering only one-half (or one-quarter) of the structure through separating it at the axis or plane of symmetry. The members situated on the axis or plane of symmetry for the part analyzed should have properties such as a, I, J, or K equal to half the values in the actual structure. The same coordinate systems shown on one-half (or one-quarter) of the structure may be used. An exception is the frame in Figure 5.16a: separation at E will result in a new node at which the vertical translation is unknown, requiring an additional coordinate.

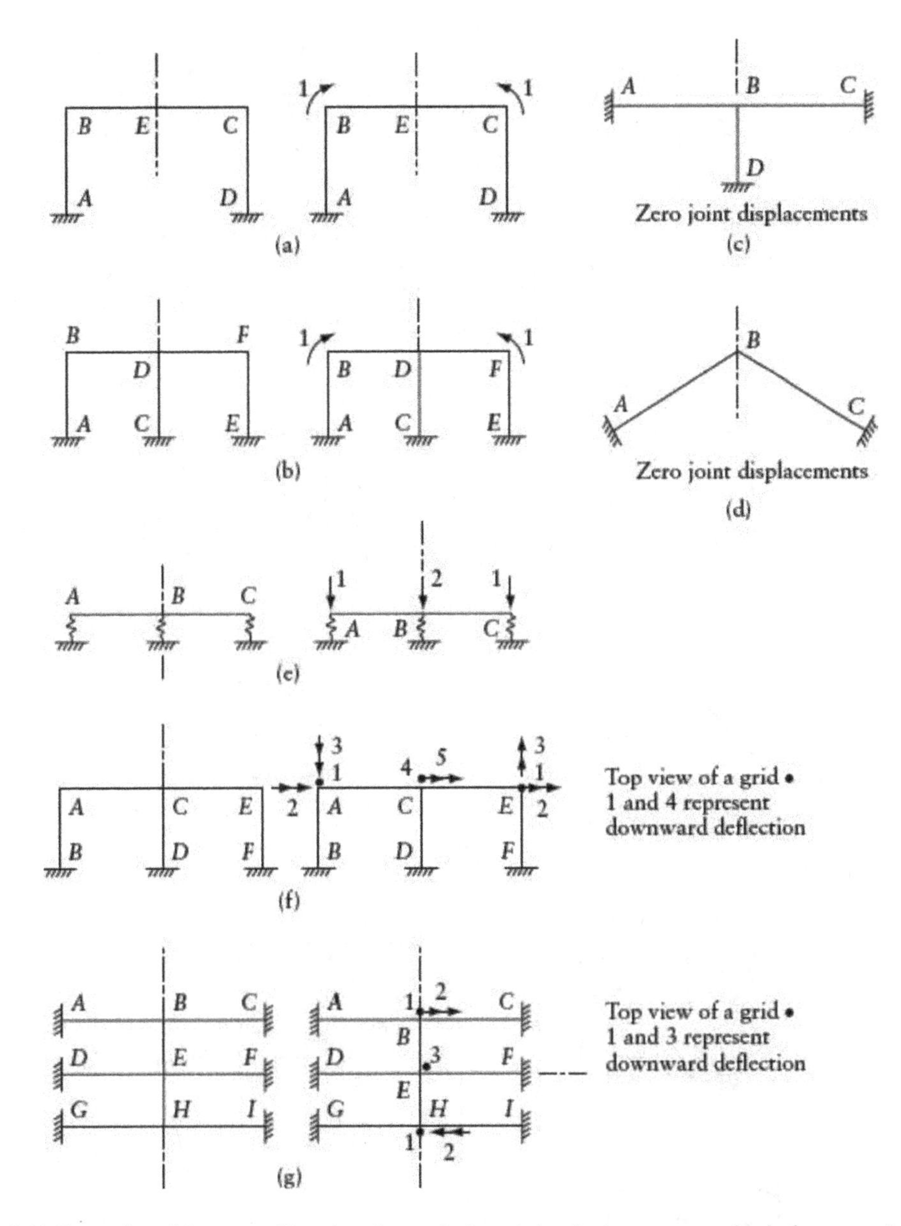

Figure 5.16 Examples of degrees of freedom for analysis and the displacement method of symmetrical plane frames and grids subjected to symmetrical loading.

When the analysis of small structures is done by hand or by a calculator, with the matrices generated by the analyst, consideration of one-half or one-quarter of the structure may represent little advantage. However, in large structures with many members and nodes, the analysis is usually performed entirely by computer, considering as small a part of the structure as possible and taking full advantage of symmetry. This is further discussed in Sections 16.4 to 16.6.

The structure shown in Figure 5.17f, representing a typical bay of a frame with an infinite number of bays, may be analyzed by the displacement method, using three degrees of freedom at each of A, B, and C: a translation in the horizontal and vertical directions and a rotation. The corresponding three displacements at A and C have the same magnitude and direction; hence, there are only six unknown displacements.

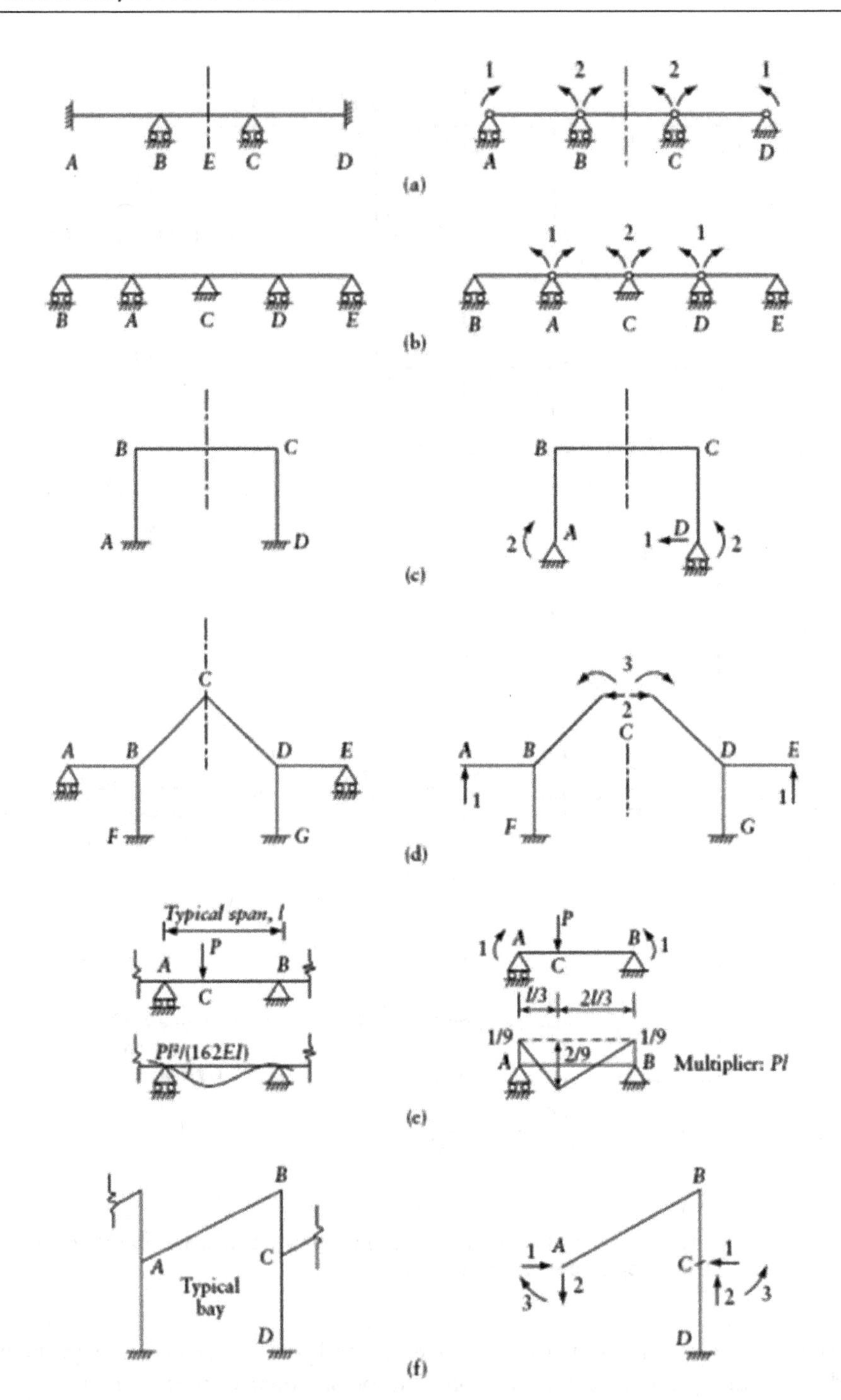

Figure 5.17 Examples of releases for analysis by the force method of symmetrical continuous beams and frames subjected to symmetrical loading. (a) to (d) Continuous beams and plane frames. (e) Typical interior span of continuous beam having infinite number of spans; bending moment diagram and deflected shape. (f) Typical bay of a plane frame having infinite bays.

EXAMPLE 5.11: SINGLE-BAY SYMMETRICAL PLANE FRAME

For the frame shown in Figure 5.18a find the bending moment diagram and the reactions R_1 and R_2 at support A. Ignore axial deformations and consider only bending deformations.

We apply the five steps of the displacement method (Section 5.4):

Step 1 Because axial deformations are ignored, only rotations can occur at B and C. A system with a single coordinate is defined in Figure 5.18b. The rotation at A and D will be allowed to occur freely in the analysis steps; thus, no coordinate to represent the rotations is needed at the two hinges. Also, the overhangs EB and FC produce statically determinate forces at B and C. Thus, by taking these forces into account, the analysis can be done for a frame without the cantilevers. The end moment $M_{BE} = (ql/2)(0.4l) = 0.2ql^2$. We define the unknown actions as:

$$\{A\} = \{M_{BA}, M_{BC}\}$$

The values of the three end moments will be sufficient to draw the bending moment diagram for the left-hand half of the frame. From symmetry, the reaction $R_1 = 1.5ql$ (half the total downward load). The reaction R_2 will be calculated by statics from M_{BA}.

Step 2 $\{F\} = \left\{0.2ql^2 - q(2l)^2/12\right\} = \left\{-0.1333ql^2\right\}$

The first term is the moment to prevent the rotation due to the load on the cantilever. The second term is the fixed-end moment for BC (Appendix B).

$$\{A_r\} = \left\{0, -q(2l)^2/12\right\} = \left\{0, -0.333ql^2\right\}$$

Step 3 $[S] = \left[\left(\frac{3EI}{l}\right)_{BA} + \left(\frac{2EI}{l}\right)_{BC}\right] = \left[EI\left(\frac{3}{1.077l} + \frac{2}{2l}\right)\right] = \left[3.785\frac{EI}{l}\right]$

$$[A_u] = \begin{bmatrix}(3EI/l)_{BA} \\ (2EI/l)_{BC}\end{bmatrix} = \begin{bmatrix}2.785EI/l \\ 1.0EI/l\end{bmatrix}$$

With $D_1 = 1$, the deflected shape of the frame will be as shown in Figure 5.18c; the terms $(3EI/l)_{BA}$ and $(2EI/l)_{BC}$ in the above calculations are taken from Appendix C (Equations C.8 and C.11).

Step 4 $\{D\} = [S]^{-1}\{-F\} = [3.785EI/l]^{-1}\left\{0.1333ql^2\right\} = \left\{35.22\times10^{-3}\frac{ql^3}{EI}\right\}$

Step 5

$$\{A\} = \{A_r\} + [A_u]\{D\} = \begin{Bmatrix}0 \\ -0.333ql^2\end{Bmatrix} + \begin{bmatrix}2.785EI/l \\ 1.0EI/l\end{bmatrix}\left\{35.22\times10^{-3}\frac{ql^3}{EI}\right\} = \begin{Bmatrix}98 \\ -298\end{Bmatrix}\frac{ql^2}{1000}$$

These are the values of the end moments used to draw the bending moment diagram in Figure 5.18d. The reaction R_2 is given by:

$$M_{BA} = -R_1(0.4l) + R_2 l$$

Substitution of $M_{BA} = 0.098\,ql^2$ and $R_1 = 1.5\,ql$ gives $R_2 = 0.698\,ql$.

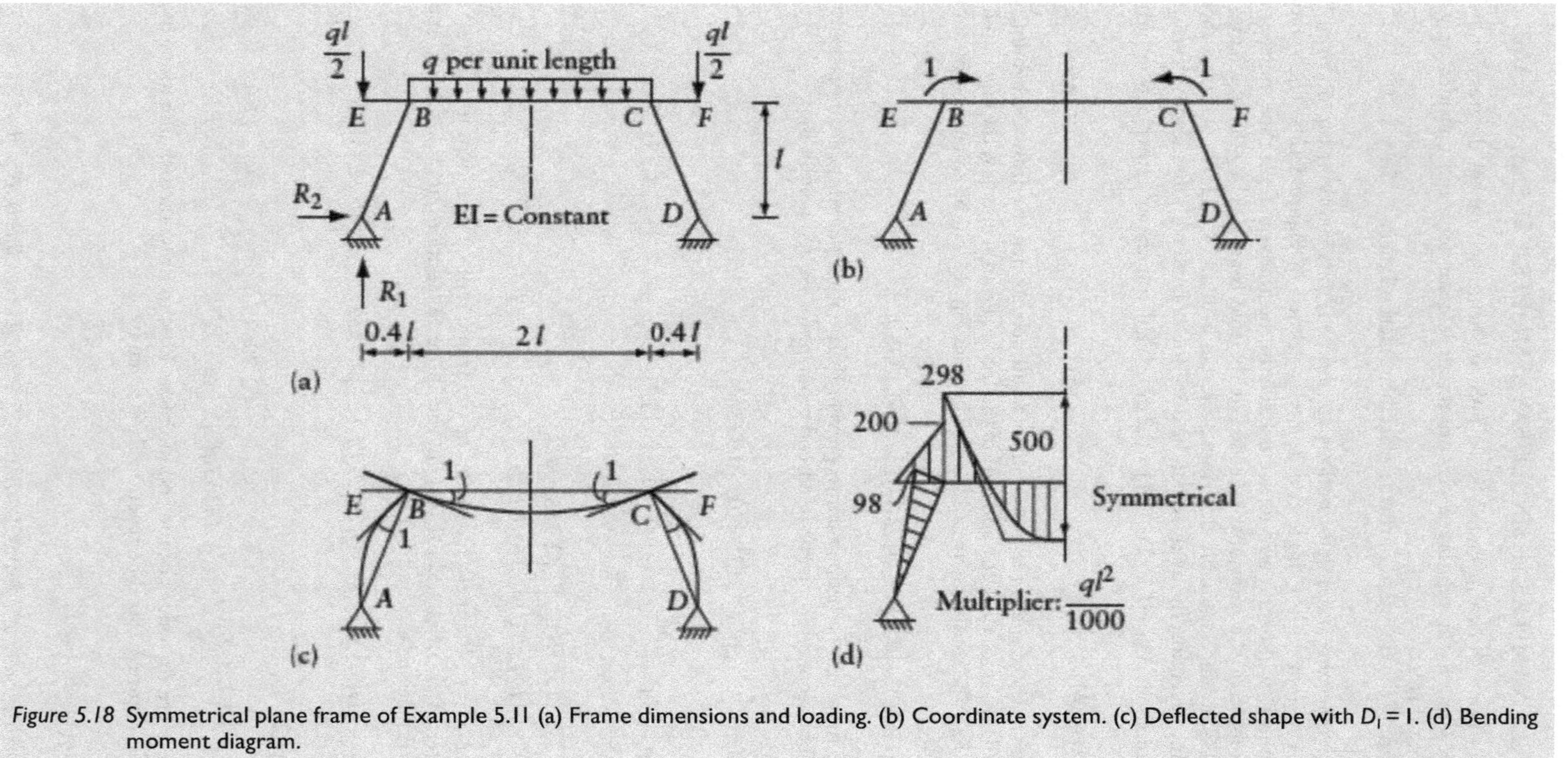

Figure 5.18 Symmetrical plane frame of Example 5.11 (a) Frame dimensions and loading. (b) Coordinate system. (c) Deflected shape with $D_1 = 1$. (d) Bending moment diagram.

EXAMPLE 5.12: A HORIZONTAL GRID SUBJECTED TO GRAVITY LOAD

Figure 5.19a shows a horizontal grid subjected to a uniformly distributed gravity load q per unit length on AB. Find the bending moment diagram for member AB. All members have the same cross section, with $GJ/EI = 0.8$.

We apply the five steps of the displacement method (Section 5.4):

Step 1 Three coordinates define the symmetrical displacement components at A and B (Figure 5.19b). To draw the bending moment diagram for AB, we need the value of M_A (or M_B). We consider the bending moment to be positive when it produces tension at the bottom face of the member. Thus, we define the required action:

$$\{A\} = \{M_A\}$$

Step 2 $\{F\} = \begin{Bmatrix} -ql/2 \\ 0 \\ -ql^2/12 \end{Bmatrix}$; $\{A_r\} = \left\{-\dfrac{ql^2}{12}\right\}$

The elements of these vectors are fixed-end forces taken from Appendix B (Equations B.7 and B.8).

Step 3

$$[S] = \begin{bmatrix} \left(\dfrac{12EI}{l^3}\right)_{AC} & & \text{Symmetrical} \\ -\left(\dfrac{6EI}{l^2}\right)_{AC} & \left(\dfrac{4EI}{l}\right)_{AC} & \\ 0 & 0 & \left(\dfrac{2EI}{l}\right)_{AB} + \left(\dfrac{GJ}{l}\right)_{AC} \end{bmatrix} = \frac{EI}{l}\begin{bmatrix} 96/l^2 & & \text{Symm.} \\ -24/l & 8 & \\ 0 & 0 & 3.6 \end{bmatrix}$$

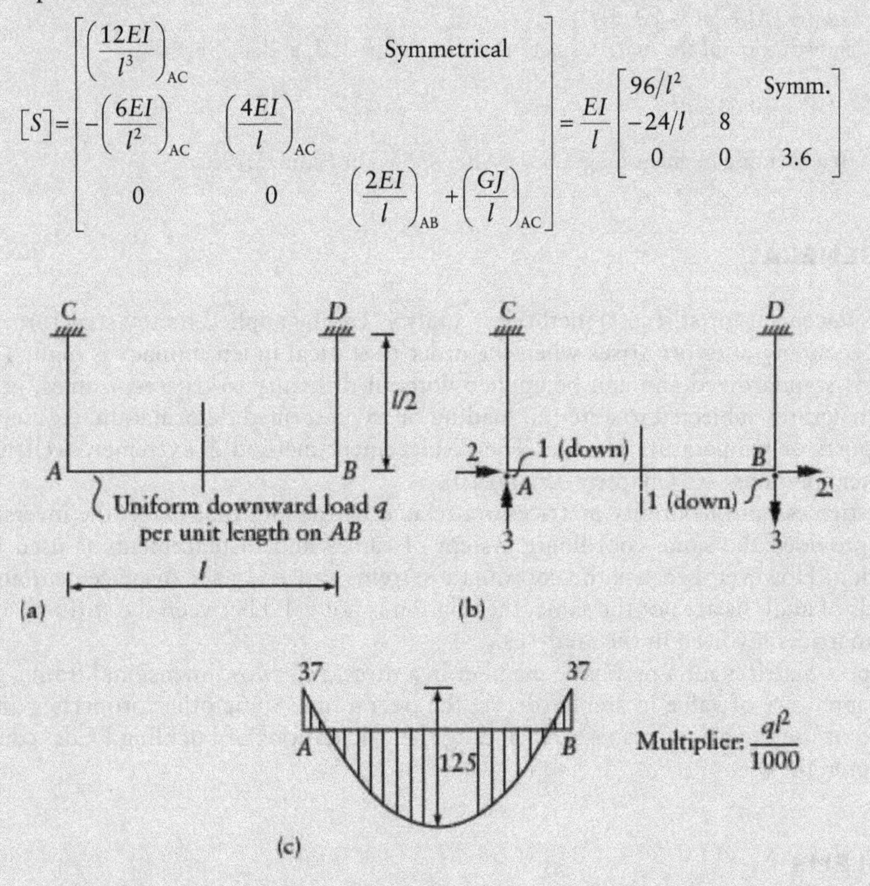

Figure 5.19 Symmetrical horizontal grid of Example 5.12. (a) Top view. (b) Coordinate system. (c) Bending moment diagram for member AB.

$$[A_u] = \begin{bmatrix} 0 & 0 & \left(\frac{2EI}{l}\right)_{AB} \end{bmatrix}$$

For the elements of the first columns of $[S]$ and $[A_u]$ a unit downward deflection, $D_1 = 1$ is introduced at each of A and B. Members AC and BD take the deflected shape shown in Appendix C (see Equations C.1 and C.2), while AB will translate downward as a rigid body (without deformation or end forces). With $D_2 = 1$, members AC and BD will take the deflected shape shown in Appendix C (see Equations C.3 and C.4); while, again, member AB will rotate as a rigid body. For the third columns of $[S]$ and $[A_u]$, rotations $D_3 = 1$ are introduced at A and B, causing AB to deflect as shown in the last figure in Appendix C (Equation C.11); members AC and BD are twisted without bending. The member end-forces given in Appendix C are used to determine the elements of $[S]$ and $[A_u]$.

Step 4 $\{D\} = [S]^{-1}\{-F\}$

Substituting $[S]$ and $\{F\}$ determined above and performing the matrix inversion and multiplication gives:

$$\{D\} = \frac{ql^3}{1000EI}\{20.8l, 62.5, 23.1\}$$

Step 5 $\{A\} = \{A_r\} + [A_u]\{D\}$

Substitution of the matrices determined in Steps 2, 3, and 4 gives:

$$\{A\} = -0.037ql^2$$

The bending moment diagram for AB is shown in Figure 5.19c.

5.14 GENERAL

The displacement (or stiffness) method of analysis can be applied to any structure, but the largest economy of effort arises when the order of statical indeterminacy is high. The procedure is standardized and can be applied without difficulty to trusses, frames, grids and other structures subjected to external loading or to prescribed deformation, e.g. settlement of supports or temperature change. The displacement method is extremely well-suited to computer solutions (see Chapters 19 and 20).

The stiffness and flexibility matrices are related by the fact that one is the inverse of the other, provided the same coordinate system of forces and displacements is used in their formation. However, because the coordinate systems chosen in the displacement and force methods of analysis are not the same, the relation is not valid between the stiffness and flexibility matrices involved in the analyses.

Stiffness matrices for a prismatic member in a three-and two-dimensional frame, given in this chapter, are of value in standardizing the operations. Some other properties discussed are also of use in various respects, including the calculations of buckling loads, considered in Chapter 10.

PROBLEMS

5.1 Use the displacement method to find the forces in the members of the truss shown. Assume the value l/aE to be the same for all members.

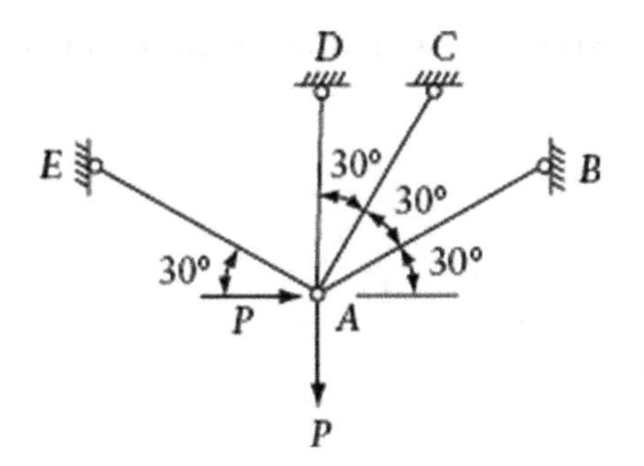

5.2 A rigid mast AB is hinged at A, free at B, and is pin-jointed to n bars of which a typical one is shown. Using the displacement method, find the force A in the ith bar if l/aE = constant for each bar. Assume that the mast and all the bars are in one plane.

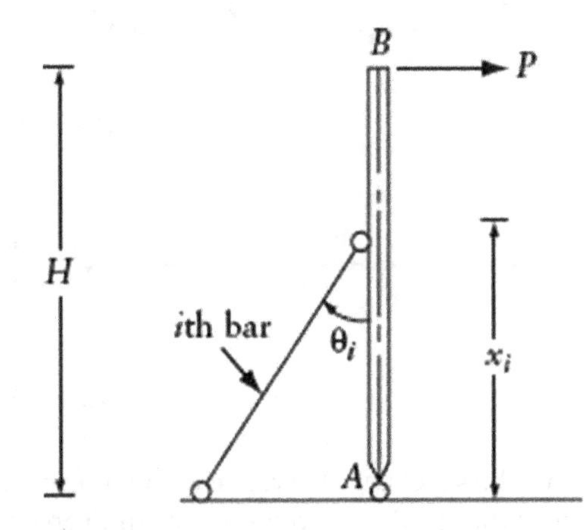

5.3 For the truss shown in the figure, write the stiffness matrix corresponding to the four coordinates indicated. Assume a and E to be the same for all members.

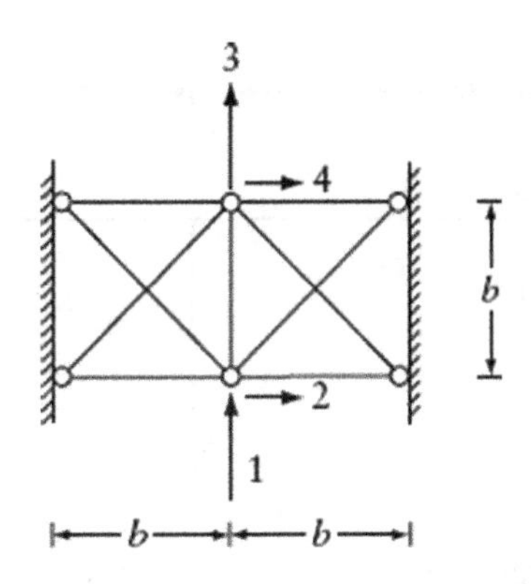

5.4 Find the forces in the members of the truss of Problem 5.3 due to $F_1 = -P$ at coordinate 1.

5.5 Member AB represents a chimney fixed at B and stayed by prestressed cables AC and AD. Find the bending moment at B and the changes in the cable forces due to a horizontal force P at A. Assume that AB, AC, and AD are in the same plane and that the cables continue to be in tension after application of P. Consider that: $(Ea)_{\text{cable}} = 10(EI/l^2)_{\text{AB}}$.

Only bending deformation of AB and axial deformation of the cables need to be considered.

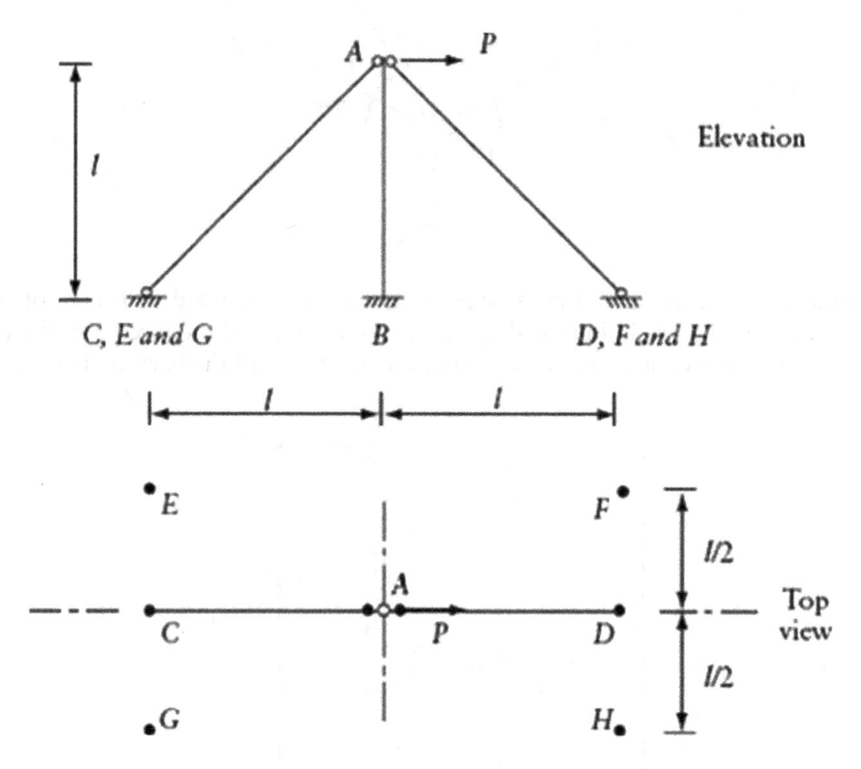

5.6 Solve Problem 5.5 with cables AC and AD replaced by four cables joining A to each of E, F, G, and H. Consider that $(Ea)_{cable}$ is the same as in Problem 5.5.

5.7 Neglecting axial deformations, find the end moments M_{BC} and M_{CF} for the frame shown. Consider an end moment to be positive when it acts in a clockwise direction on the member end.

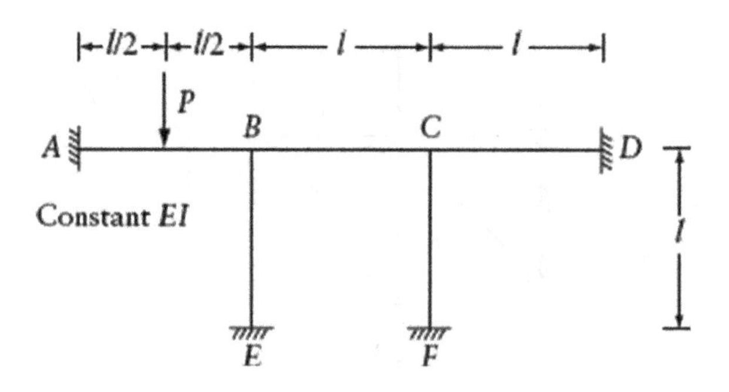

5.8 Solve Problem 5.7 if support E moves downward a distance Δ and the load P is not acting.

5.9 Use the displacement method to find the bending moment and shearing force diagrams for the frame in the figure. Ignore axial deformations.

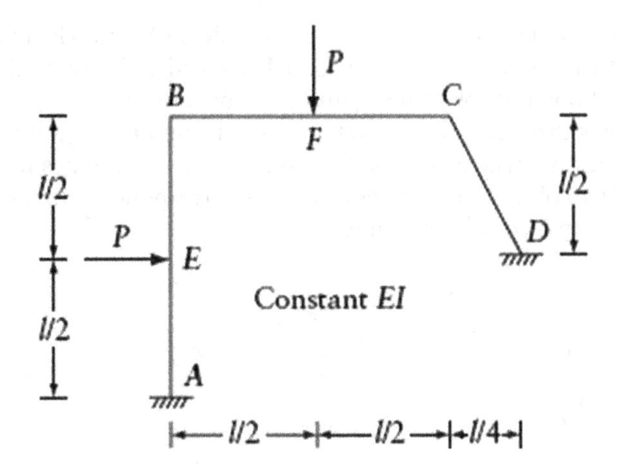

5.10 For the grid shown in plan in the figure, calculate the displacements corresponding to the three degrees of freedom at joint B. Assume $GJ/EI = 0.8$ for AB and BC. Draw the bending moment diagram for BC.

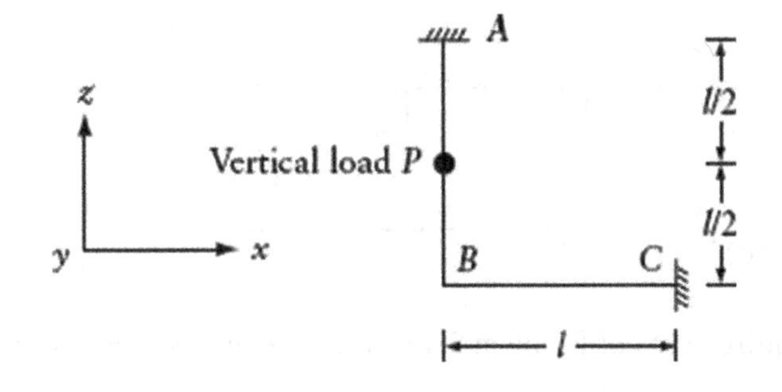

5.11 Considering three degrees of freedom: a downward deflection, and rotations represented by vectors in the x and z directions at each of the joints I, J, K, and L of the grid in the figure. Write the first three columns of the corresponding stiffness matrix. Number the twelve coordinates in the order in which they are mentioned above, that is, the first three at I followed by three at J, and so on. Take $GJ/EI = 0.5$ for all members.

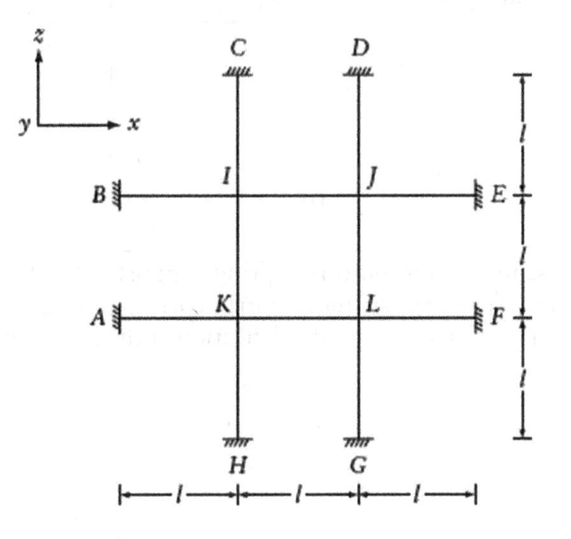

5.12 Ignoring torsion, find the bending moment in girders BE and AF of the grid of Problem 5.11 subjected to a vertical load P at joint I. Consider the vertical deflections at the joints as the unknown displacements and use Appendix D.

5.13 The figure shows three coordinate systems for a beam of constant flexural rigidity EI. Write the stiffness matrix corresponding to the four coordinates in (a). Condense this stiffness matrix to obtain the stiffness matrices corresponding to the three coordinates in (b) and to the two coordinates in (c).

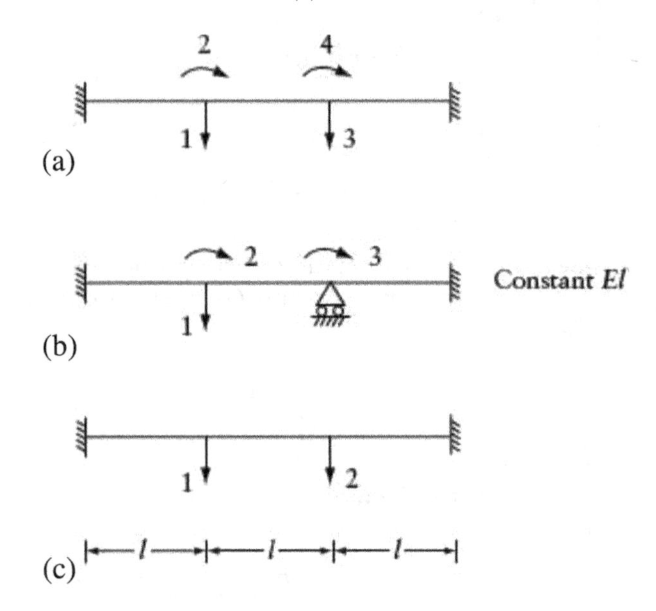

5.14 Apply the requirements of Problem 5.13 to the member shown in the figure.

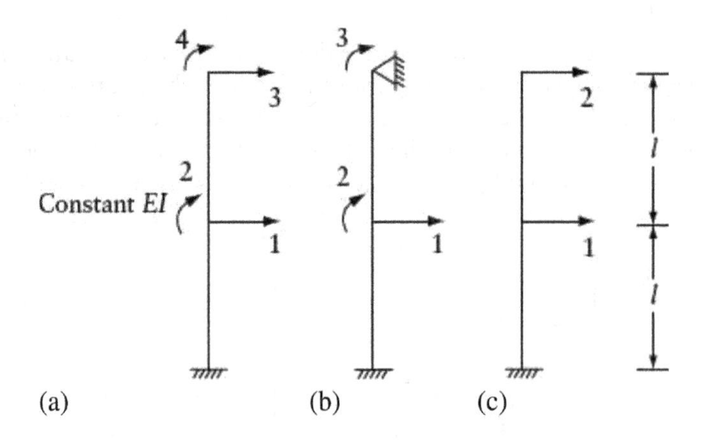

5.15 The figure represents a beam on three spring supports of stiffness $K_1 = K_2 = K_3 = EI/l$. Using Appendix D, derive the stiffness matrix corresponding to the three coordinates in the figure. Use this matrix to find the deflection at the three coordinates due to loads $\{F\} = \{3P, P, 0\}$.

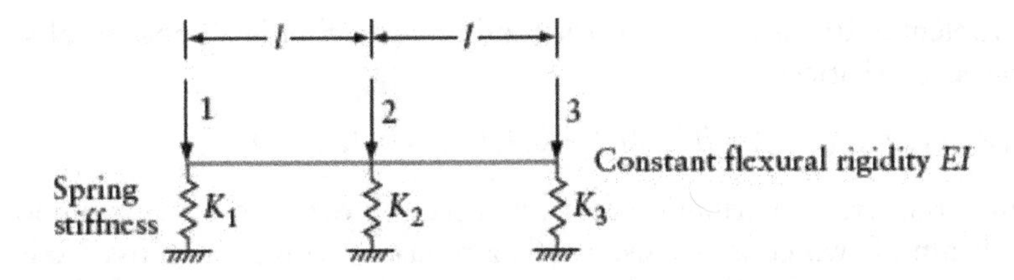

5.16 If the stiffness of any two of the spring supports in Problem 5.15 is made equal to zero, the stiffness matrix becomes singular. Verify this and explain why.

5.17 Ignoring axial deformations and using Appendix C, write the stiffness matrix for the frame in the figure corresponding to the four coordinates in (a). Condense this matrix to find the 2×2 stiffness matrix corresponding to the coordinates in (b).

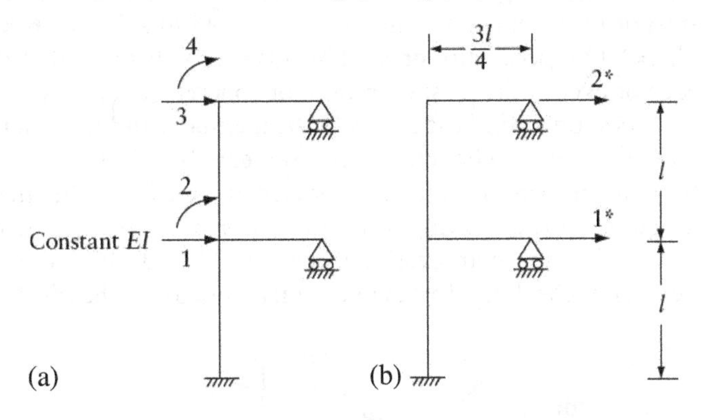

5.18 A simple beam of length l and rectangular cross section of width b and depth d is subjected to a rise of temperature which is constant over the length of the beam but varies over the depth of the section. The temperature rise at the top is T and varies linearly to zero at mid-depth; the rise of temperature is zero for the lower half of the section. Determine the stress distribution at any section, the change in length of the centroidal axis, and the deflection at mid-span. Modulus of elasticity is E and coefficient of thermal expansion is α.

5.19 If the beam of Problem 5.18 is continuous over two spans, each of length l, with one support hinged and the other two on rollers, obtain the bending moment diagrams, the reactions, and the stress distribution at the central support due to the same rise of temperature.

5.20 Find the bending moment diagram, the reactions, and the deflection at the middle of span AB due to prestressing of the continuous beam shown. Assume the prestressing force P is constant and the tendon profile is a second-degree parabola in each span. Draw the shearing force diagram.

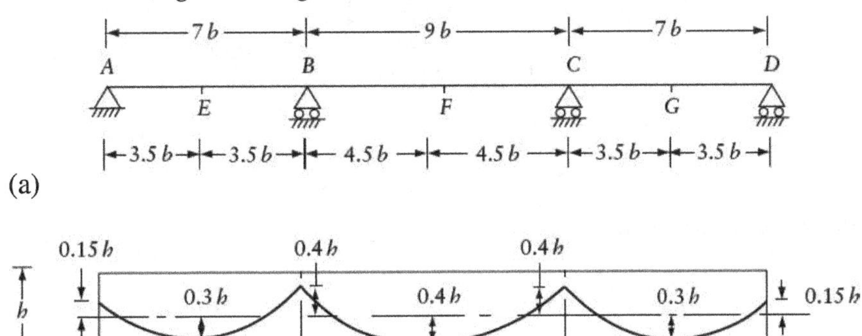

5.21 Solve Problem 5.20 assuming that the tendon has a profile composed of straight lines with the eccentricities:

$$\{e_A, e_E, e_B, e_F, e_C, e_G, e_D\} = h\{0, 0.3, -0.4, 0.4, -0.4, 0.3, 0\}$$

5.22 Determine the stress distribution at a cross section over the interior support of a continuous beam of two equal spans. The beam has a rectangular cross section subjected to a rise of temperature, T, varying over the height of the section, h, according to the equation: $T = T_{top}(0.5 - \mu)^5$, where T_{top} is the temperature rise at the top fiber; $\mu = y/h$, with y being the distance measured downward from the centroid to any fiber. Give the answer in terms of E, α, T_{top}, and μ, where E and α are the modulus of elasticity and the thermal expansion coefficient of the material. What is the value of stress at the extreme tension fiber? Using SI units, assume $E = 30$ GPa, $\alpha = 10^{-5}$ per degree Celsius, and $T_{top} = 30$ degrees Celsius; or using Imperial units, assume $E = 4300$ ksi; $\alpha = 0.6 \times 10^{-5}$ per degree Fahrenheit; $T_{top} = 50$ degrees Fahrenheit. The answers will show that the stress is high enough to reach or exceed the tensile strength of concrete; thus, cracking could occur.

5.23 For the space truss shown, find the forces in the members and the reaction components due to an external force at O having the components $\{P_x, P_y, P_z\} = P\{1.0, 2.0, 3.0\}$.

Hint: Although the truss is statically indeterminate, it can be analyzed using the equations of equilibrium by taking advantage of symmetry; e.g. P_x produces tensile forces in OA and OD and compressive forces in OB and OC having equal absolute value. As an exercise, check the forces in the members using the displacement method.

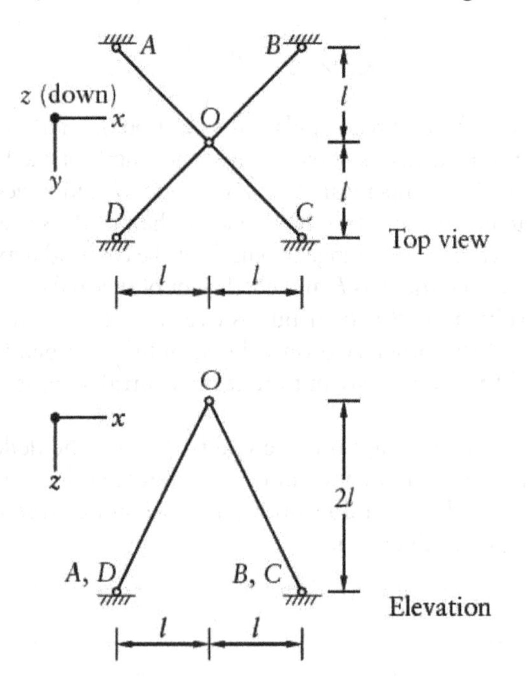

NOTE

1. See Reference 1, Chapter 4.

Chapter 6

Time-dependant displacements in structural concrete

6.1 INTRODUCTION

The present chapter applies linear analysis to compute strains, stresses, and displacements in concrete frames, considering cracking and the time-dependent effects of creep and shrinkage of concrete and relaxation of prestressed reinforcement. Virtual work is employed (unit load theory) to determine the displacement at a coordinate from the normal strain and curvature, ignoring shear deformation. The structure is loaded and prestressed at t_0; the load is sustained up to time t (the end of service life of the structure). The immediate and long-term displacements at t_0 and t are determined.

Linear analysis of the frame, made of homogeneous elastic material, gives the internal forces N and M at sections; where N=normal force at centroid and M=bending moment induced by applied load combined with effective prestressing. In general, the time-dependent behavior of the material changes the stress distribution; but the stress resultants N and M remain unchanged when the structure is statically determinate; changes in the stress resultants generally occur in other structures. The change in values of N and M is ignored in the sectional analysis in the current chapter. However, when the change in the values of N and M is significant, iterative analysis may be used with new values of N and M.

Consider structures composed of parts, cast, or loaded in stages; the time-dependent changes in internal forces and displacements can be determined by repetitive use of linear computer programs (Sections 6.4 and 6.5). For example, consider a prestressed floor having rigidly connected supports, constructed in an earlier stage. As another example, consider a bridge deck composed of segments, precast or cast-in-situ, made of concrete having different ages.

6.2 LONG-TERM DISPLACEMENTS IN STRUCTURAL CONCRETE

Consider a structure consisting of beam-like members. At time t_0, the structure is subjected to quasi-permenant load, equal to the dead load plus a fraction of the design live load. It is required to calculate the long-term displacement (translation or rotation) at a section at time t (end of the structure service life). Assume that during construction, cracking has occured by heavier loading, such as shoring. It is required to calculate the displacements at time t considering cracking, creep, and shrinkage of concrete and relaxation of prestressed reinforcement. Consider only normal stress σ and assume that the values of N and M at any section (Figure 6.1) remain constant (unaffected by creep, shrinkage, or relaxation).

Input data for the analysis consists of section geometry, creep and aging coefficients, free shrinkage strain, tensile strength of concrete, elasticity modulus of concrete at time t_0,

DOI: 10.1201/9780429286858-6

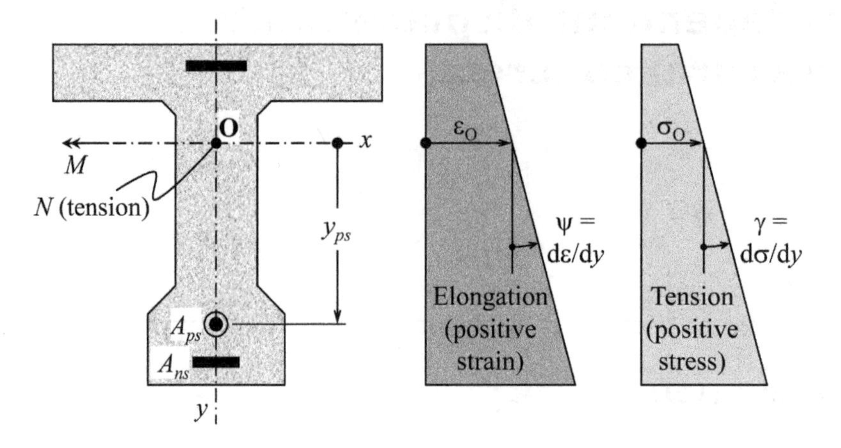

Figure 6.1 Positive sign convention. *y* is the vertical principal axis. Reference point O is arbitrarily located on the *y*-axis. *x* is the horizontal axis through O.

and intrinsic relaxation of prestressing reinforcement. The sustained load and the highest transient load – that may cause cracking – and the initial prestressing are part of the input. The analysis gives the deflection at time t, the stresses in the concrete, the prestressed, and non-prestressed reinforcements at time t. The results also include the time-dependent loss of prestressing and its effect on long-term deflection and stresses.

6.2.1 Creep of concrete

Concrete stress σ introduced at age t_0 and sustained to age t induces immediate strain and creep given by:

$$\varepsilon(t,t_0) = \frac{\sigma}{E_c(t_0)}\left[1+\phi(t,t_{.0})\right] \tag{6.1}$$

Equation 6.1 is linear stress–strain relationship; the stress value, σ is constant in the period $(t{-}t_0)$; at time t, the strain = σ multiplied by a constant. When the stress σ – introduced in the period t_0 to t – varies with time (as the relaxation, Figure 6.2), the total strain is:

$$\varepsilon(t,t_0) = \frac{\sigma}{E_c(t_0)}\left[1+\chi\phi(t,t_0)\right] = \frac{\sigma}{\bar{E}_c(t,t_0)} \tag{6.2}$$

$$\bar{E}_c(t,t_0) = \frac{E_c(t_0)}{1+\chi\phi(t,t_0)} \tag{6.3}$$

$E_c(t_0)$ = modulus of elasticity of concrete at age t_0; $\varphi(t,t_0)$ and $\chi(t,t_0)$ = creep and aging coefficients, respectively; $\bar{E}_c(t,t_0)$ = age-adjusted elasticity modulus of concrete. The benefit of the aging coefficient, χ is retaining the linear stress–strain relationship, Equation 6.2.

The value of $\varphi(t,t_0)$ depends upon the age of concrete at loading t_0 and age t for which the creep is calculated. For example, when t_0 is one month and t is infinity, $\varphi(t,t_0) = 2$ to 4 depending on the quality of concrete, the ambient temperature, and humidity as well as the dimensions of the section considered. The *fib* Model Code[1] and ACI Committee 209[2] give expressions and graphs for the value $\varphi(t,t_0)$ to be used in calculating the strain parameters at

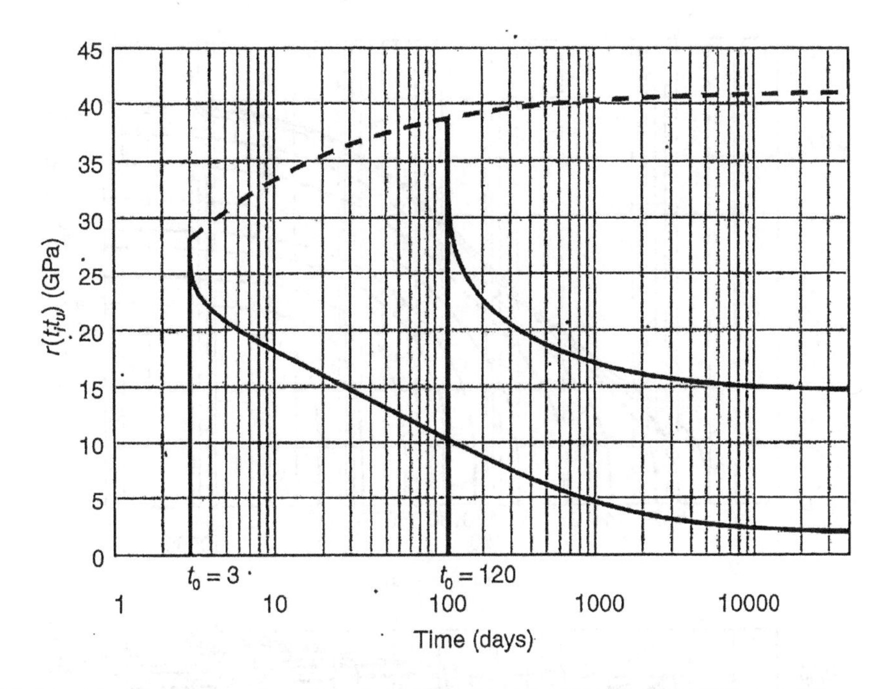

Figure 6.2 Example of relaxation function, $r(t, t_0)$ with time: f_{ck}=40 MPa; RH=50%; h_0=400 mm; s=0.25. Based on equations of MC-2010. The dashed curve represents the variation of $E_c(t_0)$.

time t. The aging coefficient, $\chi(t,t_0)$ generally varies between 0.6 and 0.9. Ghali et al. 2012[3] calculate $\chi(t,t_0)$ with a step-by-step procedure, and provide graphs and computer programs for φ and χ.

The aging coefficient may be taken from a table or a graph (Ghali et al. 2012). Since χ is always used as a multiplier to φ, which is rarely accurately determined, high accuracy in the derivation of χ is hardly justified. For a specified age t_0, the value of $\chi(t,t_0)$ is almost constant when $(t-t_0) \geq$ one year. For long-term displacements, $\chi(t,t_0)$ may be expressed in terms of t_0 in days:

$$\chi(t,t_0) \approx \frac{\sqrt{t_0}}{1+\sqrt{t_0}} \tag{6.4}$$

The error in Equation 6.4 is ±10 percent when $t_0 > 28$ days. A step-by-step procedure gives the relaxation function:

$$r(t,t_0) = \sigma(t)/\varepsilon_c; \quad \sigma_c(t) = \varepsilon_c\, r(t,t_0) \tag{6.5}$$

where $\sigma(t)$ = the stress remaining at time t induced by imposed strain ε_c, introduced at t_0 and sustained. Then, $\chi(t,t_0)$ can be expressed in terms of $\varphi(t,t_0)$ and $E_c(t_0)$ as:

$$\chi(t,t_0) = \frac{1}{1-r(t,t_0)/E_c(t_0)} - \frac{1}{\phi(t,t_0)} \tag{6.6}$$

Ghali et al. (2012,) employ a step-by-step procedure to derive the relaxation function $r(t, t_0)$. Figure 6.2 shows the relaxation function $r(t,t_0)$ and the variation of ε_c with t_0, based on equations of CEB-FIP MC 2010. Figure 6.3 shows the creep and aging coefficients for

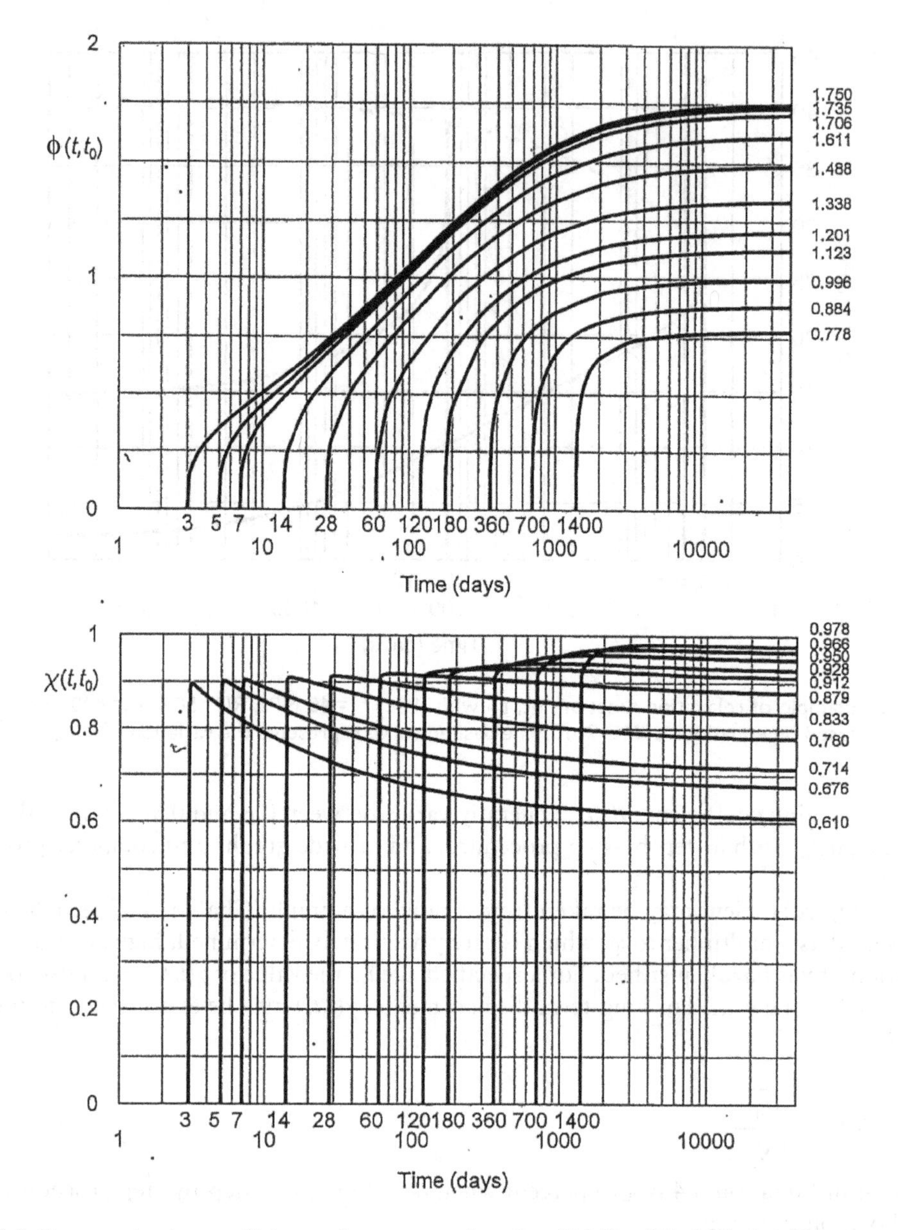

Figure 6.3 Creep and aging coefficients of concrete when f_{ck} = 40 MPa; RH = 80%; h_0 = 100 mm; s = 0.25. (Ghali et al. 2012)

f_{ck} = 40 MPa; RH = 80%; h_0 = 100 mm; s = 0.25. Ghali et al. 2012 and Chapter 20 of the present text include computer program CREEP to calculate $r(t,t_0)$, $\varphi(t,t_0)$, $\chi(t, t_0)$ and the free shrinkage as functions of f_{ck}, h_0, RH, s, t_1, and t_2; where f_{ck} = the characteristic compressive strength (in MPa) of cylinders 150 mm in diameter and 300 mm high, stored in water at 20±2°Celsius and tested at the age of 28 days; h_0 = the notional size (in mm) = 2 times the cross-sectional area of a member divided by its perimeter in contact with the atmosphere; RH = relative humidity (percent); t_1 and t_2 = ages of concrete at loading and at end of a period in which the load is sustained; s = coefficient that depends upon the type of cement, taken equal to 0.25 for $f_{ck} < 60$ MPa and 0.20 for $f_{ck} \geq 60$ MPa.

6.2.2 Relaxation of prestressing

The effect of creep on prestressing steel is commonly determined by a relaxation test. A tendon is stretched and maintained at a constant length and temperature; the loss in the tendon tensile stress, referred to as *intrinsic relaxation*, $\Delta\sigma_{pr}$ is measured over a long period. Because the tendon length is not constant in a concrete member, in analysis, a reduced relaxation, $\Delta\bar{\sigma}_{pr}\ (=\chi_r\,\Delta\sigma_{pr})$ is used, where χ_r = relaxation reduction factor (≈ 0.8). The values of χ_r can be expressed as integral or approximated by Equation 6.7 (Ghali et al. 2012):

$$\chi_r = e^{\Omega\left(-6.7+5.3\,\lambda_p\right)} \tag{6.7}$$

Ω = ratio of (total change in prestress – intrinsic relaxation) to the initial prestress; λ_p = ratio of prestress value at t_0, σ_{p0} to the characteristic tensile strength.

The intrinsic relaxation of stress-relieved wires or strands is (Magura, Sozen, and Siess 1964[4]):

$$\frac{\Delta\sigma_{pr}}{\sigma_{p0}} = -\frac{\log(\tau - t_0)}{10}\left(\frac{\sigma_{p0}}{f_{py}} - 0.55\right) \tag{6.8}$$

τ = any time instant; the time difference ($\tau - t_0$) is expressed in years; σ_{p0} = initial prestress value at time t_0; f_{py} = yield stress of prestressing steel = stress at a strain of 0.001. The value of the total prestress change is generally not known aprior, because it depends upon the reduced relaxation; iteration is here required: the total loss is calculated using an estimated value of the reduction factor, for example $\chi_r = 0.7$, which is later adjusted, if necessary. Ghali et al 2012 give graphs of χ_r versus Ω and their derivation.

6.2.3 Immediate strain parameters

Consider a beam cross section, having a vertical symmetry axis (principal y-axis; Figure 6.1), and a horizontal x-axis through an arbitrary reference point O. Assuming linear stress–strain relationship and that plane cross section remains plane, Equations 6.9 to 6.11 apply:

$$\varepsilon = \varepsilon_O + \psi\, y; \quad \sigma = \sigma_O + \gamma\, y \tag{6.9}$$

$$\int \sigma \, dA = N; \quad \int \sigma y \, dA = M \tag{6.10}$$

$$E\int \varepsilon \, dA = N; \quad E\int \varepsilon y \, dA = M \tag{6.11}$$

ε and σ = strain and stress at any point with vertical coordinate y; $\psi = d\varepsilon/dy$; $\gamma = d\sigma/dy$; N and M = values of the stress resultants at any instant. N is located at the reference point O; M = moment about a horizontal x-axis through O. Substituting Equation 6.9 into Equations 6.10 and 6.11 and solving for the strain parameters $\{\varepsilon_O, \psi\}$ gives:

$$\begin{Bmatrix} \varepsilon_O \\ \psi \end{Bmatrix} = \frac{1}{E\left(AI - B^2\right)} \begin{bmatrix} I & -B \\ -B & A \end{bmatrix} \begin{Bmatrix} N \\ M \end{Bmatrix} \tag{6.12}$$

$$\begin{Bmatrix} \sigma_O \\ \gamma \end{Bmatrix} = E\begin{Bmatrix} \varepsilon_O \\ \psi \end{Bmatrix} = \frac{1}{\left(AI - B^2\right)} \begin{bmatrix} I & -B \\ -B & A \end{bmatrix} \begin{Bmatrix} N \\ M \end{Bmatrix} \tag{6.13}$$

The sectional area properties $\{A, B, I\}$ of transformed non-cracked section are: $A = \int \mathrm{d}A$ = area; $B = \int y\,\mathrm{d}A$ = moment of area about x-axis; $I = \int y^2\,\mathrm{d}A$ = moment of inertia about x-axis.

When the reference point O is at the centroid of the transformed section, $B = 0$ and Equation 6.12 becomes:

$$\varepsilon_O = N/(EA); \quad \psi = M/(EI)$$

However, these simple equations are not suitable for analysis of time-dependent strain because the position of the centroid of the transformed section is also time-dependent.

The transformed uncracked section (in state 1) consists of the net concrete area plus $(E_{ns}/E_c)A_{ns}$ without $(E_{ps}/E_c)A_{ps}$ (only for post-tensioned reinforcement, due to lack of strain compatibility between the tendons and surrounding concrete); where A_{ns} and A_{ps} = areas of non-prestressed and prestressed reinforcements, respectively; E_{ns} and E_{ps} = elasticity moduli of non-prestressed and prestressed reinforcements, respectively. The values of N and M include the effect of initial prestressing P and Py_{ps}; where P = absolute value of prestressing force and y_{ps} is its eccentricity, with respect to the x-axis (see Figure 6.1).

The modulus of elasticity of concrete depends on its age; also, $\{A, B, I\}$ are time-dependent. Using the values of $\{A, B, I\}$ and $E_c(t_0)$ at time t_0 in Equation 6.12 gives the immediate strain parameters $\{\varepsilon_O(t_0), \psi(t_0)\}$.

6.2.4 Long-term strain parameters

Creep and shrinkage at any fiber in the period $(t-t_0)$, is artificially prevented by a restraining stress $\sigma_{res}(t, t_0)$:

$$\sigma_{res}(t,t_0) = -\bar{E}_c\left\{\phi\left[\varepsilon_O(t_0) + y\cdot\psi(t_0)\right] + \varepsilon_{cs}(t,t_0)\right\} \tag{6.14}$$

The resultants of $\sigma_{res}(t, t_0) = -\{\Delta N, \Delta M\}$. To remove the artificial restraint, apply $\{\Delta N, \Delta M\}$, whose values are given by:

$$\begin{Bmatrix}\Delta N\\ \Delta M\end{Bmatrix} = \bar{E}_c\,\phi\begin{bmatrix}A_c & B_c\\ B_c & I_c\end{bmatrix}\begin{Bmatrix}\varepsilon_O(t_0)\\ \psi(t_0)\end{Bmatrix} + \bar{E}_c\,\varepsilon_{cs}\begin{Bmatrix}A_c\\ B_c\end{Bmatrix} - \Delta\bar{\sigma}_{pr}\begin{Bmatrix}A_{ps}\\ A_{ps}\,y_{ps}\end{Bmatrix} \tag{6.15}$$

The first, second, and last terms on the right-hand side of Equation 6.15 are the resultants of the artificial stress-restraining: creep, shrinkage, and relaxation, respectively, with sign reversed. Subscripts c and ps refer to concrete and prestressed reinforcement, respectively.

Creep and shrinkage of concrete and relaxation of prestressed reinforcement induce increments $\Delta\varepsilon_O(t,t_0)$ and $\Delta\psi(t,t_0)$ given by Equation 6.12, substituting $\{N, M\}$ by $\{\Delta N, \Delta M\}$, and using the age-adjusted transformed section properties $\bar{E}_c$, $\bar{A}$, $\bar{B}$, and $\bar{I}$:

$$\begin{Bmatrix}\Delta\varepsilon_O(t,t_0)\\ \Delta\psi(t,t_0)\end{Bmatrix} = \frac{1}{\bar{E}_c\left(\bar{A}\bar{I} - \bar{B}^2\right)}\begin{bmatrix}\bar{I} & -\bar{B}\\ -\bar{B} & \bar{A}\end{bmatrix}\begin{Bmatrix}\Delta N\\ \Delta M\end{Bmatrix} \tag{6.16}$$

The properties of the age-adjusted transformed section are calculated in the same way as A, B, and I using $\bar{E}_c$ instead of E_c. However, the age-adjusted transformed section includes the term $(E_{ps}/\bar{E}_c)A_{ps}$, ignoring the lack of strain compatibility in the period $(t-t_0)$. The strain parameters for the section at time t are given by:

$$\varepsilon_O(t) = \varepsilon_O(t_0) + \Delta\varepsilon_O(t,t_0); \quad \psi(t) = \psi(t_0) + \Delta\psi(t,t_0) \tag{6.17}$$

The stress in concrete at time t at any fiber – at distance y from O – is calculated by:

$$\begin{aligned} \sigma_c(t) &= E_c(t_0)\left[\varepsilon_O(t_0) + y\cdot\psi(t_0)\right] \\ &\quad + \sigma_{res}(t,t_0) + \bar{E}_c(t,t_0)\left[\Delta\varepsilon_O(t,t_0) + y\cdot\Delta\psi(t,t_0)\right] \end{aligned} \tag{6.18}$$

The stress in prestressed steel at time t is:

$$\sigma_{ps}(t) = \sigma_{ps}(t_0) + E_{ps}\left[\Delta\varepsilon_O(t,t_0) + y_{ps}\cdot\Delta\psi(t,t_0)\right] + \Delta\bar{\sigma}_{pr}(t,t_0) \tag{6.19}$$

With post-tensioning, the first term on the right-hand side of Equation 6.19 $= P/A_{ps}$; there is no bond between the post-tensioned tendons and adjacent concrete. Thus, the strain change in a tendon is smeared along its length. While with pre-tensioned strands, the bond between tendons and adjacent concrete is retained with $\sigma_{ps}(t_0) = E_{ps}\ [\varepsilon_O(t_0) + y_{ps}\ \psi(t_0)]$. Stress in non-prestressed steel at time t, $\sigma_{ns}(t)$ can be calculated similar to pre-tensioned reinforcement (Equation 6.19) by replacing $\{E_{ps}, y_{ps}\}$ by $\{E_{ns}, y_{ns}\}$ and eliminating the last term on the right-hand side, $\Delta\bar{\sigma}_{pr}(t,t_0)$.

To calculate deflection due to daily temperature rise, T, determine the strain change $\{\Delta\varepsilon_O, \Delta\psi\}$:

$$\begin{Bmatrix} \Delta\varepsilon_O \\ \Delta\psi \end{Bmatrix} = \left[\left(AI - B^2\right)\right]^{-1} \begin{bmatrix} I & -B \\ -B & A \end{bmatrix} \begin{Bmatrix} \int \alpha_t T\,dA \\ \int \alpha_t T y\,dA \end{Bmatrix} \tag{6.20}$$

α_t = coefficient of thermal expansion (degree^{-1}).

6.2.5 Cracking

As an approximation, assume that prestressing is introduced at t_0; assume that cracking occurred also at t_0. The heaviest load – likely to induce cracking – and the prestressing are introduced simultaneously at t_0. By linear analysis of non-cracked structure, idealized as an assembly of rigidly connected short prismatic beam elements, determine the values of N and M at beam ends. Identify the cracked sections, where stress at the extreme fiber $\geq$ (1.0 or, conservatively, $1/\sqrt{2}$) times the tensile strength, f_{ct}.

Apply the quasi-permanent load and initial prestressing; determine, by linear analysis of non-cracked structure, the values of N and M at each beam end. At each cracked section, calculate the strain parameters at time t in states 1 and 2 (Equation 6.12); in state 1, cracking is ignored; in state 2, the section consists of concrete in the compression zone plus the reinforcements. Interpolation of the strain parameters of a section in state 1, $\{\varepsilon_O, \psi\}_1$, and state 2, $\{\varepsilon_O, \psi\}_2$, is done according to *fib* Model Code 2010. The interpolation coefficient, ξ, is given by:

$$\xi = 1 - 0.5\,(M_{cr}/M)^2 = 1 - 0.5\,(N_{cr}/N)^2$$
$$= 1 - 0.5\,(f_{ct}/\sigma_{max})^2 \geq 0.5 \tag{6.21}$$

where $\{N, M\}_{cr}$ combined induce cracking at an extreme tension fiber; $\{N, M\}$ = internal forces (assumed constants in the period $(t\text{-}t_0)$); σ_{max} = calculated tensile stress at extreme fiber when cracking is ignored. The values of $\{N, M\}_{cr}$ = fraction of $\{N, M\}$, such that $N_{cr}/N = M_{cr}/M$. Deformations are calculated from the interpolated strain parameters:

$$\varepsilon_O = (1-\xi)\varepsilon_{O1} + \xi\varepsilon_{O2}; \quad \psi = (1-\xi)\psi_1 + \xi\psi_2 \tag{6.22}$$

Equations 6.21 and 6.22 apply when (M/M_{cr}) or $(N/N_{cr}) \geq 1.0$ or $1/\sqrt{2}$ (when the maximum concrete stress $\geq$ tensile strength or, conservatively, $1/\sqrt{2}$ (tensile strength)).

6.2.6 Strain–displacement relationship

Calculation of translation or rotation from the strain parameters $\{\varepsilon_O, \psi\}$ is a geometry problem, solved in different ways. Virtual work or the *unit load theorem* (Section 7.7) is employed. Ignoring deformations due to shearing force and twisting moment, displacement D_i at coordinate i is given by:

$$D_i = \int N_{ui}\,\varepsilon_O\,dl + \int M_{ui}\,\psi\,dl \tag{6.23}$$

N_{ui} and M_{ui} = normal force and bending moment due to a unit virtual force applied at coordinate i in the direction of the required displacement. Consider downward deflection v_C of a horizontal beam (Figure 6.4); deflection at C measured from chord AB is related to the curvature ψ by:

$$v_C = \int_A^B M_u\,\psi\,dl \tag{6.24}$$

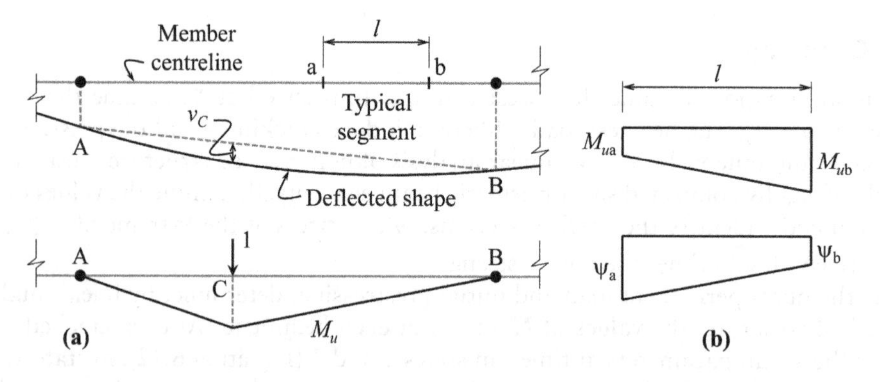

Figure 6.4 Deflection-curvature relation. (a) Deflection variation over a grid line AB. (b) Linear variation of M_u and ψ over a typical segment of AB.

M_u = internal bending moment due to a unit downward force at C (Figure 6.4a). The integral can be evaluated numerically as the following sum:

$$v_C = \sum \frac{l}{6} \left(2M_{ua}\,\psi_a + 2M_{ub}\,\psi_b + M_{ua}\,\psi_b + M_{ub}\,\psi_a \right) \tag{6.25}$$

ψ and M_u vary linearly over length of a typical segment (Figure 6.4b); subscripts a and b refer to segment ends.

6.3 LONG-TERM DEFLECTION OF REINFORCED CONCRETE FLOORS

Concrete slabs with drop panels or beams may be idealized as horizontal grids of rigidly connected short prismatic members (analysis with computer program, EGRID, Chapter 20), subjected to gravity loads and prestressing forces; plane grids (PLANEG, Chapter 20) can be used for analysis of floors of constant thickness without prestressing. Elastic analysis of torsionless grid gives the internal forces at member ends, M and N. These forces are used in the equations provided in Section 6.2 to calculate strain parameters at t_0 and t at different sections, considering cracking, creep, and shrinkage of concrete and relaxation of prestressing reinforcement.

Gravity dead load and fraction of the live load are applied at time t_0 and sustained up to the end of life of the structure (time t). The internal forces due to the sustained gravity loads are assumed unchanged by cracking, creep, shrinkage, or relaxation. Use of torsionless grid analysis errs on the conservative side for the calculation of long-term deflections. The procedure is illustrated in Example 6.2 where PLANEG is used in the analysis of a trapezoidal slab.

EXAMPLE 6.1: SIMPLY SUPPORTED POST-TENSIONED MEMBER

Figure 6.5 shows the mid-span cross section of a simply supported beam, representing a one-way slab with span l = 18 m and width = 1.0 m; it has a parabolic tendon profile with sag at mid-span = 145 mm. The slab thickness, h = 400 mm reinforced with non-prestressed flexural reinforcement at top and bottom with area = 800 mm². Determine the short- and long-term deflections at mid-span due to quasi-permanent load = 11.5 kPa, sustained in the time $(t\text{-}t_0)$. Given data: moduli of elasticity of concrete, reinforcing steel, and prestressed reinforcement are: $E_c(t_0)$ = 26.6 GPa, E_{ns} = 200 GPa, and E_{ps} = 190 GPa, respectively; free shrinkage strain, $\varepsilon_{cs}(t,t_0) = -300 \times 10^{-6}$;

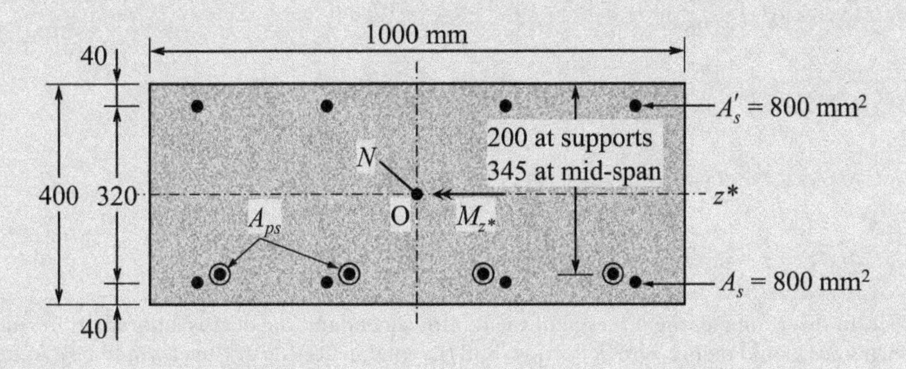

Figure 6.5 Section of a simply supported prestressed slab strip, Example 6.1.

Table 6.1 Results of sectional analyses of Example 6.1 at the supports and at mid-span

Strain parameter	*Section at support*	*Section at mid-span*
$\psi(t_0)$, m^{-1}	0.00	0.367×10^{-3}
$\psi(t)$, m^{-1}	0.00	1.820×10^{-3}
$\varepsilon_O(t_0)$	-260.5×10^{-6}	-259.7×10^{-6}
$\varepsilon_O(t)$	-919.0×10^{-6}	-939.7×10^{-6}

$\phi(t,t_0) = 2.0$; $\chi(t,t_0) = 0.8$; reduced relaxation, $\Delta\bar{\sigma}_{pr}(t,t_0) = -103$ MPa. Assume the same initial prestressing, P over all tendon length; set $f_{pc} = P/A_g$; where f_{pc} = compressive stress is induced by effective prestress on the gross area, A_g (= 0.4 m^2). The f_{pc}-value balancing 87 percent of the applied load = 7.0 MPa; corresponding value of P = 2800 kN; A_{ps} = 1800 mm^2; total ducts area = 6360 mm^2. What is the change in length of the member at time t at the reference point, O?

Computer program TDA is used to perform the sectional analysis (Section 6.2) at mid-span and at supports; the creep and aging coefficients (φ and χ) can be calculated from the CREEP program (Chapter 20). At t_0, the stress at bottom fiber at mid-span = −4.76 MPa $< f_{ct}$; thus, no cracking takes place. Table 6.1 shows results of TDA program at the supports and at the mid-span.

Assuming a parabolic variation of curvature along the span, deflection at mid-span is calculated as:

$$D_{\text{mid-span}} = \frac{l^2}{96}\left(\psi_{s1} + 10\psi_c + \psi_{s2}\right) \tag{6.26}$$

With linear variation of axial strain between sections at O, ε_O, the change in member length can be calculated by:

$$\Delta l = \frac{l}{4}\left(\varepsilon_{O-s1} + 2\varepsilon_{O-c} + \varepsilon_{O-s2}\right) \tag{6.27}$$

The subscripts $s1$ and $s2$ in Equations 6.26 and 6.27 refer to the sections at the supports; the subscript c refers to the mid-span section. Substituting the values in Table 6.1 (calculated by TDA program) in Equations 6.26 and 6.27 gives:

$$D_{\text{mid-span}}(t_0) = \frac{18^2}{96}\left(10 \times 0.367 \times 10^{-3}\right) = 12.4 \times 10^{-3} \text{ m}$$

$$D_{\text{mid-span}}(t) = \frac{18^2}{96}\left(10 \times 1.820 \times 10^{-3}\right) = 61.4 \times 10^{-3} \text{ m}$$

$$\Delta l(t_0) = \frac{18}{4} \times 2\left(-0.2605 - 0.2597\right) \times 10^{-3} = -4.68 \times 10^{-3} \text{ m}$$

$$\Delta l(t) = \frac{18}{4} \times 2\left(-0.9190 - 0.9397\right) \times 10^{-3} = -16.7 \times 10^{-3} \text{ m}$$

If, in this example, the presence of the reinforcement and the duct is ignored, uniform shrinkage would induce zero deflection, and the total deflection at time t would be = 74.5 mm.

EXAMPLE 6.2: NON-PRESTRESSED SLAB IDEALIZED AS PLANE GRID

Compute the long-term deflection at the middle of the free edge of the 200 mm trapezoidal slab in Figure 6.6a. Given data: $E_c(t_0)=26$ GPa; $E_s=200$ GPa; tensile strength, $f_{ct}=1.9$ MPa; free shrinkage, $\varepsilon_{cs}(t,t_0)=-300\times10^{-6}$; $\phi(t,t_0)=2.0$; $\chi(t,t_0)=0.8$. Quasi-permanent load, $q_p=6.5$ kPa. Flexural reinforcement ratio, $\rho=0.46\%$ in x- and z-directions; top reinforcement at the fixed edge, $\rho'=0.5\%$. Assume that the maximum load producing cracking, $q_{max}=7.7$ kPa.

Use computer program Plane Grid (PLANEG; Chapter 20) with three degrees-of-freedom per node. The moments at member ends per unit width due to $q_{max}=7.7$ kPa and $q_p=6.5$ kPa, are listed in columns 2 and 4 of Table 6.2. Cracking occurs at member ends at nodes: 3, 4, and B; the interpolation coefficient, ξ (Equation 6.21) is listed in column 3 of Table 6.2. For load, $q_p=6.5$ kPa, the curvatures at time t_0 are listed in columns 5 and 7 of Table 6.2 and at time t in columns 6 and 8 for states 1 (uncracked sections) and 2 (fully-cracked sections), respectively; the interpolated curvatures are calculated by Equation 6.22 and are listed in columns 9 and 10 of Table 6.2. The deflection v_B is calculated from curvatures listed in Table 6.3 over grid line AB. Deflections at node B at times t_0 and t are 15.5 and 29.9 mm, respectively (Table 6.3).

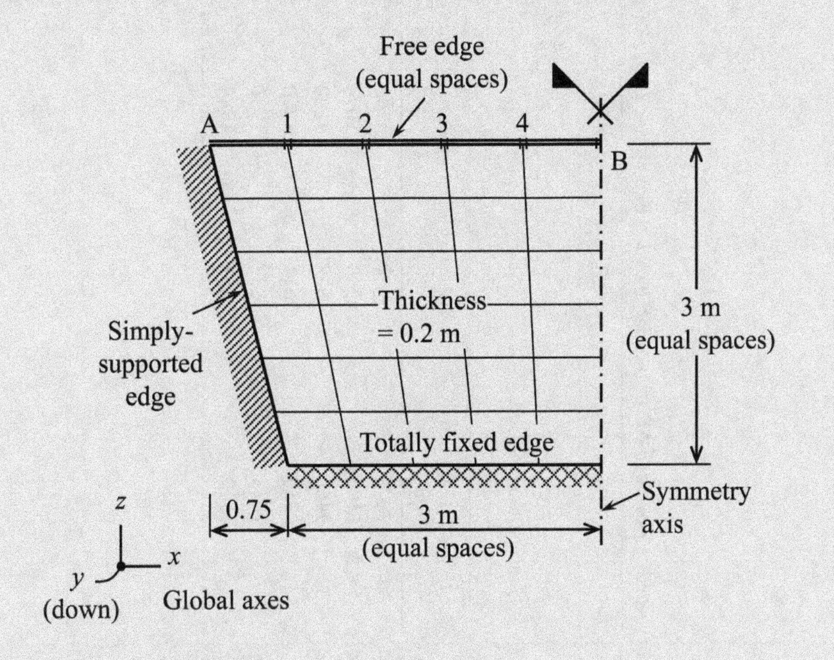

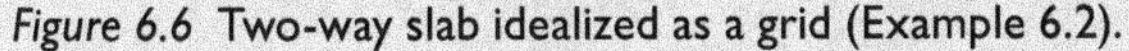

Figure 6.6 Two-way slab idealized as a grid (Example 6.2).

6.4 USE OF LINEAR COMPUTER PROGRAMS

The procedure in this section, based on the displacement method (Chapter 5), applies when there is no cracking. Linear analysis computer programs (Chapter 20) – in which the strain is proportional to the stress and superposition of displacements, stresses, and internal forces is allowed – are used to determine long-term displacements and stresses in reinforced concrete structures (e.g. bridge decks composed of segments, precast or cast-in-situ made of concrete of different ages, or of concrete and cable stays). Other examples are precast members,

Table 6.2 Immediate and long-term curvatures along edge AB of the slab panel in Example 6.2 (Fig. 6.6a)

(1)	(2)	(3)	(4)	(5)	(6)	(7)	(8)	(9)	(10)
				Curvatures due to quasi-permanent load					
	Maximum load, q_{max} = 7.7 kPa		*Quasi-permanent load, q_p = 6.5 kPa*	*State 1, uncracked*		*State 2, cracked*		*Interpolation between states 1 and 2*	
Section	M_{max} (N)	ξ	$M_{service}$ (N)	$\psi_1(t_0) \times 10^{-6}$ (m^{-1})	$\psi_1(t) \times 10^{-6}$ (m^{-1})	$\psi_2(t_0) \times 10^{-6}$ (m^{-1})	$\psi_2(t) \times 10^{-6}$ (m^{-1})	$\psi(t_0) \times 10^{-6}$ (m^{-1})	$\psi(t) \times 10^{-6}$ (m^{-1})
A	0	Uncracked	0	0	136	0	136	0	136
1	7233	Uncracked	6106	340	1090	340	1090	340	1090
2	11210	Uncracked	9463	528	1620	528	1620	528	1620
3	12982	0.52	10959	611	1850	4430	7740	2610	4940
4	13544	0.56	11433	637	1930	4620	7990	2880	5340
B	13638	0.57	11513	642	1940	4650	8030	2920	5400

Table 6.3 Calculation of deflection at middle of free edge (Point B, Example 6.2)

(1)	(2)	(3)	(4)	(5)	(6)	(7)	(8)	(9)	(10)
Segment	l (m)	M_{ua} (m)	M_{ub} (m)	$\psi_a(t_0)$ (10^{-6} m^{-1})	$\psi_b(t_0)$ (10^{-6} m^{-1})	$\int_{segment} M_u\ \psi(t_0)\ dl$ (10^{-3} m)	$\psi_a(t)$ (10^{-6} m^{-1})	$\psi_b(t)$ (10^{-6} m^{-1})	$\int_{segment} M_u\ \psi(t)\ dl$ (10^{-3} m)
A-1	0.75	0	0.375	0	340	0.03	136	1090	0.11
1-2	0.75	0.375	0.750	340	528	0.19	1090	1620	0.59
2-3	0.75	0.750	1.125	528	2610	1.15	1620	4940	2.38
3-4	0.75	1.125	1.500	2610	2880	2.71	4940	5340	5.07
4-B	0.75	1.500	1.875	2880	2920	3.67	5340	5400	6.80
	$v_A(t_0)=0;\ (v_B - v_A)\ (t_0) = v_B\ (t_0) = 2\sum \int_{segment} M_u\ \psi(t_0)\ dl =$					15.5	$v_A(t)=0;\ v_B\ (t)=2\sum \int_{segment} M_u\ \psi(t)\ dl =$		29.9

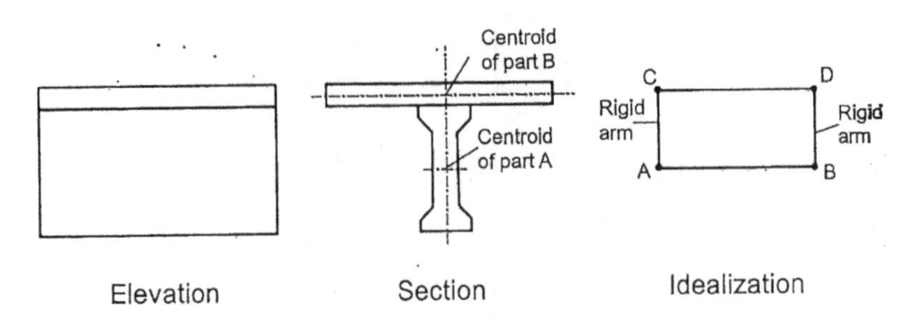

Figure 6.7 Idealization of composite members as two bars of different material properties.

erected with or without the use of temporary supports, are made continuous with cast-in-situ joints or with post-tensioned tendons. In all these cases, the time-dependent analysis can be done by superposition of the results of linear computer programs.

6.4.1 Assumptions and limitations

Any member is made of homogeneous material. The longitudinal axis of a member runs through a cross-sectional centroid. Because the cross section is considered homogeneous, no transformed cross-sectional properties are required; the location of the centroids of cross sections does not vary with time. Ignoring the presence of the reinforcement reduces the rigidity and generally increases the absolute value of the displacements induced by an externally applied load.

Figure 6.7 shows idealization of a concrete member whose cross section is composed of precast web and cast-in-situ flange. The two parts are treated as homogeneous members connected by rigid arms joining their centroidal axes.

6.4.2 Two computer runs

Consider a structure subjected at time t_0 to gravity load sustained up to time t; prestressing is also introduced at t_0. The structure is idealized as rigidly connected short members. At any time, the strain is assumed to vary linearly over the length of the members. The loading is concentrated at the nodes. It is required to determine the long-term displacements and stresses at time t, considering the effects of creep and shrinkage of concrete and relaxation of prestressed reinforcement during the period $(t-t_0)$. Two computer runs are required:

Computer run 1: The idealized structure is analyzed for the immediate strains and stresses at time t_0. The elasticity modulus of concrete members is $E_c(t_0)$. The cross-sectional area of the concrete member is equal to the gross concrete section less A_{ns} less the area of the prestressing duct(s). The gravity load is idealized as concentrated forces at the nodes. The prestressing is presented by restraining force, $A_{r\ \text{prestress}} = -P$ at node 1 and $+P$ at node 2; where P = prestressing absolute load value. In the displacement method of analysis, the member end forces are calculated as:

$$\{A\} = \{A_r\} + [A_u]\{D^*\} \tag{6.28}$$

where $\{D^*\}$ = vector of displacements at the two member ends with respect to its local directions; these local displacements can be calculated by transformation of the global displacements (Section 19.7); $[A_u]$ = member end forces due to separate unit values of the displacements D_1^*, D_2^*, ...; thus, $[A_u]$ represents the member stiffness matrix $\left[S^*\right]$. Computer run 1 gives the instantaneous displacements, $\{D(t_0)\}$, and the member end forces, $\{A(t_0)\}$. Setting $\left[A_u\right]=\left[S^*\right]$ in Equation 6.28 and defining the last term on the right-hand side as:

$$\left\{A_D^*\left(t_0\right)\right\}=\left[S^*\right]\left\{D^*\right\}_{\text{Run 1}}=\left\{A\left(t_0\right)\right\}_{\text{Run 1}}-\left\{A_r\left(t_0\right)\right\}_{\text{Run 1}} \tag{6.29}$$

where $\left[S^*\right]$ = stiffness matrix of the member with respect to its local coordinates (Figure 19.2); $\left\{D^*\right\}_{\text{Run 1}}$ = nodal displacement in local coordinates; $\left\{A\left(t_0\right)\right\}_{\text{Run 1}}$ = member end forces resulting from Computer run 1; $\left\{A_r\left(t_0\right)\right\}_{\text{Run 1}}$ = restraining end forces, used as input in Computer run 1. The force vector $\left\{A_D^*\left(t_0\right)\right\}$ represents the self-equilibrating forces induced by introduction of the displacements $\left\{D^*\right\}_{\text{Run 1}}$ at the member ends.

Computer run 2: In this run, the structure is idealized with the age-adjusted modulus of elasticity of concrete, $\bar{E}_c\left(t,t_0\right)$ (Equation 6.3). The vector of fixed-end forces $\left\{A_r\left(t,t_0\right)\right\}$ for any member comprises a set of forces in equilibrium. Calculation of the elements of the vector $\left\{A_r\left(t,t_0\right)\right\}$ is discussed below, considering the separate effect of each of creep, shrinkage, and relaxation.

Fixed-end forces due to creep are:

$$\left\{A_r^*\left(t,t_0\right)\right\}_{\text{creep}}=-\frac{\bar{E}_c\left(t,t_0\right)}{E_c\left(t_0\right)}\left\{A_D^*\left(t_0\right)\right\}\phi\left(t,t_0\right) \tag{6.30}$$

Fixed-end forces due to shrinkage are:

$$\left\{A_{r\,\text{axial}}^*\left(t,t_0\right)\right\}_{\text{shrinkage}}=-\bar{E}_c\left(t,t_0\right)A_c^*\,\varepsilon_{cs}\left(t,t_0\right)\begin{cases}+1 & \text{at first node}\\ -1 & \text{at second node}\end{cases} \tag{6.31}$$

Fixed-end forces due to reduced relaxation are:

$$\left\{A_{r\,\text{axial}}^*\left(t,t_0\right)\right\}_{\text{relaxation}}=A_{ps}^*\,\Delta\bar{\sigma}_{pr}\left(t,t_0\right)\begin{cases}-1 & \text{at first node}\\ +1 & \text{at second node}\end{cases} \tag{6.32}$$

6.5 MULTI-STAGE CONSTRUCTION

Consider a framed structure composed of members cast, prestressed, or loaded in stages; each of these is treated as an event occurring at a specific instant. Introduction or removal of a support is considered an event. The subscript 0 refers to the effect of a typical event occurring at the instant t_0; the analysis is for the changes in strain, stress, displacement, and internal forces in a typical period $(t–t_0)$. Two computer runs for each event are superimposed to give the final results.

EXAMPLE 6.3: PROPPED CANTILEVER

The cantilever AB in Figure 6.8a is subjected at time t_0 to a uniform load q. At time t_1, a simple support is introduced at B, thus, preventing the increase in deflection at B due to creep. Determine the change in the end forces at A and B between time t_1 and a later time t_2. Given data: $\phi(t_1, t_0) = 0.9$; $\phi(t_2, t_0) = 2.6$; $\phi(t_2, t_1) = 2.45$; $\chi(t_2, t_1) = 0.8$. Ignore the difference between $E_c(t_0)$ and $E_c(t_1)$.

The computer program PLANEF is used here, but the results can be checked by hand computation. Table 6.4 shows the input data and the results of Computer run 1, analyzing the immediate effect of the load introduced at time t_0. Each of $E_c(t_0)$, q and l are considered equal to unity; the support conditions are those of a cantilever encastré at A and free at B (Fig. 6.8a); the end forces for a totally fixed beam subjected to uniform load are entered as the load input:

$$\{A_r(t_0)\} = \{0, -0.5ql, -0.0833ql^2, 0, -0.5ql, 0.0833ql^2\}$$

The result of this computer run includes the member end forces immediately after load application:

$$\{A(t_0)\} = \{0, -ql, -0.5ql^2, 0, 0, 0\}$$

As expected, these are the forces at the ends of a cantilever. Apply Equation 6.29 to obtain:

$$\{A_D(t_0)\} = \{0, -0.5ql, -0.4167ql^2, 0, 0.5ql, -0.0833ql^2\}$$

These are the changes in end forces produced by varying the nodal displacements from null, when the nodal displacements are prevented, to the values $\{D^*\}$ included in the results of Computer run 1. Creep freely increases these displacements in the period t_0 to t_1. The hypothetical end forces that can prevent further increase in the period t_1 to t_2 are (Equation 6.30):

$$\{A_r(t_2,t_1)\} = -\frac{\bar{E}_c(t_2,t_1)}{E_c(t_0)}\{A_D(t_0)\}\left[\phi(t_2,t_0) - \phi(t_1,t_0)\right]$$

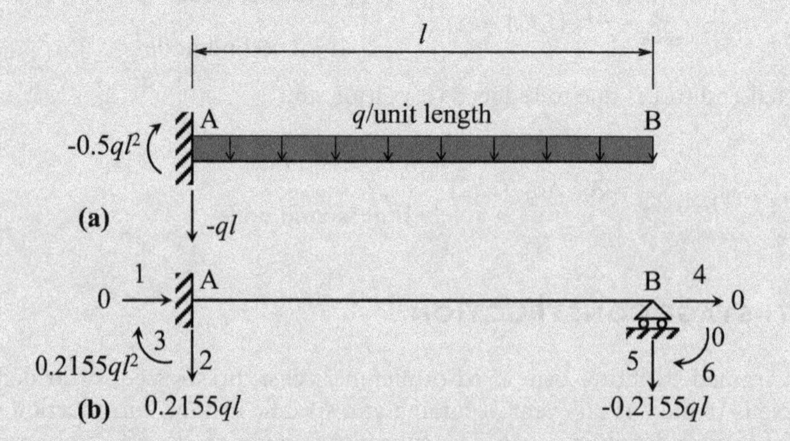

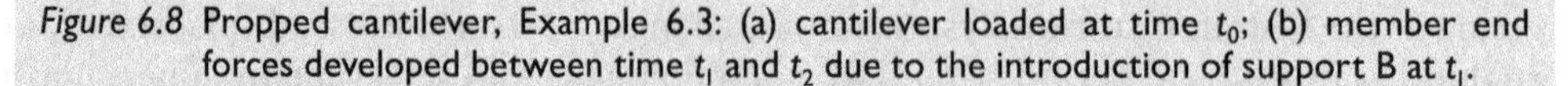
Figure 6.8 Propped cantilever, Example 6.3: (a) cantilever loaded at time t_0; (b) member end forces developed between time t_1 and t_2 due to the introduction of support B at t_1.

Table 6.4 Input and results of Computer run 1 with program PLANEF, Example 6.3

Number of joints = 2   Number of members = 1

Number of joints with prescribed displacement(s) = 2

Number of load cases = 1   Elasticity modulus = 1.0   Poisson's ratio = 0.2

Nodal coordinates

Node	x	y
1	0.0	0.0
2	1.0	0.0

Element information

Element	1st node	2nd node	a	I	a_r
1	1	2	1.0	1.0	1.0E+06

Support conditions

Node	Restraint indicators			Prescribed displacements		
	u	v	θ	u	v	θ
1	0	0	0	0.0	0.0	0.0
2	1	1	1	0.0	0.0	0.0

Forces applied at the nodes

Load case	Node	F_1	F_2	F_3
10	0	0.0	0.0	0.0

Member end forces with nodal displacement restrained

Ld. Case	Member	A_{r1}	A_{r2}	A_{r3}	A_{r4}	A_{r5}	A_{r6}
1	1	0.0	−0.5	−0.0833	0.0	−0.5	0.0833
10	10	0.0	0.0	0.0	0.0	0.0	0.0

(*Continued*)

Table 6.4 (Continued) Input and results of Computer run 1 with program PLANEF, Example 6.3

Analysis results						
Load case no. 1						
Nodal displacements						
Node	u	v	θ			
1	0.0	0.0	0.0			
2	0.0	0.0	0.1667			
Forces at the supported nodes						
Node	F_x	F_y	M_z			
1	0.0	−1.0	−0.5			
2	0.0	0.0	0.0			
Member end forces						
Memb.	F_1	F_2	F_3	F_4	F_5	F_6
1	0.0	−1.0	−0.5	0.0	0.0	0.0

The age-adjusted elasticity modulus is (Equation 6.3):

$$\bar{E}_c(t_2,t_1)=\frac{E_c(t_1)}{1+\chi\phi(t_2,t_1)}=\frac{E_c(t_1)}{1+0.8(2.45)}=0.3378\,E_c(t_1)$$

Substitution in Equation 6.30 gives a set of self-equilibrating end forces to be used as load input data in Computer run 2:

$$\{A_r(t_2,t_1)\}=-0.3378\,(2.6-0.9)\,\{A_D(t_0)\}$$

$$=\{0,\,0.2872\,ql,\,0.2393\,ql^2,\,0,\,-0.2872\,ql,\,0.0479\,ql^2\}$$

Table 6.5 includes the input data and the analysis results of Computer run 2. We note that the age-adjusted elasticity modulus is used and the support conditions are those of end encastré at A and simply supported at B. The required changes in member end forces between times t_1 and t_2 are a set of self-equilibrating forces (Fig. 6.8b), which are copied here:

$$\{A(t_2,t_1)\}=\{0,\,0.2155\,ql,\,0.2154\,ql^2,\,0,\,-0.2155\,ql,\,0\}$$

EXAMPLE 6.4: CANTILEVER CONSTRUCTION METHOD

The girder ABC (Figure 6.9a) is constructed as two separate cantilevers subjected at time t_0 to a uniform load q/unit length, representing the self-weight. At time t_0, the two cantilevers are made continuous at B by a cast-in-situ joint. Determine the changes in member end forces for AB between t_0 and a later time t. Use the same creep and aging coefficients as in Example 6.3 and ignore the difference between $E_c(t_0)$ and $E_c(t)$.

Because of symmetry, the computer analysis needs to be done for half the structure only (say, part AB). Computer run 1 and calculation of $\{A_r(t,t_0)\}$ – for use as load input in Computer run 2 – are the same as in Example 6.3 (Table 6.4).

Parts of the input and the results of Computer run 2 are presented in Figure 6.9, rather than in a table. Figure 6.9b shows $\{A_r(t,t_0)\}$; these are the forces that can artificially prevent the changes due to creep in the displacements at ends A and B of the cantilever AB. Figure 6.9c shows the results of Computer run 2; the computer applies the forces $\{A_r(t,t_0)\}$ in a reversed direction and determines the corresponding changes in member end forces and superposes them on $\{A_r(t,t_0)\}$. Figure 6.9d shows the sum of the forces in Figure 6.9c and 6.8a; this gives the forces on member AB at time t. The bending moment diagram at time t is shown in Figure 6.9e.

EXAMPLE 6.5: COMPOSITE SPACE TRUSS

Figure 6.10a depicts a cross section of a concrete floor slab supported by structural steel members. The structure is idealized as a space truss shown in pictorial view, elevation and top views in Figures 6.10b, c, and d. The truss has a span of 36.0 m; but for symmetry, half the span is analyzed. Consider that the half truss is subjected at time t_1 to downward forces: P at each of nodes 1, 2, 10, and 11 and $2P$ at each of nodes 4, 5, 7, and 8; where $P=40$ kN. Find the deflection at mid-span at time t_1 and the change in deflection, at the

Table 6.5 Input (abbreviated) and results of Computer run 2 using program PLANEF, Example 6.3

Number of joints = 2 Number of members = 1
Number of joints with prescribed displacement(s) = 2
Number of load cases = 1 Elasticity modulus = 0.3378 Poisson's ratio = 0.2

Nodal coordinates

Same as in Table 6.4.

Element information

Same as in Table 6.4.

Support conditions

Node	Restraint indicators			Prescribed displacements		
	u	v	θ	u	v	θ
1	0	0	0	0.0	0.0	0.0
2	1	0	1	0.0	0.0	0.0

Forces applied at the nodes

Same as in Table 6.4.

Member end forces with nodal displacement restrained

Ld. Case	Member	A_{r1}	A_{r2}	A_{r3}	A_{r4}	A_{r5}	A_{r6}
1	1	0.0	0.2872	0.2393	0.0	−0.2872	0.0479
10	10	0.0	0.0	0.0	0.0	0.0	0.0

Analysis results

Load case no. 1

Nodal displacements

Node	u	v	θ
1	0.0	0.0	0.0
2	0.0	0.0	−0.35451E-01

Forces at the supported nodes

Node	F_1	F_2	F_3
1	0.0	0.21535	0.21535
2	0.0	−0.21535	0.0

Member end forces

Member	F_1	F_2	F_3	F_4	F_5	F_6
1	0.0	0.21535	0.21535	0.0	−0.21535	0.0

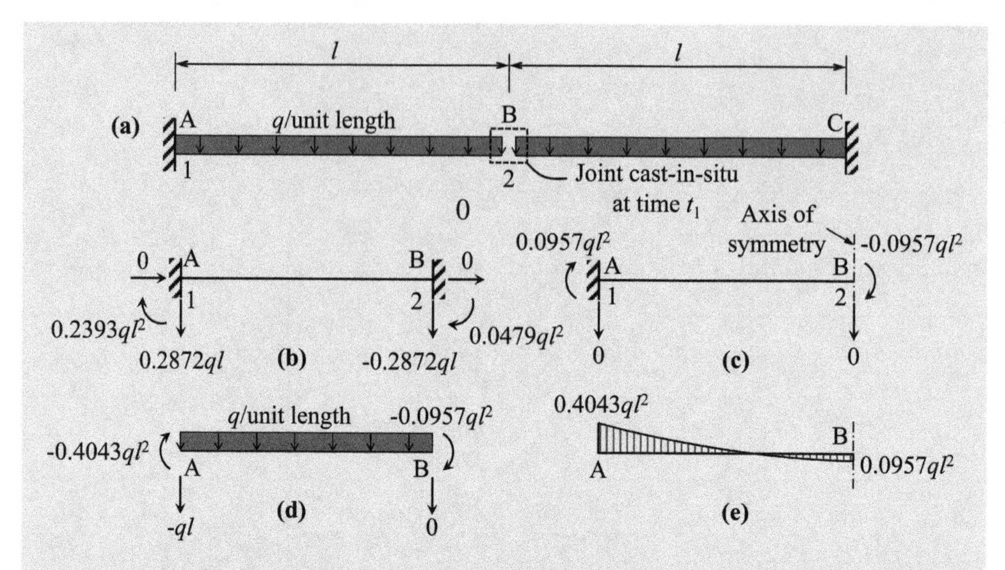

Figure 6.9 Cantilever construction, Example 6.4: (a) girder ABC constructed as two separate cantilevers subjected to uniform load at time t_0; (b) member end forces $\{A_r(t,t_0)\}$ calculated in Example 6.3; (c) changes in member end forces in Computer run 2; (d) superposition of member end forces in Figures 6.8a and 6.9c to give $\{A(t)\}$; (e) bending moment diagram at time t_2.

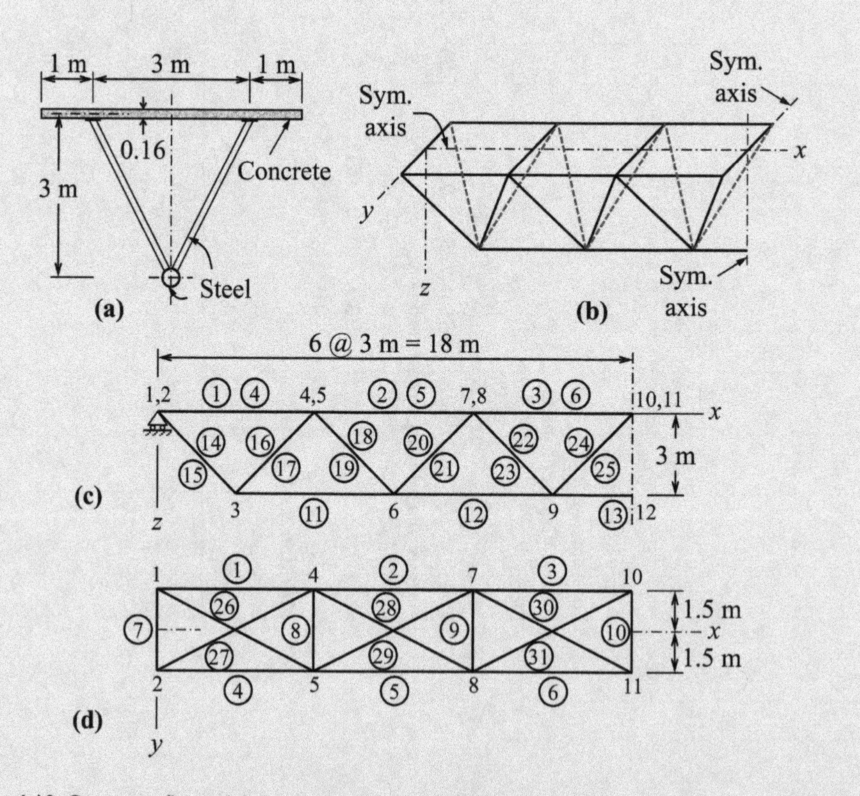

Figure 6.10 Concrete floor slab on structural steel members idealized as a space truss (Example 6.5): (a) cross section; (b) pictorial view with the diagonal members in the x-y plane omitted for clarity; (c) elevation; (d) top view.

same location, occurring between time t_1 and a later time t_2 due to creep and shrinkage of concrete. Given data: for concrete, $E_c(t_1)=25$ GPa; $\phi(t_2, t_1)=2.25$; $\chi(t_2, t_1)=0.8$; $\varepsilon_{cs}=-400 \times 10^{-6}$; for structural steel, $E_s=200$ GPa. The material for members 1 to 6 is concrete; all other members are structural steel. The cross-sectional areas of members are:

For each of members 1 to 6,	the cross-sectional area = 0.4 m^2
For each of members 11 to 13,	the cross-sectional area = 9100 mm^2
For each of members 14 to 25,	the cross-sectional area = 2300 mm^2
For the remaining members,	the cross-sectional area = 1200 mm^2

Light steel members running along lines 1–10 and 2–11 may be necessary during construction; these are ignored here.

The computer program SPACET (space truss, Chapter 20) is used in two runs. In Computer run 1, the modulus of elasticity is $E_c(t_1)=25$ GPa; a transformed cross-sectional area $= A_s\, E_s/E_c(t_1)$ is entered for the steel members of the truss. Table 6.6 shows the results, which include the deflection at mid-span (nodes 10 or 11) at time $t_1=55.8$ mm.

The age-adjusted elasticity modulus is (Equation 6.3):

$$\bar{E}_c(t_2,t_1)=\frac{25\text{ GPa}}{1+0.8(2.25)}=8.929\text{ GPa}$$

This modulus is used in Computer run 2 and a transformed cross-sectional area $= A_s\; E_s/\bar{E}_c$ is entered for the steel members. The load data are the two axial end forces $\{A_r(t_2,t_1)\}_{\text{creep}}$ calculated by Equation 6.28 for each of the concrete members (1 to 6):

$$\{A_r(t_2,t_1)\}_{\text{creep}}=-\frac{\bar{E}_c(t_2,t_1)}{E_c(t_1)}\{A_D(t_0)\}\,\phi(t_2,t_1)$$

The values of $\{A_D(t_0)\}$ are calculated by Equation 6.29 using the results of Computer run 1 and noting that $\{A_r(t_0)\}=\{0\}$ for all members. The artificial restraining forces are calculated below for member 1 as example:

$$\{A_D(t_0)\}_{\text{member 1}}=\{192.09,-192.09\}\text{ kN}$$

$$\{A_r(t_2,t_1)\}_{\text{creep, member 1}}=-\frac{8.929}{25.0}(2.25)\{192.09,-192.09\}$$

$$=\{-154.4,154.4\}\text{ kN}$$

The restraining forces for shrinkage are the same for any of the concrete members (1 to 6). Equation 6.31 gives:

$$\{A_r(t_2,t_1)\}_{\text{shrinkage, members 1 to 6}}=8.929\text{ GPa}\left(-400\times10^{-6}\right)0.4\{1,-1\}$$

$$=\{-1428.6,1428.6\}\text{ kN}$$

Table 6.7 gives abbreviated input and results of Computer run 2. Because this structure is statically determinate externally, creep and shrinkage do not affect the reactions (omitted in Table 6.7). The changes in displacements due to creep and shrinkage in the period t_1 to t_2 are given in Table 6.7, including the change of mid-span deflection of 27.5 mm (node 10 or 11). The changes in member end forces are given in Table 6.7 only for the members where the change is non-zero.

Table 6.6 Example 6.5, Space truss. Abbreviated results of Computer run 1: immediate displacements and forces at time t_1

Nodal displacements			
Node	u	v	w
1	0.83225E-3	0.60029E-3	0.44022E-8
2	0.83225E-3	−0.60029E-3	0.44022E-8
...			
10	0.41362E-9	−0.27428E-3	0.55758E-1
11	0.41362E-9	0.27428E-3	0.55758E-1
12	−0.23736E-8	0.00000E+0	0.00000E+0
Forces at the supported nodes			
Node	F_x	F_y	M_z
1	0.00000E+0	0.00000E+0	−0.24000E+6
2	0.00000E+0	0.00000E+0	−0.24000E+6
9	0.00000E+0	−0.45097E-9	0.00000E+0
10	−0.72000E+6	0.00000E+0	0.00000E+0
11	−0.72000E+6	0.00000E+0	0.00000E+0
12	0.14400E+7	0.00000E+0	0.00000E+0
Member end forces			
Member	F_1	F_2	
1	0.19209E+6	−0.19209E+6	
2	0.51887E+6	−0.51887E+6	
3	0.67612E+6	−0.67612E+6	
4	0.19209E+6	−0.19209E+6	
5	0.51887E+6	−0.51887E+6	
6	0.67612E+6	−0.67612E+6	
7	0.96046E+6	−0.96046E+6	
...			
31	0.43425E+4	−0.43425E+4	

6.6 GENERAL

Two procedures are presented in this chapter: the sectional method and linear computer analyses. The former procedure calculates the strain parameters at t_0 and in the period $(t\text{-}t_0)$ due to known internal forces (N, M). This procedure uses the cross-sectional properties of the concrete and the transformed sections; it accounts for the reinforcement bars and prestressing reinforcement locations and cracking. Computer program TDA performs the analysis and the deformations are calculated by the Unit Load Theory (Section 7.7).

Linear computer analyses/programs apply to homogeneous members, ignoring cracking. Displacements are calculated in two separate runs: the first gives results at time t_0, while the second calculates the values in the period $(t\text{-}t_0)$. Ignoring the presence of the reinforcement reduces the rigidity and increases the absolute value of the displacements.

Table 6.7 Example 6.5, space truss. Abbreviated input and results of Computer run 2. Analysis of changes in displacements and internal forces between time t_1 and t_2

Elasticity modulus = 0.89286E+10

Member end forces with nodal displacement restrained

Ld. Case	Member	A_{r1}	A_{r2}
1	1	−0.1544E+6	0.1544E+6
1	2	−0.4170E+6	0.4170E+6
1	3	−0.5433E+6	0.5433E+6
1	4	−0.1544E+6	0.1544E+6
1	5	−0.4170E+6	0.4170E+6
1	6	−0.5433E+6	0.5433E+6
1	1	−0.1429E+7	0.1429E+7
1	2	−0.1429E+7	0.1429E+7
1	3	−0.1429E+7	0.1429E+7
1	4	−0.1429E+7	0.1429E+7
1	5	−0.1429E+7	0.1429E+7
1	6	−0.1429E+7	0.1429E+7

Analysis results

Nodal displacements

Node	u	v	w
1	0.87131E-2	−0.20020E-3	0.28363E-22
2	0.87131E-2	0.20020E-3	−0.13222E-22
...			
10	0.29462E-8	−0.48412E-3	0.27301E-1
11	0.29462E-8	0.48412E-3	0.27543E-1
12	−0.28656E-23	0.00000E+0	0.00000E+0

Member end forces

Member	F_1^*	F_2^*
1	−0.64064E+5	0.64064E+5
2	−0.72448E+5	0.72448E+5
3	−0.77460E+5	0.77460E+5
4	−0.64064E+5	0.64064E+5
5	−0.72448E+5	0.72448E+5
6	−0.77460E+5	0.77460E+5
7	−0.32032E+5	0.32032E+5
8	−0.68256E+5	0.68256E+5
9	−0.74954E+5	0.74954E+5
10	−0.38730E+5	0.38730E+5
...		
26	0.71626E+5	−0.71626E+5
27	0.71626E+5	−0.71626E+5
28	0.80999E+5	−0.80999E+5
29	0.80999E+5	−0.80999E+5
30	0.86603E+5	−0.86603E+5
31	0.86603E+5	−0.86603E+5

REFERENCES

1. Fédération Internationale du Béton, *fib, CEB-FIP, Model Code for Concrete Structures*, Ernst & Sohn, Vol. 1, 2010, 317 pp.
2. ACI 209.2R-08, *Guide for Modeling and Calculation of Shrinkage and Creep in Hardened Concrete*, American Concrete Institute, Farmington Hills, MI, 2008, 48 pp.
3. Ghali, A., Favre, R. and Elbadry, M., *Concrete Structures: Stresses and Deformations – Analysis and Design for Serviceability*, 4th ed., CRC Press, Boca Raton, FL, 2012, 637 pp.
4. Magura, D., Sozen, M.A. and Siess, C.P., "A Study of Stress Relaxation in Prestressing Reinforcement," *PCI J.*, 9(2), (1964), pp. 13–57.

PROBLEMS

6.1 The line AB represents the centroidal axis of a concrete cantilever. At time t_1, the cantilever is subjected to its own weight, $q = 25$ kN/m and a prestressing force $P(t_1) = 200$ kN, introduced by the steel cable AC. Calculate the changes in deflection at the tip of the cantilever and in the force in the cable in the period t_1 to a later time t_2, caused by creep and shrinkage of concrete and relaxation of prestressed steel. Ignore cracking and presence of reinforcement in AB. Given data: for concrete, $E_c(t_1) = 25$ GPa; $\phi(t_2, t_1) = 2$; $\chi = 0.8$; $\varepsilon_{cs} = -300 \times 10^{-6}$; $\Delta\bar{\sigma}_{pr} = -50$ MPa. Cross-sectional area properties for AB: $A_c = 1.0$ m^2; $I = 0.1$ m^4. For the cable, $A_s = 250$ mm^2; $E_s = 200$ GPa.

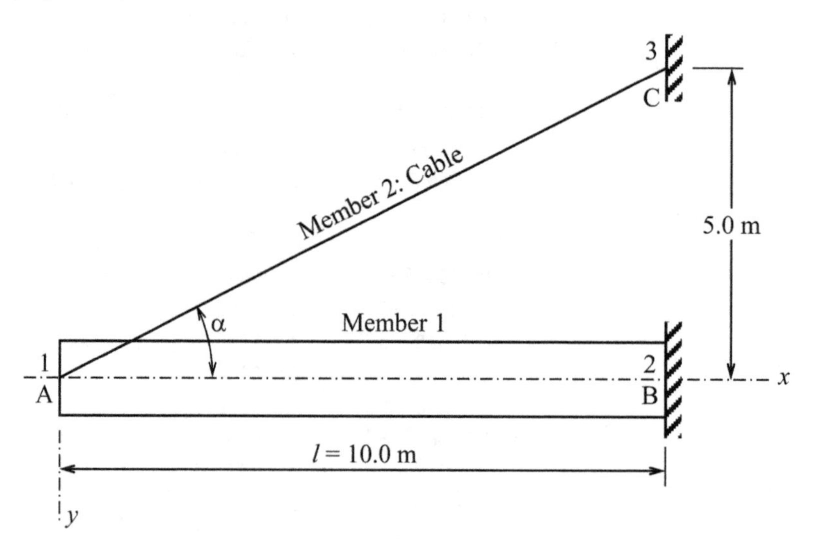

6.2 Find the mid-span deflection at time t and the immediate deflection in the beam of Example 6.1, using computer program PLANEF, treating the beam cross section as homogenous. Ignore A_{ns}, A_{ps}, and A_{duct} in calculating the cross-sectional area properties. Alternatively, use hand calculations instead of computer analysis.

6.3 Find the deflection at mid-span and member end moments at time t of the post-tensioned frame shown. At time t_0 prestressing force, P and gravity load q are applied on member BC; where $P = 2640$ kN and $q = 26$ kN/m. Member AC is non-prestressed. The prestressing tendon has parabolic profile and eccentricities $= -0.072$ and 0.528 m at B and C, respectively. Given data: $E_c(t_0) = 25$ GPa; $\phi(t,t_0) = 2$; $\chi(t,t_0) = 0.8$; $\varepsilon_{cs}(t,t_0) = -300 \times 10^{-6}$; $(A_{ps}\Delta\bar{\sigma}_{pr}) = -0.05\ P$. Use computer program PLANEF (two runs); assume that the cross sections of members are homogenous having the following area properties: Member AB, area = 0.16 m^2; second moment of area = 2.13 × 10^{-3} m^4. Member BC, area = 0.936 m^2; second moment of area = 55.86 × 10^{-3} m^4.

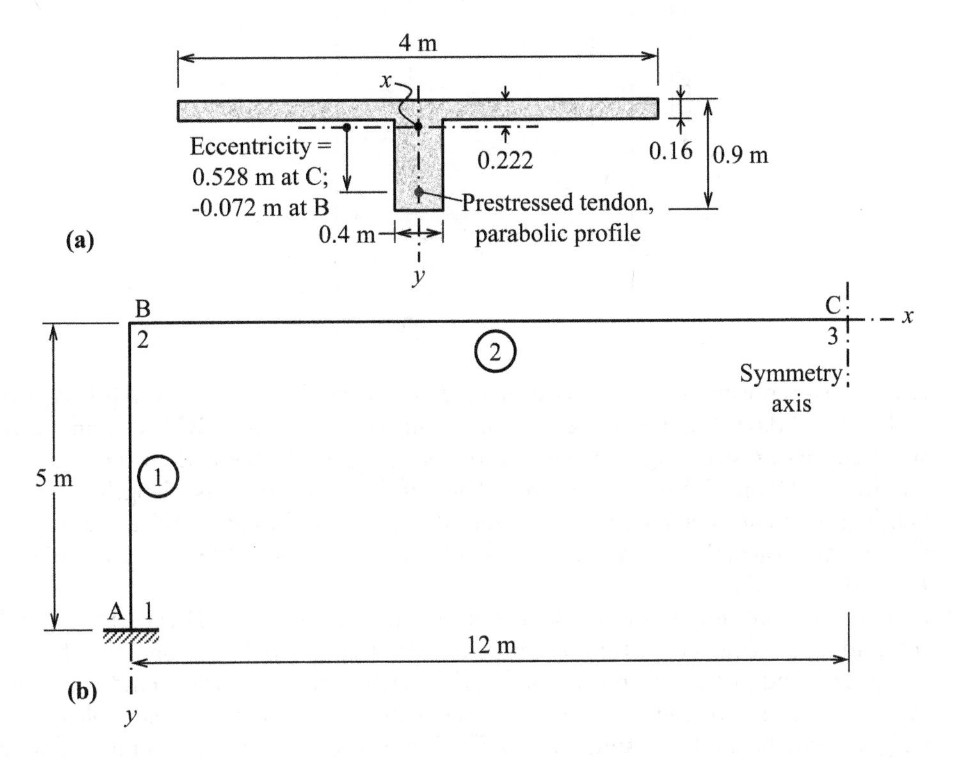

6.4 The figure shows a simply supported beam without prestressing, with: span, $l = 8$ m; $E_c(t_0) = 30$ GPa; $E_{ns} = 200$ GPa; $\phi(t,t_0) = 3.0$; $\chi(t,t_0) = 0.8$. The reference point O is chosen at the centroid of the transformed section at t_0; thus, $B(t_0) = 0$. Calculate the curvature and mid-span deflection at time t for the case where: $N = 0$, $M = 0$, and uniform shrinkage, $\varepsilon_{cs} = -400 \times 10^{-6}$.

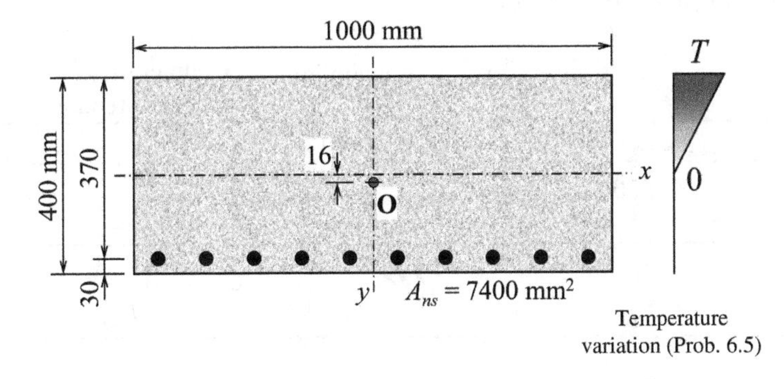

6.5 Consider the beam of Problem 6.4 with a rise of temperature with triangular distribution: $T = 30$°C at the top and zero at mid-height.

6.6 Calculate the stresses in steel and concrete at time t_0 and t due to $N = -3200$ kN at the center of the shown column cross section. Given data: $E_c(t_0) = 30$ GPa; $E_{ns} = 200$ GPa; $\phi(t,t_0) = 3.0$; $\chi(t,t_0) = 0.8$; $\varepsilon_{cs} = -400 \times 10^{-6}$.

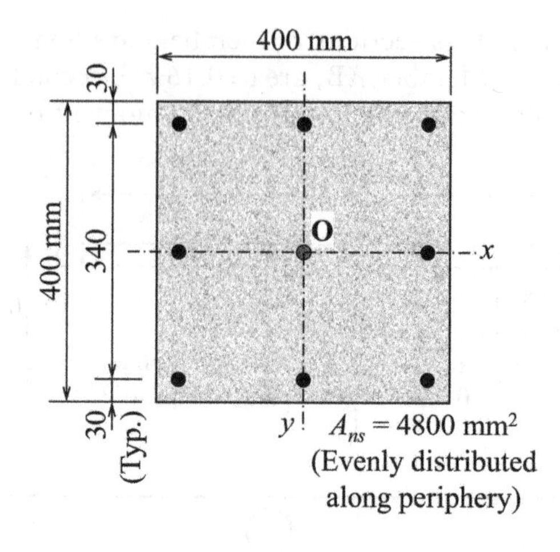

6.7 A two-span continuous prestressed beam ABC is cast in two stages: AB is cast first, and at age 7 days it is prestressed and its forms removed; span BC is cast in a second stage and its prestressing and removal of forms are performed when the ages of AB and BC are 60 and 7 days, respectively. Each of the two spans has a length $= l$. Find the bending moment diagram at time infinity due to the self-weight of the beam, q only. Given data: $\phi(\infty,7)=2.7$; $\chi(\infty,7)=0.74$; $\phi(60,7)=1.1$; $\phi(\infty,60)=2.3$; $\chi(\infty,60)=0.78$; $E_c(60)/E_c(7)=1.26$.

6.8 A two-span continuous beam ABC is built in two stages. Part AD is cast first, and its scaffolding is removed at time t_0 immediately after prestressing. Shortly after, part DC is cast and at time t_1, prestressed and its scaffolding removed. Find the bending moment diagram for the beam at a much later time t_2 due to prestressing plus the self-weight of the beam, q per unit length. The initial prestress creates an upward load of intensity $= 0.75q$ per unit length; the prestress loss (assumed to be the total amount in the period t_1 to t_2) $= 15\%$ of the initial value.

Assume that the time is measured from the day of casting of part AD and that the prestress for DC is applied at time t_1 when the age of DC is t_0. Given data: $t_0 = 7$ days; $t_1 = 60$ days; $t_2 = \infty$; $\phi(t_1,t_0) = 1.1$; $\chi(t_1,t_0) = 0.79$; $\phi(t_2,t_0) = 2.7$; $\chi(t_2,t_0) = 0.74$; $\phi(t_2,t_1) = 2.3$; $\chi(t_2,t_1) = 0.78$; $E_c(t_1)/E_c(t_0) = 1.26$.

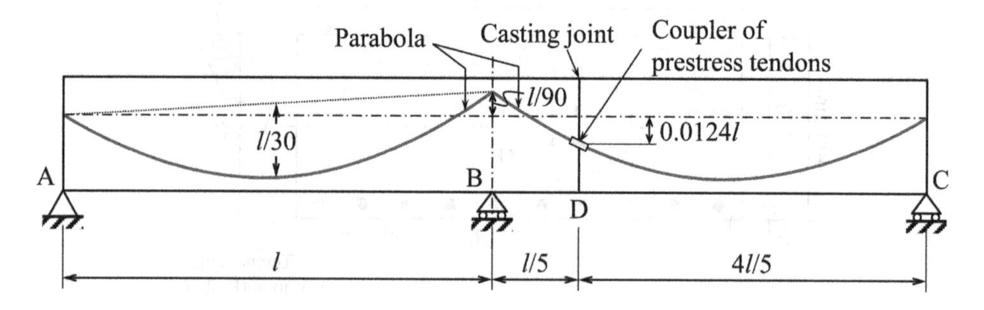

Chapter 7

Strain energy and virtual work

7.1 INTRODUCTION

We have already seen that the knowledge of the magnitude of displacements in a structure is necessary in the analysis of statically indeterminate structures, and in the preceding two chapters we used, for the purpose, either the displacements due to forces or the forces induced by imposed displacements. Displacements are, of course, also of interest in design; in fact, in some cases, the consideration of deflections under design loads may be the controlling factor in proportioning of members.

Calculation of displacements of structures made of materials obeying Hooke's law requires the knowledge of the modulus of elasticity in tension and compression (usually identical) E and of the shear modulus G. When the stress–strain relation is nonlinear, it is necessary to develop an expression relating forces and deformations, in terms of stress and strain, axial load and extension, or moment and curvature. In this chapter, we deal mainly with linear analysis. Virtual work is used to calculate displacement from strains.

7.2 STRAIN ENERGY

Work done by applied forces on an elastic structure is completely stored as strain energy, provided that no work is lost in the form of kinetic energy causing vibration of the structure, or of heat energy causing a rise in its temperature. In other words, the load must be applied gradually, and the stresses must not exceed the elastic limit of the material. When the structure is gradually unloaded, the internal energy is recovered, causing the structure to regain its original shape. Therefore, the external work W and the internal energy U are equal to one another:

$$W = U \tag{7.1}$$

This relation can be used to calculate deflections or forces, but we must first consider the method of calculating the internal strain energy.

Consider a small element of a linear elastic structure in the form of a prism of cross-sectional area $\mathrm{d}a$ and length $\mathrm{d}l$. The area $\mathrm{d}a$ can be subjected to either a normal stress σ (Figure 7.1a), or to a shear stress τ (Figure 7.1b). Assume that the left-hand end B of the element is fixed while the right-hand end C is free. The displacement of C under the two types of stress is then:

$$\Delta_1 = \frac{\sigma}{E}\mathrm{d}l \quad \text{and} \quad \Delta_2 = \frac{\tau}{G}\mathrm{d}l$$

DOI: 10.1201/9780429286858-7

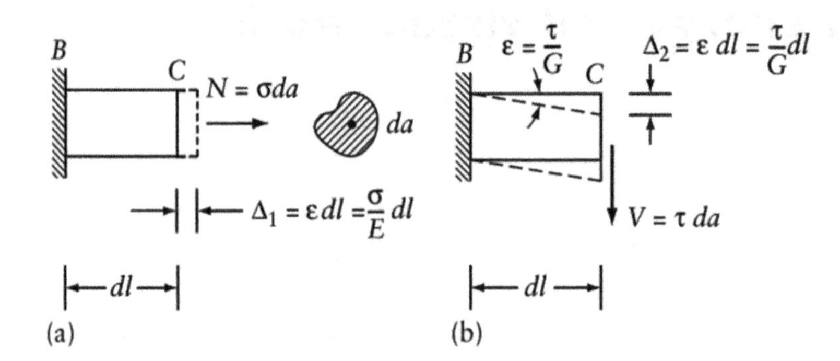

Figure 7.1 Deformation of an element due to (a) normal stress, and (b) shearing stress.

where E = modulus of elasticity in tension or compression; G = modulus of elasticity in shear. When the forces σ da and τ da, which cause the above displacements, are applied gradually, the energy stored in the two elements is:

$$dU_1 = \frac{1}{2}(\sigma da)\Delta_1 = \frac{1}{2}\frac{\sigma^2}{E} dl\, da$$

$$dU_2 = \frac{1}{2}(\tau da)\Delta_2 = \frac{1}{2}\frac{\tau^2}{G} dl\, da$$

Using ε as a general symbol for strain, the above equations can be put in the general form:

$$dU = \frac{1}{2}\sigma\varepsilon dv \tag{7.2}$$

where dv = dl, da = volume of the element, and σ represents a generalized stress, that is, either a normal or a shearing stress.

The strain ε in Equation 7.2 is either due to a normal stress and has magnitude $\varepsilon = \sigma/E$, or due to a shearing stress, in which case $\varepsilon = \tau/G$. But G and E are related by:

$$G = \frac{E}{2(1+\nu)} \tag{7.3}$$

where ν = Poisson's ratio, so that the strain due to the shearing stress can be expressed as:

$$\varepsilon = 2(\tau / E)(1+\nu)$$

The increase in strain energy in any elastic element of volume dv due to a change in strain from $\varepsilon = 0$ to $\varepsilon = \varepsilon_f$ is:

$$dU = dv \int_0^{\varepsilon_f} \sigma d\varepsilon \tag{7.4}$$

where the integral $\int_0^{\varepsilon_f} \sigma d\varepsilon$ is called the *strain energy density* and is equal to the area under the stress–strain curve for the material (Figure 7.2a). If the material obeys Hooke's law, the stress–strain curve is a straight line (Figure 7.2b), and the strain energy density is (1/2) $\sigma_f\, \varepsilon_f$.

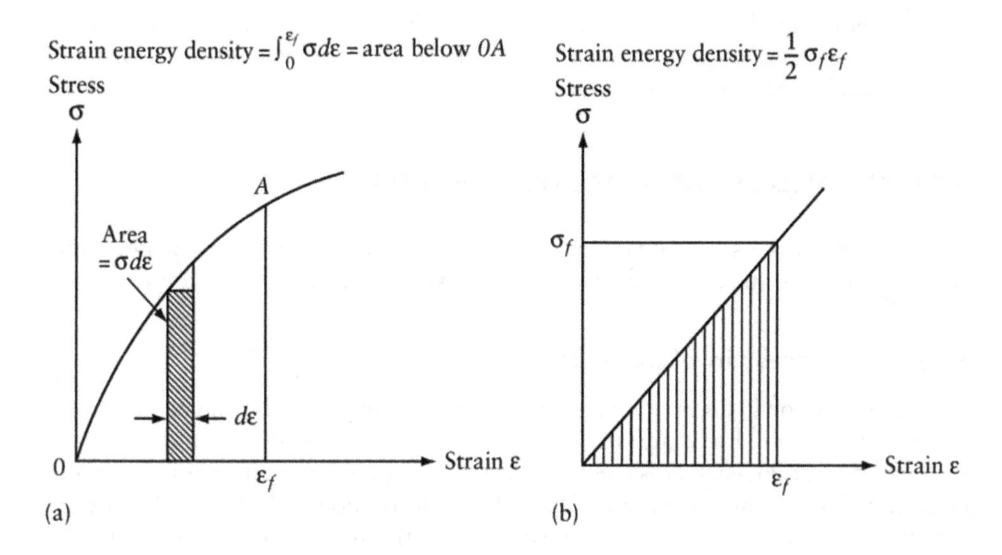

Figure 7.2 Stress–strain relations: (a) nonlinear, and (b) linear.

Any structure can be considered to consist of small elements of the type shown in Figure 7.3 subjected to normal stresses σ_x, σ_y, σ_z and to shearing stresses τ_{xy}, τ_{xz}, τ_{yz}, with resulting strains ε_x, ε_y, ε_z, γ_{xy}, γ_{xz}, and γ_{yz}, where the subscripts x, y, and z refer to rectangular Cartesian coordinate axes. The total strain energy in a linear structure is then:

$$U = \frac{1}{2}\sum_{m=1}^{6}\int_{v} \sigma_m \varepsilon_m \, dv \tag{7.5}$$

where m refers to the type of stress and to the corresponding strain. This means that the integration has to be carried out over the volume of the structure for each type of stress separately.

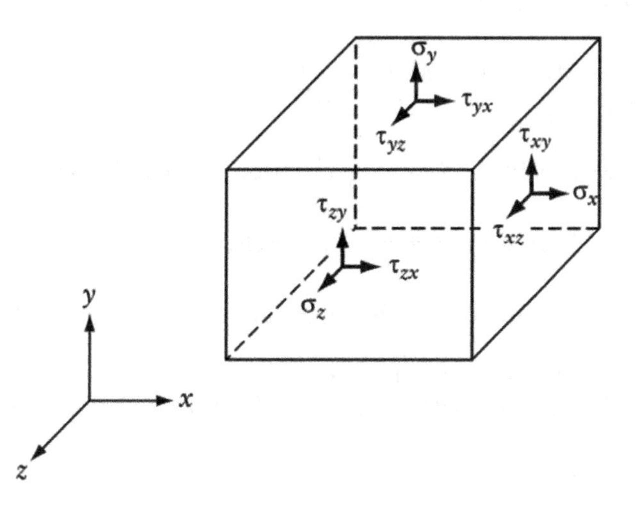

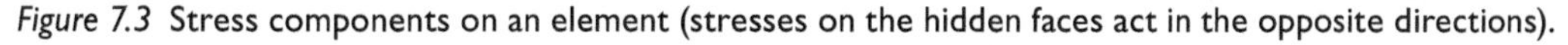

Figure 7.3 Stress components on an element (stresses on the hidden faces act in the opposite directions).

Gauss numerical method is commonly used in evaluating the integrals (see Chapter 13). Frame structures, composed of beam elements are considered in Sections 7.3.1 to 7.3.4. Other elastic bodies are considered in Section 7.4.

7.3 NORMAL AND SHEAR STRESSES IN BEAMS

The strain energy U is a scalar quantity of units of force × length. U can be determined as the sum of the energy due to normal and shear stresses, determined separately.

7.3.1 Normal stress in plane sections

The beam cross section in Figure 7.4a is subjected to a normal force, N, and bending moments M_x and M_y. It is required to determine the normal stress variation over the section. In the deformed configuration, each cross section, originally a plane, remains plane and normal to the longitudinal axis. This assumption is attributed to Bernoulli (1700–1782). Considering linearly elastic material with modulus of elasticity E, the strain ε in the longitudinal direction and the stress σ at any point (x, y) can be expressed as:

$$\varepsilon = \varepsilon_O + \psi_x y + \psi_y x; \quad \sigma = \sigma_O + E\psi_x y + E\psi_y x \tag{7.6}$$

where ε_O and σ_O $(= E\,\varepsilon_O)$ = strain and stress at the centroid O; $\psi_x = \partial\varepsilon/\partial y$ = the curvature about x-axis; $\psi_y = \partial\varepsilon/\partial x$ = the curvature about y-axis; $E\psi_x = E\partial\varepsilon/\partial y$; $E\psi_y = E\partial\varepsilon/\partial x$. The stress resultants are:

$$N = \int \sigma\, da; \quad M_x = \int \sigma y\, da; \quad M_y = \int \sigma x\, da \tag{7.7}$$

Substitution of Equation 7.6 into 7.7 and solution of the resulting equations give the parameters σ_O, $E\psi_x$, and $E\psi_y$:

$$\sigma_O = \frac{N}{a}; \quad E\psi_x = \frac{M_x I_y - M_y I_{xy}}{I_x I_y - I_{xy}^2}; \quad E\psi_y = \frac{M_y I_x - M_x I_{xy}}{I_x I_y - I_{xy}^2} \tag{7.8}$$

Thus, the stress at any point is:

$$\sigma = \frac{N}{a} + \left(\frac{M_x I_y - M_y I_{xy}}{I_x I_y - I_{xy}^2}\right) y + \left(\frac{M_y I_x - M_x I_{xy}}{I_x I_y - I_{xy}^2}\right) x \tag{7.9}$$

where $a = \int da$ = area of section; I_x and $I_y = \int y^2\, da$, and $\int x^2\, da$ = second moments of area (moments of inertia) about x and y axis, respectively; $I_{xy} = \int xy\, da$ = product of inertia. The derivation of Equation 7.9 involves the first moments of area $\int y\, da$ and $\int x\, da$ about x and y axis, respectively. However, both integrals vanish because x and y axes pass through the centroid O. When x and y are centroidal principal axes, I_{xy} also vanishes and Equation 7.9 becomes:

$$\sigma = \frac{N}{a} + \frac{M_x}{I_x} y + \frac{M_y}{I_y} x \tag{7.10}$$

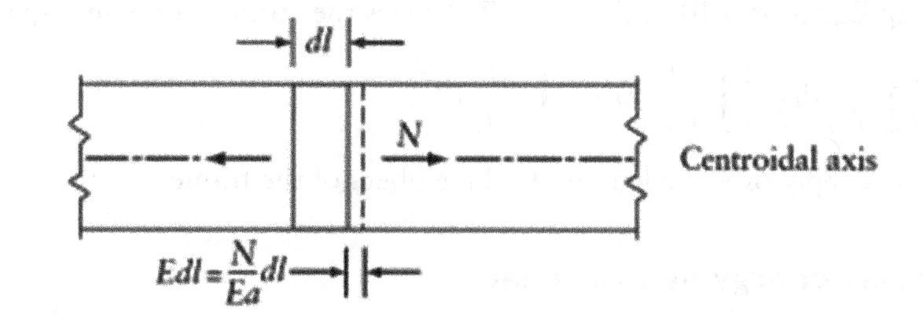

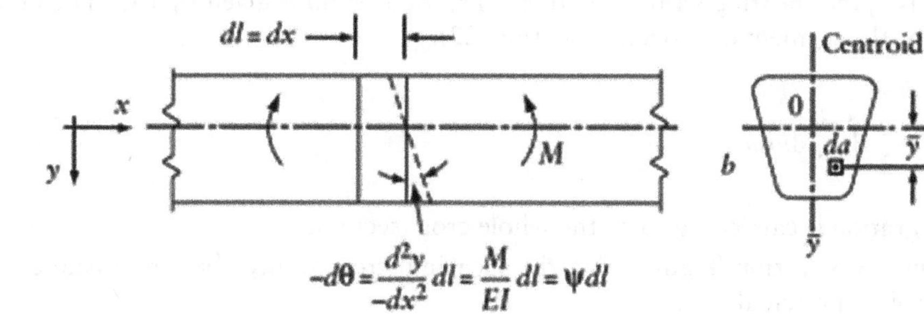

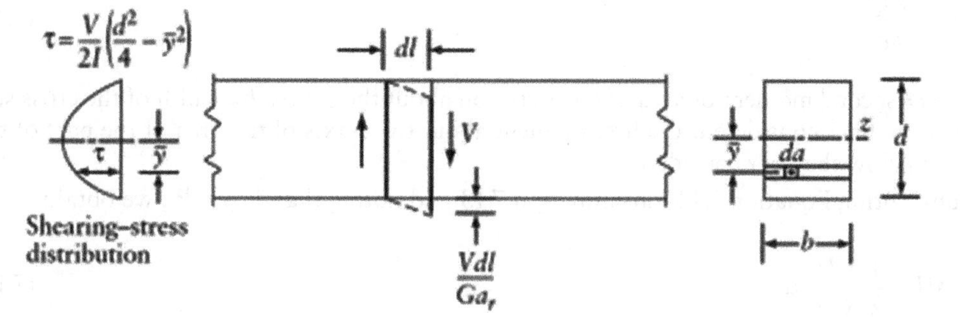

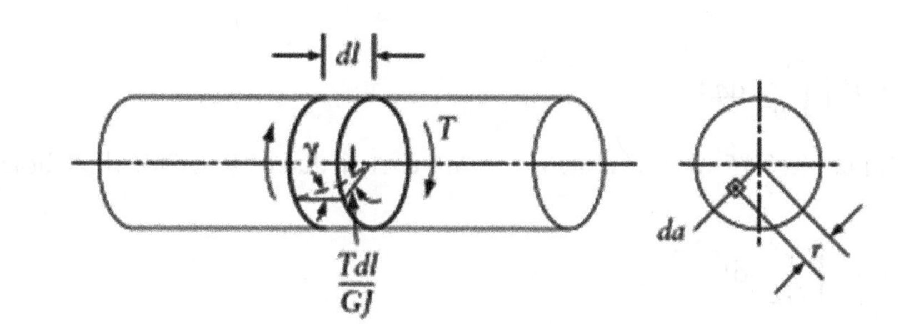

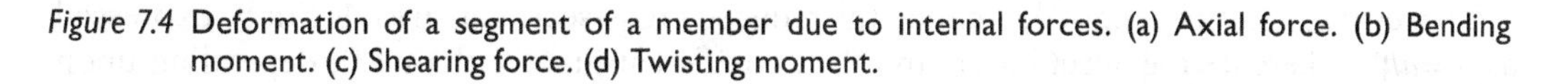

Figure 7.4 Deformation of a segment of a member due to internal forces. (a) Axial force. (b) Bending moment. (c) Shearing force. (d) Twisting moment.

When the section is symmetrical about axis x or y, $I_{xy} = 0$, and the two axes are principal. When x and y are centroidal non-principal axes, σ can be calculated by Equation 7.9.

The strain energy due to normal stress (Equation 7.2):

$$U = \frac{1}{2}\int \frac{\sigma^2}{E} dv = \frac{1}{2}\int \frac{\sigma^2}{E} da\, dl \tag{7.11}$$

Substituting Equation 7.10 in Equation 7.11 gives the strain energy in a space frame:

$$U = \frac{1}{2}\int \frac{N^2}{Ea} dl + \frac{1}{2}\int \frac{M_x^2}{EI_x} dl + \frac{1}{2}\int \frac{M_y^2}{EI_y} dl \tag{7.12}$$

The integrals apply over the length of all members of the frame.

7.3.2 Strain energy due to shear

Consider the segment dl of Figure 7.4c, subjected to a shearing force V. If the shearing stress induced is τ, the shearing strain ε in an element of area da is given by τ/G. Then the strain energy for the segment dl is (from Equation 7.2):

$$\Delta U = \frac{1}{2}\int \frac{\tau^2}{G} dl\, da \tag{7.13}$$

The integration is carried out over the whole cross section.

For any cross section (Figure 7.4c), the shearing stress at any fiber at a distance $\bar{y}$ below the centroidal principal z-axis is:

$$\tau = \frac{VQ}{Ib} \tag{7.14}$$

where I = second moment of area of cross section about the z-axis; b = width of the cross section at the fiber considered; Q = first moment about the z-axis of the area of the part of the section below the fiber considered.

Substituting Equation 7.14 into Equation 7.13 and noting that $da = b\, d\bar{y}$, we obtain:

$$\Delta U = \frac{1}{2}\frac{V^2}{Ga_r} dl \tag{7.15}$$

where

$$a_r = I^2\left(\int \frac{Q^2}{b^2} da\right)^{-1} \tag{7.16}$$

From Equation 7.15 we can write the strain energy due to shear for the whole structure as

$$U = \frac{1}{2}\int \frac{V^2}{Ga_r} dl \tag{7.17}$$

where the integration is carried out over the entire length of each member of the structure.

The term a_r in units (length)^2, is called the reduced cross-sectional area. It can be expressed as $a_r = a/\beta$; where a = the actual area and β = a coefficient greater than 1.0, depending upon the geometrical shape of the cross section. Values of β are 1.2 and 10/9 for rectangular and circular cross sections, respectively. For a rolled steel I-section, $a_r \simeq$ area of the web. For a thin-walled hollow circular cross section, $a_r \simeq a/2$. For a thin-walled hollow rectangular section subjected to vertical shearing force, $a_r \simeq$ area of the two vertical sides of the section.

We may wish to verify the value of a_r for a rectangular section of width b and depth d (Figure 7.4c). The value of Q at any fiber is (the first moment of the area $b\,[(d/2) - \bar{y}]$ about z-axis):

$$Q = \frac{b}{2}\left(\frac{d^2}{4} - \bar{y}^2\right) \tag{7.18}$$

The shearing stress distribution is parabolic (Figure 7.4c). Substituting this equation in Equation 7.16, with $I = bd^3/12$ and $da = b\ d\bar{y}$, and evaluating the integral between $\bar{y} = -d/2$ and $d/2$, gives $a_r = 5bd/6 = a/1.2$.

7.3.3 Strain energy due to torsion

Figure 7.4d shows a segment dl of a circular bar subjected to a twisting moment T. The shearing stress at any point distance r from the center is $\tau = (Tr/J)$; where J = polar moment of inertia. The corresponding strain is $\varepsilon = (\tau/G)$. From Equation 7.2, the strain energy in the segment dl is:

$$\Delta U = \frac{1}{2}\int \frac{T^2 r^2}{GJ^2}\, da\, dl = \frac{1}{2}\frac{T^2\, dl}{GJ^2}\int r^2\, da$$

The integral $\int r^2\, da = J$ is the polar moment of inertia. Therefore, $\Delta U = (1/2)\left(T^2\, dl/(GJ)\right)$. Thus, for the whole structure the strain energy due to torsion is:

$$U = \frac{1}{2}\int \frac{T^2\, dl}{GJ} \tag{7.19}$$

This equation can be used for members with cross sections other than circular, but in this case, J is a torsion constant [in units $(\text{length})^4$] which depends on the shape of the cross section. Expressions for J for some structural sections are given in Appendix F.

7.3.4 Total strain energy

In a structure in which all the four types of internal forces shown in Figure 7.4 are present, the values of energy obtained by Equations 7.12, 7.17, and 7.19 are added to give the total strain energy:

$$U = \frac{1}{2}\int \frac{N^2\, dl}{Ea} + \frac{1}{2}\int \frac{M^2\, dl}{EI} + \frac{1}{2}\int \frac{V^2\, dl}{Ga_r} + \frac{1}{2}\int \frac{T^2\, dl}{GJ} \tag{7.20}$$

The integration is carried out along the whole length of each member of the structure. We should note that each integral involves a product of an internal force N, M, V, or T acting on a segment dl and of the relative displacement of the cross section at the two ends of the segment. These displacements are $Ndl/(Ea)$, $Mdl/(EI)$, $Vdl/(Ga_r)$, and $Tdl/(GJ)$ (see Figure 7.4).

7.4 BASIC EQUATIONS OF ELASTICITY

The stresses and strains in an elastic body are related by Hooke's law, which can be written in the generalized form:

$$\{\sigma\} = [d]\cdot\{\varepsilon\} \tag{7.21}$$

where $\{\sigma\}$ and $\{\varepsilon\}$ = generalized stress and strain vectors respectively; and $[d]$ = a square symmetrical matrix referred to as the *elasticity matrix*.

The strain components are defined as derivatives of the displacement component by the generalized equation:

$$\{\varepsilon\} = [\partial]\{f\} \tag{7.22}$$

where $[\partial]$ = a matrix of the differential operator; $\{f\}$ = a vector of functions describing the displacement field.

The symbols $\{\sigma\}$ and $\{\varepsilon\}$ will be used to represent stress or strain components in one-, two-, or three-dimensional bodies. The displacement field $\{f\}$ will have one, two, and three components: u, v, and w in the direction of orthogonal axes x, y, and z. The differential operator matrix $[\partial]$ will represent derivatives with respect to one, two, or three of the variables x, y, and z.

For a bar subjected to an axial force (Figure 7.4a), each of $\{\sigma\}$ and $\{\varepsilon\}$ has one component and $[d]$ has one element equal to E, the modulus of elasticity. The strain ε is equal to du/dx; where u = displacement along the beam axis and x = distance measured in the same direction. Thus, we can use Equations 7.21 and 7.22 for the uniaxial stress state, with the symbols having the following meanings:

$$\{\sigma\} \equiv \sigma; \quad \{\varepsilon\} \equiv \varepsilon; \quad [d] \equiv E; \quad \{f\} \equiv u; \quad \{\partial\} \equiv \mathrm{d}/\mathrm{d}x \tag{7.23}$$

We shall also use the symbol $\{\sigma\}$ to represent a vector of stress resultants. For the bar considered above, we can take $\{\sigma\} \equiv N$, the axial force on the bar cross section, and $[d] \equiv Ea$; where a = cross-sectional area. Again, Equations 7.21 and 7.22 apply, with the symbols having the following meanings:

$$\{\sigma\} \equiv N; \quad \{\varepsilon\} \equiv \varepsilon; \quad [d] \equiv Ea; \quad \{f\} \equiv u; \quad \{\partial\} \equiv \mathrm{d}/\mathrm{d}x \tag{7.24}$$

The generalized Equations 7.21 and 7.22 apply to a bar in bending (Figure 7.4b), with the symbols having the following meanings:

$$\{\sigma\} \equiv M; \quad \{\varepsilon\} \equiv \psi; \quad [d] \equiv EI; \quad \{f\} \equiv v; \quad \{\partial\} \equiv \mathrm{d}^2/\mathrm{d}x^2 \tag{7.25}$$

where M = bending moment; ψ = curvature; I = second moment of area about the centroidal axis; and v = displacement in the y-direction.

The product $\{\sigma\}^T \{\varepsilon\}$ integrated over the volume of an element appears in the strain energy Equation 7.4 and later in the virtual work Equation 7.44, both of which will be frequently used. When $\{\sigma\}$ represents stress resultants over a cross section of a bar, the integral over the volume has to be replaced by an integral over the length (see Equation 7.20). For plates in bending, we shall use $\{\sigma\}$ to represent bending and twisting moments $\{M_x, M_y, M_{xy}\}$ and the integral will be over the area. In the following subsections we shall apply the generalized Equations 7.21 and 7.22 in three stress states.

7.4.1 Plane stress and plane strain

Consider a plate subjected to in-plane forces (Figure 7.5). At any point, the stress, strain, and displacement components are:

$$\{\sigma\} = \{\sigma_x, \sigma_y, \tau_{xy}\}; \quad \{\varepsilon\} = \{\varepsilon_x, \varepsilon_y, \gamma_{xy}\}; \quad \{f\} = \{u, v\} \tag{7.26}$$

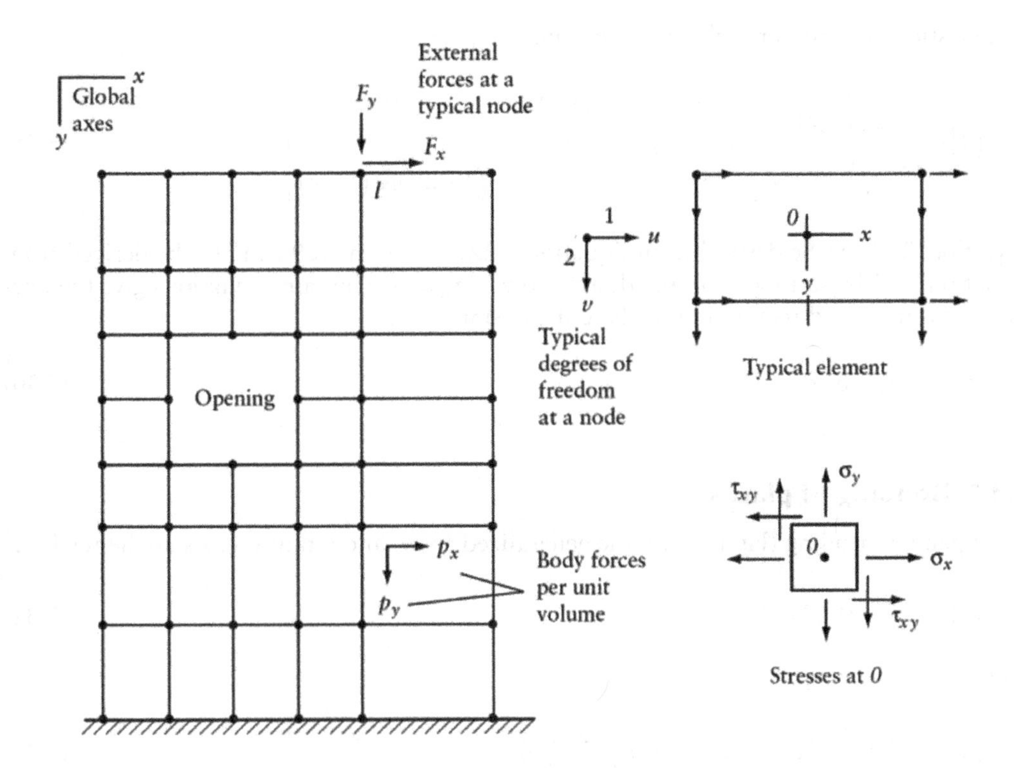

Figure 7.5 Example of a finite-element model for the analysis of stresses in a wall subjected to in-plane forces.

The strains are defined as derivatives of $\{f\}$ by the generalized Equation 7.22, with the differential operator matrix:

$$[\partial]=\begin{bmatrix}\partial/\partial x & 0\\ 0 & \partial/\partial y\\ \partial/\partial y & \partial/\partial x\end{bmatrix} \tag{7.27}$$

The stress and strain vectors are related by generalized Hooke's law (Equation 7.21) with the elasticity matrix $[d]$ given by Equation 7.28 or 7.29.

When strain in the z direction is free to occur, $\sigma_z = 0$ and we have the state of plane stress. Deep beams and shear walls are examples of structures in a state of plane stress. When strain in the z direction cannot occur, $\varepsilon_z = 0$ and we have the state of plane strain. The state of plane strain occurs in structures which have a constant cross section perpendicular to the z direction and also have the dimension in the z direction much larger than those in the x and y directions. Concrete gravity dams and earth embankments are examples of structures in this category. The analysis of these structures may be performed for a slice of unit thickness in a state of plane strain.

For an isotropic material, the elasticity matrix in a plane-stress state is:

$$[d]=\frac{E}{1-\nu^2}\begin{bmatrix}1 & \nu & 0\\ \nu & 1 & 0\\ 0 & 0 & (1-\nu)/2\end{bmatrix} \tag{7.28}$$

The elasticity matrix for a plane-strain state is

$$[d] = \frac{E(1-\nu)}{(1+\nu)(1-2\nu)} \begin{bmatrix} 1 & \nu/(1-\nu) & 0 \\ \nu/(1-\nu) & 1 & 0 \\ 0 & 0 & (1-2\nu)/2(1-\nu) \end{bmatrix} \tag{7.29}$$

Equation 7.28 can be derived from Equation 7.22. Equation 7.29 can also be derived from Equation 7.22 by setting $\varepsilon_z = 0$ (in addition to $\tau_{xz} = \tau_{yz} = 0$). The same equations give the normal stress in the z direction in the plane strain state:

$$\sigma_z = \nu(\sigma_x + \sigma_y) \tag{7.30}$$

7.4.2 Bending of plates

For a plate in bending (Figure 7.6), the generalized stress and strain vectors are defined as:

$$\{\sigma\} = \{M_x, M_y, M_{xy}\} \tag{7.31}$$

and

$$\{\varepsilon\} = \{-\partial^2 w/\partial x^2, -\partial^2 w/\partial y^2, 2\partial^2 w/(\partial x\, \partial y)\} \tag{7.32}$$

One component of body force and one component of displacement exist:

$$\{p\} \equiv \{q\} \tag{7.33}$$

$$\{f\} \equiv \{w\} \tag{7.34}$$

where q = force in the z direction per unit area and w = deflection in the same direction. The generalized Equations 7.21 and 7.22 apply to a plate in bending, with:

$$[\partial] = \begin{Bmatrix} -\partial^2/\partial x^2 \\ -\partial^2/\partial y^2 \\ 2\partial^2/(\partial x\, \partial y) \end{Bmatrix} \tag{7.35}$$

For an orthotropic plate in bending, the elasticity matrix is:

$$[d] = \frac{h^3}{12} \begin{bmatrix} E_x/(1-\nu_x \nu_y) & \nu_x E_y/(1-\nu_x \nu_y) & 0 \\ \nu_x E_y/(1-\nu_x \nu_y) & E_y/(1-\nu_x \nu_y) & 0 \\ 0 & 0 & G \end{bmatrix} \tag{7.36}$$

where E_x, E_y = moduli of elasticity in tension or in compression in the x and y directions; ν_x and ν_y = Poisson's ratios; and h = plate thickness. The shear modulus of elasticity:

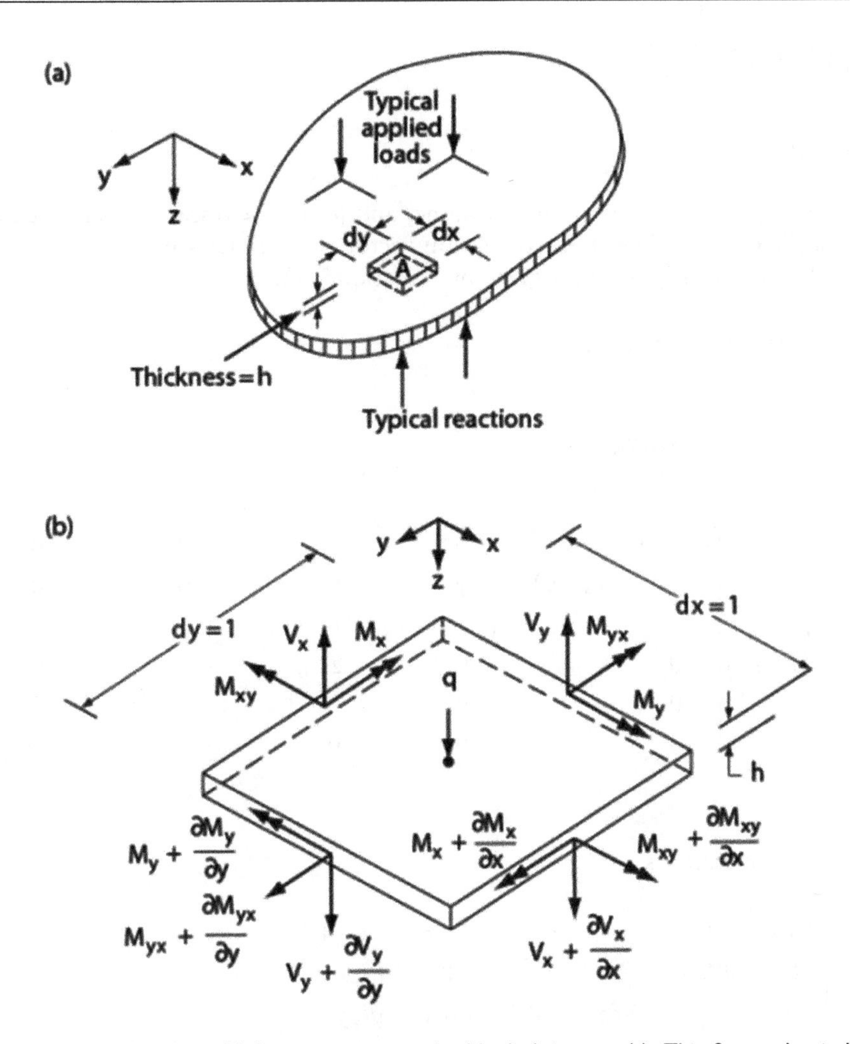

Figure 7.6 (a) Plate in bending. (b) Forces acting on the block A in part (a). This figure also indicates the positive directions of the stress resultants.

$$G = \sqrt{E_x E_y} \Big/ \left[2\left(1 + \sqrt{\nu_x \nu_y}\right) \right]$$

When the plate is isotropic, we set $E = E_x = E_y$ and $\nu = \nu_x = \nu_y$ in Equation 7.36:

$$[d] = \frac{Eh^3}{12\left(1-\nu^2\right)} \begin{bmatrix} 1 & \nu & 0 \\ \nu & 1 & 0 \\ 0 & 0 & (1-\nu)/2 \end{bmatrix} \tag{7.37}$$

7.4.3 Three-dimensional solid

For a three-dimensional body, the generalized Equations 7.21 and 7.22 apply again, with the vectors $\{p\}$ and $\{f\}$ having the following meaning:

$$\{p\} = \{p_x, p_y, p_z\} \tag{7.38}$$

$$\{f\} = \{u, v, w\} \tag{7.39}$$

where $\{p_x, p_y, p_z\}$ = body forces per unit volume, and $\{u, v, w\}$ = translations in the x, y, and z directions (Figure 7.3). The stress and strain vectors $\{\sigma\}$ and $\{\varepsilon\}$ are defined by Equations 7.40 and 7.41, and the $[d]$ matrix is given by Equation 7.42.

$$\{\sigma\} = \{\sigma_x, \sigma_y, \sigma_z, \tau_{xy}, \tau_{xz}, \tau_{yz}\} \tag{7.40}$$

$$\{\varepsilon\} = \{\varepsilon_x, \varepsilon_y, \varepsilon_z, \gamma_{xy}, \gamma_{xz}, \gamma_{yz}\} \tag{7.41}$$

$$[d] = \frac{E}{(1+\nu)(1-2\nu)} \begin{bmatrix} (1-\nu) & \nu & \nu & 0 & 0 & 0 \\ \nu & (1-\nu) & \nu & 0 & 0 & 0 \\ \nu & \nu & (1-\nu) & 0 & 0 & 0 \\ 0 & 0 & 0 & (1-2\nu)/2 & 0 & 0 \\ 0 & 0 & 0 & 0 & (1-2\nu)/2 & 0 \\ 0 & 0 & 0 & 0 & 0 & (1-2\nu)/2 \end{bmatrix} \tag{7.42}$$

The differential operator is:

$$[\partial]^{\mathrm{T}} = \begin{bmatrix} \partial/\partial x & 0 & 0 & \partial/\partial y & \partial/\partial z & 0 \\ 0 & \partial/\partial y & 0 & \partial/\partial x & 0 & \partial/\partial z \\ 0 & 0 & \partial/\partial z & 0 & \partial/\partial x & \partial/\partial y \end{bmatrix} \tag{7.43}$$

7.5 DIFFERENTIAL EQUATIONS FOR DEFORMATIONS

7.5.1 Beam in bending

In the majority of cases, the deflection of a beam is primarily due to bending. Thus, it is not unreasonable to ignore the contribution of shear to deflection and to obtain the elastic deflection of a beam by solving the differential equation of the deflected shape (also referred to as elastic line). In this section, we shall derive the appropriate differential equations and, in subsequent sections, deal with their solution.

Let us consider the beam of Figure 7.7a subjected to arbitrary lateral and axial loads. Making the usual assumptions in the theory of bending, that plane transverse sections remain plane, and that the material obeys Hooke's law, we can show that (see Equation 7.25)

$$M = -EI \frac{\mathrm{d}^2 y}{\mathrm{d}x^2} \tag{7.44}$$

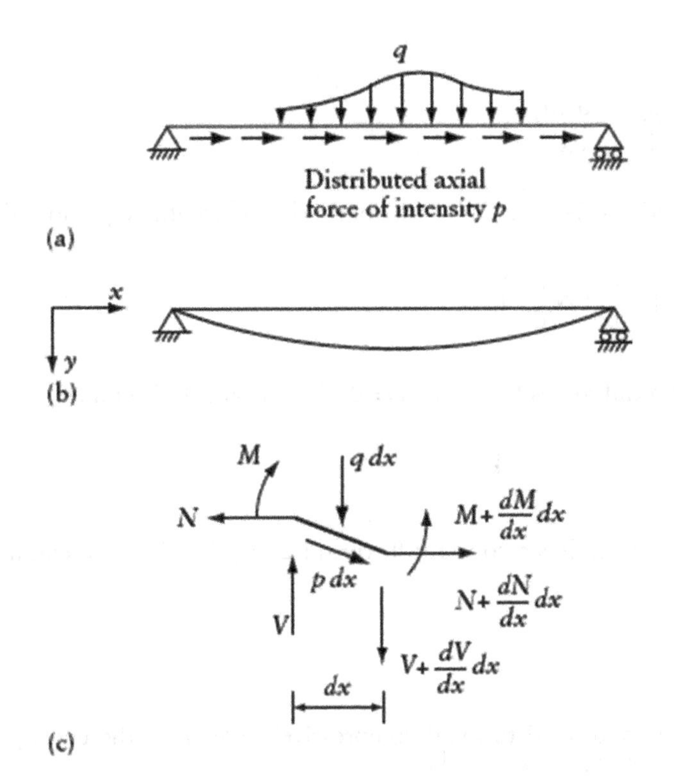

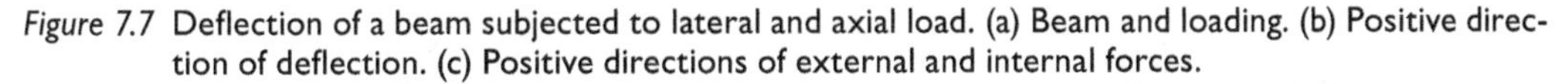
Figure 7.7 Deflection of a beam subjected to lateral and axial load. (a) Beam and loading. (b) Positive direction of deflection. (c) Positive directions of external and internal forces.

where y is the deflection and M is the bending moment at any section x (see Figure 7.7b); EI is the flexural rigidity, which may vary with x. The positive direction of y is indicated in Figure 7.7b and the bending moment is considered positive if it causes tensile stress at the bottom face of the beam.

An element of the beam of length dx is in equilibrium in a deflected position under the forces shown in Figure 7.7c. Summing the forces in the x and y directions, we obtain:

$$\left.\begin{aligned} &\frac{dn}{dx} = -p \\ &\text{and} \\ &q = -\frac{dV}{dx} \end{aligned}\right\} \tag{7.45}$$

where N is the thrust, V is the shear, and p and q are the intensity, respectively, of the axial and lateral distributed loads. The positive directions of q, p, y, M, V, and N are as indicated in Figure 7.7c.

Taking moments about the right-hand edge of the element:

$$V\,dx - N\frac{dy}{dx}\,dx - \frac{dM}{dx}\,dx = 0 \tag{7.46}$$

whence

$$\frac{dV}{dx} - \frac{d}{dx}\left(N\frac{dy}{dx}\right) - \frac{d^2M}{dx^2} = 0 \tag{7.47}$$

Substituting Equations 7.44 and 7.45, we obtain the differential equation of the elastic line:

$$\frac{d^2}{dx^2}\left(EI\,\frac{d^2y}{dx^2}\right) - \frac{d}{dx}\left(N\frac{dy}{dx}\right) = q \tag{7.48}$$

In the absence of axial forces (i.e. when $N=0$), Equation 7.48 becomes:

$$\frac{d^2}{dx^2}\left(EI\,\frac{d^2y}{dx^2}\right) = q \tag{7.49a}$$

If, in addition, the beam has a constant flexural rigidity EI, the differential equation of the elastic line is:

$$\frac{d^4y}{dx^4} = \frac{q}{EI} \tag{7.50}$$

In a beam–column subjected to axial compressive forces P at the ends, $p=0$ and $N=-P$, whence Equation 7.48 becomes:

$$\frac{d^2}{dx^2}\left(EI\,\frac{d^2y}{dx^2}\right) + P\,\frac{d^2y}{dx^2} = q \tag{7.51}$$

All the equations developed in this section can be applied with a slight modification to a beam on an elastic foundation, that is, to a beam which, in addition to the forces already mentioned, receives transverse reaction forces proportional at every point to the deflection of the beam. Let the intensity of the distributed reaction be:

$$\bar{q} = -k\,y$$

where k is the foundation modulus with dimensions of force per (length)^2. The modulus represents the intensity of the reaction produced by the foundation on a unit length of the beam due to unit deflection. The positive direction of reaction $\bar{q}$ is downward. We can therefore use Equations 7.44 to 7.51 for a beam on an elastic foundation if the term q is replaced by the resultant lateral load of intensity $q^* = (q + \bar{q}) = (q - ky)$. For example, Equation 7.49 gives the differential equation of a beam with a variable EI on elastic foundation, subjected to lateral load q with no axial forces, as:

$$\frac{d^2}{dx^2}\left(EI\,\frac{d^2y}{dx^2}\right) = q - k\,y \tag{7.49}$$

Any one of the differential Equations 7.44, 7.48, 7.49, 7.50, or 7.51 has to be solved to yield the lateral deflection y. Direct integration is suitable only in a limited number of cases, considered in standard books on strength of materials. In other cases, a solution of the differential equations by other means is necessary. In the following sections, we consider the method of elastic weights, the numerical method of finite differences, and solutions by series.

7.5.2 Plates in bending

The rectangular element in Figure 7.8, and used in the analysis of plates in bending, has twelve degrees of freedom (three at each corner), defined as:

$$\{D^*\} = \{(w,\theta_x,\theta_y)_1 \mid (w,\theta_x,\theta_y)_2 \mid (w,\theta_x,\theta_y)_3 \mid (w,\theta_x,\theta_y)_4\} \tag{7.52}$$

Here, the displacement $f \equiv w$ is the deflection at any point. The rotations θ are treated as derivatives of w:

$$\theta_x = \partial w/\partial y; \quad \theta_y = -\partial w/\partial x \tag{7.53}$$

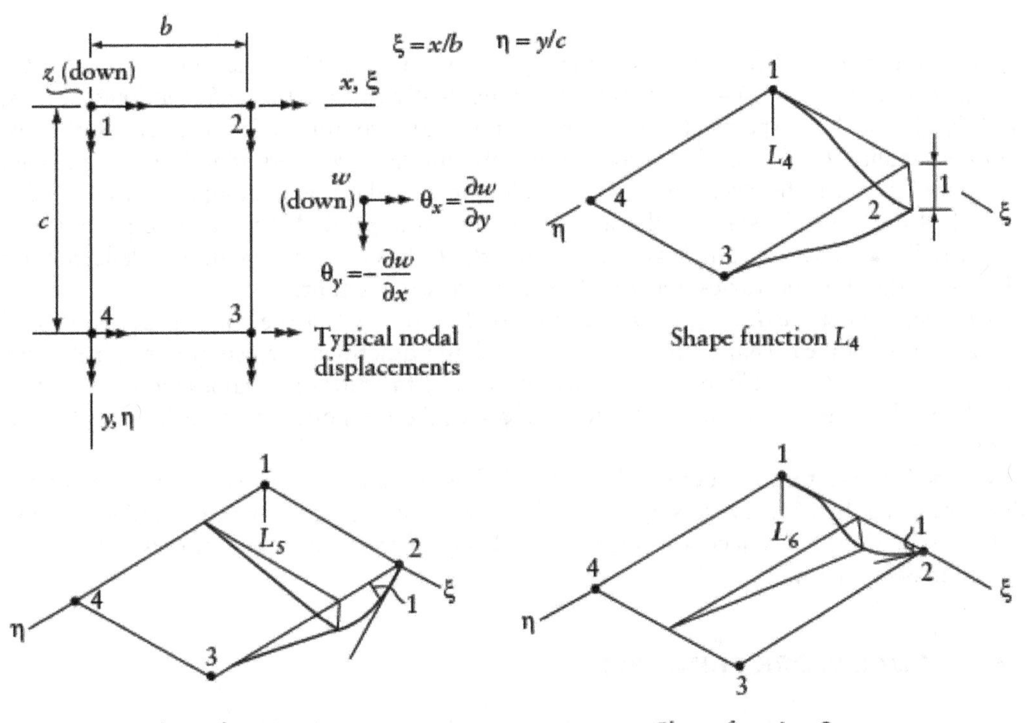

Figure 7.8 Plate-bending element.

The shape functions for the rectangular bending element (Figure 7.8) are

$$[L] = \begin{bmatrix} (1-\xi)(1-\eta) - \frac{1}{b}(L_3 + L_6) + \frac{1}{c}(L_2 + L_{11}) & 1 \\ c\eta(\eta-1)^2(1-\xi) & 2 \\ -b\xi(\xi-1)^2(1-\eta) & 3 \\ \xi(1-\eta) + \frac{1}{b}(L_3 + L_6) + \frac{1}{c}(L_5 + L_8) & 4 \\ c\eta(\eta-1)^2\xi & 5 \\ -b\xi^2(\xi-1)(1-\eta) & 6 \\ \xi\eta + \frac{1}{b}(L_9 + L_{12}) - \frac{1}{c}(L_5 + L_8) & 7 \\ c\eta^2(\eta-1)\xi & 8 \\ -b\xi^2(\xi-1)\eta & 9 \\ (1-\xi)\eta - \frac{1}{b}(L_9 + L_{12}) - \frac{1}{c}(L_2 + L_{11}) & 10 \\ c\eta^2(\eta-1)(1-\xi) & 11 \\ -b\xi(\xi-1)^2\eta & 12 \end{bmatrix} \tag{7.54}$$

where b and c are lengths of element sides; $\xi = x/b$ and $\eta = y/c$. The functions L_2, L_3, L_5, L_6, L_8, L_9, L_{11}, and L_{12} are shape functions given explicitly in Equation 7.54 on lines 2, 3, 5, 6, 8, 9, 11, and 12, respectively; they correspond to unit rotations. Pictorial views of three deflected shapes, L_4, L_5, and L_6, corresponding to unit displacements at node 2, are included in Figure 7.8. It can be seen that L_6 is zero along three edges, while along the fourth edge (1-2) the function is the same as the shape function for a beam. Along any line ξ = constant, L_6 varies linearly, as shown in Figure 7.8. Similarly, L_5 has the same shape as a deflected bar along the edge 3-2 and varies linearly along any line η = constant.

The shape functions L_1, L_4, L_7, L_{10}, corresponding to $w = 1$ at the corners, are expressed as the sum of bilinear shape functions and the shape functions corresponding to rotations at the nodes equal to $\pm(1/b)$ or $\pm(1/c)$. Each of the four functions has values along two edges which are the same as the shape functions of a bar. At any point, the sum $L_1 + L_4 + L_7 + L_{10}$ is equal to unity.

We should note that the deflected surface defined by any of the twelve shape functions does not generally have a zero slope normal to the element edges. This can produce incompatibility of slopes in adjacent elements; the effects of these incompatibilities will be discussed in Section 13.4.

7.6 VIRTUAL WORK PRINCIPLE

This principle relates a system of forces in equilibrium to a compatible system of displacements in a linear or nonlinear structure. The name of the principle is derived from the fact that a fictitious (virtual) system of forces in equilibrium or of small virtual displacements is applied to the structure and related to the actual displacements or actual forces, respectively. Any system of virtual forces or displacements can be used, but it is necessary that the condition of equilibrium of the virtual forces or of compatibility of the virtual displacements be satisfied. This means that the virtual displacements can be any geometrically possible infinitesimal displacements: they must be continuous within the structure boundary and must satisfy the boundary conditions. With an appropriate choice of virtual forces or displacements, the principle of virtual work can be used to compute displacements or forces.

Let us consider a structure deformed by the effect of external applied forces and of environmental causes, such as temperature variation or shrinkage. Let the *actual* total strain at any point be ε, and the corresponding (actual) displacements at n chosen coordinates be $D_1, D_2, \ldots, D_n$. Suppose now that before these actual loads and deformations have been introduced, the structure was subjected to a system of *virtual* forces $F_1, F_2, \ldots, F_n$ at the coordinates 1, 2, ..., n causing a stress σ at any point. The system of virtual forces is in equilibrium, but it need not correspond to the actual displacements $\{D\}$. The principle of virtual work states that the product of the actual displacements and the corresponding virtual forces (which is the virtual complementary work) is equal to the product of the actual internal displacements and the corresponding virtual internal forces (which is the virtual complementary energy). Thus,

Virtual complementary work = Virtual complementary energy

This can be expressed in a general form:

$$\sum_{i=1}^{n} F_i D_i = \int_v \{\sigma\}^{\mathrm{T}} \{\varepsilon\} \, dv \tag{7.55}$$

where σ = a stress corresponding to virtual forces F, and ε = a real strain compatible with the real displacements $\{D\}$. The integration is carried out over the volume of the structure and the summation is for all the virtual forces $\{F\}$. Equation 7.55 states that the values of the complementary work of the virtual external forces and of the complementary energy of the virtual internal forces while *moving* along the real displacements are equal. In other words,

$$\sum_{i=1}^{n} \begin{pmatrix} \text{virtual force} \\ \text{at } i \end{pmatrix} \begin{pmatrix} \text{actual displacement} \\ \text{at } i \end{pmatrix} = \int_v \begin{pmatrix} \text{virtual internal} \\ \text{forces} \end{pmatrix} \begin{pmatrix} \text{actual internal} \\ \text{displacements} \end{pmatrix} dv$$

The principle of virtual work in this form is used in Section 8.2 to calculate the displacement at any coordinate from the strains due to known actual internal forces. The principle can also be used to determine the external force at a coordinate from the internal forces. In the latter case, the structure is assumed to acquire virtual displacements $\{D\}$ compatible with virtual strain pattern ε at any point. The product of the actual external forces $\{F\}$ and the virtual displacements $\{D\}$ is equal to the product of the actual internal forces and the virtual internal displacements compatible with $\{D\}$. This relation can be written:

Virtual work = Virtual strain energy

which is also expressed by Equation 7.55:

$$\sum_{i=1}^{n} F_i D_i = \int_v \{\sigma\}^{\mathrm{T}} \{\varepsilon\} \, dv$$

but with σ being the actual stress corresponding to the actual forces $\{F\}$ and ε being the virtual strain compatible with the virtual displacements $\{D\}$. In this case, Equation 7.55 states that the external and internal virtual work of the real forces while moving along the virtual displacements is the same. The same equation can be written in words as follows:

$$\sum_{i=1}^{n}\begin{pmatrix}\text{real force}\\ \text{at } i\end{pmatrix}\begin{pmatrix}\text{virtual displacement}\\ \text{at } i\end{pmatrix}=\int_v \begin{pmatrix}\text{real internal}\\ \text{forces}\end{pmatrix}\begin{pmatrix}\text{virtual internal}\\ \text{displacements}\end{pmatrix}\mathrm{d}v$$

When the principle of virtual work is used for the calculation of a displacement (or a force) the virtual loads (or the virtual displacements) are chosen in such a manner that Equation 7.55 directly gives the desired quantity. This is achieved by the so-called Unit Load Theorem or unit-displacement theorem for the calculation of displacement and force, respectively.

7.7 UNIT-LOAD AND UNIT-DISPLACEMENT THEOREMS

When the principle of virtual work is used to calculate the displacement D_j at a coordinate j, the system of virtual forces $\{F\}$ is chosen so as to consist only of a unit force at the coordinate j. Equation 7.55 becomes:

$$1\times D_j=\int_v \{\sigma_{uj}\}^{\mathrm{T}}\{\varepsilon\}\mathrm{d}v$$

or

$$D_j=\int_v \{\sigma_{uj}\}^{\mathrm{T}}\{\varepsilon\}\mathrm{d}v \tag{7.56}$$

where σ_{uj} = virtual stress corresponding to a unit virtual force at j, and ε = real strain due to the actual loading. This equation is known as the *unit-load theorem*, and is the general form (not limited by linearity) of Equation 7.65 developed later for linear elastic framed structures.

We should observe that the principle of virtual work as used in the unit-load theorem achieves a transformation of an actual geometrical problem into a fictitious equilibrium problem.

The principle of virtual work can also be used to determine the force at a coordinate j if the distribution of the real stresses or of the internal forces is known. The structure is assumed to acquire a virtual displacement D_j at the coordinate j, but the displacement at the points of application of all the other forces remains unaltered. The corresponding compatible internal displacements are now determined. The external and internal virtual work of the real forces while moving along the virtual displacements is the same: that is, we can write from Equation 7.55:

$$F_j\times D_j=\int_v \{\sigma\}^{\mathrm{T}}\{\varepsilon\}\mathrm{d}v \tag{7.57}$$

Thus, $\{\sigma\}$ are the actual stress components due to the system of real loads, and $\{\varepsilon\}$ are the virtual strain components compatible with the configuration of the virtual displacements.

In a linear elastic structure, the strain component ε at any point is proportional to the magnitude of displacement at j, so that:

$$\varepsilon=\varepsilon_{uj}\,D_j \tag{7.58}$$

where ε_{uj} = strain compatible with a unit displacement at j, with no other displacement at the points of application of other forces. Equation 7.57 becomes:

$$F_j = \int_v \{\sigma\}^T \{\varepsilon_{uj}\} \, dv \tag{7.59}$$

This equation is known as the *unit-displacement theorem.*

The unit-displacement theorem is valid for linear elastic structures only, owing to the limitation of Equation 7.58. No such limitation is imposed on the unit-load theorem because it is always possible to find a statically determinate virtual system of forces in equilibrium, which results in linear relations between virtual external and internal forces, even though the material does not obey Hooke's law.

The unit-displacement theorem is the basis of calculation of the stiffness properties of structural elements used in the finite-element method of analysis (see Chapter 13), where a continuous structure (e.g. a plate or a three-dimensional body) is idealized into elements with fictitious boundaries (e.g. triangles or tetrahedra). In one of the procedures used to obtain the stiffness of an element, a stress or displacement distribution within the element is assumed. The unit-displacement theorem is then used to determine the forces corresponding to a unit displacement at specified coordinates on the element.

Although the unit-displacement theorem is valid strictly only for linear elastic structures, it is used in nonlinear finite-element analysis, which usually involves a series of linear analyses, resulting in errors corrected by iterations.

7.7.1 Symmetry of flexibility and stiffness matrices

An element f_{ij} of a flexibility matrix of a linearly elastic structure is the displacement at coordinate i due to unit force at j. Application of Equation 7.56, using the stress–strain relationship $\{\varepsilon\} = [e]\{\sigma\}$, gives:

$$f_{ij} = \int_v \{\sigma_{uj}\}^T [e] \{\sigma_{ui}\} \, dv \tag{7.60}$$

where $[e]$ = a symmetrical matrix representing the flexibility of the material of an elemental volume dv. The interchange of i and j in this equation gives the displacement at coordinate j due to a unit force at i:

$$f_{ji} = \int_v \{\sigma_{ui}\}^T [e] \{\sigma_{uj}\} \, dv \tag{7.61}$$

Because the right-hand side of Equation 7.61 and its transpose are equal, and the transpose is the same as the right-hand side of Equation 7.60, we conclude that $f_{ij} = f_{ji}$. This proves that the flexibility matrix is symmetrical. Its inverse, the stiffness matrix ($[S] = [f]^{-1}$) is also symmetrical.

7.8 GENERAL

The concept of strain energy is important in structural analysis, and it is useful to express the strain energy due to any type of stress in a general form amenable to matrix treatment. It is possible then to consider, at the same time, components of strain energy due to axial force, bending moment, shear, and torsion.

Complementary energy has no physical meaning, but it is of value in helping to understand some of the energy equations. The same applies to the analogous concept of complementary work.

The principle of virtual work relates a system of forces in equilibrium to a compatible system of displacements in any structure, linear or nonlinear. In analysis, we apply virtual forces or virtual displacements and use the equality of the complementary work of virtual external forces and the complementary energy of the virtual internal forces moving along the real displacements. Alternatively, we utilize the equality of external and internal virtual work of the real forces moving along the virtual displacements. Unit-load and unit-displacement theorems offer a convenient formulation. It should be noted that the latter theorem is applicable only to linear structures.

We should note that the principle of virtual work makes it possible to transform an actual geometrical problem into a fictitious equilibrium problem or an actual equilibrium problem into a fictitious geometrical problem, and there exist circumstances when either transformation is desirable. Chapter 8 discusses further the applications of virtual work for calculation of displacements. The problems at the end of Chapter 8 are related to the material presented in both Chapters 7 and 8.

Chapter 8

Virtual work applications

8.1 INTRODUCTION

In Chapter 7, we adopted the principle of virtual work and used it (in Section 7.6) to derive the unit-load theorem, which is applicable to linear or nonlinear structures of any shape. In this chapter, we shall further consider the method of virtual work, first with reference to trusses and later to beams and frames. The present chapter will also deal with evaluation of integrals for calculation of displacements by the method of virtual work.

To begin with, we shall use the unit-load theorem to calculate displacements due to external applied loading on linear framed structures and we shall also present the theory for this particular application.

8.2 DISPLACEMENT CALCULATION BY VIRTUAL WORK

Consider the linear elastic structure shown in Figure 8.1 subjected to a system of forces F_1, F_2, ..., F_n, causing the stress resultants N, M, V, and T at any section. The magnitude of external and internal work is the same, so that from Equations 7.1 and 7.20:

$$\frac{1}{2}\sum_{i=1}^{n} F_i D_i = \frac{1}{2}\int \frac{N^2\,dl}{Ea} + \frac{1}{2}\int \frac{M^2\,dl}{EI} + \frac{1}{2}\int \frac{V^2\,dl}{Ga_r} + \frac{1}{2}\int \frac{T^2\,dl}{GJ} \tag{8.1}$$

where D_i = displacement at the location and in the direction of F_i; N, M, V, and T = stress resultants at any section due to the $\{F\}$ system.

Suppose that at the time when the forces $\{F\}$ are applied to the structure, there is already a virtual force Q_j acting at the location and in the direction of a coordinate j (Figure 8.1). This force induces at any section internal forces N_{Qj}, M_{Qj}, V_{Qj}, and T_{Qj}. The magnitude of internal and external work during the application of the $\{F\}$ system of loads is again the same, so that:

$$\frac{1}{2}\sum_{i=1}^{n} F_i D_i + Q_j D_j = \frac{1}{2}\left[\int \frac{N^2}{Ea}dl + \int \frac{M^2}{EI}dl + \int \frac{V^2}{Ga_r}dl + \int \frac{T^2}{GJ}dl\right]$$
$$+\left[\int \frac{N_{Qj}N}{Ea}\,dl + \int \frac{M_{Qj}M}{EI}dl + \int \frac{V_{Qj}V}{Ga_r}\,dl + \int \frac{T_{Qj}T}{GJ}dl\right] \tag{8.2}$$

where D_j = displacement at j due to the $\{F\}$ system in the direction of the virtual force Q_j. The second term on each side of Equation 8.2 represents the work due to the force Q_j while moving along the displacement by the $\{F\}$ system. The coefficient 1/2 does not appear in these

DOI: 10.1201/9780429286858-8

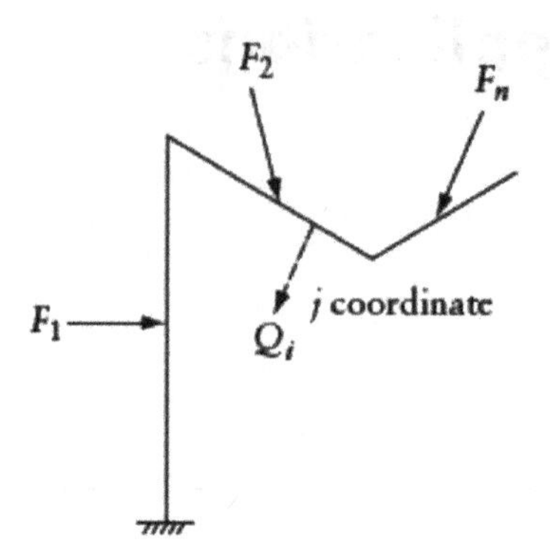

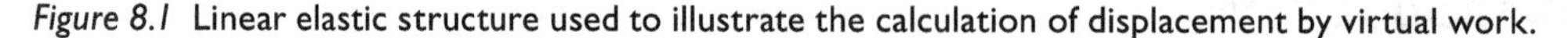
Figure 8.1 Linear elastic structure used to illustrate the calculation of displacement by virtual work.

terms because the load Q_j and the corresponding internal forces act at their full value along the entire displacement by the $\{F\}$ system.

Subtracting Equation 8.1 from 8.2, we find:

$$Q_j D_j = \int \frac{N_{Qj} N}{Ea} dl + \int \frac{M_{Qj} M}{EI} dl + \int \frac{V_{Qj} V}{G a_r} dl + \int \frac{T_{Qj} T}{GJ} dl \tag{8.3}$$

To determine the deflection at any location and in any direction due to the $\{F\}$ system, we divide Equation 8.3 by Q_j. Hence, the displacement at j is:

$$D_j = \int \frac{N_{uj} N}{Ea} dl + \int \frac{M_{uj} M}{EI} dl + \int \frac{V_{uj} V}{G a_r} dl + \int \frac{T_{uj} T}{GJ} dl \tag{8.4}$$

where

$$N_{uj} = \frac{N_{Qj}}{Q_j}; \; M_{uj} = \frac{M_{Qj}}{Q_j}; \; V_{uj} = \frac{V_{Qj}}{Q_j}; \; T_{uj} = \frac{T_{Qj}}{Q_j}$$

These are the values of the internal forces at any section due to a unit virtual force ($Q_j = 1$) applied at the coordinate j where the displacement is required.

Referring back to Equation 7.56, we can see that Equation 8.3 is, in fact, a particular case of the unit-load theorem applicable to linear framed structures, subjected to external loads only (that is, with no environmental effects). The only internal force in a truss member is an axial force, which is constant at all its sections. Thus, the displacement at coordinate j in a truss can be calculated by replacing the first integral in Equation 8.4 by a summation and dropping the remaining integrals:

$$D_j = \sum N_{uj} \frac{Nl}{Ea} \tag{8.5}$$

where the summation is for all members of the truss; l, E, and a = length, modulus of elasticity, and cross-sectional area of a typical member. For the displacement due to change in length – Δ due to temperature or shrinkage – apply Equation 8.5 with the term $Nl/(Ea)$ replaced by Δ.

In order to use Equation 8.4 for the determination of the displacement at any section, the internal forces at all sections of the structure must be determined due to: (i) the actual loads and (ii) a unit virtual force. The latter is a fictitious force or a *dummy load* introduced solely for the purpose of the analysis. Specifically, if the required displacement is a translation, the fictitious load is a concentrated unit force acting at the point and in the direction of the required deflection. If the required displacement is a rotation, the unit force is a couple acting in the same direction and at the same location as the rotation. If the relative translation of two points is to be found, two unit loads are applied in opposite directions at the given points along the line joining them. Similarly, if a relative rotation is required, two unit couples are applied in opposite directions at the two points.

The internal forces N_{uj}, M_{uj}, V_{uj}, and T_{uj} are forces per unit virtual force. If the displacement to be calculated is a translation and the Newton-meter system is used, then N_{uj}, M_{uj}, V_{uj}, and T_{uj} have, respectively, the dimensions: N/N, N-m/N, N/N, and N-m/N. When the virtual force is a couple, N_{uj}, M_{uj}, V_{uj}, and T_{uj} have, respectively, the dimensions: N/N-m, N-m/N-m, N/N-m, and N-m/N-m. A check on the units in Equation 8.4, when used to determine translation or rotation, should easily verify the above statements.

Each of the four terms on the right-hand side of Equation 8.4 represents the contribution of one type of internal forces to the displacement D_j. In the majority of practical cases, not all the four types of the internal forces are present, so that some of the j terms in Equation 8.4 may not be required. Furthermore, some of the terms may contribute very little compared to the others and may, therefore, be neglected. For example, in frames in which members are subjected to lateral loads, the effect of axial forces and shear is very small compared to bending. This is, however, not the case in members with a high depth-to-length ratio or with a certain shape of cross sections, when the displacement due to shear represents a significant percentage of the total.

The internal forces at a section of a space frame are generally composed of six components: N, V_{y^*}, V_{z^*}, M_{y^*}, M_{z^*}, and T, shown in Figure 8.2 in the directions of local orthogonal axes x^*, y^*, and z^* of a typical member. The x^* axis is normal to the cross section through the centroid; y^* and z^* are centroidal principal axes in the plane of the section.

When Equation 8.4 is applied to a space frame, the second term, representing the contribution of bending deformation, is to be replaced by two terms:

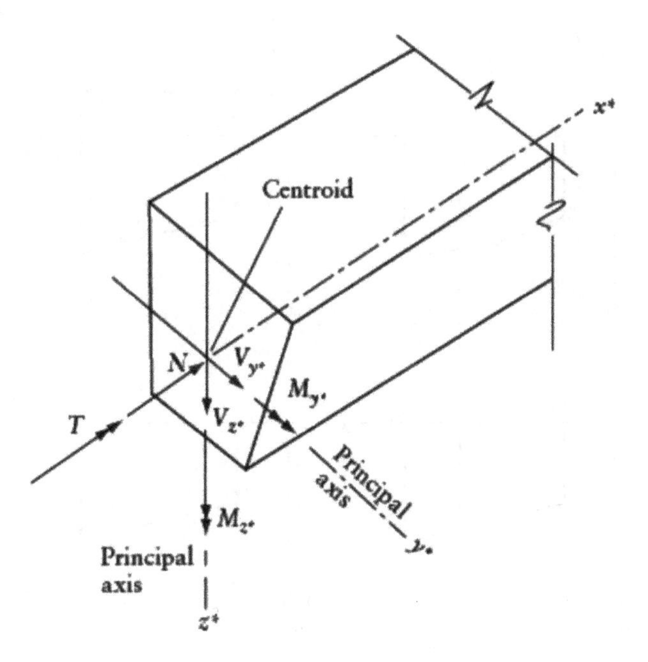

Figure 8.2 Internal forces in a section in a member of a space frame: a normal force N, two shearing forces V_{y*} and V_{z*}, two bending moments M_{y*} and M_{z*}, and a twisting moment T.

$$\int \left(M_{uj}\, M/EI\right)_{y^*} \mathrm{d}l \text{ and } \int \left(M_{uj}\, M/EI\right)_{z^*} \mathrm{d}l$$

The subscripts y^* or z^* indicate that the M and I values inside the brackets are about the principal axis y^* or z^*, respectively. Similarly, the third term in Equation 8.4, representing the contribution of shear deformation, is to be replaced by two terms:

$$\int \left(V_{uj}\, V/Ga_r\right)_{y^*} \mathrm{d}l \text{ and } \int \left(V_{uj}\, V/Ga_r\right)_{z^*} \mathrm{d}l$$

In Equation 8.4, the terms $(N/Ea) = \varepsilon_O$ and $(M/EI) = \psi$ represent, respectively, the centroidal normal strain and the curvature due to N and M on a cross section (Figure 6.1). When D_j is required for the combined effects of N and M together with thermal expansion, the terms (N/Ea) and (M/EI) are to be replaced, respectively, by $[(N/Ea) + \varepsilon_{OT}]$ and $[(M/EI) + \psi_T]$, where ε_{OT} and ψ_T=centroidal strain and curvature due to thermal expansion (Equations 8.32 and 8.33). When the structure is statically indeterminate, N and M must include the statically indeterminate axial force and moment due to temperature (Section 8.8).

8.2.1 Definite integral of product of two functions

Consider functions $y_1(x)$ and $y_2(x)$ of which y_1 is arbitrary function and y_2 is one straight line between x_1 and x_2. The value of the product of the functions is:

$$\int_{x_1}^{x_2} y_1(x)\, y_2(x)\,\mathrm{d}x = a_{y1}\, \bar{y}_2 \tag{8.6}$$

where a_{y1} = area of diagram of y_1 between x_1 and x_2; $\bar{y}_2$ = the value of y_2 at the centroid of Figure 8.3a. In the special case when each of the two functions is a straight line between x_1 and x_2 (Figure 8.3b):

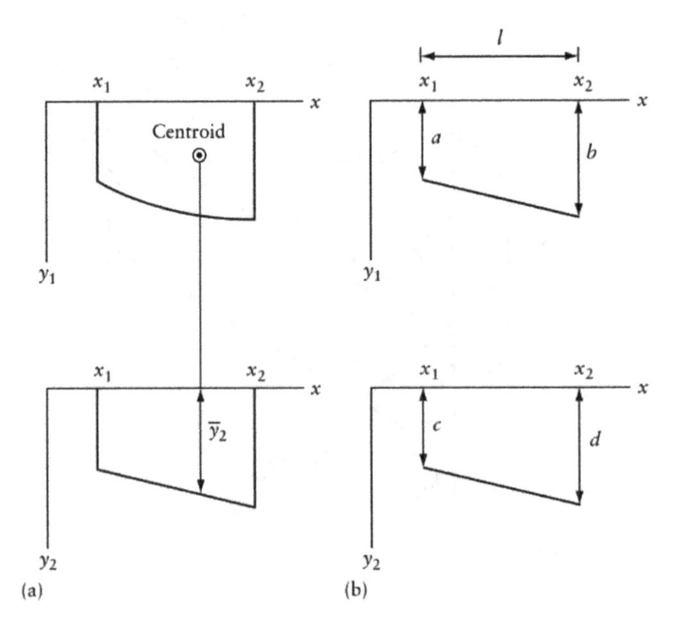

Figure 8.3 Evaluation of the integral of the product of two functions $y_1(x)$ and $y_2(x)$. (a) General case: y_1 is arbitrary function and y_2 is a straight line between x_1 and x_2. (b) Special case: each of y_1 and y_2 is a straight line between x_1 and x_2.

$$\int_{x_1}^{x_2} y_1(x)\, y_2(x)\mathrm{d}x = \frac{l}{6}(2ac + 2bd + ad + bc) \tag{8.7}$$

where $l = x_2 - x_1$ and a to d are ordinates defined in Figure 8.3b.

EXAMPLE 8.1: LONG-TERM DEFLECTION OF A CRACKED CONCRETE BEAM

The long-term curvature of a simply supported beam over its span is shown in Figure 8.4. Calculate the long-term deflection at mid-span.

Deflection at mid-span (Equations 8.4, 8.6 and Figure 8.4):

$$D = \int \psi\, M_u\, \mathrm{d}l = 10^{-6} \left\{ \begin{aligned} &\frac{0.95 \times 2}{6}(113 \times 0.475 + 2 \times 883 \times 0.475) \\ &+2(4081 \times 1.2375 + 4567 \times 1.4281) \end{aligned} \right\}$$

$$= 0.0234 \text{ m}$$

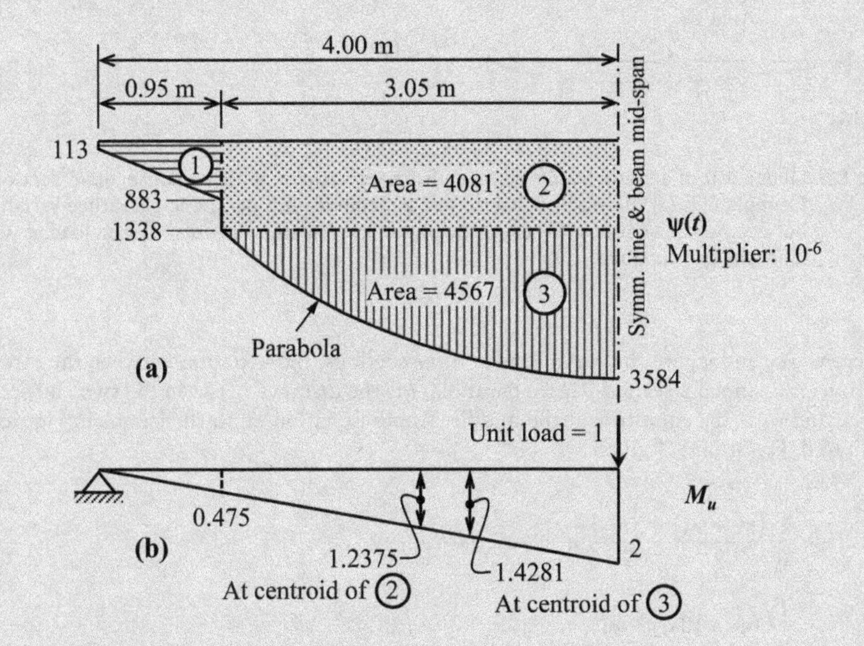

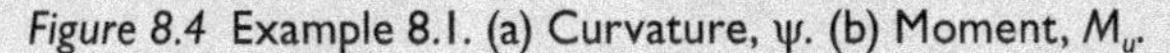

Figure 8.4 Example 8.1. (a) Curvature, ψ. (b) Moment, M_u.

EXAMPLE 8.2: MID-SPAN DEFLECTION OF CONTINUOUS BEAM OR PLANE FRAME MEMBER

Show that the deflection at the center of a straight member with respect to the chord is given (Figure 8.5a) by:

$$D_2 = \frac{l^2}{96}(\psi_1 + 10\,\psi_2 + \psi_3) \tag{a}$$

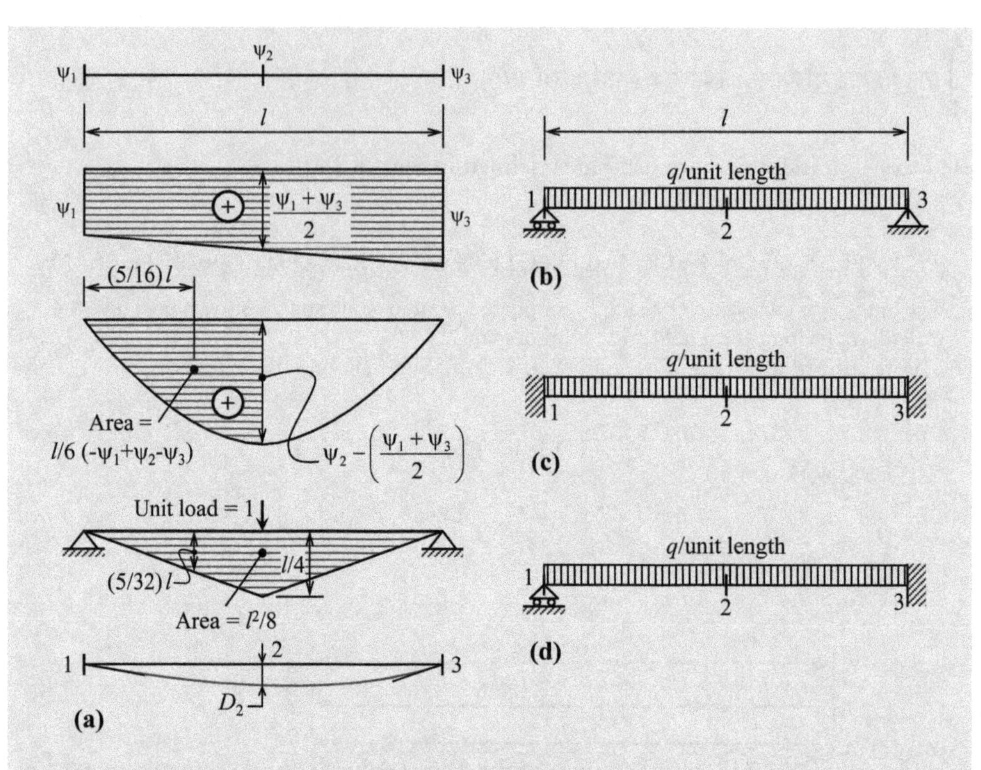

Figure 8.5 Deflection of a plane frame member from curvatures at three consecutive sections, Example 8.2. (a) Mid-span deflection of a member with parabolic curvature variation between three sections. (b) to (d) Beams subjected to evenly distributed load, q with different end supports.

where ψ_1, ψ_2 and ψ_3 are the curvatures at three sections; the variation between the three sections is assumed a second-degree parabola; l is the distance between the two end sections; and ψ_2 is the curvature at the middle. Apply Equation (a) to the beams in Figures 8.5b to d. Equation 8.4 gives:

$$D_2 = \frac{l^2}{8}\frac{(\psi_1 + \psi_3)}{2} + \frac{5l}{32}\left(\frac{l}{3}\right)(-\psi_1 + 2\psi_2 - \psi_3)$$

$$= \frac{l^2}{96}(\psi_1 + 10\psi_2 + \psi_3)$$

Using Appendix G to evaluate the second term on the right-hand side of Equation 8.4 for the two areas shown in Figure 8.5a gives the mid-span deflection at C:

$$D_2 = \frac{l^2}{16}(\psi_1 + \psi_3) + \frac{5l^2}{96}(-\psi_1 - \psi_3 + 2\psi_2) = \frac{l^2}{96}(\psi_1 + 10\psi_2 + \psi_3)$$

For the beam in Figure 8.5b to d, the curvatures at the three sections, in terms of M/EI and the mid-span deflections, are listed in Table 8.1.

Table 8.1 Section curvatures ($\psi = M/EI$; Appendix B) and mid-span deflections, Example 8.2

Case of Figure 8.5	ψ_1	ψ_2	ψ_3	D_2 *(Equation (a))*
(b)	0	$\frac{ql^2}{8EI}$	0	$\frac{5ql^4}{384EI}$
(c)	$\frac{-ql^2}{12EI}$	$\frac{ql^2}{24EI}$	$\frac{-ql^2}{12EI}$	$\frac{ql^4}{384EI}$
(d)	0	$\frac{ql^2}{16EI}$	$\frac{-ql^2}{8EI}$	$\frac{ql^4}{192EI}$

8.3 DISPLACEMENTS REQUIRED IN THE FORCE METHOD

The five steps in the analysis of statically indeterminate structures by the force method (Section 4.6) include calculation of displacements of the released structure due to the specified loads (step 2) and displacements of the released structure due to unit values of the redundant forces (flexibility coefficients, step 3). Virtual work can be used to determine these displacements.

Consider the linear analysis of a statically indeterminate framed structure by the force method (Section 4.6). When the number of the statically indeterminate forces is n, the displacements $\{D\}_{n\times 1}$ can be determined by Equation 8.4, with $j = 1, 2, \ldots, n$; N_{uj}, M_{uj}, V_{uj}, and T_{uj} are internal forces in the released structure due to a force $F_j = 1$; N, M, V, and T are internal forces in the released structure subjected to the specified loads.

Any coefficient, f_{ij}, required to generate the flexibility matrix, $[f]_{n\times n}$, can be determined by application of Equation 8.4. In this application, f_{ij} is the displacement at coordinate i due to a load $F_j = 1$ on the released structure; the corresponding internal forces at any section are N_{uj}, M_{uj}, V_{uj}, and T_{uj}. Apply a unit virtual load, $F_i = 1$, at coordinate i on the released structure, producing at any section the internal forces N_{ui}, M_{ui}, V_{ui}, and T_{ui}. Application of Equation 8.4 gives the flexibility coefficient:

$$f_{ij} = \int \frac{N_{ui}\,N_{uj}}{Ea}\,dl + \int \frac{M_{ui}\,M_{uj}}{EI}\,dl + \int \frac{V_{ui}\,V_{uj}}{Ga_r}\,dl + \int \frac{T_{ui}\,T_{uj}}{GJ}\,dl \tag{8.8}$$

For a plane or a space truss, the flexibility coefficient:

$$f_{ij} = \sum N_{ui}\,N_{uj}\,\frac{l}{Ea} \tag{8.9}$$

where the summation is for all members of the truss; l, E, and a are, respectively, the length, the modulus of elasticity, and the cross-sectional area of a typical member.

EXAMPLE 8.3: PLANE FRAME WITH ANTI-SYMMETRICAL LOADING

Calculate the horizontal displacement at node C of the frame in Figure 8.6a with and without the brace. Consider bending deformations of ABCD and axial deformation of the tie; assume $(Ea)_{\text{Tie}} = 20\,(EI)_{\text{ABCD}}/l^2$.

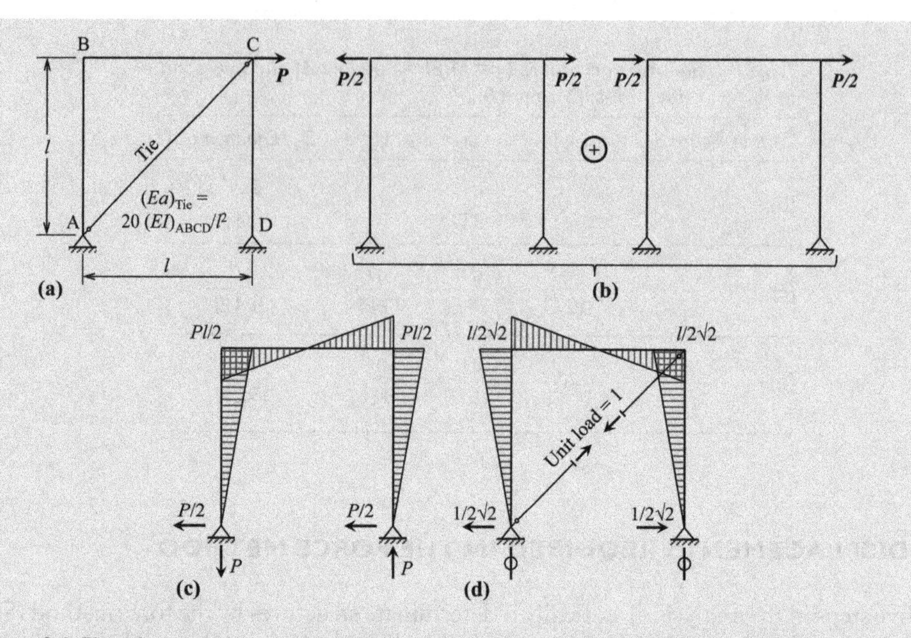

Figure 8.6 Plane frame with or without a tie, Example 8.3. (a) Geometry of braced frame. (b) Load representedasthesumofsymmetricalandanti-symmetricalloads.(c)Bendingmomentofaframe without tie. (d) Moment due to unit and opposite forces applied on a cut tie.

Any load on a symmetrical structure can be replaced by the sum of a symmetrical and anti-symmetrical loading as indicated in Figure 8.6b. The reactions in the anti-symmetrical case are statically determinate because of symmetry. The symmetrical case (first part of Figure 8.6b) has no bending moment and no reactions. For the frame without tie, the horizontal displacement in Figure 8.6c:

$$D_C = \int \frac{M\,M_u}{EI}\,dl = \frac{2}{EI}\left(\frac{Pl^2}{4}\times\frac{l}{3}+\frac{Pl^2}{8}\times\frac{l}{3}\right) = 0.25\frac{Pl^3}{EI}$$

The frame with tie is once statically indeterminate; release the structure by cutting the tie; the flexibility f_{11} is (Figure 8.6d).

$$f_{11} = \left[\int \frac{M_u^2}{EI}\,dl\right]_{\text{ABCD}} + \left(\frac{N_u^2\,l}{Ea}\right)_{\text{Tie}}$$

$$= \left(\frac{1}{\sqrt{2}}\right)^2\left(0.25\frac{Pl^3}{EI}\right) + \frac{\sqrt{2}\,l^3}{20EI} = 0.1957\frac{l^3}{EI}$$

Displacement of the released structure at coordinate 1:

$$D_1 = \int \frac{M\,M_u}{EI}\,dl = \frac{-1}{\sqrt{2}}\left(0.25\frac{Pl^3}{EI}\right) = -0.1768\frac{Pl^3}{EI}$$

$$F_1 = \frac{-D_1}{f_{11}} = \frac{0.1768}{0.1957}P = 0.9033P$$

For the frame with tie,

$$D_C = \frac{0.25Pl^3}{EI}\left(1-\frac{0.9033}{\sqrt{2}}\right) = 0.0903\frac{Pl^3}{EI}$$

8.4 DISPLACEMENT OF STATICALLY INDETERMINATE STRUCTURES

As shown in Chapter 7, the principle of virtual work is applicable to any structure, whether determinate or indeterminate. However, in the latter case, the internal forces induced by the real loading in all parts of the structure must be known. This requires the solution of the statically indeterminate structure by any of the methods discussed in Chapters 4 and 5.

Furthermore, we require the internal forces N_{uj}, M_{uj}, V_{uj} and T_{uj} due to a unit virtual load applied at j. These forces can be determined for any released structure satisfying the requirement of equilibrium with the unit virtual load at j. Thus, it is generally sufficient to determine the internal forces due to a unit virtual load at j acting on a released stable statically determinate structure obtained by the removal of arbitrarily chosen redundants. This is so because the principle of virtual work relates a compatible system of deformations of the actual structure to a virtual system of forces in equilibrium which need not correspond to the actual system of forces (see Section 7.6).

As an example, we can apply the above procedure to the frame of Figure 8.7a in order to find the horizontal displacement at C, D_4. Bending deformations only need be considered. The frame has a constant flexural rigidity EI. The bending moment diagram was obtained in Example 5.2 and is shown again in Figure 8.7b. A unit virtual load is now applied at coordinate 4 to a statically determinate system obtained by cutting the frame just to the left of C (thus forming two cantilevers), Figure 8.7c. The bending moment diagram for M_{u4} is shown in Figure 8.7d.

Applying Equation 8.4 and considering bending only,

$$D_4 = \int \frac{M_{u4}\, M}{EI} \mathrm{d}l \tag{b}$$

This integral needs to be evaluated for the part CD only because M_{u4} is zero in the remainder of the frame. Using Appendix G, we find

$$D_4 = \frac{1}{EI}\left[\frac{1}{2}\frac{(l/2)}{6}(2\times 0.271\,Pl - 0.334\,Pl)\right] = 0.0087\,\frac{Pl^3}{EI}$$

Suppose now that we want to find the vertical deflection at F, denoted by D_5. We assume that the virtual unit load acts on the three-hinged frame of Figure 8.7e. The corresponding bending moment diagram M_{u5} is shown in Figure 8.7f. By the same argument as used for D_4, we obtain

$$D_5 = \int \frac{M_{u5}\, M}{EI} \mathrm{d}l \tag{c}$$

The integral has to be evaluated for the part BC only because M_{u5} is zero elsewhere. Using the table in Appendix G, we find

$$D_4 = \frac{1}{EI}\left[\frac{1}{4}\frac{(l/2)}{6}(2\times 0.276 + 0.385)\,Pl + \frac{1}{4}\frac{(l/2)}{6}(2\times 0.276 - 0.334)\,Pl\right]$$

$$= 0.0240\,\frac{Pl^3}{EI}$$

It may be instructive to calculate D_4 and D_5 using some other choice of the virtual system of forces.

Application of the unit load on a released structure, instead of the actual structure, does not change the results, but greatly simplifies the calculations. To prove this, assume that $F_5 = 1$

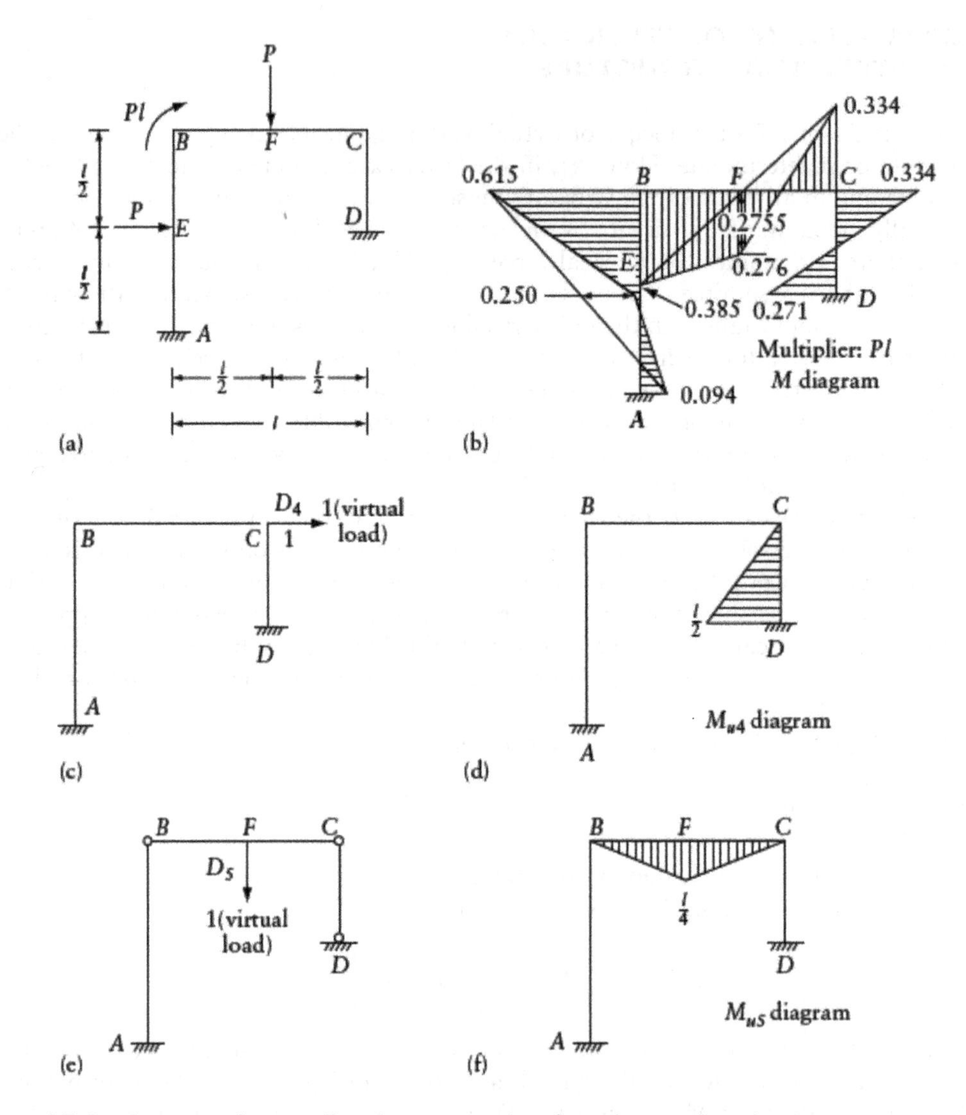

Figure 8.7 Displacement of a statically indeterminate plane frame by virtual work.

is applied on the actual structure to obtain $(M_{u5})_{\text{alternative}}$ to replace $(M_{u5})_s$ when Equation (c) is used to calculate D_5; where $(M_{u5})_s$ is the statically determinate bending moment depicted in Figure 8.7f and used in the above calculation. We will show that:

$$D_5 = \int (M_{u5})_s \frac{M}{EI} \mathrm{d}l \tag{d}$$

or

$$D_5 = \int (M_{u5})_{\text{alternative}} \frac{M}{EI} \mathrm{d}l \tag{e}$$

Each of (M/EI) and M_{u5} is a straight line over BF; thus, Equation 8.7 gives the value of the product of the two functions integrated over BF $= 19.52 \times 10^{-3}$ (Pl^3/EI); similarly, the integral over FC $= 4.54 \times 10^{-3}$ (Pl^3/EI); the sum $= D_5 = 24.1 \times 10^{-3}$ (Pl^3/EI).

Any ordinate of the diagram of $(M_{u5})_{\text{alternative}}$ (not shown) may be expressed as:

$$\left(M_{u5}\right)_{\text{alternative}} = \left(M_{u5}\right)_s + F_6\, M_{u6} + F_7\, M_{u7} + F_8\, M_{u8} \tag{f}$$

where F_6, F_7, and F_8 are, respectively, the ordinates of $(M_{u5})_{\text{alternative}}$ at B, C, and D; M_{u6}, M_{u7}, or M_{u8} is the value of the bending moment at any section due to a pair of opposite unit couples applied on the released structure (Figure 8.7e) at B, C, or D, respectively. Substitution of Equation (f) in Equation (e) gives:

$$D_5 = \int \left(M_{u5}\right)_s \frac{M}{EI} \mathrm{d}l + F_6 \int \left(M_{u6}\right) \frac{M}{EI} \mathrm{d}l + F_7 \int \left(M_{u7}\right) \frac{M}{EI} \mathrm{d}l + F_8 \int \left(M_{u8}\right) \frac{M}{EI} \mathrm{d}l \tag{g}$$

Any of the last three integrals in this equation is zero because it represents the relative rotation of the two sections adjacent to B, C, or D in the actual structure (subjected to M in Figure 8.7b). Since the actual structure is continuous at these locations, the relative rotations are zero. Thus, calculations of D_5 by Equation (d) or (e) must give the same result.

EXAMPLE 8.4: MID-SPAN DEFLECTION OF A FLOOR GRID

Figure 8.8a is a plan view of a post-tensioned slab panel of span = 10 m in x and z-directions; AB is a free edge; AE, BF and EF are symmetry planes; at AE and BF the rotation

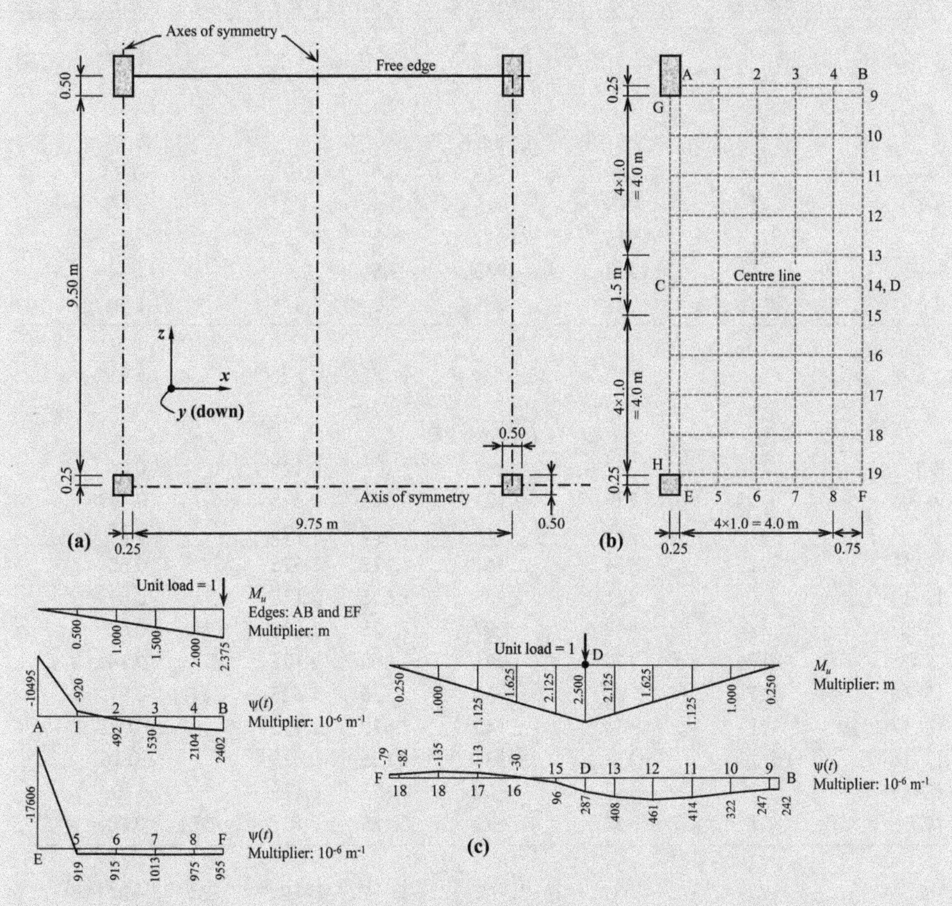

Figure 8.8 Post-tensioned slab panel, Example 8.4. (a) Plan geometry showing supporting columns and boundaries of the slab panel. (b) Grid idealization for half panel. (c) Curvatures and M_u-diagrams for edges: AB, EF, and BF.

$\theta_z = 0$; rotation, $\theta_x = 0$ at EF. Figure 8.8c gives the curvatures along edges AB, EF, and BF; it is required to use the unit force method to calculate the deflection at node D.

Deflections D_B, D_F, and $(D_D - D_{BF})$ are calculated in Table 8.2 using Equation 8.7:

$$D = \int M_u \, \psi(t) \, dl = \frac{l}{6}\left(2\,\psi_a\, M_{ua} + 2\,\psi_b\, M_{ub} + \psi_a\, M_{ub} + \psi_b\, M_{ua}\right)$$

Table 8.2 Curvatures at edges of a post-tensioned flat slab analyzed as an eccentric grid, Example 8.4 (Figure 8.8c)

Segment	l (m)	$\psi_a(t)$ (10^{-6} m^{-1})	$\psi_b(t)$ (10^{-6} m^{-1})	M_{ua} (m)	M_{ub} (m)	$\int_{segment} M_u \psi(t) dl$ (10^{-3} m)
			(a) Edge AB			
A-1	1	−10495	−920	0	0.5	−1.03
1–2	1	−920	492	0.5	1.0	−0.10
2–3	1	492	1530	1.0	1.5	1.3
3–4	1	1530	2104	1.5	2.0	3.20
4-B	0.75	2104	2402	2.0	2.375	3.70
			$D_B(t) = 2\sum \int_{segment} M_u\, \psi(t)\, dl =$			14.2
			(b) Edge EF			
E-5	1	−17606	919	0	0.5	−1.31
5–6	1	919	915	0.5	1.0	0.69
6–7	1	915	1013	1.0	1.5	1.21
7–8	1	1013	975	1.5	2.0	1.74
8-F	0.75	975	955	2.0	2.375	1.58
			$D_F(t) = 2\sum \int_{segment} M_u\, \psi(t)\, dl =$			7.81
			(c) Edge BF			
B-9	0.25	242	247	0	0.125	0.00
9–10	0.25	247	322	0.125	0.625	0.11
10–11	1	322	414	0.625	1.125	0.33
11–12	1	414	461	1.125	1.625	0.60
12–13	1	461	408	1.625	2.125	0.81
13-D	1	408	287	2.125	2.500	0.60
D-15	0.75	287	96	2.500	2.125	0.34
15–16	0.75	96	−30	2.125	1.625	0.07
16–17	1	−30	−113	1.625	1.125	−0.02
17–18	1	−113	−135	1.125	0.625	−0.16
18–19	1	−135	−82	0.625	0.125	−0.04
19-F	1	−82	−79	0.125	0	0.00
			$(D_D - D_{BF})(t) = \sum \int_{segment} M_u\, \psi(t)\, dl =$			2.63

$D_B = 14.2$ mm; $D_F = 7.81$ mm; $(D_D - D_{BF}) = 2.63$ mm
The vertical deflection at node D is calculated (Table 8.2) as:

$$D_D(t) = \frac{D_B + D_F}{2} + (D_D - D_{BF})$$

$$= \frac{14.2 + 7.81}{2} + 2.63 = 13.64 \text{ mm}$$

8.5 TRUSS DEFLECTION

In a plane or space truss composed of m pin-jointed members, with loads applied solely at joints, the only internal forces present are axial, so that Equation 8.4 can be written:

$$D_j = \sum_{i=1}^{m} \int \frac{N_{uij}\, N_i}{Ea}\, dl \tag{8.10}$$

Generally, the cross section of any member is constant along its length. Equation 8.10 can therefore be written:

$$D_j = \sum_{i=1}^{m} \frac{N_{uij}\, N_i}{E_i\, a_i}\, l_i \tag{8.11}$$

where m is the number of members; N_{uij} is the axial force in member i due to a virtual unit load at j, and $N_i l_i/(E_i a_i)$ is the change in length of a member caused by the real loads, assuming that the material obeys Hooke's law.

Equation 8.11 can also be written in the form:

$$D_j = \sum_{i=1}^{m} N_{uij}\, \Delta_i \tag{8.12}$$

where Δ_i is the real change in length of the ith member. This form is used when the deflection due to causes other than applied loading is required, for example, in the case of a change in temperature of some members.

When the material obeys Hooke's law:

$$\Delta_i = \Delta_{iT} + \left(\frac{Nl}{aE}\right)_i \tag{8.13}$$

where Δ_i is the free (unrestrained) elongation of the member due to temperature; N_i is the axial force in the member due to all effects (including the effect of temperature when the truss is statically indeterminate). The unrestrained elongation of the member due to a rise of temperature T degrees is:

$$\Delta_{iT} = \alpha T l_i \tag{8.14}$$

where l_i is the length of the member and α is the coefficient of thermal expansion. Equation 8.12 is valid for linear and nonlinear trusses.

EXAMPLE 8.5: EFFECT OF TEMPERATURE VARIATION ON A TRUSS

Use the unit-load theorem to calculate the downward displacement at C for the truss shown in Figure 8.9a due to a drop of temperature T degrees in members AB and AD only. Assume $\alpha T = 0.003$, where α is coefficient of thermal expansion.

In a statically determinate truss, a change in temperature produces elongation or shortening of members, but induces no forces. Thus, only members AB and AD change length by the amounts: $\Delta_{AB} = \Delta_{AD} = -\alpha T\left(\sqrt{2}\, l\right)$. A unit force introduced at node C produces the axial forces $\{N_u\}$ given in Figure 8.9b; in AB and AD, the axial forces are: $N_{AB} = N_{AD} = -1/\sqrt{2}$. Application of Equation 8.12 gives the deflection at node C:

$$D_C = \sum N_{ui}\,\Delta_i = 2\left(\frac{-1}{\sqrt{2}}\right)\left(-\alpha T\sqrt{2}\, l\right) = 2\alpha T l = 0.006\, l$$

The deflection can also be calculated from the geometry of Figure 8.9c.

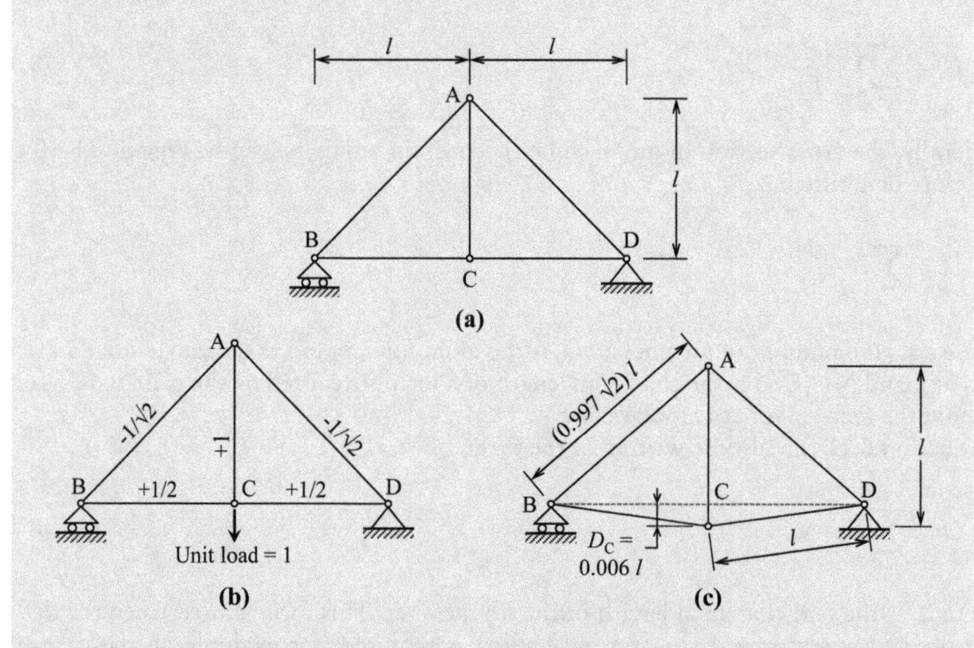

Figure 8.9 Deflection of a plane truss due to drop of temperature of members AB and AD, Example 8.5. (a) Truss dimensions. (b) Axial forces $\{N_u\}$ due to a unit downward load at C. (c) Deflected shape of the truss.

EXAMPLE 8.6: PLANE TRUSS

The *plane truss* of Figure 8.10a is subjected to two equal loads, P at E and D. The cross-sectional area of the members labeled 1, 2, 3, 4, and 5 is a and that of members 6 and 7 is $1.25a$. Determine the horizontal displacement, D_1, at joint C and the relative movement, D_2, of joints B and E.

The internal forces in the members due to the actual loading are calculated by statics and are given in Figure 8.10b. The internal forces due to a unit virtual load at the coordinates D_1 and D_2 are then determined (Figures 8.10c and d). The values obtained can be checked by considering the equilibrium of the joints. We have taken axial forces as positive when tensile, which is the common practice. However, the final result is not affected by the sign convention used.

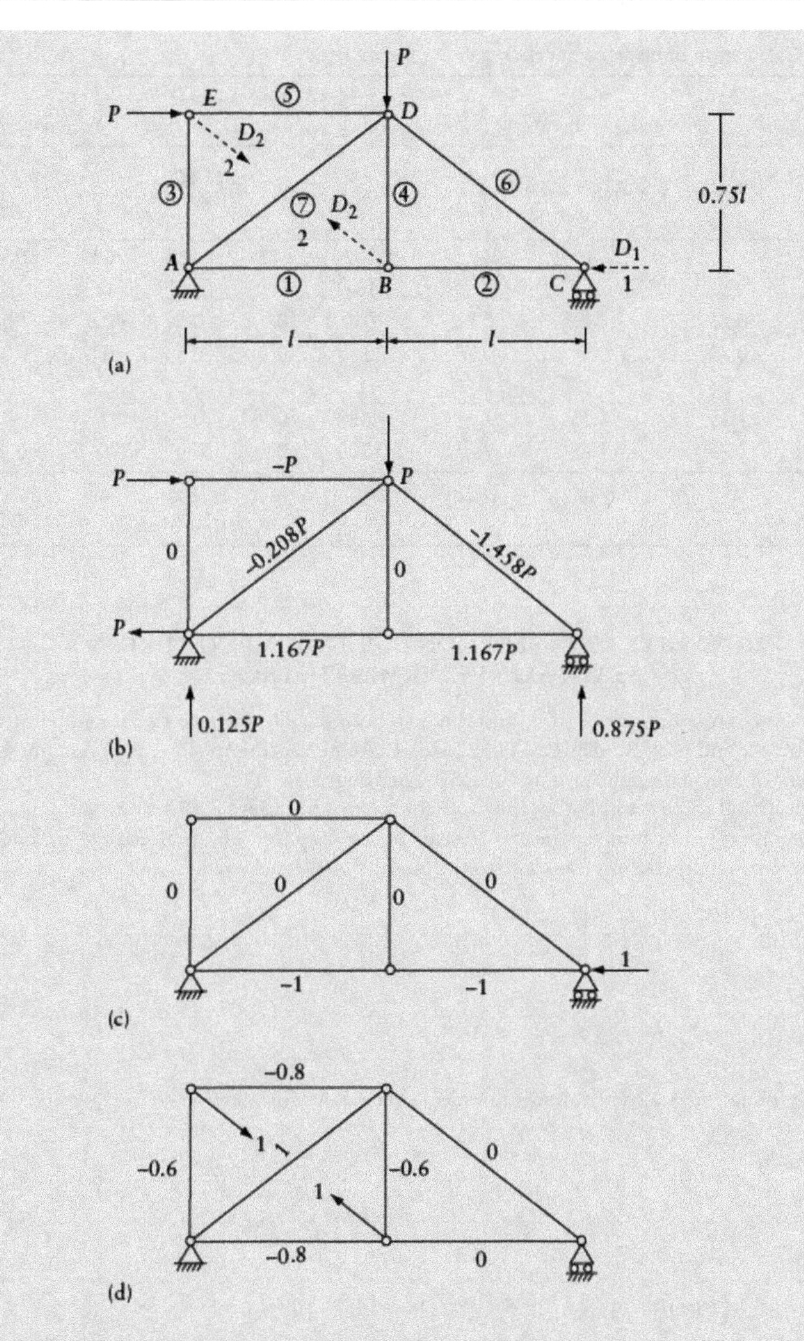

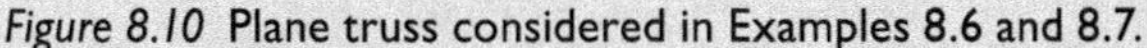
Figure 8.10 Plane truss considered in Examples 8.6 and 8.7.

The displacements D_1 and D_2 are calculated by Equation 8.11. It is convenient to use a tabular form, as shown in Table 8.3. The table is self-explanatory, the displacement D_1 or D_2 being equal to the sum of the appropriate column of $N_u Nl/(Ea)$. Thus, $D_1 = -2.334$ $Pl/(Ea)$ and $D_2 = -0.341$ $Pl/(Ea)$.

The negative sign of D_1 indicates that the displacement is in a direction opposite to the direction of the virtual load in Figure 8.10c. This means that the horizontal translation of joint C is to the right. Likewise, the negative sign of D_2 means that the relative movement of B and E is opposite to the directions of the virtual forces assumed, that is, separation.

Table 8.3 Calculation of displacements D_1 and D_2, Example 8.6

	Properties of member			Actual loading	Calculation of D_1		Calculation of D_2	
Member	Length, l	Cross-sectional area, a	$\frac{l}{Ea}$	N	N_u	$\frac{N_u N l}{Ea}$	N_u	$\frac{N_u N l}{Ea}$
1	1	1	1	1.167	−1	−1.167	−0.8	−0.933
2	1	1	1	1.167	−1	−1.167	0	0
3	0.75	1	0.75	0	0	0	−0.6	0
4	0.75	1	0.75	0	0	0	−0.6	0
5	1	1	1	−1	0	0	−0.8	0.8
6	1.25	1.25	1	−1.458	0	0	0	0
7	1.25	1.25	1	−0.208	0	0	1	−0.208
Multiplier	l	a	$l/(Ea)$	P	—	$Pl/(Ea)$ −2.334	—	$Pl/(Ea)$ −0.341

EXAMPLE 8.7: DEFLECTION DUE TO TEMPERATURE: STATICALLY DETERMINATE TRUSS

For the same truss of Figure 8.10a, find the displacement D_2 due to a rise in temperature of members 5 and 6 by 30 degrees. (The loads P are not acting in this case.) Assume the coefficient of thermal expansion $\alpha = 0.6 \times 10^{-5}$ per degree.

A unit virtual force is applied in the same manner as in Figure 8.10d, so that the forces in the members, indicated on the figure, can be used again. The real change in length occurs in two members only, so that from Equation 8.14:

$$\Delta_5 = 0.6 \times 10^{-5} \times 30 \times l = 18 \times 10^{-5}\, l$$

and

$$\Delta_6 = 0.6 \times 10^{-5} \times 30 \times 1.25 l = 22.5 \times 10^{-5}\, l$$

Applying Equation 8.12 with the summation carried out for members 5 and 6 only:

$$D_2 = \sum_{i=5,6} N_{ui2}\, \Delta_i$$

or

$$D_2 = -0.8\left(18 \times 10^{-5}\, l\right) + 0 \times \left(22.5 \times 10^{-5}\, l\right) = -14.4 \times 10^{-5}\, l$$

It is clear that the same method of calculation can be used if Δ_i is due to any other cause, for example, lack of fit (see Section 4.4).

8.6 EQUIVALENT JOINT LOADING

In the analysis of structures by the force method, we have to know the displacements at a number of coordinates (usually chosen at the joints) due to several loading arrangements.

This requires that the loads be applied at joints only, but any loads acting between joints can be replaced by equivalent loads acting at the joints. The equivalent loads are chosen so that the resulting displacements at the joints are the same as the displacements due to the actual loading. The displacements at points other than the joints will not necessarily be equal to the displacements due to the actual loading.

Consider the beam in Figure 8.11a for which the displacements at the coordinates 1 and 2 at B (Figure 8.11b) are required. We consider B as a joint between members AB and BC. In Figure 8.11c, the displacements at joints B and C are restrained and the fixed-end forces due to the actual loading on the restrained members are determined. The formulas in Appendix B may be used for this purpose. The fixed-end forces at each joint are then totaled and reversed on the actual structure, as in Figure 8.11d. These reversed forces are statically equivalent to the actual loading on the structure and produce the same displacement at the coordinates 1 and 2 as the actual loading. The rotation at C is also equal to the rotation due to the actual loading, but this is not true for the displacements at the other joints A and D. This statement can be proved as follows.

The displacements at 1 and 2 and the rotation at C can be obtained by superposition of the displacements due to the loadings under the conditions shown in Figures 8.11c and d. However, the forces in Figure 8.11c produce no displacements at the restrained joints B and C. Removal of the three restraining forces is equivalent to the application of the forces in Figure 8.11d, and it follows that, due to these forces, the structure will undergo displacements at joints B and C equal to the displacements due to the actual loading.

From the above, we can see that restraints have to be introduced only at the coordinates where the deformation is required. Sometimes, additional restraints may be introduced at

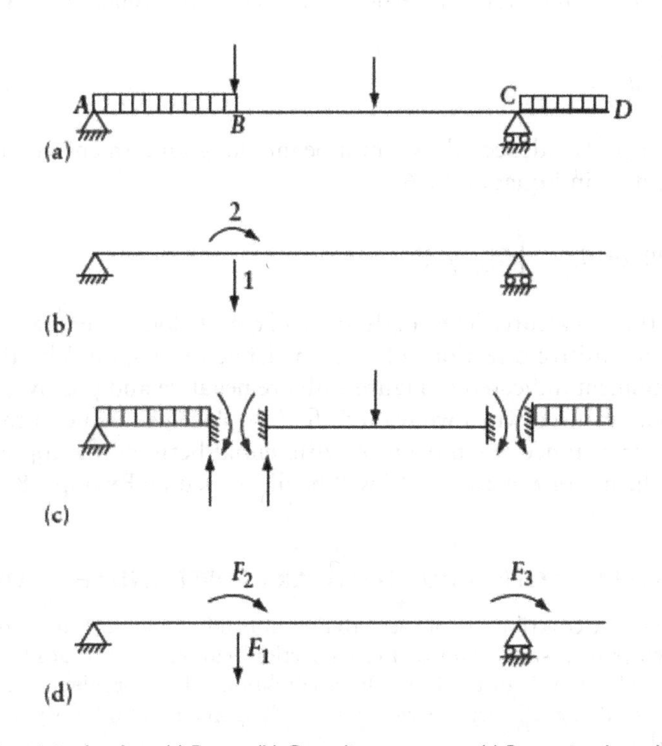

Figure 8.11 Equivalent joint loading. (a) Beam. (b) Coordinate system. (c) Restrained condition. (d) Restraining forces reversed on actual structure.

other convenient locations (such as C in the structure considered above) in order to facilitate the calculation of the fixed-end forces on the structure.

It is apparent that the use of the equivalent joint loading results in the same reactions at supports as the actual loading. The internal forces at the ends of the members caused by the equivalent joint loading, when added to the fixed-end forces caused by the actual loading, give the end-forces in the actual condition.

The advantage of the use of equivalent loads concentrated at the joints instead of the actual loading is that the diagrams of the stress resultants become straight lines. As a result, the evaluation of the integrals of the type involved in Equation 8.4 can be done by Equation 8.7.

8.7 DEFLECTION OF BEAMS AND FRAMES

The main internal forces in beams and frames are bending moments and shearing forces. Axial forces and twisting moments are either absent or they make little contribution to the lateral deflections and rotations. For this reason, in most cases, the terms in Equation 8.4 representing the contribution of the axial force and torsion can be omitted when lateral deflections or rotations are calculated. It follows that the displacement in beams subjected to bending moment and shear is given by:

$$D_j = \int \frac{M_{uj}\ M}{EI} dl + \int \frac{V_{uj}\ V}{G a_r} dl \tag{8.15}$$

Furthermore, the cross section of beams generally used in practice is such that contribution of shear to deflection is small and can be neglected. Thus, the deflection is given by:

$$D_j = \int \frac{M_{uj}\ M}{EI}\ dl \tag{8.16}$$

The change in slope of the deflected axis of a beam along an element of length dl is $d\theta = -(M/EI)\ dl$. Substituting in Equation 8.16:

$$D_j = -\int M_{uj}\ d\theta \quad \text{or} \quad D_j = \int M_{uj}\ \psi\ dl \tag{8.17}$$

where $\psi = M/EI$ = the curvature. The angle $\theta = dy/dx$ and $d\theta/dx = d^2y/dx^2$, where y is the deflection. With the positive directions of x and y defined in Figure 7.4b, the angle change and the bending moment indicated in Figure 7.4b are negative and positive, respectively.

Equation 8.17 can be used when we want to find displacements due to causes other than external loading, for instance, a temperature differential between the top and bottom surfaces of a beam. The use of Equation 8.17 will be illustrated by Example 8.9.

EXAMPLE 8.8: DEFLECTION DUE TO SHEAR IN DEEP AND SHALLOW BEAMS

Find the ratio of the contribution of shear to that of bending moment in the total deflection at the center of a steel beam of I-cross section, carrying a uniformly distributed load over a simple span (Figure 8.12a). Other conditions of the problem are: the second moment of area = I; $a_r \simeq a_w$ = area of web; $G/E = 0.4$; span = l; and intensity of loading = q per unit length.

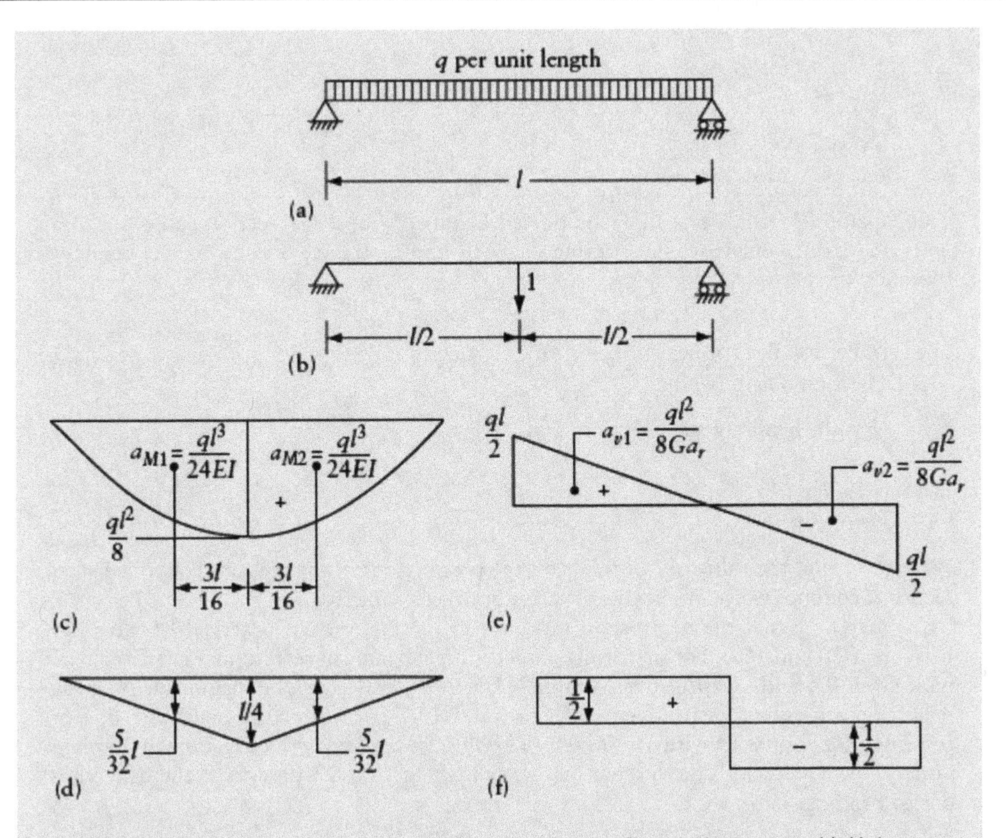

Figure 8.12 **Beam considered in Example 8.8. (a) Beam. (b) Coordinate system. (c) *M* diagram. (d) M_{u1} diagram. (e) *V* diagram. (f) V_{u1} diagram.**

Figures 8.12c and e show the bending moment and shearing force diagrams due to the real loading. A unit virtual load is applied at coordinate 1 (Figure 8.12b) where the deflection is required. The corresponding plots of the bending moment M_{u1} and the shearing force V_{u1} are shown in Figures 8.12d and f.

The M and V diagrams are now divided into two parts such that the corresponding portions of the M_{u1} and V_{u1} diagrams are straight lines. The ordinates $\overline{M}_{uj}$ and $\overline{V}_{uj}$, corresponding to the centroids of the two parts of the M and V diagrams, are then determined. The total deflection at the center is:

$$D_1 = \int \frac{M_{u1} M}{EI} \, dl + \int \frac{V_{u1} V}{G a_r} \, dl \tag{8.18}$$

Of this, the deflection due to bending is:

$$\int \frac{M_{u1} M}{EI} \, dl = \sum a_M \overline{M}_{u1} = 2\left(\frac{q l^3}{24EI}\right)\left(\frac{5l}{32}\right) = \frac{5}{384} \frac{q l^4}{EI}$$

and the deflection due to shear is:

$$\int \frac{V_{u1} V}{G a_r} \, dl = \sum a_V \overline{V}_{u1} = 2\left(\frac{q l^2}{8 G a_r}\right)\left(\frac{1}{2}\right) = \frac{q l^2}{8 G a_r}$$

Hence,

$$\frac{\text{Deflection due to shear}}{\text{Deflection due to bending}} = 9.6\left(\frac{E}{G}\right)\left(\frac{I}{l^2\, a_r}\right) \tag{8.19}$$

This equation is valid for simply supported beams of any cross section subjected to a uniform load. In our case, substituting $G=0.4E$ and $a_r=a_w$, we find for a steel beam of I-section

$$\frac{\text{Deflection due to shear}}{\text{Deflection due to bending}} = 24\left(\frac{I}{l^2\, a_w}\right) = c\left(\frac{h}{l}\right)^2 \tag{8.20}$$

Where h is the height of the I-section, and:

$$c = \frac{24\, I}{h^2\, a_w} \tag{8.21}$$

We can see that the value of c depends on the proportions of the section. For rolled steel sections, commonly used in beams, c varies between 7 and 20.

We may note that the depth/span ratio, h/l in the majority of practical I-beams lies between 1/10 and 1/20. For uniformly loaded simple beams of rectangular cross section, with $G/E=0.4$ and a depth/span ratio, $h/l=1/5$, 1/10, and 1/15, the magnitude of shear deflection represents, respectively, 9.6, 2.4, and 1.07 percent of the deflection due to bending (Equation 8.19 with $a_r=5/6\ bh$; $I=bh^3/12$; where b = width of rectangle). In plate girders, the deflections due to shear can be as high as 15 to 25 percent of the deflection due to bending.

EXAMPLE 8.9: DEFLECTION DUE TO TEMPERATURE GRADIENT

Find the deflection at the center of a simply supported beam of length l and depth h (Figure 8.13a) caused by a rise in temperature which varies linearly between the top and bottom of the beam (Figure 8.13c). The coefficient of thermal expansion is α per degree.

Consider an element ABCD of length dl, shown in Figure 8.13c. It is convenient to assume AB to be fixed in position. The rise of temperature will then cause a displacement of CD to C′D′. The angular rotation of CD with respect to AB is:

$$d\theta = \frac{\alpha\, t\, dl}{h} \tag{8.22}$$

Figure 8.13d shows the plot of the bending moment M_{u1} due to a unit virtual load at coordinate 1 (Figure 8.13b), corresponding to the vertical deflection at the center of the span. Substituting Equation 8.22 in Equation 8.17:

$$D_1 = -\frac{\alpha\, t}{h}\int M_{u1}\, dl$$

The integral $\int M_{u1}\, dl$ is the area under the M_{u1}, diagram, that is:

$$\int M_{u1}\, dl = \frac{l^2}{8}$$

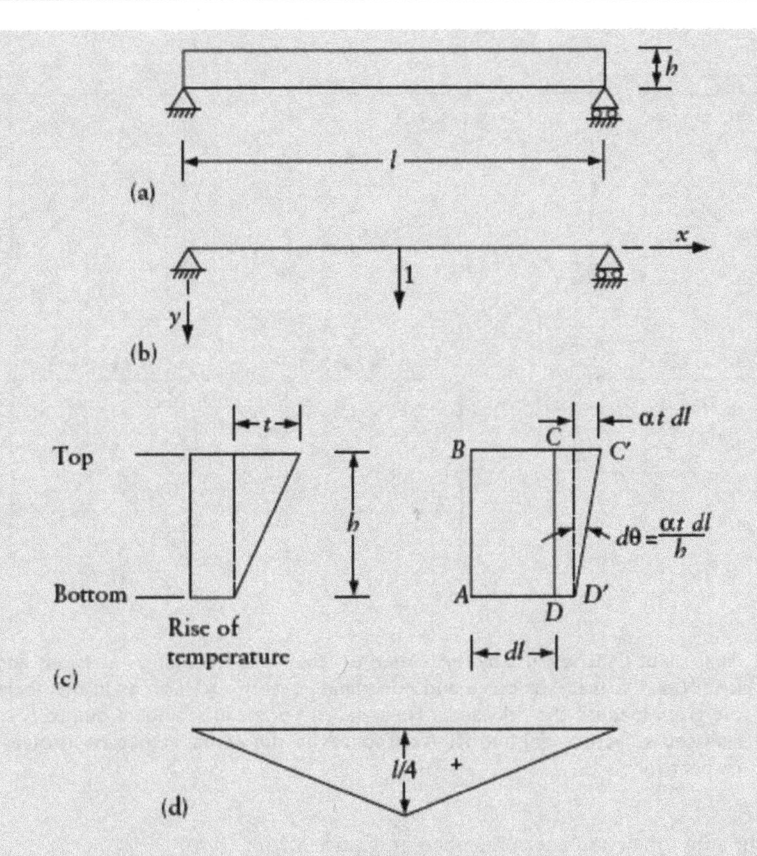

Figure 8.13 Beam considered in Example 8.9. (a) Beam. (b) Coordinate system. (c) Deformation of an element of length dl. (d) M_{u1} diagram.

Whence,

$$D_1 = -\frac{\alpha\, t\, l^2}{8h}$$

The minus sign indicates that the deflection is in a direction opposite to the coordinate 1, that is, upward (as expected).

EXAMPLE 8.10: PLANE TRUSS: ANALYSIS BY THE FORCE METHOD

Find the forces in the members of the truss shown in Figure 8.14a due to: (*i*) the force P and (*ii*) a rise of temperature T degrees in all members. The modulus of elasticity E, the cross-sectional area a and the thermal expansion coefficient α are the same for all members.

We follow the five steps of the force method (Section 4.6). In Step 1, the structure is released by cutting a member; the two arrows in Figure 8.14b define a system of one coordinate. In Step 2, we calculate the displacement D_1 at the coordinate for each load case nonzero using Equation 8.11 or 8.12.

$$(D_1)_{\text{Case i}} = \sum\left(\frac{N_{u1}\, N_s\, l}{Ea}\right)_i = \frac{1}{Ea}\left[P\sqrt{2}(1)l\sqrt{2} - P\left(-\frac{1}{\sqrt{2}}\right)l\right] = 2.707\frac{Pl}{Ea}$$

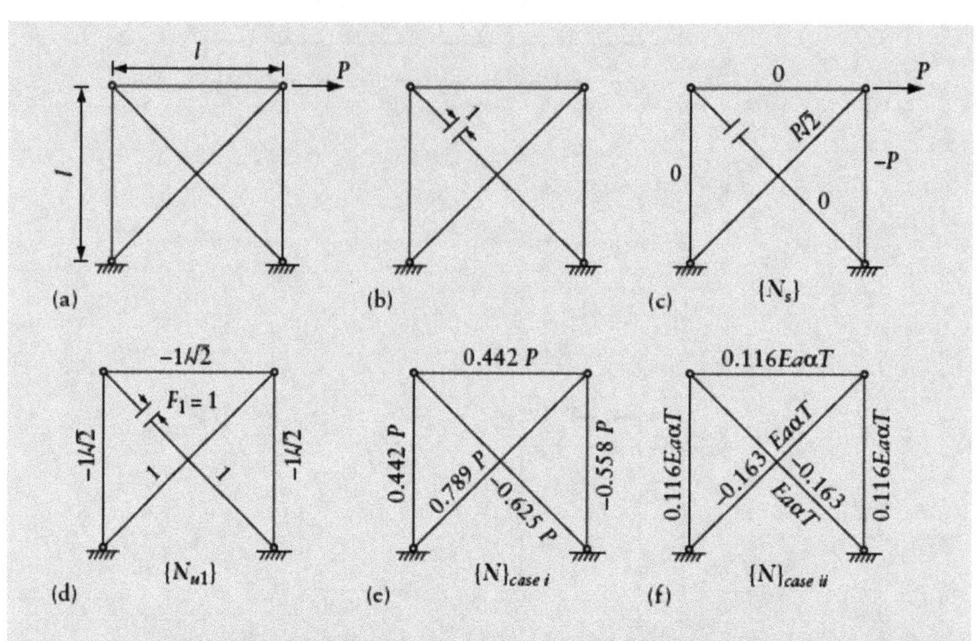

Figure 8.14 Analysis of a plane truss by force method, Example 8.10. (a) Dimensions and applied load. (b) Released structure and coordinate system. (c) Forces in members due to the given load on the released structure. (d) Forces in members due to $F_1 = 1$ on the released structure. (e) and (f) Axial forces in the actual structure in cases i and ii, respectively.

This is the sum for two nonzero members in Figure 8.14c:

$$(D_1)_{\text{Case ii}} = \sum (N_{u1}\,\Delta_t)_i \text{ with}$$

$$\Delta_{ti} = \alpha T\, l_i = \alpha T\left[3\left(-\frac{1}{\sqrt{2}}\right)l + 2(1)\sqrt{2}\,l\right] = 0.7071\alpha T\, l$$

Figure 8.14d shows three members of length l and $N_{u1} = \dfrac{-1}{\sqrt{2}}$ and two diagonal members of length $l\sqrt{2}$ and $N_{u1} = 1$. The subscript i refers to any member and the summations are for all members.

In Step 3, we determine the flexibility matrix $[f]_{1\times1}$ (Equation 8.9):

$$f_{11} = \sum\left(\frac{N_{u1}^2\, l}{Ea}\right)_i = \frac{1}{Ea}\left[3\left(-\frac{1}{\sqrt{2}}\right)^2 l + 2(1)^2\, l\sqrt{2}\right] = 4.328\frac{l}{Ea}$$

In Step 4, we solve the geometry (compatibility) equation (see Equation 4.9):

$$\left[F_{1\,\text{Case i}} \quad F_{1\,\text{Case ii}}\right] = \left[f_{11}\right]^{-1}\left[-D_{1\,\text{Case i}} \quad -D_{1\,\text{Case ii}}\right]$$

$$\left[F_{1\,\text{Case i}} \quad F_{1\,\text{Case ii}}\right] = \left[4.328\frac{l}{Ea}\right]^{-1}\left[-2.707\frac{Pl}{Ea} \quad -0.7071\alpha T\, l\right]$$

$$= \left[-0.628P \quad -0.163\,Ea\alpha T\right]$$

Step 5 is done by inspection. Superposition of $\{N_s\}$ and $(-0.625P)\{N_{u1}\}$ (Figures 8.14c and d) gives the forces in the actual structure in case *i* (Figure 8.14e). For case *ii*, the forces in the actual structure are simply equal to $(-0.163\ E\ a\ \alpha\ T)\{N_{u1}\}$ (see Figure 8.14f).

EXAMPLE 8.11: SPACE TRUSS: ANALYSIS BY FORCE METHOD

For the space truss and loading in Problem 2.8, find the forces in the members when a member is added to join nodes A and C. Assume constant cross-sectional area, a and elasticity modulus, E for all members. Use the answers to Problem 2.8 in the analysis.

The addition of member AC makes the truss statically indeterminate to the first degree; a released structure is obtained by cutting AC. Let coordinate 1 and D_1 represent the relative translation of the cut sections; positive D_1 indicates that the two sections approach each other.

$$D_1 = \sum_{i=1}^{m} \left(\frac{N_s N_u l}{Ea} \right)_i$$

m is equal to the number of members, including AC ($m = 13$). N_{si} are the forces in the ith member due to the applied forces on the released structure (forces equal to P at A and C); N_{ui} is the force in the ith member due to equal and opposite unit forces applied along the member at its cut section. For member AC, $N_{s\,13} = 0$ and $N_{u\,13} = 1.0$. The axial forces $\{N_s\}$ and $\{N_u\}$ are (see answers to Problem 2.8):

$$\{N_s\} = P\{0, 0.707, 0, 0.707, -0.707, 0.707, -0.707, 0.707, 1.0, -1.0, 1.0, -1.0, 0\}$$

$$\{N_u\} = \{0, -0.707, 0, -0.707, 0.707, -0.707, 0.707, -0.707, -1.0, 1.0, -1.0, 1.0, 1.0\}$$

The lengths of members are:

$$\{l\} = l\{1.0, 1.0, 1.0, 1.0, 1.0, 1.0, 1.0, 1.0, \sqrt{2}, \sqrt{2}, \sqrt{2}, \sqrt{2}, \sqrt{2}\}$$

Ea = constant for all members:

$$D_1 = \sum_{i=1}^{13} \left(\frac{N_s N_u l}{Ea} \right)_i = \frac{-8.657\,Pl}{Ea}$$

The displacement D_1 due to $F_1 = 1$ is:

$$f_{11} = \sum_{i=1}^{13} \left(\frac{N_u^2 l}{Ea} \right)_i = \frac{10.071 l}{Ea}$$

Solution of Equation 4.4 gives the force F_1, which closes the gap (make $\Delta_1 = 0$):

$$F_1 = -f_{11}^{-1}\, D_1$$

$$F_1 = -\left(\frac{10.071\, l}{Ea} \right)^{-1} \left(\frac{-8.657\, Pl}{Ea} \right) = 0.8596\, P$$

The forces in the members are given by Equation 4.8; here the required actions $\{A\} \equiv \{N\}$, the forces in the members are:

$$\{N\} = \{N_s\} + \{N_u\}F_1$$

$$\{N\} = 10^{-3}P\{0,\ 99.3,\ 0,\ 99.3,\ -99.3,\ 99.3,\ -99.3,\ 99.3,\ 140.4,\ -140.4,\ 140.4,\ -140.4,\ 859.6\}$$

EXAMPLE 8.12: PLANE FRAME: CONTRIBUTIONS OF AXIAL AND SHEAR DEFORMATIONS TO DISPLACEMENTS

The rigid frame of Figure 8.15a is made of a steel I-section which has the following properties: $a = 134 \times 10^{-6}\, l^2$; $a_r = a_{web} = 65 \times 10^{-6}\, l^2$ and $I = 53 \times 10^{-9}\, l^4$; $G = 0.4E$. Using computer program PLANEF, determine the contributions in Figure 8.15b.

By entering a large value (e.g. 1.0E6) for a or a_r, we ignore axial or shear deformation, respectively. The answers obtained, in terms of P, l, and E are:

	D_1 (P/El)	D_2 (P/El)	D_3 (P/El^2)
Considering all three deformations, {D} =	0.32652E8	−0.37385E6	0.33019E8
Considering axial deformations, {D} =	0.74629E4	0	0
Considering shear deformations, {D} =	0.19218E5	0.19214E5	0
Considering bending deformations only, {D} =	0.32625E8	−0.39306E5	0.33019E8

From the answers, we conclude that in the analyzed frame, the displacements are induced by bending deformation; the contributions of axial and shear deformations are negligible.

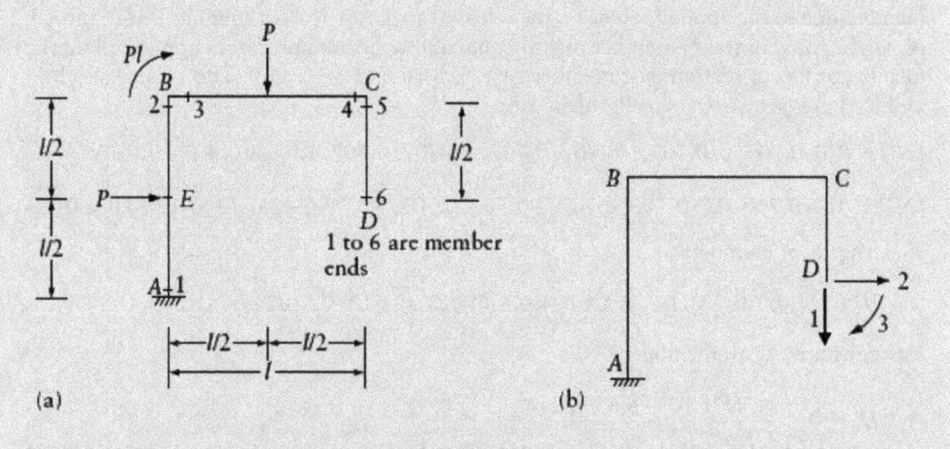

Figure 8.15 Frame considered in Example 8.12. (a) Frame geometry and loading. (b) Coordinate system.

8.8 EFFECT OF TEMPERATURE VARIATION

Analysis of changes in stresses and internal forces in structures, due to a variation in temperature or due to shrinkage or creep, can be done in the same way. The distribution of temperature over the cross section of members is generally nonlinear, as shown in Figure 8.16a and b for a bridge girder exposed to the radiation of the sun. In a cross section composed of different materials, such as concrete and steel, the components tend to contract or expand differently because of shrinkage and creep. However, contraction and expansion cannot occur freely, and changes in stresses occur. In the following, we consider the effect of temperature rise varying nonlinearly over the cross section of members of a framed structure. The temperature rise is assumed constant over the length of individual members.

In a statically determinate frame, no stresses are produced when the temperature variation is linear; in this case the thermal expansion occurs freely, without restraint. This results in changes in length or in curvature of the members but produces no changes in the reactions

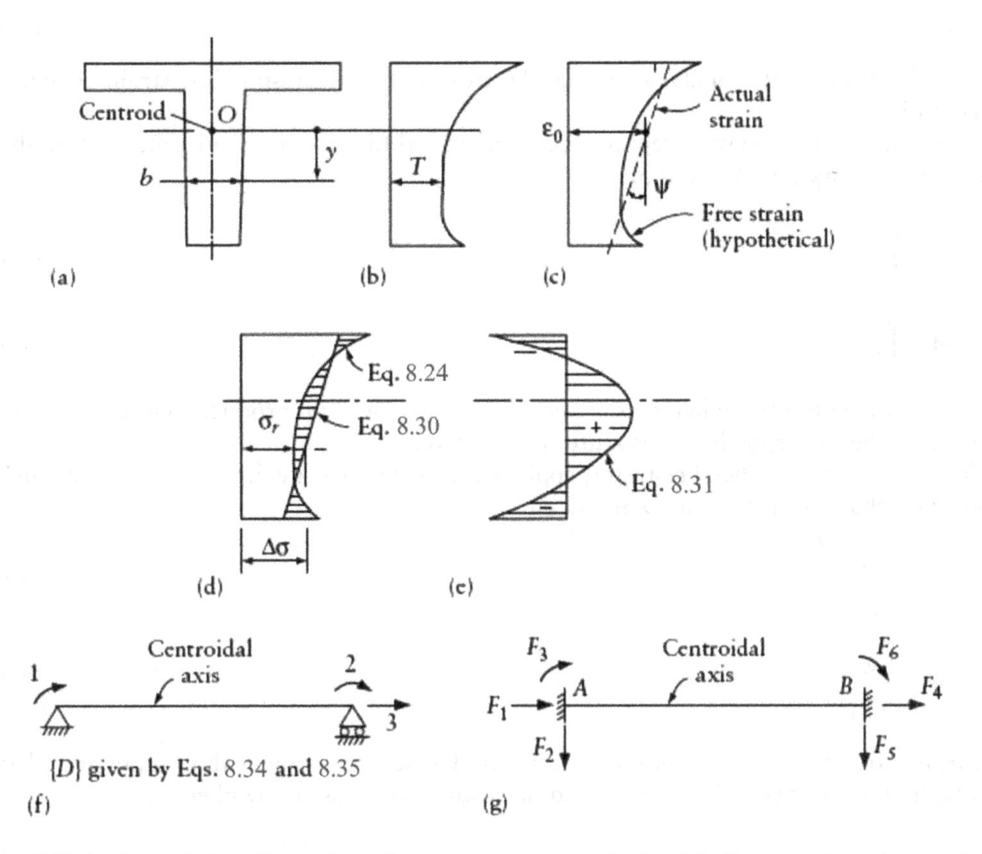

Figure 8.16 Analysis of the effects of nonlinear temperature variation. (a) Cross section of a member. (b) Distribution of temperature rise. (c) Strain distribution. (d) Stresses σ_r and $\Delta\sigma$. (e) Self-equilibrating stresses. (f) Displacements due to a temperature rise in a simple beam. (g) Fixed-end forces due to a temperature rise.

or in the internal forces. However, when the temperature variation is nonlinear, each fiber, being attached to adjacent fibers, is not free to undergo the full expansion and this induces stresses. These stresses must be self-equilibrating in an individual cross section as long as the structure is statically determinate. The *self-equilibrating stresses* caused by nonlinear temperature (or shrinkage) variation over the cross section of a statically determinate frame are sometimes referred to as the *eigenstresses.*

If the structure is statically indeterminate, the elongations and the rotations at the member ends may be restrained or prevented. This results in changes in the reactions and in the internal forces which can be determined by an analysis using the force or the displacement method. We should note that the reactions produced by temperature must represent a set of forces in equilibrium.

Let us now analyze the self-equilibrating stresses in a statically determinate member, e.g. a simple beam of homogeneous material, subjected to a nonlinear rise in temperature (Figure 8.16a). The hypothetical strain which would occur in each fiber if it were free to expand is:

$$\varepsilon_f = \alpha T \tag{8.23}$$

where α is the coefficient of thermal expansion and $T = T(y)$ is the temperature rise in any fiber at a distance y below the centroid O. If the expansion is artificially prevented, the stress in the restrained condition will be:

$$\sigma_r = -E\varepsilon_f \tag{8.24}$$

where E is the modulus of elasticity. Tensile stress and the corresponding strain are considered positive.

The resultant of σ_r may be represented by a normal force N at O and a moment M about the horizontal axis at O, given by:

$$N = \int \sigma_r \, \mathrm{d}a \tag{8.25}$$

$$M = \int \sigma_r \, y \, \mathrm{d}a \tag{8.26}$$

N is considered positive when tensile and M is positive when it produces tension in the bottom fiber; the corresponding curvature ψ is positive.

To eliminate the artificial restraint, apply N and M in opposite directions, resulting in the following changes in strain at O and in curvature:

$$\varepsilon_O = -\frac{N}{Ea} \tag{8.27}$$

$$\psi = -\frac{M}{EI} \tag{8.28}$$

where a and I are the area of the cross section and its second moment about a horizontal axis through O, respectively. The corresponding strain and stress at any fiber are:

$$\varepsilon = \varepsilon_O + y\,\psi \tag{8.29}$$

$$\Delta\sigma = E\left(\varepsilon_O + y\,\psi\right) \tag{8.30}$$

The addition of σ_r to $\Delta\sigma$ gives the self-equilibrating stress due to temperature:

$$\sigma_s = E\left(-\alpha T + \varepsilon_O + y\,\psi\right) \tag{8.31}$$

The stress σ_s must have a zero resultant because its components σ_r and $\Delta\sigma$ have equal and opposite resultants. The distribution of the self-equilibrating stress is shown in Figure 8.16e; the ordinates of this graph are equal to the ordinates between the curve σ_r and the straight line $\Delta\sigma$ in Figure 8.16d.

The changes in axial strain and curvature due to temperature are derived from Equations 8.23 to 8.28:

$$\varepsilon_O = \frac{\alpha}{a}\int T\,b\,\mathrm{d}y \tag{8.32}$$

$$\psi = \frac{\alpha}{I}\int T\,b\,y\,\mathrm{d}y \tag{8.33}$$

where $b = b(y)$ is the width of the section. The actual strain distribution over the depth of the section is presented in Figure 8.16c by a dashed line defined by the values ε_O and ψ. The two values may be used to calculate the displacements at the coordinates in Figure 8.16f (see Appendix A):

$$D_1 = -D_2 = \psi \frac{l}{2} \tag{8.34}$$

$$D_3 = \varepsilon_O \, l \tag{8.35}$$

When using Equation 8.34 we should note that, according to the sign convention adopted, ψ in Figure 8.16c is negative.

If the structure is statically indeterminate, the displacements $\{D\}$, such as those given above, may be used in the force method for the analysis of statically indeterminate reactions and internal forces.

When the analysis is by the displacement method, the values ε_O and ψ (Equations 8.32 and 8.33) can be used to determine the internal forces in a member in the restrained condition and the corresponding member end-forces (Figure 8.16g):

$$N = -E a \varepsilon_O \tag{8.36}$$

$$M = -E I \psi \tag{8.37}$$

$$\{F\} = E\{a\varepsilon_O, 0, -I\psi, -a\varepsilon_O, 0, I\psi\} \tag{8.38}$$

In the special case when the rise in temperature varies linearly from T_{top} to T_{bot} at top and bottom fibers in a member of constant cross section, the fixed-end forces (Figure 8.16g) may be calculated by Equation 8.38, with $\varepsilon_O = \alpha T_O$ and $\psi = \alpha\,(T_{bot} - T_{top})/h$; here, T_O is the temperature at the cross-section centroid and h is the section depth.

The forces F_1 and F_3 are along the centroidal axis and the other forces are along centroidal principal axes of the member cross section. The six forces are self-equilibrating. The restraining forces at the ends of individual members meeting at a joint should be transformed in the directions of the global axes and summed to give the external restraining forces which will artificially prevent the joint displacements of the structure. (The assemblage of end-forces is discussed further in Section 19.9.) In the restrained condition, the stress in any fiber may be calculated by Equation 8.24.

When the temperature rise varies from section to section or when the member has a variable cross section, ε_O and ψ will vary over the length of the member. Equations 8.36 and 8.37 may be applied at any section to give the variables N and M in the restrained condition; the member forces are given by:

$$\{F\} = E\{\{a\varepsilon_O, 0, -I\psi\}_A, \{-a\varepsilon_O, 0, I\psi\}_B\} \tag{8.39}$$

The subscripts A and B refer to the member ends (Figure 8.16g). For equilibrium, a distributed axial load p and a transverse load q must exist. The load intensities (force per length) are given by:

$$p = E\frac{\mathrm{d}(a\varepsilon_O)}{\mathrm{d}x} \tag{8.40}$$

$$q = E\frac{\mathrm{d}^2(I\psi)}{\mathrm{d}x^2} \tag{8.41}$$

Positive p is in the direction A to B and positive q is downwards; x is the distance from A to any section. Equations 8.40 and 8.41 can be derived by considering the equilibrium of a small length of the beam separated by two sections dx apart.

The restraining forces given by Equations 8.39 to 8.41 represent a system in equilibrium. The displacements due to temperature can be analyzed by considering the effect of these restraining forces applied in reversed directions. In the restrained condition, the displacement at *all* sections is zero and the internal forces are given by Equations 8.36 and 8.37. These internal forces must be superimposed on the internal forces resulting from the application of the reversed self-equilibrating restraining forces in order to give the total internal forces due to temperature.

The self-equilibrating stresses and the statically indeterminate forces caused by temperature change are proportional to the modulus of elasticity, E. Some materials, such as concrete, exhibit creep (increase in strain), when subjected to sustained stress. The thermal effects will be overestimated when the value of E used in the analysis is based on the relation of stress to instantaneous strain, ignoring creep (see Section 6.2.1).

Inevitable cracking of reinforced concrete members reduces the effective values of a and I. If cracking is ignored in the values of a and I used in analysis of thermal effects, the results can be greatly overestimated.

8.8.1 Computer analysis for effect of temperature

We use the computer program PLANEF to analyze the effect of temperature variation in a plane frame. We idealize the structure as assembly of prismatic beam elements; the temperature variation over the cross section varies nonlinearly; but the same variation applies at all sections of the element. Artificially prevent the longitudinal thermal expansion by stress $\sigma_r = -E\,\alpha T$, which has resultants:

$$\text{Normal force} = -E\alpha \int T\,\mathrm{d}a; \quad \text{moment} = -E\alpha \int T\,y\,\mathrm{d}a$$

where E = elasticity modulus; α = coefficient of longitudinal thermal expansion; da = elemental area. T is positive for temperature rise; x and y are centroidal principal axes; ε is positive for elongation; ψ (curvature) = dε/dy. Eliminate the artificial restraint by application of the stress resultants in opposite directions to induce planar strain, ε and stress = $E\varepsilon$:

$$\varepsilon = \varepsilon_O + \psi y \tag{h}$$

$$\varepsilon_O = \frac{\alpha}{a} \int T\,\mathrm{d}a \tag{i}$$

$$\psi = \frac{\alpha}{I} \int T\,y\,\mathrm{d}a \tag{j}$$

The sum of $\sigma_r + E\varepsilon$ is a self-equilibrating stress:

$$\sigma_s = E\left(\varepsilon_O + \psi y\right) + \sigma_r \tag{k}$$

The computer program PLANEF gives the nodal displacements, the reactions and the member end forces induced by temperature variation by entering $\{A_r\}$, the member end forces with displacements restrained (Figure 8.17b):

$$\{A_r\} = E\left\{\left(a\varepsilon_O, 0, -I\psi\right)_{\text{End }1}, \left(-a\varepsilon_O, 0, I\psi\right)_{\text{End }2}\right\} \tag{8.42}$$

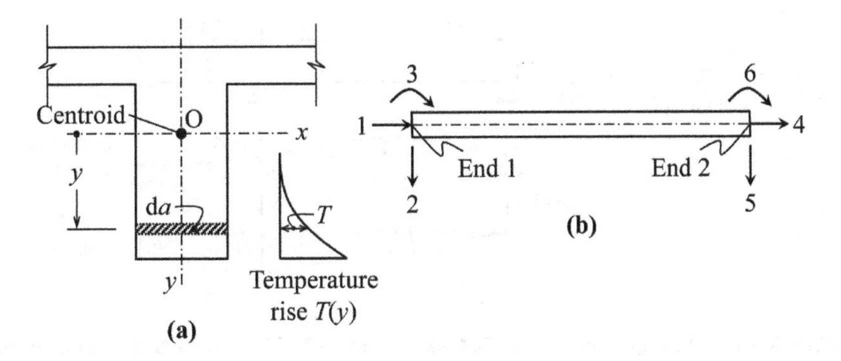

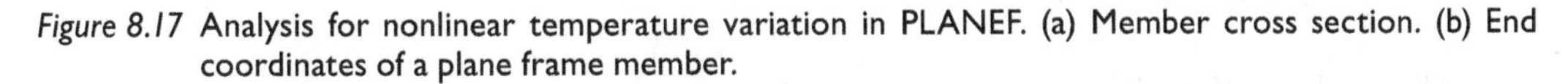

Figure 8.17 Analysis for nonlinear temperature variation in PLANEF. (a) Member cross section. (b) End coordinates of a plane frame member.

EXAMPLE 8.13: NONLINEAR TEMPERATURE VARIATION

Use computer program PLANEF to determine the mid-span defection, the reactions, and bending moments in: (a) Simple beam of span = 40 m; (b) Beam continuous over three 40 m spans, due to the rise of temperature shown (Figure 8.18). Other data: $E = 30 \times 10^9$ N/m^2; $\alpha = 12 \times 10^{-6}$ per degree Celsius:

$$\varepsilon_O = \frac{\alpha}{a}\int T\,\mathrm{d}a = \frac{12\times10^{-6}}{1.12}(9.981) = 106.9\times10^{-6}$$

$$\psi = \frac{\alpha}{I}\int T\,y\,\mathrm{d}a = \frac{12\times10^{-6}}{0.2725}(-3.969) = -174.8\times10^{-6}\ \mathrm{m}^{-1}$$

$$\{A_r\} = 30\times10^{3}\Big\{\big[1.12(106.9),\,0,\,-0.2725(-174.8)\big]_{\mathrm{End}\,1},$$

$$\big[-1.12(106.9),\,0,\,0.2725(-174.8)\big]_{\mathrm{End}\,2}\Big\}$$

$$= 10^{6}\,\{3.592,\,0,\,1.429,\,-3.592,\,0,\,-1.429\}$$

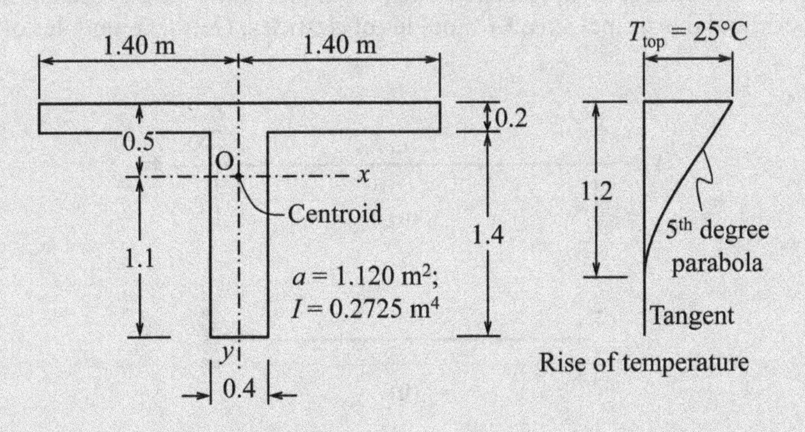

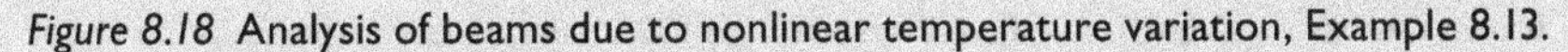

Figure 8.18 Analysis of beams due to nonlinear temperature variation, Example 8.13.

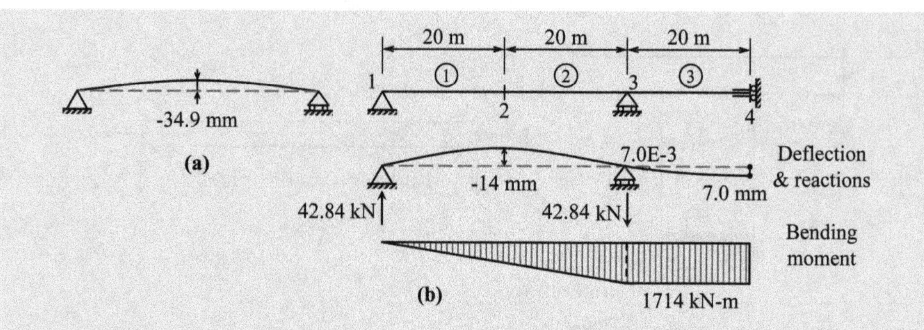

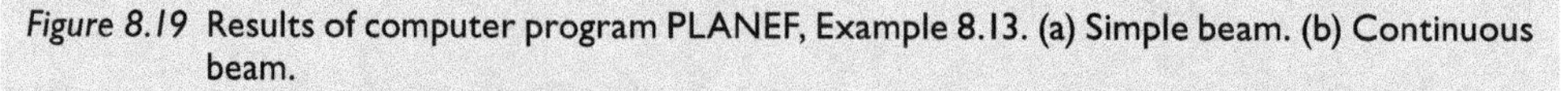

Figure 8.19 Results of computer program PLANEF, Example 8.13. (a) Simple beam. (b) Continuous beam.

The results of PLANEF are:

(a) *Simple beam:* Deflection at mid-span = –34.9 mm. Reactions and bending moment are zero; this is the case for any statically determinate structure.
(b) *Continuous beam:* Deflections at middle of edge and intermediate spans = –14.0 and +7.0 mm, respectively. Reactions and bending moment diagram are shown in Figure 8.19b.

In the statically determinate structure in Figure 8.19a, the reactions and the member end forces, $\{A\}$, are zero; but the curvature ψ occurs freely causing upward mid-span deflection = –34.9 mm. In the statically indeterminate structure (Figure 8.19b), the reactions and the corresponding moment induce a downward deflection at middle of the central span $= 1714 \times 10^3 \left(40^2/8\right) / \left[30 \times 10^9 \left(0.2725\right)\right] = 41.9$ mm; the net deflection = 41.9–34.9 = 7 mm. The bending moment and the deflection shape in Figure 8.19b are not compatible.

8.9 STIFFNESS MATRIX OF PLANE FRAME MEMBER CONSIDERING SHEAR, BENDING, AND AXIAL DEFORMATIONS

Derive the stiffness matrix of a prismatic beam (Figure 8.20a) in terms of a, a_r, I, l, E, and G; where a = cross-sectional area; a_r = effective shear area (Section 7.3.2); I = second moment of area about centroidal principal axis; E = modulus of elasticity; G = shear modulus of elasticity.

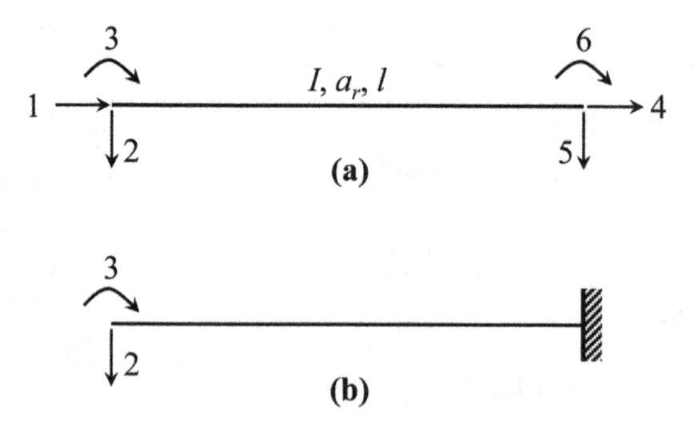

Figure 8.20 Plane frame member. (a) End coordinate system. (b) Coordinates corresponding to shear and bending deformations.

The flexibility matrix corresponding to coordinates 2 and 3 in Figure 8.20b is:

$$[f]=\begin{bmatrix}\dfrac{l^3}{3EI}+\dfrac{l}{Ga_r} & -\dfrac{l^2}{2EI}\\ -\dfrac{l^2}{2EI} & \dfrac{l}{EI}\end{bmatrix} \tag{8.43}$$

$l/(Ga_r)$ = shear deflection due to $F_2=1$. Inversion of $[f]$ gives the stiffness matrix:

$$[f]^{-1}=\begin{bmatrix}\dfrac{12EI}{(1+\eta)\,l^3} & \text{Symm.}\\ \dfrac{6EI}{(1+\eta)\,l^2} & \dfrac{EI}{l}\dfrac{4+\eta}{1+\eta}\end{bmatrix}=\begin{bmatrix}S_{22} & S_{23}\\ S_{32} & S_{33}\end{bmatrix}$$

where

$$\eta=\frac{12EI}{l^2Ga_r} \tag{8.44}$$

For a prismatic beam:

$$\begin{bmatrix}S_{22} & S_{23}\\ S_{32} & S_{33}\end{bmatrix}=\begin{bmatrix}S_{55} & S_{56}\\ S_{65} & S_{66}\end{bmatrix}\quad;\quad S_{11}=S_{44}=\frac{Ea}{l}\quad;\quad S_{41}=-\frac{Ea}{l}$$

The six elements on each column of $[S]$ are self-equilibrating. Thus, the stiffness matrix of plane frame member considering shear deformation is:

$$[S]=\begin{bmatrix}\dfrac{Ea}{l} & & & & & \text{Symm.}\\ 0 & \dfrac{12EI}{(1+\eta)\,l^3} & & & & \\ 0 & \dfrac{6EI}{(1+\eta)\,l^2} & \dfrac{(4+\eta)EI}{(1+\eta)\,l} & & & \\ -\dfrac{Ea}{l} & 0 & 0 & \dfrac{Ea}{l} & & \\ 0 & -\dfrac{12EI}{(1+\eta)\,l^3} & -\dfrac{6EI}{(1+\eta)\,l^2} & 0 & \dfrac{12EI}{(1+\eta)\,l^3} & \\ 0 & \dfrac{6EI}{(1+\eta)\,l^2} & \dfrac{(2-\eta)EI}{(1+\eta)\,l} & 0 & -\dfrac{6EI}{(1+\eta)\,l^2} & \dfrac{(4+\eta)EI}{(1+\eta)\,l}\end{bmatrix} \tag{8.45}$$

If we set $\eta=0$ in Equation 8.45, it gives the stiffness matrix with shear deformation ignored (Equation 5.19)

EXAMPLE 8.14: DEFLECTION DUE TO SHEAR DEFORMATION BY COMPUTER

Using computer program PLANEF (Chapter 20), find mid-span deflection of a simple beam subjected to uniform load q/unit length. Assume a rectangular cross section with span = l; width = b; depth, $h=l/5$; $G=0.4E$.

For the PLANEF input, setting l and b=unity, the cross-sectional properties are: $a=0.2$; $I=(2/3)\times10^{-3}$; $a_r=(5/6)\ a=0.1667$. Considering shear deformation, PLANEF gives: $D=21.406\times(2/3)\times10^{-3}\ ql^4/(EI)=14.271\times10^{-3}\ ql^4/(EI)$. Considering bending deformation only, $D=\dfrac{5}{384}\dfrac{ql^4}{EI}=13.021\times10^{-3}\dfrac{ql^4}{EI}$.

$$\frac{\text{Deflection due to shear}}{\text{Deflection due to bending}}=\frac{14.271-13.021}{13.021}=9.6\%$$

8.10 TRANSFORMATION OF STIFFNESS AND FLEXIBILITY MATRICES

Consider a coordinate system on a linear structure defining the location and direction of forces $\{F\}$ and displacements $\{D\}$, and let the corresponding stiffness matrix be $[S]$ and the flexibility matrix $[f]$. Another system of coordinates is defined for the same structure referring to forces $\{F^*\}$ and displacements $\{D^*\}$, with the stiffness and flexibility matrices $[S^*]$ and $[f^*]$, respectively. We assume that the two systems of forces $\{F\}$ and $\{F^*\}$ are equivalent to each other. The displacements $\{D\}$ would be the same due to forces $\{F\}$ or due to their equivalents $\{F^*\}$. Similarly, the displacements $\{D^*\}$ can be produced by forces $\{F\}$ or $\{F^*\}$. Also, the systems $\{F\}$ and $\{F^*\}$ do the same work to produce the displacement $\{D\}$ or $\{D^*\}$.

If the displacements or forces at the two systems of coordinates are related by:

$$\{D\}=[H]\{D^*\} \tag{8.46a}$$

or:

$$\{F^*\}=[H]^T\{F\} \tag{8.46b}$$

then the stiffness matrix $[S]$ can be transferred to $[S^*]$ by the equation:

$$[S^*]=[H]^T[S][H] \tag{8.47}$$

Equation 8.46a is a geometry relationship between $\{D^*\}$ and $\{D\}$. The elements on the ith column of $[H]$ are equal to the D-displacements when $D_i^*=1$, while $D_j^*=0$, with $j\neq i$. Thus, $[H]$ can be generated and Equation 8.47 is valid only when the D^*-displacements can be varied arbitrarily and independently. For the proof of Equation 8.47, the work done by the forces $\{F\}$ to produce the displacements $\{D\}$ is:

$$W=\frac{1}{2}\{D\}^T[S]\{D\}$$

Substitution for $\{D\}$ using Equation 8.46a, we obtain:

$$W=\frac{1}{2}\{D^*\}^T[H]^T[S][H]\{D^*\} \tag{l}$$

The forces $\{F^*\}$ do the same work to produce the displacements $\{D^*\}$. Thus:

$$W=\frac{1}{2}\{D^*\}^T[S^*]\{D^*\} \tag{m}$$

Comparing Equation (l) with (m) gives the relationship between $[S^*]$ and $[S]$ in Equation 8.47. Application of Equation 8.47 is demonstrated in Example 8.15.

To derive an equation for the transformation of flexibility matrix (Equation 8.49), assume:

$$\{D^*\} = [C]\{D\} \tag{8.48a}$$

$$\{F\} = [C]^T \{F^*\} \tag{8.48b}$$

Equation 8.48a and the transformation Equation 8.49 are valid only when each D-displacement can be varied arbitrarily and independently. The elements on the ith column of [C] are equal to $\{D^*\}$ when $D_i = 1$ while $D_j = 0$ with $j \neq i$.

The work done by the forces $\{F\}$ to produce displacements $\{D\}$ is:

$$W = \frac{1}{2}\{F\}^T [f]\{F\}$$

Substituting for $\{F\}$ using Equation 8.48b, we obtain:

$$W = \frac{1}{2}\{F^*\}^T [C][f^*][C]^T \{F^*\} \tag{n}$$

The same work is done by the forces $\{F^*\}$ to produce displacements $\{D^*\}$; thus:

$$W = \frac{1}{2}\{F^*\}^T [f^*]\{F^*\} \tag{o}$$

Comparison of Equations (n) and (o) indicates that:

$$[f^*] = [C][f][C]^T$$

Transpose both sides of this equation and note that the flexibility matrices are symmetrical to obtain:

$$[f^*] = [C]^T [f][C] \tag{8.49}$$

8.11 STIFFNESS MATRIX OF PLANE FRAME MEMBER WITH RESPECT TO ECCENTRIC COORDINATES

Stiffness matrix with respect to the eccentric coordinates in Figure 8.21 is:

$$[S] = \begin{bmatrix} \frac{Ea}{l} & & & & & \text{Symmetrical} \\ 0 & \frac{12EI}{l^3} & & & & \\ -e\frac{Ea}{l} & \frac{6EI}{l^2} & e^2\frac{Ea}{l}+\frac{4EI}{l} & & & \\ -\frac{Ea}{l} & 0 & e\frac{Ea}{l} & \frac{Ea}{l} & & \\ 0 & -\frac{12EI}{l^3} & -\frac{6EI}{l^2} & 0 & \frac{12EI}{l^3} & \\ e\frac{Ea}{l} & \frac{6EI}{l^2} & -e^2\frac{Ea}{l}+\frac{2EI}{l} & -e\frac{Ea}{l} & -\frac{6EI}{l^2} & e^2\frac{Ea}{l}+\frac{4EI}{l} \end{bmatrix} \tag{8.50}$$

Setting $e = 0$, Equation 8.50 becomes the same as Equation 5.19.

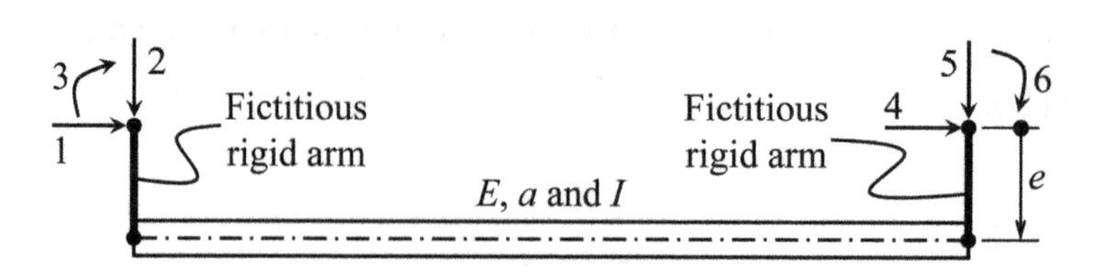

Figure 8.21 Eccentric coordinates of a plane frame member.

EXAMPLE 8.15: TRANSFORMATION OF STIFFNESS MATRIX WITH RESPECT TO ECCENTRIC COORDINATES

Find the stiffness matrix of a prismatic member with respect to the six coordinates 1 to 6 (Figure 8.22). Ignore shear deformations.

$$\{D^*\}=[T]\{D\} \quad ; \quad [T]=\begin{bmatrix}[t] & [0]\\ [0] & [t]\end{bmatrix} \quad ; \quad [t]=\begin{bmatrix}1 & 0 & -e\\ 0 & 1 & 0\\ 0 & 0 & 1\end{bmatrix}$$

The stiffness matrix corresponding to coordinates 1* to 6* is (Equation 5.19):

$$[S^*]=\begin{bmatrix}1/15 & & & & & \text{Symm.}\\ 0 & 1/1500 & & & & \\ 0 & 1/300 & 1/45 & & & \\ -2/30 & 0 & 0 & 1/15 & & \\ 0 & -1/1500 & -1/45 & 0 & 1/1500 & \\ 0 & 1/300 & 1/90 & 0 & -1/300 & 1/45\end{bmatrix}$$

The required stiffness matrix is:

$$[S]=[T]^T[S^*][T]=\begin{bmatrix}1/15 & & & & & \text{Symm.}\\ 0 & 1/1500 & & & & \\ -1/15 & 1/300 & 4/45 & & & \\ -1/15 & 0 & 1/15 & 1/15 & & \\ 0 & -1/1500 & -1/45 & 0 & 1/1500 & \\ 1/15 & 1/300 & -1/18 & -1/15 & -1/300 & 4/45\end{bmatrix}$$

The forces on each column can be verified by one run of computer program PLANEF.

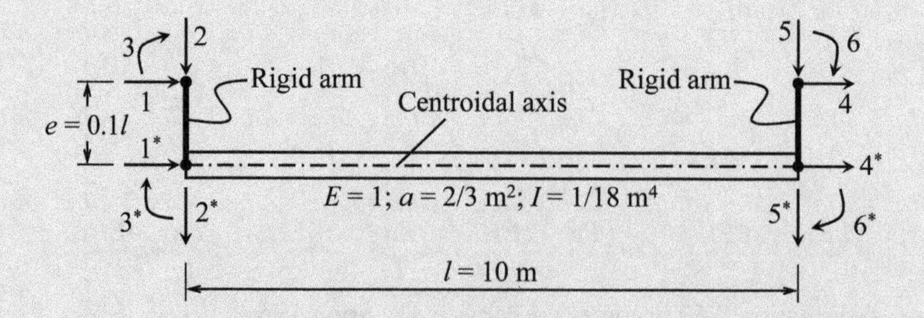

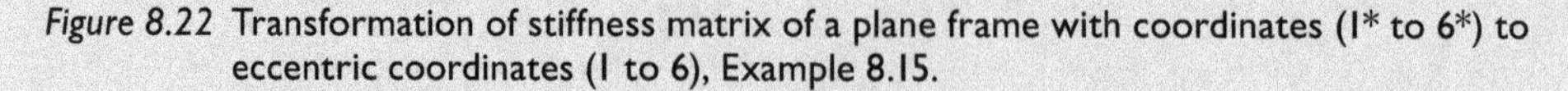

Figure 8.22 Transformation of stiffness matrix of a plane frame with coordinates (1* to 6*) to eccentric coordinates (1 to 6), Example 8.15.

8.12 GENERAL

The energy principles are based on the law of conservation of energy which requires that the work done by external forces on an elastic structure be stored in the form of strain energy which is completely recovered when the load is removed. Energy law, applied in linear analysis, serves a useful purpose in transformation of information given from one form into another.

PROBLEMS

8.1 Using the method of virtual work, find D_1, the vertical deflection of joint H, and D_2, the relative translation in the direction of OB of joints O and B of the truss shown in the figure. The changes in length of the bars (in. or cm) are indicated in the figure.

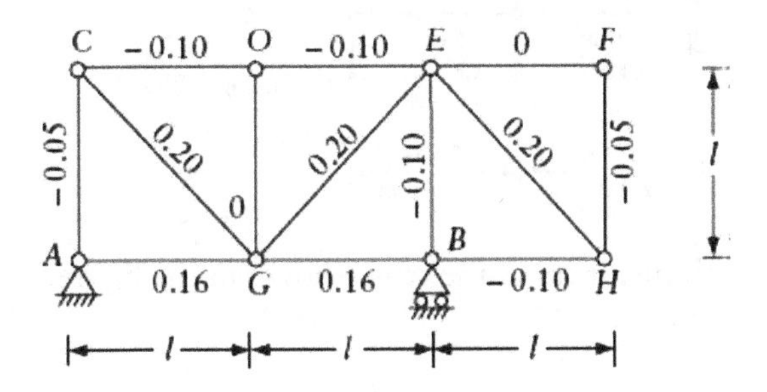

8.2 Find the displacement along the line of action of the force P for the space truss shown in the figure. The value of Ea is the same for all members.

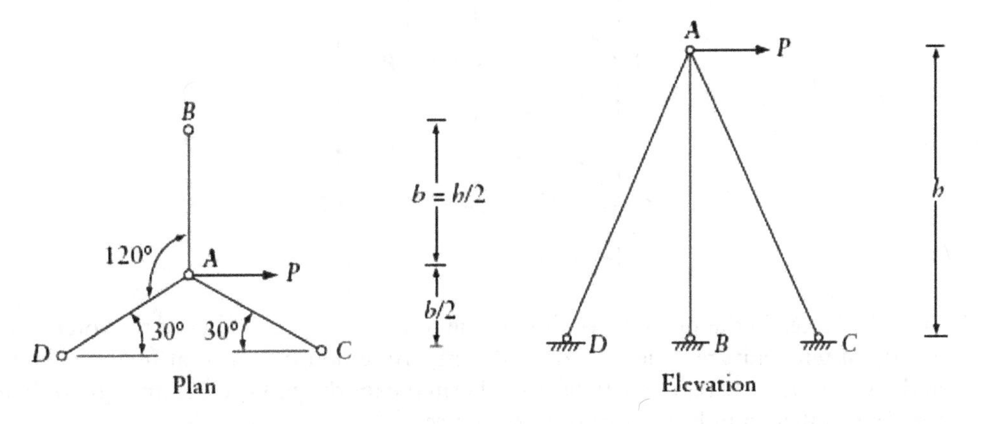

8.3 For the plane truss shown in the figure, find: (a) the vertical deflection at E due to the given loads, (b) the camber at E in the unloaded truss if member EF is shortened by $b/2000$, and (c) the forces in all members if a member is added between C and F and the truss is subjected to the given loads. Consider Ea = constant for all members.

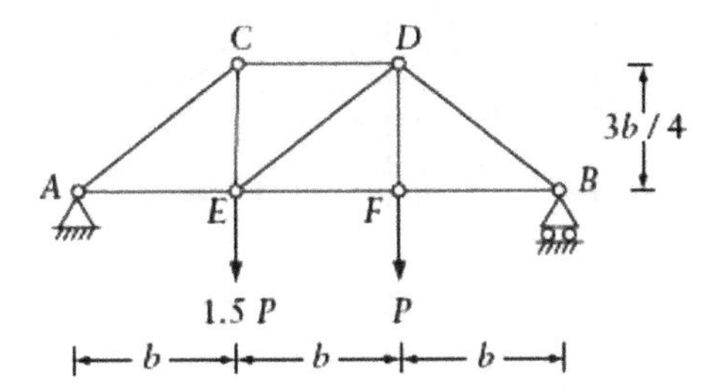

8.4 For the plane truss shown in the figure, find: (a) the deflection at C due to the load *P*, (b) the vertical deflection at C in the unloaded truss if members DE and EC are each shortened by 3 mm, and (c) the forces in all members if a member of cross-sectional area of 2400 mm² is added between B and D, and the truss is subjected to the load *P*. Assume $E = 200$ GN/m² and the cross-sectional area of the members as indicated in the figure.

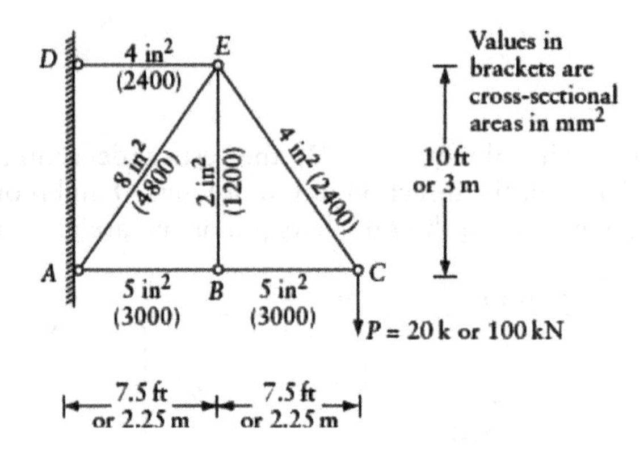

8.5 Find the forces in all members of the truss shown in the figure. Assume l/Ea to be the same for all members.

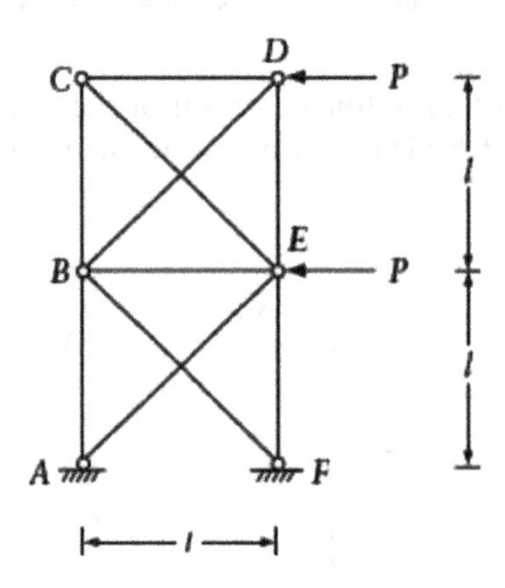

8.6 Find the forces in the plane truss shown due to the combined effect of the force *P* and a drop of temperature *T* degrees of ABC only. Assume Ea = constant for all members and $\alpha T = P/(aE)$, where α = coefficient of thermal expansion. Give the answers only for the six members numbered 1 to 6 in the figure.

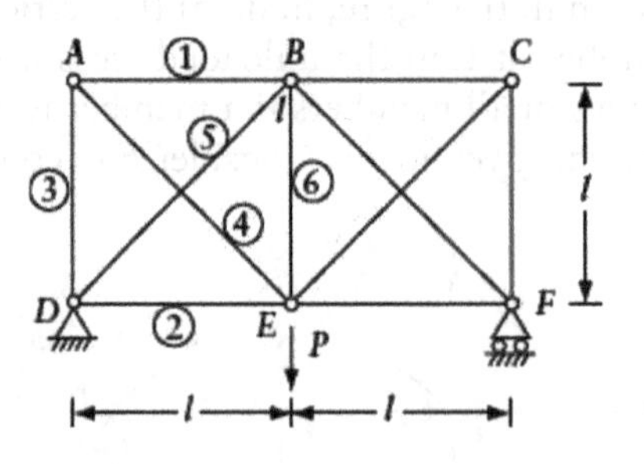

8.7 Find the fixed-end moments in the beam shown.

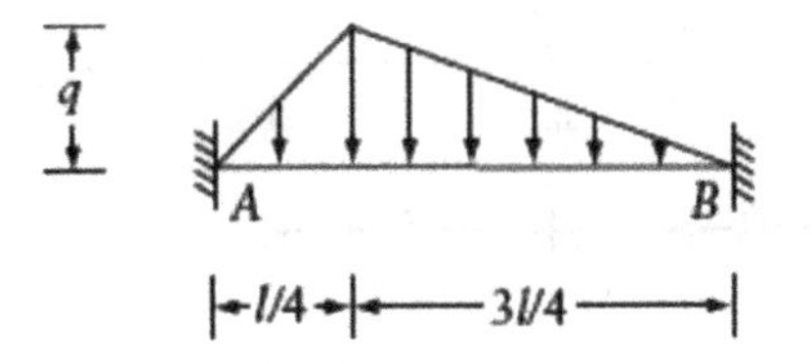

8.8 Find the fixed-end moments in the beam shown.

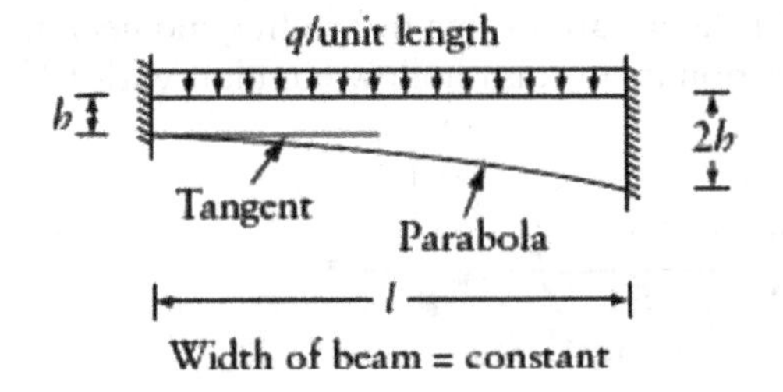

8.9 Find the bending moment diagram for the tied arch shown, considering only the bending deformation in the arch and only the axial deformation in the tie. What is the force in the tie? $(Ea)_{\text{Tie}} = (43/b^2)\,(EI)_{\text{Arch}}$.

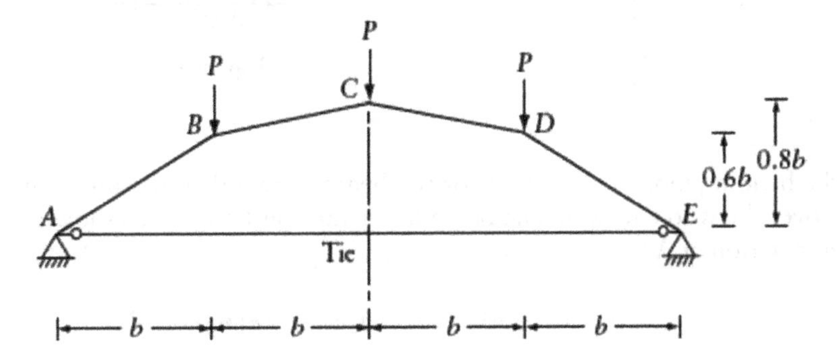

8.10 Considering the deformations due to bending only, find the vertical deflection at A for the frame shown in the figure.

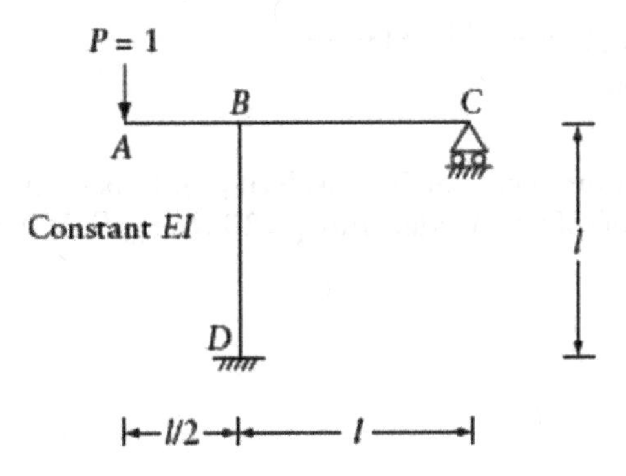

8.11 Find the vertical deflection at D and the angular rotation at A for the beam in the figure. Consider bending deformation only.

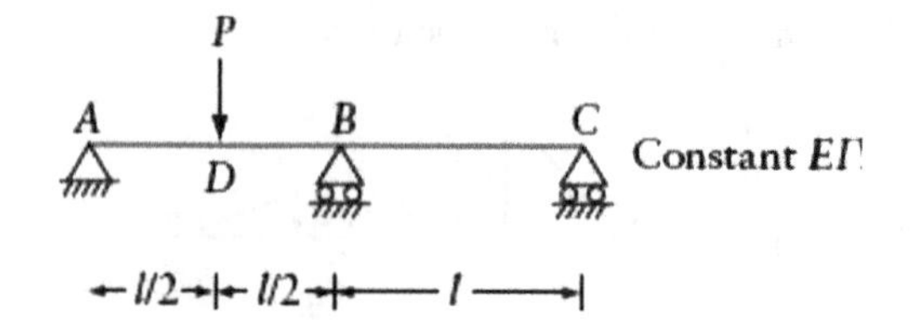

8.12 The figure represents a pole serving as support of two cables subjected to a tensile force P. Determine the translations D_1, D_2, and D_3 at E in the positive directions of x, y, and z axes, assuming: (a) both cables are supported, and (b) only the cable at E is supported. Consider deformations due to bending moment and twisting moment and assume the cross section is constant hollow circular, with $EI = 1.3GJ$.

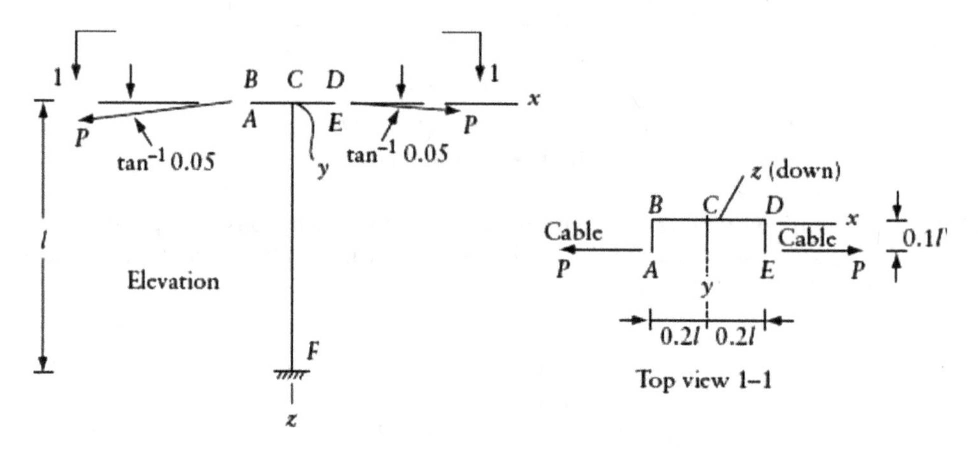

8.13 Find the bending moment diagram for the beam shown due to a uniform load q/unit length over the whole length. The beam has a varying EI value as shown. What is the vertical reaction at A?

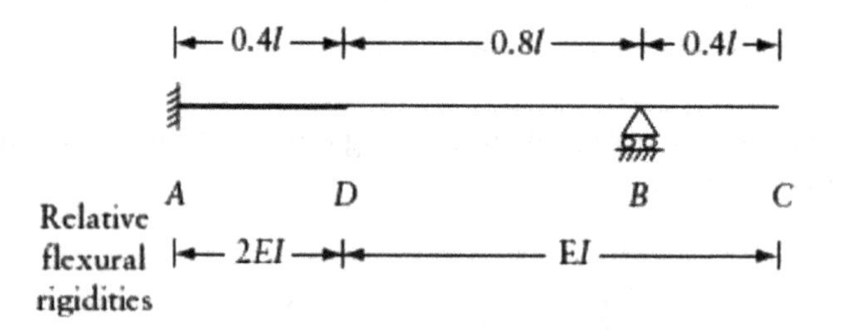

8.14 Find the reaction components at B considering only bending deformation of AB and axial deformation of CD. Consider $(Ea)_{CD} = 20\ (EI)_{AB}/l^2$. What is the deflection at A?

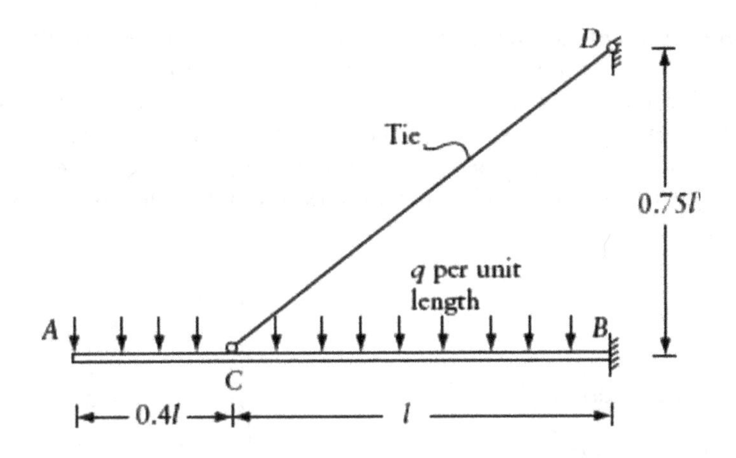

8.15 The figure shows a top view of a curved member of a horizontal grid. Determine the fixed-end forces due to the self-weight of the member, q/unit length. Angle $\theta = 1$ radian. Assume constant cross section, with $GJ/EI = 0.8$. Give the answer as components in the directions of the coordinates shown. Consider only deformations due to bending and torsion.

Hint: The only statically indeterminate force at the central section is a bending moment; the torsional moment and the shearing force are zero because of symmetry.

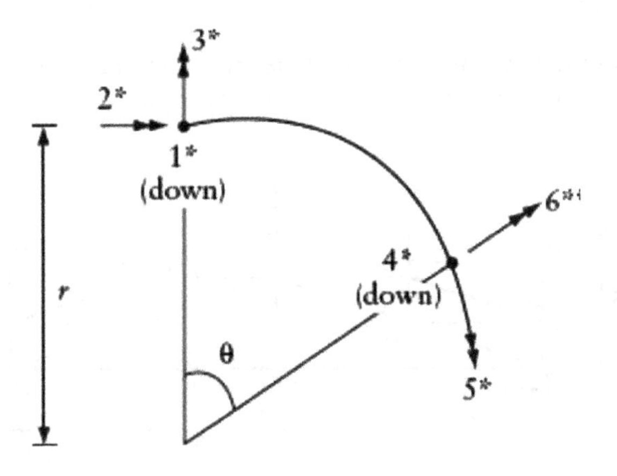

8.16 The figure represents half a symmetrical and symmetrically loaded three-span cable-stayed bridge. Find the changes in the cable forces and the bending moment diagram for AC and BD due to a uniform load q/unit length, covering the whole central span. Consider only axial deformation in the cables and bending deformation in AC and BD. Assume that the initial tension in the cable is sufficient to remain stretched under the effect of all subsequent loading. Ignore the effect of the initial tension on cable stiffness. Consider EI = constant for AC and BD and $(Ea)_{cable} = 2000EI/l^2$.

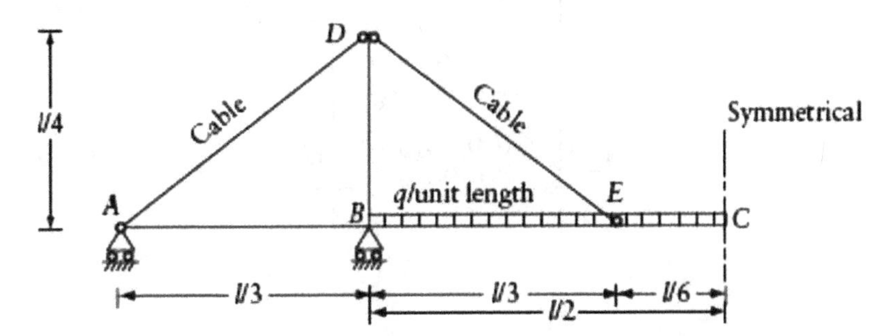

8.17 The figure represents a chimney ABC stayed by four cables BD, BE, BF, and BG. Find the reaction components at A due to a force P in the x direction at C. What is the displacement in the x direction at C? Assume that the cables have sufficient initial tension to remain stretched after deformation; ignore the effect of the initial tension on cable stiffness and the self-weight of the cables. Consider only bending deformation of the chimney and axial deformation of the cables; assume $(Ea)_{\text{cable}} = (20/l^2)\,(EI)_{\text{chimney}}$.

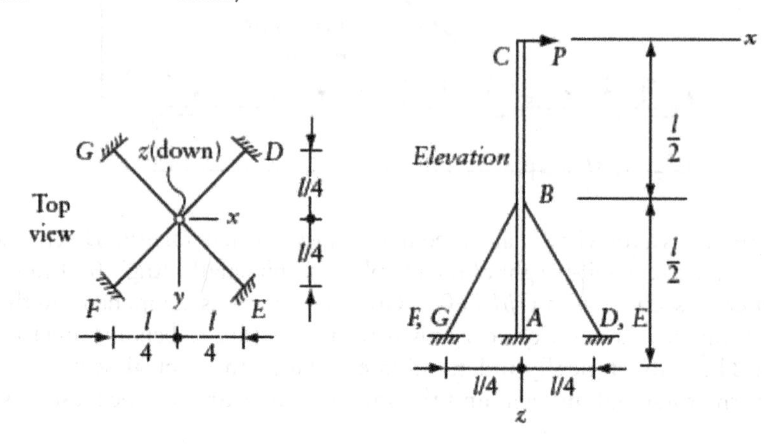

8.18 Find the forces in the members of the plane truss shown. All members have the same axial rigidity Ea except member AB, whose axial rigidity is $Ea/2$.

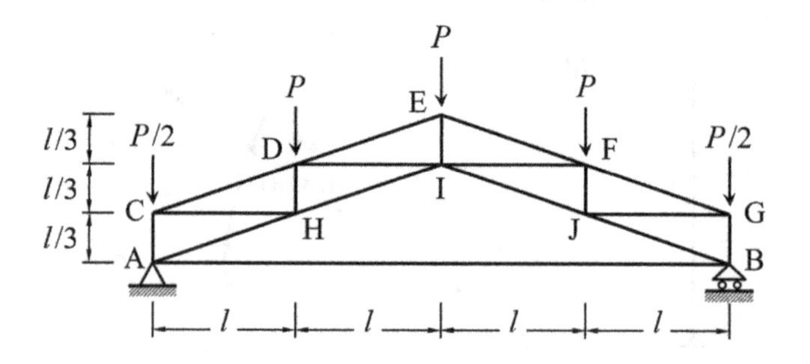

8.19 Find the horizontal displacement at C of the frame shown. Consider only bending deformation of the frame ABCD and the axial deformations of AC and BD. Assume $(Ea)_{\text{AC or BD}} = 20\,(EI)_{\text{ABCD}}/l^2$.

Hint: See Example 8.3.

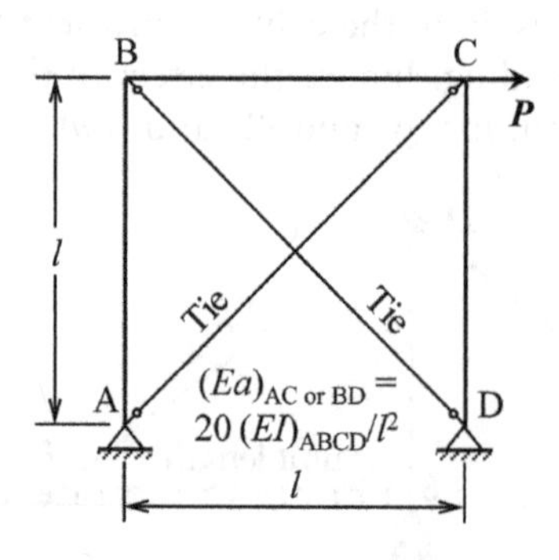

8.20 Use the unit displacement method to find the force F_1 in terms of the displacement D_1 in the beam in the figure. Assume the following approximate equation for the deflection:

$$y = -\frac{D_1}{2}\sin\frac{\pi x}{l}$$

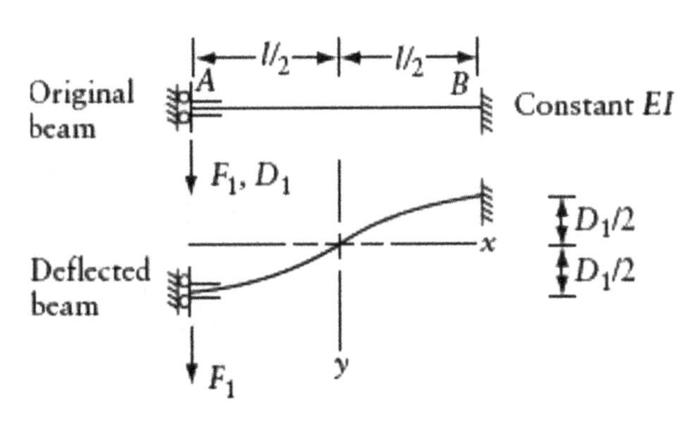

Chapter 9

Application of displacement method

Moment distribution

9.1 INTRODUCTION

Application of the displacement method has led to the classical procedure known as *moment distribution*. In the moment distribution method, for analysis of plane frames considering only bending deformations, the joint displacements are first assumed restrained. The effect of joint displacements is then introduced by successive iterations, which can be continued to any desired precision. Thus, the moment distribution is a displacement method of analysis. There is, however, a fundamental difference: in moment distribution generally, no equations are solved to find the joint displacements; instead, these displacements are allowed to take place in succession, and their effect on the end moments is introduced as a series of successive converging corrections. This absence of the need to solve simultaneous equations has made the moment distribution method an extremely popular one, especially when the calculations are done by a simple calculator.

The moment distribution procedure yields bending moments, and it is the value of the moments that are generally needed for design; thus, we avoid the procedure of first finding the joint displacements and then calculating the moments. A further advantage of the moment distribution procedure is that it is easily remembered and easily applied.

Despite the use of computers, which can rapidly analyze frames with numerous joints and members, it is useful to analyze such frames (or their isolated parts) by hand, using moment distribution, in preliminary design or to check the computer results. Performing such a check is of great importance.

9.2 END-ROTATIONAL STIFFNESS AND CARRYOVER MOMENT

Consider the prismatic frame member in Figure 9.1a, with the four coordinates representing end moments and shear forces. The elements of columns 2 and 4 of the stiffness matrix represent the end-forces when the beam is deflected as shown in Figures 9.1b and c, respectively. Thus, the stiffness matrix of the member in Figure 9.1a, may be written in the form:

$$[S]=\begin{bmatrix} (S_{AB}+S_{BA}+2t)/l^2 & (S_{AB}+t)/l & -(S_{AB}+S_{BA}+2t)/l^2 & (S_{BA}+t)/l \\ (S_{AB}+t)/l & S_{AB} & -(S_{AB}+t)/l & t \\ -(S_{AB}+S_{BA}+2t)/l^2 & -(S_{AB}+t)/l & (S_{AB}+S_{BA}+2t)/l^2 & -(S_{BA}+t)/l \\ (S_{BA}+t)/l & t & -(S_{BA}+t)/l & S_{BA} \end{bmatrix} \tag{9.1}$$

where the member end moment S_{AB} $(= S_{22})$, referred to as the *rotational stiffness* of end A of member AB (Figure 9.1b), is the moment required to produce unit rotation of end A while

DOI: 10.1201/9780429286858-9

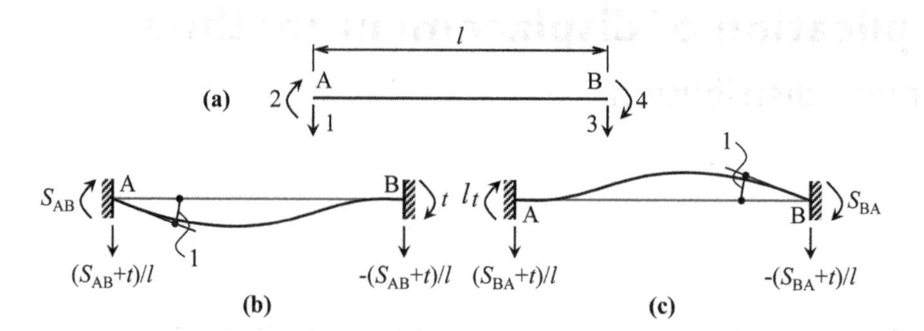

Figure 9.1 Member end-forces produced by unit rotations at the ends of a prismatic member. (a) Plane frame member with four coordinates. (b) and (c) Forces on columns 2 and 4, respectively.

end B is fixed; the corresponding moment t at end B is referred to as the *carryover moment*. The ratio $C_{AB} = t/S_{AB}$ is referred to as the *carryover factor* (COF) from A to B. Similarly, with respect to Figure 9.1c, the rotational stiffness of end B of member AB, S_{BA} is the moment at B required to produce unit rotation at B when end A is fixed; the corresponding carryover moment at A is also equal to t. The carryover factor from B to A is $C_{BA} = t/S_{BA}$.

For a prismatic member, the end-rotational stiffness, the carryover moment, and the carryover factor are:

$$S_{AB} = S_{BA} = \frac{4EI}{l}; \quad t = \frac{2EI}{l}; \quad C_{BA} = \frac{1}{2} \tag{9.2}$$

The end-rotational stiffnesses and the carryover factors will be used in the next section. Substitution of Equation 9.2 in 9.1 gives the stiffness matrix for a prismatic member (Equation 5.20).

9.3 PROCESS OF MOMENT DISTRIBUTION

The moment distribution method of analysis of framed structures was introduced by Hardy Cross[1] and was extended by others to the cases of structures subjected to high axial forces, and to axisymmetrical circular plates and shells of revolution.[2] Here, we deal mainly with plane frames supported in such a way that the only possible joint displacements are rotations without translation.

Consider the beam ABC of Figure 9.2a, encastré at A and C and continuous over support B. Let us assume first that the rotation of joint B is prevented by a restraining external couple acting at B. Due to the lateral loads on the members AB and BC, with the end rotations prevented at all ends, fixed-end moments (FEM) result at the ends (A, B), and (B, C).

Arbitrary values are assigned to these moments in Figure 9.2b, using the convention that a positive sign indicates a clockwise end moment. The restraining moment required to prevent the rotation of joint B is equal to the algebraic sum of the fixed-end moments of the members meeting at B, that is –50.

So far, the procedure has been identical with the general displacement method considered in Chapter 5. However, we now recognize that the joint B is, in fact, not restrained, and we allow it to rotate by removing the restraining moment of –50. The same effect is achieved

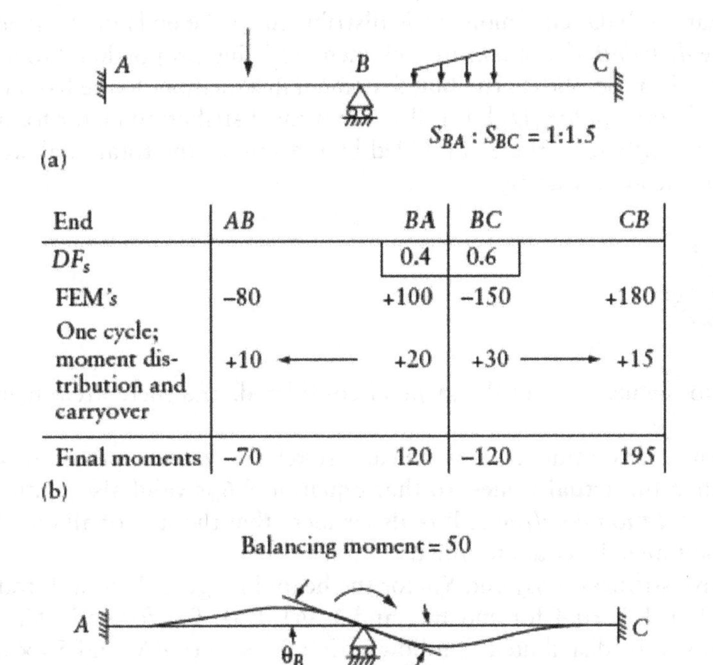

End	AB	BA	BC	CB
DF_s		0.4	0.6	
FEM's	−80	+100	−150	+180
One cycle; moment distribution and carryover	+10 ←	+20	+30 →	+15
Final moments	−70	120	−120	195

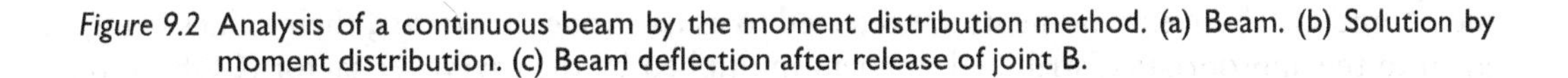

Figure 9.2 Analysis of a continuous beam by the moment distribution method. (a) Beam. (b) Solution by moment distribution. (c) Beam deflection after release of joint B.

by applying to the joint B an external couple $M = +50$ (Figure 9.2c), that is, a moment equal and opposite to the algebraic sum of the fixed-end moments at the joint. This is known as the *balancing moment.* Its application causes the ends of the members BA and BC meeting at B to rotate through the same angle θ_B; hence, end moments M_{BA} and M_{BC} develop. For equilibrium of the moments acting on joint B:

$$M = M_{BA} + M_{BC} \tag{9.3}$$

The end moments M_{BA} and M_{BC} can be expressed in terms of the rotational stiffnesses of end B of members AB and BC, viz. S_{BA} and S_{BC}, thus:

$$M_{BA} = \theta_B S_{BA} \quad \text{and} \quad M_{BC} = \theta_B S_{BC} \tag{9.4}$$

From Equations 9.3 and 9.4, the end moments can be expressed as a fraction of the balancing moment:

$$\left.\begin{aligned} M_{BA} &= \left(\frac{S_{BA}}{S_{BA} + S_{BC}}\right) M \\ M_{BC} &= \left(\frac{S_{BC}}{S_{BA} + S_{BC}}\right) M \end{aligned}\right\} \tag{9.5}$$

This means that the balancing moment is distributed to the ends of the members meeting at the joint, the *distributed moment* in each member being proportional to its relative rotational stiffness. The ratio of the distributed moment in a member to the balancing moment is called *the distribution factor (DF)*. It follows that the distribution factor for an end is equal to the rotational stiffness of the end divided by the sum of the rotational stiffnesses of the ends meeting at the joint, that is:

$$(DF)_i = \frac{S_i}{\sum_{j=1}^{n} S_j} \tag{9.6}$$

where i refers to the near end of the member considered, and there are n members meeting at the joint.

It is clear that to determine the DFs we can use relative values of the end-rotational stiffnesses rather than the actual values, so that Equation 9.6 is valid also, with S representing the *relative end-rotational stiffness*. It is also evident that the sum of all the DFs of the ends meeting at a joint must be equal to unity.

Let the relative stiffnesses S_{BA} and S_{BC} for the beam in Figure 9.2a be 1 and 1.5. The DFs are therefore: 1/(1+1.5) = 0.4 for end BA, and 1.5/(1+1.5) = 0.6 for end BC. The balancing moment of +50 will be distributed as follows: 50 × 0.4 = 20 to BA, and 50 × 0.6 = 30 to BC. The distributed moments are recorded in a table in Figure 9.2b.

The rotation of joint B produced in the previous step induces end moments at the far fixed ends A and C. These end moments are referred to as *carryover moments*, their values being equal to the appropriate distributed moment multiplied by the carryover factor (COF): C_{BA} from B to A, and C_{BC} from B to C. The value of a carryover factor depends upon the variation in the cross section of the member; for a prismatic member, COF = ½. When the far end of a member is pinned, the COF is, of course, zero.

The carryover moments in the beam considered are recorded in Figure 9.2b on the assumption that $C_{BA} = C_{BC} = ½$, that is, the cross section is taken to be constant within each span. The two arrows in the table pointing away from joint B indicate that the rotation at B (or moment distribution at B) causes the carryover moment of the value indicated at the head of the arrow. The use of the arrows makes it easier to follow a proper sequence of operations and also facilitates checking.

The process of moment distribution followed by carrying over is referred to as one *cycle*. In the problem considered, no further cycles are required as there is no out-of-balance moment. The final end moments are obtained by adding the end moments in the restrained condition (FEMs) to the moments caused by the rotation of joint B in the cycle in Figure 9.2b. However, if rotation can occur at more than one joint, further cycles of distribution and carryover have to be performed, as shown in the following example.

EXAMPLE 9.1: PLANE FRAME: JOINT ROTATIONS WITHOUT TRANSLATIONS

Use the method of moment distribution to analyze the frame in Figure 9.3a. If the axial deformations are ignored, the only possible joint displacements are rotations at B and C. We assume the frame to have prismatic members so that the carryover factor in all cases is ½. Then, the rotational end stiffness of any member is $S = 4EI/l$, where EI = flexural rigidity of the cross section and l = length of the member. It follows that the relative rotational stiffnesses can be taken as $K = I/l$. The relative K values for all the members are shown in Figure 9.3a. The applied load is assumed to be of such a magnitude as to produce the FEMs given in Figure 9.3b.

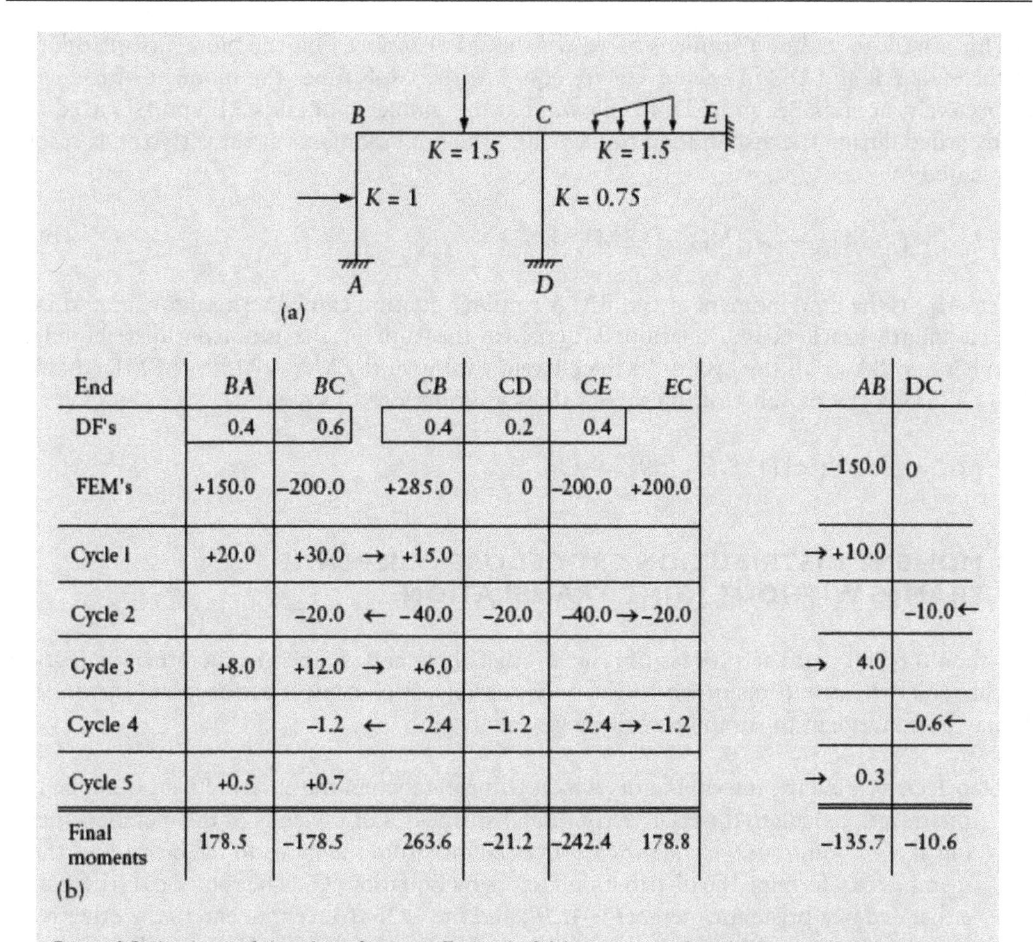

End	BA	BC	CB	CD	CE	EC	AB	DC
DF's	0.4	0.6	0.4	0.2	0.4			
FEM's	+150.0	-200.0	+285.0	0	-200.0	+200.0	-150.0	0
Cycle 1	+20.0	+30.0 →	+15.0				→ +10.0	
Cycle 2		-20.0 ←	-40.0	-20.0	-40.0 →	-20.0		-10.0 ←
Cycle 3	+8.0	+12.0 →	+6.0				→ 4.0	
Cycle 4		-1.2 ←	-2.4	-1.2	-2.4 →	-1.2		-0.6 ←
Cycle 5	+0.5	+0.7					→ 0.3	
Final moments	178.5	-178.5	263.6	-21.2	-242.4	178.8	-135.7	-10.6

Figure 9.3 Analysis of the plane frame in Example 9.1 by moment distribution without joint translation. (a) Plane frame without joint translation. (b) Moment distribution.

The DFs are calculated by Equation 9.6 and indicated in Figure 9.3b. The first cycle of moment distribution and carryover is done by allowing joint B to rotate; joint C continues to be restrained. In the second cycle, joint C is allowed to rotate, joint B now being restrained. The balancing moment for this cycle is equal to minus the algebraic sum of the FEMs at CB, CD, and CE plus the carryover moment caused by the rotation of joint B in the previous cycle, that is $-(285+0-200+15)=-100$. The second cycle is terminated by the carryover of moments to the far ends of the three members meeting at C, as shown by arrows. It is evident that cycle 2 has induced an imbalance at joint B, that is, if joint B is now released, a further rotation will occur. The effect of this release is followed through in the same way as in cycle 1, but with the balancing moment equal to minus the moment carried over to BC from the previous cycle.

The carryover in the third cycle results in an unbalanced moment at joint C. In order to remove the external constraint, joint C must be balanced again. The process is repeated until the unbalanced moments at all the joints are so small as to be considered negligible. The final moments are obtained by adding the FEMs to the end moments produced by the joint displacement allowed in *all* the cycles. Because we reach a final position of equilibrium under the applied loading, the sum of the final end moments at any joint must be zero. This fact can be used as a check on the arithmetic in the distribution process.

No distribution is carried out at the fixed ends A, D, and E because the ends of the members at these points can be imagined to be attached to a body of infinite rigidity. Thus, the

DF for a built-in end of a frame is zero. We should also note that the moments introduced at the ends AB and DC in each cycle are equal to the COF times the moment introduced, respectively, at ends BA and CD. It follows that the moments at ends AB and DC need not be recorded during the distribution process, and the final moments at these two ends can be calculated by:

$$M_{AB} = (\text{FEM})_{AB} + C_{BA}[M_{BA} - (\text{FEM})_{BA}] \tag{9.7}$$

where M_{BA} is the final moment at end BA. A similar equation can be written for the end DC.

The square brackets in Equation 9.7 enclose the sum of the moments distributed to member end BA in all the cycles. In the current example: $(\text{FEM})_{AB} = -150$; $(\text{FEM})_{BA} = +150$; $M_{BA} = 178.5$; $C_{BA} = ½$. Substituting these values in Equation 9.7 gives:

$$M_{AB} = -150 + (½)(178.5 - 150) = -135.75$$

9.4 MOMENT DISTRIBUTION PROCEDURE FOR PLANE FRAMES WITHOUT JOINT TRANSLATION

We should recall that the process of moment distribution described in the previous section applies solely to structures in which the only possible displacement at the joints is rotation. It may be convenient to summarize the steps involved.

Step 1 Determine the internal joints which will rotate when the external load is applied to the frame. Calculate the relative rotational stiffnesses of the ends of the members meeting at these joints, as well as the carryover factors from the joints to the far ends of these members. Determine the distribution factors by Equation 9.6. The rotational stiffness of either end of a prismatic member is $4EI/l$, and the COF from either end to the other is ½. If one end of a prismatic member is hinged, the rotational end stiffness of the other end is $3EI/l$, and, of course, no moment is carried over to the hinged end. In a frame with all members prismatic, the relative rotational end stiffness can be taken as $K = I/l$, and when one end is hinged, the rotational stiffness at the other end is $(3/4)\ K = (3/4)\ I/l$.

Step 2 With all joint rotations restrained, determine the fixed-end moments due to the lateral loading on all the members.

Step 3 Select the joints to be released in the first cycle. It may be convenient to take these as alternate internal joints. (For example, in the frame of Figure 9.4, we can release in the first

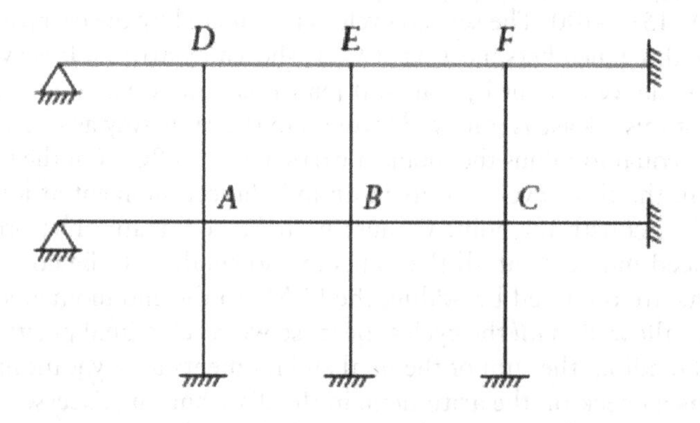

Figure 9.4 Plane frame without joint translation.

cycle either A, C, and E or D, B, and F.) Calculate the balancing moment at the selected joints; this is equal to minus the algebraic sum of the fixed-end moments. If an external clockwise couple acts at any joint, its value is simply added to the balancing moment.

Step 4 Distribute the balancing moments to the ends of the members meeting at the released joints. The distributed moment is equal to the DF multiplied by the balancing moment. The distributed moments are then multiplied by the COFs to give the carryover moments at the far ends. Thus, the first cycle is terminated.

Step 5 Release the remaining internal joints, while further rotation is prevented at the joints released in the first cycle. The balancing moment at any joint is equal to minus the algebraic sum of the FEMs and of the end moments carried over in the first cycle. The balancing moments are distributed and moments are carried over to the far ends in the same way as in Step 3. This completes the second cycle.

Step 6 The joints released in Step 3 are released again, while the rotation of the other joints is prevented. The balancing moment at a joint is equal to minus the algebraic sum of the end moments carried over to the ends meeting at the joint in the previous cycle.

Step 7 Repeat Step 6 several times, for the two sets of joints in turn, until the balancing moments become negligible.

Step 8 Sum the end moments recorded in each of Steps 2 to 7 to obtain the final end moments. The reactions or stress resultants, if required, may then be calculated by simple equations of statics.

If a frame has an overhanging part, its effect is replaced by a force and a couple acting at the joint of the overhang with the rest of the structure. This is illustrated in the following example.

EXAMPLE 9.2: CONTINUOUS BEAM

Obtain the bending moment diagram for the continuous beam of Figure 9.5a.

The effect of the cantilever AB on the rest of the beam is the same as that of a downward force, qb and an anticlockwise couple of $0.6qb^2$ acting at B, as shown in Figure 9.5b. The end B is thus free to rotate; the distribution has to be carried out at joints C, D, and E. During the distribution, joint B (Figure 9.5b) will be free to rotate so that B is considered to be a hinge. No moment is therefore carried over from C to B, and the relative end-rotational stiffness of (3/4) $K = (3/4)\ I/l$. Since I is constant throughout, it may be taken as unity. The relative rotational stiffnesses of the ends at the three joints C, D, and E are recorded in Figure 9.5c, together with the COFs and the DFs.

Apply the external applied loads on the beam BCDEF (Figure 9.5b), with the rotation at B allowed and prevented at C, D, and F. In this state, member BC has the end moment $(-0.6qb^2)$ at B and $C_{BC}\ (-0.6qb^2)$ at C; where C_{BC} = the carryover factor from B to C = ½ (Section 9.2); member DE has the end moment $[-2.4q(2b)^2/12]$ at D and $[2.4q(2b)^2/12]$ at E (Equation B.7, Appendix B); the remaining end moments are nil because members CD and EF have no load.

The moment distribution is performed at joint D in one cycle and at joints C and E in the following cycle. The moments produced at ends CB, EF, and FE are not recorded; the final moments at these ends are calculated from the values of the final moments at the other end of the respective members. First, we write $M_{CB} = -M_{CD} = -0.285qb^2$ and $M_{EF} = -M_{ED} = -0.532qb^2$. Then, by an equation similar to Equation 9.7:

$$M_{FE} = (\text{FEM})_{FE} + C_{EF}[M_{EF} - (\text{FEM})_{EF}]$$

The FEMs at the two ends of member EF are zero and $C_{EF} = ½$; therefore, $M_{FE} = (½)\ M_{EF} = -0.266\ qb^2$. The final bending moment diagram of the beam is shown in Figure 9.5d.

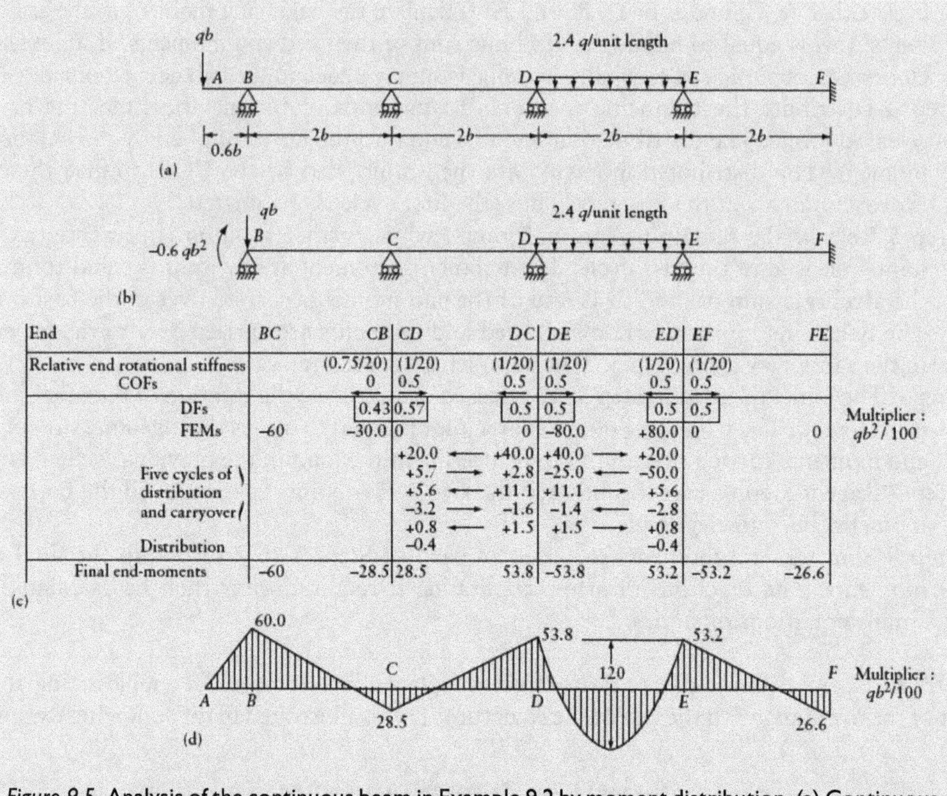

End	BC	CB	CD	DC	DE	ED	EF	FE
Relative end rotational stiffness		(0.75/20)	(1/20)	(1/20)	(1/20)	(1/20)	(1/20)	
COFs		0	0.5	0.5	0.5	0.5	0.5	
DFs		0.43	0.57	0.5	0.5	0.5	0.5	
FEMs	−60	−30.0	0	0	−80.0	+80.0	0	0
Five cycles of distribution and carryover			+20.0	+40.0	+40.0	+20.0		
			+5.7	+2.8	−25.0	−50.0		
			+5.6	+11.1	+11.1	+5.6		
			−3.2	−1.6	−1.4	−2.8		
			+0.8	+1.5	+1.5	+0.8		
Distribution			−0.4			−0.4		
Final end-moments	−60	−28.5	28.5	53.8	−53.8	53.2	−53.2	−26.6

Figure 9.5 Analysis of the continuous beam in Example 9.2 by moment distribution. (a) Continuous beam. (b) Replacement of the actual load on overhang by equivalent loading at B. (c) Moment distribution. (d) Bending moment diagram.

9.5 ADJUSTED END-ROTATIONAL STIFFNESSES

The process of moment distribution can be made shorter in certain cases if adjusted end-rotational stiffnesses are used instead of the usual stiffnesses. Expressions will be derived for these adjusted end-rotational stiffnesses of prismatic members.

The end-rotational stiffness S_{AB} was defined in Section 9.2 as the value of the moment required at A to rotate the beam end A through a unit angle while the far end B is fixed. Similarly, S_{BA} is the end moment to produce a unit rotation at B while end A is fixed. The deflected shapes of the beam corresponding to these two conditions are shown in Figure 9.6a. The moment t at the fixed end has the same value for the two deflected configurations. The rotated ends in Figure 9.6a are sketched with a roller support, but they can also be represented as in Figures 9.1b and c. Both figures indicate the same conditions, that is, a unit rotation without transverse translation of the end. The axial and shear deformations are ignored for the present purposes. For a prismatic beam, $S_{AB} = S_{BA} = 4EI/l$ and $t = 2EI/l$.

The special cases, which we shall now consider are: (a) when rotation is applied at one end of a member whose far end is not fixed but hinged (Figure 9.6b); (b) when such a member is subjected to symmetrical or anti-symmetrical end moments and rotations (Figures 9.6c and d). The adjusted end-rotational stiffnesses in these cases will be denoted by S with a subscript indicating the beam end and a superscript indicating the conditions at the far end as defined in Figure 9.6. Let us express the adjusted end-rotational stiffnesses $S_{AB}^{①}$, $S_{AB}^{②}$, and $S_{AB}^{③}$ in terms of the stiffnesses, S_{AB}, S_{BA}, and t for the same beam.

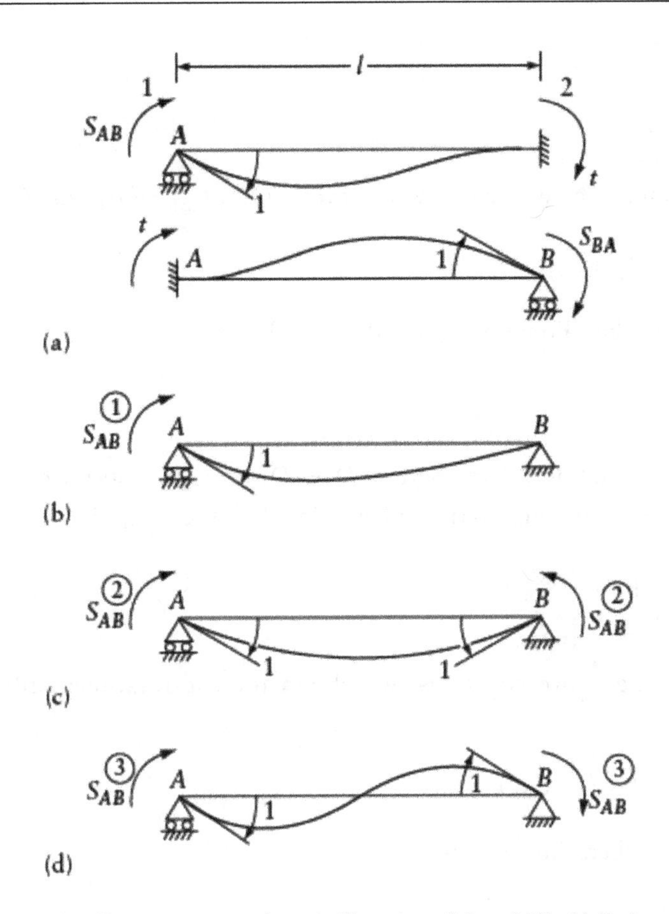

Figure 9.6 End-rotational stiffnesses in special cases (Equations 9.6 to 9.13). (a) End moments caused by a unit rotation at one end while the other end is fixed. (b) End moment caused by a unit rotation at end A while end B is hinged. (c) End moments caused by symmetrical unit rotations at ends A and B. (d) End moments caused by anti-symmetrical unit rotations at ends A and B.

The forces and displacements along the coordinates 1 and 2 in Figure 9.6a are related by:

$$[S]\,\{D\} = \{F\} \tag{9.8}$$

where

$$[S] = \begin{bmatrix} S_{AB} & t \\ t & S_{BA} \end{bmatrix}$$

and where $\{D\}$ are the end-rotations and $\{F\}$ the end moments.

Putting $D_1 = 1$ and $F_2 = 0$ in Equation 9.8 represents the conditions in Figure 9.6b. The force F_1 in this case will be equal to the adjusted stiffness $S_{AB}^{①}$. Thus:

$$\begin{bmatrix} S_{AB} & t \\ t & S_{BA} \end{bmatrix} \begin{Bmatrix} 1 \\ D_2 \end{Bmatrix} = \begin{Bmatrix} S_{AB}^{①} \\ 0 \end{Bmatrix}$$

Solving:

$$S_{AB}^{①} = S_{AB} - \frac{t^2}{S_{BA}} \tag{9.9}$$

The same equation can be written in terms of the COFs, $C_{AB} = t/S_{AB}$, and $C_{BA} = t/S_{BA}$, thus:

$$S_{AB}^{①} = S_{AB}\left(1 - C_{AB}\, C_{BA}\right) \tag{9.10}$$

For a prismatic member, Equations 9.9 and 9.10 reduce to:

$$S_{AB}^{①} = \frac{3EI}{l} \tag{9.11}$$

Referring again to Equation 9.8 and putting $D_1 = -D_2 = 1$, represents the conditions in Figure 9.6c. The force $F_1 = -F_2$ is equal to the end-rotational stiffness $S_{AB}^{②}$. Thus:

$$\begin{bmatrix} S_{AB} & t \\ t & S_{BA} \end{bmatrix} \begin{Bmatrix} 1 \\ -1 \end{Bmatrix} = \begin{Bmatrix} S_{AB}^{②} \\ -S_{AB}^{②} \end{Bmatrix}$$

Solving either of the above equations, we obtain the end-rotational stiffness in case of symmetry:

$$S_{AB}^{②} = S_{AB} - t = S_{AB}\left(1 - C_{AB}\right) \tag{9.12}$$

For a prismatic member, this reduces to:

$$S_{AB}^{②} = \frac{2EI}{l} \tag{9.13}$$

Similarly, if we put $D_1 = D_2 = 1$, Equation 9.8 represents the conditions in Figure 9.6d. The end-forces are $F_1 = F_2 = S_{AB}^{③}$, and the end-rotational stiffness in the anti-symmetrical case is:

$$S_{AB}^{③} = S_{AB} + t = S_{AB}\left(1 + C_{AB}\right) \tag{9.14}$$

For a prismatic member, this reduces to:

$$S_{AB}^{③} = \frac{6EI}{l} \tag{9.15}$$

9.6 ADJUSTED FIXED-END MOMENTS

Figure 9.7a represents a straight member with two fixed ends subjected to transverse loading. Let the end moments for this beam be M_{AB} and M_{BA} and let the vertical reactions be F_A and F_B. The beam is assumed to be prismatic with end-rotational stiffnesses S_{AB} and S_{BA}, the carryover moment t, and the carryover factors C_{AB} and C_{BA}. Let us now find the end moments for an identical beam subjected to the same transverse loading but with the support conditions of Figure 9.7b. In this figure, the end A is hinged; thus, rotation at A is free to take place and there is only one end moment $M_{BA}^{①}$ to be determined.

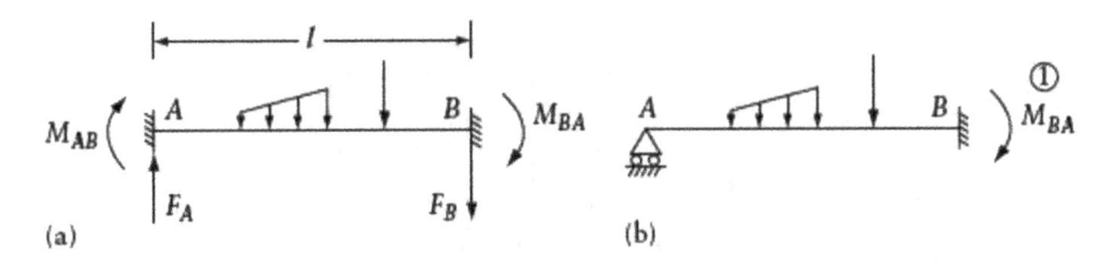

Figure 9.7 Adjusted fixed-end moments. (a) Two ends encastré. (b) Free rotation at A.

To shorten the process of moment distribution, in some cases we can use the adjusted fixed-end moment $M_{BA}^{①}$ together with the adjusted end-rotational stiffnesses. Consider first the beam in Figure 9.7a with the end-displacements restrained; the end moments are M_{AB} and M_{BA}. Now, allow the end A to rotate by an amount such that a moment $-M_{AB}$ is developed at end A. The corresponding moment developed at B is $-C_{AB}\, M_{AB}$. Superposing the end moments developed by the rotation will give a zero moment at A, and this represents the condition of the beam in Figure 9.7b. The adjusted FEM at B when end A is hinged is therefore:

$$M_{BA}^{①} = M_{BA} - C_{AB}\, M_{AB} \tag{9.16}$$

where M_{AB} and M_{BA} are the FEMs of the same beam with the two ends encastré, and C_{AB} is the COF from A to B. For a prismatic beam, Equation 9.16 becomes:

$$M_{BA}^{①} = M_{BA} - \frac{M_{AB}}{2} \tag{9.17}$$

This equation can be used to derive Equations B.22, B.25, and B.28 of Appendix B from other equations in the appendix.

EXAMPLE 9.3: PLANE FRAME SYMMETRY AND ANTI-SYMMETRY

Find the bending moment in the symmetrical frame shown in Figure 9.8a by replacing the loading by equivalent symmetrical and anti-symmetrical loadings.

Any load on a symmetrical structure can be replaced by the sum of a symmetrical and an anti-symmetrical loading, as indicated in Figures 9.8b and c. The same concept can be used in computer analysis of large structures which exceeds the capacity of the available computer; by virtue of the structural symmetry, while the loading is not symmetrical, it is possible to find the solution by superposition of two or more analyses performed on a repetitive part of the structure.

We shall now introduce this concept for a relatively simple frame, using the adjusted stiffnesses and FEMs, although this does not achieve much saving in calculations. With symmetry or anti-symmetry of loading, the moment distribution need be done for one-half of the frame only. Figures 9.8d and e deal with the symmetrical and anti-symmetrical cases, respectively. The end-rotational stiffnesses for the members meeting at B are calculated using Equations 9.11, 9.13, and 9.15, and we find for the symmetrical case:

$$S_{BA} : S_{BE} : S_{BC} = 3K_{BA} : 4K_{BE} : 2K_{BC}$$

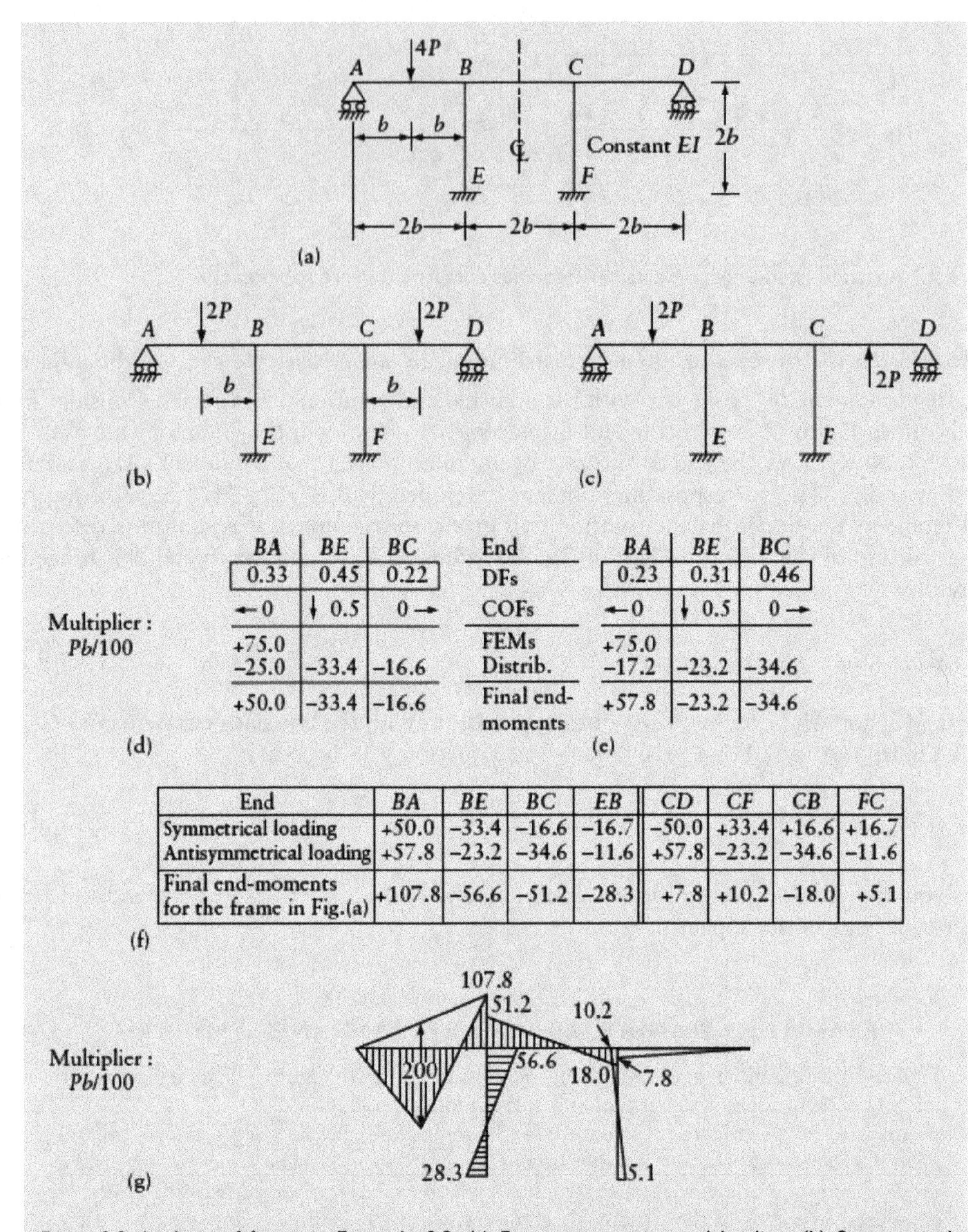

End	BA	BE	BC	EB	CD	CF	CB	FC
Symmetrical loading	+50.0	−33.4	−16.6	−16.7	−50.0	+33.4	+16.6	+16.7
Antisymmetrical loading	+57.8	−23.2	−34.6	−11.6	+57.8	−23.2	−34.6	−11.6
Final end-moments for the frame in Fig.(a)	+107.8	−56.6	−51.2	−28.3	+7.8	+10.2	−18.0	+5.1

Figure 9.8 Analysis of frame in Example 9.3. (a) Frame properties and loading. (b) Symmetrical loading. (c) Anti-symmetrical loading. (d) Moment distribution for the symmetrical case. (e) Moment distribution for the anti-symmetrical case. (f) Summation of end moments calculated in parts (d) and (e). (g) Bending moment diagram for the frame in part (a).

and for the anti-symmetrical case:

$$S_{BA} : S_{BE} : S_{BC} = 3K_{BA} : 4K_{BE} : 6K_{BC}$$

where $K = I/l$. In this frame, K is the same for all members. The DFs are calculated in the usual way and are given in Figures 9.8d and e.

The FEM in end BA (for a beam with one end hinged) is obtained by the use of Appendix B and Equation 9.16; thus, $(FEM)_{BA} = 0.75Pb$ for both cases. The COF, $C_{BE} = 0.5$. No moments are carried over from B to C or from B to A. Thus, only one cycle of moment distribution is required at B, as shown in Figures 9.8d and e.

It is important to note that in the symmetrical case, the end moments in the right-hand half of the frame are equal in magnitude and opposite in sign to the end moments in the left-hand half, while in the anti-symmetrical case, they are equal and of the same sign in the two halves. The summation of the end moments in the two cases is carried out in the table in Figure 9.8f. This gives the end moments of the frame in Figure 9.8a. The corresponding bending moment diagram is shown in Figure 9.8g.

9.7 GENERAL

The moment distribution is used in this chapter for the analysis of plane frames considering bending deformations only. Solution of simultaneous equations is avoided in structures not involving translation of the joints. The amount of computation can be further reduced for symmetrical frames subjected to symmetrical or anti-symmetrical loading (and any loading can be resolved into such components) by using adjusted end-rotational stiffnesses and FEMs.

REFERENCES

1. Hardy Cross, "Analysis of Continuous Frames by Distributing Fixed-End Moments," *Trans. ASCE.*, Paper 1793 (96), 1932.
2. Gere, J.M., *Moment Distribution*, Van Nostrand, New York, 1963, pp. 247–268.

PROBLEMS

9.1 and 9.2 Considering bending deformations only, find the bending moment diagram and the reactions (if any) for the plane frame shown.

Hint: Where possible, take advantage of the symmetry of the structure or of both the structure and the loading. In the latter case, cutting a section at an axis of symmetry gives a released structure with two unknown internal forces, instead of three (the third one is known to be nil).

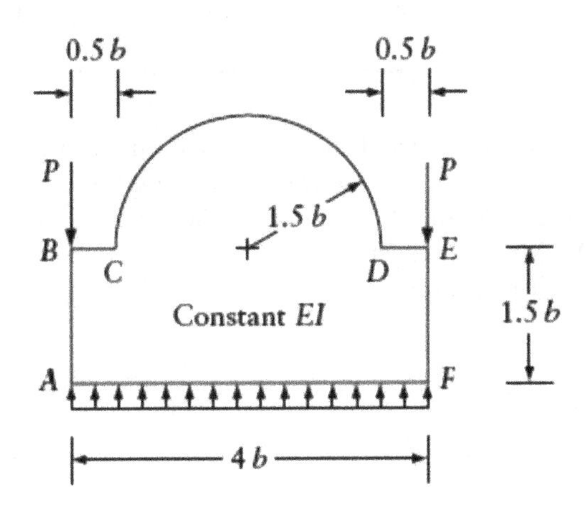

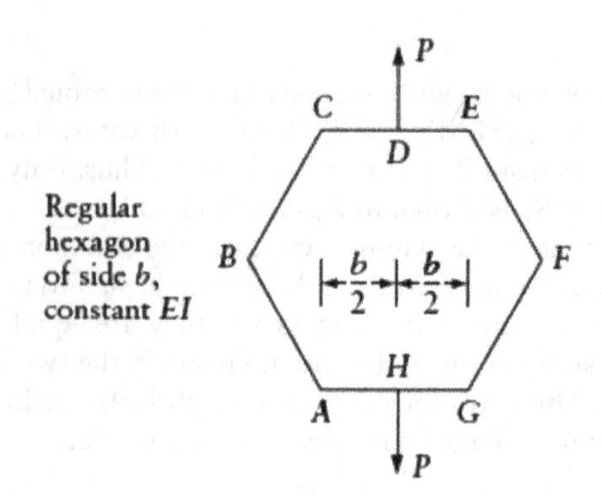

9.3 Find the forces at the ends A and B of the gable frame shown, subjected to a uniform vertical load of intensity q per unit length of horizontal projection. The frame is assumed to be encastré at A and B. Consider bending deformations only.

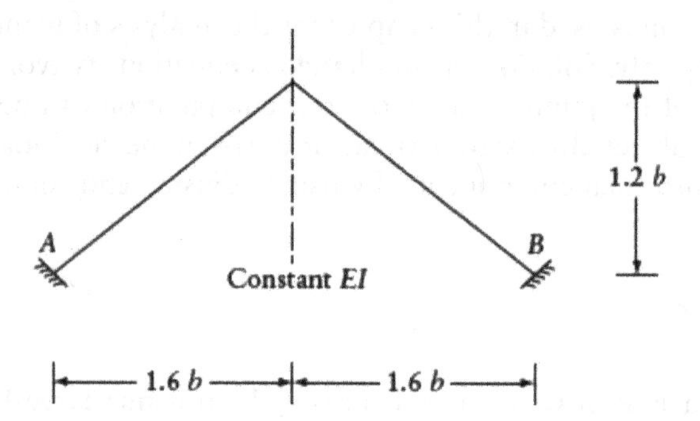

9.4 Find the bending moment diagram for the frame shown in the figure. Find the three reaction components at A.

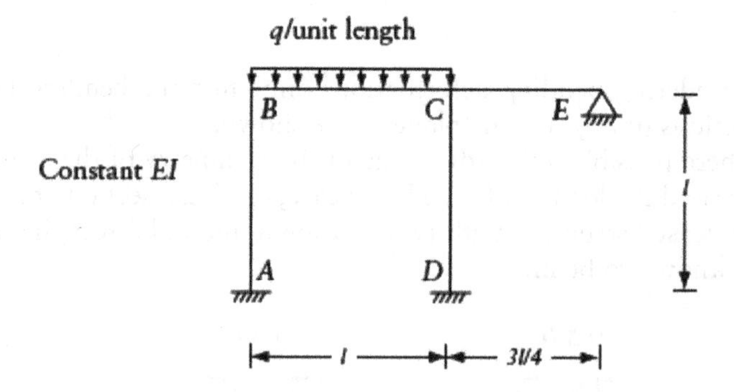

9.5 The beam in the figure is subjected to a dead load and a live load of intensities q and $2q$ per unit length, respectively. Draw the curve of maximum moment.

Hint: Solve for the following four cases of loading:

(i) D.L. on AD with L.L. on AB
(ii) D.L. on AD with L.L. on AC

(iii) D.L. on AD with L.L. on BC
(iv) D.L. on AD with L.L. on CD.

Draw the four bending moment diagrams in one figure to the same scale. The curve which has the maximum ordinates at any portion of the beam is the required curve.

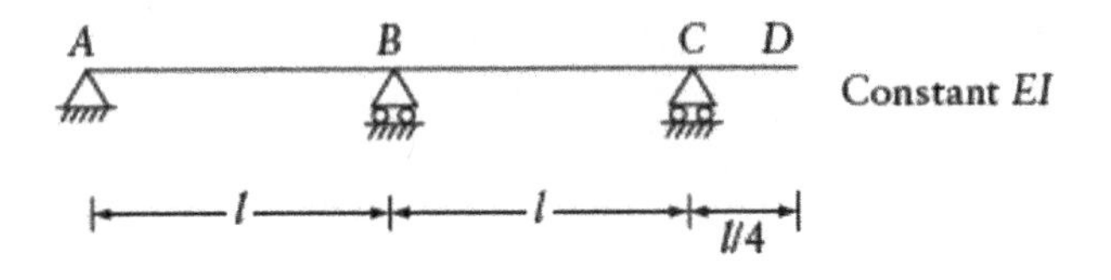

Chapter 10

Effects of axial forces on flexural stiffness

10.1 INTRODUCTION

The forces required to cause a unit rotation or translation in the transverse direction at one end of a member decrease in the presence of axial compressive force and an increase in the presence of axial tensile force. The presence of relatively high axial forces in slender members changes significantly their n bending stiffness. We refer to this as the *beam-column effect* (or, casually, the *P-delta effect*).

In the following sections, we discuss the beam-column effect in plane frames and. calculate the critical buckling loads for members and frames. The P-delta effect of gravity load on a laterally deformed structure can significantly influence its earthquake response beyond elasticity (see Section 17.18.1).

10.2 STIFFNESS OF A PRISMATIC MEMBER SUBJECTED TO AN AXIAL FORCE

We recall that the force and displacement methods of analysis considered in the previous chapters are for linear structures for which the principle of superposition holds. In such structures, the deformations are proportional to the applied loads; superposition of set of effects applies to displacements and internal forces.

In Section 3.6, we saw that the superposition of deflections does not apply in the case of a strut subjected to axial compression together with a transverse load because of the additional moment caused by the change in geometry of the member by the loading. However, in presence of unaltered axial force, the displacements induced by a system of loads can be superimposed. Thus, the axial force becomes a parameter, affecting the stiffness or the flexibility, just like I. Once these have been determined, the methods of analysis of linear structures apply.

In the analysis of rigid frames with very slender members, the axial forces are generally unknown at the outset of the analysis. We estimate a set of axial forces and employ them to determine the stiffness (or flexibility) of the members; then we analyze the frame as a linear structure. If the results show that the axial forces obtained by this analysis differ greatly from the assumed values, we use the calculated values to find new stiffness (or flexibility) and repeat the analysis. This simplified analysis, applicable for many structures, does not fully consider the change of geometry due to loading. The nodes are in equilibrium in their initial locations – not in their real displaced positions in the presence of applied load. For very flexible systems, equilibrium equations use trial displacement values and improve the accuracy by iterations (see Chapter 18).

Let us now consider the effects of an axial compressive and axial tensile force on the stiffness of a prismatic member.

DOI: 10.1201/9780429286858-10

10.3 EFFECT OF AXIAL COMPRESSION

The differential equation governing the deflection y of a prismatic member AB subjected to a compressive force P and any end restraint (Figure 10.1a and Equation 10.7) is:

$$\frac{d^4y}{dx^4} + \frac{P}{EI}\frac{d^2y}{dx^2} = \frac{q}{EI} \tag{10.1}$$

where q is the intensity of transverse loading. When $q = 0$, the general solution of Equation 10.1 is:

$$y = A_1 \sin u\frac{x}{l} + A_2 \cos u\frac{x}{l} + A_3\, x + A_4 \tag{10.2}$$

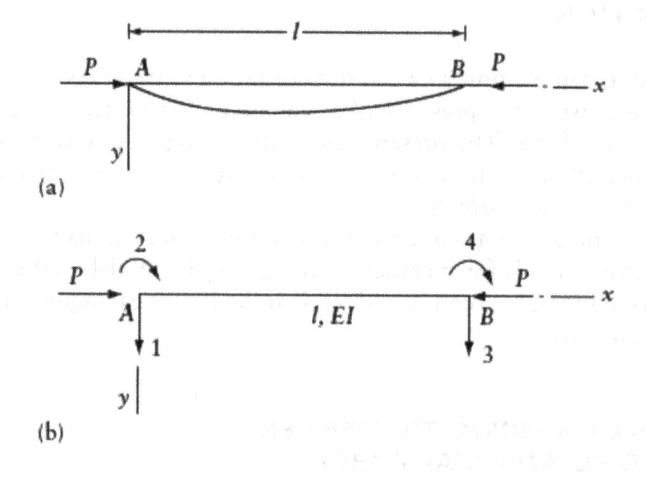

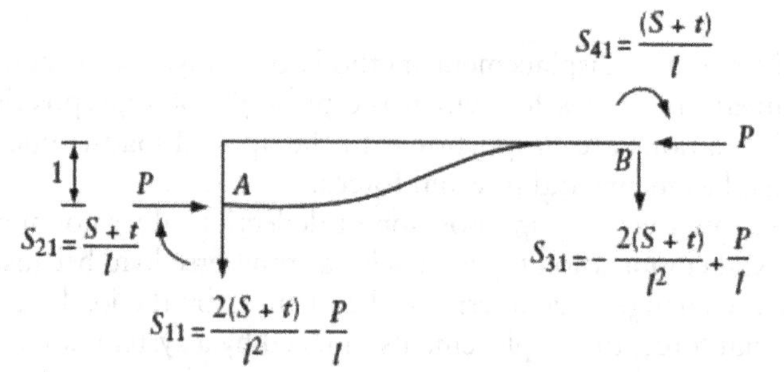

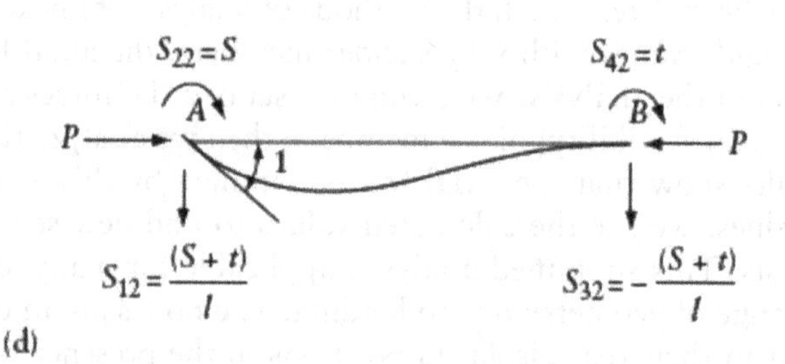

Figure 10.1 Stiffness of a strut. (a) Deflection of a strut assumed to be subjected to end-forces and end displacements at the coordinates indicated in part (b). (b) Coordinate system corresponding to the stiffness matrix of a strut in Equation 10.14. (c) End-forces corresponding to $D_1 = 1$, while $D_2 = D_3 = D_4 = 0$. (d) End-forces corresponding to $D_2 = 1$, while $D_1 = D_3 = D_4 = 0$.

where

$$u = l\sqrt{\frac{P}{EI}} \tag{10.3}$$

A_1, A_2, A_3, and A_4 are the integration constants to be determined from the boundary conditions.

We use Equation 10.2 to derive the stiffness matrix of an axially compressed member corresponding to the coordinates 1, 2, 3, and 4 in Figure 10.1b. The displacements $\{D\}$ at the four coordinates:

$$D_1 = (y)_{x=0};\quad D_2 = \left(\frac{dy}{dx}\right)_{x=0};\quad D_3 = (y)_{x=l};\quad D_4 = \left(\frac{dy}{dx}\right)_{x=l}$$

They relate to the constants $\{A\}$ by the equation:

$$\begin{Bmatrix} D_1 \\ D_2 \\ D_3 \\ D_4 \end{Bmatrix} = \begin{bmatrix} 0 & 1 & 0 & 1 \\ \frac{u}{l} & 0 & 1 & 0 \\ s & c & l & 1 \\ \frac{u}{l}c & -\frac{u}{l}s & 1 & 0 \end{bmatrix} \begin{Bmatrix} A_1 \\ A_2 \\ A_3 \\ A_4 \end{Bmatrix} \tag{10.4}$$

$s = \sin u$ and $c = \cos u$.

$$\{D\} = [B]\,\{A\} \tag{10.5}$$

$[B]$ is the 4×4 matrix in Equation 10.4.

The forces $\{F\}$ at the four coordinates are the shear and bending moment at $x = 0$ and $x = l$; thus:

$$\begin{Bmatrix} F_1 \\ F_2 \\ F_3 \\ F_4 \end{Bmatrix} = \begin{Bmatrix} (-V)_{x=0} \\ (M)_{x=0} \\ (V)_{x=l} \\ (-M)_{x=l} \end{Bmatrix} = EI \begin{Bmatrix} \left(\frac{d^3y}{dx^3} + \frac{u^2}{l^2}\frac{dy}{dx}\right)_{x=0} \\ \left(-\frac{d^2y}{dx^2}\right)_{x=0} \\ \left(-\frac{d^3y}{dx^3} - \frac{u^2}{l^2}\frac{dy}{dx}\right)_{x=l} \\ \left(\frac{d^2y}{dx^2}\right)_{x=l} \end{Bmatrix} \tag{10.6}$$

Equation 10.1 expresses the bending moment M in terms of the deflections. Using this equation with Equations 7.46 and 10.3, the shear can be expressed as:

$$V = \frac{dM}{dx} - P\frac{dy}{dx} = EI\left(-\frac{d^3y}{dx^3} - \frac{u^2}{l^2}\frac{dy}{dx}\right) \tag{10.7}$$

Differentiating Equation 10.2 and substituting in Equation 10.6 give:

$$\begin{Bmatrix} F_1 \\ F_2 \\ F_3 \\ F_4 \end{Bmatrix} = EI \begin{bmatrix} 0 & 0 & \frac{u^2}{l^2} & 0 \\ 0 & \frac{u^2}{l^2} & 0 & 0 \\ 0 & 0 & -\frac{u^2}{l^2} & 0 \\ -\frac{su^2}{l^2} & -\frac{cu^2}{l^2} & 0 & 0 \end{bmatrix} \begin{Bmatrix} A_1 \\ A_2 \\ A_3 \\ A_4 \end{Bmatrix} \tag{10.8}$$

or:

$$\{F\} = [C]\{A\} \tag{10.9}$$

$[C] = EI$ times the 4×4 matrix in Equation 10.8.

Solving for $\{A\}$ from Equation 10.5 and substituting in Equation 10.9:

$$\{F\} = [C]\,[B]^{-1}\{D\} \tag{10.10}$$

Putting:

$$[S] = [C][B]^{-1} \tag{10.11}$$

Equation 10.10 takes the form:

$$\{F\} = [S]\{D\} \tag{10.12}$$

where $[S]$ is the required stiffness matrix. The inverse of $[B]$ is:

$$[B]^{-1} = \frac{1}{2-2c-us} \begin{bmatrix} -s & \frac{l}{u}(1-c-us) & s & -\frac{l}{u}(1-c) \\ (1-c) & \frac{l}{u}(s-uc) & -(1-c) & \frac{l}{u}(u-s) \\ \frac{u}{l}s & (1-c) & -\frac{u}{l}s & (1-c) \\ (1-c-us) & -\frac{l}{u}(s-uc) & (1-c) & -\frac{l}{u}(u-s) \end{bmatrix} \tag{10.13}$$

Substituting Equation 10.13 into Equation 10.11, we obtain the stiffness of a strut corresponding to the coordinates in Figure 10.1b:

$$[S] = EI \begin{bmatrix} \frac{u^3 s}{l^3(2-2c-us)} & & & \text{Symmetrical} \\ \frac{u^2(1-c)}{l^2(2-2c-us)} & \frac{u(s-uc)}{l(2-2c-us)} & & \\ -\frac{u^3 s}{l^3(2-2c-us)} & -\frac{u^2(1-c)}{l^2(2-2c-us)} & \frac{u^3 s}{l^3(2-2c-us)} & \\ \frac{u^2(1-c)}{l^2(2-2c-us)} & \frac{u(u-s)}{l(2-2c-us)} & -\frac{u^2(1-c)}{l^2(2-2c-us)} & \frac{u(s-uc)}{l(2-2c-us)} \end{bmatrix} \tag{10.14}$$

When the axial force P vanishes, $u \rightarrow 0$ and the above stiffness matrix becomes identical with the stiffness matrix in Equation 5.20. Thus:

$$\lim_{u \to 0}[S] = \begin{bmatrix} \frac{12EI}{l^3} & & & \text{Symmetrical} \\ \frac{6EI}{l^2} & \frac{4EI}{l} & & \\ -\frac{12EI}{l^3} & -\frac{6EI}{l^2} & \frac{12EI}{l^3} & \\ \frac{6EI}{l^2} & \frac{2EI}{l} & -\frac{6EI}{l^2} & \frac{4EI}{l} \end{bmatrix} \tag{10.15}$$

The elements in the first and second column of $[S]$ in Equation 10.14 are the forces necessary to hold the member in the deflected configuration shown in Figures 10.1c and d. The two end moments in Figure 10.1d are the end-rotational stiffness S ($= S_{AB} = S_{BA}$) and the carryover moment t, which are needed for moment distribution. Thus, the end-rotational stiffness and the carryover moment for a prismatic member subjected to an axial compressive force P (Figure 10.1d) are:

$$S = \frac{u(s-uc)}{(2-2c-us)} \frac{EI}{l} \tag{10.16}$$

and:

$$t = \frac{u(u-s)}{(2-2c-us)} \frac{EI}{l} \tag{10.17}$$

The carryover factor C ($= C_{AB} = C_{BA}$) is:

$$C = \frac{t}{S} = \frac{u-s}{s-uc} \tag{10.18}$$

When u tends to zero, S, t, and C tend to $4EI/l$, $2EI/l$, and ½, respectively. It can be seen thus that the axial compressive force reduces the value of S and increases C. When the force P reaches the critical buckling value, S becomes zero, which means that the displacement D_2 can be caused by an infinitely small value of the force F_2, and the value of C approaches infinity. From Equation 10.16, S is zero when $s = uc$ or $\tan u = u$. The smallest value of u to satisfy this equation is $u = l\sqrt{P_{cr}/(EI)} = 4.4934$. Thus, the buckling load of the strut with the end conditions of Figure 10.1d is $P_{cr} = 20.19\ (EI/l^2)$. The critical buckling load corresponding to any end conditions can be derived from the stiffness matrix in Equation 10.14.

The deflected shape in Figure 10.1c can be achieved in two stages. First, a unit downward translation is introduced at A while the end-rotations are allowed to occur freely; this condition can be represented by a straight line (a chord) joining A and B. Second, clockwise end-rotations, through angles equal to $1/l$, at A and B produce the deflected shape in Figure 10.1c and require the end moments $M_{AB} = S_{21} = (S+t)/l$ and $M_{BA} = S_{41} = (S+t)/l$. The vertical forces S_{11} and S_{31} can now be determined by considering the equilibrium of the system of forces in Figure 10.1c. Thus, we express the elements in the first column of the stiffness matrix corresponding to the coordinates in Figure 10.1b in terms of S and t. The end moments S and t in Figure 10.1d are, respectively, equal to S_{22} and S_{42}; again, considering equilibrium of the forces in this figure, $S_{12} = (S+t)/l = -S_{32}$. This gives the elements in the second column of [S] in terms of S and t. In a similar way, we derive the elements in the remaining two columns of [S] can be derived. Thus, the stiffness matrix for a prismatic member subjected to a compressive axial force (Figure 10.1b) is:

$$[S] = \begin{bmatrix} \frac{2(S+t)}{l^2} - \frac{P}{l} & & & \text{Symmetrical} \\ \frac{S+t}{l} & S & & \\ -\frac{2(S+t)}{l^2} + \frac{P}{l} & -\frac{(S+t)}{l} & \frac{2(S+t)}{l^2} - \frac{P}{l} & \\ \frac{S+t}{l} & t & -\frac{(S+t)}{l} & S \end{bmatrix} \tag{10.19}$$

If we substitute $P = u^2EI/l^2$ in the above equation, it becomes apparent that all the elements of the stiffness matrix are a function of the dimensionless parameter u, the member length l and the flexural rigidity EI. Figure 10.2 shows the values of the end-rotational stiffness S and the carryover moment t in terms of the dimensionless parameter, $u = l\sqrt{P/(EI)}$. Table 10.1 lists the numerical data. All these values apply to prismatic members only. The negative values of S correspond to P values higher than the critical buckling load, so that the negative sign indicates that a restraining moment holds the end rotation at a value not exceeding unity.

10.4 EFFECT OF AXIAL TENSION

The stiffness of a prismatic member in tension can be derived from the expressions for a member in compression by replacing P by $(-P)$. Accordingly, the notation $u = l\sqrt{P/(EI)}$ is to be replaced by $l\sqrt{-P/(EI)} = iu$, where $i = \sqrt{-1}$. Applying this to Equations 10.16 to 10.18 and making use of the fact that:

$$\sinh u = -i \sin iu \text{ and } \cosh u = \cos iu$$

we obtain the end-rotational stiffness, carryover moment, and carryover factor for a prismatic member subjected to an axial tensile force P:

$$S = \frac{u(u\cosh u - \sinh u)}{(2 - 2\cosh u + u\sinh u)} \frac{EI}{l} \tag{10.20}$$

$$t = \frac{u(\sinh u - u)}{(2 - 2\cosh u + u\sinh u)} \frac{EI}{l} \tag{10.21}$$

and

$$C = \frac{t}{S} = \frac{\sinh u - u}{u\cosh u - \sinh u} \tag{10.22}$$

where $u = l\sqrt{P/(EI)}$, P being the absolute value of the axial tensile force. The values of S and t calculated by the above equations are included in Table 10.1 and are plotted against u in Figure 10.2.

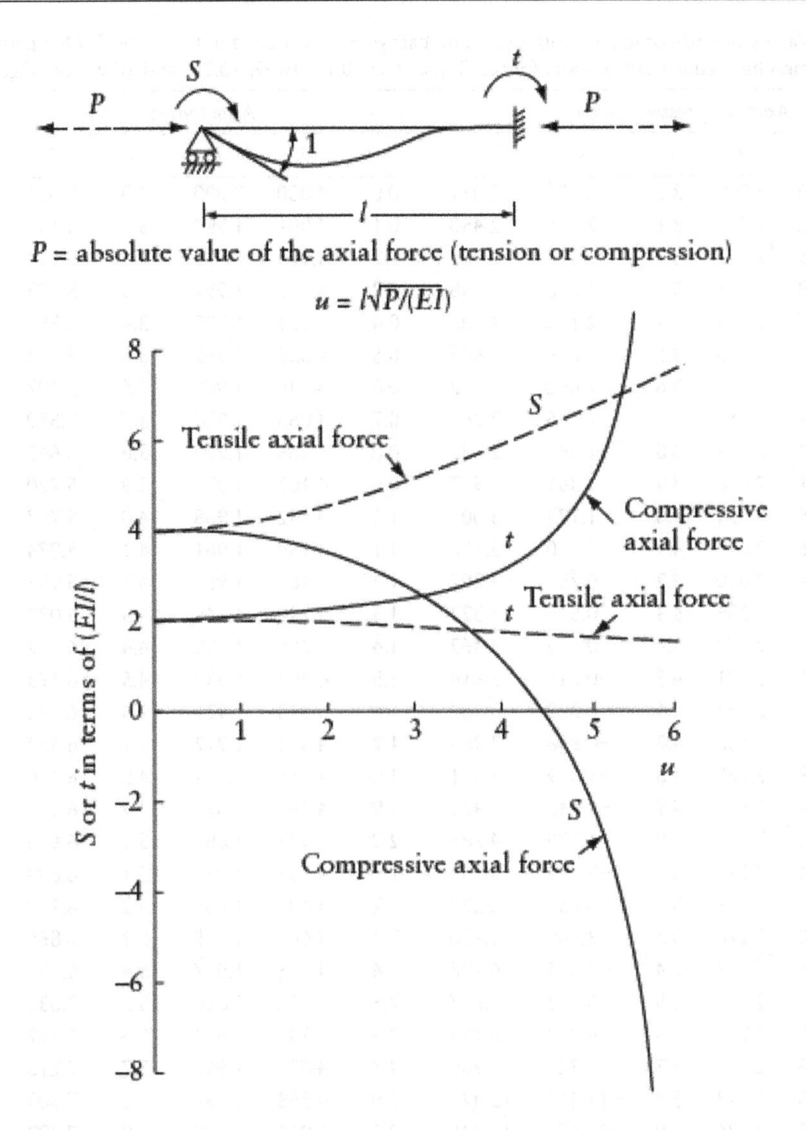

Figure 10.2 End-rotational stiffness and carryover moment for a prismatic member subjected to an axial force.

Considering equilibrium, we can readily see that the stiffness matrix corresponding to the coordinates in Figure 10.1b for a member in tension is:

$$[S]=\begin{bmatrix} \frac{2(S+t)}{l^2}+\frac{P}{l} & & & \text{Symm.} \\ \frac{S+t}{l} & S & & \\ -\frac{2(S+t)}{l^2}-\frac{P}{l} & -\frac{(S+t)}{l} & \frac{2(S+t)}{l^2}+\frac{P}{l} & \\ \frac{S+t}{l} & t & -\frac{(S+t)}{l} & S \end{bmatrix} \tag{10.23}$$

Table 10.1 Values of end-rotational stiffness **S** and carryover moment **t** in terms of ***EI/l*** for prismatic members subjected to axial forces (Equations 10.16, 10.17, 10.20, and 10.21; see Figure 10.2)

Axial compressive force P						*Axial tensile force P*					
u•	*S*	*t*	*u*	*S*	*t*	*u*	*S*	*t*	*u*	*S*	*t*
0.0	4.000	2.000	3.0	2.624	2.411	0.0	4.000	2.000	3.0	5.081	1.766
0.1	3.998	2.000	3.1	2.515	2.450	0.1	4.001	1.999	3.1	5.147	1.754
0.2	3.995	2.001	3.2	2.399	2.492	0.2	4.005	1.999	3.2	5.214	1.742
0.3	3.988	2.003	3.3	2.276	2.538	0.3	4.012	1.997	3.3	5.283	1.730
0.4	3.977	2.005	3.4	2.146	2.588	0.4	4.021	1.995	3.4	5.353	1.718
0.5	3.967	2.008	3.5	2.008	2.642	0.5	4.033	1.992	3.5	5.424	1.706
0.6	3.952	2.012	3.6	1.862	2.702	0.6	4.048	1.988	3.6	5.497	1.694
0.7	3.934	2.016	3.7	1.706	2.767	0.7	4.065	1.984	3.7	5.570	1.683
0.8	3.914	2.022	3.8	1.540	2.838	0.8	4.085	1.979	3.8	5.645	1.671
0.9	3.891	2.028	3.9	1.363	2.917	0.9	4.107	1.974	3.9	5.720	1.659
1.0	3.865	2.034	4.0	1.173	3.004	1.0	4.132	1.968	4.0	5.797	1.648
1.1	3.836	2.042	4.1	0.970	3.100	1.1	4.159	1.961	4.1	5.874	1.636
1.2	3.804	2.050	4.2	0.751	3.207	1.2	4.188	1.954	4.2	5.953	1.625
1.3	3.769	2.059	4.3	0.515	3.327	1.3	4.220	1.946	4.3	6.032	1.614
1.4	3.732	2.070	4.4	0.259	3.462	1.4	4.255	1.938	4.4	6.112	1.603
1.5	3.691	2.081	4.5	−0.019	3.614	1.5	4.292	1.930	4.5	6.193	1.592
1.6	3.647	2.093	4.6	−0.323	3.787	1.6	4.330	1.921	4.6	6.275	1.581
1.7	3.599	2.106	4.7	−0.658	3.984	1.7	4.372	1.912	4.7	6.357	1.571
1.8	3.548	2.120	4.8	−1.029	4.211	1.8	4.415	1.902	4.8	6.440	1.561
1.9	3.494	2.135	4.9	−1.443	4.475	1.9	4.460	1.892	4.9	6.524	1.551
2.0	3.436	2.152	5.0	−1.909	4.785	2.0	4.508	1.881	5.0	6.608	1.541
2.1	3.374	2.170	5.1	−2.439	5.151	2.1	4.557	1.871	5.1	6.693	1.531
2.2	3.309	2.189	5.2	−3.052	5.592	2.2	4.608	1.860	5.2	6.779	1.521
2.3	3.240	2.210	5.3	−3.769	6.130	2.3	4.661	1.849	5.3	6.865	1.512
2.4	3.166	2.233	5.4	−4.625	6.798	2.4	4.716	1.837	5.4	6.952	1.503
2.5	3.088	2.257	5.5	5.673	7.647	2.5	4.773	1.826	5.5	7.039	1.494
2.6	3.005	2.283	5.6	−6.992	8.759	2.6	4.831	1.814	5.6	7.127	1.485
2.7	2.918	2.312	5.7	−8.721	10.269	2.7	4.891	1.802	5.7	7.215	1.476
2.8	2.825	2.342	5.8	−11.111	12.428	2.8	4.953	1.791	5.8	7.303	1.468
2.9	2.728	2.376	5.9	−14.671	15.745	2.9	5.016	1.779	5.9	7.392	1.460

• $u = l\sqrt{P/(EI)}$. Enter the table with the dimensionless parameter *u* and read the corresponding values of *S*/(*EI*/*l*) and *t*/(*EI*/*l*).

10.5 LINEAR ANALYSIS CONSIDERING EFFECT OF AXIAL FORCE

The flexural stiffness of a member is appreciably affected by an axial force, *P*, when the parameter $u = l\sqrt{P/(EI)}$ > say, 1.5. We can employ the linear force and displacement method for analysis of structures in which the axial forces affect their stiffness, by assuming a set of axial forces to be present in all analysis steps. The *u*-parameter for each member is regarded as known property, just like *E*, *I*, and *l*. The computer program PDELTA (Chapter 20) applies the displacement method first, ignoring the effect of axial forces on stiffness (by setting $u = 0$). Then, by iteration, the computer improves the accuracy of the values of *u*.

10.6 FIXED-END MOMENTS FOR A PRISMATIC MEMBER SUBJECTED TO AN AXIAL FORCE

Consider a straight member subjected to transverse loading and an axial force shown in Figure 10.3. The end-displacement along the beam axis can take place freely but the end rotation is prevented, so that fixed-end moments are induced. The presence of an axial compressive force causes an increase in the fixed-end moments whereas a tensile force results in a decrease. We consider two cases of loading: a uniform load and a concentrated load.

10.6.1 Uniform load

The deflection of the prismatic strut AB in Figure 10.3a carrying a transverse load of constant intensity q is governed by the differential Equation 10.1, for which the solution is:

$$y = G_1 \sin u\frac{x}{l} + G_2 \cos u\frac{x}{l} + G_3\, x + G_4 + \frac{qx^2}{2P} \tag{10.24}$$

The integration constants {G} are determined from the boundary conditions $y=0$ and $(dy/dx)=0$ at $x=0$ and $x=l$; these four conditions can be put in the form:

$$[B]\begin{Bmatrix} G_1 \\ G_2 \\ G_3 \\ G_4 \end{Bmatrix} + \frac{q}{P}\begin{Bmatrix} 0 \\ 0 \\ l^2/2 \\ l \end{Bmatrix} = \{0\}$$

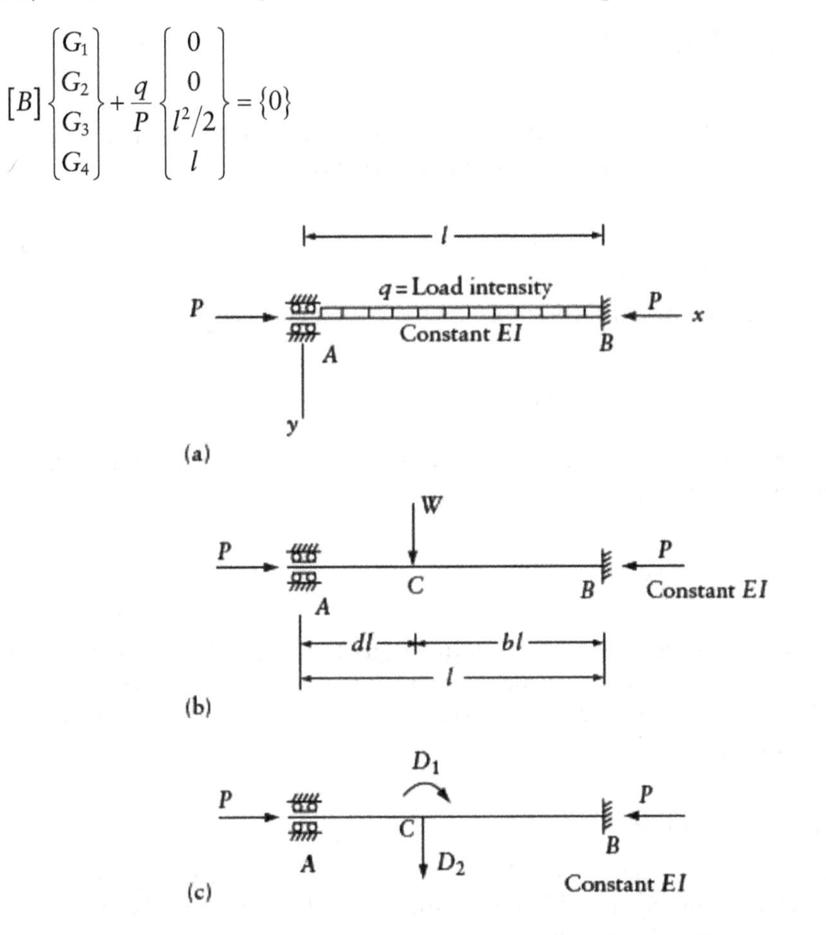

Figure 10.3 Straight prismatic member subjected to an axial force and transverse loading. (a) Loading corresponding to the FEM Equations 10.26 or 10.28. (b) Loading corresponding to the FEM Equation 10.29 and Table 10.3. (c) Degrees of freedom (D_1 and D_2) considered for the analysis of the strut in part (b).

where $[B]$ has the same meaning as in Equation 10.5. Solving for the integration constants, we obtain:

$$\begin{Bmatrix} G_1 \\ G_2 \\ G_3 \\ G_4 \end{Bmatrix} = -\frac{ql^2}{u^2 EI}[B]^{-1}\begin{Bmatrix} 0 \\ 0 \\ l^2/2 \\ l \end{Bmatrix} \tag{10.25}$$

where $[B]^{-1}$ is that given by Equation 10.13.

Considering the end moments to be positive if clockwise and making use of symmetry of the member, the two fixed-end moments are related to the deflection by:

$$M_{AB} = -M_{BA} = -EI\left(\frac{d^2y}{dx^2}\right)_{x=0}$$

Differentiating twice Equation 10.24 and substituting in the above equation, we obtain:

$$M_{AB} = -M_{BA} = EI\left(\frac{u^2}{l^2}G_2 - \frac{q}{P}\right)$$

With G_2 from Equation 10.25 substituted in the above equation, the fixed-end moments due to a uniform load on a member subjected to an axial compression are:

$$M_{AB} = -M_{BA} = -ql^2\,\frac{1}{u^2}\left(1 - \frac{u}{2}\cot\frac{u}{2}\right) \tag{10.26}$$

If the axial force is tensile, the expression for fixed-end moments becomes:

$$M_{AB} = -M_{BA} = -ql^2\,\frac{1}{u^2}\left(\frac{u}{2}\coth\frac{u}{2} - 1\right) \tag{10.27}$$

The fixed-end moments due to a uniform transverse load given by Equations 10.26 and 10.27 can be expressed in the form:

$$M_{AB} = -M_{BA} = -\frac{EI}{2l(S+t)}ql^2 \tag{10.28}$$

where the values of S and t are determined as a function of $u = l\sqrt{P/(EI)}$ (P being the absolute value of the axial compressive or tensile force) by Equations 10.16 and 10.17, Equations 10.20 and 10.21, or by Table 10.1. When the axial force P vanishes, $u \to 0$, and Equations 10.26 to 10.28 give $M_{AB} = -M_{BA} = -ql^2/12$.

10.6.2 Concentrated load

Consider the prismatic strut AB in Figure 10.3b carrying a transverse load W at a distance dl and bl from the left- and right-hand ends, respectively. The strut can be treated as an assemblage of two members AC and CB with the two degrees of freedom indicated in Figure 10.3c, for which the stiffness matrix can be derived from the stiffness of the individual members (Equation 10.19). Thus:

$$[S]=\begin{bmatrix} (S_d+S_b) & \text{Symmetrical} \\ \left\{\frac{(S_b+t_b)}{bl}-\frac{(S_d+t_d)}{dl}\right\} & \left\{\frac{(2S_d+t_d)}{d^2l^2}+\frac{(2S_b+t_b)}{b^2l^2}-\frac{P}{dbl}\right\} \end{bmatrix}$$

where the subscripts d and b refer to members AC and BC, respectively. Equations 10.16 and 10.17 give the values of S and t for the two members.

The displacements at the two coordinates are:

$$\begin{Bmatrix} D_1 \\ D_2 \end{Bmatrix}=[S]^{-1}\begin{Bmatrix} 0 \\ W \end{Bmatrix}$$

and the fixed-end moment at A (considered positive if clockwise) is given by:

$$M_{AB}=t_d\,D_1-(S_d+t_d)\frac{D_2}{dl}$$

The above procedure applies for members subjected to axial tension. Thus, the fixed-end moment in member AB subjected to an axial compression (Figure 10.3b) is:

$$M_{AB}=-Wl\,\frac{(bu\cos u-\sin u+\sin du+\sin bu-u\cos bu+du)}{u(2-2\cos u-u\sin u)} \tag{10.29}$$

and when the axial force P is tension:

$$M_{AB}=-Wl\,\frac{(bu\cosh u-\sinh u+\sinh du+\sinh bu-u\cosh bu+du)}{u(2-2\cosh u-u\sinh u)} \tag{10.30}$$

Equations 10.29 and 10.30 are used to calculate influence coefficients of the fixed-end moment M_{AB} in Table 10.3 by fixing a value for u and varying d (and $b=1-d$). The same table applies for the fixed-end moments at the right-hand end M_{BA} by considering d in the table to represent b and changing the sign of the moment given, for example, for $d=0.3$, $b=0.7$, $P=0$, and $u=0$. Table 10.3 gives (on first row): $M_{AB}=-0.1470$ Wl and $M_{BA}=0.0630$ Wl. Equations B.3 and B.4, Appendix B, would of course give the same answers.

The computer program PDELTA gives the displacements and the member end moments due to distributed or concentrated load, as demonstrated in Examples 10.1.

EXAMPLE 10.1: FIXED-END MOMENTS BY COMPUTER

Find the bending moments and displacements at the ends and at the middle of the members in Figures 10.3a and b. Place point C at the middle of AB; take $l=10.0$ m; $I=40\times10^{-6}$ m^4; $P=320\times10^3$ N. Thus, $u=2.0$. The computer gives:

For the member in Figure 10.3a:

Take $q=1$; $\{A_r\}=\{0.0, -2.5, -2.08333, 0.0, -2.5, 2.08333\}$. The computer gives: $\{M_A, M_C, M_B\}=\{-8.912, 4.746, -8.912\}$ $q=\{-0.08912, 0.04746, -0.08912\}$ ql^2. Deflection at C, $D=3.617\times10^{-6}$ $q=2.894\times10^{-3}$ ql^4/EI. We can verify that the static bending moment, $ql^2/8$ is approximately equal to $[M_C-(M_A \text{ or } M_B)-PD]$.

For the member in Figure 10.3b:

Take $W=1$. The computer gives: $\{M_A, M_C, M_B\}=\{-1.3568, 1.3568, -1.3568\}$ $W=\{-0.1357, 0.1357, -0.1357\}$ Wl. Deflection at C, $D=0.72348\times10^{-6}$

$W = 5.788 \times 10^{-3}\ Wl^3/EI$. We can verify that the static bending moment, $Wl/4$ is approximately equal to $[M_C - (M_A \text{ or } M_B) - PD]$.

EXAMPLE 10.2: PLANE FRAME BY DISPLACEMENT METHOD

Analyze the frame in Figure 10.4 by the displacement method. Compare the member-end moments with the values by the computer program PDELTA, setting $b = 2$ m; $I = 10^{-5}$ m^4; $E = 200 \times 10^9$ N/m^2; and $q = 4 \times 10^3$ N/m.

The approximate values of the axial forces are: $P_{AB} = 0$, $P_{BC} = 0.05qb$ tensile, and $P_{BD} = 7qb$ compressive; the corresponding u values are: $u_{AB} = 0$, $u_{BC} = 0$ and $u_{BD} = 1.34$. For member BD, $S = 3.75EI/l$ and $t = 2.06EI/l$. With these values, Table 10.2 gives the end-forces corresponding to unit end displacements.

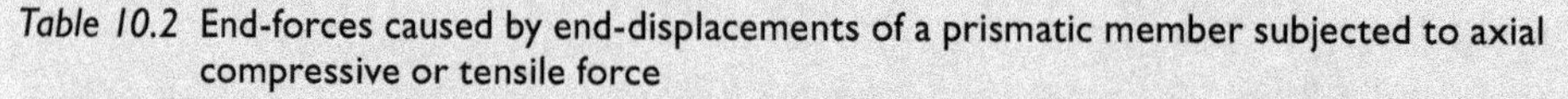

Table 10.2 End-forces caused by end-displacements of a prismatic member subjected to axial compressive or tensile force

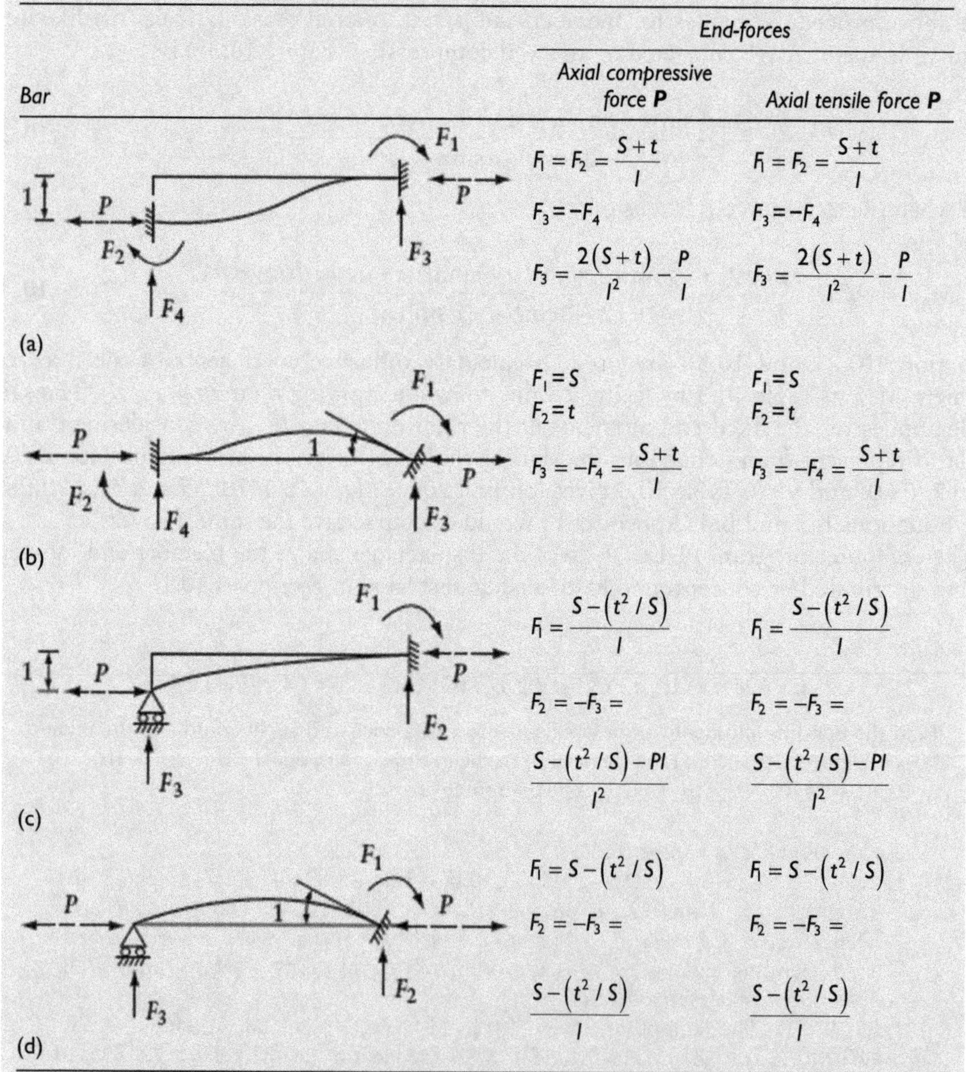

Bar	End-forces: Axial compressive force **P**	End-forces: Axial tensile force **P**
(a)	$F_1 = F_2 = \frac{S+t}{l}$; $F_3 = -F_4$; $F_3 = \frac{2(S+t)}{l^2} - \frac{P}{l}$	$F_1 = F_2 = \frac{S+t}{l}$; $F_3 = -F_4$; $F_3 = \frac{2(S+t)}{l^2} - \frac{P}{l}$
(b)	$F_1 = S$; $F_2 = t$; $F_3 = -F_4 = \frac{S+t}{l}$	$F_1 = S$; $F_2 = t$; $F_3 = -F_4 = \frac{S+t}{l}$
(c)	$F_1 = \frac{S-(t^2/S)}{l}$; $F_2 = -F_3 = \frac{S-(t^2/S)-Pl}{l^2}$	$F_1 = \frac{S-(t^2/S)}{l}$; $F_2 = -F_3 = \frac{S-(t^2/S)-Pl}{l^2}$
(d)	$F_1 = S-(t^2/S)$; $F_2 = -F_3 = \frac{S-(t^2/S)}{l}$	$F_1 = S-(t^2/S)$; $F_2 = -F_3 = \frac{S-(t^2/S)}{l}$

Note: S and t are given in terms of $u = l\sqrt{P/(EI)}$ in Table 10.1. EI is flexural rigidity and l is the length of the member. P may be expressed as $P = u^2EI/l^2$.

Table 10.3 Influence coefficient of the fixed-end moment* M_{AB} due to a unit transverse load on a prismatic beam subjected to an axial force M_{AB} = −coefficient × Wl (refer to Figure 10.3b, calculations by Equations 10.29 and 10.30)

Force **P**; $u = l\sqrt{\frac{P}{EI}}$	Value of **d**										
	0	*0.1*	*0.2*	*0.3*	*0.4*	*0.5*	*0.6*	*0.7*	*0.8*	*0.9*	*1.0*
0	0	0.0810	0.1280	0.1470	0.1440	0.1250	0.0960	0.0630	0.0320	0.0090	0
Compressive											
1.0	0	0.0815	0.1294	0.1493	0.1467	0.1276	0.0981	0.0644	0.0327	0.0091	0
2.0	0	0.0831	0.1342	0.1569	0.1558	0.1365	0.1054	0.0692	0.0350	0.0097	0
3.0	0	0.0863	0.1438	0.1725	0.1747	0.1552	0.1208	0.0796	0.0402	0.0111	0
4.0	0	0.0922	0.1624	0.2037	0.2136	0.1946	0.1540	0.1022	0.0515	0.0141	0
5.0	0	0.1057	0.2070	0.2820	0.3150	0.3009	0.2459	0.1662	0.0843	0.0229	0
6.0	0	0.1943	0.5218	0.8689	1.1150	1.1750	1.0288	0.7282	0.3789	0.1039	0
Tensile											
1.0	0	0.0805	0.1265	0.1447	0.1413	0.1224	0.0939	0.0616	0.0313	0.0088	0
2.0	0	0.0791	0.1226	0.1386	0.1341	0.1155	0.0883	0.0579	0.0295	0.0083	0
3.0	0	0.0770	0.1168	0.1297	0.1239	0.1058	0.0806	0.0529	0.0271	0.0077	0
4.0	0	0.0745	0.1100	0.1196	0.1125	0.0951	0.0722	0.0474	0.0244	0.0070	0
5.0	0	0.0718	0.1029	0.1093	0.1012	0.0848	0.0641	0.0423	0.0220	0.0064	0
6.0	0	0.0690	0.0959	0.0996	0.0908	0.0754	0.0569	0.0377	0.0198	0.0059	0

* A fixed-end moment is positive when clockwise. To find the fixed-end moment at the right-hand end, enter the value of *b* in lieu of *d* and change the sign of the moment. *P* is the absolute value of the axial compressive or tensile force.

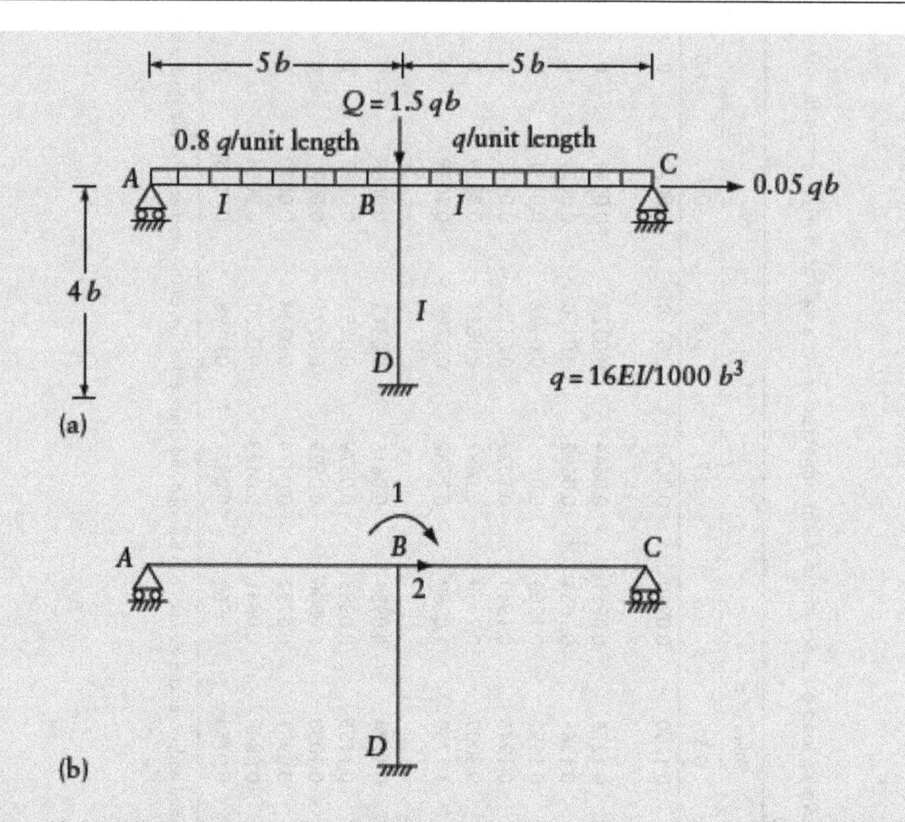

Figure 10.4 Analysis of the frame of Example 10.2 by the general displacement method taking into account the beam–column effect. (a) Frame properties and loading. (b) Coordinate system.

The stiffness matrix of the structure corresponding to the coordinates 1 and 2 in Figure 10.4b is:

$$[S]=\begin{bmatrix} 3\left(\frac{EI}{l}\right)_{BA}+3\left(\frac{EI}{l}\right)_{BC}+3.75\left(\frac{EI}{l}\right)_{BD} & \text{Symmetrical} \\ -(3.75+2.06)\left(\frac{EI}{l^2}\right)_{BD} & 2(3.75+2.06)\left(\frac{EI}{l^3}\right)_{BD}-\left(\frac{P}{l}\right)_{BD} \end{bmatrix}$$

The fixed-end moments at end B of members BA, BC, BD, and DB are $2.5qb^2$, $-3.125qb^2$, 0, and 0, respectively. The restraining forces at coordinates 1 and 2 to prevent displacements at B are:

$$\{F\}=\begin{Bmatrix} -0.625\,qb^2 \\ -0.05\,qb \end{Bmatrix}$$

The displacements at the coordinates are (Equation 5.3):

$$\{D\}=[S]^{-1}\{-F\}$$

Substituting the values of I, l, and P $(= u^2[EI/l^2])$ into $[S]$ and inverting it, then substituting in Equation 5.3, we obtain:

$$\{D\}=\frac{qb^2}{EI}\begin{Bmatrix} 0.58 \\ 1.71\,b \end{Bmatrix}$$

We determine by Equation 5.5:

$$\{A\} = \{A_r\} + [A_u]\,\{D\}$$

The end moments M_{BA}, M_{BC}, M_{BD}, and M_{DB} are:

$$\begin{Bmatrix} M_{BA} \\ M_{BC} \\ M_{BD} \\ M_{DB} \end{Bmatrix} = qb^2 \begin{Bmatrix} 2.50 \\ -3.125 \\ 0 \\ 0 \end{Bmatrix} + \begin{bmatrix} 3.00\left(\frac{EI}{l}\right)_{BA} & 0 \\ 3.00\left(\frac{EI}{l}\right)_{BC} & 0 \\ 3.00\left(\frac{EI}{l}\right)_{BD} & -(3.75+2.06)\left(\frac{EI}{l^2}\right)_{BD} \\ 3.00\left(\frac{EI}{l}\right)_{DB} & -(3.75+2.06)\left(\frac{EI}{l^2}\right)_{DB} \end{bmatrix} \{D\}$$

We find the member end moment by substituting the values of $\{D\}$:

$$\begin{Bmatrix} M_{BA} \\ M_{BC} \\ M_{BD} \\ M_{DB} \end{Bmatrix} = \frac{qb^2}{100} \begin{Bmatrix} 285 \\ -277 \\ -8 \\ -32 \end{Bmatrix}$$

The computer program PDELTA gives: $\{M_{BA}, M_{BC}, M_{DB}\} = qb^2\,\{2.77, -2.73, -0.29\}$.

10.7 ELASTIC STABILITY OF FRAMES

In design based on a definite load factor, we need failure loads. Broadly speaking, failure can occur by yielding of the material at a sufficient number of locations to form a mechanism (see Chapters 14 and 15) or by buckling due to axial compression without the stresses exceeding the elastic limit.

In Section 10.3, we considered buckling of individual members. In the present section, we shall deal with buckling of rigidly jointed plane frames.

Consider a plane frame subjected to a system of forces $\{Q\}$ (Figure 10.5a) causing axial compression in some of the members. The buckling, or critical loading, $\alpha\{Q\}$ is defined by the value of the scalar α at which the structure can be given small displacements without application of disturbing forces. In other words, when the buckling loading $\alpha\{Q\}$ acts, it is possible to maintain the structure in a displaced configuration without additional loading, as shown in Figure 10.5b.

If, for the frame considered, an estimate is made of the value of α and the corresponding values of the axial forces $\{P(\alpha)\}$ are computed, it becomes possible to find the stiffness matrix $[S(\alpha)]$ of the structure corresponding to any chosen coordinate system. The elements of this matrix are functions of the value α. Any system of forces $\{F\}$ and displacements $\{D\}$ at the coordinates are related by the equation:

$$[S(\alpha)]\,\{D\} = \{F\} \tag{10.31}$$

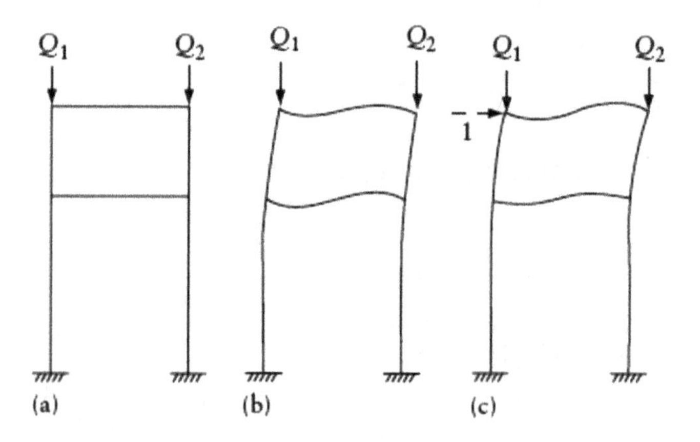

Figure 10.5 Buckling of a plane frame.

If α corresponds to the critical buckling value, then it is possible to let the structure acquire some small displacements $\{\delta D\}$ without application of forces and the last equation thus becomes:

$$[S(\alpha)]\,\{\delta D\} = \{0\} \tag{10.32}$$

For a non-trivial solution to exist, the stiffness matrix $[S(\alpha)]$ must be singular – that is, the determinant:

$$[S(\alpha)] = 0 \tag{10.33}$$

The collapse loading is that which has the smallest value of α, which satisfies Equation 10.33. In general, there is more than one value of α, which satisfies this equation and each value has associated with it values of $\{\delta D\}$ of arbitrary magnitude but in definite proportions defining an associated critical mode.

To solve Equation 10.33 for α, we calculate the value of the determinant for a set of values of α and determine by interpolation the value corresponding to a zero value of the determinant. Once α has been calculated, the associated displacement vector, if required, can be obtained in a way similar to that used in the calculation of eigenvectors.

The above procedure may involve a large amount of numerical work so that access to a computer is necessary. However, in most structures, it is possible to guess the form of the critical mode associated with the lowest critical load. For example, it can be assumed that when the buckling load is reached in the frame of Figure 10.5a, a small disturbing force at coordinate 1 will cause buckling in sidesway (Figure 10.5c). At this stage, the force at 1 required to produce a displacement at this coordinate is zero, that is $S_{11} = 0$. We use this condition to determine the critical load. A value of α lower than the critical value is estimated and the stiffness coefficient $S_{11}(\alpha)$ corresponding to coordinate 1 is determined. As α is increased, the value of S_{11} decreases, and it vanishes when the critical value of α is attained. If a set of values of α is assumed and plotted against S_{11}, the value of α corresponding to $S_{11} = 0$ can be readily determined.

Figure 10.6 gives the critical buckling loads for a straight prismatic member with various end conditions; we may use these to establish lower and upper bounds to the buckling load of a frame; hence, they are of help in estimating the approximate values of α.

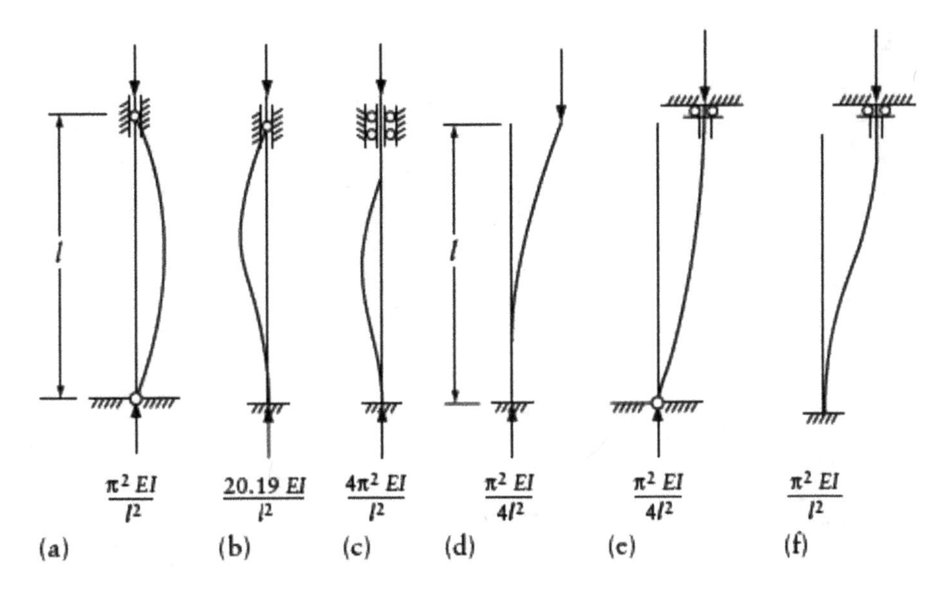

Figure 10.6 Critical buckling load for a prismatic member.

EXAMPLE 10.3: BUCKLING IN PORTAL FRAME

Find the value of Q that causes buckling of the frame in Figure 10.7a.

Table 10.2 gives the stiffness matrix for the three coordinates in Figure 10.7b:

$$[S] = \begin{bmatrix} \left(\dfrac{S - t^2/S - u^2\, EI/l}{l^2}\right)_{BA} + \left(\dfrac{S - t^2/S - u^2\, EI/l}{l^2}\right)_{CD} & & \text{Symmetrical} \\ -\left(\dfrac{S - t^2/S}{l}\right)_{BA} & \left(S - t^2/S\right)_{BA} + 4\left(\dfrac{EI}{l}\right)_{BC} & \\ -\left(\dfrac{S - t^2/S}{l}\right)_{CD} & 2\left(\dfrac{EI}{l}\right)_{BC} & \left(S - t^2/S\right)_{CD} + 4\left(\dfrac{EI}{l}\right)_{BC} \end{bmatrix} \quad \text{(a)}$$

Element S_{11} is derived by the equation for F_2, case (c), Table 10.2, substituting $u^2(EI/l^2)$ for P; S_{21} and S_{31} are derived from the equation for F_1 of the same case. Elements S_{22} and S_{32} are derived by the appropriate equations from cases (d) and (b), Table 10.2, noting that member BC has zero axial forces thus, $S_{BC} = (4EI/l)_{BC}$ and $t_{BC} = (2EI/l)_{BC}$. S_{33} is also derived using equations from cases (d) and (b), Table 10.2.

The lowest buckling load for this frame corresponds to a sidesway by a small disturbing force at coordinate 1. It can be shown that the stiffness S_{11}^* corresponding to coordinate 1* (Figure 10.7c) is given by (see Section 5.12, Equation 5.55):

$$S_{11}^* = S_{11} - \frac{S_{21}^2\, S_{31} - 2\, S_{21}\, S_{31}\, S_{22} + S_{31}^2\, S_{22}}{S_{22}\, S_{33} - S_{32}^2} \qquad (10.34)$$

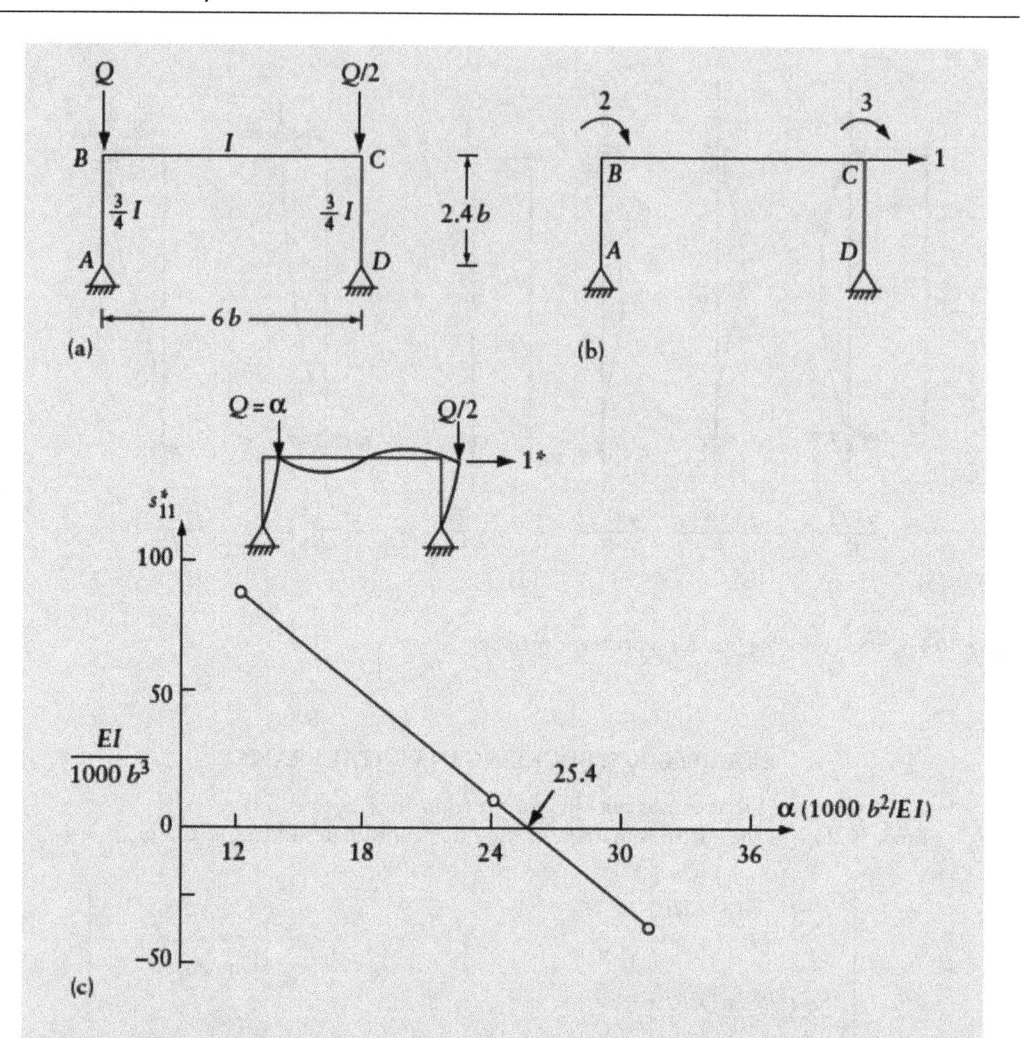

Figure 10.7 Determination of the critical load on the frame of Example 10.3 by consideration of stiffness reduction. (a) Frame dimensions and loading. (b) Coordinate system; the corresponding stiffness matrix in presence of the Q and Q/2 forces is given in Equation (a). (c) Variation of stiffness S^*_{11} with α.

where S_{ij} are elements of the matrix in Equation (a). Equation 10.32 can, in fact, be used for condensation of any 3 × 3 stiffness matrix when the forces at coordinates 2 and 3 are zero.

Let the buckling load occur when $Q=\alpha$. Considering that the rotation of end C of member CD is partially restrained, it can be concluded that the buckling load is lower than the value $(\pi^2 EI/4l^2)_{BA}$ which corresponds to the conditions in Figure 10.6e. Thus, an upper bound for α is:

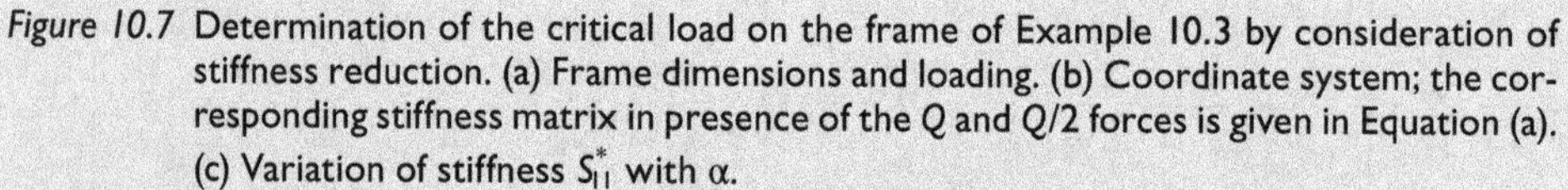

$$\alpha < \left(\frac{\pi^2 EI}{4l^2}\right)_{BA} = 0.32\frac{EI}{b^2};\quad Q < 0.32\frac{EI}{b^2}$$

The values of α are assumed, the corresponding u values for BA and CD are determined, and the corresponding S and t values are taken from Table 10.1:

α (b^2/EI)	u_{BA}	$\dfrac{S_{BA}}{(EI/l)_{BA}}$	$\dfrac{t_{BA}}{(EI/l)_{BA}}$	u_{CD}	$\dfrac{S_{CD}}{(EI/l)_{CD}}$	$\dfrac{t_{CD}}{(EI/l)_{CD}}$
0.12	0.960	3.876	2.031	0.679	3.938	2.015
0.24	1.357	3.747	2.066	0.960	3.876	2.031
0.31	1.543	3.672	2.086	1.091	3.839	2.041

Substituting into Equations (a) and 10.32, we obtain the following value of S_{11}^*:

α (b^2/EI)	$S_{11}^*\left(b^3/EI\right)$
0.12	0.0888
0.24	0.0092
0.31	−0.0376

The above values are plotted in Figure 10.7c from which it is seen that $S_{11}^* = 0$ when $\alpha = 0.254EI/b^2$. Thus, the buckling load $Q_{cr} = 0.254EI/b^2$.

Computer program PDELTA (Chapter 20) gives:

For $\alpha = 0.24$ (EI/b^2), $S_{11}^* = 0.00915$ (b^3/EI)
For $\alpha = 0.26$ (EI/b^2), $S_{11}^* = -0.00412$ (b^3/EI)

For S_{11}^*, linear interpolation gives:

$$\alpha = \left[0.24 + \frac{0.00915\left(0.26-0.24\right)}{0.00915-\left(-0.00412\right)}\right]\frac{b^2}{EI} = 0.254\,\frac{b^2}{EI}$$

10.8 ELASTIC STABILITY OF FRAMES: GENERAL SOLUTION

The smallest value of α that satisfies Equation 10.32 can be obtained by trial without the determinant Equation 10.33, which is not suitable when the number of degrees of freedom, n, is large. We assume a value of α and determine $[S(\alpha)]$, considering the axial forces in the members. Find the force F_i to produce a small displacement δD_i; where i is a selected coordinate; element ii of the stiffness matrix is:

$$S_{ii}\left(\alpha\right) = F_i/\delta D_i \tag{10.35}$$

As α is increased, S_{ii} (α) becomes a small value or a negative. Linear interpolation gives the value of α for which:

$$S_{ii}\left(\alpha_{\text{buckling}}\right) = 0 \tag{10.36}$$

The choice of i assumes a buckling mode that may not correspond to the smallest α; occasionally, it may be necessary to try another choice. The computer program PDELTA determines $F_i\,(\alpha)$ by iteration.

10.9 EIGENVALUE PROBLEM

Determining the smallest buckling load and the corresponding buckling mode of a frame is an eigenvalue problem that can be solved by iterative procedure. For this purpose, we search for the smallest eigenvalue and the corresponding eigenvector to satisfy:

$$\left[S(\alpha)\right]_{n\times n}\{\phi\}_k=\lambda_k\left[I\right]_{n\times n}\{\phi\}_k \tag{10.37}$$

where $[I]$ is identity matrix; $[S(\alpha)]_{n\times n}$ is the stiffness matrix accounting for the axial forces in the member based on a trial value of α; k refers to the first eigen pair; and n is the number of degrees of freedom. All the coordinates that are likely to produce the smallest α should be included. Equation 10.37 has the same form as the structural dynamic equation of free vibration:

$$\left[S\right]_{n\times n}\{\phi\}_k=\lambda_k\left[m\right]_{n\times n}\{\phi\}_k \tag{10.38}$$

But in the equation of motion, $[I]$ in Equation 10.37 is replaced by $[m]$ and the stiffness matrix is not dependent on a trial value, α. For practical application of the iterative procedure, we need a computer program that generates $[S(\alpha)]$, then applies an iterative procedure such as the method used in the program EIGEN (Chapter 20). An alternative technique may be necessary when the stiffness matrix is large.

EXAMPLE 10.4

Use the computer program PDELTA (Chapter 20) to determine the value of $Q=\alpha\,(EI/l^2)$ which causes buckling of the plane frame in Figure 10.8. The second moments of the cross-sectional area of columns and beams are I and $2I$, respectively. Consider bending deformations only.

Select coordinate i as the horizontal displacement at G.

Trial 1: $\alpha=2.0$. Introduce a horizontal displacement $\delta D_i=0.001l$ at G. Program PDELTA gives the force $F_i=6.524\times10^{-3}\,(EI/l^3)$ and Equation 10.35 gives the stiffness:

$$S_{ii}(\alpha)=\frac{6.524\times10^{-3}}{0.001}\left(\frac{EI}{l^3}\right)=6.524\,\frac{EI}{l^3}$$

Trial 2: $\alpha=3.0$. Introduce a horizontal displacement $\delta D_i=0.001l$ at G. PDELTA gives the force $F_i=-7.538\times10^{-3}\,(EI/l^3)$ and Equation 10.35 gives the stiffness:

$$S_{ii}(\alpha)=\frac{-7.538\times10^{-3}}{0.001}\left(\frac{EI}{l^3}\right)=-7.538\,\frac{EI}{l^3}$$

Linear interpolation gives: $\alpha=2.0+\dfrac{6.524}{6.524-(-7.538)}=2.46$

The forces that produce buckling are equal to $Q_{cr}=2.46\,(EI/l^2)$. Now, we attempt to find a smaller value of Q_{cr}. Select coordinate i as the horizontal displacement at K and repeat the analysis.

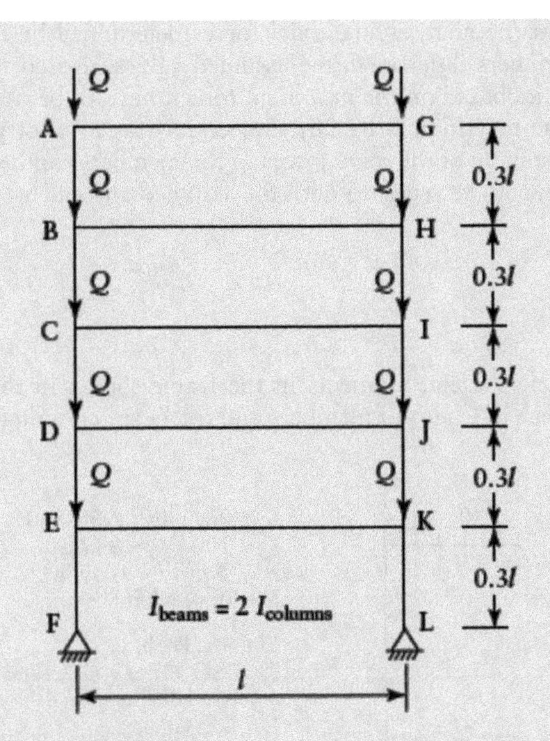

Figure 10.8 Plane frame of Example 10.4.

With $\alpha = 2.0$; $\delta D_i = 0.001$; $F_i = 28.71 \times 10^{-3}$; $S_{ii} = 28.71\ (EI/l^3)$
With $\alpha = 3.0$; $\delta D_i = 0.001$; $F_i = -9.848 \times 10^{-3}$; $S_{ii} = -9.848\ (EI/l^3)$

Linear interpolation gives: $\alpha = 2.0 + \dfrac{6.524}{28.71 - (-9.848)} = 2.74$

The more accurate answer of the trial values is $Q_{cr} = 2.46\ (EI/l^2)$.

10.10 GENERAL

By inspection of the numerical values of S and t in Table 10.1, we can see that the stiffness of a member is appreciably affected by an axial force P only when $u = l\sqrt{P/(EI)}$ is relatively large, say 1.5.

In the plane structures considered in this chapter, the deflections and bending moments are assumed to be in the plane of the structure, while displacements normal to this plane are assumed to be prevented. For space structures, similar methods of analysis can be developed, but they become complicated by the possibility of torsional-flexural buckling.

The force or displacement methods of analysis for linear structures can be used for structures in which the axial forces in the members affect their stiffness, provided that a set of axial forces is assumed to be present in the members in all the steps of the calculations. These forces are regarded as known parameters of the members, just like E, I, and l. Tables 10.1 and 10.2, based on a dimensionless parameter, can be used for the calculation of the stiffness of prismatic members subjected to axial forces.

If, after carrying out the analysis, the axial forces determined by considering the static equilibrium of the members differ from the assumed values, revised values of the member stiffnesses, or flexibilities based on the new axial forces, have to be used and the analysis is repeated. However, the repetition is usually unnecessary, as in most practical problems, a reasonably accurate estimate of the axial forces in the members can be made. If a computer is used, the calculation may be repeated until the assumed and calculated axial forces agree to any desired degree.

PROBLEMS

10.1 or 10.2 Determine the end moments in the frame shown in the figure, taking into account the change in length of members and the beam-column effect.

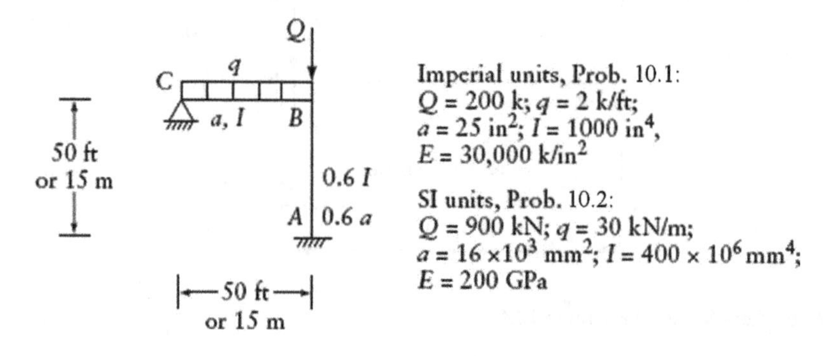

10.3 or 10.4 Apply the requirements of Problem 10.1 to the frame shown.

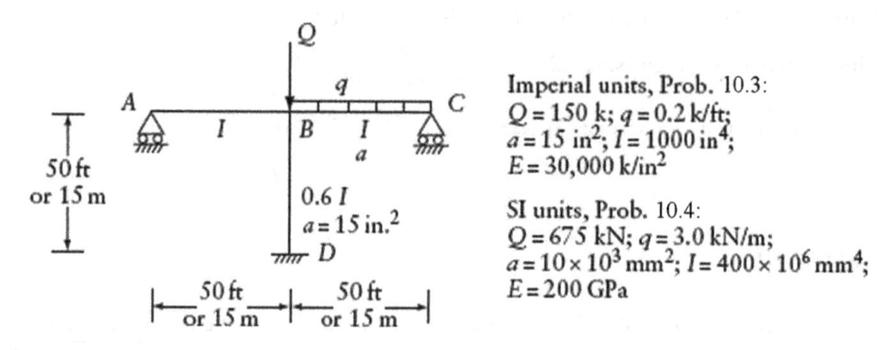

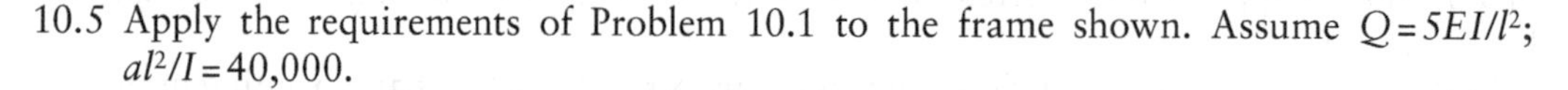

10.5 Apply the requirements of Problem 10.1 to the frame shown. Assume $Q = 5EI/l^2$; $al^2/I = 40{,}000$.

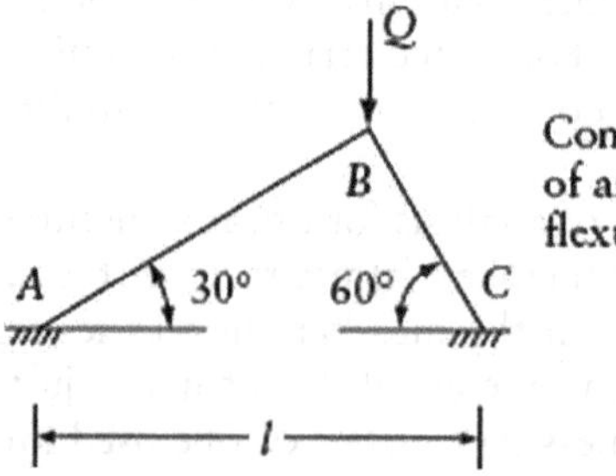

10.6 Prove Equations 10.27 and 10.28.

10.7 Verify the critical buckling loads given in Figure 10.6 for a prismatic member with different end conditions.

Hint: Determine S_{11} corresponding to a rotational or a translational coordinate 1 and verify that $S_{11}=0$, assuming that the member is subjected to an axial compressive force equal to the critical value given in the figure.

10.8 Obtain the bending moment diagram for beam ABC shown in the figure. The beam is fixed at A, on a roller at C, and on a spring support of stiffness $= 3EI/l^3$ at B.

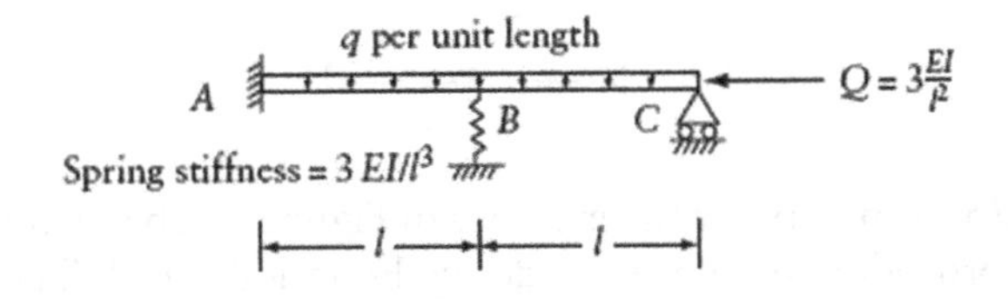

10.9 Solve Example 10.3, with the roller support at C changed to a hinged support.

10.10 Find the value of the force Q that causes buckling of the frame of Problem 10.5. Assume $I_{AB}=1.73\ I_{BC}$. Ignore axial deformations.

10.11 Find the value of Q that causes buckling of the beam in Problem 10.8 without the distributed load.

10.12 Show that the critical load Q associated with side-sway mode for the frame in the figure is given by the equation:

$$(S+t)^2 = \left(S + 6\frac{EI_1}{l}\right)(2S + 2t - Qh)$$

where S and t are respectively the end-rotational stiffness and carryover moment for BA (S and t are defined in Figure 10.2).

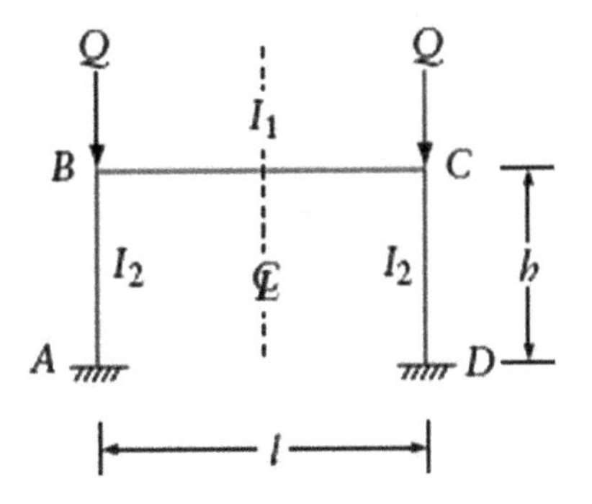

10.13 Solve Example 10.4 by moment distribution.

10.14 Determine the value of Q that causes instability of the structure shown in the figure. Assume E = constant.

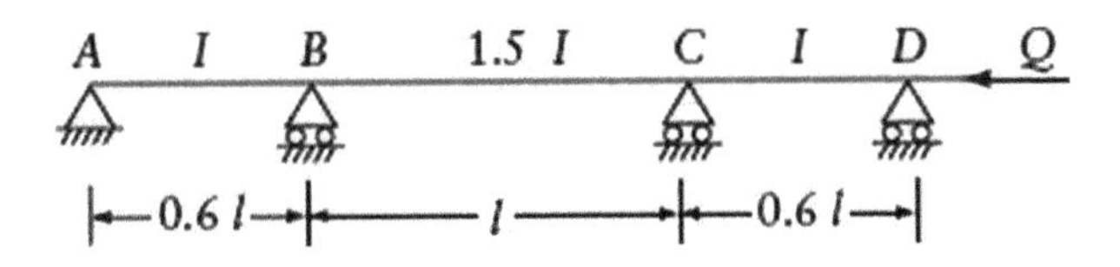

10.15 Apply the requirements of Problem 10.14 to the structure shown.

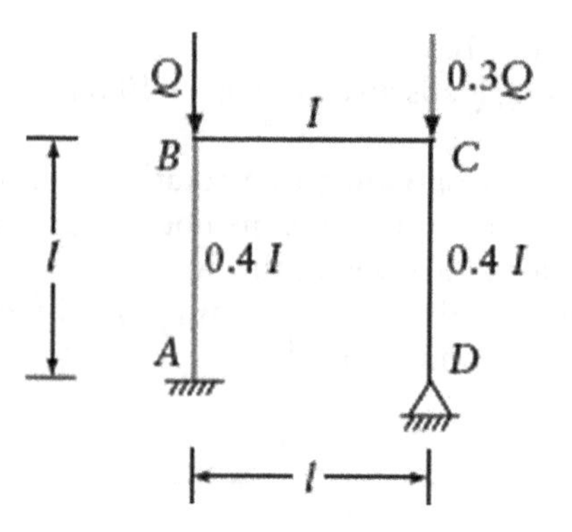

10.16 Derive the stiffness matrix for the member in Figure 10.1b when $P=0$ (Equation 5.6), following the procedure employed to derive Equation 10.14. The solution of the differential Equation 10.1 governing the deflection when P and q are zero is: $y=A_1+A_2 x+A_3 x^2+A_4 x^3$.

10.17 Solve Example 10.6 using Equation 10.43.

10.18 Find the critical buckling force Q_{cr} in the presence of a horizontal distributed load, $q=1.0\ EI/l^3$ on the beam of Figure 10.8d. The answer obtained by program PDELTA is: $Q_{cr}=2.4\ EI/l^2$. Horizontal forces $\{P\}=ql\ \{0.1, 0.2, 0.2, 0.2, 0.2, 0.1\}$ spaced at $0.2l$ are assumed as a substitute of the distributed load.

10.19 Use Equation 10.43, combined with computer program PDELTA (Chapter 20) to verify the answers of any or each of Problems 10.3, 10.5, 10.14, and 10.15.

Chapter 11

Analysis of shear wall structures

11.1 INTRODUCTION

In buildings, it is important to ensure adequate stiffness to resist lateral forces induced by wind, seismic, or blast effects. These forces can develop high stresses and produce sway movement or vibration, thereby causing discomfort to the occupants. Concrete walls, which have high in-plane stiffness, placed at convenient locations, are often economically used to provide the necessary resistance to horizontal forces. This type of wall is called a shear wall. The walls may be placed in the form of assemblies surrounding lift shafts or stairwells; this box-type structure is efficient in resisting horizontal forces. Columns, of course, also resist horizontal forces; their contribution depends on their stiffness relative to the shear walls. The objective of the analysis for horizontal forces is to determine in what proportion are the external loads at each floor level distributed among shear walls and the columns (Figure 11.1).

The horizontal forces are usually assumed to act at floor levels. The stiffness of the floors in the horizontal direction is very large compared with the stiffness of shear walls or columns. For this reason, it is common to assume that each floor diaphragm is displaced in its horizontal plane as a rigid body. This rigid-body movement can be defined by translations along horizontal perpendicular axes and a rotation about a vertical axis at an arbitrary point in the floor (Figure 11.1c). The assumption of rigid-body in-plane behavior is important in that it reduces considerably the degree of kinematic indeterminacy.

A major simplification of the problem is achieved if the analysis can be limited to a plane structure composed of shear walls and frames subjected to horizontal forces in their plane. This is possible when the building is laid out in a symmetrical rectangular grid pattern so that the structure can be assumed to be made up of two sets of parallel frames acting in perpendicular directions (Figure 11.1b). Very often, even in an irregular building, an idealized plane structure is used to obtain an approximate solution. The majority of published papers on the analysis of shear wall structures deal with this type of two-dimensional problem.

The unsymmetrical arrangement of shear walls in the building in Figure 11.1c causes the diaphragms to rotate and translate under the action of symmetrical horizontal forces. The shear walls in this case are subjected to twisting moments which cannot be calculated if the analysis is limited to an idealized plane structure. In the following sections, the analysis of an idealized plane shear wall structure is treated on the basis of certain simplifying assumptions.

11.2 STIFFNESS OF A SHEAR WALL ELEMENT

In the analysis to follow, we shall treat shear walls as vertical deep beams transmitting loads to the foundations. The effect of shear deformations in these walls is of greater importance

DOI: 10.1201/9780429286858-11

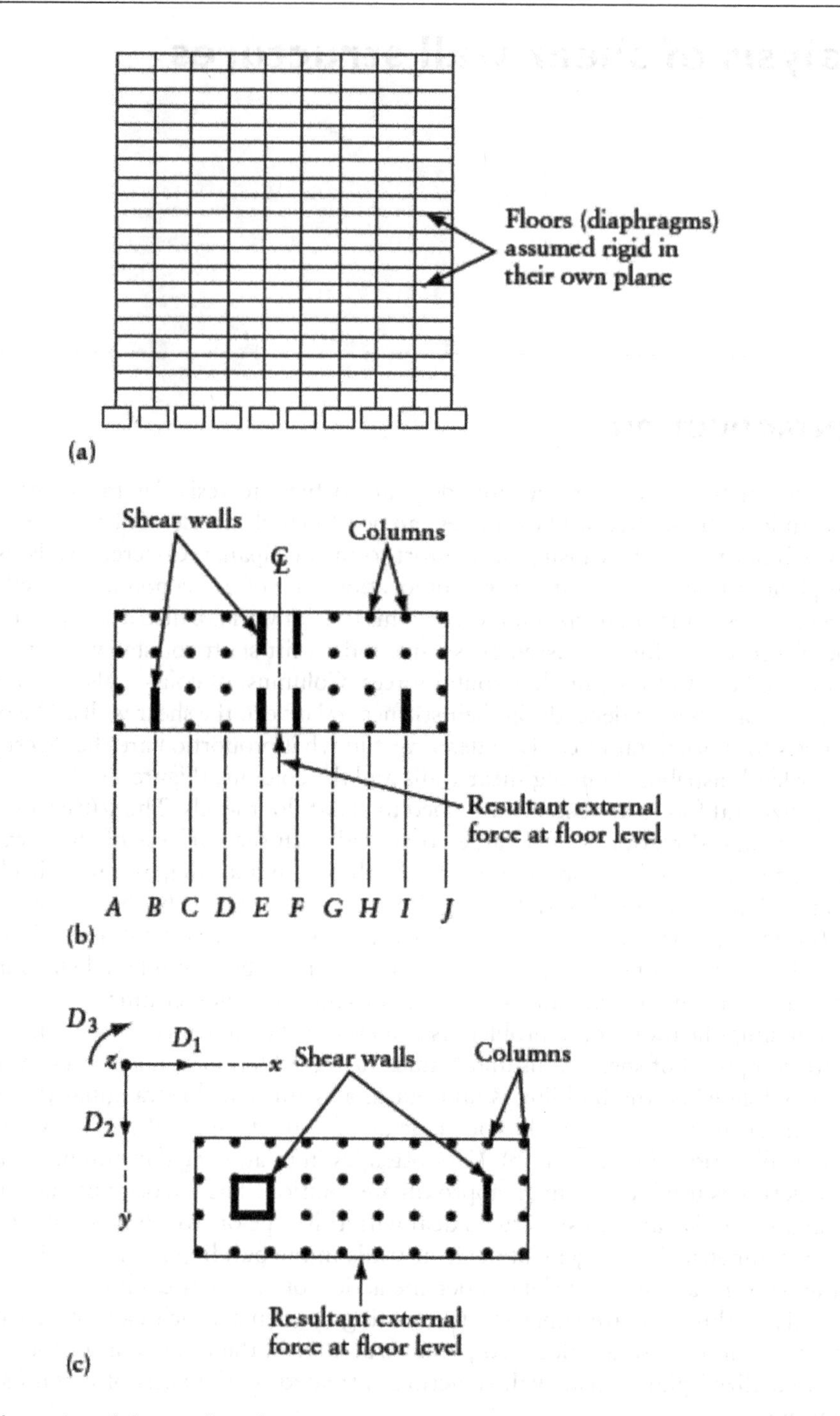

Figure 11.1 Illustration of shear walls and some assumptions involved in their analysis. (a) Elevation of a multistory structure. (b) Plan for a regular, symmetric building. (c) Plan of unsymmetrical building (rigid-body translation of the floor is defined by displacements at the coordinates D_1, D_2, and D_3).

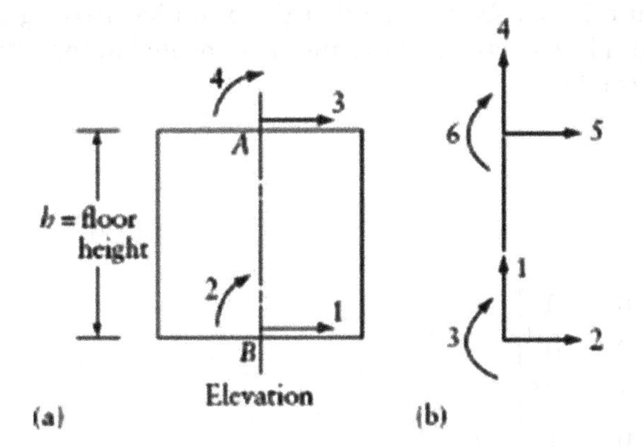

Figure 11.2 Flexibility and stiffness of a member considering shear, bending, and axial deformations. (a) Shear wall element. (b) Coordinates corresponding to the flexibility or stiffness matrix in Equation 11.5.

than in conventional beams, where the span-to-depth ratio is much larger. The stiffness matrix of an element of a shear wall between two consecutive floors (Figures 11.2a and b), considering shear deformation, is derived.

Consider the member AB in Figure 11.2; in some cases, it is necessary to consider the axial deformation so that the stiffness matrix has to be written for the six coordinates shown in Figure 11.2b. The stiffness matrix corresponding to these coordinates taking into account bending, shear and axial deformation is:

$$[S]=\begin{bmatrix} \frac{Ea}{h} & & & & & \\ & \frac{12EI}{(1+\alpha)h^3} & & & & \\ & \frac{6EI}{(1+\alpha)h^2} & \frac{(4+\alpha)EI}{(1+\alpha)h} & & & \\ -\frac{Ea}{h} & & & \frac{Ea}{h} & & \\ & \frac{-12EI}{(1+\alpha)h^3} & \frac{-6EI}{(1+\alpha)h^2} & & \frac{12EI}{(1+\alpha)h^3} & \\ & \frac{6EI}{(1+\alpha)h^2} & \frac{(2-\alpha)EI}{(1+\alpha)h} & & \frac{-6EI}{(1+\alpha)h^2} & \frac{(4+\alpha)EI}{(1+\alpha)h} \end{bmatrix} \quad \text{Symmetrical; Elements not shown are zero} \tag{11.1}$$

where

$$\alpha = \frac{12\,EI}{h^2 G a_r} \tag{11.2}$$

E = modulus of elasticity; a and I = area and moment of inertia of the cross section perpendicular to the axis, respectively; G = shear modulus of elasticity; a_r = effective shear area (see Section 7.3.2); and h = floor height. Substituting $\alpha = 0$, Equation 11.1 becomes the same as Equation 5.19 with the shear deformations ignored.

11.3 STIFFNESS MATRIX OF A BEAM WITH RIGID END PARTS

Shear walls are usually connected by beams, and, for purposes of analysis, we have to find the stiffness of such a beam corresponding to coordinates at the wall axis. Consider the beam AB of Figure 11.3a. We assume that the beam has two rigid parts AA' and B'B (Figure 11.3b). The displacements $\{D^*\}$ at A and B are related to the displacements $\{D\}$ at A' and B' by geometry as follows:

$$\{D\} = [H]\,\{D^*\} \tag{11.3}$$

where

$$[H] = \begin{bmatrix} 1 & \delta l & 0 & 0 \\ 0 & 1 & 0 & 0 \\ 0 & 0 & 1 & -\beta l \\ 0 & 0 & 0 & 1 \end{bmatrix} \tag{11.4}$$

The elements in the first and second column of $[H]$, which are the $\{D\}$ displacements due to $D_1^* = 1$ and $D_2^* = 1$, respectively, can be checked by examining Figure 11.3c.

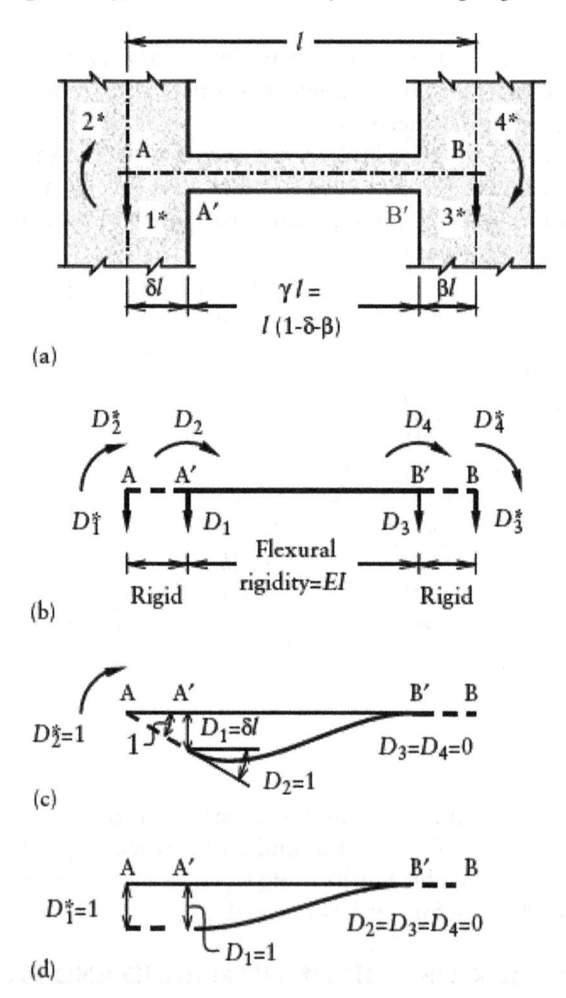

Figure 11.3 Coordinate system corresponding to the stiffness matrices in Equations 11.8 and 11.10 of a beam between shear walls. (a) Elevation of a beam between shear walls. (b) Coordinate systems. (c) and (d) Deflected configurations corresponding to $D_2^* = 1$ and $D_1^* = 1$, respectively.

If shear deformations are to be considered, the stiffness matrix [S] of the beam corresponding to the {D} coordinates is:

$$[S]=\frac{1}{1+\alpha}\begin{bmatrix} \frac{12EI}{(\gamma l)^3} & & & \text{Symmetrical; Elements not shown are zero} \\ \frac{6EI}{(\gamma l)^2} & (4+\alpha)\frac{EI}{\gamma l} & & \\ -\frac{12EI}{(\gamma l)^3} & -\frac{6EI}{(\gamma l)^2} & \frac{12EI}{(\gamma l)^3} & \\ \frac{6EI}{(\gamma l)^2} & (2-\alpha)\frac{EI}{\gamma l} & -\frac{6EI}{(\gamma l)^2} & (4+\alpha)\frac{EI}{\gamma l} \end{bmatrix} \tag{11.5}$$

where $\gamma = (1\text{-}\delta\text{-}\beta)$; δ and β = ratios of the lengths of rigid parts of the beam to the total length l.

From Equation 8.47, the stiffness matrix corresponding to the $\{D^*\}$ coordinates is given by:

$$\left[S^*\right]=[H]^{\mathrm{T}}[S][H] \tag{11.6}$$

Substituting Equations 11.4 and 11.5 into 11.6 gives:

$$\left[S^*\right]=\frac{EI}{1+\alpha}\begin{bmatrix} \frac{12}{(\gamma l)^3} & & & \text{Symmetrical; Elements not shown are zero} \\ \frac{6}{(\gamma l)^2}+\frac{12\delta}{\gamma^3 l^2} & \frac{4+\alpha}{\gamma l}+\frac{12\delta}{\gamma^2 l}+\frac{12\delta^2}{\gamma^3 l} & & \\ -\frac{12}{(\gamma l)^3} & \frac{-6}{(\gamma l)^2}-\frac{12\delta}{\gamma^3 l^2} & \frac{12}{(\gamma l)^3} & \\ \frac{6}{(\gamma l)^2}+\frac{12\beta}{\gamma^3 l^2} & \frac{2-\alpha}{\gamma l}+\frac{6\delta+6\beta}{\gamma^2 l}+\frac{12\delta\beta}{\gamma^3 l} & \frac{-6}{(\gamma l)^2}-\frac{12\beta}{\gamma^3 l^2} & \frac{4+\alpha}{\gamma l}+\frac{12\beta}{\gamma^2 l}+\frac{12\beta^2}{\gamma^3 l} \end{bmatrix} \tag{11.7}$$

EXAMPLE 11.1: USE OF COMPUTER PROGRAM PLANEF (CHAPTER 20)

Using computer program PLANEF (Chapter 20), verify the elements on the first two columns of $[S^*]$ for the beam AB in Figure 11.3b. Assume: E = 1.0; G = E/2.4 (Poisson's ratio, ν = 0.2); l = 8.0 m; δ = 0.1; β = 0.2; γ = 0.7; assume α = 0.0588. For the part A'B', assume a rectangular section 0.25 × 0.8 m² for which: a = 0.2 m²; $I = 10.667 \times 10^{-3}$ m⁴; a_r = 0.16667 m². We analyze a structure having 4 nodes and 3 members. For the rigid parts, we assume each of a, I, and a_r equal to 1.0E6.

The prescribed displacements for the first column of $[S^*]$ are zero for displacements at the end, except $D_1^*=1$. For the second column, the prescribed displacements are zero, except $D_2^*=1$. The computer program gives the first two columns of $[S^*]$:

$$\left[S^*\right]=\begin{bmatrix} 0.6884\times10^{-3} & 0.2478\times10^{-2} & \cdots & \cdots \\ 0.2478\times10^{-2} & 0.1083\times10^{-1} & \cdots & \cdots \\ -0.6884\times10^{-3} & -0.2478\times10^{-2} & \cdots & \cdots \\ 0.3029\times10^{-2} & 0.9000\times10^{-2} & \cdots & \cdots \end{bmatrix}$$

Equation 11.7 gives the same answers.

11.4 ANALYSIS OF A PLANE FRAME WITH SHEAR WALLS

Consider the structure shown in Figure 11.1b, composed of frames parallel to the axis of symmetry. Some of these frames include shear walls. Because of symmetry in structure and in loading, the diaphragms translate without rotation. With the diaphragms assumed rigid in their own planes, all the frames sway by the same amount D^* at a given floor level, as shown in Figure 11.4a.

The stiffness matrix $[S^*]_i$ (of the order $n \times n$, where n = number of floors), corresponding to the $\{D^*\}$ coordinates is calculated for each plane frame. The matrices are then added to obtain the stiffness $[S^*]$ of the entire structure:

$$\left[S^*\right] = \sum_{i=1}^{m} \left[S^*\right]_i$$

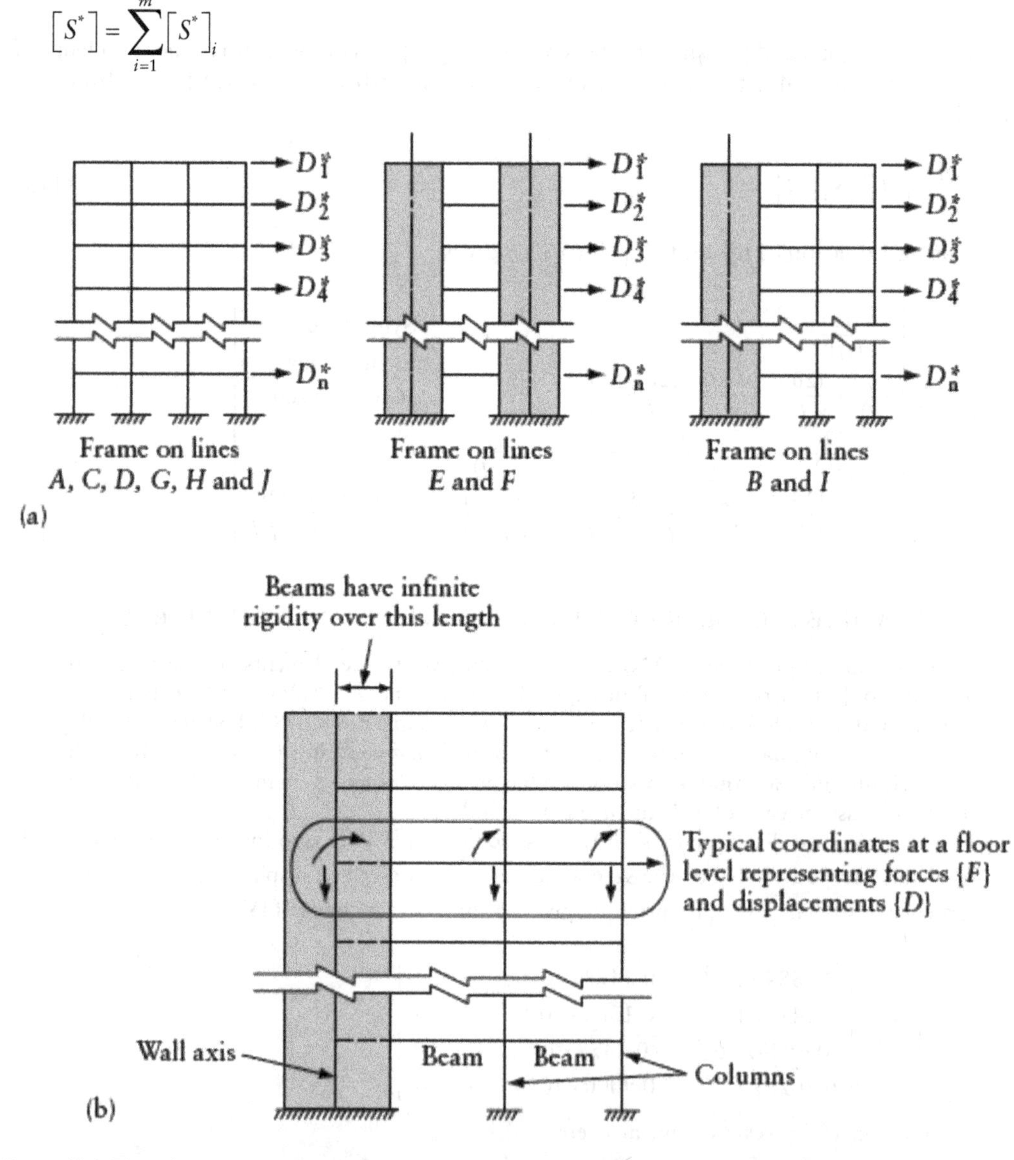

Figure 11.4 Plane frames considered in the analysis of the symmetrical three-dimensional structure of Figure 11.1b. (a) Frames parallel to the line of symmetry in the building of Figure 11.1b. (b) Coordinate system corresponding to the stiffness matrix $[S]_i$ for the frame on lines B and I.

where m is the number of the frames.

The sway at the floor levels is calculated by:

$$\left[S^*\right]_{n\times n}\left\{D^*\right\}_{n\times 1}=\left\{F^*\right\}_{n\times 1} \tag{11.8}$$

where $\{F^*\}$ are the resultant horizontal forces at floor levels.

In order to determine $[S^*]_i$ for any frame, say, the frame on line B or I (Figure 11.4a), coordinates are taken at the frame joints as shown in Figure 11.4b. These represent rotation and vertical displacement at each joint and sway of the floor as a whole. The corresponding stiffness matrix $[S]_i$ (of the order $7n \times 7n$, in this case) is first derived using the stiffness of the shear wall and beam attached to it (as obtained in Sections 11.2 and 11.3). The stiffness matrix $[S]_i$ is then condensed into matrix $[S^*]_i$ corresponding to coordinates for the sidesway at floor level (see Section 5.12). The elements of $[S^*]_i$ are forces at floor levels corresponding to unit horizontal displacements at alternate floors with the rotations and vertical joint displacements allowed to take place.

After solving Equation 11.8 for $\{D^*\}$, the horizontal forces at floor levels for each plane frame are determined by:

$$\left[S^*\right]_i\left\{D^*\right\}=\left\{F^*\right\}_i \tag{11.9}$$

When the horizontal forces $\{F^*\}_i$ are applied at the floors of the ith frame, without forces at the other coordinates in Figure 11.4b, the displacements at all the other coordinates in this figure can be calculated. From these, the stress resultants in any element can be determined.

Clough et al.[1] give details of an analysis involving the same assumptions as in the above approach and describe an appropriate computer program. It may be interesting to give here the results of their analysis of a 20-story structure with the plan arrangement shown in Figure 11.5a, subjected to wind loading in the direction of the x-axis. Story heights are 10 ft (3.05 m), except for the ground floor which is 15 ft (4.58 m) high. All columns and shear walls are fixed at the base. The properties of the structural members are listed in Table 11.1.

For this structure, two types of frames need to be considered: frame A with five bays, with all columns considered to be of zero width in the x direction; and frame B of three bays: 26, 36, and 26 ft (7.93, 10.98, and 7.93 m) with two shear wall columns 20 ft (6.10 m) wide.

Figure 11.5b, taken from Reference 1, shows the distribution of shear force between the columns and shear walls; we can see that the major part of the lateral resistance is provided by the shear walls.

Figure 11.5c, also reproduced from Reference 1, shows the bending moment in the shear wall and in an inner column of frame A. These results clearly demonstrate the different behavior of columns and shear walls: in the words of Clough et al., "*the shear wall is basically a cantilever column, with frame action modifying its moment diagram only slightly, whereas the single column shows essentially pure frame action.*" The effect of discontinuity in column stiffness between the tenth and eleventh floors is apparent in Figures 11.5b and c.

11.5 SIMPLIFIED APPROXIMATE ANALYSIS OF A BUILDING AS A PLANE STRUCTURE

The preceding example (see Figure 11.5c) shows that the columns have a point of inflection within the height of each story, while the deflection of the walls is similar to that of a cantilever. This is so because the rotation of the column ends is elastically restrained by the beams. When the walls have a very high I-value compared to that of the beams, which is the case

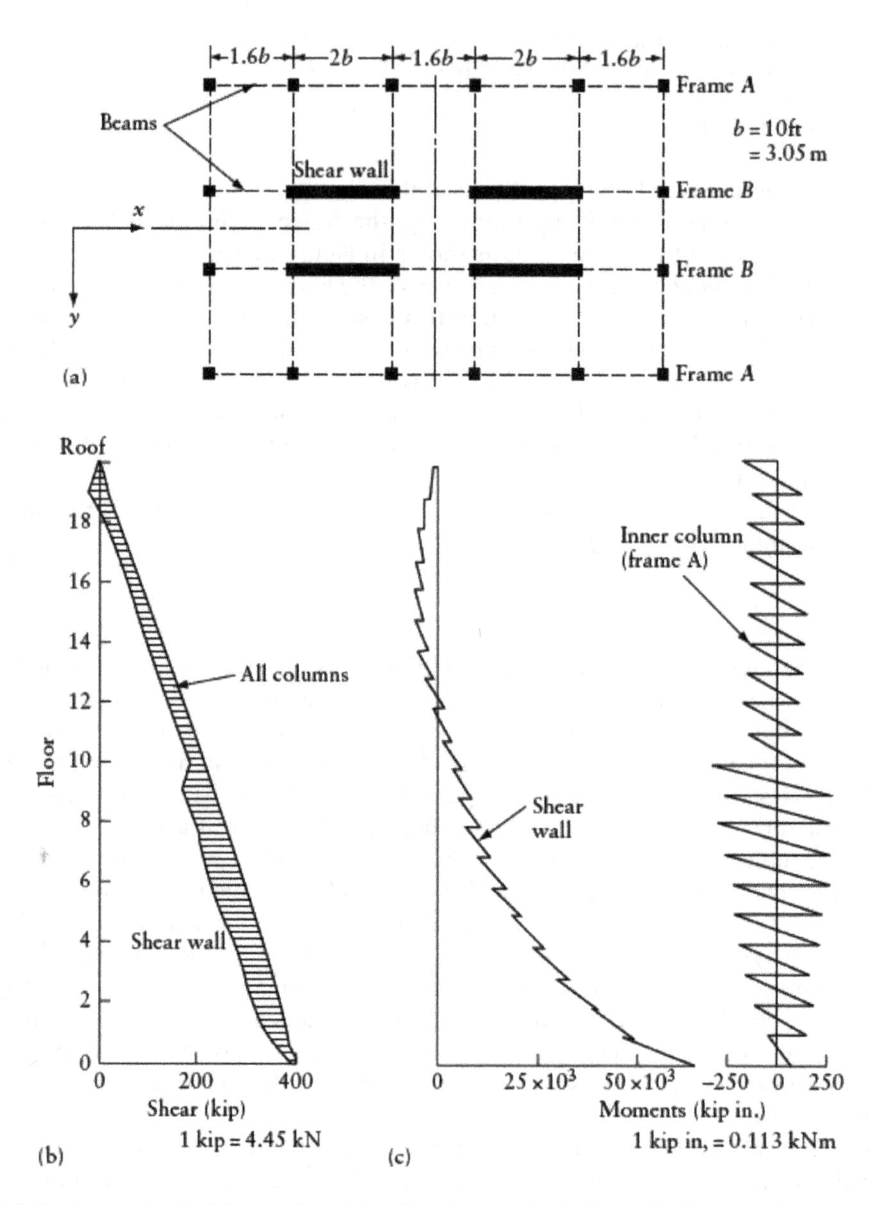

Figure 11.5 Example of a building analyzed by Clough et al.[1] (a) Plan. (b) Shearing force distribution. (c) Moments in vertical members.

Table 11.1 Member properties for the example of Figure 11.5

	Columns		*Shear Walls*		*Girders*	
	I *(in.⁴)*	*Area (in.²)*	***I*** *(in.⁴)*	*Area (in.²)*	***I*** *(in.⁴)*	*Area (in.²)*
Stories 11–20	3437	221.5	1793×10^4	3155	6875	—
Stories 1–10	9604	313.5	2150×10^4	3396	11295	—

1 in.^{2} = 645 mm^2; 1 in.^{4} = 416,000 mm^4.

in practice, the beams cannot significantly prevent the rotation at the floor levels associated with the deflection in the form of a cantilever.

This behavior leads to the suggestion that structures of the type shown in Figures 11.1b or 11.5a, under the action of horizontal forces, can be idealized into a structure composed of the two systems indicated in Figure 11.6a. One of these is a shear wall, which has I-value in any story equal to the sum of the I-values of all the walls; the second system is an equivalent

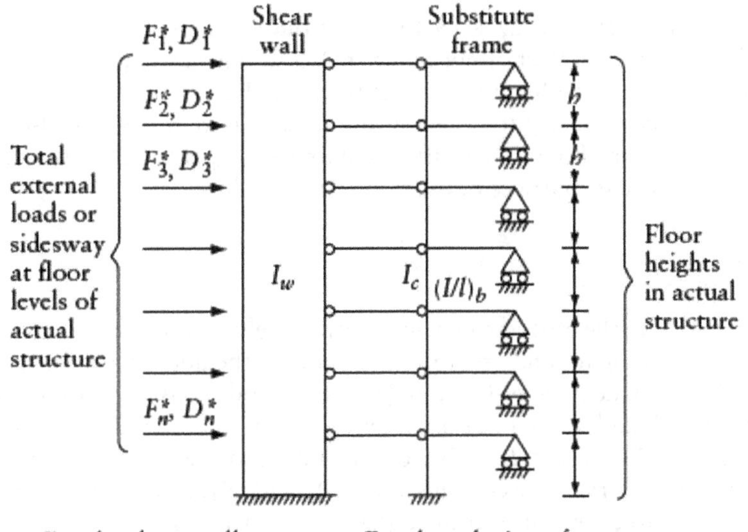

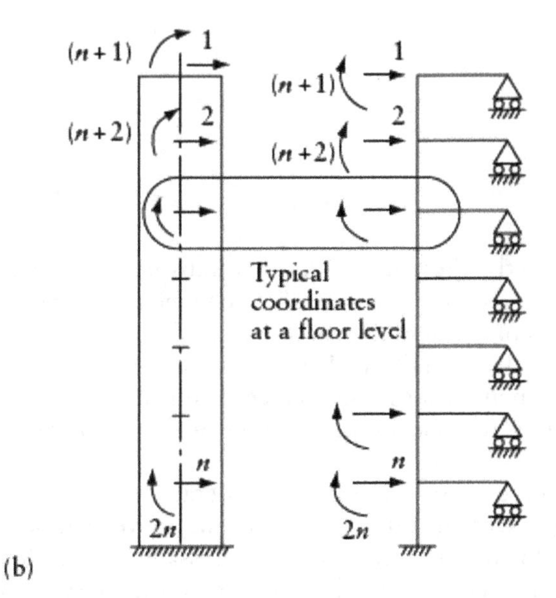

Figure 11.6 Simplified analysis of a building frame of the type shown in Figures 11.1b or 11.5a. (a) Idealized structure. (b) Coordinates corresponding to stiffness matrices $[S_w]$ and $[S_r]$.

column rigidly jointed to the beams. The I_c value for the equivalent column is the sum of the I-values for all the columns in a story. The $(I/l)_b$ value for any of the beams is equal to four times the sum of (I/l) values for all the beams running in the x-direction. The two systems, connected by inextensible link members, are assumed to resist the full external horizontal forces at floor levels. Further, the axial deformations of all the members are ignored. The shear deformations of the wall or the columns may or may not be included in the analysis. If they are, the reduced (effective) area is the sum of the reduced areas of the walls or columns in a story.

Replacement of the actual frames by the substitute frame in Figure 11.6a implies the assumption that, under the effect of horizontal forces at floor levels, the beams deflect with a point of inflection (zero bending moment) at their middle.

The idealized structure in Figure 11.6a is assumed to have n degrees-of-freedom representing the sidesway of the floors. The stiffness matrix $[S^*]_{n\times n}$ of this structure is obtained by summation of the stiffness matrices of the two systems. Thus:

$$\left[S^*\right] = \left[S^*\right]_w + \left[S^*\right]_r \tag{11.10}$$

where $[S^*]_w$ and $[S^*]_r$ are the stiffness matrices, respectively, of the shear wall and the substitute frame, corresponding to n horizontal coordinates at the floor levels. For the determination of $[S^*]_w$ or $[S^*]_r$, two degrees-of-freedom (a rotation and sidesway) are considered at each floor level for the wall and at each beam-column connection in the substitute frame. A stiffness matrix $[S]_w$ or $[S]_r$ of order $2n \times 2n$ corresponding to the coordinates in Figure 11.6b is written, then these matrices are condensed to $[S^*]_w$ and $[S^*]_r$ which relate horizontal forces to sidesway with the rotations unrestrained (see Section 5.12).

The sidesway at floor levels of the actual structure is then calculated by solving Equation 11.8, where the displacements $\{D^*\}$ represent the horizontal translation at floor levels of all columns or shear walls in the building. The external forces $\{F^*\} = \{F^*\}_w + \{F^*\}_r$, where $\{F^*\}_w$ and $\{F^*\}_r$ are, respectively, the forces resisted by walls and by the substitute frame, can be calculated from:

$$\left\{F^*\right\}_w = \left[S^*\right]_w \left\{D^*\right\} \tag{11.11}$$

and

$$\left\{F^*\right\}_r = \left[S^*\right]_r \left\{D^*\right\} \tag{11.12}$$

These forces are then applied to the shear wall and to the substitute frame, and the member end moments are determined in each system. If these end moments are apportioned to the walls, columns, and beams of the actual structure according to their (EI/h) or (EI/l) values, approximate values of the actual member end moments can be obtained. The apportionment in this manner may result in unbalanced moments at some of the joints. Improved values can be rapidly reached by performing one or two cycles of moment distribution.

It is important to note that if the shear walls differ considerably from one wall to another or if there are variations in cross section at different levels, the above method of calculation can lead to erroneous results. In such a case, it may be necessary to consider an idealized structure composed of more than one wall attached by links to the substitute frame and to derive the stiffness of each separately; the stiffness of the idealized structure is then obtained by summation.

EXAMPLE 11.2: STRUCTURES WITH FOUR AND WITH TWENTY STORIES

Use computer program PLANEF (Chapter 20) to find the approximate values of the end moments in a column and a shear wall in a structure which has the same plan as in Figure 11.5a, and has four stories of equal height $h = b$. The frame is subjected to a horizontal force in the x direction of magnitude $P/2$ at roof and P at each of the other floor levels. The properties of members are: for any column, $I = 17 \times 10^{-6}b^4$; for any beam, $I = 34 \times 10^{-6}b^4$; and for any wall, $I = 87 \times 10^{-3}b^4$. Take $E = 2.3G$. The area of wall cross section is $222 \times 10^{-3}b^2$. Consider shear deformation in the walls only.

The wall and the substitute frame of the idealized structure are shown in Figure 11.7a. In the actual structure there are 16 columns, 4 walls, 12 beams of length $1.6b$, and 4 beams of length $2b$. The properties of members of the idealized structure are:

$$I_c = \sum I_{ci} = 16 \times 17 \times 10^{-6}\, b^4 = 272 \times 10^{-6}\, b^4$$

$$(I/l)_b = 4\sum (I/l)_{bi} = 4 \times 34 \times 10^{-6}\, b^4 \left(\frac{12}{1.6b} + \frac{4}{2.0b}\right) = 1292 \times 10^{-6}\, b^3$$

$$I_w = \sum I_{wi} = 4 \times 87 \times 10^{-3}\, b^4 = 348 \times 10^{-3}\, b^4$$

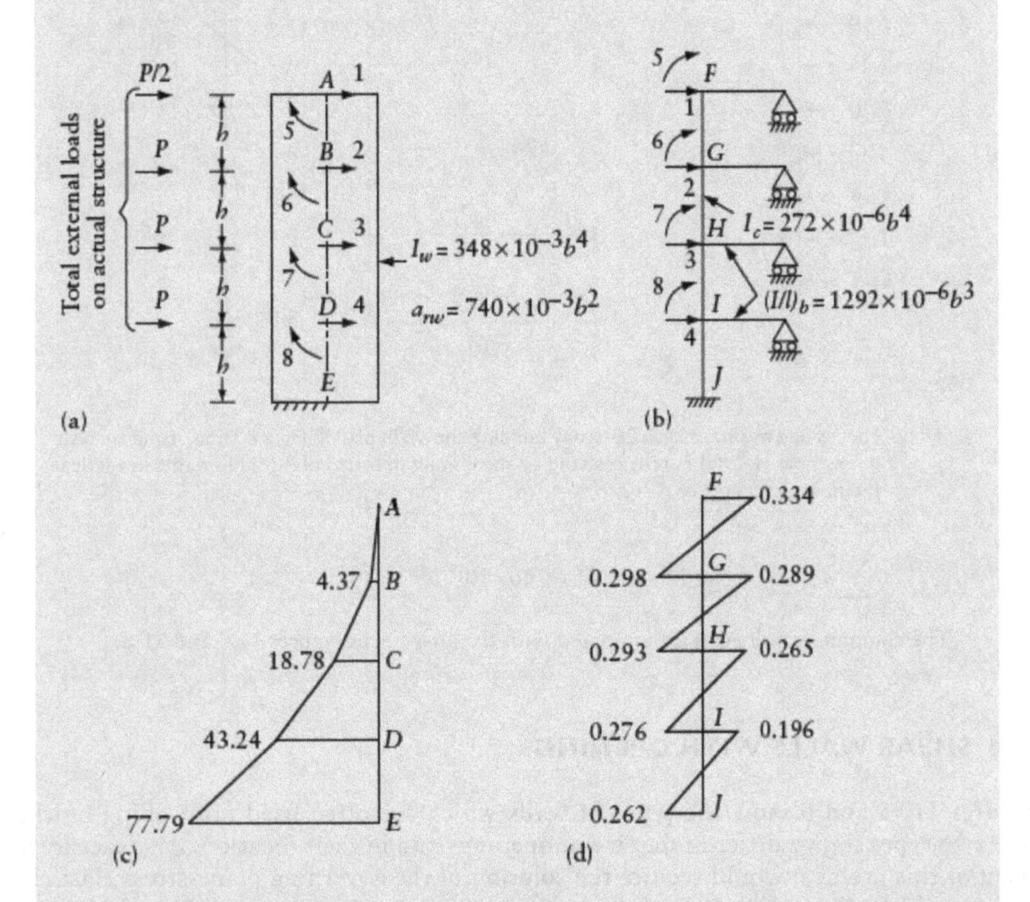

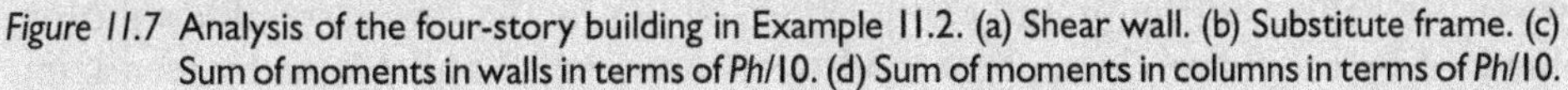
Figure 11.7 Analysis of the four-story building in Example 11.2. (a) Shear wall. (b) Substitute frame. (c) Sum of moments in walls in terms of $Ph/10$. (d) Sum of moments in columns in terms of $Ph/10$.

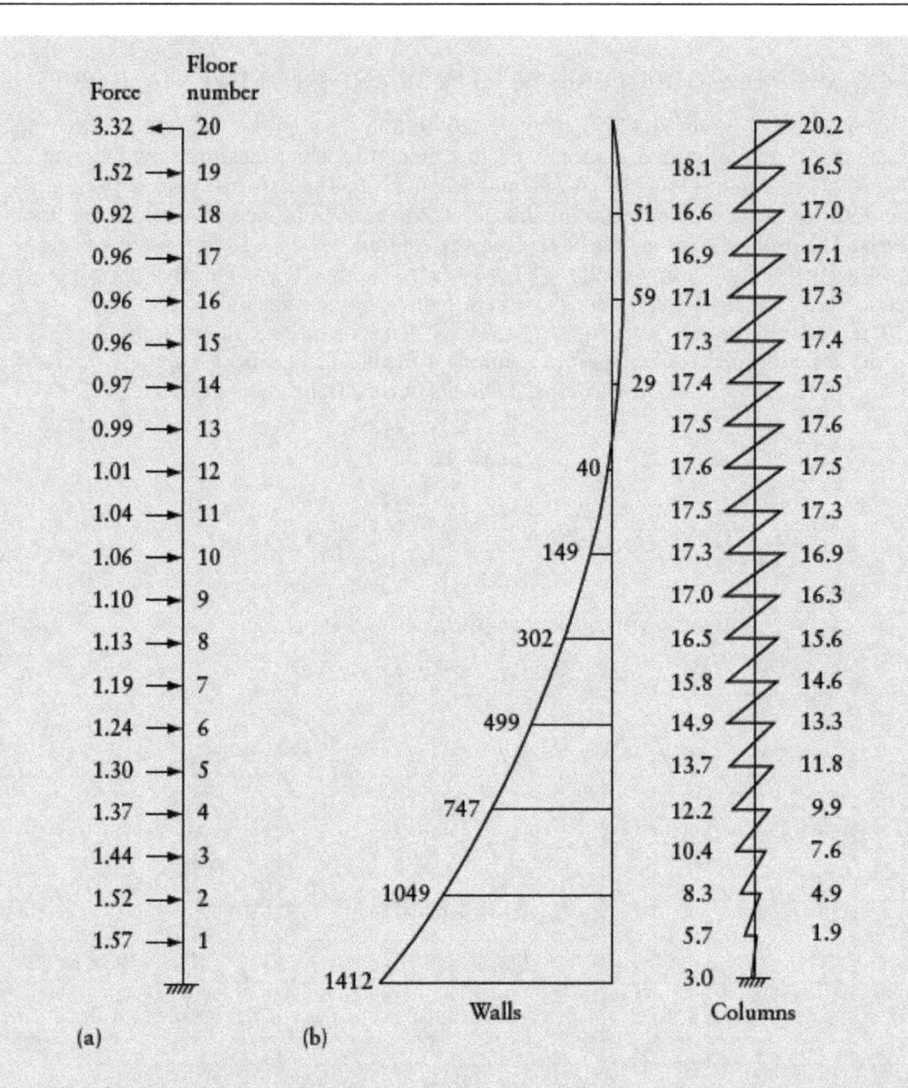

Figure 11.8 Forces and moments in a 20-story building shown in plan in Figure 11.5a, using analysis in Example 11.2. (a) Forces resisted by shear walls in terms of *P*. (b) Moments in vertical members in terms of *Ph*/10.

$$a_{rw} = \sum a_{rwi} = 4 \times \frac{5}{6} \times 222 \times 10^{-3}\, b^2 = 740 \times 10^{-3}\, b^2$$

The computer results can be compared with the answers in Figures 11.7 and 11.8.

11.6 SHEAR WALLS WITH OPENINGS

Figures 11.9a and b show the types of walls which are often used in dwelling blocks. The two types shown differ in the size of openings and in their location. An exact treatment of this problem would require the solution of the governing plane-stress elasticity equations, but this is difficult and cannot be used in practice. A reasonable solution can be obtained by the finite-element method (see Chapters 12 and 13) or by idealizing the wall into different types of latticed frames composed of small elements; however, the calculation generally requires the solution of a large number of equations. These methods are used in the analysis of walls with openings arranged in any pattern, and they give a better picture of the stress distribution than the much more simplified analysis described below.

In the simplified analysis, walls with a row of openings of the type shown in Figures 11.9a and b are idealized to a frame composed of two wide columns connected by beams with end parts infinitely rigid. The stiffness of the elements forming such a structure and the method of analysis are given in Sections 11.2 and 11.3. MacLeod[2] showed by model testing that the idealization of a wall of this type by a frame gives a good estimate of stiffness (corresponding to sidesway) for most practical cases. It seems, therefore, that the finite-element idealization offers little advantage in this respect.

The symmetrical wall in Figure 11.9a can be analyzed using a suitable frame composed of one column rigidly connected to beams (see Figure 11.7b). Each beam has a rigid part near its connection with the column. The stiffness matrix of such a beam with one end hinged, corresponding to the coordinates in Figure 11.9d can be easily derived from Equation 11.10 and is:

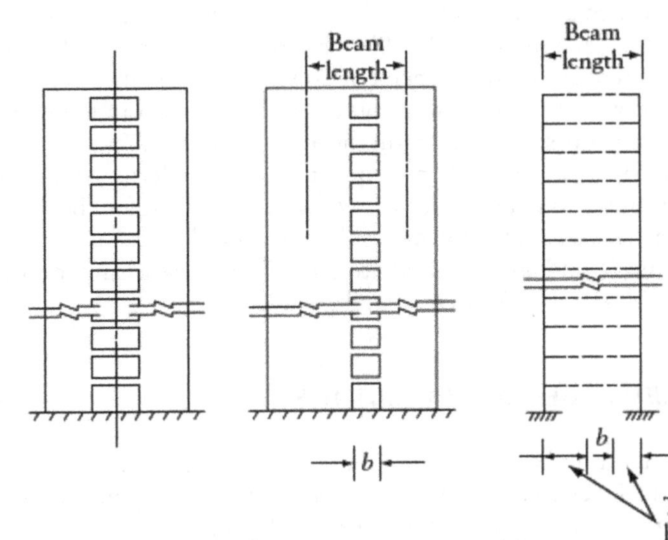

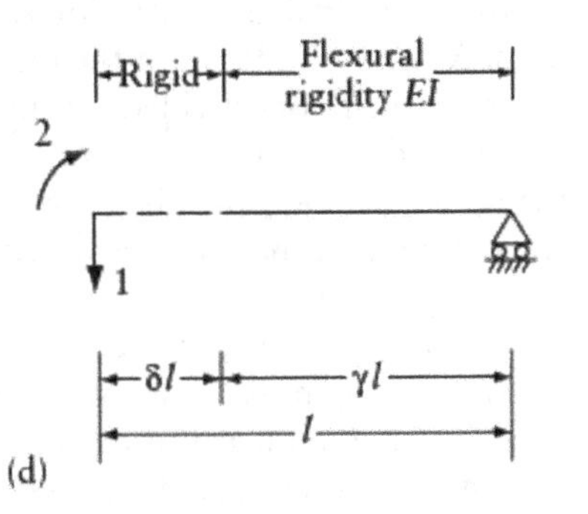

Figure 11.9 Idealized structure for the analysis of a wall with a row of openings. (a) Symmetrical wall. (b) Wall with a row of openings. (c) Idealized structure for the analysis of the wall in part (b). (d) Coordinates corresponding to the stiffness matrix $\bar{S}$ given by Equation 11.13.

$$\left[\bar{S}\right] = \begin{bmatrix} S_{11}^{*} - \dfrac{S_{14}^{*\,2}}{S_{44}^{*}} & \text{Symm.} \\ S_{21}^{*} - \dfrac{S_{24}^{*}\,S_{41}^{*}}{S_{44}^{*}} & S_{21}^{*} - \dfrac{S_{24}^{*\,2}}{S_{44}^{*}} \end{bmatrix} \tag{11.13}$$

where S_{ij}^{*} are elements of the stiffness matrix in Equation 11.7 with $\beta = 0$.

11.7 THREE-DIMENSIONAL ANALYSIS

A joint in a three-dimensional framed structure has, in general, six degrees-of-freedom: three rotations and three translations in the x, y, and z directions (Figure 11.10a). The assumption that the diaphragms in a multistory building are rigid constrains three of the displacements (D_1, D_2, and D_3) to be the same at all joints in one floor. Even with this important simplification, the analysis is complicated in the case of three-dimensional structures incorporating members having an arbitrary orientation in space.

This section deals with the case of a structure formed of shear walls in a random arrangement. Any two or more walls which are monolithic will be referred to as a *wall assembly*. A typical wall assembly is shown in plan in Figure 11.10b. The structure is analyzed to determine the forces resisted by different shear walls when horizontal forces in any direction are applied at floor levels. In addition to the rigid-diaphragm assumption used in the previous sections, we assume here that the floors do not restrain the joint rotations about the x and y axes (D_4 and D_5 in Figure 11.10a). This assumption is equivalent to considering that the diaphragm has a small flexural rigidity compared with the walls and can therefore be ignored. With this additional assumption, horizontal forces result in no axial forces in the walls; thus, the vertical displacements (D_6 in Figure 11.10a) are zero. With these assumptions, the analysis is done in Section 14.7.1 and 14.7.2 of Ghali and Neville.[3]

11.8 OUTRIGGER-BRACED HIGH-RISE BUILDINGS

The drift of a tall building caused by horizontal forces due to wind or earthquake can be reduced by the provision of outriggers. Figure 11.11a represents plan of a typical floor of a tall building in which the core is composed of two halves; each consists of concrete walls housing the elevator shafts. The core is the main lateral force-resistant component. The maximum horizontal displacement (the drift) is limited by codes for the stability of the building and for the comfort of its occupants. Also, the codes limit the interstory drift ratio, defined as the difference of drift in two consecutive floors divided by the vertical distance between them. The sum of the moments at the ends of a column at a floor level is a couple transferred, in the opposite direction, to the floor; the floor must be designed for the flexural and shear stress caused by the transfer. The moments transferred between the columns and the floors are mainly dependent on the interstory drift ratio.

The outriggers are relatively stiff horizontal members, of depth commonly equal to one-story height, connecting the core to the exterior columns, inducing in them tensile and compressive forces, forming stabilizing couples that reduce the maximum drift and the interstory drift ratio. The outriggers can be at one level or at two levels (Figure 11.11b), or at up to four levels in very tall buildings. For lateral loads in the west-east direction of the building in Figure 11.11, the columns at the east and the west extremities of the outriggers are subjected to tensile and compressive forces, respectively; the corresponding internal

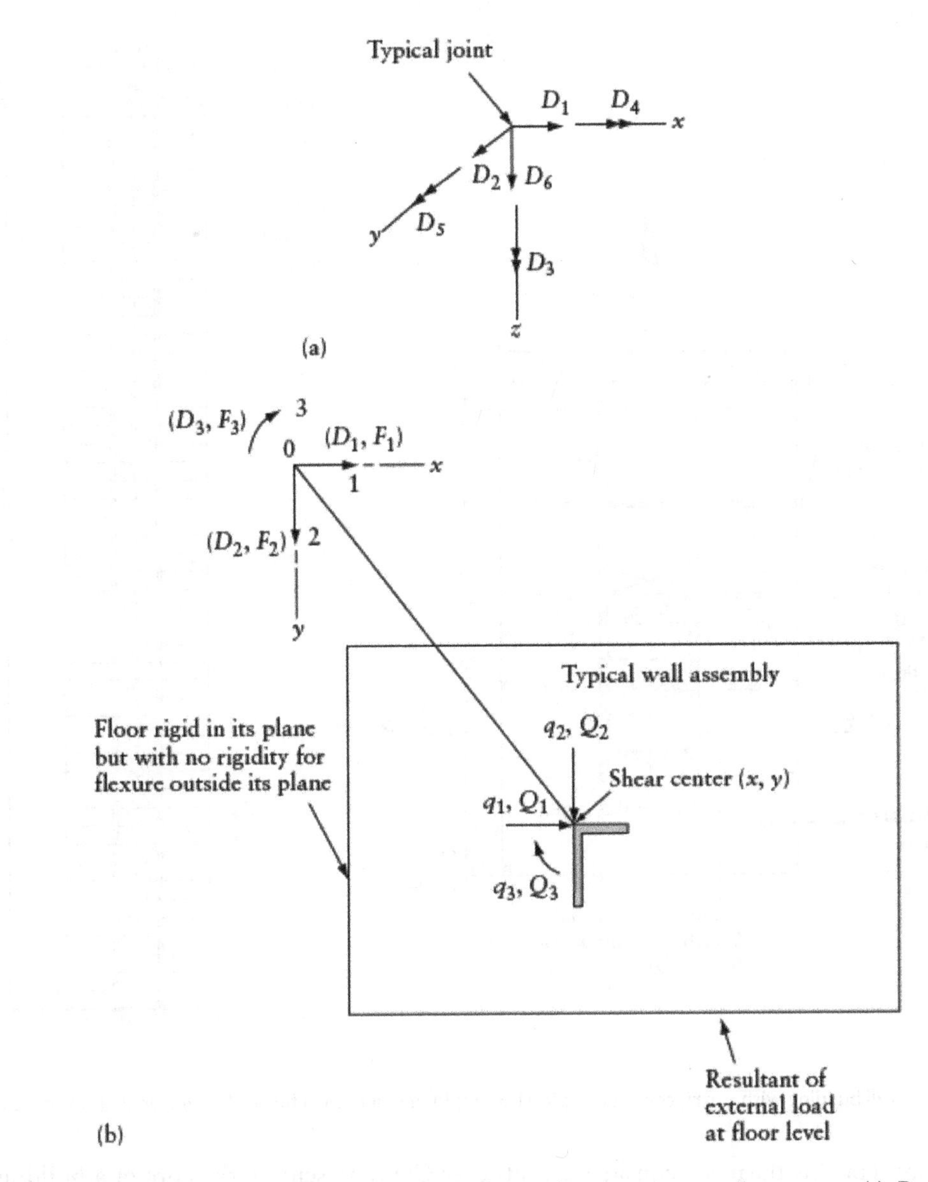

Figure 11.10 Coordinate system for the analysis of a single-story shear wall structure. (a) Degrees-of-freedom of a typical joint in a building frame. (b) Coordinate system.

forces act downward and upward at the east and the west extremities, respectively, of the outriggers. A stiff spandrel girder or truss on the periphery can also be provided to connect the outer ends of the outriggers and also to mobilize the columns on the north and south ends of the building to stiffen the core.

11.8.1 Location of the outriggers

The outriggers are walls that impair the use of the floor area; thus, they are usually located in a tall building in the floor(s) of mechanical equipment. However, it is useful to study at what level an outrigger is most effective in reducing the drift and the bending moment

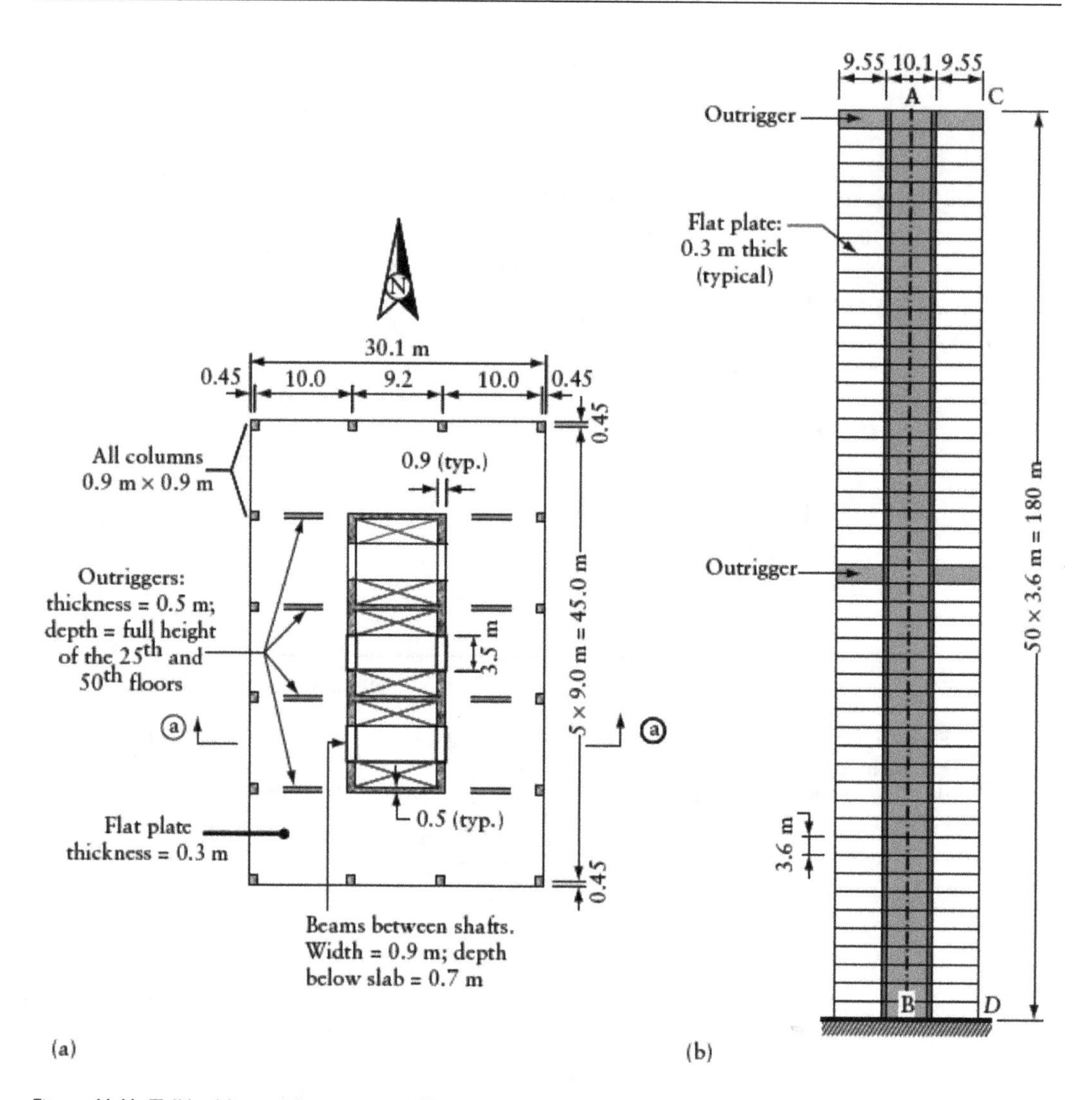

Figure 11.11 Tall building with outriggers (Example 11.3). (a) Plan of a typical floor. (b) Sectional elevation a-a.

in the core. For this purpose, consider a cantilever CD representing the core of a building subjected to uniform load q/unit length (Figure 11.12a). A rigid outrigger AB elastically restrains the rotation of the cantilever at E by columns extending from A and B to the base, at the same level as D. The elasticity modulus E is assumed to be the same for all members. Constant I and a are assumed for the core and for the columns, respectively; the flexural rigidity of the columns is ignored, and only their axial deformation and the bending deformation of the core are considered. The floors connecting the core to the exterior columns also restrain the free rotation and translation of the core; the existence of these floors is ignored here, but it is assumed that their presence excludes buckling of the exterior columns. We analyze the structure by the force method, releasing a single redundant F_1 representing the moment connecting AB to the cantilever at E; the redundant F_1 is a clockwise couple on AB and anticlockwise on the core. The incompatible rotation of the released core at E is (Equation A.28):

$$D_1 = -\frac{ql^3}{6EI}\left(3\xi - 3\xi^2 + \xi^3\right) \tag{11.14}$$

where $\xi = x/l$. The relative rotation of the released core relative to AB at E due to $F_1 = 1$ is (Equation A.31):

$$f_{11} = \frac{\xi l}{EI} + \frac{2\xi l}{Eab^2} \tag{11.15}$$

$$f_{11} = \frac{\xi l}{\beta EI} \tag{11.16}$$

where

$$\beta = ab^2 / \left(ab^2 + 2I\right) \tag{11.17}$$

Thus, the redundant F_1 (= $-D_1/f_{11}$) can be expressed as:

$$F_1 = \frac{ql^2}{6}\beta\left(3 - 3\xi + \xi^2\right) \tag{11.18}$$

The drift at the top and the bending moment at the base of the core are (Equations A.27 and A.30):

$$D_{top} = \frac{ql^4}{8EI} - \frac{F_1 l^2}{EI}\xi(1 - 0.5\xi) \quad ; \quad M_{base} = \frac{ql^2}{2} - F_1 \tag{11.19}$$

When the outrigger is at the top, $\xi = 1.0$ and:

$$F_1 = \frac{ql^2\beta}{6} \quad ; \quad D_{top} = \frac{ql^4}{EI}\left(\frac{1}{8} - \frac{\beta}{12}\right) \quad ; \quad M_{base} = ql^2\left(0.5 - \frac{\beta}{6}\right) \tag{11.20}$$

The location of the outrigger that minimizes D_{top} is at $\xi = 0.54$ giving:

$$F_1 = 0.279\beta ql^2 \quad ; \quad D_{top} = \frac{ql^4}{EI}\left(\frac{1}{8} - 0.110\beta\right) \quad ; \quad M_{base} = ql^2\left(0.5 - 0.279\beta\right) \tag{11.21}$$

The bending moment diagram of the core with $\xi = 0.54$ and $\beta = 0.8$ is shown in Figure 11.12b. It can be seen that the outrigger reduces D_{top} (from $0.125ql^4/EI$ to $0.037ql^4/EI$) and M_{base} (from $0.500ql^2$ to $0.278ql^2$) without affecting the shear at the base. Figure 11.12c shows the bending moment diagram of the core with rigid outriggers at the top and mid-height; the corresponding drift at the top is also given in the figure.

Considering the flexibility of the outrigger in the above equations will only increase f_{11} (Equation 11.15) to become $\bar{f}_{11} = f_{11} + b(1-\delta)^3/(12EI_o)$, where EI_o is the flexural rigidity of the outrigger, δ is a ratio (= the length of the rigid part of AB divided by $\overline{AB}$ (Figure 11.3a). For practical values of I_o/I, considering the flexibility of the outriggers will have a small effect on the values given in Figures 11.12b and c.

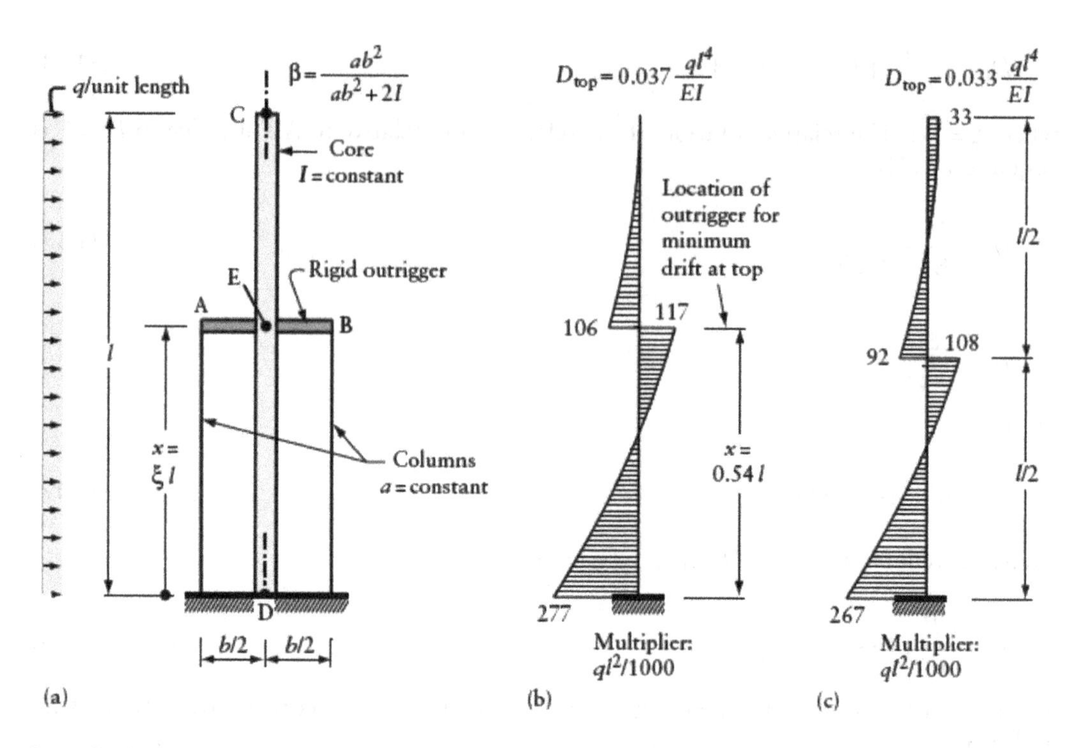

Figure 11.12 Core wall with one and two rigid outriggers. (a) Simplified model considering only bending deformation of the core and axial deformation of the columns. (b) Bending moment in core with one outrigger at $x = 0.54l$, with $\beta = 0.8$. (c) Bending moment in core with outriggers at top and at mid-height, with $\beta = 0.8$.

EXAMPLE 11.3: CONCRETE BUILDING WITH TWO OUTRIGGERS SUBJECTED TO WIND LOAD

The 50-story reinforced concrete building shown in Figure 11.11 is analyzed for the effect of wind load in the east-west direction. For simplicity, uniform load q per unit height of the building is considered. The building is idealized as a plane frame subjected to a uniform load of intensity q (Figure 11.13a). Line AB represents the centroidal axis of the core; at each floor level, AB is connected to rigid elements representing the parts within the shaft width. Members on line CD represent the six columns on the east façade. The two columns, away from the corners, on the north and the south façades can contribute to resisting the wind load by framing with the corner columns; this relatively small contribution is here ignored for simplicity of presentation.

In calculating the cross-sectional properties of the members given in Figure 11.13a, the gross concrete sections are considered for the core (AB) and for the edge columns on CD; to account for cracking, half the gross concrete sections are considered for the floor slabs. Within the width of the core, the horizontal members are considered rigid. A single value of the elasticity modulus is assumed for all members.

The reactions at B and D and the bending moments in AB and CD are given in Figures 11.13a, b, and c, respectively. The applied moment of the wind load about the base is $500 \times 10^{-3}ql^2$. This is partly resisted by the anticlockwise moment reactions at B, D, and F; the total moment value = $(0.322+0.308+280)\ 10^{-3}ql^2 = 281 \times 10^{-3}ql^2$, representing 56 percent of the applied moment. The remainder of the applied moment is resisted by the couple induced by the upward and downward vertical reactions at B and D, respectively

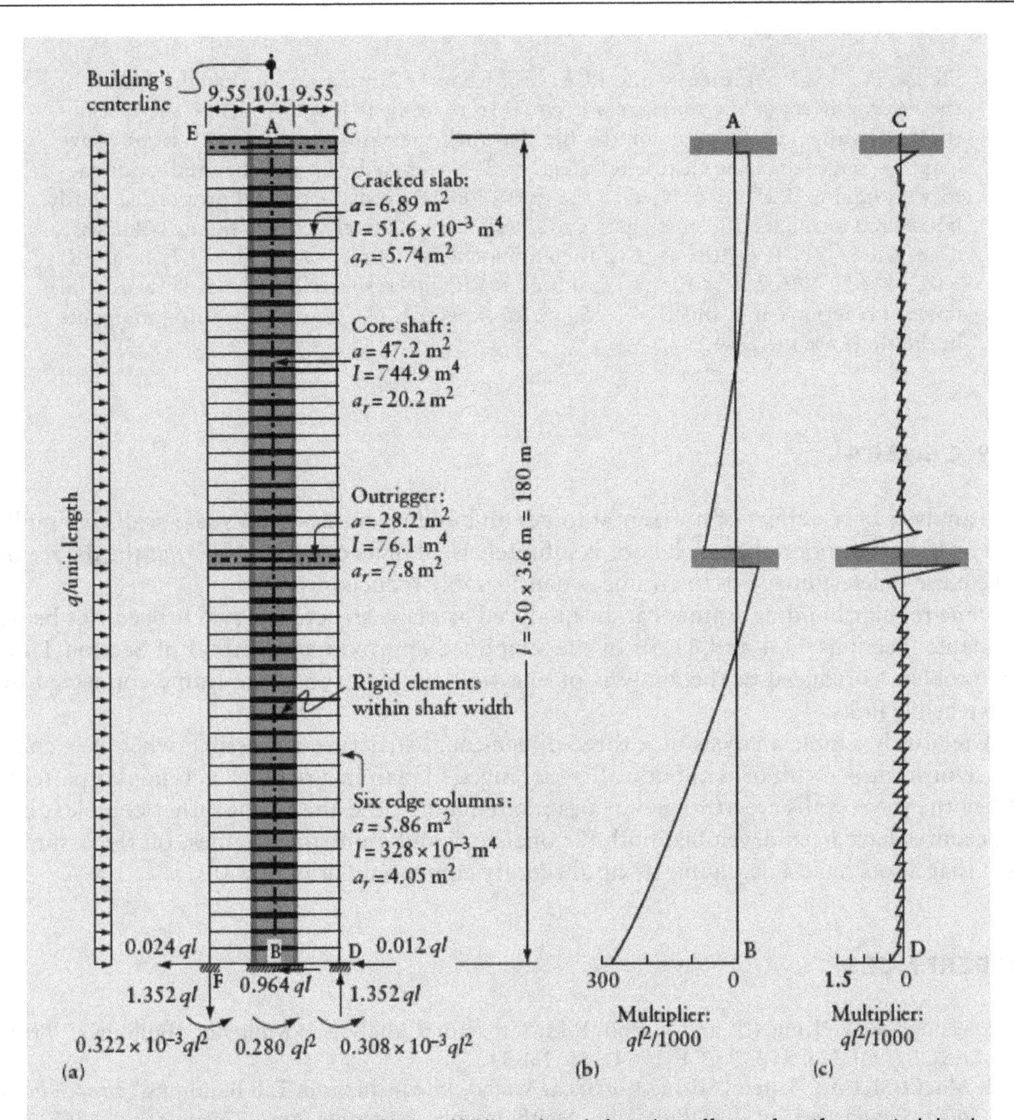

Figure 11.13 The tall building of Figure 11.11 analyzed for the effect of uniform wind load (Example 11.3). (a) Plane frame idealization of the right-hand half of the building. (b) Bending moment in the core. (c) Total bending moment in the six columns on the east edge.

(Figure 11.13a); the resisting moment of the vertical reaction components = 1.35 $ql \times 0.1622\ l = 219 \times 10^{-3} ql^2$.

If the wind load were resisted totally by the core, the horizontal displacement at the top would be $125 \times 10^{-3} ql^4/(EI)$, with $I = 744.9$ m⁴ = the second moment of area of the core about its centroidal axis. Because of the interaction of the core with other members, the drift at the top is reduced to $D_{top} = 44 \times 10^{-3} ql^4/(EI)$. Repetition of the analysis with the cross-sectional area properties of the outriggers reduced by a factor of 0.5 (e.g. to account for cracking) gives results that are not significantly different.

In the above elastic analysis, a reduction factor of 0.5 is applied to the gross concrete cross-sectional area properties to obtain effective values that account for cracking of the slabs. Codes differ in their requirements for the reduction factors that account for the variation in concrete cracking of the structural components; different reduction factors are required for the slabs, the core, and the columns. The results in

Figure 11.13 are obtained (using PLANEF, Chapter 20) assuming that the length of the flexible parts of the outriggers is equal to their clear length (outside the faces of the core walls). To account for the high-strain penetration in the wall, an effective length – longer than the clear length (e.g. by 20%) – may be used. When reduction factors of 0.5 and 0.25 are applied to the gross concrete cross-sectional properties of all horizontal members, the outriggers and the slabs respectively, the following results are obtained: $D_{top} = 47 \times 10^{-3}ql^4/(EI)$; the anticlockwise moment reactions at F, B, and D = $0.328 \times 10^{-3}ql^2$, $296 \times 10^{-3}ql^2$, and $0.303 \times 10^{-3}ql^2$ respectively; the downward and upward reactions at F and D = $1.253ql$. As expected, D_{top} and the bending moments in the shaft are increased.

11.9 GENERAL

The analysis of the effect of horizontal forces on building frames with shear walls is simplified by the assumption that each floor is infinitely rigid in its own plane so that the degree of kinematic indeterminacy of the frame is considerably reduced.

Some regular building frames can be analyzed as plane structures, two procedures being available (Sections 11.4 and 11.5). In the simplified approximate method of Section 11.5, the problem is reduced to the analysis of one wall and one substitute frame connected by inextensible links.

A relatively simple analysis of a three-dimensional structure is possible when it is composed of frames – with or without walls – arranged in plan in a regular rectangular pattern. When the shear walls are arranged in a random manner, the analysis is rather complex, but procedures have been developed both for one-story and multistory frames, on the assumption that floors have a negligible flexural rigidity compared with the walls.

REFERENCES

1. Clough, R.W., King, I.P. and Wilson, E.L., "Structural Analysis of Multistory Buildings," *Proc. ASCE*, (90), No. ST3, Part 1 (1964), pp. 19–34.
2. MacLeod, I.A., "Lateral Stiffness of Shear Walls with Openings in Tall Buildings," *Proceedings of a Symposium on Tall Buildings*, Southampton, Pergamon Press, New York, 1967, pp. 223–244.
3. Ghali, A. and Neville, A.M., *Structural Analysis: A Unified Classical and Matrix Approach*, 7th ed., CRC Press, Taylor & Francis Group, Boca Raton, FL, 2017, 933 pp.

PROBLEMS

11.1 Using computer program PLANEF verify columns 3 and 4 of $[S^*]$ in Equation 11.7.
Hint: Follow the procedure in procedure Example 11.1.

11.2 Solve Example 11.1 with the properties of members in the top two stories as follows: for any column, $I = 17 \times 10^{-6}b^4$; for any beam, $I = 24.3 \times 10^{-6}b^4$; for any wall, $I = 58 \times 10^{-3}b^4$; and area of wall cross section = $146 \times 10^{-3}b^2$. All other data is unchanged.

11.3 Find the bending moment at section A-A and the end moment in the beam at the lower floor in the symmetrical wall with openings shown in the figure. Neglect shear deformation in the beams and axial deformations in all elements. Take $E = 2.3G$. The wall has a constant thickness. To idealize the shear wall, use a substitute frame similar to that shown in Figure 11.7b. The substitute frame has one vertical column connected to horizontal beams. The top three beams are of length $3b/4$ and the lower three are of length b.

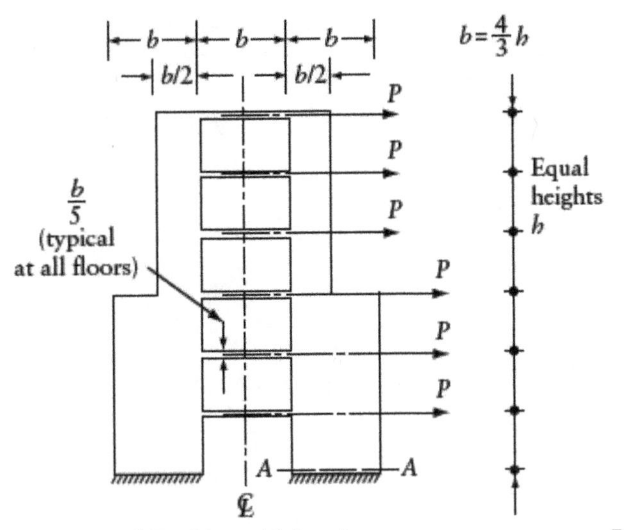

11.4 Analyze the structure of Problem 11.3 using computer program PLANEF (Chapter 20). Assume $b = 5$ m; $h = 3.75$ m; thickness of shear walls and width of beams = 0.25 m; $E = 1.0$; Poisson's ratio, $\nu = 0.2$; $G = E/2.4$. Ignore axial deformation of members of the idealized plane frame shown.

Chapter 12

Methods of finite differences and finite-elements

12.1 INTRODUCTION

Numerical solution of differential equation relating beam deflection and applied loading, in finite difference form, gives the deflections at equally spaced nodes. Then we use the calculated deflections to determine the shearing force, the bending moment, and the support reactions. The procedure applies to beams, slabs, and shells; but in the present chapter, we use it only for beams on elastic foundations. The same equation that applies to beams on elastic foundations applies also to circular cylindrical walls subjected to axisymmetrical loading. The computer program CTW (Cylindrical Tank Walls) performs the analysis (Chapter 20). The major part of the present chapter concerns finite-element analysis.

The finite-element method is widely used in structural analysis. The method is also used in a wide range of physical problems including heat transfer, seepage, flow of fluids, and electrical and magnetic potential. In the finite-element method, a continuum is idealized as an assemblage of finite-elements with specified nodes. The infinite number of degrees-of-freedom of the continuum is replaced by specified unknowns at the nodes.

In essence, the analysis of a structure by the finite-element method is an application of the displacement method. In frames, trusses, and grids, the elements are bars connected at the nodes; these elements are considered to be one-dimensional. Two-dimensional or three-dimensional finite-elements are used in the analysis of walls, slabs, shells, and mass structures. The finite-elements can have many shapes with nodes at the corners or on the sides (Figure 12.1). The unknown displacements are nodal translations or rotations or derivatives of these.

12.2 REPRESENTATION OF DERIVATIVES BY FINITE DIFFERENCES

Figure 12.2 represents a function $y=f(x)$, which for our purposes can, for example, be the deflection of a beam. Consider equally spaced abscissae x_{i-1}, x_i, and x_{i+1} and the corresponding ordinates y_{i-1}, y_i, and y_{i+1}. The derivative (or the slope) of the curve at point $x_{i-1/2}$ midway between i and $i-1$ can be approximated by:

$$\left(\frac{dy}{dx}\right)_{i-\frac{1}{2}} \cong \frac{1}{\lambda}\left(y_i - y_{i-1}\right) \tag{12.1}$$

where λ is the spacing of the abscissae. Similarly, the slope of the curve midway between i and i+1 is:

$$\left(\frac{dy}{dx}\right)_{i+\frac{1}{2}} \cong \frac{1}{\lambda}\left(y_{i+1} - y_i\right) \tag{12.2}$$

DOI: 10.1201/9780429286858-12

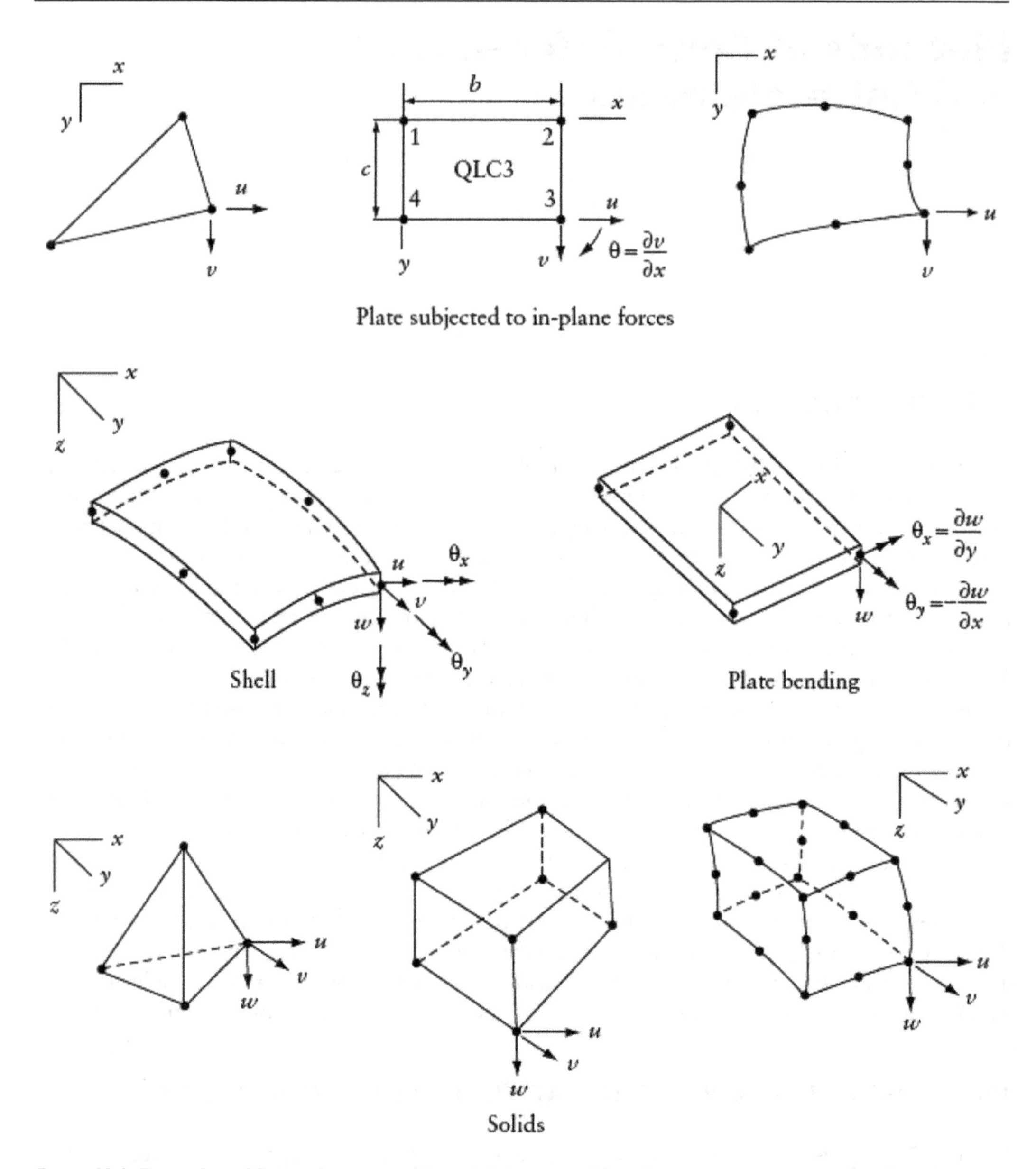

Figure 12.1 Examples of finite-elements with nodal degrees of freedom shown at a typical node only.

The second derivative at i (which is the rate of change of slope) is approximately equal to the difference between the slope at $i+1/2$ and at $i-1/2$ divided by λ thus:

$$\left(\frac{\mathrm{d}^2y}{\mathrm{d}x^2}\right)_i \cong \frac{1}{\lambda}\left[\left(\frac{\mathrm{d}y}{\mathrm{d}x}\right)_{i+\frac{1}{2}} - \left(\frac{\mathrm{d}y}{\mathrm{d}x}\right)_{i-\frac{1}{2}}\right]$$

Substituting from Equations 12.1 and 12.2:

$$\left(\frac{\mathrm{d}^2y}{\mathrm{d}x^2}\right)_i \cong \frac{1}{\lambda^2}\left(y_{i+1} - 2y_i + y_{i-1}\right) \tag{12.3}$$

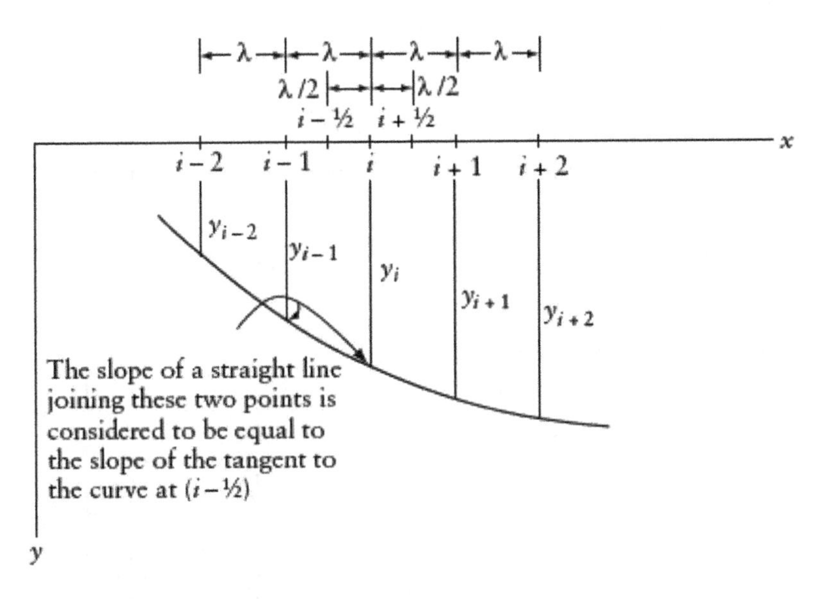

Figure 12.2 Graph of function $y=f(x)$.

In the above expressions, we have used central differences because the derivative of the function in each case was expressed in terms of the values of the function at points located symmetrically with respect to the point considered. The process can be repeated to calculate higher derivatives, in which case the values of y at a greater number of equally spaced points are required. This is done in the finite difference pattern of coefficients as shown in Figure 12.3.

The first derivative at i can also be expressed in terms of y_{i-1} and y_{i+1} with interval 2λ. Similarly, the third derivative at i can be expressed from the difference of the second derivatives at i+1 and i–1 with interval 2λ. The resulting coefficients are given in the last two rows of Figure 12.3.

Other finite difference expressions can be obtained by considering forward or backward differences in which the derivative at any point is expressed in terms of the value of the function at points in ascending or descending order with respect to the point under consideration. The central differences are more accurate than either forward or backward differences and they will be used in this chapter.

12.3 BEAM ON ELASTIC FOUNDATION

The differential equation for a beam on elastic foundation in (Figure 12.4):

$$\frac{d^2}{dx^2}\left(EI\frac{d^2y}{dx^2}\right)=q-ky \tag{12.4}$$

q = transverse load per unit length; y = transverse deflection; k = foundation modulus = foundation reaction per unit length per unit deflection (force/length²). The term between brackets in Equation 12.4 = – (the bending moment) expressed in finite difference form as:

$$M_i \cong -\frac{EI_i}{\lambda^2}\left(y_{i+1}-2y_i+y_{i-1}\right) \tag{12.5}$$

The second derivative of moment is:

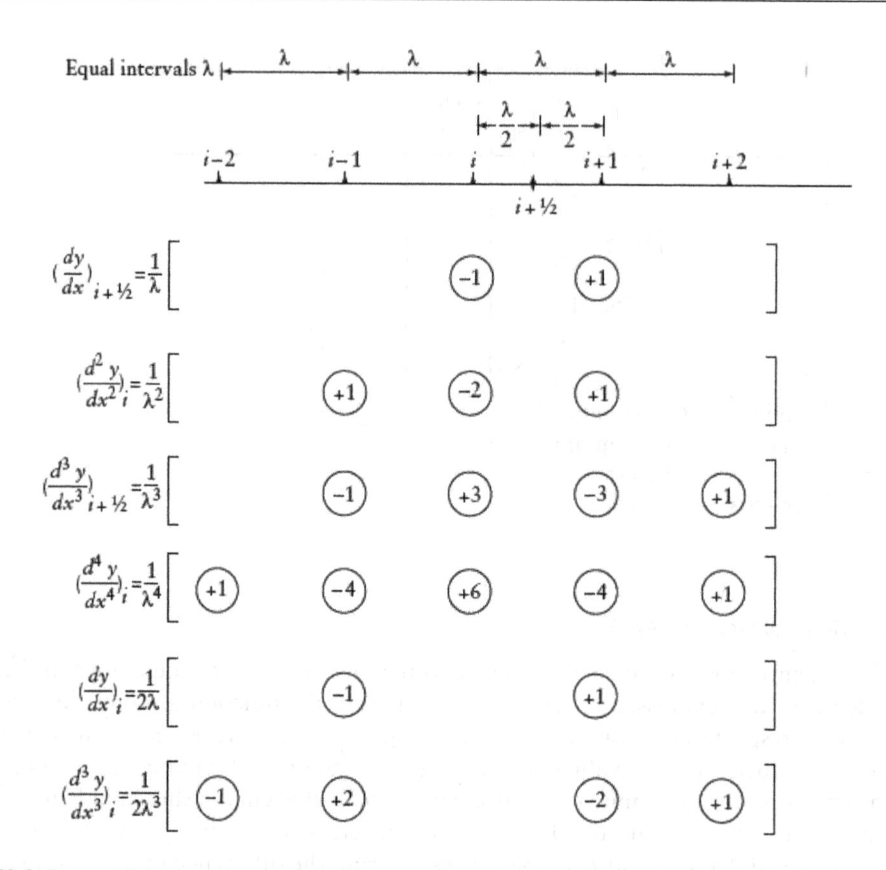

Figure 12.3 Finite difference pattern of coefficients, using central differences.

$$\frac{d^2}{dx^2}\left(EI\frac{d^2y}{dx^2}\right) = -\left(\frac{d^2M}{dx^2}\right) \cong -\frac{1}{\lambda^2}\left(M_{i+1} - 2M_i + M_{i-1}\right) \tag{12.6}$$

Combining Equations 12.4, 12.5 and 12.6 gives a finite difference equation applied at a general node i:

$$\frac{E}{\lambda^3}\begin{bmatrix} I_{i-1} & -2\left(I_{i-1}+I_i\right) & \left(I_{i-1}+4I_i+I_{i+1}+\dfrac{k_i\lambda^4}{E}\right) & -2\left(I_i+I_{i+1}\right) & I_{i+1}\end{bmatrix} \times \begin{Bmatrix} y_{i-2} \\ y_{i-1} \\ y_i \\ y_{i+1} \\ y_{i+2}\end{Bmatrix} \cong Q_i \tag{12.7}$$

I_i and k_i = moment of inertia of the beam and the foundation modulus at node i; Q_i = equivalent concentrated load accounting for the variation of applied load q over two consecutive spacings (assumed to be equal in Figure 12.5b; $\lambda_r = \lambda_l = \lambda$).

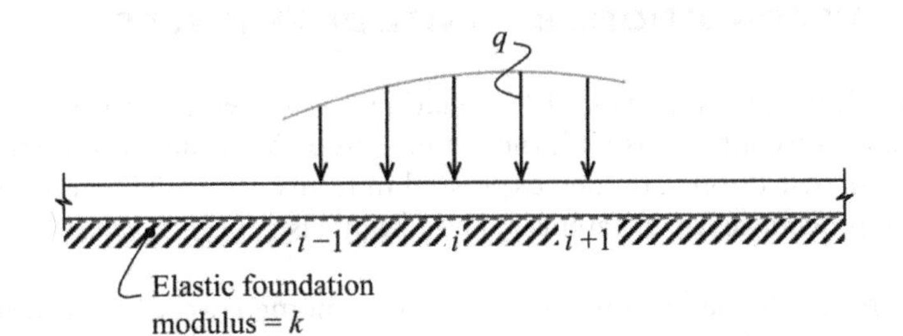

Figure 12.4 Beam on elastic foundation.

$$Q_i = \frac{1}{\lambda}\left(\int_0^{\lambda} q x_1 \, dx + \int_0^{\lambda} q x_2 \, dx\right) \tag{12.8}$$

$$Q_i = \frac{1}{\lambda}\left(-M_{i-1} + 2M_i - M_{i+1}\right) \tag{12.9}$$

Figure 12.5 presents finite difference equations relating deflection of beam over elastic foundation to equivalent applied load, Q_i.

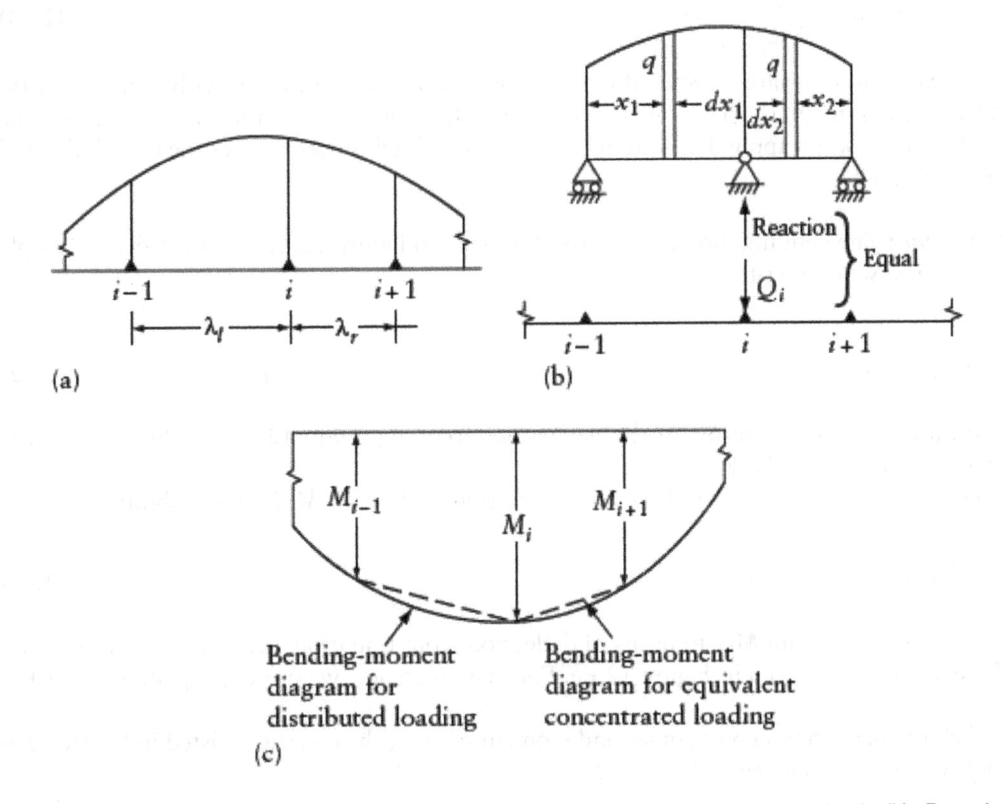

Figure 12.5 Bending moment using equivalent concentrated loading. (a) Applied load. (b) Equivalent concentrated load at *i*. (c) Loadings (a) and (b) induce the same bending moment at the nodes.

12.4 BOUNDARY CONDITIONS BY FINITE DIFFERENCES

When the finite difference Equation 12.7 relating the deflection to external loading is applied at or near a discontinuous end, the deflections at fictitious nodes outside the beam are included. These deflections are then expressed in terms of the deflections at other nodes on the beam, the procedure for different end conditions is:

(a) ***Simple support:*** The deflection and the bending moment are zero. Referring to Figure 12.6b, $y_{i-1}=0$ and $M_{i-1}=0$.

These two conditions are satisfied if the beam is considered to be continuous with another similar beam with similar loading acting in the opposite direction. Therefore, at a fictitious point i–2 (not shown in the figure), the deflection $y_{i-2}=-y_i$ and the finite difference Equation 12.7, when applied at a point i adjacent to a simple support, takes the form of Equation 12.10 in Figure 12.6b.

$$y_{i-1}=0 \quad \text{and} \quad \left(\frac{d^2y}{dx^2}\right)_{i-1}=0 \tag{12.10}$$

(b) ***Fixed end***: The deflection and the slope are zero. Referring to Figure 12.6c:

$$y_{i-1}=0 \quad \text{and} \quad \left(\frac{dy}{dx}\right)_{i-1}=0 \tag{12.11}$$

These two conditions are satisfied if the beam is considered continuous with a similar beam loaded in the same manner as the actual beam. Therefore, $y_{i-2}=y_i$ and the finite difference Equation 12.7, when applied at a point i adjacent to a fixed support, takes the form indicated in Figure 12.6c.

(c) ***Free end***: The bending moment is zero. Referring to Figure 12.6d and considering simple statics, we can write:

$$M_{i+1}=-Q_i\,\lambda \tag{12.12}$$

Substituting for M_{i+1} in terms of the deflections from Equation 12.5, we obtain the finite difference equation in Figure 12.6d.

Referring to Figure 12.6e and applying Equation 12.9, with $M_{i-1}=0$, we obtain:

$$-\frac{1}{\lambda}\left(-2M_i+M_{i+1}\right)=Q_i \tag{12.13}$$

Substituting for M_i and M_{i+1} in terms of deflection from Equation 12.5, we obtain the finite difference equation given in Figure 12.6e. For easy reference, we present Equations 12.10 to 12.13 in Figure 12.6.

When the beam has a constant second moment of area, the equations listed in Figure 12.6 simplify to the form given in Figure 12.7.

	Position of node i	Coefficient of the deflection in terms of E/λ^3					Right-hand side	Equation number
		y_{i-2}	y_{i-1}	y_i^*	y_{i+1}	y_{i+2}		
(a)	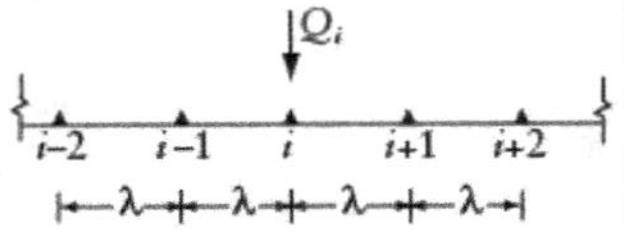 Q_i; $i-2$, $i-1$, i, $i+1$, $i+2$; λ, λ, λ, λ	I_{i-1}	$-2(I_{i-1}+I_i)$	$(I_{i-1}+4I_i+I_{i+1})$	$-2(I_i+I_{i+1})$	I_{i+1}	$=Q_i$	12.7
(b)	Hinged support; 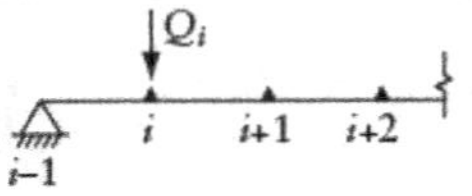 Q_i; $i-1$, i, $i+1$, $i+2$	–	–	$(4I_i+I_{i+1})$	$-2(I_i+I_{i+1})$	I_{i+1}	$=Q_i$	12.10
(c)	Fixed support; 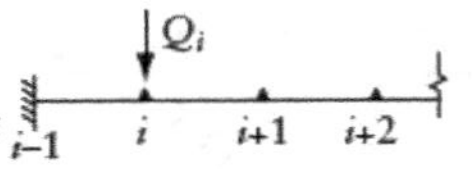 Q_i; $i-1$, i, $i+1$, $i+2$	–	–	$2I_{i-1}+4I_i+I_{i+1}$	$-2(I_i+I_{i+1})$	I_{i+1}	$=Q_i$	12.11
(d)	Free end; 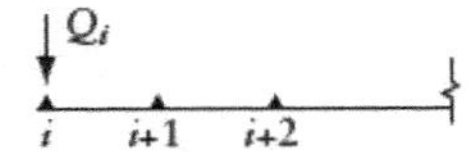 Q_i; i, $i+1$, $i+2$	–	–	I_{i+1}	$-2I_{i+1}$	I_{i+1}	$=Q_i$	12.12
(e)	Free end; 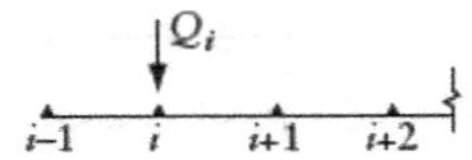 Q_i; $i-1$, i, $i+1$, $i+2$	–	$-2I_i$	$(4I_i+I_{i+1})$	$-2(I_i+I_{i+1})$	I_{i+1}	$=Q_i$	12.13

*For a beam on elastic foundation of modulus k_i (force/length^2), add $k_i\lambda/2$ to the coefficient of y_i in Eq.12.12 a and $k_i\lambda$ in other equations. If the beam is on elastic spring at i of stiffness k_i (force/length), add k_i to the coefficient of y_i in all equations.

Figure 12.6 Finite difference equations relating beam deflection to applied load.

	Position of node i	Coefficient of the deflection in terms of EI/λ^3					Right-hand side	Equation number
		y_{i-2}	y_{i-1}	y_i^*	y_{i+1}	y_{i+1}		
(a)	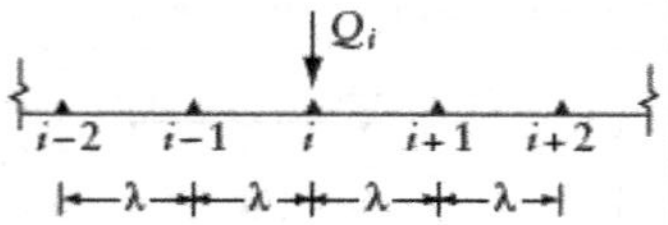 Q_i; $i-2$, $i-1$, i, $i+1$, $i+2$; λ, λ, λ, λ	1	−4	6	−4	1	$= Q_i$	12.7
(b)	Hinged support; 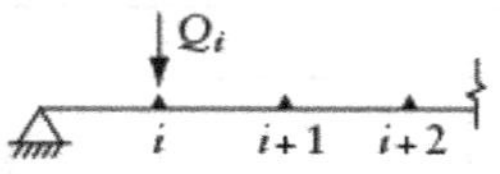 Q_i; i, $i+1$, $i+2$	–	–	5	−4	1	$= Q_i$	12.10
(c)	Fixed support; 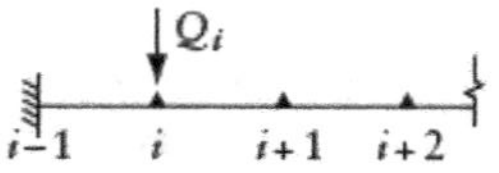 Q_i; $i-1$, i, $i+1$, $i+2$	–	–	7	−4	1	$= Q_i$	12.11
(d)	Free end; 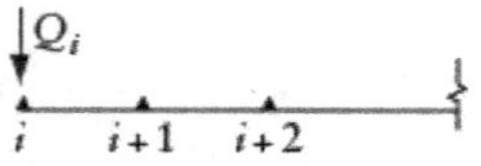 Q_i; i, $i+1$, $i+2$	–	–	1	−2	1	$= Q_i$	12.12
(e)	Free end; 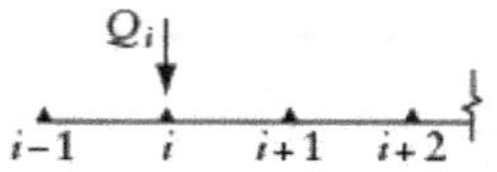 Q_i; $i-1$, i, $i+1$, $i+2$	–	−2	5	−4	1	$= Q_i$	12.13

*For a beam on elastic foundation of modulus k_i (force/length^2), add $k_i\lambda/2$ to the coefficient of y_i in Eq. 12.12 and $k_i\lambda$ in other equations. If the beam is on elastic spring at i of stiffness K_i (force/length), add K_i to the coefficient of y_i in all equations.

Figure 12.7 Finite difference equations relating beam deflection to applied load, when I is constant.

12.5 STRESS RESULTANTS AND REACTION: FINITE DIFFERENCE RELATIONSHIP TO DEFLECTION

The shearing force midway between nodes i and i+1 is:

$$V_{i+\frac{1}{2}} \cong \frac{1}{\lambda}\left(M_{i+1} - M_i\right) \tag{12.14}$$

Substituting for M_i, from Equation 12.5,

$$V_{i+\frac{1}{2}} \cong \frac{E}{\lambda^3}\left[I_i\, y_{i-1} - \left(2I_i + I_{i+1}\right)y_i + \left(I_i + 2I_{i+1}\right)y_{i+1} - I_{i+1}\, y_{i+2}\right] \tag{12.15}$$

The same value represents also the shear at any point between the nodes i and i+1 when no load acts between those nodes.

The reaction at an intermediate support i in a continuous beam is given in Figure 12.8a as:

$$R_i = Q_i - \frac{E}{\lambda^3}\left[I_{i-1}\, y_{i-2} - 2\left(I_{i-1} + I_i\right)y_{i-1} + \left(I_{i-1} + 4I_i + I_{i+1}\right)y_i - 2\left(I_i + I_{i+1}\right)y_{i+1} + I_{i+1}\, y_{i+2}\right] \tag{12.16}$$

where Q_i is the equivalent concentrated load acting directly above the support.

The reaction at a hinged end i (Figure 12.8b) is:

$$R_i = Q_i + \frac{M_{i+1}}{\lambda}$$

Substituting for M_{i+1} from Equation 12.5:

$$R_i \cong Q_i - \frac{EI_{i+1}}{\lambda^3}\left(y_i - 2\, y_{i+1} + y_{i+2}\right) \tag{12.17}$$

The reaction at a totally fixed end is:

$$R_i = Q_i + \frac{1}{\lambda}\left(M_{i+1} - M_i\right) \tag{12.18}$$

The moment M_i at a fixed end i where rotation is prevented but transverse displacement y_i can occur (Figure 12.8c) is:

$$M_i \cong \frac{EI_i}{\lambda^2}\left(2\, y_i - 2\, y_{i+1}\right) \tag{12.19}$$

The bending moment M_i is, of course, numerically equal to the fixed-end moment. A positive M_i indicates a clockwise moment for a left-hand end of the beam. Substituting for M_{i+1} and M_i in Equation 12.18 in terms of deflection, we obtain the reaction at a totally fixed-end i:

$$R_i = Q_i + \frac{E}{\lambda^3}\left[-2\left(I_i + I_{i+1}\right)y_i + 2\left(I_i + I_{i+1}\right)y_{i+1} - I_{i+1}\, y_{i+2}\right] \tag{12.20}$$

The various equations are collected in Figure 12.8, together with the corresponding equations for the case when I is constant. When there is no settlement at the support, $y_i = 0$ in the appropriate equation.

	Type of support		Coefficient of the deflection in terms of E/λ^3 when I is variable, or in terms of EI/λ^3 when I is constant					Right-hand side	Equation number
			y_{i-2}	y_{i-1}	y_i^*	y_{i+1}	y_{i+2}		
(a)	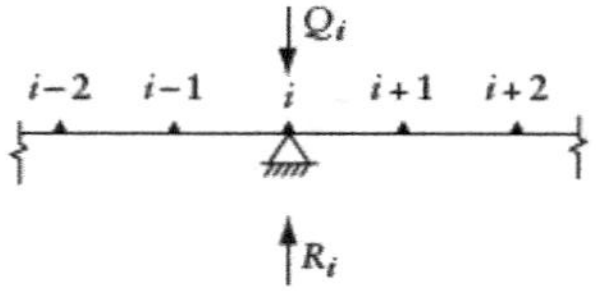 $i-2$ $i-1$ i $i+1$ $i+2$ Q_i R_i Intermediate support	Variable I	$-I_{i-1}$	$2(I_{i-1}+I_i)$	$-(I_{i-1}+4I_i+I_{i+1})$	$2(I_i+I_{i+1})$	$-I_{i+1}$	$=(R_i-Q_i)$	12.16
		Constant I	–1	4	–6	4	–1	$=(R_i-Q_i)$	12.16
(b)	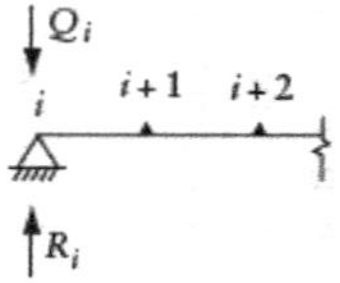 Q_i i $i+1$ $i+2$ R_i Hinged end	Variable I	–	–	$-I_{j+1}$	$2I_{j+1}$	$-I_{j+1}$	$=(R_i-Q_i)$	12.17
		Constant I	–	–	–1	2	–1	$=(R_i-Q_i)$	12.17
(c)	Q_i i $i+1$ $i+2$ R_i Fixed-end moment M_i (given by Eq. 12.19) 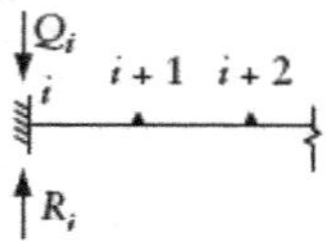 Rotation at i is prevented	Variable I	–	–	$-2(I_i+I_{i+1})$	$2(I_i+I_{i+1})$	$-I_{i+1}$	$=(R_i-Q_i)$	12.20
		Constant I	–	–	–3	4	–1	$=(R_i-Q_i)$	12.20

*For a beam on elastic foundation of modulus k_i (force/length²), add to the coefficient of y_i the value $(-k_i\lambda)$ in the first two equations and $(-k_i\lambda/2)$ in the other equations of this table.

Figure 12.8 Finite difference equations relating beam deflection to reaction.

12.6 AXISYMMETRICAL CIRCULAR CYLINDRICAL SHELL: IDEALIZATION AS BEAM ON ELASTIC FOUNDATION

Consider a thin-walled elastic cylinder subjected to any axisymmetrical radial loading. Because of symmetry, any section of the shell perpendicular to the cylinder axis will remain circular, while the radius r will undergo a change $\Delta r = y$. We need therefore to consider the deformation of only one strip parallel to the generatrix of the cylinder (Figure 12.9a). Let the width of the strip be unity.

The radial displacement y must be accompanied by a circumferential (or hoop) force (Figure 12.9b) whose magnitude per unit length of the generatrix is:

$$N = \frac{Eh}{r} y \tag{12.21}$$

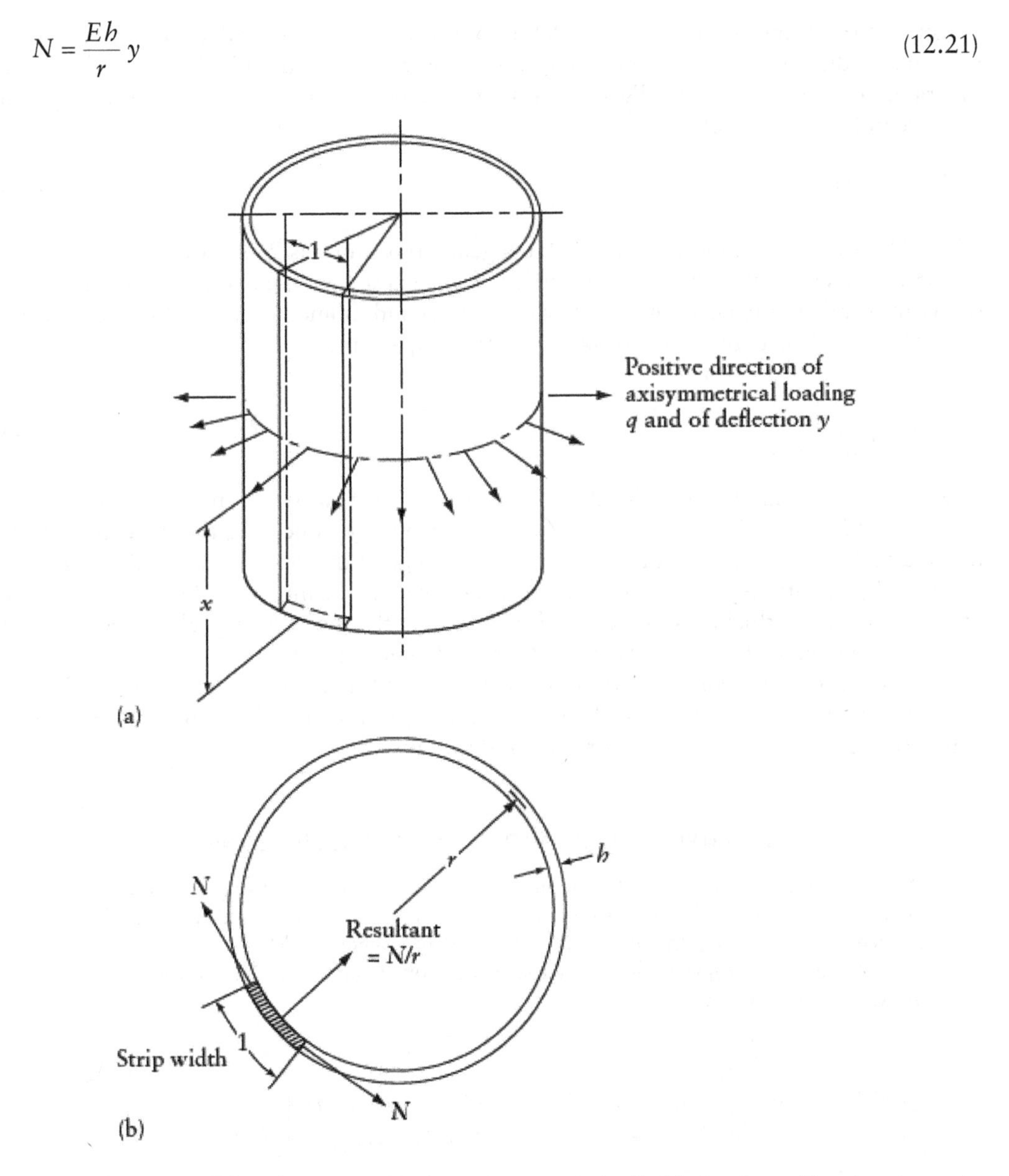

Figure 12.9 A strip parallel to the generatrix of an axisymmetrical cylindrical wall considered analogous to a beam on an elastic foundation. (a) Pictorial view. (b) Cross section.

where h is the thickness of the cylinder and E is the modulus of elasticity. The hoop forces are considered positive when tensile. The radial deflection and loading are positive when outward.

The resultant of the hoop forces N on the two edges of the strip acts in the radial direction opposing the deflection and its value per unit length of the strip is:

$$-\frac{N}{r} = -\frac{Eh}{r^2}y \tag{12.22}$$

Hence, the strip may be regarded as a beam on an elastic foundation whose modulus is:

$$k = \frac{Eh}{r^2} \tag{12.23}$$

Because of the axial symmetry of the deformation of the wall, the edges of any strip must remain in radial planes and lateral extension or contraction (caused by bending of the strip in a radial plane) is prevented. This restraining influence is equivalent to a bending moment in a circumferential direction.

$$M_\phi = \nu M \tag{12.24}$$

where M is the bending moment parallel to a generatrix and ν is Poisson's ratio of the material. It can be shown that the stiffening effect of M_ϕ on the bending deformation of the beam strip can be taken into account by increasing the second moment of area of the strip by the ratio $1/(1-\nu^2)$; hence, the flexural rigidity of the strip of unit width is:

$$EI = \frac{Eh^3}{12\left(1-\nu^2\right)} \tag{12.25}$$

The differential equation for the deflection of a beam resting on an elastic foundation (Equation 12.4) can thus be used for the cylinder with y indicating the deflection in the radial direction and x the distance from the lower edge of the wall (Figure 12.9a); EI and k are given by Equations 12.25 and 12.23, respectively; q is the intensity of radial pressure on the wall. When the thickness of the wall is variable, both EI and k vary. However, this does not cause an appreciable difficulty in the finite difference's solution.

The finite difference equations derived for beams and listed in Figures 12.6 to 12.8 can be used for circular cylindrical shells subjected to axisymmetrical loading, the equivalent concentrated load Q_i being taken for a strip of unit width.

EXAMPLE 12.1: BEAM ON ELASTIC FOUNDATION

Determine the deflection, bending moment, and end support reactions for the beam of Figure 12.10a, which has a constant EI. The beam is resting on an elastic foundation between the supports with a modulus $k = 0.1024EI/\lambda^4$, where $\lambda = l/4$.

Applying the appropriate finite difference equations from Figure 12.7 at each point of division, we can write:

$$\frac{EI}{\lambda^3}\begin{bmatrix} (5+0.1024) & -4 & 1 \\ -4 & (6+0.1024) & -4 \\ 1 & -4 & (5+0.1024) \end{bmatrix}\begin{Bmatrix} y_1 \\ y_2 \\ y_3 \end{Bmatrix} = q\lambda\begin{Bmatrix} 1 \\ 1 \\ 1 \end{Bmatrix}$$

for which the solution is:

$$y_1 = y_3 = 1.928\,\frac{q\lambda^4}{EI} \quad \text{and} \quad y_2 = 2.692\,\frac{q\lambda^4}{EI}$$

or

$$y_1 = y_3 = 7.532 \times 10^{-3}\, ql^4/EI \quad \text{and} \quad y_2 = 10.514 \times 10^{-3}\, ql^4/EI$$

We could have made use of symmetry of the structure by putting $y_1 = y_3$; hence, only two simultaneous equations would have required solving.

Applying Equation 12.5 at points 1 and 2, we obtain:

$$M_1 = -\frac{EI}{\lambda^2} \times q\frac{\lambda^4}{EI}(0 - 2 \times 1.928 + 2.692) = 1.164\,q\lambda^2 = 0.0728\,ql^2$$

and

$$M_2 = -\frac{EI}{\lambda^2} \times q\frac{\lambda^4}{EI}(1.928 - 2 \times 2.692 + 1.928) = 1.528\,q\lambda^2 = 0.0955\,ql^2$$

The bending moment diagram for the beam is plotted in Figure 12.10c. The reaction at 0 is given by Equation 12.17a (Figure 12.8b) as:

$$\frac{EI}{\lambda^3}(2y_1 - y_2) = (R_0 - Q_0)$$

Substituting for the y values calculated previously and $Q_0 = q\lambda/2 = ql/8$, we find:

$$R_0 = R_4 = 0.416\,ql$$

The exact values using an analytical solution[1] are: $y_1 = y_3 = 0.00732ql^4/EI$; $y_2 = 0.0102ql^4/EI$; $M_1 = M_3 = 0.0745ql^2$; $M_2 = 0.0978ql^2$; and $R_0 = R_4 = 0.414ql$.

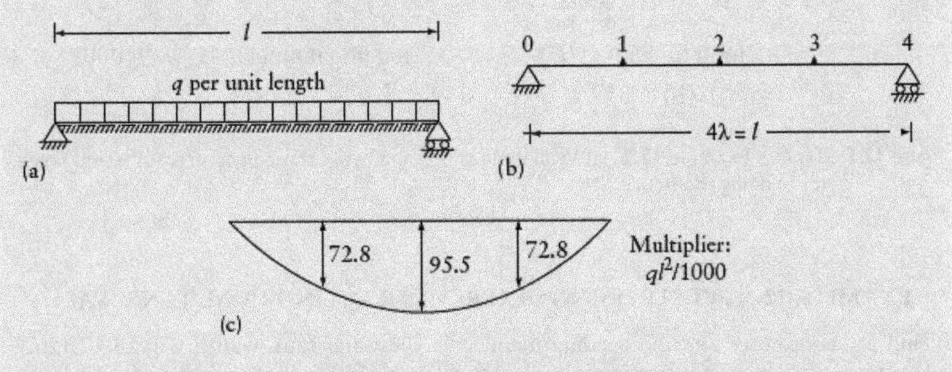

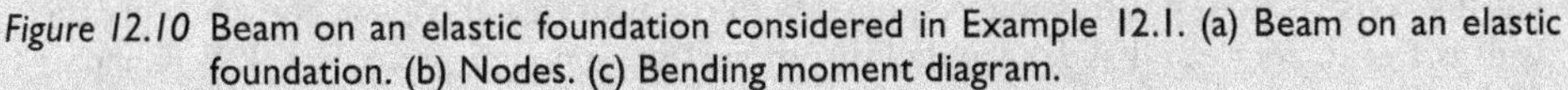

Figure 12.10 Beam on an elastic foundation considered in Example 12.1. (a) Beam on an elastic foundation. (b) Nodes. (c) Bending moment diagram.

EXAMPLE 12.2: CIRCULAR CYLINDRICAL TANK WALL

Using computer program CTW (Chapter 20), determine the hoop force and the bending moment for the wall of the concrete tank in Figure 12.11a. Assume Poisson's ratio, $\nu = 0.2$. The results are presented in Figures 12.11b and c.

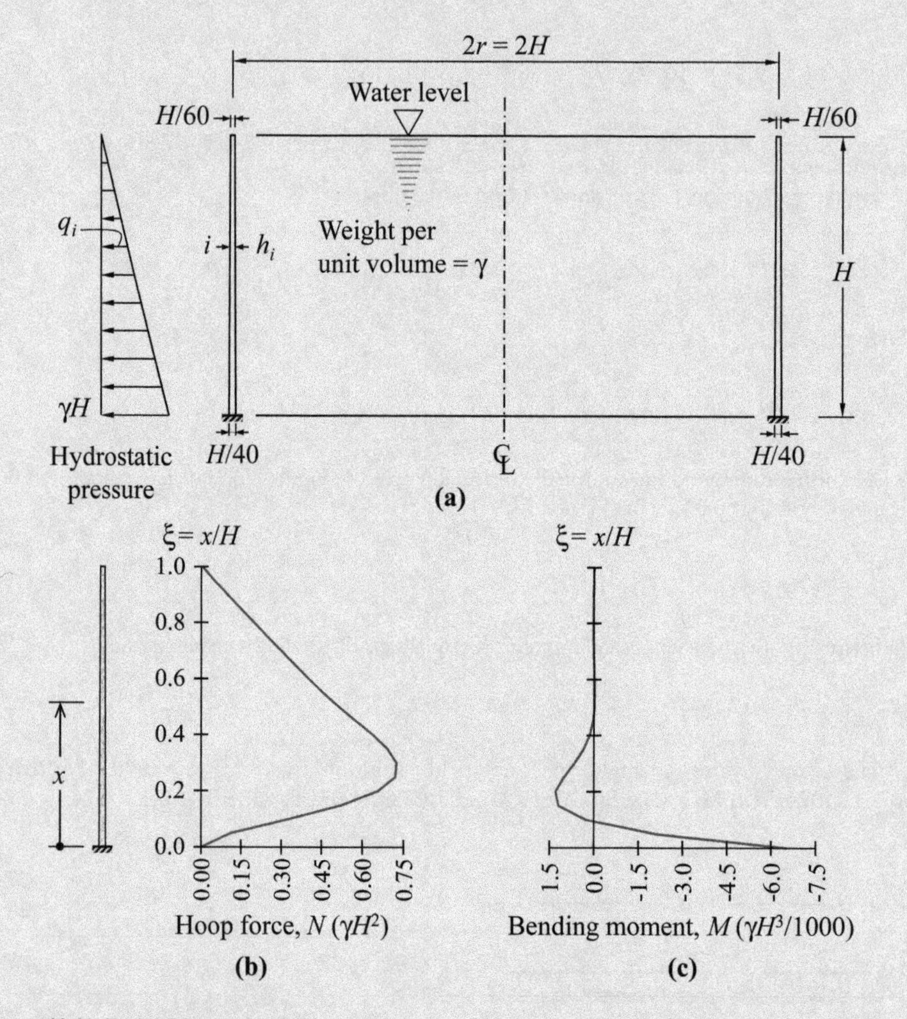

Figure 12.11 Tank of Example 12.2. (a) Wall dimensions and hydrostatic pressure. (b) Hoop force. (c) Bending moment.

EXAMPLE 12.3: PRESTRESSING OF CIRCULAR CYLINDRICAL TANK WALL

Find the hoop force and the bending moment in the water tank wall in Figure 12.12a. During construction, prestressing is applied with the wall free at top; at bottom edge, the wall is free to rotate and slide in radial direction. Use computer program CTW (Chapter 20) to determine the hoop force and the bending moment due to six circumferential post-tendons. Data: Force per tendon $= 2.19 \times 10^6$ N; corresponding inward radial pressure is shown in Figure 12.12b. Assume $E = 26$ GPa and Poisson's ratio, $\nu = 0.2$. Assume that the pressure induced by the prestressing is distributed over a depth = thickness of wall (= 0.25 m).

The prestressing is designed to produce initial compression equal to the effect of pressure shown in Figure 12.12b. In this example, the trapezoid in Figure 12.2b is divided into six equal segments, each has an area = the tendon force = 2.19×10^6 N; the top segment is further divided into two parts. The prestressed tendons are located at centroids of those segments. For any of the bottom five typical segments:

$$P/\text{tendon} = 2.19 \text{ MN} = \frac{N_i + N_j}{2}\left(\xi_j - \xi_i\right) H \tag{a}$$

$$\Delta x_{CG-i} = \frac{N_i + 2N_j}{N_i + N_j}\left(\xi_j - \xi_i\right)\frac{H}{3} \tag{b}$$

Subscripts i and j in Equations (a) and (b) refer to bottom and top ends of a typical segment; N = hoop force value; $\xi = x/H$; x = distance measured from bottom edge of wall; H = overall height of tank wall = 12 m. Successive application of Equation (a) from the bottom edge of the wall, with N_j (10^6 N/m) = 1.89–1.59 ξ_j, gives locations of top edges of the segments.

When the tank is full, the hydrostatic pressure will be equal to zero at the top and 1.18×10^6 N/m^2 at the bottom. Figures 12.12c and d show the effect of the initial prestressing obtained by computer program CTW when the tank is empty. Two graphs are plotted in Figure 12.12d for the cases when the prestress forces are applied as point loads or distributed over a height = 0.25 m; as expected, distributing the prestress forces slightly reduces the bending moment values. Further reduction of M might be achieved by provision of more tendons with smaller forces.

The initial prestress is the force exerted by the tendons on the tank wall; it does not account for the effect of friction or anchor setting. The required bending moment and hoop force are those when the tank is empty, but creep, shrinkage, and relaxation of the prestressed tendons will change the calculated values[2]. In design, consider the effects of the two cases (when the tank is full and empty) on the values of the bending moment and the hoop force.

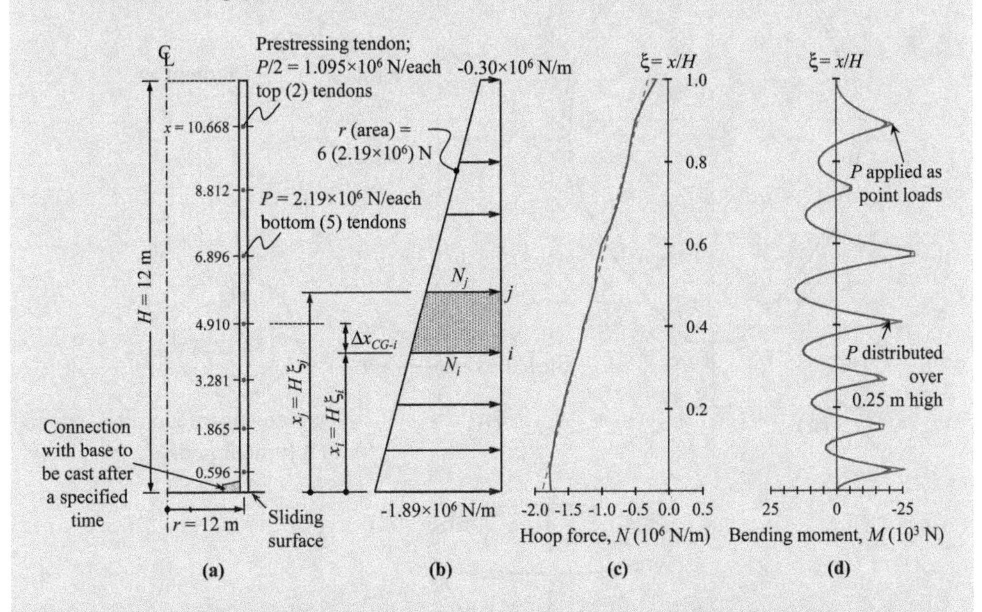

Figure 12.12 Circular cylindrical tank wall, Example 12.3. (a) Dimensions. (b) Target hoop forces; radial pressure inducing hoop force = sum of forces in the tendons. (c) Hoop force. (d) Bending moment.

12.7 FINITE-ELEMENT ANALYSIS OF SHELLS OF REVOLUTION

Walls, bases, and covers of concrete silos – having the shape of axisymmetrically loaded circular shells – are analyzed as continuous structures. The shell is idealized as assemblage of finite-elements as shown in Figure 12.13. The relatively simple element used here is adequate for use in design. Shell of Revolution (SOR; Chapter 20) is a computer program that performs the analysis. The equations used in SOR are based on assumed deformed shape functions of finite-elements.

12.7.1 Nodal displacements and nodal forces

Figure 12.14a depicts a typical element subjected to axisymmetrical pressure represented by forces $\{F^*\}$ = static equivalent load, q multiplied by the length of nodal lines 1 and 2:

$$\begin{Bmatrix} F_1^* \\ F_2^* \\ F_3^* \end{Bmatrix} = 2\pi r_1 \begin{Bmatrix} q_1^* \\ q_2^* \\ q_3^* \end{Bmatrix}; \quad \begin{Bmatrix} F_4^* \\ F_5^* \\ F_6^* \end{Bmatrix} = 2\pi r_2 \begin{Bmatrix} q_4^* \\ q_5^* \\ q_6^* \end{Bmatrix} \tag{12.26}$$

The element nodal forces and displacements in local directions in Figure 12.14a are related by geometry to forces and displacements in global directions (Figure 12.14b):

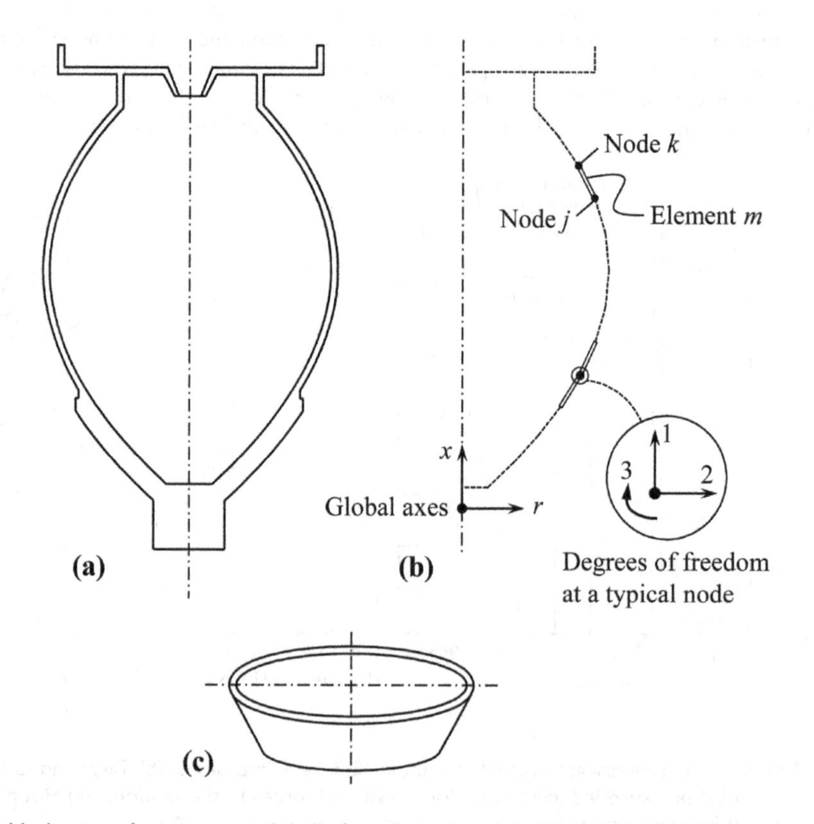

Figure 12.13 Idealization of axisymmetrical shell of revolution by conical shell element. (a) Vertical section of egg-shaped digestor for sewage water treatment. (b) Finite-element idealization. (c) Typical shell element.

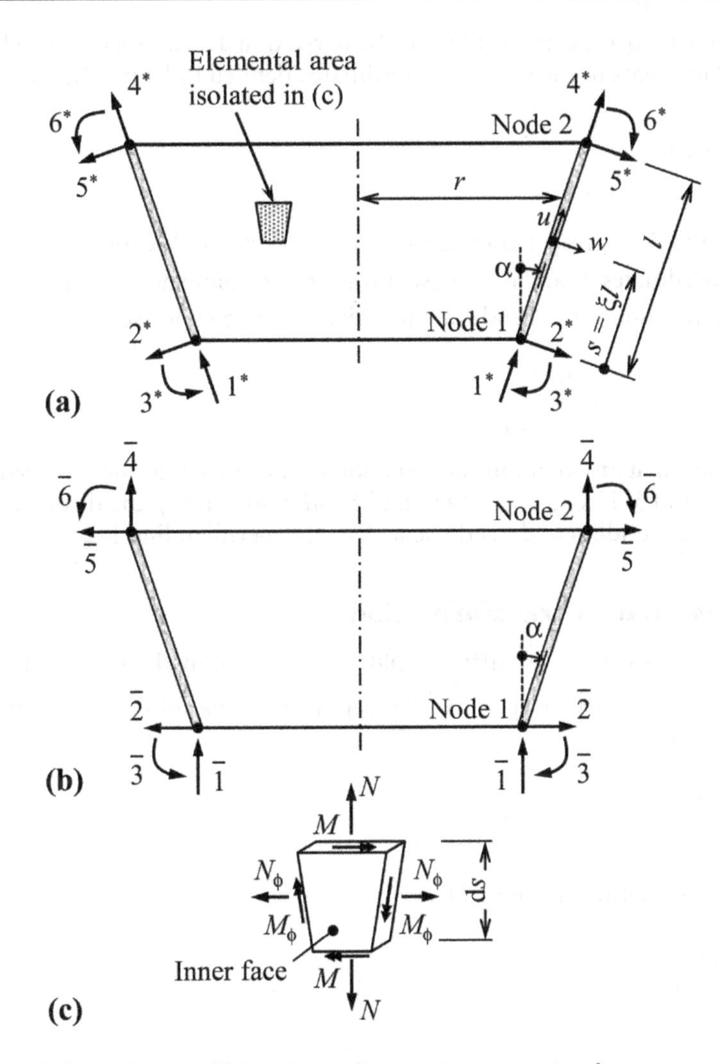

Figure 12.14 Typical finite-element. (a) Local coordinates; sign convention for u, w, s, and α. (b) Degrees of freedom in directions of global axes; order of numbering of coordinates at the nodes of a typical element. (c) Pictorial view of an elemental area showing the positive sign convention for N, M, N_ϕ, and M_ϕ.

$$\{D^*\} = [T]\{\bar{D}\}; \quad \{\bar{F}\} = [T]^{\mathrm{T}}\{F^*\} \tag{12.27}$$

where

$$[T] = \begin{bmatrix} [t] & [0] \\ [0] & [t] \end{bmatrix}; \quad [t] = \begin{bmatrix} \cos\alpha & \sin\alpha & 0 \\ -\sin\alpha & \cos\alpha & 0 \\ 0 & 0 & 1 \end{bmatrix} \tag{12.28}$$

The angle α and its positive sign convention are defined in Figures 12.14a and b.

The finite-element stiffness matrix $[S^*]$, to be derived in 12.7.5, relates the element nodal forces and displacements in the six local coordinates defined in Figure 12.14a:

$$\left[S^*\right]\left\{D^*\right\}=\left\{F^*\right\} \tag{12.29}$$

Coordinates 2^* and 5^* represent translations or forces in the direction of a normal to the cone surface. Coordinates 3^* and 6^* represent rotations or moments in radial planes. D_3^* and D_6^*, representing rotations at the ends of a meridian, can be expressed as:

$$D_3^*=\left(\frac{dw}{ds}\right)_{\xi=0}, \quad D_6^*=\left(\frac{dw}{ds}\right)_{\xi=1} \tag{12.30}$$

where w = translation at any point in the direction of the normal to the shell (same direction as coordinates 2^* and 5^* in Figure 12.14a); and $\xi = s/l$, with s being the distance from node 1 to any point on the meridian and l is the length of the meridian line 1–2.

12.7.2 Stiffness matrix transformation

Equation 12.31 transforms the stiffness matrix $[S^*]$ of individual element with respect to local coordinates to stiffness matrix $\left[\bar{S}\right]$ referring to coordinates in global directions (Figures 12.14a and b).

$$\left[\bar{S}\right]=\left[T\right]^T\left[S^*\right]\left[T\right] \tag{12.31}$$

where $\left[\bar{S}\right]$ = matrix relating $\left\{\bar{F}\right\}$ and $\left\{\bar{D}\right\}$,

$$\left[\bar{S}\right]\left\{\bar{D}\right\}=\left\{\bar{F}\right\} \tag{12.32}$$

12.7.3 Displacement interpolation

The displacements $\{u, w\}$ are assumed to be related to the nodal displacements by:

$$\begin{bmatrix} u \\ w \end{bmatrix}=\begin{bmatrix} 1-\xi & 0 & 0 & \xi & 0 & 0 \\ 0 & L_1 & L_2 & 0 & L_3 & L_4 \end{bmatrix}\left\{D^*\right\} \tag{12.33}$$

where u = translation of any point in direction of the meridian (node 1 to node 2 in Figure 12.14a); u is assumed to vary linearly between D_1^* and D_4^*. L_1 to L_4 are shape functions of ξ interpolating the deflection w between its nodal values to give the value at any point on the meridian line 1–2.

$$\left[L_1 \quad L_2 \quad L_3 \quad L_4\right]=\left[1-3\xi^2+2\xi^3 \quad l\xi(\xi-1)^2 \quad \xi^2(3-2\xi) \quad l\xi^2(\xi-1)\right] \tag{12.34}$$

12.7.4 Stress resultants

The stresses in meridian and circumference directions are related to strains by Equations 12.35 to 12.37:

$$\{\sigma\} = [d_e]\{\varepsilon\} \tag{12.35}$$

$$\{\sigma\} = \begin{Bmatrix} N \\ N_\phi \\ M \\ M_\phi \end{Bmatrix}; \quad \{\varepsilon\} = \begin{Bmatrix} du/ds \\ (w\cos\alpha + u\sin\alpha)/r \\ -d^2w/ds^2 \\ -(\sin\alpha/r)(dw/ds) \end{Bmatrix} \tag{12.36}$$

$$[d_e] = \frac{Eh}{1-\nu^2}\begin{bmatrix} 1 & \nu & 0 & 0 \\ \nu & 1 & 0 & 0 \\ 0 & 0 & h^2/12 & \nu h^2/12 \\ 0 & 0 & \nu h^2/12 & h^2/12 \end{bmatrix} \tag{12.37}$$

where $\{\sigma\}$ and $\{\varepsilon\}$ = generalized stress and strain vectors defined by Equation 12.36; $[d_e]$ = element's elasticity matrix (Equation 12.37); N and M = the normal force and the moment in the meridian direction per unit length; N_ϕ and M_ϕ = the normal force and the moment in the circumferential (hoop) direction per unit length (Figure 12.14c); h = element thickness; E = modulus of elasticity; ν = Poisson's ratio. The positive sign conventions of the stress resultants are shown in Figure 12.14c. Substitution of Equation 12.33 in 12.36 gives the relationship of $\{\varepsilon\}$ to nodal displacements, $\{D^*\}$:

$$\{\varepsilon\} = [B]\{D^*\} \tag{12.38}$$

$$[B] = \begin{bmatrix} -\frac{1}{l} & 0 & 0 & \frac{1}{l} & 0 & 0 \\ \frac{\sin\alpha}{r}(1-\xi) & \frac{\cos\alpha}{r}L_1 & \frac{\cos\alpha}{r}L_2 & \frac{\sin\alpha}{r}\xi & \frac{\cos\alpha}{r}L_3 & \frac{\cos\alpha}{r}L_4 \\ 0 & \frac{1}{l^2}(6-12\xi) & \frac{1}{l}(4-6\xi) & 0 & \frac{1}{l^2}(12\xi-6) & \frac{1}{l}(2-6\xi) \\ 0 & \frac{\sin\alpha}{rl}(6\xi-6\xi^2) & \frac{\sin\alpha}{r}(4\xi-3\xi^2-1) & 0 & \frac{\sin\alpha}{rl}(6\xi^2-6\xi) & \frac{\sin\alpha}{r}(2\xi-3\xi^2) \end{bmatrix} \tag{12.39}$$

The stress resultants are given by:

$$\{\sigma\} = [d_e][B]\{D^*\} + \{\sigma_r\} \tag{12.40}$$

$\{\sigma_r\}$ = stress resultants with nodal displacements prevented; $\{\sigma_r\} = \{0\}$ when the analysis is for the effect of volume change due to temperature, shrinkage, or swelling. The first term on the right-hand side of Equation 12.40 is derived from Equations 12.35 and 12.38. Substituting $\xi = 0.5$ and $r = (r_1 + r_2)/2$ gives $\{\sigma\}$ midway between nodal lines of the element (computer program SOR results).

12.7.5 Element stiffness

The element stiffness matrix with respect to its local coordinate is given by Equations 12.41 and 12.42:

$$\left[S^*\right] = 2\pi l \int_0^1 r[B]^{\mathrm{T}} [d_e][B]\, \mathrm{d}\xi \tag{12.41}$$

$$r = (1-\xi)\, r_1 + \xi\, r_2 \tag{12.42}$$

r_1 and r_2 = radii at nodes 1 and 2. The value of the integral in Equation 12.41 = $(g_1 + g_2)$; where g_1 and g_2 = values of the integral at Gauss' sampling points: $\xi = \left(3-\sqrt{3}\right)/6$ and $\left(3+\sqrt{3}\right)/6$. Two sampling points give sufficient accuracy.

12.7.6 Effect of temperature

Consider the effect of rise of temperature varying linearly through the thickness between T_i and T_o at inner and outer faces of the finite-element in Figure 12.14a. The stress vector, $\{\sigma_r\}$ is calculated as:

$$\{\sigma_r\} = -[d_e]\begin{Bmatrix} \alpha(T_o - T_i)/2 \\ \alpha(T_o + T_i)/2 \\ \alpha(T_o - T_i)/h \\ \alpha(T_o - T_i)/h \end{Bmatrix} \tag{12.43}$$

$\{\sigma_r\}$ = generalized stress that would develop if the thermal expansion is arbitrarily restrained by forces $\left\{F_r^*\right\}$ at the six coordinates in Figure 12.14a; where

$$\left\{F_r^*\right\} = 2\pi l \int_0^1 r[B]^{\mathrm{T}} \{\sigma_r\}\, \mathrm{d}\xi \tag{12.44}$$

The equivalent restraining forces of individual element in global directions are:

$$\left\{\bar{F}_r\right\} = [T]^{\mathrm{T}} \left\{F_r^*\right\} \tag{12.45}$$

The element restraining forces are then assembled to give the restraining forces of the structure to be applied in reversed direction to eliminate the artificial restraint and give nodal displacements due to temperature. When the nodal displacements are determined, Equation 12.40 gives the generalized thermal stress resultants in individual elements.

12.7.7 Nodal forces due to distributed loads

The distributed load in Figure 12.15a is represented by statically equivalent nodal forces in Figure 12.15b:

$$\left\{F^*\right\} = \frac{\pi l}{3}\begin{Bmatrix} 2q_{t1} r_1 + q_{t2} r_2 \\ 2q_{n1} r_1 + q_{n2} r_2 \\ 0 \\ q_{t1} r_1 + 2q_{t2} r_2 \\ q_{n1} r_1 + 2q_{n2} r_2 \\ 0 \end{Bmatrix} \tag{12.46}$$

$\{F^*\}$ = equivalent nodal force per unit length in local coordinates (Figures 12.15c and d). q_t and q_n = intensity of tangential and normal loads on an individual element; it is assumed that the product $(q\ r)$ varies linearly over the length l. The subscripts 1 and 2 refer to nodes 1 and 2.

12.7.8 Assemblage of stiffness matrices and nodal forces

The displacements at the nodes, $\{D\}$ in global directions are determined by solution of equilibrium equation:

$$[S]\{D\} = \{F\} \tag{12.47}$$

$[S]$ = stiffness matrix of the structure; $\{F\}$ = a vector of nodal forces in global directions = the sum of applied nodal forces and the forces restraining thermal expansion in reversed direction. Before solving Equation 12.47, $[S]$ must be adjusted to satisfy prescribed conditions at specified nodes; for example, zero displacement at support.

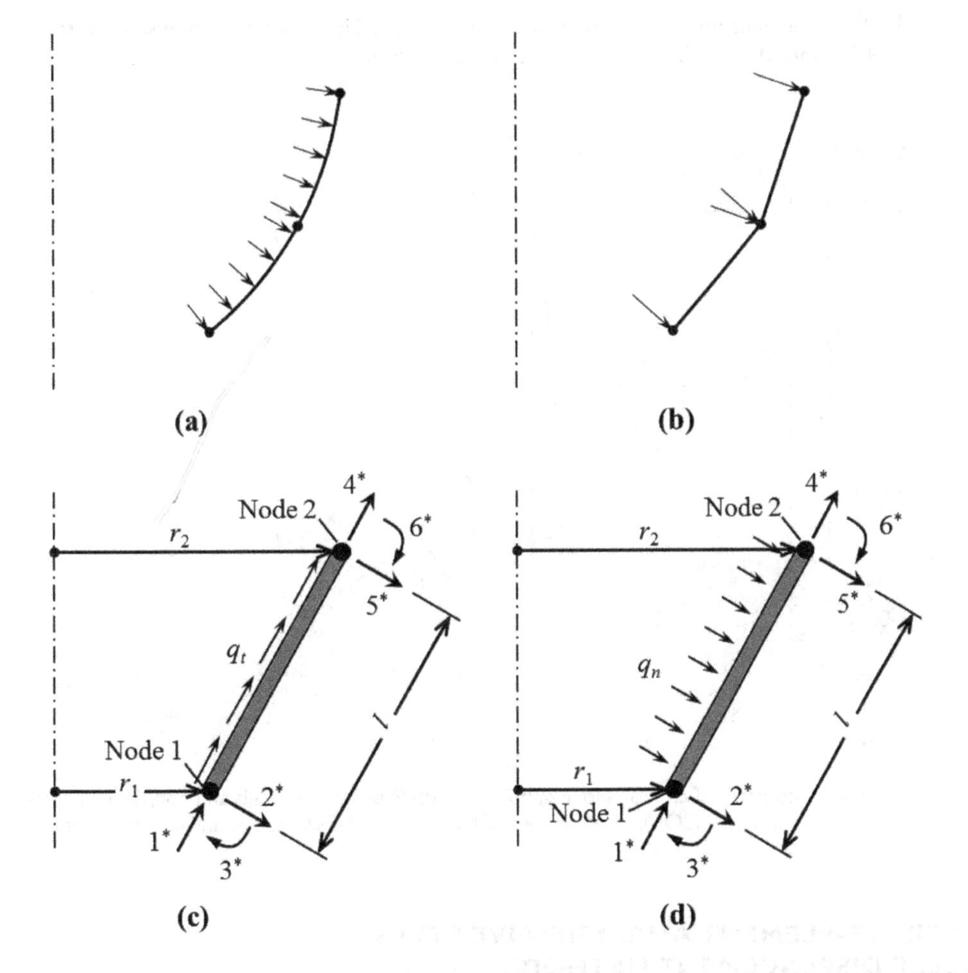

Figure 12.15 Nodal forces equivalent to a distributed load on the surface of an axisymmetrical finite-element. (a) Actual load on a shell with a curved meridian. (b) Idealization as conical elements loaded only at the nodes. (c and d) Equivalent nodal forces in local coordinates.

EXAMPLE 12.4: STRESS RESULTANTS IN A CIRCULAR SILO: FINITE-ELEMENT ANALYSIS

Determine N_ϕ and M for a 30 ft diameter concrete silo 140 ft high with an 8 in. wall and a conical hopper 20 ft deep. The wall is continuous with a 10 in. thick × 40 ft high cylindrical wall (Figure 12.16a). Tangential pressure, q_t and normal pressure, q_n applied on the silo and hopper walls are shown in Figure 12.16b and are calculated by Equations (c) and (d), respectively; the y-values in Equations (c) and (d) should be in feet. Other data: modulus of elasticity, $E = 3420$ ksi; Poisson's ratio, $\nu = 0.2$; assume total fixity at the base.

$$q_t\,(\text{psi}) = \begin{cases} 347.22\times10^{-3}y - 17.097\left(1-e^{-y/49.24}\right) & 0 < y\,(\text{ft}) < 95.337 \\ 2.647 + 62.796\times10^{-3}\left(y-95.337\right) & y\,(\text{ft}) > 95.337 \end{cases} \tag{c}$$

$$q_n\,(\text{psi}) = \begin{cases} 10.429\left(1-e^{-y/49.24}\right) & 0 < y\,(\text{ft}) < 95.337 \\ 10.703 + 0.254\left(y-95.337\right) & y\,(\text{ft}) > 95.337 \end{cases} \tag{d}$$

The finite-element model is shown in Figure 12.16c. Figure 12.16d shows the variations of N_ϕ and M, calculated by computer program SOR.

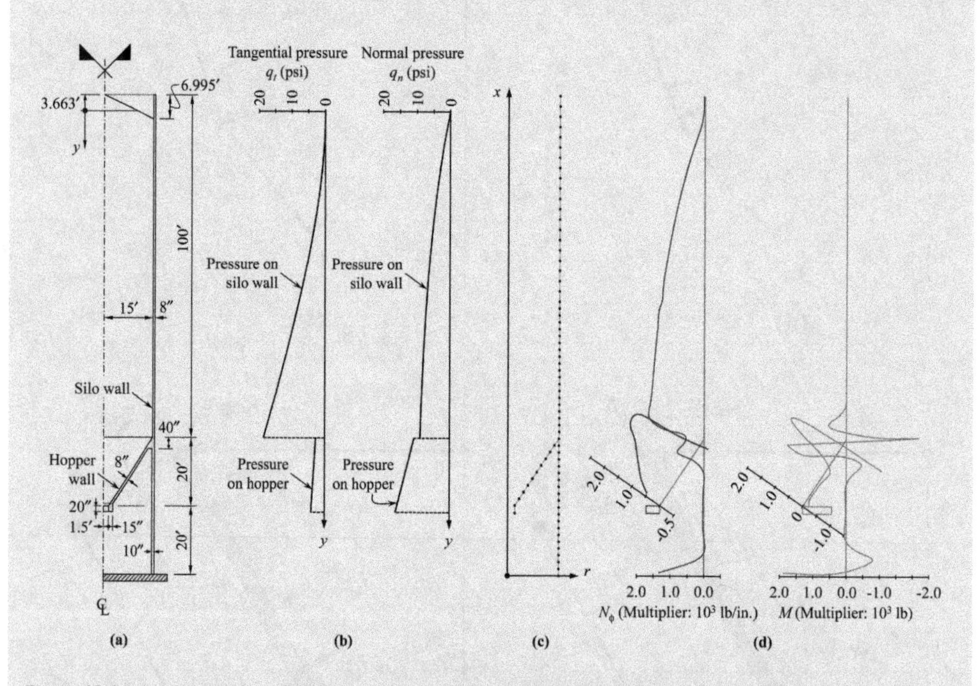

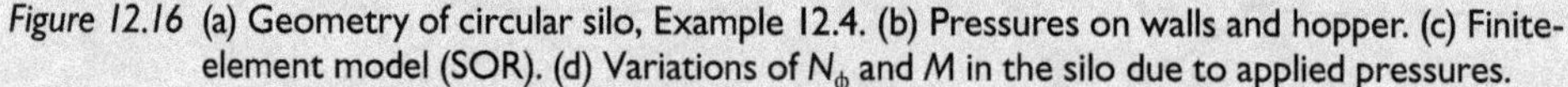

Figure 12.16 (a) Geometry of circular silo, Example 12.4. (b) Pressures on walls and hopper. (c) Finite-element model (SOR). (d) Variations of N_ϕ and M in the silo due to applied pressures.

12.8 FINITE-ELEMENT ANALYSIS: FIVE STEPS (OF DISPLACEMENT METHOD)

In essence, the analysis of a structure by the finite-element method is an application of the five steps of the displacement method summarized in Section 5.4. This analysis is explained below by reference to a plate subjected to in-plane forces (Figure 12.17), which is idealized as an assemblage of rectangular finite-elements. Each element has four nodes with two degrees-of-freedom per node, that is, translations u and v in the x and y directions, respectively. The purpose of the analysis is to determine the stress components $\{\sigma\}=\{\sigma_x, \sigma_y, \tau_{xy}\}$ at O, the center of each element. The external loads are nodal forces $\{F_x, F_y\}_i$ at any node i and body forces with intensities per unit volume of $\{p_x, p_y\}_m$ distributed over any element m. Other loadings can also be considered, such as the effects of temperature variation or of shrinkage (or swelling).

The five steps in the analysis are as follows:

Step 1 Define the unknown degrees-of-freedom by two coordinates u and v at each node. The actions to be determined for any element m are $\{A\}_m=\{\sigma\}_m=\{\sigma_x, \sigma_y, \tau_{xy}\}_m$.
Step 2 With the loading applied, determine the restraining forces $\{F\}$ to prevent the displacements at all coordinates. Also, for any element, determine $\{A_r\}_m=\{\sigma_r\}_m$, which represents the values of the actions (the stresses) with the nodal displacements prevented.

The stresses $\{\sigma_r\}$ are produced only when effects of temperature are considered; $\{\sigma_r\}$ due to body forces is commonly ignored. The vector $\{F\}$ is considered equal to the sum of two vectors:

$$\{F\} = \{F_a\} + \{F_b\} \tag{12.48}$$

The vector $\{F_a\}$ is composed of the external nodal forces reversed in sign; $\{F_b\}$ is generated by assemblage of $\{F_b^*\}_m$ for individual elements. The vector $\{F_b^*\}_m$ is composed of forces at the nodes of element m in equilibrium with the external forces on the element

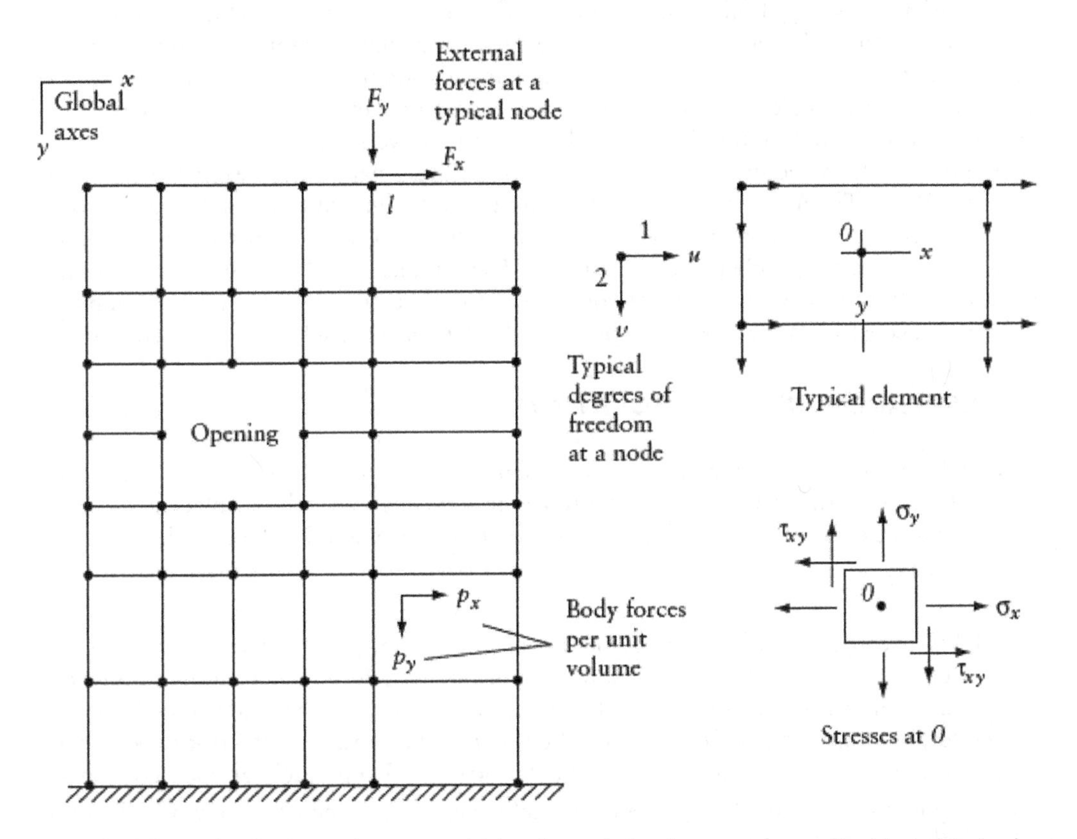

Figure 12.17 Example of a finite-element model for the analysis of stresses in a wall subjected to in-plane forces.

body away from the nodes; in the case of temperature variation, $\{F_b^*\}_m$ represents a system of nodal forces in equilibrium producing stresses $\{\sigma_r\}$. (Equation 5.42 gives the nodal forces due to temperature variation for a member of a plane frame; for other finite-elements, see Section 12.12.1.)

Step 3 Generate the structure stiffness matrix $[S]$ by assemblage of the stiffness matrices $[S]_m$ of individual elements. Also, generate $[A_u]_m = [\sigma_u]_m$, which represents the stress components at O in any element due to unit displacement introduced separately at the element nodal coordinates. For the example, considered in Figure 12.17, $[\sigma_u]_m$ will be a 3×8 matrix.

Step 4 Solve the equilibrium equations:

$$[S]\{D\} = -\{F\} \tag{12.49}$$

This gives the structure nodal displacements $\{D\}$. In the example considered, the number of elements in $\{D\}$ is twice the number of nodes.

Step 5 Calculate the required stress components for each element:

$$\{A\}_m = \{A_r\}_m + [A_u]_m \{D\}_m \tag{12.50}$$

or

$$\{\sigma\}_m = \{\sigma_r\}_m + [\sigma_u]_m \{D\}_m \tag{12.51}$$

The values $\{D\}_m$ are the nodal displacements for the element m; in the example considered (Figure 12.17), $\{D\}_m$ has eight values (subset of the structure displacement vector $\{D\}$).

Ignoring $\{\sigma_r\}$ caused by body forces (steps 2 and 5) produces an error which diminishes as the size of the finite-elements is reduced. However, when the elements are bars (in framed structures), $\{\sigma_r\} \equiv \{A_r\}$, and the other matrices for individual members can be determined exactly; for this reason, $\{\sigma_r\}$ is commonly not ignored and the exact answers can be obtained without the need to reduce the size of the elements for convergence.

Assemblage of the structure load vector and of the stiffness matrix may be done by Equations 19.38 and 19.35. The nonzero elements of $[S]$ are generally limited to a band adjacent to the diagonal. This property, combined with the symmetry of $[S]$, is used to conserve computer storage and reduce the number of computations. These topics and the methods of solution of Equation 12.49 to satisfy displacement constraints are discussed in Chapter 19. Examples of displacement constraints are a zero or a prescribed value for the displacement at a support.

12.9 DISPLACEMENT INTERPOLATION

In the derivation of element matrices, interpolation functions are required to define the deformed shape of the element. For a finite-element of any type, the displacement at any point within the element can be related to the nodal displacement by the equation:

$$\{f\} = [L]\{D^*\} \tag{12.52}$$

where $\{f\}$ is a vector of displacement components at any point; $\{D^*\}$ is a vector of nodal displacements; and $[L]$ is a matrix of functions of coordinates defining the position of the point considered within the element (e.g. x and y or ξ and η in two-dimensional finite-elements).

The interpolation functions $[L]$ are also called *shape functions*; they describe the deformed shape of the element due to unit displacements introduced separately at each coordinate. Any interpolation function L_i represents the deformed shape when $D_i^* = 1$ while the other nodal displacements are zero.

The accuracy of a finite-element depends upon the choice of the shape functions. These should satisfy conditions which will ensure convergence to correct answers when a finer finite-element mesh is used. The derivation of the shape functions and the conditions which should be satisfied are discussed for plate-bending elements in Section 7.5.2 and will be discussed in the following sub-sections for other types of finite-elements.

12.9.1 Straight bar element

For the axial deformation of a bar (Figure 12.18), $\{f\} \equiv \{u\}$ is the translation in the x direction at any section and $\{D^*\} = \{u_1, u_2\}$ is the translation at the two ends. The matrix $[L]$ may be composed of two linear interpolation functions:

$$[L] = [1 - \xi, \quad \xi] \tag{12.53}$$

where $\xi = x/l$, and l is the bar length.

The shape functions which can be used for a bar in bending are given in Figure 12.19. For this element, $\{f\} \equiv \{v\}$ is the transverse deflection in the y direction at any section; the nodal displacements are defined as:

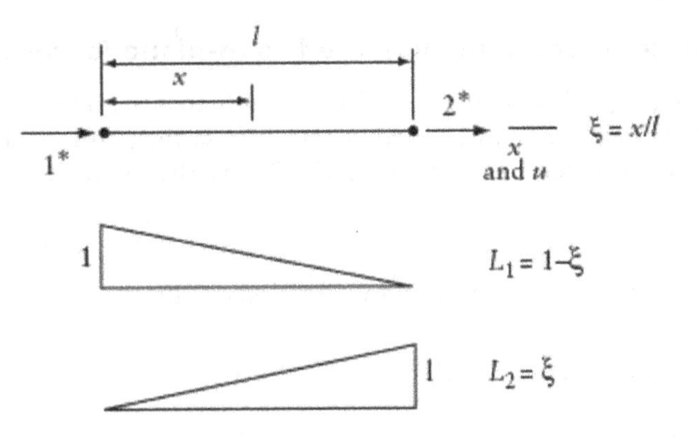

Figure 12.18 Linear interpolation functions. Shape functions for a bar subjected to an axial force.

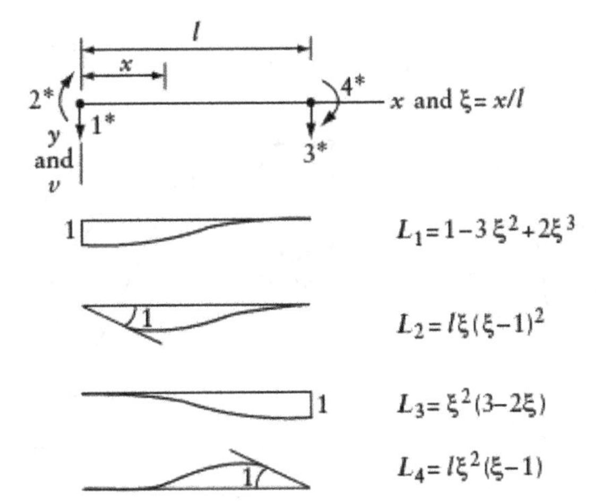

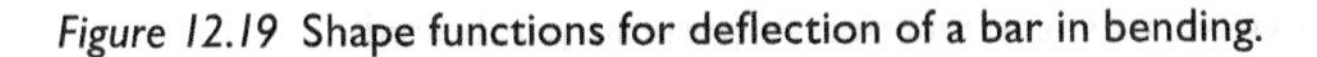

Figure 12.19 Shape functions for deflection of a bar in bending.

$$\{D^*\} = \left\{ v_{x=0}, \quad \left(\frac{dv}{dx}\right)_{x=0}, \quad v_{x=l}, \quad \left(\frac{dv}{dx}\right)_{x=l} \right\} \tag{12.54}$$

The shape functions $[L]$ can be the four cubic polynomials:

$$[L] = \left[1 - 3\xi^2 + 2\xi^3 \quad l\xi(\xi-1)^2 \quad \xi^2(3-2\xi) \quad l\xi^2(\xi-1)\right] \tag{12.55}$$

Each shape function L_i in Equations 12.53 and 12.55 satisfies the requirement that $D_i^* = 1$ with other displacements zero. This requirement is sufficient to derive the functions. The shape functions in Equations 12.53 and 12.55 correspond to the true deformed shapes of a prismatic bar (with shear deformation ignored). The same shape functions may be used to derive matrices for nonprismatic bars. It can be shown that the shape functions in Figures 12.18 and 12.19 are the same as the influence lines for the nodal forces, reversed in sign.

The shape function L_1 in Figure 12.19 can be considered to be equal to the sum of the straight lines $1 - \xi$ and $(L_2 + L_4)/l$; the latter term represents the deflected shape corresponding to clockwise rotations, each equal to $1/l$ at the two ends. By similar reasoning, we can verify that $L_3 = \xi - (L_2 + L_4)/l$.

12.9.2 Quadrilateral element subjected to in-plane forces

Figure 12.20 shows a quadrilateral plane element with corner nodes and two degrees-of-freedom per node. The element may be used in plane-stress and plane-strain analyses. In this case, the symbols in the generalized Equation 12.52 have the following meaning:

$$[f] = \{u, v\} \tag{12.56}$$

where u and v are translations in the x and y directions, and:

$$\{D^*\} = \{u_1, v_1, u_2, v_2, u_3, v_3, u_4, v_4\} \tag{12.57}$$

where u_i and v_i are translations at node i in the x and y directions. The $[L]$ matrix for the element in Figure 12.20 is:

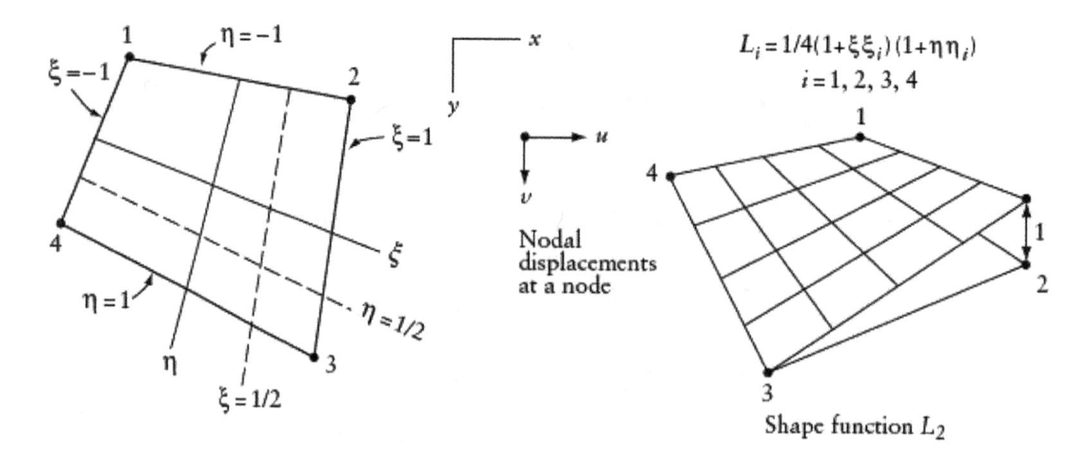

Figure 12.20 Plane-stress or plane-strain quadrilateral element. Natural coordinates ξ and η define the location of any point. Pictorial view of a shape function (hyperbolic paraboloid).

$$[L] = \begin{bmatrix} L_1 & 0 & L_2 & 0 & L_3 & 0 & L_4 & 0 \\ 0 & L_1 & 0 & L_2 & 0 & L_3 & 0 & L_4 \end{bmatrix} \tag{12.58}$$

The function L_i is a sum of bilinear functions in the natural coordinates ξ and η, defined in Figure 12.20. The value of L_i is unity at node i and zero at the other three nodes; any of the four shape functions may be expressed as:

$$L_i = \frac{1}{4}(1+\xi\xi_i)(1+\eta\eta_i) \quad \text{with} \quad i = 1, 2, 3, 4 \tag{12.59}$$

If the value of L_i is plotted perpendicular to the surface of the element, a hypersurface is obtained. Along the lines ξ = constant or η = constant, the surface follows straight lines. The shape function L_2 is plotted in pictorial view in Figure 12.20; at node 2, $\xi_i = 1$ and $\eta_i = -1$, and the function L_2 defining the hypersurface is obtained by substitution of the two values ξ and η in Equation 12.59.

Equation 12.59 represents one of a family of functions used for interpolation in the *iso-parametric elements*, discussed in Section 13.7. We may note that the sum of the values of the four L_i functions at any point is unity.

12.10 STIFFNESS AND STRESS MATRICES FOR DISPLACEMENT-BASED ELEMENTS

Equation 12.52 defines the element displacement field in terms of the nodal displacement. By appropriate differentiation, we can derive the strain (Equation 7.22):

$$\{\varepsilon\} = [\partial][L]\{D^*\} \tag{12.60}$$

Thus, the strain at any point in a displacement-based element is:

$$\{\varepsilon\} = [B]\{D^*\} \tag{12.61}$$

where

$$[B] = [\partial][L] \tag{12.62}$$

The matrix $[B]$ may be referred to as the nodal displacement-strain transformation matrix. Any column j of $[B]$ represents the strain components $\{\varepsilon_{uj}\}$ due to $D_j^* = 1$. Use of Hooke's law (Equation 7.21) gives the stress at any point in a displacement-based element:

$$\{\sigma\} = [d][B]\{D^*\} \tag{12.63}$$

or

$$\{\sigma\} = [\sigma_u]\{D^*\} \tag{12.64}$$

where $[\sigma_u]$ is the stress matrix for the element:

$$[\sigma_u] = [d][B] \tag{12.65}$$

The elements of any column j of $[\sigma_u]$ are the stress components at any point due to $D_j^* = 1$.

An element S_{ij}^* of the stiffness matrix is the force at coordinate i corresponding to unit displacement at j; S_{ij}^* can be determined by the unit-displacement theorem (Equation 7.59):

$$S_{ij}^* = \int_v \{\sigma_{uj}\}^{\mathrm{T}} \{\varepsilon_{ui}\} \, dv \tag{12.66}$$

where $\{\sigma_{uj}\}$ represents the "actual" stresses at any point due to unit displacement at j; $\{\varepsilon_{ui}\}$ represents the strains at the same point corresponding to unit virtual displacement at i; and dv is an elemental volume.* The integral over the volume is replaced by an integral over the length in the case of a bar and over the area in the case of a plate. For this purpose, the symbols $\{\sigma\}$ and $\{\varepsilon\}$ in Equation 12.66 represent generalized stress and strain, respectively. For example, in a bar, $\{\sigma\}$ represents internal forces at a section; in a plate in bending, $\{\varepsilon\}$ represents curvatures (Equation 7.32).

Using the shape functions $[L]$ to determine the actual stresses and the virtual strains via Equations 12.62 and 12.65, substitution in Equation 12.66 gives any element of the stiffness matrix:

$$S_{ij}^* = \int_v \{B\}_j^{\mathrm{T}} [d] \{B\}_i \, dv \tag{12.67}$$

where $\{B\}_i$ and $\{B\}_j$ are the ith and j^{th} columns of $[B]$. The stiffness matrix of a finite-element is given by:

$$\left[S^*\right] = \int_v [B]_j^{\mathrm{T}} [d] [B]_i \, dv \tag{12.68}$$

In Equations 12.66 and 12.67, we are accepting an assumed displacement field, namely L_j, as actual. However, in general, the assumed shape is different from the actual. What we are doing then is tantamount to imposing the assumed configuration by the application of small distributed forces on the element body in addition to the nodal forces. The distributed forces have the effect of changing the actual configuration to the assumed shape; these forces, not accounted for, cause the stiffness calculated by Equation 12.68 to be an overestimate. In other words, a finite-element analysis in which the element stiffness matrices are derived by the above procedure is expected to give smaller displacements than the actual ones.

12.11 ELEMENT LOAD VECTORS

The vector of restraining forces, to be used in the equilibrium Equation 12.49, includes a component $\{F_b\}$ representing the forces in equilibrium with the external loads applied on the body of the elements away from the nodes (Equation 12.48). Considering a single element, the equilibrant at node j to the body forces can be determined by:

$$F_{bi} = -\int_v \{L_i\}^{\mathrm{T}} \{p\} \, dv \tag{12.69}$$

where $\{p\}$ represents the magnitudes per unit volume of forces applied in the same directions as the displacements $\{f\}$.

Equation 12.69 can be explained by the principle of virtual work (Equation 7.55). Here, the body forces and the nodal equilibrants form one system; the element subjected only to those nodal forces which produce the displacement configuration L_j represents the second system. The work of the forces of the first system during displacements by the second system is equal to the work of the second system during displacements by the first system. Now, the second quantity is zero because the second system has forces at the nodes only and the nodal displacements in the first system are all zero.

By the use of Equation 12.69, we are treating the shape function L_j as the influence line (or influence surface) of the nodal force at j, reversed in sign (see Section 4.8.1). We should remember that an approximation is involved in Equation 12.69 by the acceptance of an assumed deflected shape as the actual displacement field.

The vector of nodal forces in equilibrium with the forces applied on the element away from the nodes is:

$$\left\{F_b^*\right\} = -\int_v [L]^{\mathrm{T}} \{p\}\, \mathrm{d}v \qquad (12.70)$$

This vector is referred to as the element consistent load vector because the same shape functions $[L]$ are used to generate $[S^*]$ and $\left\{F_b^*\right\}$. The superscript * is used here to refer to local coordinates of an individual element.

When the external forces are applied to the surface of the element, the integral in Equation 12.70 should be taken over the area of the element. When concentrated forces act, the integral is replaced by a summation of the forces multiplied by the values of $[L]^{\mathrm{T}}$ at the load positions.

12.11.1 Analysis of effects of temperature variation

When an element is subjected to volume change due to temperature variation (or to shrinkage), with the displacements restrained, the stresses at any point are given by (Equation 7.21)

$$[\sigma_r] = -[d][\varepsilon_0] \qquad (12.71)$$

where $\{\varepsilon_0\}$ represents the strains that would exist if the change in volume were free to occur. In a two-dimensional plane-stress or plane-strain state, a rise of temperature of T degrees produces the free strain (Equation 7.26),

$$\{\varepsilon_0\} = \alpha T \begin{Bmatrix} 1 \\ 1 \\ 0 \end{Bmatrix} \qquad (12.72)$$

where α is the coefficient of thermal expansion.

For an element subjected to volume change, the consistent vector of restraining forces is given by:

$$\left\{F_b^*\right\} = -\int_v [B]^{\mathrm{T}} [d]\{\varepsilon_0\}\, \mathrm{d}v \qquad (12.73)$$

Again, the unit-displacement theory may be used to derive Equation 12.73. With the actual stress being $\{\sigma_r\} = -[d]\ \{\varepsilon_0\}$ and the virtual strain being $\{\varepsilon_{uj}\} = \{B\}_j$, Equation 7.59 gives the jth element of the consistent load vector.

In most cases, the integrals involved in generating the stiffness matrix and the load vectors for individual elements are evaluated numerically using Gaussian quadrature (see Section 13.12).

EXAMPLE 12.5: AXIAL FORCES AND DISPLACEMENTS OF A BAR OF VARIABLE CROSS SECTION

Generate the stiffness matrix and the vector of restraining forces due to a uniform rise in temperature of T degrees for the bar shown in Figure 12.18. Assume that the cross section varies as $a = a_0(2 - \xi)$, where a_0 is constant. The coefficient of thermal expansion is α and the modulus of elasticity is E.

Using the shape functions of Equation 12.53, the $[B]$ matrix is (Equation 12.62)

$$[B] = \frac{\mathrm{d}}{\mathrm{d}x}[1-\xi, \xi] = \frac{\mathrm{d}}{l\,\mathrm{d}\xi}[1-\xi, \xi] = \frac{1}{l}[-1, 1]$$

The elasticity matrix in this case has one element $[d] \equiv [E]$. Substitution in Equation 12.68 gives the stiffness matrix:

$$[S^*] = l\int_0^1 \frac{1}{l}\begin{Bmatrix}-1\\1\end{Bmatrix}[E]\frac{1}{l}[-1, 1]a_0(2-\xi)\,\mathrm{d}\xi$$

or

$$[S^*] = \frac{Ea_0}{l}\begin{bmatrix}1 & -1\\-1 & 1\end{bmatrix}\left[2\xi - \frac{\xi^2}{2}\right]_0^1 = \frac{1.5\,Ea_0}{l}\begin{bmatrix}1 & -1\\-1 & 1\end{bmatrix}$$

If the effect of the temperature change is not restrained, the strain is $\varepsilon_0 = \alpha T$. The nodal forces to restrain nodal displacements (Equation 12.73) are then:

$$\{F_b\}_m = -l\int_0^1 \frac{1}{l}\begin{Bmatrix}-1\\1\end{Bmatrix}E\,\alpha\,T a_0(2-\xi)\,\mathrm{d}\xi = 1.5\,E\,\alpha\,T a_0\begin{Bmatrix}-1\\1\end{Bmatrix}$$

The same results would be obtained if the member were treated as a prismatic bar with a constant cross-sectional area equal to the average of the values at the two ends. The exact answer for the stiffness matrix is the same as above with the constant 1.5 replaced by $(1/\ln 2) = 1.443$ (obtained by considering the true deformed shape). As expected, the use of assumed shape functions resulted in an overestimate of stiffness.

EXAMPLE 12.6: STIFFNESS MATRIX OF A BEAM IN FLEXURE WITH VARIABLE CROSS SECTION

Determine element S_{12}^* of the stiffness matrix for the bar shown in Figure 12.19, assuming the second moment of the cross-sectional area to vary as $I = I_0(1+\xi)$, where I_0 is constant. Consider bending deformations only; E = constant. Also, generate the vector of nodal equilibrants of a uniform load q per unit length covering the entire length.

For a beam in bending, $[d] = [EI]$, $\{\sigma\} \equiv \{M\}$ and $\{\varepsilon\} \equiv -\mathrm{d}^2v/\mathrm{d}x^2$. Using the shape functions in Figure 12.19, the $[B]$ matrix is (Equation 12.62):

$$[B] = -\frac{\mathrm{d}^2}{\mathrm{d}x^2}\left[1-3\xi^2+2\xi^3 \quad l\xi(\xi-1)^2 \quad \xi^2(3-2\xi) \quad l\xi^2(\xi-1)\right]$$

$$= -\frac{1}{l^2}\left[-6+12\xi \quad l(6\xi-4) \quad 6-12\xi \quad l(6\xi-2)\right]$$

The required element of the stiffness matrix (Equation 12.67) is:

$$S_{12}^{*}=\int_{0}^{1}\left(\frac{6-12\xi}{l^{2}}\right)EI_{0}(1+\xi)\left(\frac{-6\xi+4}{l}\right)l\,\mathrm{d}\xi=8\frac{EI_{0}}{l^{2}}$$

The exact answer can be calculated, giving $S_{12}^{*}=7.72EI_{0}/l^{2}$. As expected, the stiffness is overestimated by the use of the assumed shape function L_2 instead of the true deflected shape due to $D_2^{*}=1$.

The entire stiffness $[S^{*}]$ derived by Equation 12.68 may be compared with the exact stiffness matrix in the answers to Problem 12.1.

The nodal forces in equilibrium with the uniform load q, with nodal displacements prevented, are (Equation 12.70)

$$\{F_b\}_m=-\int_{0}^{1}\begin{Bmatrix}1-3\xi^{2}+2\xi^{3}\\ l\xi(\xi-1)^{2}\\ \xi^{2}(3-2\xi)\\ l\xi^{2}(\xi-1)\end{Bmatrix}ql\,\mathrm{d}\xi=\begin{Bmatrix}-ql/2\\ -ql^{2}/12\\ -ql/2\\ ql^{2}/12\end{Bmatrix}$$

These are the same forces as for a prismatic beam; again, an approximation is involved in accepting the deflected shapes of a prismatic bar for a bar with a variable I.

EXAMPLE 12.7: STIFFNESS MATRIX OF A RECTANGULAR PLATE SUBJECTED TO IN-PLANE FORCES

Determine element S_{22}^{*} of the stiffness matrix for a rectangular plane-stress element of constant thickness h (Figure 12.1). The shape functions of this element, as will be derived in Example 13.3, are:

$$[L]=\begin{bmatrix}L_1 & 0 & 0 & L_4 & 0 & 0 & L_7 & 0 & 0 & L_{10} & 0 & 0\\ 0 & L_2 & L_3 & 0 & L_5 & L_6 & 0 & L_8 & L_9 & 0 & L_{11} & L_{12}\end{bmatrix}$$

$$L_1=(1-\xi)(1-\eta)\qquad L_2=(1-3\xi^{2}+2\xi^{3})(1-\eta)\qquad L_3=b\xi(\xi-1)^{2}(1-\eta)$$

$$L_4=\xi(1-\eta)\qquad L_5=\xi^{2}(3-2\xi)(1-\eta)\qquad L_6=b\xi^{2}(\xi-1)(1-\eta)$$

$$L_7=\xi\eta\qquad L_8=\xi(3-2\xi)\eta\qquad L_9=b\xi^{2}(\xi-1)\eta$$

$$L_{10}=(1-\xi)\eta\qquad L_{11}=(1-3\xi^{2}+2\xi^{3})\eta\qquad L_{12}=b\xi(\xi-1)^{2}\eta$$

where $\xi=x/b$; $\eta=y/c$ (Figure 12.1). Determine also the stresses at any point due to $D_2^{*}=1$.

Due to $D_2^{*}=1$, the displacements at any point are given by Equation 12.52 where $\{f\}=\{u, v\}$,

$$\begin{Bmatrix}u\\ v\end{Bmatrix}=\begin{Bmatrix}0\\ L_2\end{Bmatrix}=\begin{Bmatrix}0\\ (1-3\xi^{2}+2\xi^{3})(1-\eta)\end{Bmatrix}$$

The strain at any point (Equations 12.62 and 7.27) is:

$$\{B\}_2=\begin{Bmatrix}0\\ (1/c)(-1+3\xi^{2}-2\xi^{3})\\ (1/b)(-6\xi+6\xi^{2})(1-\eta)\end{Bmatrix}$$

The required element of the stiffness matrix is given by Equation 12.67 as:

$$S^*_{22} = bch \int_0^1 \int_0^1 \{B\}_2^T [d] \{B\}_2 \, d\eta \, d\xi$$

Substituting for $[d]$ from Equation 7.28 and performing the integral gives:

$$S^*_{22} = \frac{Eh}{1-\nu^2} \left[\frac{13}{35}\left(\frac{b}{c}\right) + \frac{1}{5}\left(\frac{c}{b}\right)(1-\nu) \right]$$

The stress at any point due to $D^*_2 = 1$ is (Equation 12.65):

$$\{\sigma_u\}_2 = [d]\{B\}_2 = \frac{Eh}{1-\nu^2} \left\{ \frac{\nu}{c}\left(-1+3\xi^2-2\xi^3\right), \frac{1}{c}\left(-1+3\xi^2-2\xi^3\right), \frac{3(1-\nu)}{b}(1-\eta)\left(-\xi+\xi^2\right) \right\}$$

12.12 GENERAL

The displacement method of analysis is applicable to structures composed of finite-elements which may be one-, two-, or three-dimensional. The analysis requires the generation of stiffness and stress matrices and of load vectors for individual elements. Exact element matrices can be generated only for bars. Procedures to generate approximate matrices for elements of any type have been presented in this chapter. Convergence to the exact solution can be ensured as a finer finite-element mesh is used.

REFERENCES

1. Hetényi, M., *Beams on Elastic Foundation*, University of Michigan Press, Ann Arbor, 1952, 60 pp.
2. Ghali, A., *Circular Storage Tanks and Silos*, 3rd ed., CRC Press, Taylor & Francis Group, Boca Raton, FL, 2014, 367 pp.

PROBLEMS

12.1 Determine any element S^*_{ij} for the bar of Example 12.6, using the shape functions in Figure 12.19. Compare the answer with an exact value as given by Equation 9.5.

12.2 Consider the member of Problem 12.1 as a cantilever fixed at the end $x=l$ and subjected to a transverse concentrated *load P* at the free end. Find the deflection at the tip of the cantilever, using the two matrices given in the answers to Problem 12.1.

12.3 For the prismatic bar shown, generate the stiffness matrix corresponding to the three coordinates indicated. Use the following shape functions: $L_1 = -½\xi(1-\xi)$; $L_2 = ½\xi(1+\xi)$; $L_3 = 1-\xi^2$ (Lagrange polynomials, Figure 13.5). Condense the stiffness matrix by elimination of node 3.

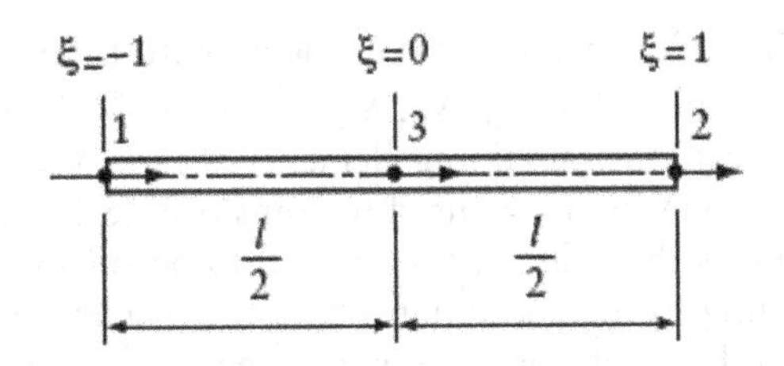

12.4 Assume that the displacement at coordinate 1 of the bar element of Problem 12.3 is prevented. A spring support is provided at coordinate 2 so that $F_2 = -Ku_2$, where F_2 and u_2 are, respectively, the force and the displacement at coordinate 2, and K is the spring constant equal to $Ea/(3l)$. What will be the displacements at coordinates 2 and 3 and the stress in the bar when it is subjected to a rise in temperature of $T = ½T_0(1+\xi)$ where T_0 is a constant? Use the shape functions given in Problem 12.3. Do you expect exact answers?

12.5 Determine S^*_{11}, S^*_{21}, S^*_{31}, S^*_{41} and S^*_{51} of the stiffness matrix of the plane-stress finite-element in Figure 12.20. Assume that the element is rectangular with sides b and c parallel to the x and y axes, respectively, and is made of an isotropic material with $\nu = 0.2$. Element thickness is constant and equal to h. The nodal displacement vector and shape functions are defined by Equations 12.57 to 12.59.

12.6 Considering symmetry and equilibrium, use the results of Problem 12.5 to generate $[S^*]$ for the element when $b = c$.

12.7 Use the results of Problem 12.5 to calculate the deflection at the middle of a beam idealized by two elements as shown. Take the beam width as h and Poisson's ratio as 0.2. Compare the result with that obtained by beam theory, considering bending and shear deformations. Note that the deflection calculated by the finite-element method is smaller than the more accurate value obtained by beam theory and that the percentage error increases with an increase in the ratio b/c.

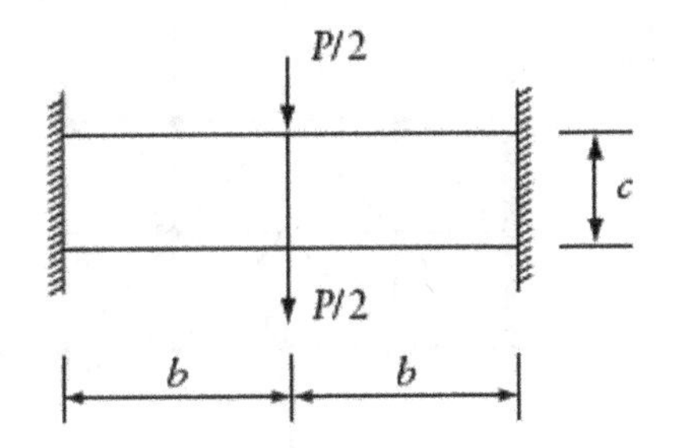

12.8 Use the displacement determined in Problem 12.7 to calculate the stresses in the top fiber at the fixed end.

12.9 Determine S^*_{11} for the rectangular plate-bending element shown in Figure 7.8. The nodal displacement vector and the shape functions are defined in Equations 16.33 to 16.35. Use the result to calculate the central deflection of a rectangular plate $2b \times 2c$ with built-in edges and subjected to a concentrated load P at mid-point. Consider an isotropic material with $\nu = 0.3$. Idealize the plate by four elements and perform the analysis for one element only, taking advantage of symmetry. Use Figure 7.8 to represent the element analyzed with node 1 at the center of the plate. Determine also M_x at node 2.

12.10 The figure shows identical plane-stress elements with three coordinate systems. For the system in (a), calculate S^*_{11}, S^*_{21}, S^*_{31}, S^*_{33}, S^*_{34}, and S^*_{44}; by considering equilibrium and symmetry, generate the remaining elements of $[S^*]$. What are the transformation matrix $[T]$ and the equation to be used to transform $[S^*]$ into $[S]$ for the coordinate system in (b)? What is the value of the angle a to be substituted in the equations to give $[S]$ corresponding to the coordinates in (c)? Consider an isotropic material with Poisson's ratio $\nu=0$. Element thickness is h; element sides are b, b, and $b\sqrt{2}$.

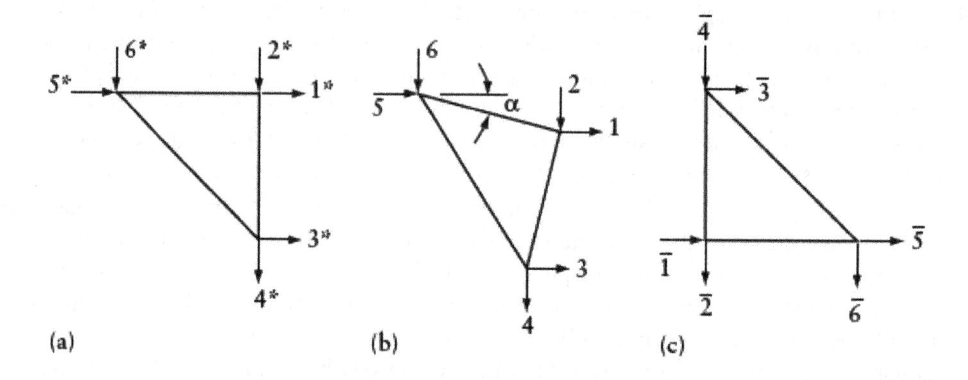

12.11 Using the answers of Problem 12.2 and taking advantage of symmetry, determine, for the beam shown, the nodal displacements and the stress in element 2-6-3. The beam is of isotropic material; $\nu=0$; beam width is h. The structure idealization is composed of isosceles right-angle triangles of the same size. The equilibrium equations can be checked by substitution of $\{D\}$ given in the answers.

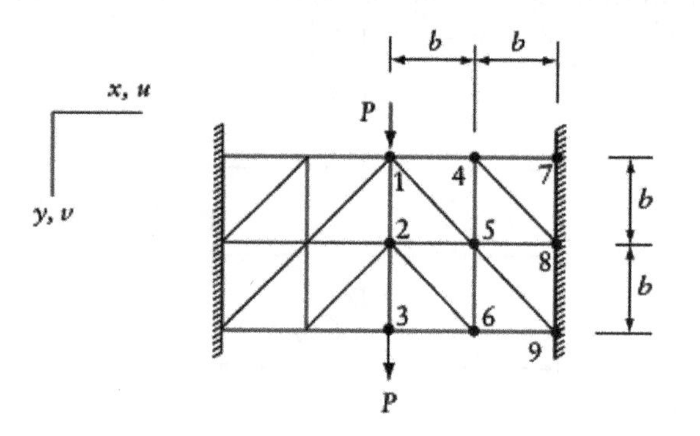

12.12 Generate columns 1, 2, and 3 of the $[B]$ matrix for element QLC3 (Figure 12.1).

NOTE

* The symbol dv representing elemental volume should not be confused with v, which represents a translational displacement.

Chapter 13

Finite-element analysis

13.1 INTRODUCTION

This chapter is a continuation of Chapter 12, introducing various types of finite-elements frequently used in practice. We derive the shape functions for few elements. The isoparametric formulation is widely used, because it allows the elements to have curved shapes. We use the Gauss quadrature method for numerical evaluation of integrals involved in generating the finite-element matrices. The idealization of spatial structures, as assemblage of flat shell elements, is frequently used in practice. The stiffness matrix of shell elements is a combination of the stiffness matrices of plane-stress and bending elements.

13.2 DERIVATION OF SHAPE FUNCTIONS

The displacement field $\{f\}$ (Equation 12.52) may be expressed as polynomials of the coordinates x and y (or ξ and η) defining the position of any point. For example, the deflection w in a plate-bending element or the translations u and v in a plane-stress or a plane-strain element may be expressed as:

$$f(x,y) = \left[1,\, x,\, y,\, x^2 \cdots\right] \{A\} = [P]\{A\} \tag{13.1}$$

where $\{A\}$ is a vector of constants, yet to be determined, and $[P]$ is a matrix of polynomial terms, the number of which equals the number of nodal degrees-of-freedom. Pascal's triangle (Figure 13.1) can be used to select the polynomial terms to be included in $[P]$. In general, the lower-degree terms are used. Examples of the polynomial terms used in several elements are given later in this section.

The nodal displacements $\{D^*\}$ can be related to the constants $\{A\}$ by substituting x and y (or ξ and η) values at the nodes in Equation 13.1 (or its derivatives). This gives:

$$\left\{D^*\right\} = [C]\{A\} \tag{13.2}$$

The elements of [C] are known values depending upon (x_i, y_i), with $i = 1, 2, \ldots$ referring to the node numbers. The undetermined constants $\{A\}$ can now be expressed in terms of $\{D^*\}$ by inversion:

$$\{A\} = [C]^{-1}\left\{D^*\right\} \tag{13.3}$$

DOI: 10.1201/9780429286858-13

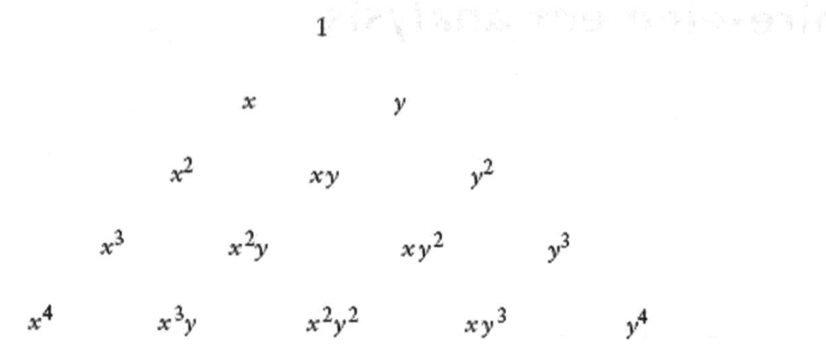

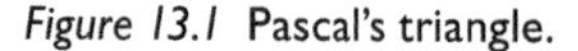
Figure 13.1 Pascal's triangle.

Substituting Equation 13.3 into 13.1, and by analogy of the resulting equation with Equation 12.52, we obtain the shape functions:

$$[L]=[P][C]^{-1} \tag{13.4}$$

As an example of polynomial selection, let us consider the bar element shown in Figure 12.18, which is subjected to an axial force. With two degrees-of-freedom, $[P]$ has only two terms:

$$[P]=[1 \quad \xi] \tag{13.5}$$

For a bar in bending (Figure 12.19):

$$[P]=\left[1 \quad \xi \quad \xi^2 \quad \xi^3\right] \tag{13.6}$$

For a plate element subjected to in-plane forces (Figure 12.20), each of the displacements u and v is associated with four nodal displacements. A polynomial with four terms is used for each of u and v:

$$[P]=[1 \quad \xi \quad \eta \quad \xi\eta] \tag{13.7}$$

For the plate-bending element in Figure 7.8, with twelve degrees-of-freedom, we express the deflection as $w=[P]\,\{A\}$, with:

$$[P]=\left[1 \quad \xi \quad \eta \quad \xi^2 \quad \xi\eta \quad \eta^2 \quad \xi^3 \quad \xi^2\eta \quad \xi\eta^2 \quad \eta^3 \quad \xi^3\eta \quad \xi\eta^3\right] \tag{13.8}$$

For elements with two or more variables, such as x and y (or x, y, and z), the terms included in $[P]$ should be invariant if the reference axes x and y (or x, y, and z) are interchanged. Thus, in Equation 13.7, we should not replace the terms $\xi\eta$ by ξ^2 or by η^2. Similarly, in Equation 13.8, we should not replace $\xi^3\eta$ or $\xi\eta^3$ by $\xi^2\eta^2$. In other words, $[P]$ includes symmetrical terms from Pascal's triangle (Figure 13.1). When this requirement is satisfied, the element does not have a "preferred" direction. In consequence, the use of such an element in the analysis of the structure shown in Figure 12.17 will give the same answers regardless of whether the global axes x and y are as shown or are rotated through 90° so that x becomes vertical.

The invariance requirement is relaxed for the rectangular plate element subjected to in-plane forces included in Figure 12.1. The element has four nodes with three nodal displacements per node: u, v, and $\partial v/\partial x$. The inclusion of $\partial v/\partial x$ (but not of other derivatives of u and v) makes the element behave differently in the x and y directions. The polynomials used for u and for v are different:

$$u(x,y) = [1 \quad x \quad y \quad xy]\{A\}_u \tag{13.9}$$

$$v(x,y) = \left[1 \quad x \quad y \quad x^2 \quad xy \quad x^3 \quad x^2y \quad x^3y\right]\{A\}_v \tag{13.10}$$

The variation of v in the x direction is cubic, while a linear variation is used for v in the y direction and for u in both the x and y directions. The shape functions for the element are given in Equation 13.11 (see Example 13.3).

This element* gives excellent accuracy when used for structures which have beam-like behavior, e.g. folded plates and box girders. For this use, the element local x-axis must be in the direction of the "beam."

EXAMPLE 13.1: BEAM IN FLEXURE

Derive the shape functions for the bar element in Figure 12.19. The deflection v is expressed as a cubic polynomial of ξ, where $\xi = x/l$ (Equations 13.1 and 13.6).

$$v = \left[1 \quad \xi \quad \xi^2 \quad \xi^3\right]\{A\} = [P]\{A\}$$

The nodal displacements defined by Equation 12.54 are substituted in the above equations to give:

$$\{D^*\} = \begin{bmatrix} 1 & 0 & 0 & 0 \\ 0 & 1/l & 0 & 0 \\ 1 & 1 & 1 & 1 \\ 0 & 1/l & 2/l & 3/l \end{bmatrix}\{A\}$$

Inversion of the square matrix in the above equation gives:

$$[C]^{-1} = \begin{bmatrix} 1 & 0 & 0 & 0 \\ 0 & l & 0 & 0 \\ -3 & -2l & 3 & -l \\ 2 & l & -2 & l \end{bmatrix}$$

The product $[P]\,[C]^{-1}$ gives the shape functions $[L]$ (Equation 12.55).

EXAMPLE 13.2: QUADRILATERAL PLATE SUBJECTED TO IN-PLANE FORCES

Derive the shape functions for the plane-stress or plane-strain quadrilateral element shown in Figure 12.20.

We have u or $v = [P]\{A\}$, with $[P]$ given in Equation 13.7. Substituting for ξ and η by their values at the four corners, we write for u:

$$\begin{Bmatrix} u_1 \\ u_2 \\ u_3 \\ u_4 \end{Bmatrix} = \begin{bmatrix} 1 & -1 & -1 & 1 \\ 1 & 1 & -1 & -1 \\ 1 & 1 & 1 & 1 \\ 1 & -1 & 1 & -1 \end{bmatrix} \{A\}$$

Inversion of [C], which is the square matrix in this equation, gives:

$$[C]^{-1} = (1/4)\begin{bmatrix} 1 & 1 & 1 & 1 \\ -1 & 1 & 1 & -1 \\ -1 & -1 & 1 & 1 \\ 1 & -1 & 1 & -1 \end{bmatrix}$$

Substitution in Equation 13.4 gives:

$$[L_1, L_2, L_3, L_4] = \frac{1}{4}[1-\xi-\eta+\xi\eta \quad 1+\xi-\eta-\xi\eta \quad 1+\xi+\eta+\xi\eta \quad 1-\xi+\eta-\xi\eta]$$

which is the same as Equation 12.59. The same shape functions apply to v.

EXAMPLE 13.3: RECTANGULAR ELEMENT QLC3

Using the polynomials in Equation 13.10, derive the shape functions for the displacement v in a rectangular plate element QLC3 subjected to in-plane forces (Figure 12.1).

The nodal displacements are defined as:

$$\{D^*\} = \left\{ \left(u, v, \frac{\partial v}{\partial x}\right)_1, \left(u, v, \frac{\partial v}{\partial x}\right)_2, \left(u, v, \frac{\partial v}{\partial x}\right)_3, \left(u, v, \frac{\partial v}{\partial x}\right)_4 \right\}$$

Using the symbols $\xi = x/b$ and $\eta = y/c$, Equation 13.10 can be written as $v = [P]\{\bar{A}\}$, with:

$$[P] = [1 \quad \xi \quad \eta \quad \xi^2 \quad \xi\eta \quad \xi^3 \quad \xi^2\eta \quad \xi^3\eta]$$

The nodal displacements associated with v are:

$$\{D_v^*\} = \{D_2^*, D_3^*, D_5^*, D_6^*, D_8^*, D_9^*, D_{11}^*, D_{12}^*\}$$

Substituting for ξ and η by their values at the nodes, we can write $\{D_v^*\} = [C]\{\bar{A}\}$ with:

$$[C] = \begin{matrix} 2 \\ 3 \\ 5 \\ 6 \\ 8 \\ 9 \\ 11 \\ 12 \end{matrix} \begin{bmatrix} 1 & 0 & 0 & 0 & 0 & 0 & 0 & 0 \\ 0 & 1/b & 0 & 0 & 0 & 0 & 0 & 0 \\ \hline 1 & 0 & 0 & 1 & 0 & 1 & 0 & 0 \\ 0 & 1/b & 0 & 2/b & 0 & 3/b & 0 & 0 \\ \hline 1 & 0 & 1 & 1 & 1 & 1 & 1 & 1 \\ 0 & 1/b & 0 & 2/b & 1/b & 3/b & 2/b & 3/b \\ \hline 1 & 0 & 1 & 0 & 0 & 0 & 0 & 0 \\ 0 & 1/b & 0 & 0 & 1/b & 0 & 0 & 0 \end{bmatrix}$$

Inversion of [C] and substitution in Equation 13.4 gives the eight shape functions associated with v: L_2, L_3, L_5, L_6, L_8, L_9, L_{11}, L_{12}. For reference, we give the complete shape functions for the element as follows:

$$\begin{array}{lll} L_1=(1-\xi)(1-\eta) & L_2=\left(1-3\xi^2+2\xi^3\right)(1-\eta) & L_3=b\xi(\xi-1)^2(1-\eta) \\ L_4=\xi(1-\eta) & L_5=\xi^2(3-2\xi)(1-\eta) & L_6=b\xi^2(\xi-1)(1-\eta) \\ L_7=\xi\eta & L_8=\xi(3-2\xi)\eta & L_9=b\xi^2(\xi-1)\eta \\ L_{10}=(1-\xi)\eta & L_{11}=\left(1-3\xi^2+2\xi^3\right)\eta & L_{12}=b\xi(\xi-1)^2\eta \end{array} \tag{13.11}$$

These functions can be used to express u and v by the equation $\{u, v\}=[L]\{D^*\}$ (see Equation 12.52), with:

$$[L]=\begin{bmatrix} L_1 & 0 & 0 & L_4 & 0 & 0 & L_7 & 0 & 0 & L_{10} & 0 & 0 \\ 0 & L_2 & L_3 & 0 & L_5 & L_6 & 0 & L_8 & L_9 & 0 & L_{11} & L_{12} \end{bmatrix} \tag{13.12}$$

Element QLC3 is biased by having nodal displacement $\partial v/\partial x$, without $\partial v/\partial y$. It is intended for the use in a structure having beam-like behavior in the x-direction.

13.3 SHELLS AS ASSEMBLAGE OF FLAT ELEMENTS

Finite-elements in the form of flat quadrilateral or triangular plates can be used to idealize a shell (Figure 13.2). In general, the elements will be subjected to in-plane forces and to bending. The element matrices derived separately in earlier sections for elements in a state of plane-stress and for bending elements can be combined for a shell element as discussed below.

Idealization of a shell, using curved elements, may be necessary if large elements are employed, particularly in double-curved shells. However, in practice, many shells, particularly those of cylindrical shape, have been analyzed successfully using triangular, quadrilateral, or rectangular elements.

Flat shell elements can be easily combined with beam elements to idealize edge beams or ribs, which are common in practice.

13.3.1 Rectangular shell element

Figure 13.3 represents a rectangular shell element with six nodal displacements at each corner, three translations $\{u, v, w\}$ and three rotations $\{\theta_x, \theta_y, \theta_z\}$. The derivation of the matrices for a plane-stress rectangular element with nodal displacements u, v, and θ_z at each corner was discussed in Section 13.2 (see Example 13.2). A rectangular bending element with nodal displacements w, θ_x, and θ_y at each corner was discussed in Section 7.5.2.

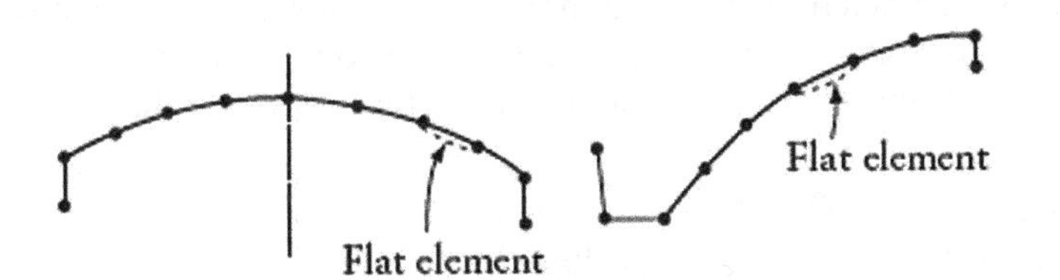

Figure 13.2 Cylindrical shells idealized as an assemblage of flat elements.

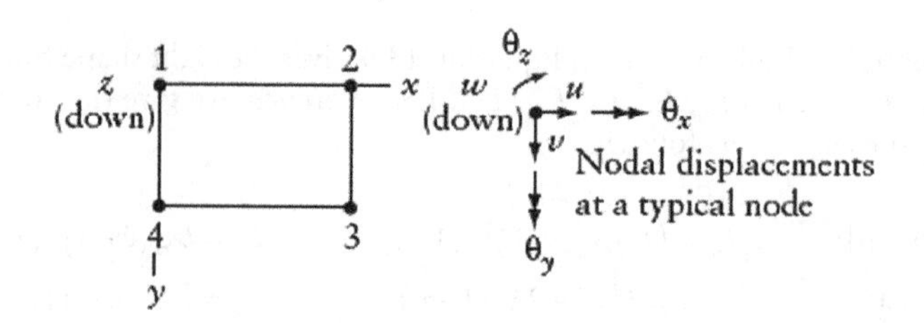

Figure 13.3 Rectangular flat shell element.

The stiffness matrix for the shell element in Figure 13.3 can be written in the form:

$$\left[S^*\right]=\begin{bmatrix}\left[S_m^*\right] & [0] \\ [0] & \left[S_b^*\right]\end{bmatrix} \tag{13.13}$$

where $\left[S_m^*\right]$ is a 12 × 12 membrane stiffness matrix relating forces to the displacements u, v, and θ_z and $\left[S_b^*\right]$ is a 12 × 12 bending stiffness matrix relating forces to the displacements w, θ_x, and θ_y. The off-diagonal submatrices in Equation 13.13 are null because the first set of displacements produces no forces in the directions of the second set and vice versa. The two sets of displacements are said to be *uncoupled.*

For convenience in coding, the degrees-of-freedom at each node are commonly arranged in the order $\{u, v, w, \theta_x, \theta_y, \theta_z\}$. The columns and rows in the 24 × 24 stiffness matrix in Equation 13.13 are rearranged so that the nodal coordinates are numbered in the following sequence: six coordinates at node 1, followed by six coordinates at node 2, and so on.

13.3.2 Fictitious stiffness coefficients

The majority of plane-stress elements have two degrees-of-freedom per node, u and v. Combining such an element with a bending element (with nodal displacements w, θ_x, and θ_y) so as to form a shell element will leave the θ_z coordinate unused. The five degrees-of-freedom: u, v, w, θ_x, and θ_y, are not sufficient to analyze a spatial structure. This can be seen by considering elements in two intersecting planes; no matter how the global axes are chosen, the structure will, in general, have three rotation components in the three global directions at the nodes on the line of intersection.

We may think of generating a stiffness matrix for the shell element with six degrees-of-freedom per node, with zero columns and rows corresponding to the θ_z coordinates. However, when elements situated in one plane meet at a node, a zero will occur at the diagonal of the stiffness matrix of the assembled structure; this will result in an error message with common computer equation solvers (Section 19.11) when attempting to divide a number by the stiffness coefficient on the diagonal. To avoid this difficulty, Zienkiewicz[1] assigns fictitious stiffness coefficients, instead of zeros, in the columns and rows corresponding to θ_z. For a triangular element with three nodes, the fictitious coefficients are:

$$[S]_{\text{fict}}=\beta\,E\begin{bmatrix}1 & -0.5 & -0.5 \\ -0.5 & 1 & -0.5 \\ -0.5 & -0.5 & 1\end{bmatrix}(\text{volume}) \tag{13.14}$$

where E is the modulus of elasticity, the multiplier at the end is the volume of the element, and β is an arbitrarily chosen coefficient. Zienkiewicz suggested a value $\beta = 0.03$ or less. We can note that the sum of the fictitious coefficients in any column is zero, which is necessary for equilibrium.

13.4 CONVERGENCE CONDITIONS

Regardless of length, the shape functions for prismatic beam elements, Equation 12.55 are exact. For monotonic convergence to the correct answer as smaller elements are used, the shape functions of elements, other than prismatic beams, should satisfy the following condition:

1. The displacements of adjacent elements along a common boundary must be identical.

It can be seen that this condition is satisfied in the plane-stress or plane-strain element in Figure 12.20 and in the plate-bending element in Figure 7.8. Along any side 1-2, the translations u and v in Figure 12.20 or the deflection w in Figure 7.8 are functions of the nodal displacements at 1 and 2 only. It thus follows that two adjacent elements sharing nodes 1 and 2 will have the same nodal displacements at the two nodes and the same u and v or w along the line 1-2.

In some cases, the first partial derivatives of the element should also be compatible. This condition needs to be satisfied in plate-bending elements but not in plane-stress or plane-strain elements.

Two adjacent elements of the type shown in Figure 7.8 have the same deflection and hence the same slope along the common edge. However, normal to the common edge, the tangents of the deflected surfaces of the two elements have the same slope only at the nodes. Away from the nodes, the tangents normal to a common edge can have slopes differing by an angle α. The quantity $(1/2) \int M_n \, \alpha \, ds$ represents, for the assembled structure, strain energy not accounted for in the process of minimizing the total potential energy; here, M_n is the resultant of stresses normal to the common edge, and ds is an elemental length of the edge.

The rule to ensure convergence is that the compatibility be satisfied for $[L]$ and its derivatives of order one less than the derivatives included in $[B]$. For a plate-bending element, $[B]$ includes second derivatives (Equations 12.62 and 7.35): $-\partial^2 w/\partial x^2$, $-\partial^2 w/\partial y^2$, $2\partial^2 w/\partial x \, \partial y$. Thus, compatibility is required for $\partial w/\partial x$ and $\partial w/\partial y$ in addition to w.

Several elements, said to be incompatible or nonconforming, such as the element in Figure 7.8, do not satisfy the requirement of compatibility of displacement derivatives, and yet some of these elements give excellent results. An explanation of this behavior is that the excess stiffness, which is a characteristic of displacement-based finite-elements, is compensated by the increase in flexibility resulting from a lack of compatibility of slopes.

Nonconforming elements converge towards the correct answer when the incompatibilities disappear as the mesh becomes finer and the strains within the element tend to be constants.

2. When the nodal displacements $\{D^*\}$ correspond to rigid-body motion, the strains $[B]$ $\{D^*\}$ must be equal to zero.

This condition is easily satisfied when the polynomial matrix $[P]$ in Equation 13.1 includes the lower order terms of Pascal's triangle. For example, for the plate-bending element (Figure 7.8), inclusion of the terms 1, x, and y would allow w to be represented by the equation of an inclined plane.

We can verify that the shape functions in Equation 7.54, for the element shown in Figure 7.8, allow translation as a rigid body by setting $w = 1$ at the four corners and $\theta_x = \theta_y = 0$ at all nodes; Equation 12.52 will then give $w = 1$, representing unit downward translation. (This is because

$L_1 + L_4 + L_7 + L_{10} = 1$, as noted earlier.) To check that the shape functions allow rigid-body rotation, let $w = 1$ at nodes 1 and 2 and $\theta_x = -1/c$ at the four nodes of the element, while all other nodal displacements are zero. It can be seen that substitution of these nodal displacements and of Equation 7.54 in Equation 12.52 gives $w = 1 - \eta$, which is the equation of the plane obtained by rotation of the element through an angle $\theta_x = -1/c$ about the edge 4-3. In a similar way, we can verify that the shape functions allow a rigid-body rotation, θ_y = constant.

3. The shape functions must allow the element to be in a state of constant strain.

This is required because, as the elements become smaller, the strains within individual elements tend to be constant. Thus, a smooth curve (or surface) representing the strain variation can be approximated by step variation.

This requirement will be satisfied when the polynomial [P] in Equation 13.1 includes the lower terms which contribute to the strain. For example, for the rectangular bending element shown in Figure 7.8, the strains are $\{\varepsilon\} = \{-\partial^2 w/\partial x^2, -\partial^2 w/\partial y^2, 2\partial^2 w/(\partial x\, \partial y)\}$; the terms x^2, xy, and y^2 of Pascal's triangle must be included in [P].

We can verify that the shape functions in Equation 7.54 allow a constant curvature, i.e. $-\partial^2 w/\partial x^2$ = constant, by setting $\theta_y = 1$ at nodes 1 and 4, and $\theta_y = -1$ at nodes 2 and 3, while the remaining nodal displacements are equal to zero. Substitution in Equation 12.52 gives $w = b\xi\,(1-\xi)$, which represents the surface of a cylinder with a constant curvature of $2/b$.

For the same element, $2\partial^2 w/(\partial x\, \partial y)$ will be constant ($= -4/bc$) when the element is twisted so that $w = 1$ at nodes 2 and 4, with the edges remaining straight. Thus, the nodal displacements will be: $w = 0$ at 1 and 3; $w = 1$ at 2 and 4; $\theta_x = 1/c$ at 1 and 4; $\theta_x = -1/c$ at 2 and 3; $\theta_y = -1/b$ at 1 and 2; and $\theta_y = 1/b$ at 3 and 4.

Requirement 2 can be considered to be a special case of requirement 3 when the constant strain is zero.

13.5 LAGRANGE INTERPOLATION

Consider a function $g(\xi)$ for which the n values $g_1, g_2, \ldots, g_n$ are known at $\xi_1, \xi_2, \ldots, \xi_n$ (Figure 13.4). The Lagrange equation gives a polynomial $g(\xi)$ which passes through the n points. The equation takes the form of a summation of n polynomials:

$$g(\xi) = \sum_{i=1}^{n} g_i\, L_i \qquad (13.15)$$

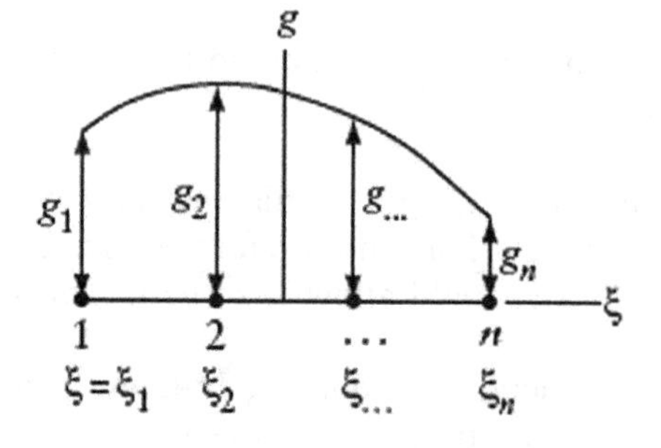

Figure 13.4 Function $g(\xi)$ with n known values $g_1, g_2, \ldots, g_n$ at $\xi = \xi_1, \xi_2, \ldots, \xi_n$.

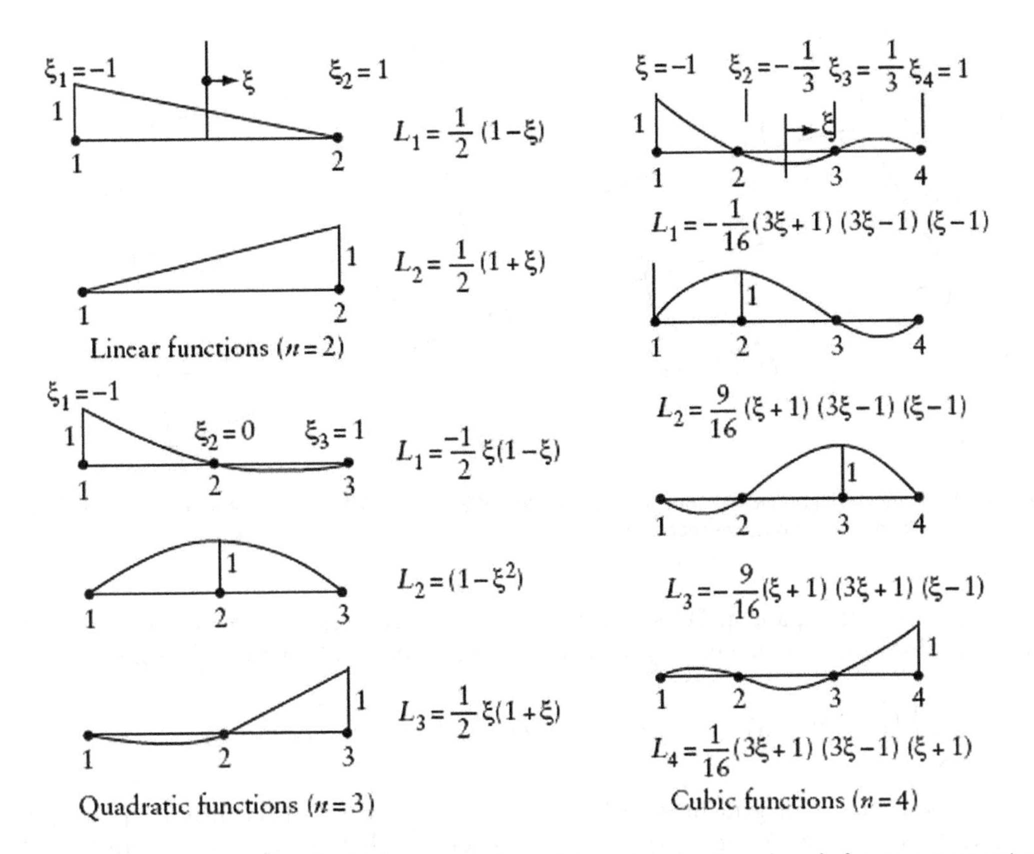

Figure 13.5 Linear, quadratic and cubic Lagrange polynomials used to give the value of a function at any point in terms of n known values, with $n=2$, 3, and 4, respectively (Equation 13.15).

where L_i is a polynomial in ξ of degree n–1. Equation 13.15 can be used to interpolate between $g_1, g_2, \ldots, g_n$ to give the g value at any intermediate ξ. Figure 13.5 shows *Lagrange interpolation functions* for $n=2$, 3, and 4. It can be seen that any function L_i has a unit value at ξ_i and a zero value at ξ_j, where $j \neq i$.

Lagrange interpolation functions are expressed as a product of n–1 terms:

$$L_i = \frac{\xi-\xi_1}{\xi_i-\xi_1}\frac{\xi-\xi_2}{\xi_i-\xi_2}\cdots\frac{\xi-\xi_{i-1}}{\xi_i-\xi_{i-1}}\frac{\xi-\xi_{i+1}}{\xi_i-\xi_{i+1}}\cdots\frac{\xi-\xi_n}{\xi_i-\xi_n} \tag{13.16}$$

which gives polynomials of order n–1.

Equation 13.16 can be used to derive any of the functions L_i in Figure 13.5, which are linear, quadratic and cubic corresponding to $n=2$, 3 and 4, respectively. It can be verified that, for any n, the sum $(L_1 + L_2 + \ldots + L_n)$ is equal to 1 for any value of ξ.

13.6 COORDINATES OF GRID NODES

Given the (x, y) coordinates of the corners of a quadrilateral (Figure 13.6a), we find the (x, y) coordinates of a grid node k, whose natural coordinates are $(\xi, \eta)_k$. Assume that:

$$L_i = \frac{1}{4}(1+\xi\xi_i)(1+\eta\eta_i) \quad ; \quad i=1, 2, 3, 4 \tag{13.17}$$

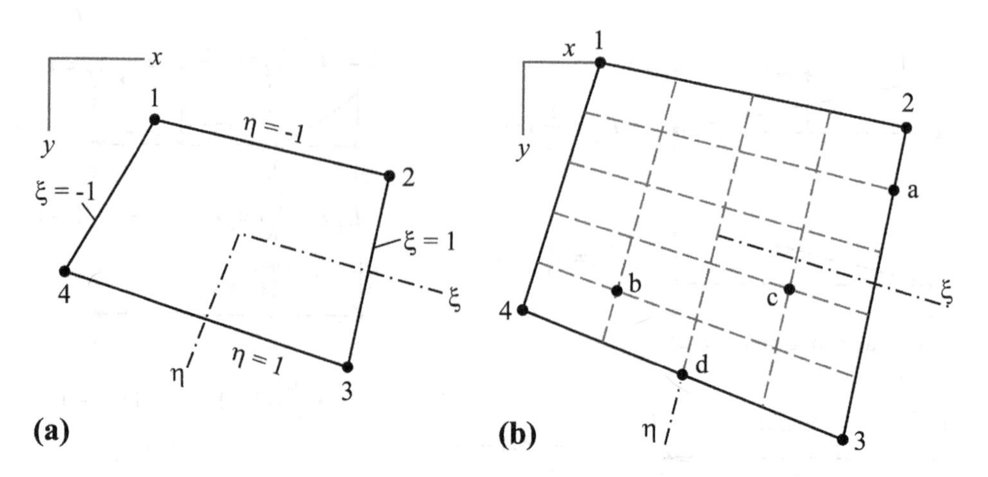

Figure 13.6 Plane grid. (a) Natural coordinates. (b) Grid of Example 13.4. n_ξ and n_η = number of grid lines in ξ and η directions, respectively.

L_i is a polynomial in ξ and η. The subscript i refers to the corners of the quadrilateral (Figure 13.6a). Substituting the values of ξ and η at the four corners in Equation 13.17 gives:

$$[L]=\frac{1}{4}\left[(1-\xi)(1-\eta)\quad(1+\xi)(1-\eta)\quad(1+\xi)(1+\eta)\quad(1-\xi)(1+\eta)\right] \tag{13.18}$$

Assume n_ξ = number of grid lines in ξ-direction; n_η = number of grid lines in η-direction (Figure 13.6b). To determine $(x, y)_k$ at $(n_\xi \times n_\eta)$ nodes, apply Equation 13.19) with $k = 1, 2, \ldots, n_\xi$ and $k = 1, 2, \ldots, n_\eta$:

$$[x_k, y_k]=[L]_k\begin{bmatrix} x_1 & y_1 \\ x_2 & y_2 \\ x_3 & y_3 \\ x_4 & y_4 \end{bmatrix} \tag{13.19}$$

EXAMPLE 13.4: APPLICATION OF LAGRANGE INTERPOLATION: COORDINATES OF GRID NOTES

Find the (x, y) coordinates of nodes a, b, c, and d of the grid in Figure 13.6b. Data: For corner nodes, $(x, y)_1 = (2.5, 0)$; $(x, y)_2 = (12.5, 2.0)$; $(x, y)_3 = (10.5, 12)$; $(x, y)_4 = (0, 8)$; $(\xi, \eta)_a = (1.0, -0.6)$; $(\xi, \eta)_b = (-0.5, 0.6)$; $(\xi, \eta)_c = (0.5, 0.2)$; $(\xi, \eta)_d = (0, 1.0)$.

We apply Equation 13.19 for the four nodes:

$$\begin{bmatrix} x_a & y_a \\ x_b & y_b \\ x_c & y_c \\ x_d & y_d \end{bmatrix}=\frac{1}{4}\begin{bmatrix} 0.0(1.6) & 2.0(1.6) & 2.0(0.4) & 0.0(0.4) \\ 1.5(0.4) & 0.5(0.4) & 0.5(1.6) & 1.5(1.6) \\ 0.5(0.8) & 1.5(0.8) & 1.5(1.2) & 0.5(1.2) \\ 1.0(0.0) & 1.0(0.0) & 1.0(2.0) & 1.0(2.0) \end{bmatrix}\begin{bmatrix} 2.5 & 0.0 \\ 12.5 & 2.0 \\ 10.5 & 12.0 \\ 0.0 & 8.0 \end{bmatrix}=\begin{bmatrix} 12.10 & 4.00 \\ 3.10 & 7.30 \\ 8.73 & 7.80 \\ 5.25 & 10.0 \end{bmatrix}$$

The computer program QMESH can give (x, y) coordinates of all nodes $(n_\xi \times n_\eta = 6 \times 5 = 30)$.

13.7 SHAPE FUNCTIONS FOR TWO- AND THREE-DIMENSIONAL ISOPARAMETRIC ELEMENTS

The use of Lagrange polynomials at the edges and linear interpolation between opposite sides enables us to write directly (by intuition) the shape functions for any corner or intermediate node on the element edges. Figure 13.7 gives the shape functions for an eight-node element; such an element is widely used in plane-stress or plane-strain analysis, because it gives accurate results. In general form, the edges are curved (second-degree parabola) and the coordinates ξ and η are curvilinear. The shape functions given in this figure can be used with equations of Section 13.8 to generate a stiffness matrix and a load vector (without complication caused by the curvilinear coordinates).

The above approach can be extended directly to three-dimensional solid elements with three degrees-of-freedom per node, u, v, and w, in the global x, y, *and* z directions. The shape functions are given in Figure 13.8 for an element with corner nodes and for an element with corner as well as mid-edge nodes. With the addition of mid-edge nodes, the accuracy is increased and the element can have curved edges.

For the three-dimensional isoparametric elements shown in Figure 13.8, the shape functions $L(\xi, \eta, \zeta)$ are used for interpolation of displacements u, v, w and coordinates x, y, z as follows:

$$u = \sum L_i \, u_i \quad ; \quad v = \sum L_i \, v_i \quad ; \quad w = \sum L_i \, w_i \tag{13.20}$$

$$x = \sum L_i \, x_i \quad ; \quad y = \sum L_i \, y_i \quad ; \quad z = \sum L_i \, z_i \tag{13.21}$$

with $i = 1$ to 8 or 1 to 20.

It is of interest to note that, by use of Lagrange polynomials, the sum of the interpolation functions at any point equals unity. This can be verified for the two- and three-dimensional elements discussed above.

13.8 STIFFNESS MATRIX AND LOAD VECTOR OF ISOPARAMETRIC ELEMENTS

Isoparametric elements can be in the form of a curved bar, a triangular or quadrilateral plate with curved edges, or a three-dimensional brick with curved edges. The quadrilateral

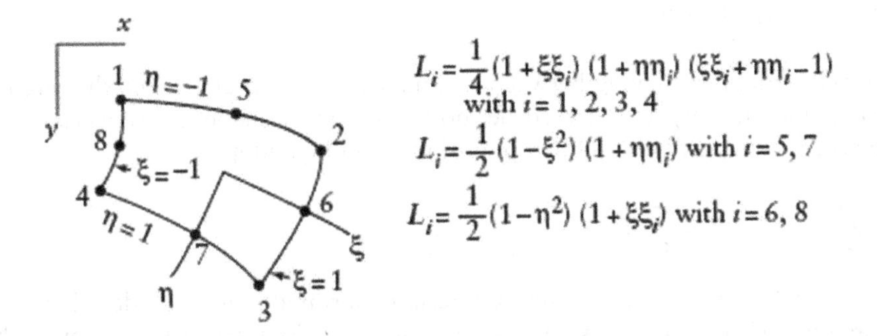

Figure 13.7 Shape functions for a two-dimensional element with corner and mid-side nodes.

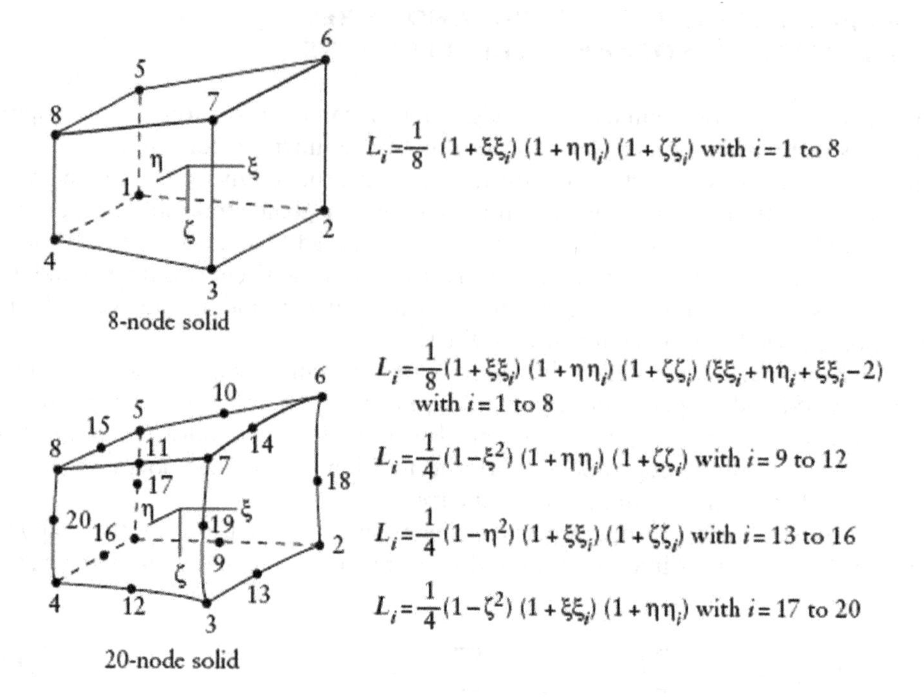

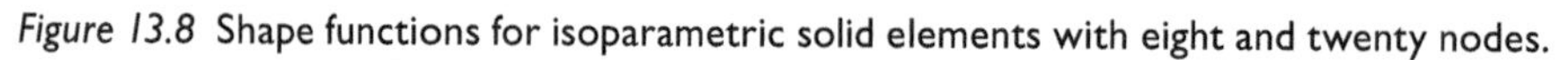

Figure 13.8 Shape functions for isoparametric solid elements with eight and twenty nodes.

plane-stress or plane-strain element in Figure 12.20 is an example of an isoparametric element. The nodal displacements for the element are:

$$\left\{D^*\right\} = \left\{u_1, v_1, u_2, v_2, u_3, v_3, u_4, v_4\right\} \tag{13.22}$$

The displacements u and v at any point are determined from the nodal displacements, using shape functions:

$$u = \sum L_i\, u_i \quad ; \quad v = \sum L_i\, v_i \tag{13.23}$$

where L_i with $i = 1, 2, 3, 4$ represents shape functions of the natural coordinates ξ and η. The shape functions for this element are given by Equation 12.59, which is repeated here:

$$L_i = \frac{1}{4}\left(1 + \xi\xi_i\right)\left(1 + \eta\eta_i\right) \tag{13.24}$$

The natural coordinates varying between −1 and 1 (Figure 12.20) define the relative position of any point with respect to the corner nodes. The same shape functions can be used to determine the (x, y) coordinates of any point in terms of ξ and η:

$$x = \sum L_i\, x_i \quad ; \quad y = \sum L_i\, y_i \tag{13.25}$$

where (x_i, y_i) with $i = 1, 2, 3, 4$ represent Cartesian coordinates at the nodes. The element is called isoparametric because the same interpolation functions are used to express the location of a point and the displacement components in terms of ξ and η.

The strain components at any point involve derivatives $\partial/\partial x$ and $\partial/\partial y$ with respect to the Cartesian coordinates. Derivatives of any variable g with respect to the natural coordinates are obtained by the chain rule:

$$\frac{\partial g}{\partial \xi}=\frac{\partial g}{\partial x}\frac{\partial x}{\partial \xi}+\frac{\partial g}{\partial y}\frac{\partial y}{\partial \xi} \quad ; \quad \frac{\partial g}{\partial \eta}=\frac{\partial g}{\partial x}\frac{\partial x}{\partial \eta}+\frac{\partial g}{\partial y}\frac{\partial y}{\partial \eta} \tag{13.26}$$

We can rewrite Equation 13.26 in a matrix form:

$$\begin{Bmatrix}\partial g/\partial \xi \\ \partial g/\partial \eta\end{Bmatrix}=[J]\begin{Bmatrix}\partial g/\partial x \\ \partial g/\partial y\end{Bmatrix} \quad ; \quad \begin{Bmatrix}\partial g/\partial x \\ \partial g/\partial y\end{Bmatrix}=[J]^{-1}\begin{Bmatrix}\partial g/\partial \xi \\ \partial g/\partial \eta\end{Bmatrix} \tag{13.27}$$

Here, $[J]$ is the *Jacobian matrix*; it serves to transform the derivatives of any variable with respect to ξ and η into derivatives with respect to x and y, and vice versa. The elements of $[J]$ are given by:

$$[J]=\begin{bmatrix}\partial x/\partial \xi & \partial y/\partial \xi \\ \partial x/\partial \eta & \partial y/\partial \eta\end{bmatrix} \tag{13.28}$$

When Equation 13.25 applies, the Jacobian can be expressed as:

$$[J]=\sum\begin{bmatrix}x_i\,(\partial L_i/\partial \xi) & y_i\,(\partial L_i/\partial \xi) \\ x_i\,(\partial L_i/\partial \eta) & y_i\,(\partial L_i/\partial \eta)\end{bmatrix} \tag{13.29}$$

The element geometry is defined by the Cartesian coordinates (x, y) at the four nodes. For a given quadrilateral, the Jacobian varies, with ξ and η defining the position of any point.

The strains at any point are given by $\{\varepsilon\}=[\partial]\{u, v\}$; the derivative operator $[\partial]$ is defined by Equation 7.27. Substitution of Equation 13.23 in 12.61 gives:

$$\{\varepsilon\}=\sum_{i=1}^{n}\left\{[B]_i\begin{Bmatrix}u_i \\ v_i\end{Bmatrix}\right\} \tag{13.30}$$

where n is the number of nodes; for the quadrilateral element in Figure 12.20, $n=4$. Here:

$$[B]_i=\begin{bmatrix}\dfrac{\partial L_i}{\partial x} & 0 \\ 0 & \dfrac{\partial L_i}{\partial y} \\ \dfrac{\partial L_i}{\partial y} & \dfrac{\partial L_i}{\partial x}\end{bmatrix} \quad ; \quad [B]=\left[[B]_1 \quad [B]_2 \quad \cdots \quad [B]_n\right] \tag{13.31}$$

When $[J]$ and its inverse are determined at a particular point, numerical values of the derivatives $(\partial L_i/\partial x)$ and $(\partial L_i/\partial y)$, required for generating $[B]_i$, are calculated from $\partial L_i/\partial \xi$ and $\partial L_i/\partial \eta$ via Equation 13.27. The integrals involved in the generation of the element stiffness matrix and of the consistent load vector are obtained numerically.

The element stress matrix relating the stresses at any point to nodal displacements is given by Equation 12.65. The element stiffness matrix is given by (Equation 12.68):

$$\left[S^*\right]=h\int_{-1}^{1}\int_{-1}^{1}[B]^{\mathrm{T}}\,[d]\,[B]\,|J|\,\mathrm{d}\xi\,\mathrm{d}\eta \tag{13.32}$$

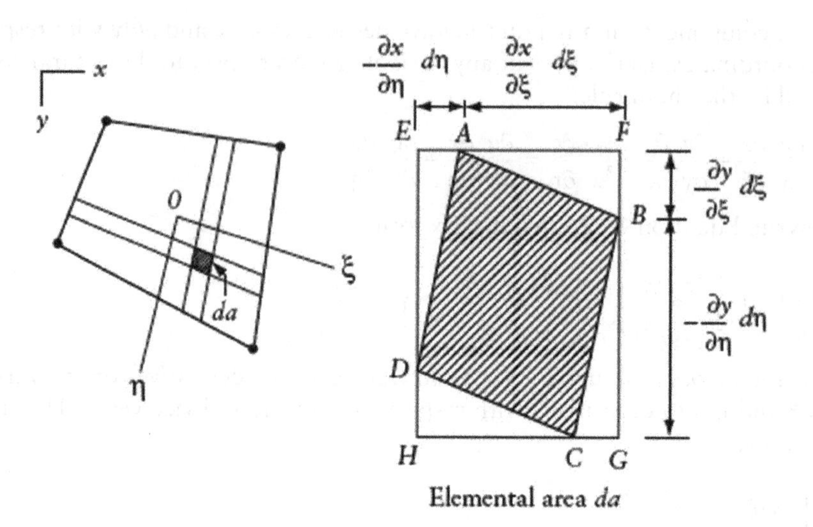

Figure 13.9 Verification of Equation 13.32.

where h is the element thickness, which is assumed constant, and:

$$|J|\,\mathrm{d}\xi\,\mathrm{d}\eta = \mathrm{d}a \tag{13.33}$$

Here, da is the elemental area (parallelogram) shown in Figure 13.9. The determinant of the Jacobian serves as a scaling factor (length^2) to transform the dimensionless product dξ dη into an elemental area. The validity of Equation 13.33 can be verified by considering the following areas:

Parallelogram ABCD = rectangle EFGH – 2(triangle EAD + triangle AFB)

If the element is subjected to a rise in temperature of T degrees and the expansion is restrained, the stress $\{\sigma_r\}$ will be the same as given by Equation 12.71, and the consistent vector of restraining forces will be given by (Equation 12.73):

$$\{F_b^*\} = -h\int_{-1}^{1}\int_{-1}^{1}[B]^{\mathrm{T}}\,[d]\,[\varepsilon_0]\,|J|\,\mathrm{d}\xi\,\mathrm{d}\eta \tag{13.34}$$

where $\{\varepsilon_0\}$ represents the change in strains if the expansion is free to occur (Equation 12.72).

For a quadrilateral element, the determinant $|J|$ at the center of the element is equal to one-quarter of its area. For a rectangle, $[J]$ is constant. In general, the determinant $|J|$ is positive, the matrix $[J]$ is nonsingular, and the matrix inversion in Equation 13.27 is possible. If two corners of the quadrilateral coincide, the element becomes triangular. At the point where they coincide, $[J]$ is singular, indicating that the strains cannot be determined at this point. When the internal angle at each of the four corners of a quadrilateral is smaller than 180°, $[J]$ is nonsingular at all points, and its determinant is positive.

Equations 13.23 and 13.25–13.34 are general for any isoparametric plane-stress or plane-strain element with two degrees-of-freedom per node. The shape functions $[L]$ differ depending on the number of nodes and the arrangement of the nodes in each element. Methods of derivation of shape functions for various isoparametric elements are given in Sections 13.5 and 13.7.

EXAMPLE 13.5: QUADRILATERAL ELEMENT: JACOBIAN MATRIX AT CENTER

Determine the Jacobian matrix at the center of the quadrilateral element shown in Figure 12.20, using the following (x, y) coordinates: node 1 (0, 0); node 2 (7, 1); node 3 (6, 9); node 4 (–2, 5). Use the determinant of the Jacobian to calculate the area of the quadrilateral.

The derivatives of the shape function in Equation 13.24 at $\xi=\eta=0$ are:

$$\frac{\partial L_i}{\partial \xi}=\frac{\xi_i}{4} \quad ; \quad \frac{\partial L_i}{\partial \eta}=\frac{\eta_i}{4}$$

The Jacobian is (Equation 13.29):

$$[J]=\begin{bmatrix} 0\left(\frac{-1}{4}\right) & 0\left(\frac{-1}{4}\right) \\ 0\left(\frac{-1}{4}\right) & 0\left(\frac{-1}{4}\right) \end{bmatrix}+\begin{bmatrix} 7\left(\frac{1}{4}\right) & 1\left(\frac{1}{4}\right) \\ 7\left(\frac{-1}{4}\right) & 1\left(\frac{-1}{4}\right) \end{bmatrix}$$

$$+\begin{bmatrix} 6\left(\frac{1}{4}\right) & 9\left(\frac{1}{4}\right) \\ 6\left(\frac{1}{4}\right) & 9\left(\frac{1}{4}\right) \end{bmatrix}+\begin{bmatrix} -2\left(\frac{-1}{4}\right) & 5\left(\frac{-1}{4}\right) \\ -2\left(\frac{1}{4}\right) & 5\left(\frac{1}{4}\right) \end{bmatrix}$$

$$=\begin{bmatrix} \frac{15}{4} & \frac{5}{4} \\ -\frac{3}{4} & \frac{13}{4} \end{bmatrix}$$

Area $=4\,|J|=4(13.125)=52.5.$

13.9 CONSISTENT LOAD VECTORS FOR RECTANGULAR PLANE ELEMENT

Equations 12.70 and 12.73 can be used to determine the consistent load vectors due to body forces or due to temperature variation for the plane-stress or plane-strain element shown in Figure 13.7. Consider a special case when the element is a rectangle subjected to a uniform load of q_x per unit volume (Figure 13.10). The consistent vector of restraining forces is (Equation 12.70):

$$\{F_b^*\}=-\frac{hbc}{4}\int_{-1}^{1}\int_{-1}^{1}\begin{bmatrix} L_1 & 0 & L_2 & 0 & \cdots & L_8 & 0 \\ 0 & L_1 & 0 & L_2 & \cdots & 0 & L_8 \end{bmatrix}\begin{Bmatrix} q_x \\ 0 \end{Bmatrix} d\xi\, d\eta \tag{13.35}$$

where b and c are the element sides and h is its thickness; L_1 to L_8 are given in Figure 13.7. The integrals in Equation 13.35 give the consistent nodal forces shown in Figure 13.10.

Note the unexpected result: the forces at corner nodes are in the opposite direction to the forces at mid-side nodes. But, as expected, the sum of the forces at the eight nodes is equal to $q_x hbc$. Apportionment of the load on the nodal forces intuitively may be sufficient when the

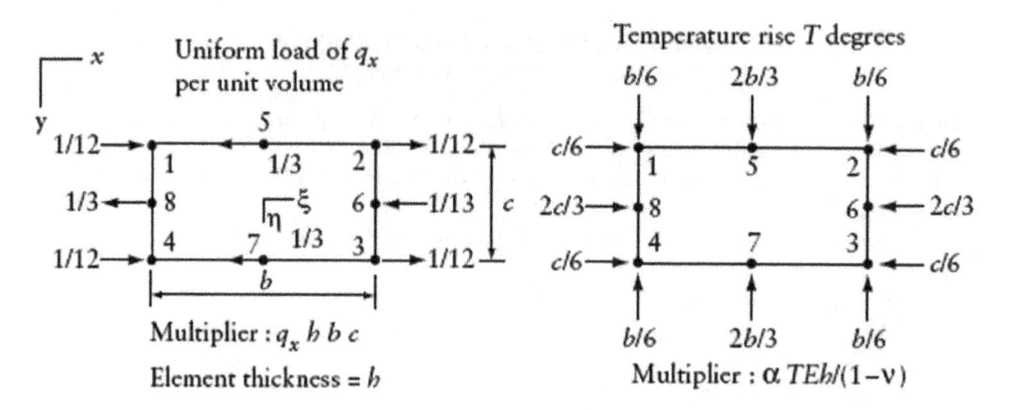

Figure 13.10 Rectangular plane-stress element. Consistent nodal forces due to a uniform load q_x and due to a rise in temperature of T degrees, with the nodal displacements prevented.

finite-element mesh is fine, but greater accuracy is obtained using the consistent load vector, particularly when the mesh is coarse.

Let the element in Figure 13.10 be subjected to a temperature rise of T degrees; if the expansion is restrained, the stresses are:

$$\{\sigma\}_r = -\left[E\alpha T/(1-\nu)\right]\{1, 1, 0\} \tag{13.36}$$

where E, ν and α are modulus of elasticity, Poisson's ratio and coefficient of thermal expansion of the material (assumed isotropic). The corresponding consistent nodal forces can be determined by substitution of Equations 13.31, 13.36, and 12.71 into 12.73, using the shape functions in Figure 13.7. The resulting nodal forces are included in Figure 13.10.

The same results can be obtained by integration, over the length of the edges, of the product of σ_{rx} or σ_{ry} and the shape functions, with the result multiplied by the thickness h. For example, the force at node 8 is:

$$F_8 = \sigma_{rx}\, h \int_{-\frac{c}{2}}^{\frac{c}{2}} (L_8)_{\xi=-1}\, dy = \sigma_{rx} \frac{hc}{2} \int_{-1}^{1} \left(1-\eta^2\right) d\eta = \frac{2}{3}\sigma_{rx}\, hc \tag{13.37}$$

Here again, the consistent load vector cannot be generated by intuition.

13.10 CONSTANT-STRAIN TRIANGLE

The triangular element shown in Figure 13.11 may be used in a plane-stress or plane-strain analysis. The element has three nodes at the corners, with two nodal displacements u and v. The element is called a constant-strain triangle because the strain, and hence the stress, within the element is constant. Figure 13.11b shows the shape of the element with $u_i = 1$ with zero displacement at other coordinates.

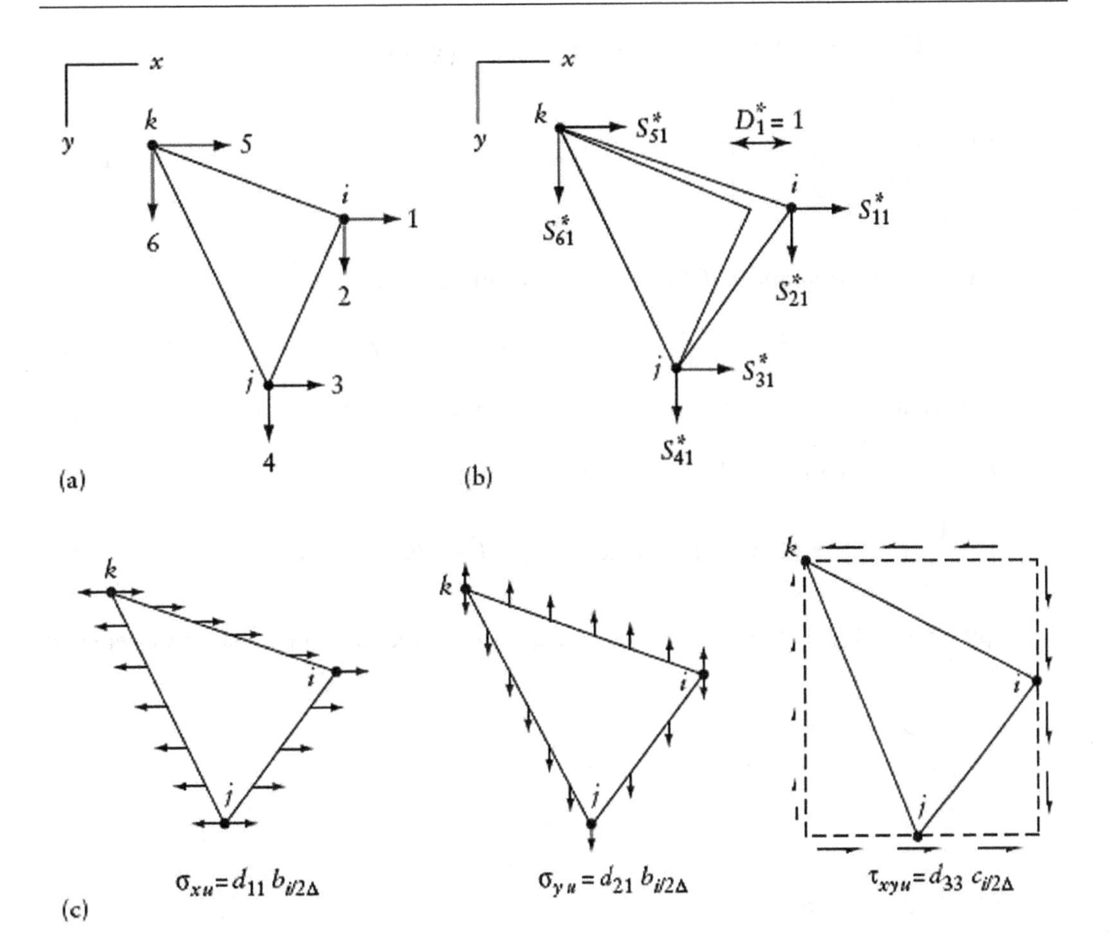

Figure 13.11 Triangular plane-stress or plane-strain element with $D_1^* = 1$ and $D^* = 0$ at other nodal coordinates. (a) Nodal coordinates. (b) and (c) Stresses on element edges lumped in equivalent forces at the nodes.

Each of u and v is associated with three of the six nodal displacements. The same polynomial may therefore be used for the two variables:

$$[P] = [1, x, y] \tag{13.38}$$

We shall now derive the shape functions associated with u, which are the same as the shape functions associated with v. At the three nodes we have:

$$\begin{Bmatrix} u_i \\ u_j \\ u_k \end{Bmatrix} = \begin{bmatrix} 1 & x_i & y_i \\ 1 & x_j & y_j \\ 1 & x_k & y_k \end{bmatrix} \{A\} \tag{13.39}$$

or

$$\{u\} = [C]\ \{A\} \tag{13.40}$$

Inversion of [C], using Cramer's rule or other methods, gives:

$$[C]^{-1} = \frac{1}{2\Delta}\begin{bmatrix} a_i & a_j & a_k \\ b_i & b_j & b_k \\ c_i & c_j & c_k \end{bmatrix} \tag{13.41}$$

where 2Δ is the determinant of $[C] = 2 \times$ area of triangle. Here:

$$a_i = x_j\, y_k - y_j\, x_k \quad ; \quad b_i = y_j - y_k \quad ; \quad c_i = x_k - x_j \tag{13.42}$$

By cyclic permutation of the subscripts i, j, and k, similar equations can be written for $\{a_j, b_j, c_j\}$ and for $\{a_k, b_k, c_k\}$.

Substitution in Equation 13.4 gives the shape functions:

$$[L] = \frac{1}{2\Delta}\left[(a_i + b_i\, x + c_i\, y) \quad (a_j + b_j\, x + c_j\, y) \quad (a_k + b_k\, x + c_k\, y) \right] \tag{13.43}$$

The displacements $\{u, v\}$ at any point may be expressed in terms of the nodal displacements (Equation 12.52):

$$\begin{Bmatrix} u \\ v \end{Bmatrix} = \begin{bmatrix} L_1 & 0 & L_2 & 0 & L_3 & 0 \\ 0 & L_1 & 0 & L_2 & 0 & L_3 \end{bmatrix} \{D^*\} \tag{13.44}$$

where

$$\{D^*\} = \{u_i, v_i, u_j, v_j, u_k, v_k\}$$

The matrix $[B]$, relating strains $\{\varepsilon\}$ to $\{D^*\}$ is (Equations 12.62 and 7.27):

$$[B] = \frac{1}{2\Delta}\begin{bmatrix} b_i & 0 & b_j & 0 & b_k & 0 \\ 0 & c_i & 0 & c_j & 0 & c_k \\ c_i & b_i & c_j & b_j & c_k & b_k \end{bmatrix} \tag{13.45}$$

The product $[d]$ $[B]$ gives the stresses due to unit nodal displacements (Equation 12.65):

$$[\sigma_u] = \frac{1}{2\Delta}\begin{bmatrix} d_{11}\, b_i & d_{21}\, c_i & d_{11}\, b_j & d_{21}\, c_j & d_{11}\, b_k & d_{21}\, c_k \\ d_{21}\, b_i & d_{22}\, c_i & d_{21}\, b_j & d_{22}\, c_j & d_{21}\, b_k & d_{22}\, c_k \\ d_{33}\, c_i & d_{33}\, b_i & d_{33}\, c_j & d_{33}\, b_j & d_{33}\, c_k & d_{33}\, c_k \end{bmatrix} \tag{13.46}$$

where d_{ij} are elements of the elasticity matrix $[d]$, given by Equations 7.28 and 7.29 for the states of plane-stress and plane-strain, respectively. It can be seen that all the elements of $[B]$ and $[\sigma_u]$ are constants, indicating constant strain and stress. If the thickness h is constant, the stiffness matrix of the element is (Equation 12.68):

$$[S^*] = h\Delta\, [B]^{\mathrm{T}}\, [d]\, [B] = h\, \Delta\, [B]^{\mathrm{T}}\, [\sigma_u] \tag{13.47}$$

that is:

$$
[S^*] = \frac{h}{4\Delta}\begin{bmatrix}
d_{11}b_i^2 + d_{33}c_i^2 & & & & & \text{Symmetrical} \\
d_{21}b_ic_i + d_{33}b_ic_i & d_{22}c_i^2 + d_{33}b_i^2 & & & & \\
d_{11}b_ib_j + d_{33}c_ic_j & d_{21}c_ib_j + d_{33}c_jb_i & d_{11}b_j^2 + d_{33}c_j^2 & & & \\
d_{21}b_ic_j + d_{33}b_jc_i & d_{22}c_ic_j + d_{33}b_ib_j & d_{21}b_jc_j + d_{33}c_jb_j & d_{22}c_j^2 + d_{33}b_j^2 & & \\
d_{11}b_ib_k + d_{33}c_ic_k & d_{21}c_ib_k + d_{33}b_ic_k & d_{11}b_jb_k + d_{33}c_jc_k & d_{21}c_jb_k + d_{33}b_jc_k & d_{11}b_k^2 + d_{33}c_k^2 & \\
d_{21}b_ic_k + d_{33}b_kc_i & d_{22}c_ic_k + d_{33}b_ib_k & d_{21}b_jc_k + d_{33}c_jb_k & d_{22}c_jc_k + d_{33}b_jb_k & d_{21}b_kc_k + d_{33}b_kc_k & d_{22}c_k^2 + d_{33}b_k^2
\end{bmatrix} \tag{13.48}
$$

If the element is subjected to uniform body forces with intensities per unit volume of $\{p\}=\{q_x, q_y\}$, the equilibrants at the nodes, when the nodal displacements are prevented, are (Equation 12.70):

$$
\{F_b^*\} = -h\iint \begin{bmatrix} L_1 & 0 & L_2 & 0 & L_3 & 0 \\ 0 & L_1 & 0 & L_2 & 0 & L_3 \end{bmatrix}^T \begin{Bmatrix} q_x \\ q_y \end{Bmatrix} \mathrm{d}x\,\mathrm{d}y \tag{13.49}
$$

Evaluation of the integrals is simplified by noting that $\iint \mathrm{d}x\ \mathrm{d}y=\Delta$; $\iint x\mathrm{d}x\ \mathrm{d}y$ is the first moment of area about the y-axis, so that $\iint y\mathrm{d}x\ \mathrm{d}y=(\Delta/3)(y_i+y_j+y_k)$. The consistent vector of forces in equilibrium with the body forces thus becomes:

$$
\{F_b^*\} = \frac{\Delta h}{3}\{q_x, q_y, q_x, q_y, q_x, q_y\} \tag{13.50}
$$

This means that one-third of the load on the triangle is assigned to each node. The same distribution could have been suggested intuitively. (However, it is not always possible to determine $\{F_b^*\}$ intuitively; see Figure 13.10.

The consistent vector of restraining forces when the element is subjected to a rise in temperature of T degrees is obtained by substitution of Equations 12.72 and 13.45 into Equation 12.73:

$$
\{F_b\}_m = -\frac{\alpha T h}{2}(d_{11}+d_{21})\{b_i, c_i, b_j, c_j, b_k, c_k\} \tag{13.51}
$$

Here, we have assumed that the material is isotropic ($d_{22}=d_{11}$).

13.11 INTERPRETATION OF NODAL FORCES

In the derivation of the stiffness matrix of individual elements, the forces distributed along the edges of the elements are replaced by equivalent forces lumped at the nodes. The equivalent forces are determined by the use of the principle of virtual work.

This approach can be seen in the example of the constant-strain triangular element shown in Figures 13.11b and c. For a unit nodal displacement, say, $D_1^* = 1$ while other nodal displacements are zero, the stresses are (Equation 13.46, first column of $[\sigma_u]$):

$$\{\sigma_{u1}\} = \frac{1}{2\Delta}\{d_{11}\, b_i,\, d_{21}\, b_i,\, d_{33}\, c_i\} \tag{13.52}$$

where d_{ij} are elements of the elasticity matrix $[d]$ given in Equations 7.28 and 7.29 for plane-stress or plane-strain states, respectively; and Δ = area of the triangle. The stresses are represented by uniform forces on the edges of the element in Figure 13.11c.

The distributed forces in Figure 13.11c are lumped at the nodes in Figure 13.11b. The force in the x or y direction at a node is equal to the sum of one-half of the distributed load on each of the two sides connected to the node. For example, the horizontal force at node k is one-half of the load on edges ki and kj; thus:

$$S_{51}^* = h\left[\frac{\sigma_x}{2}\left[(y_j - y_k) - (y_i - y_k)\right] + \frac{\tau_{xy}}{2}\left[(x_j - x_k) - (x_i - x_k)\right]\right] \tag{13.53}$$

We can now verify that this is the same as element S_{51}^* of the stiffness matrix in Equation 13.48. Any other element of the stiffness matrix can be verified in a similar way.

Equation 13.53 means that the virtual work of force S_{51}^* during nodal displacement $D_1^* = 1$ is the same as the work done by the forces distributed over edges ki and kj during their corresponding virtual displacements (which, in this example, vary linearly).

The consistent load vector also represents the distributed forces lumped at the nodes. If the element shown in Figure 13.11b is subjected to a rise in temperature of T degrees and the element expansion is restrained, the stresses in an isotropic material will be (Equations 12.71 and 12.72):

$$\{\sigma_r\} = -[d]\{\varepsilon_0\} = -\alpha T\left(d_{11} + d_{21}\right)\begin{Bmatrix}1\\1\\0\end{Bmatrix} \tag{13.54}$$

where α is the coefficient of thermal expansion.

Under the conditions of restraint, uniform forces act on the edges so as to produce $\sigma_x = \sigma_y = -\alpha T(d_{11} + d_{21})$. Lumping one-half of the distributed load on any edge at the two end nodes gives the nodal forces of Equation13.51.

13.12 TRIANGULAR PLANE-STRESS AND PLANE-STRAIN ELEMENTS

Consider the triangular element 1-2-3 shown in Figure 13.12a, which has an area Δ. Joining any point (x, y) inside the triangle to its three corners divides Δ into three areas, A_1, A_2, and A_3. The position of the point can be defined by the dimensionless parameters α_1, α_2, and α_3, where:

$$\alpha_1 = A_1/\Delta \quad ; \quad \alpha_2 = A_2/\Delta \quad ; \quad \alpha_3 = A_3/\Delta \tag{13.55}$$

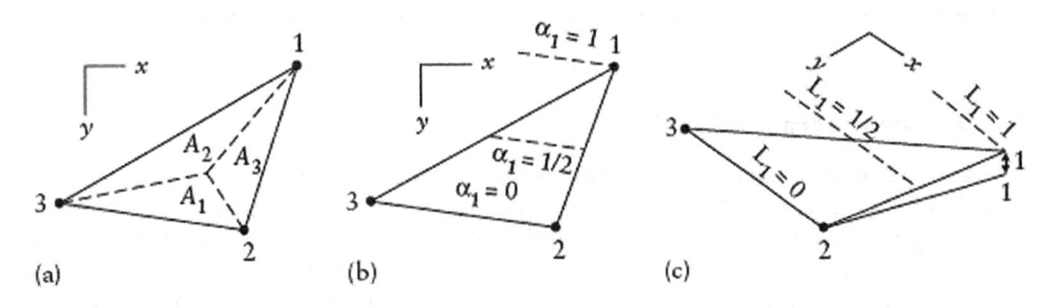

Figure 13.12 Constant-strain triangle. (a) Area coordinates $\{\alpha_1, \alpha_2, \alpha_3\} = (1/\Delta)\ \{A_1, A_2, A_3\}$. (b) Lines of equal α_1. (c) Pictorial view of shape function $L_1 = \alpha_1$.

and

$$\alpha_1 + \alpha_2 + \alpha_3 = 1 \tag{13.56}$$

The parameters α_1, α_2, and α_3 are called *area coordinates* or *areal coordinates*. Any two of them are sufficient to define the position of a point within the element. Lines parallel to the sides of the triangle are lines of equal α (Figure 13.12b).

Figure 13.12c is a pictorial view of the shape function L_1 for the constant-strain triangle discussed in Section 13.10. The element has three corner nodes with degrees-of-freedom u and v per node, representing translations in the x and y directions. The ordinates L_1, which are plotted perpendicular to the surface of the triangle, represent the variation in u or v when the nodal displacement u_1 or v_1 equals unity. If we plot lines of equal L_1, they will be identical to the lines of equal α_1, indicating that α_1 is equal to L_1. Thus, the shape functions α_1, α_2, and α_3 can serve as shape functions L_1, L_2, and L_3 for the constant-strain triangle.

Therefore, we can express the displacement at any point by:

$$u = \sum L_i\, u_i \quad ; \quad v = \sum L_i\, v_i \tag{13.57}$$

It can be shown (by linear interpolation) that:

$$x = \sum L_i\, x_i \quad ; \quad y = \sum L_i\, y_i \tag{13.58}$$

Equations 13.57 and 13.58 are the same as Equations 13.23 and 13.25, indicating that the constant-strain triangle is an isoparametric element in which the area coordinates serve as shape functions:

$$L_1 = \alpha_1 \quad ; \quad L_2 = \alpha_2 \quad ; \quad L_3 = \alpha_3 \tag{13.59}$$

For a triangle with *straight* edges (Figure 13.12a), the area coordinates can be expressed in terms of x and y by combining Equations 13.56, 13.58, and 13.59, giving:

$$\{\alpha_1, \alpha_2, \alpha_3\} = \frac{1}{2\Delta}\{a_1 + b_1\, x + c_1\, y, \quad a_2 + b_2\, x + c_2\, y, \quad a_3 + b_3\, x + c_3\, y\} \tag{13.60}$$

where

$$a_1 = x_2\, y_3 - y_2\, x_3 \quad ; \quad b_1 = y_2 - y_3 \quad ; \quad c_1 = x_3 - x_2 \tag{13.61}$$

By cyclic permutation of the subscripts 1, 2, and 3, similar equations can be written for a_2, b_2, c_2 and a_3, b_3, c_3.

13.12.1 Linear-strain triangle

An isoparametric triangular element with corner and mid-side nodes is shown in Figure 13.13a. This element is widely used in practice for plane-stress or plane-strain analysis because it gives accurate results. The nodal displacements are u_i and v_i with $i = 1, 2, \ldots, 6$. Equations 13.57 and 13.58 apply to this element, with the shape functions L_1 to L_6 derived by superposition for various shapes, in a way similar to that used for the quadrilateral element in Figure 13.7.

The shape functions for the mid-side nodes in Figure 13.13a are:

$$L_4 = 4\alpha_1\alpha_2 \quad ; \quad L_5 = 4\alpha_2\alpha_3 \quad ; \quad L_6 = 4\alpha_3\alpha_1 \tag{13.62}$$

A pictorial view of L_4, plotted perpendicular to the plane of the element, is shown in Figure 13.13b. The function is quadratic along edge 1-2, but linear along lines α_1 or α_2 = constant.

The shape functions for corner nodes are obtained by an appropriate combination of L_4, L_5, and L_6 with the shape functions for the three-node element. For example, combining the shape function of Figure 13.12 with $-L_4/2$ and $-L_6/2$ gives shape function L_1 for the six-node element. The function obtained in this way has a unit value at 1 and a zero value at all other nodes, as it should. The shape functions for the three corner nodes for the element in Figure 13.13a are:

$$L_1 = \alpha_1(2\alpha_1 - 1) \quad ; \quad L_2 = \alpha_2(2\alpha_2 - 1) \quad ; \quad L_3 = \alpha_3(2\alpha_3 - 1) \tag{13.63}$$

It can be seen from Equations 13.62 and 13.63 that the shape functions are quadratic, but the first derivatives, which give the strains, are linear. The six-node triangular element is sometimes referred to as a *linear-strain* triangle.

The six-node element can have curved sides. The x, y coordinates of any point are given by Equation 13.58, using the shape functions in Equations 13.62 and 13.63. The lines of equal a will not be straight in this case, and Equation 13.60 does not apply.

The stiffness matrix and the consistent load vectors for the six-node isoparametric triangular element can be generated using the equations given in Section 13.8 for a quadrilateral element, replacing ξ and η by α_1 and α_2, respectively. The third parameter α_3 is eliminated from the shape functions by substituting $\alpha_3 = 1 - \alpha_1 - \alpha_2$.

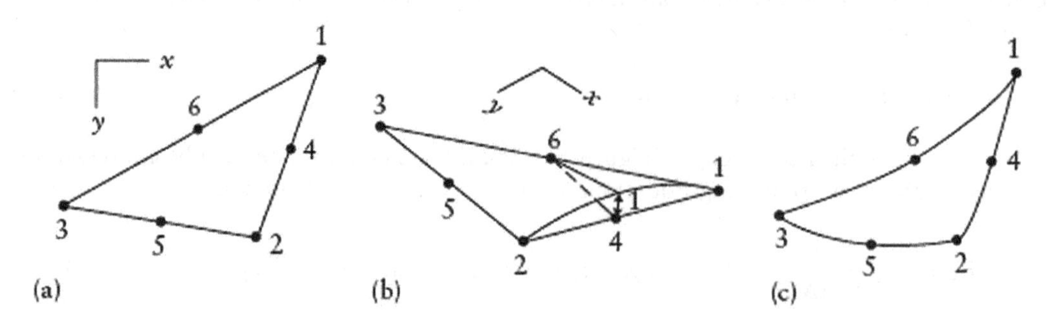

Figure 13.13 Triangular isoparametric element for plane-stress or plane-strain analysis. Two degrees-of-freedom per node: u and v in x and y directions. (a) Node numbering. (b) Pictorial view of shape function L_4. (c) General quadratic triangle.

The integrals involved in the derivation of the matrices of triangular elements can be put in the form:

$$\int_a f(\alpha_1, \alpha_2)\, da = \int_0^1 \int_0^{1-\alpha_1} f(\alpha_1, \alpha_2)\, |J|\, d\alpha_2\, d\alpha_1 \tag{13.64}$$

where $f(\alpha_1, \alpha_2)$ is a function of α_1 and α_2, and $|J|$ is the determinant of the Jacobian. The inner integral on the right-hand side represents the value of the integral over a line of constant α_1; over such a line, the parameter α_2 varies between 0 and $1-\alpha_1$.

The Jacobian can be determined by Equation 13.29, replacing ξ and η as mentioned above. We can verify that, when the sides of the triangle are straight, $[J]$ is constant and its determinant is equal to twice the area.

The integral on the right-hand side of Equation 13.64 is commonly evaluated numerically, using Equation 13.75. However, in the case of a triangle with *straight* edges, the determinant $|J|$ can be taken out of the integral and the following closed-form equation can be used to evaluate the functions in the form:

$$\int_a \alpha_1^{n_1}\, \alpha_2^{n_2}\, \alpha_3^{n_3}\, da = 2\Delta \frac{n_1!\, n_2!\, n_3!}{(n_1 + n_2 + n_3 + 2)!} \tag{13.65}$$

where n_1, n_2, and n_3 are any integers; the symbol "!" indicates a factorial.

13.13 TRIANGULAR PLATE-BENDING ELEMENTS

A desirable triangular element which can be used for the analysis of plates in bending and for shell structures (when combined with plane-stress elements) should have three degrees-of-freedom per node, namely w, θ_x, and θ_y, representing deflection in a global z direction and two rotations θ_x and θ_y (Figure 13.14a). With nine degrees-of-freedom, the deflection w should be expressed by a polynomial with nine terms. Pascal's triangle (Figure 13.1) indicates that a complete quadratic polynomial has six terms and a cubic polynomial has ten. Leaving out one of the terms x^2y or xy^2 gives an element for which the outcome is dependent upon the choice of x and y axes.

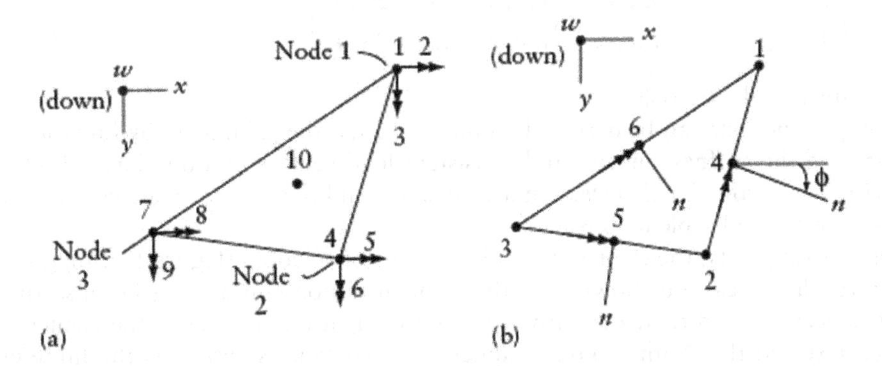

Figure 13.14 Triangular bending elements. (a) Number of degrees-of-freedom = 10: w, θ_x, θ_y at each corner and w at centroid. (b) Number of degrees-of-freedom = 6: w at corners and $\partial w/\partial n$ at mid-side nodes.

It is possible to express w using area coordinates and a polynomial of nine terms (which are not a simple selection). This gives a non-conforming element[2] which is widely used in practice because its results are accurate.

The two triangular elements, shown in Figures 13.14b and a, have six and ten degrees-of-freedom, respectively. The deflection w for the two elements can be represented by complete quadratic and cubic polynomials, respectively, containing the top six or ten terms of Pascal's triangle (Figure 13.1).

The six degrees-of-freedom for the element in Figure 13.14b. are:

$$\{D^*\} = \left\{ w_1, w_2, w_3, \left(\frac{\partial w}{\partial n}\right)_4, \left(\frac{\partial w}{\partial n}\right)_5, \left(\frac{\partial w}{\partial n}\right)_6 \right\} \tag{13.66}$$

where $\partial w/\partial n$ is the slope of a tangent normal to the sides half-way between the nodes. The positive direction of the normal vector is pointing either inwards or outwards, so that the angle ϕ, measured in the clockwise direction from the global x axis to the normal, is smaller than π. Thus, for the triangle in Figure 13.14b, the normal at node 6 points inwards. However, for an adjacent triangle (not shown) sharing side 1-3, the normal points outwards; the two triangles share the same degree-of-freedom $\partial w/\partial n$, which is indicated in a positive direction (Figure 13.14b). The derivative $\partial w/\partial n$ can be expressed as:

$$\frac{\partial w}{\partial n} = \frac{\partial w}{\partial x}\cos\phi + \frac{\partial w}{\partial y}\sin\phi \tag{13.67}$$

The deflection is expressed by the upper six terms of Pascal's triangle (Figure 13.1):

$$w = \begin{bmatrix} 1 & x & y & x^2 & xy & y^2 \end{bmatrix}\{A\} = [P]\{A\} \tag{13.68}$$

Substituting for x and y by their values at the nodes, the nodal displacements can be expressed by $[D^*] = [C]\{A\}$, with:

$$[C] = \begin{bmatrix} 1 & x_1 & y_1 & x_1^2 & x_1 y_1 & y_1^2 \\ 1 & x_2 & y_2 & x_2^2 & x_2 y_2 & y_2^2 \\ 1 & x_3 & y_3 & x_3^2 & x_3 y_3 & y_3^2 \\ 0 & c_4 & s_4 & 2x_4 c_4 & y_4 c_4 + x_4 s_4 & 2y_4 s_4 \\ 0 & c_5 & s_5 & 2x_5 c_5 & y_5 c_5 + x_5 s_5 & 2y_5 s_5 \\ 0 & c_6 & s_6 & 2x_6 c_6 & y_6 c_6 + x_6 s_6 & 2y_6 s_6 \end{bmatrix} \tag{13.69}$$

where $s_i = \sin\phi_i$ and $c_i = \cos\phi_i$.

The shape functions can be derived by Equation 13.4, the $[B]$ matrix by Equations 12.62 and 7.35, and the stiffness matrix and consistent load vectors by Equations 12.62, 12.70, and 12.73. The strains for this element are constant and the element matrices are relatively simple to derive (see Problem 13.6).

Despite its simplicity, the element with six degrees-of-freedom (Figure 13.14b) gives fairly accurate results. It can be shown that the element is non-conforming because the inter-element deflections and their derivatives are not compatible. However, the element passes the patch test and the results converge quickly to the exact solution as the finite-element mesh is refined.[3]

The element in Figure 13.14a has ten degrees-of-freedom $\{D^*\}$, as shown. The tenth displacement is a downward deflection at the centroid. The deflection can be expressed as

$w = [L]\{D^*\}$, where $[L]$ is composed of ten shape functions given in terms of area coordinates α_1, α_2, and α_3 (Equation 13.55) as follows:

$$L_1 = \alpha_1^2 (\alpha_1 + 3\alpha_2 + 3\alpha_3) - 7\alpha_1 \alpha_2 \alpha_3$$

$$L_2 = \alpha_1^2 (b_2 \alpha_3 - b_3 \alpha_2) + (b_3 - b_2)\alpha_1 \alpha_2 \alpha_3$$

$$L_3 = \alpha_1^2 (c_2 \alpha_3 - c_3 \alpha_2) + (c_3 - c_2)\alpha_1 \alpha_2 \alpha_3 \tag{13.70}$$

$$L_{10} = 27\alpha_1 \alpha_2 \alpha_3$$

where b_i and c_i are defined by Equation 13.61. The equation for L_1 can be used for L_4 and L_7 by cyclic permutation of the subscripts 1, 2, and 3. Similarly, the equation for L_2 can be used for L_5 and L_8, and the equation for L_3 can be used for L_6 and L_9. We can verify that the ten shape functions and their derivatives take unit or zero values at the nodes, as they should. This involves the derivatives $\partial\alpha_i/\partial x$ and $\partial\alpha_i/\partial y$ which are, respectively, equal to $b_i/2\Delta$ and $c_i/2\Delta$, where Δ is the area of the triangle (see Equation 13.60).

The inter-element displacement compatibility for $\partial w/\partial n$ is violated with the shape functions in Equation 13.70, but the values of w of two adjacent elements are identical along a common side. This element does not give good accuracy.

However, the accuracy of the element shown in Figure 13.14a can be improved by imposing constraints in the solution of the equilibrium equations of the assembled structure. The constraints specify continuity of normal slopes $\partial w/\partial n$ at mid-points of element sides. With such treatment, the accuracy of this element will be better than that of the element shown in Figure 13.14b.

13.14 NUMERICAL INTEGRATION

The Gauss method of integration is used extensively in generating the matrices of finite-elements. The method is given below without derivation[4] for one-, two-, and three-dimensional integration.

A definite integral of a one-dimensional function $g(\xi)$ (Figure 13.15) can be calculated by:

$$\int_{-1}^{1} g(\xi)\, d\xi \cong \sum_{i=1}^{n} W_i\, g_i \tag{13.71}$$

where g_i represents the values of the function at n sampling points $\xi_1, \xi_2, \ldots, \xi_n$ and W_i represents the weight factors. The sampling points are sometimes referred to as *Gauss points*. The values ξ_i and W_i are given to 15 decimal places in Table 13.1 for n between 1 and 6.[†] The values of ξ_i and W_i used in computer programs should include as many digits as possible.

The location of the sampling points, specified by the values of ξ_i in the table, is chosen to achieve the maximum accuracy for any given n. The sampling points are located symmetrically with respect to the origin.

If the function is a polynomial, $g = a_0 + a_1\,\xi + \ldots + a_m\,\xi^m$, Equation 13.71 is exact when the number of sampling points is $n \geq (m+1)/2$. In other words, n Gauss points are sufficient to integrate exactly a polynomial of order $2n$-1. For example, we can verify that using $n = 2$ (Figure 13.15) gives exact integrals for each of the functions $g = a_0 + a_1\,\xi$, $g = a_0 + a_1\,\xi + a_2\,\xi^2$ and $g = a_0 + a_1\,\xi + a_2\,\xi^2 + a_3\,\xi^3$. The two sampling points are at $\xi = \pm 0.577\ldots$ and the weight

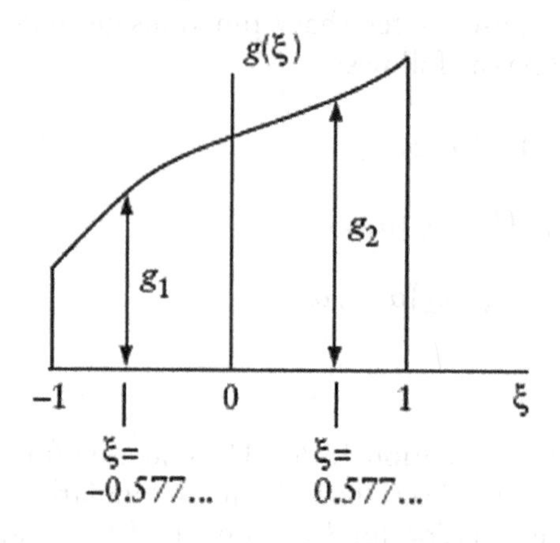

Figure 13.15 Gauss numerical integration of function g(ξ) by Equation 13.71. Example for a number of sampling points $n=2$.

Table 13.1 Location of sampling points and weight factors for Gauss numerical integration by Equations 13.71 to 13.73

n	ξ_i, η_i or ζ_i	W_i
1	0.000 000 000 000 000	2.000 000 000 000 000
2	±0.577 350 269 189 626	1.000 000 000 000 000
3	0.000 000 000 000 000	0.888 888 888 888 888
	±0.774 596 669 241 483	0.555 555 555 555 556
4	±0.339 981 043 584 856	0.652 145 154 862 546
	±0.861 136 311 594 053	0.347 858 485 137 454
	0.000 000 000 000 000	0.568 888 888 888 889
5	±0.538 469 310 105 683	0.478 628 670 499 366
	±0.906 179 845 938 664	0.236 926 885 056 189
	±0.238 619 186 083 197	0.467 913 934 572 691
6	±0.661 209 386 466 265	0.360 761 573 048 139
	±0.932 469 514 203 152	0.171 324 492 379 170

factors are equal to unity. We should note that the sum of the weight factors in Equation 13.71 is 2.0.

A definite integral for a two-dimensional function $g(\xi, \eta)$ can be calculated by:

$$\int_{-1}^{1}\int_{-1}^{1} g(\xi, \eta)\, d\xi\, d\eta \cong \sum_{j}^{n_\eta}\sum_{i}^{n_\xi} W_j\, W_i\, g(\xi_i, \eta_j) \tag{13.72}$$

where n_ξ and n_η are the number of sampling points on ξ and η axes, respectively. Figure 13.16 shows two examples of sampling points. In most cases, n_ξ and n_η are the same (equal to n); then $n \times n$ Gauss points integrate exactly a polynomial $g(\xi, \eta)$ of order $2n - 1$.

Equation 13.72 can be derived from Equation 13.71 by integrating first with respect to ξ and then with respect to η. For 3 × 2 Gauss points ($n_\xi = 3$, $n_\eta = 2$), Equation 13.72 can be rewritten (Figure 13.16) as:

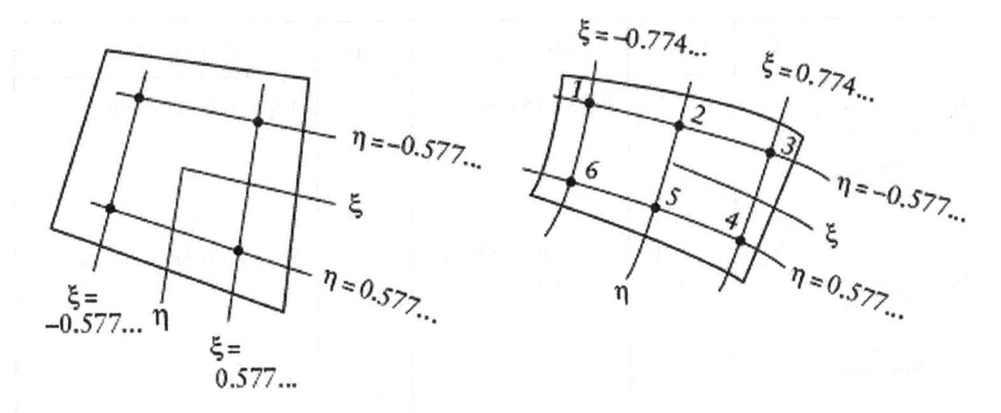

Figure 13.16 Examples of sampling points for two-dimensional Gauss numerical integration.

$$\int_{-1}^{1}\int_{-1}^{1} g(\xi,\eta)\,d\xi\,d\eta \cong (1.0)(0.555\cdots)g_1+(1.0)(0.888\cdots)g_2+(1.0)(0.555\cdots)g_3 \\ +(1.0)(0.555\cdots)g_4+(1.0)(0.888\cdots)g_5+(1.0)(0.555\cdots)g_6 \tag{13.73}$$

We can note that the sum of the multipliers of g_1 to g_6 is 4.0.

For a function in three dimensions, $g(\xi, \eta, \zeta)$, the integral is:

$$\int_{-1}^{1}\int_{-1}^{1}\int_{-1}^{1} g(\xi,\eta,\zeta)\,d\xi\,d\eta\,d\zeta \cong \sum_{k}^{n_\zeta}\sum_{j}^{n_\eta}\sum_{i}^{n_\xi} W_k\,W_j\,W_i\,g(\xi_i,\eta_j,\zeta_k) \tag{13.74}$$

Equations 13.71, 13.72, and 13.74 can be used for one-, two-, and three-dimensional elements with curvilinear axes.

When the element is triangular and the area coordinates α_1, α_2, and α_3 are used (Figure 13.12a), any variable over the area of the element can be expressed as a function $f(\alpha_1, \alpha_2)$, in which α_3 is eliminated by substituting $\alpha_3 = 1 - \alpha_1 - \alpha_2$. The integral $\int f\,da$ over the area of the triangle is expressed by Equation 13.64 and can be evaluated numerically using:

$$\int_{0}^{1}\int_{0}^{1-\alpha_1} g(\alpha_1,\alpha_2)\,d\alpha_2\,d\alpha_1 \cong \sum_{i=1}^{n} W_i\,g_i \tag{13.75}$$

In this equation, the symbol $g(\alpha_1, \alpha_2)$ stands for the product $f(\alpha_1, \alpha_2)\,|J|$ in Equation 13.64, where $|J|$ is the determinant of the Jacobian.

The location of the sampling points and the corresponding weight factors for use in Equation 13.75 are given in Figure 13.17.[6] These values give the exact integrals when g is linear, quadratic, cubic, and quintic, as indicated.

We should note that the sum of the weight factors for all the sampling points of any triangle is equal to one-half. For a triangle with straight edges, $|J| = 2\Delta$, where Δ is the area. If we set $f(\alpha_1, \alpha_2) = 1$, the function $g(a_1, a_2) = 2\Delta$, and the value of the integral given by Equation 13.75 is equal to $(2\Delta)\Sigma\, W_i$, $= \Delta$, as it should.

The number of sampling points, in the numerical integration to generate the stiffness matrices of finite-elements, influences the accuracy and the convergence. The appropriate choice of sampling points is discussed in more detail in references devoted specifically to the finite-element method.[7]

	Point	α1orα2	Point	Weight factor
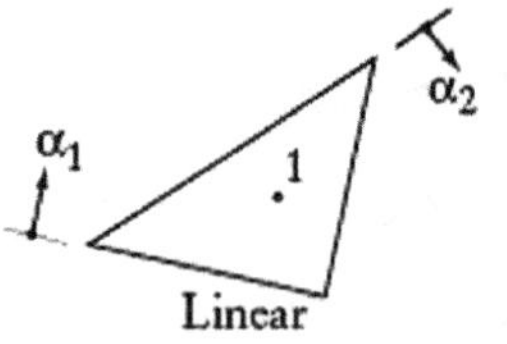 Linear	1	0.333 333 333 3	1	$W_1 = 0.5000000000$
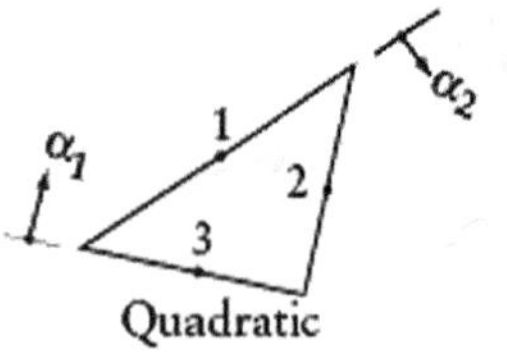 Quadratic	3 1,2	0.000 000 000 0 0.500 000 000 0	1 2,3	$W_{1,2,3} = 0.1666666667$
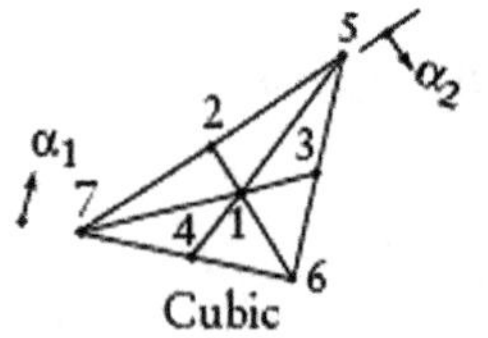 Cubic	4,6,7 1 2,3 5	0.000 000 000 0 0.333 333 333 3 0.500 000 000 0 1.000 000 000 0	2,5,7 1 3,4 6	$W_1 = 0.2250000000$ $W_{2,3,4} = 0.0666666667$ $W_{5,6,7} = 0.0250000000$
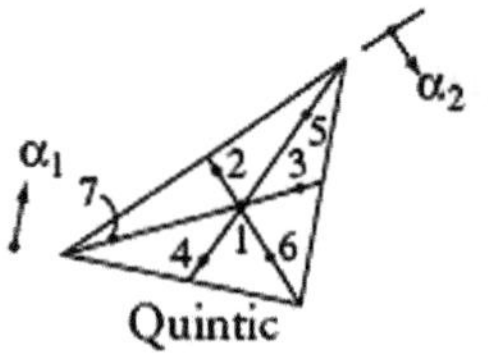 Quintic	4 6,7 1 2,3 5	0.059 715 871 8 0.101 286 507 3 0.333 333 333 3 0.470 142 064 1 0.797 426 985 4	2 5,7 1 3,4 6	$W_1 = 0.1125000000$ $W_{2,3,4} = 0.0661970764$ $W_{5,6,7} = 0.0629695903$

Figure 13.17 Numerical integration of function $g(\alpha_1, \alpha_2, \alpha_3)$ over the area of a triangle. Values of α_1, α_2 and α_3 defining sampling points and weight factors W_i for use in Equation 13.75 ($\alpha_3 = 1 - \alpha_1 - \alpha_2$).

A small number of points is generally used in order to reduce computation. Furthermore, the use of a small number of points tends to reduce the stiffness, thus compensating for the excess in stiffness associated with displacement-based finite-elements (see Section 12.10). However, the number of sampling points cannot be reduced without limit. Some elements have a *spurious mechanism* when a specified pattern of sampling points is used in the integration for the derivation of the stiffness matrix. This mechanism occurs when the element can deform in such a way that the strains at the sampling points are zero.

Two-by-two sampling points are frequently used in quadrilateral plane linear elements with corner nodes only or in quadratic elements with corner and mid-side nodes. Three or four sampling points in each direction are used when the elements are elongated or in cubic[1] elements.

EXAMPLE 13.6: STIFFNESS MATRIX OF QUADRILATERAL ELEMENT IN PLANE-STRESS STATE

Determine coefficient S^*_{11} of the stiffness matrix of the quadrilateral element of Example 13.5 and Figure 12.20. The element is of isotropic material; Poisson's ratio, $\nu = 0.2$. Also find the consistent restraining force at coordinate 1 when the element is subjected to a constant rise of temperature of T degrees. Use only one Gauss integration point.

The displacement field corresponding to $D^*_{11} = 1$ is $\{u, v\} = \{L_1, 0\}$, where:

$$L_1 = \frac{1}{4}(1-\xi)(1-\eta)$$

Derivatives of L_1 with respect to x and y at the center ($\xi = \eta = 0$) are (using Equation 13.27, with $[J]$ from Example 13.5):

$$\begin{Bmatrix} \partial L_1/\partial x \\ \partial L_1/\partial y \end{Bmatrix} = \begin{bmatrix} 15/4 & 5/4 \\ -3/4 & 13/4 \end{bmatrix}^{-1} \begin{Bmatrix} -(1-\eta)/4 \\ -(1-\xi)/4 \end{Bmatrix} = \frac{1}{210} \begin{Bmatrix} -8 \\ -18 \end{Bmatrix}$$

When $D^* = 1$, the strains at the same point are (Equation 13.31):

$$\{B\}_1 = \frac{1}{210}\{-8, 0, -18\}$$

The value of the product in the stiffness matrix of Equation 13.32 at the center of the element (Equation 7.28) is:

$$g_1 = h\,\{B\}_1^{\mathrm{T}}\,[d]\{B\}_1\,|J|$$

$$= \frac{1}{(210)^2}\{-8, 0, -18\}\frac{Eh}{1-(0.2)^2}\begin{bmatrix} 1 & 0.2 & 0 \\ 0.2 & 1 & 0 \\ 0 & 0 & (1-0.2)/2 \end{bmatrix}\begin{Bmatrix} -8 \\ 0 \\ -18 \end{Bmatrix}|J| = 4.57\times 10^{-3}\ Eh\,|J|$$

The determinant of the Jacobian is 13.125 (see Example 13.5). Substitution in Equation 13.72 with $W_1 = 2$ (from Table 13.1) gives:

$$S_{11}^* = 2\times 2\left(4.57\times 10^{-3}\right) Eh\,(13.125) = 0.240\ Eh$$

The free strains due to the temperature rise are αT {1, 1, 0}. Substitution in Equation 13.34 gives:

$$F_{b1}^* = 2\times 2\,(13.125)(-h)\left(\frac{1}{210}\right)\{-8, 0, -18\}\times\frac{Eh}{1-(0.2)^2}\begin{bmatrix} 1 & 0.2 & 0 \\ 0.2 & 1 & 0 \\ 0 & 0 & (1-0.2)/2 \end{bmatrix}\begin{Bmatrix} 1 \\ 1 \\ 0 \end{Bmatrix}\alpha T$$

$$= 2.5\,\alpha\,T\,h$$

EXAMPLE 13.7: TRIANGULAR ELEMENT WITH PARABOLIC EDGES: JACOBIAN MATRIX

Determine the Jacobian at $(\alpha_1, \alpha_2) = (1/3, 1/3)$ in a triangular isoparametric element with corner and mid-side nodes, where α_1 and α_2 are area coordinates. The (x, y) coordinates of the nodes are: $(x_1, y_1) = (0,0)$; $(x_2, y_2) = (1,1)$; $(x_3, y_3) = (0,1)$; $(x_4, y_4) = \left(\sqrt{2}/2, 1-\left[\sqrt{2}/2\right]\right)$; $(x_5, y_5) = (0.5, 1)$; $(x_6, y_6) = (0, 0.5)$. These correspond to a quarter of a circle with its center at node 3 and a radius of unity. By numerical integration of Equation 13.75, determine the area of the element, using only one sampling point.

The shape functions L_1 to L_6 are given by Equations 13.62 and 13.63. We eliminate α_3 by substituting $\alpha_3 = 1 - \alpha_1 - \alpha_2$ and then determine the derivatives $\partial L_i/\partial\alpha_1$ and $\partial L_i/\partial\alpha_2$ at $(\alpha_1, \alpha_2) = (1/3, 1/3)$. The Jacobian at this point (from Equation 13.29, replacing ξ and η by α_1 and α_2) is:

$$[J]=\begin{bmatrix}0\left(\frac{1}{3}\right) & 0\left(\frac{1}{3}\right)\\ 0\,(0) & 0\,(0)\end{bmatrix}+\begin{bmatrix}1\,(0) & 1\,(0)\\ 1\left(\frac{1}{3}\right) & 1\left(\frac{1}{3}\right)\end{bmatrix}$$

$$+\begin{bmatrix}0\left(\frac{-1}{3}\right) & 1\left(\frac{-1}{3}\right)\\ 0\left(\frac{-1}{3}\right) & 1\left(\frac{-1}{3}\right)\end{bmatrix}+\begin{bmatrix}\frac{\sqrt{2}}{2}\left(\frac{4}{3}\right) & \left(1-\frac{\sqrt{2}}{2}\right)\left(\frac{4}{3}\right)\\ \frac{\sqrt{2}}{2}\left(\frac{4}{3}\right) & \left(1-\frac{\sqrt{2}}{2}\right)\left(\frac{4}{3}\right)\end{bmatrix}$$

$$+\begin{bmatrix}0.5\left(\frac{-4}{3}\right) & 1\left(\frac{-4}{3}\right)\\ 0.5\,(0) & 1\,(0)\end{bmatrix}+\begin{bmatrix}0\,(0) & 0.5\,(0)\\ 0\left(\frac{-4}{3}\right) & 0.5\left(\frac{-4}{3}\right)\end{bmatrix}=\begin{bmatrix}0.2761 & -1.2761\\ 1.2761 & -0.2761\end{bmatrix}$$

The area of the element (from Equation 13.75 with W_1 from Figure 13.17) is:

$$\text{area}=\int_0^1\int_0^{1-\alpha_2}|J|\,d\alpha_2\,d\alpha_1=0.5\,|J|=0.7761$$

13.15 GENERAL

In this chapter and in Chapter 12, we have discussed the main concepts and techniques widely used in the analysis of structures by the finite-element method. This was possible because the finite-element method is an application of the displacement method, which is extensively discussed in earlier chapters and also in Chapter 19.

REFERENCES

1. Zienkiewicz, O.C., *The Finite Element Method in Engineering Science*, 4th ed., McGraw-Hill, London, 1989.
2. Bazeley, G.P., Cheung, Y.K., Irons, B.M. and Zienkiewicz, O.C., " Triangular Elements in Plate Bending: Conforming and Non-Conforming Solutions," *Proceedings of the Conference on Matrix Methods in Structural Mechanics*, Air Force Institute of Technology, Wright Patterson Base, Ohio, 1965.
3. Gallagher, R.H., *Finite Element Analysis Fundamentals*, Prentice-Hall, Englewood Cliffs, NJ, 1975, p. 350.
4. Irons, B.M. and Shrive, N.G., *Numerical Methods in Engineering and Applied Science*, Ellis Horwood, Chichester, 1987.
5. Kopal, Z., *Numerical Analysis*, 2nd ed., Chapman & Hall, 1961.
6. Cowper, G.R., "Gaussian Quadrature Formulas for Triangles", *Int. J. Num. Math. Eng.*, 7, (1973), pp. 405–408.
7. Irons, B.M. and Ahmad, S., *Techniques of Finite Elements*, Ellis Horwood, Chichester, 1980, pp. 149–162.

PROBLEMS

13.1 Use the shape functions in Figure 13.7 to verify the consistent vectors of restraining nodal forces given in Figure 13.10.

13.2 Using Lagrange polynomials (Figure 13.5), construct shape functions L_1, L_2, L_{10}, and L_{11} for the element shown. Using these shape functions and considering symmetry and equilibrium, determine the consistent vector of nodal forces in the y direction when the element is subjected to a uniform load of q_y per unit volume.

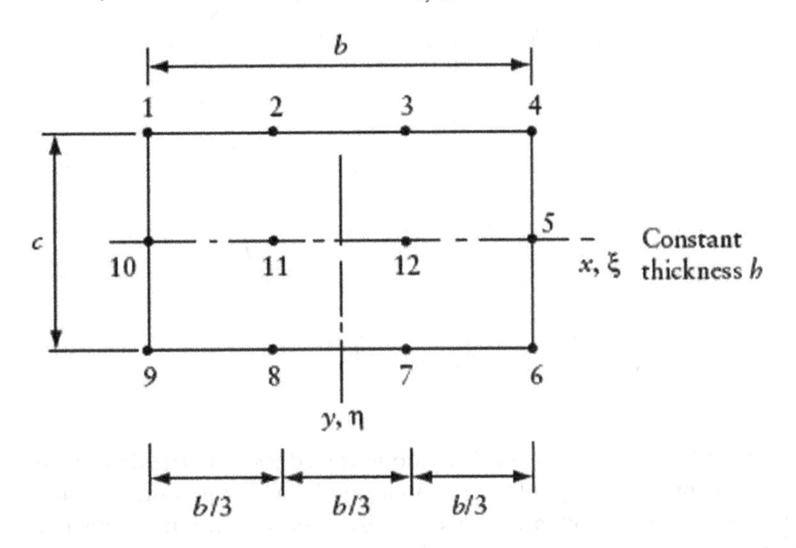

13.3 Determine the strains at nodes 1, 2, and 3 in the linear-strain triangle (Figure 13.13) due to $D_1^* \equiv u_1 \equiv 1$ and $D_2^* \equiv v_1 \equiv 1$.

13.4 Show that the shape functions of Equations 13.62 and 13.63 give linear strains in the element shown in Figure 13.13. The answers to Problem 13.3 give the strains at the three corners of the triangle. By linear interpolation, using area coordinates, determine the strains at any point in the element. These give the first two columns of the $[B]$ matrix.

13.5 Use the answers to Problem 13.4 to determine S_{11}^* and S_{12}^* of the linear-strain triangle (Figure 13.13).

13.6 Derive $[B]$ and $[S^*]$ for the triangular bending element with equal sides l shown. Assume a constant thickness h and an isotropic material with $\nu = 0.3$.

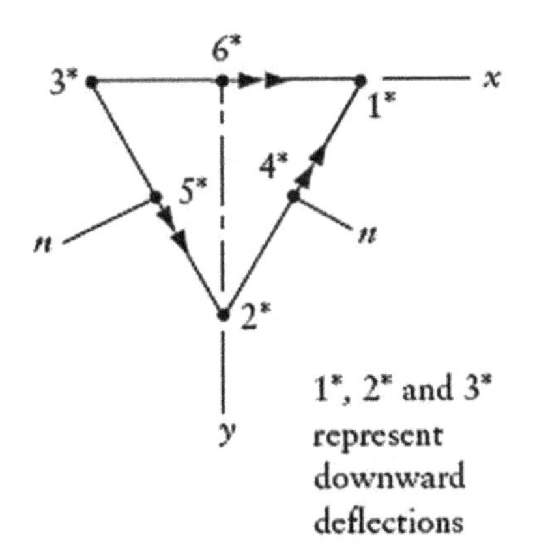

13.7 The hexagonal horizontal plate shown is simply supported at the corners and carries a downward concentrated load P at the center. Calculate the deflection at this point and the moments at A by idealizing the plate as an assemblage of six triangular elements identical to the element of Problem 13.6. Check the value of M_x by taking the moment of the forces situated to the left or to the right of the y axis.

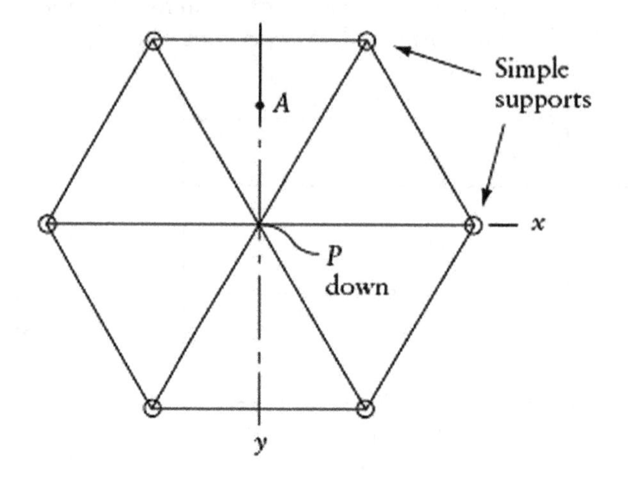

13.8 If the plate in Problem 13.7 is built-in along the edges, find the deflection at the center and the moments M_x, M_y, and M_{xy} at A caused by a change in temperature of $T/2$ and $-T/2$ degrees at the bottom and top surfaces, respectively, with a linear variation over the thickness. Take the coefficient of thermal expansion as α.

NOTES

* The rectangular element considered here is a special case of the quadrilateral element referred to as QLC3. See Reference 1. The combination of QLC3 with the rectangular bending element shown in Figure 7.8 gives a good shell element with three translations and three rotations per node for the analysis of spatial structures (See 13.3.1).

† Values of ξ_i and W_i for n between 1 and 10 can be found on p. 79 of Reference 4. See also Reference 5.

Chapter 14

Plastic analysis of plane frames

14.1 INTRODUCTION

An elastic analysis of a structure is important in order to study its performance, especially with respect to serviceability, under the service loading for which the structure is designed. However, if the load is increased until yielding occurs at some locations, the structure undergoes elastic–plastic deformations and, on further increase, a fully plastic condition is reached, at which a sufficient number of *plastic hinges* are formed to transform the structure into a mechanism. This mechanism would collapse under any additional loading. A study of the mechanism of failure and the knowledge of the magnitude of the collapse load are necessary to determine the load factor in analysis. Alternatively, if the load factor is specified, the structure can be designed so that its collapse load is equal to, or higher than, the product of the load factor and the service loading.

Design of structures based on the plastic approach (referred to as *limit design)* is accepted by various codes of practice, particularly for steel construction. The material is assumed to deform in the idealized manner shown in Figure 14.1. The strain and stress are proportional to one another up to the yield stress, at which the strain increases indefinitely without any further increase in stress. This type of stress–strain relation is not very different from that existing in mild steel. However, there exists a reserve of strength due to strain hardening, but this will not be allowed for in the analysis in this chapter.

We shall now consider the principles of plastic analysis of plane frames in which buckling instability is prevented and fatigue or brittle failure is not considered possible. In most cases, the calculation of the collapse load involves trial and error, which may become tedious in large structures. The companion computer programs of the present book include computer program PLASTICF (Chapter 20). The computer increases the values of the loads gradually until collapse; it determines the load levels that develop successive plastic hinges, the corresponding nodal displacements and member end forces. Section 14.6 explains the procedure with example.

14.2 ULTIMATE MOMENT

Consider a beam whose cross section has an I-shape as shown in Figure 14.2a. Let the beam be subjected to bending moment, $M_x = M$. If the bending moment is small, the stress and the strain vary linearly across the section, as shown in Figure 14.2b. When the moment is increased, yield stress is attained at the extreme fibers (Figure 14.2c), and with a further

DOI: 10.1201/9780429286858-14

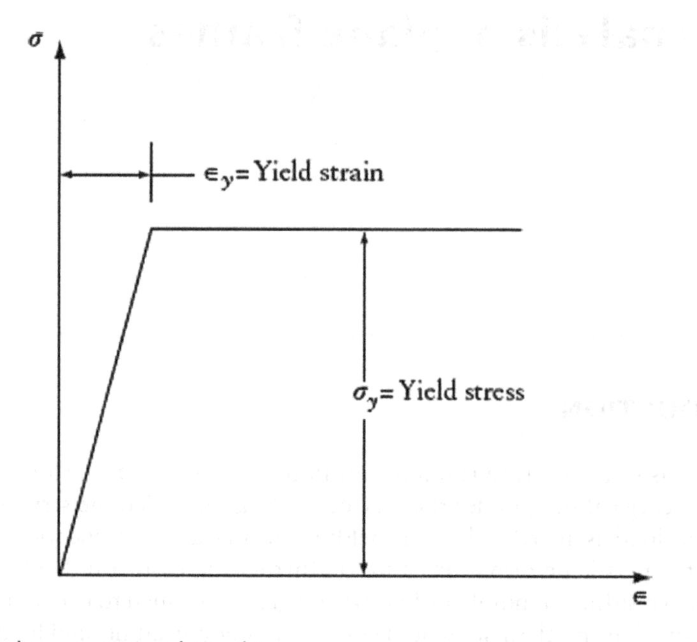

Figure 14.1 Idealized stress–strain relationship.

increase, the yield stress develops across the whole section, as shown in Figure 14.2d. If the bending moment continues to increase, yield will spread from the outer fibers inward until the two zones of yield meet (Figure 14.2e); the cross section in this state is said to *be fully plastic.*

The value of the ultimate moment in the fully plastic condition can be calculated in terms of the yield stress σ_y. Since the axial force is zero in the case considered, the neutral axis in the fully plastic condition divides the section into two equal areas, and the resultant tension and compression are each equal to ($a\ \sigma_y/2$), forming a couple equal to the ultimate moment:

$$M_p = \frac{1}{2} a \sigma_y \left(\bar{y}_c + \bar{y}_t \right) \tag{14.1}$$

where $\bar{y}_c$ and $\bar{y}_t$ are, respectively, the distances of the centroid of the compression and tension areas from the neutral axis in the fully plastic condition.

The maximum moment which a section can carry without exceeding the yield stress is $M_y = \sigma_y Z$, where Z is the section modulus. The ratio $\alpha = M_p/M_y$ depends on the shape of the cross section and is referred to as the *shape factor*; it is always greater than unity.

For a rectangular section of breadth b and depth d, $Z = bd^2/6$, $M_p = \sigma_y bd^2/4$; hence, $\alpha = 1.5$. For a solid circular cross section $\alpha = 1.7$, while for I-beams and channels α varies within a small range of 1.15–1.17.

14.3 PLASTIC BEHAVIOR OF A SIMPLE BEAM

To consider displacements, let us assume an idealized relation between the bending moment and curvature at a section, as shown in Figure 14.3. If a load P at the mid-span of a simple beam (Figure 14.4a) is increased until the bending at the mid-span cross section reaches the fully plastic moment M_p, a plastic hinge is formed at this section and collapse will occur under any further load increase. According to the assumed idealized relation, the curvature,

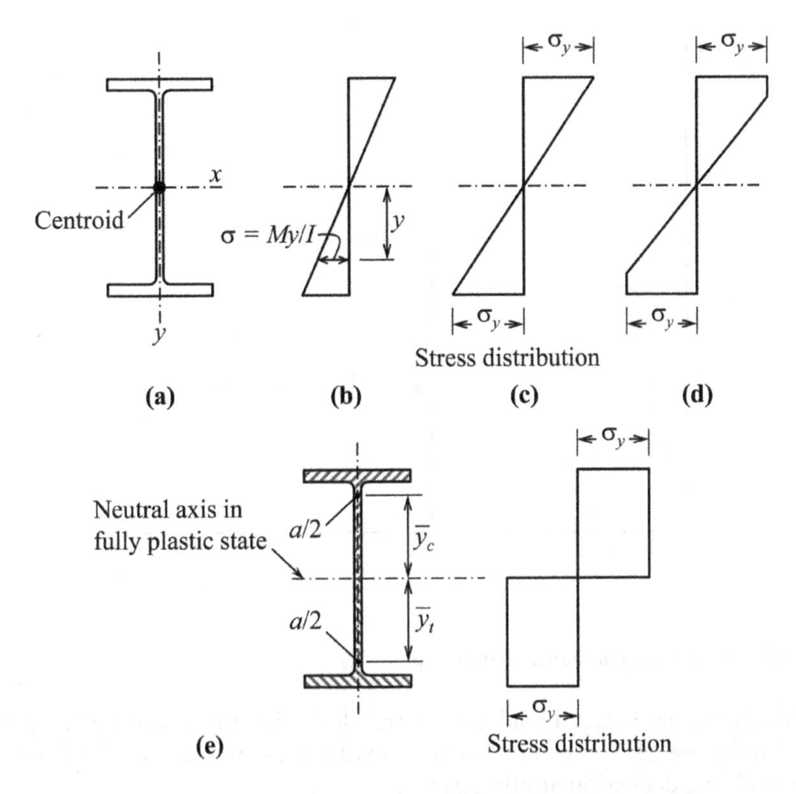

Figure 14.2 Stress distribution in a symmetrical cross section subjected to a bending moment of increasing magnitude. (a) Beam cross section. (b) Elastic state. (c) Plastic at top and bottom fibers. (d) Partial plasticity at top and bottom fibers. (e) Fully plastic state.

and hence the rotation, at the plastic hinge increase at a constant load; so does deflection. The collapse load P_c can be easily calculated from statics:

$$P_c = 4M_p/l \tag{14.2}$$

The bending moment at sections other than mid-span is less than M_p, and by virtue of the assumed idealized relations, the beam remains elastic away from this section. The deflected configurations of the beam in the elastic and plastic stages are shown in Figure 14.4b. The increase in deflection during collapse is caused by the rotation at the central hinge without a concurrent change in curvature of the two halves of the beam. Figure 14.4c represents the change in deflection during collapse; this is a straight line for each half of the beam. The same figure shows also the collapse mechanism of the beam.

The collapse load of the beam (and this applies also to statically indeterminate structures) can be calculated by equating the external and internal work during a virtual movement of the collapse mechanism. Let each half of the beam in Figure 14.4c acquire a virtual rotation θ, so that the corresponding rotation at the hinge is 2θ, and the downward displacement of the load P_c is $l\theta/2$. Equating the work done by P_c to the work of the moment M_p at the plastic hinge, we obtain:

$$P_c \frac{l\,\theta}{2} = 2M_p\,\theta \tag{14.3}$$

which gives the same result as Equation 14.2.

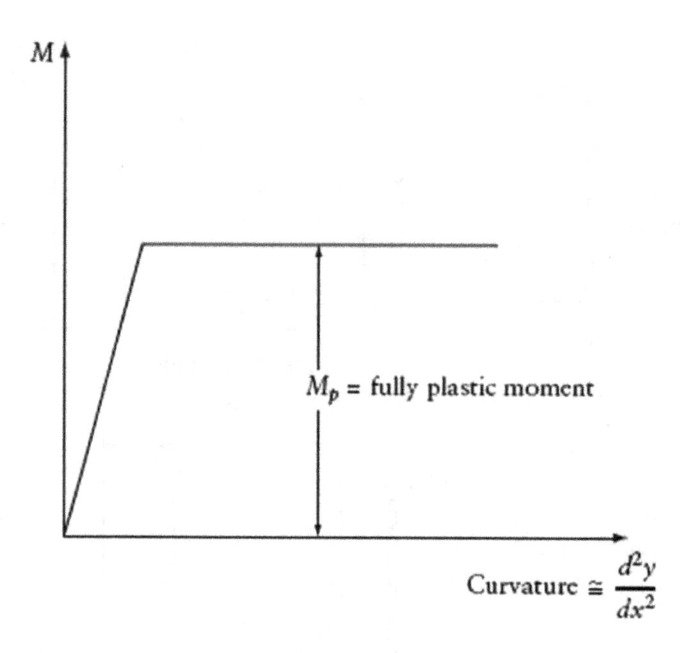

Figure 14.3 Idealized moment–curvature relationship.

The idealized relation between load and central deflection for this beam is represented by the line OFM in Figure 14.5. When the collapse load corresponding to point F in Figure 14.5 is reached, the elastic deflection at mid-span is:

$$D_F = P_c \frac{l^3}{48EI} \tag{14.4}$$

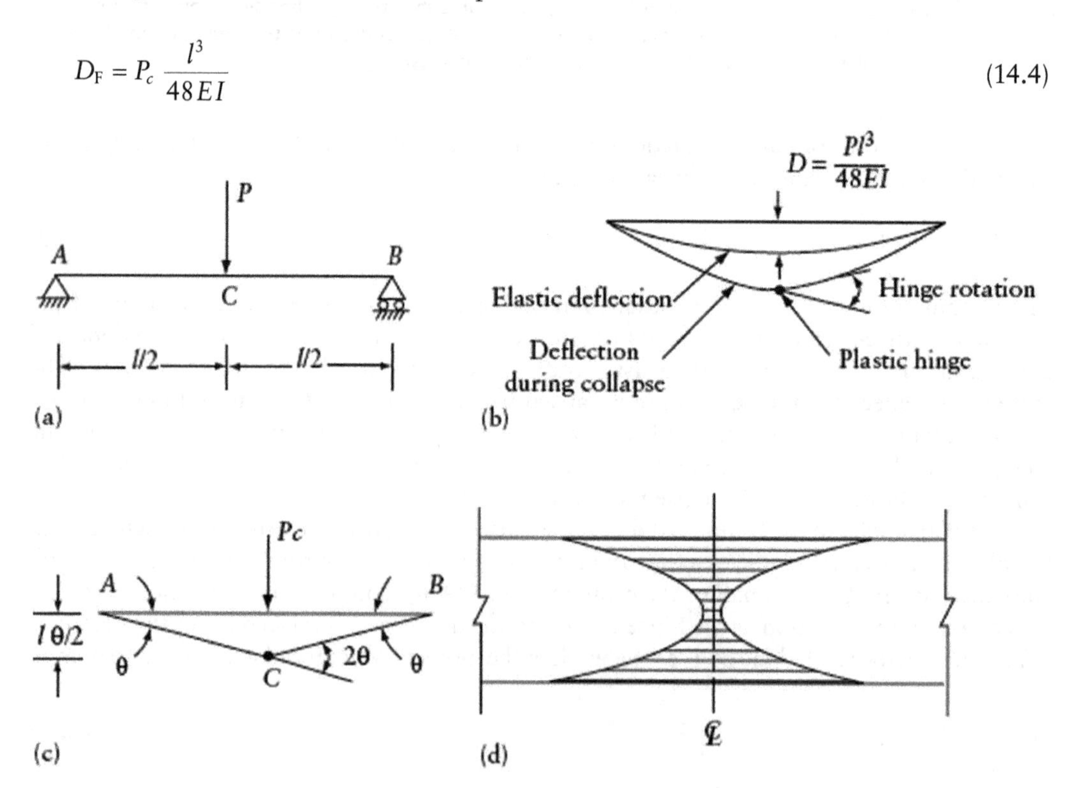

Figure 14.4 Plastic behavior of a simple beam. (a) Beam. (b) Deflection lines. (c) Change in deflection during collapse. (d) Elevation of beam showing yielding near the mid-span section.

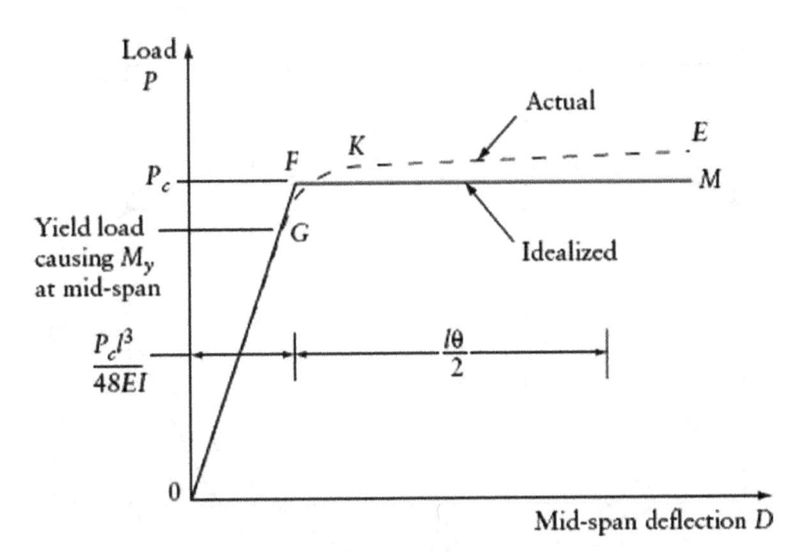

Figure 14.5 Load-deflection relationship for the beam in Figure 14.4.

However, the actual load-deflection relation follows the dashed curve GKE. When the yield moment M_y (= $\sigma_y Z$) is reached at the mid-span section, the upper or the lower fiber, or both, the elastic behavior comes to an end. If the load is increased further, yield spreads inward at this section and also laterally to other nearby sections. Figure 14.4d illustrates this spread of yield. After M_y has been reached, the deflection increases at a greater rate per unit increase of load until M_p is reached, as indicated by the curve GK in Figure 14.5. In practice, rolled steel sections continue to show a small rise in the load-deflection curve during collapse (line KE); this is due to strain-hardening, which is generally not considered in ordinary plastic analysis.

14.4 ULTIMATE STRENGTH OF FIXED-ENDED AND CONTINUOUS BEAMS AND FRAMES

Consider a prismatic fixed-ended beam subjected to a uniform load of intensity q (Figure 14.6a). The resulting bending moments are $M_A = -M_C = -ql^2/12$ and $|M_B| = ql^2/24$. When the load intensity is increased to q_1 such that the moments at the supports reach the fully plastic moment $M_p = q_1 l^2/12$, hinges are formed at A and C. If q is further increased, the moment at the supports will remain constant at M_p; free rotation will take place there so that the deflection due to the load in excess of q_1 will be the same as in a simply supported beam. The collapse will occur at a load intensity q_c which produces the moment at mid-span of magnitude M_p, so that a third hinge is formed at B. The bending moment diagrams due to load intensity q for the cases $q = q_1$, $q_1 < q < q_c$, and $q = q_c$ are shown in Figure 14.6b and the collapse mechanism in Figure 14.6c. The collapse load q_c is calculated by the virtual-work equation:

$$M_p\left(\theta + 2\theta + \theta\right) = 2\left(\frac{q_c\, l}{2}\right)\frac{\theta l}{4} \tag{14.5}$$

where θ, 2θ, and θ are the virtual rotations at the plastic hinges A, B, and C respectively, and $\theta l/4$ is the corresponding downward displacement of the resultant load on one-half of the beam. Equation 14.5 gives the intensity of the collapse load:

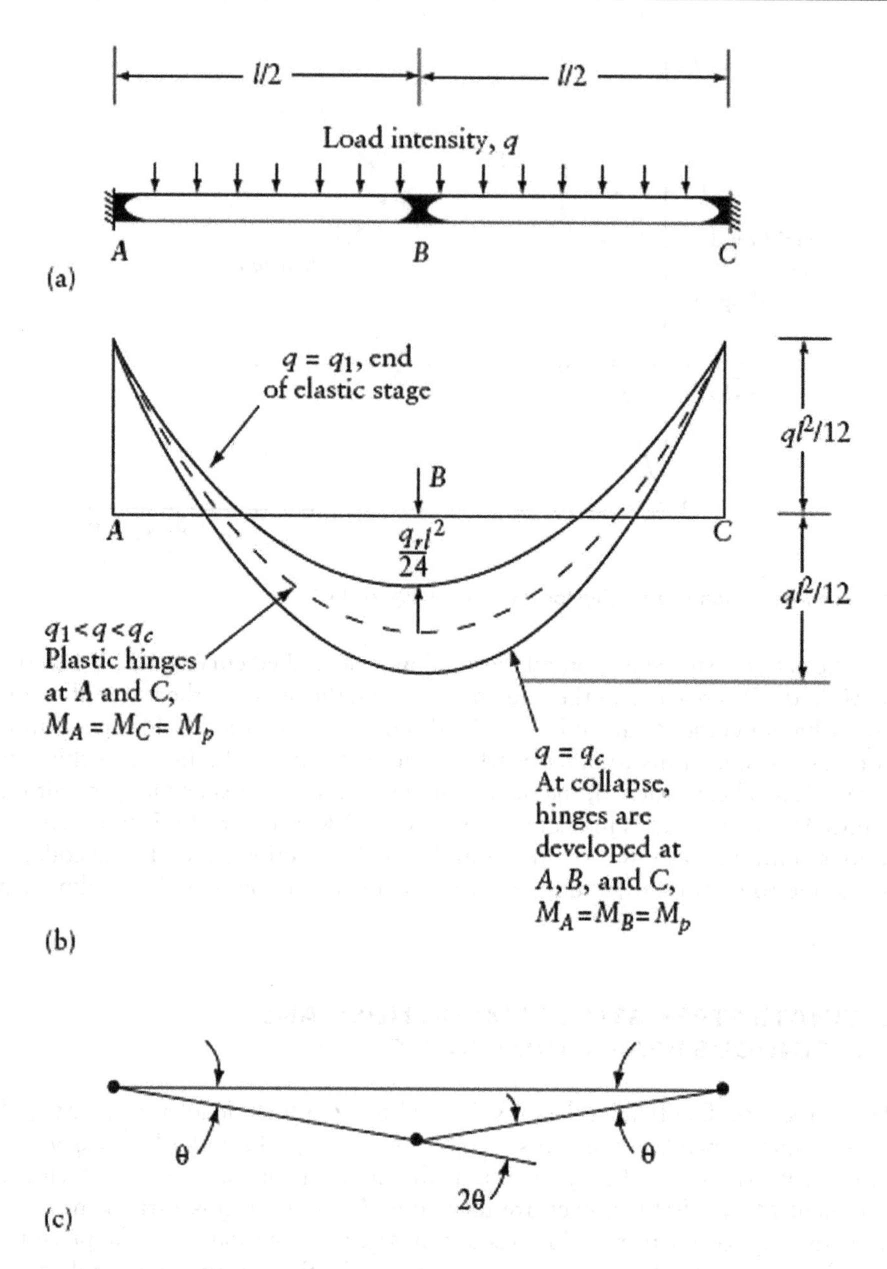

Figure 14.6 Collapse of a beam with fixed ends under a uniformly distributed load. (a) Beam. (b) Bending moment diagrams for three load intensities. (c) Collapse mechanism.

$$q_c = \frac{16\, M_p}{l^2} \tag{14.6}$$

For continuous beams and frames, the values of the collapse loads corresponding to all possible mechanisms are determined; the actual collapse load is the smallest one of these. For the continuous beam in Figure 14.7, employ virtual work equation to determine collapse loads, P_{c1} and P_{c2}, corresponding to mechanisms 1 and 2 in Figures 14.7b and c:

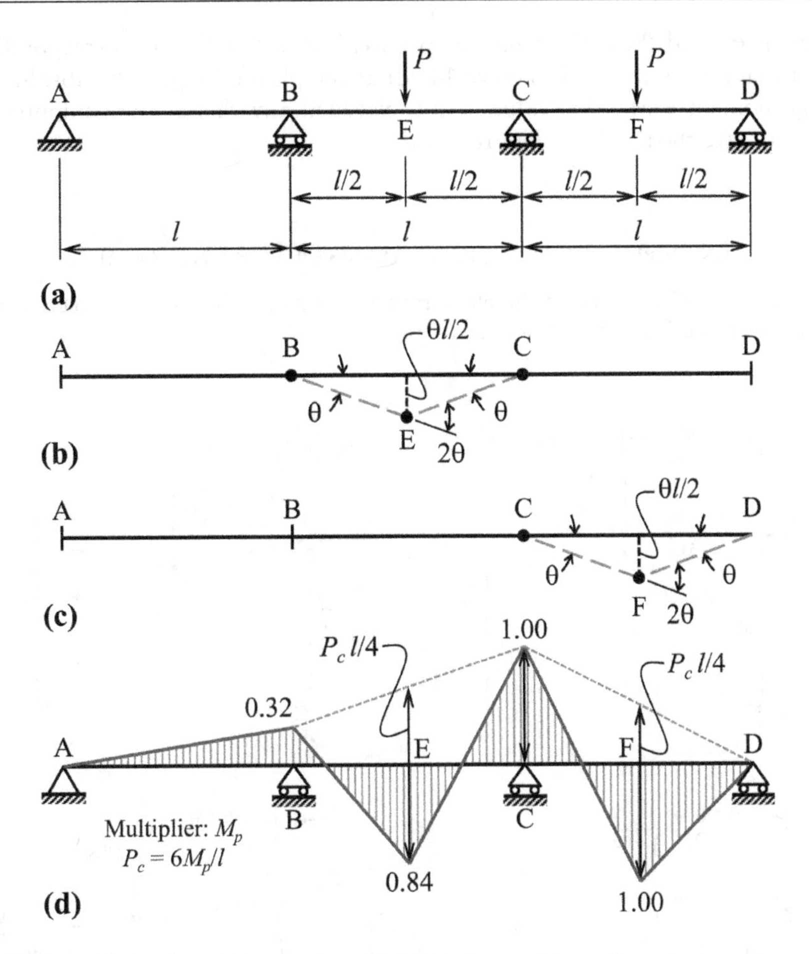

Figure 14.7 Plastic analysis of a continuous beam. (a) Continuous beam of constant section and of plastic moment of resistance M_p. (b) Collapse mechanism 1. (c) Collapse mechanism 2. (d) Bending moment diagram at collapse.

$$P_{c1}\left(\frac{\theta l}{2}\right) = M_p\left(\theta + 2\theta + \theta\right) \text{ (See Figure 14.7b)}$$

whence:

$$P_{c1} = 8M_p/l \tag{14.7}$$

and

$$P_{c2}\left(\frac{\theta l}{2}\right) = M_p\left(\theta + 2\theta\right) \text{ (See Figure 14.7c)}$$

Whence

$$P_{c2} = 6M_p/l \tag{14.8}$$

The smaller of P_{c1} and P_{c2} is the true collapse load: $P_c = 6M_p/l$. The corresponding bending moment diagram is shown in Figure 14.7d; at the plastic hinges, the absolute value of the bending moment = M_p; this value is not exceeded anywhere. The computer program PLASTICF can give the results in Figure 14.7d.

EXAMPLE 14.1: PLASTIC ANALYSIS OF PORTAL FRAME

Determine the collapse load of the plane frame in Figure 14.8a using PLASTICF; calibrate the results using virtual work equation.

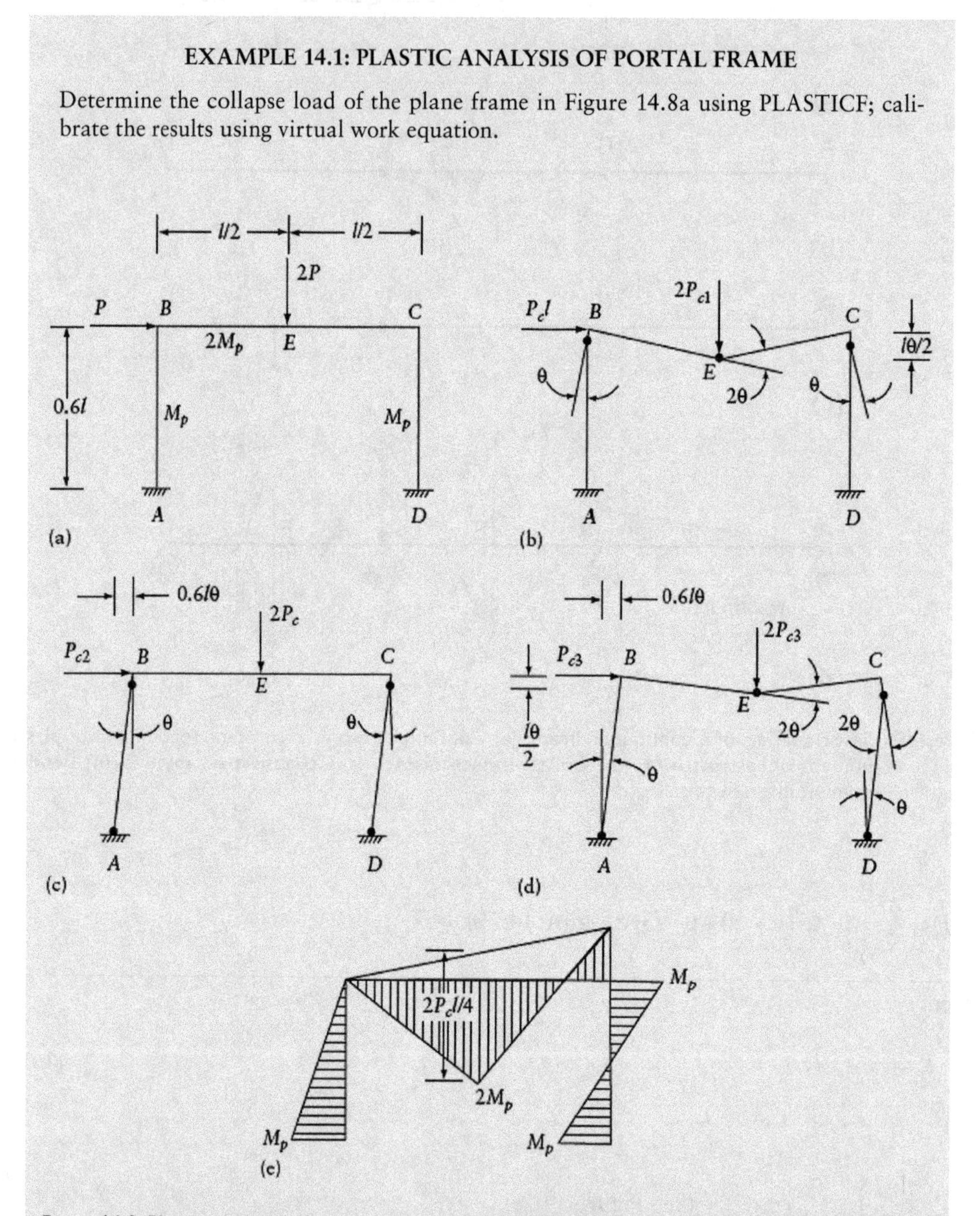

Figure 14.8 Plastic analysis of a rectangular portal frame. (a) Frame loading and properties. (b) Mechanism corresponding to collapse load P_{c1}. (c) Mechanism corresponding to collapse load P_{c2}. (d) Mechanism corresponding to collapse load P_{c3}. (e) Bending moment diagram at collapse.

Let us determine the collapse load for the frame shown in Figure 14.8a, assuming the plastic moment of resistance = $2M_p$ for the beam BC, and M_p for the columns. There are only three possible collapse mechanisms, which are shown in Figures 14.8b, c, and d.

A virtual-work equation for each of these mechanisms gives:

$$M_p(\theta+\theta)+2M_p(2\theta)=2P_{c1}\left(\frac{\theta l}{2}\right) \quad \text{(See Figure 14.8b)}$$

or:

$$P_{c1}=6M_p/l \tag{14.9}$$

$$M_p(\theta+\theta+\theta+\theta)=P_{c2}(0.6\theta l) \quad \text{(See Figure 14.8c)}$$

or:

$$P_{c2}=6.67M_p/l \tag{14.10}$$

and:

$$M_p(\theta+\theta+2\theta)+2M_p(2\theta)=P_{c3}(0.6\theta l)+2P_{c3}\left(\frac{\theta l}{2}\right) \quad \text{(See Figure 14.8d)}$$

or:

$$P_{c3}=5M_p/l \tag{14.11}$$

The collapse load is the smallest of P_{c1}, P_{c2}, and P_{c3}; thus, $P_c = 5\ M_p/l$, and failure of the frame will occur with the mechanism of Figure 14.8d. The corresponding bending moment diagram in Figure 14.8e has an ordinate M_p at the plastic hinges A, C, and D and $2M_p$ at E, with the plastic moment of resistance exceeded nowhere. A check on the calculations can be made by verifying that the bending moment diagram in Figure 14.8e satisfies static equilibrium.

EXAMPLE 14.2: PLASTIC ANALYSIS OF TWO-BAY FRAME

Use computer program PLASTICF to determine the collapse load of the plane frame in Figure 14.9a; calibrate the results by virtual work equation. Assume that all members have constant E, a, and I. Figure 14.9a indicates the relative values of plastic moment resistance.

Figure 14.9b presents the successive development of plastic hinges with incremental increase of P_c. Collapse occurs at $P_c = 1.067\ M_p/l$, in 8-load increments and 6-plastic hinges. The figure shows the collapse mechanism and the virtual work equation confirming the computer results. Figure 14.9c shows the bending moment diagram for the collapse mechanism.

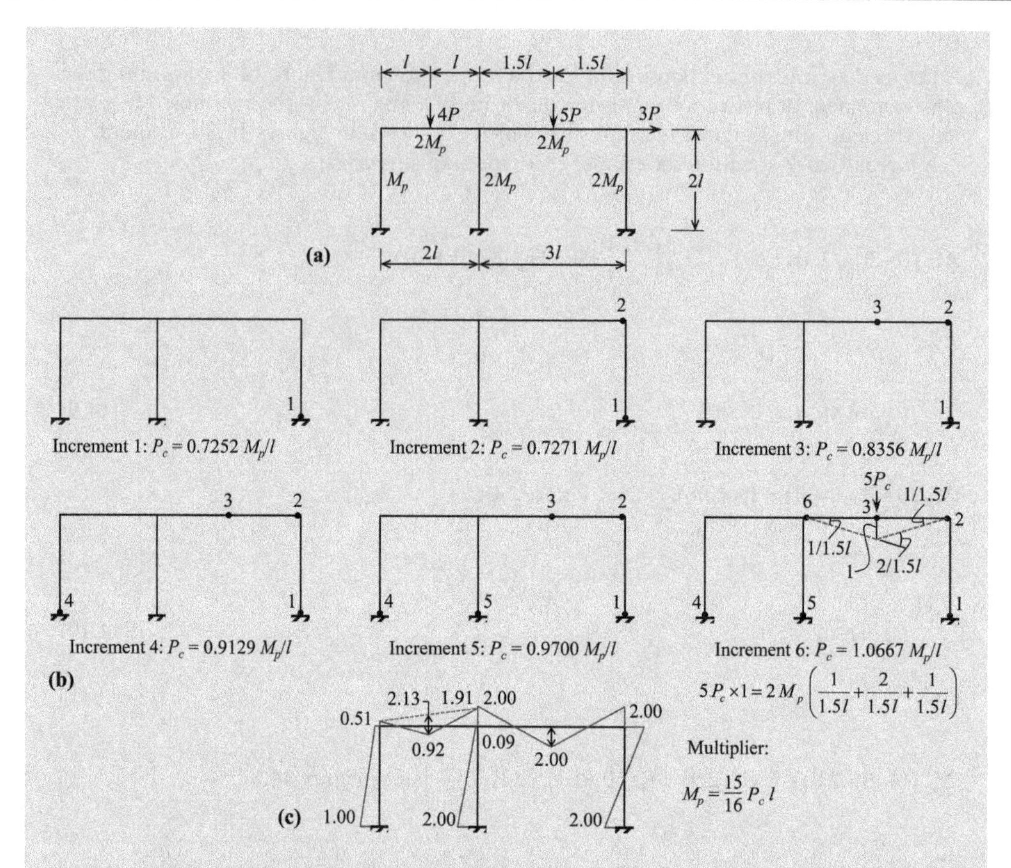

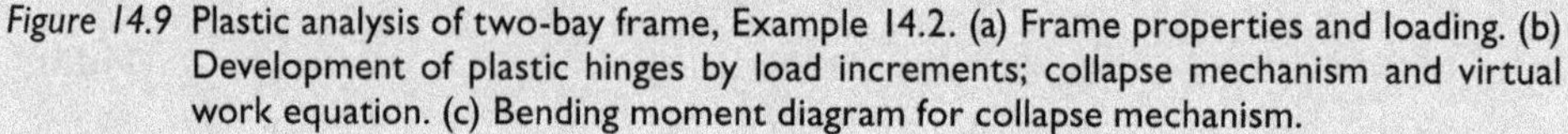
Figure 14.9 Plastic analysis of two-bay frame, Example 14.2. (a) Frame properties and loading. (b) Development of plastic hinges by load increments; collapse mechanism and virtual work equation. (c) Bending moment diagram for collapse mechanism.

EXAMPLE 14.3: PLASTIC ANALYSIS OF GABLE FRAME

Use computer program PLASTICF to determine the collapse load of the plane frame in Figure 14.10a; calibrate the results by virtual work equation. Assume that all members have constant E, a, I, and M_p.

Figure 14.10b presents the successive development of plastic hinges with incremental increase of P_c. Collapse occurs at $P_c = 0.5556\ M_p/b$ in 4-load increments and 4-plastic hinges. The figure shows the collapse mechanism and the virtual work equation confirming the computer results. Figure 14.10c shows the bending moment diagram for the collapse mechanism.

14.5 LOCATION OF PLASTIC HINGE UNDER DISTRIBUTED LOAD

Consider the frame analyzed in Example 14.1, but with a vertical load $4P$ distributed over the beam BC, as shown in Figure 14.11a; the horizontal load P is unchanged. In this case, the position of the maximum positive bending moment in BC is not known, so that the location of the plastic hinge has to be determined.

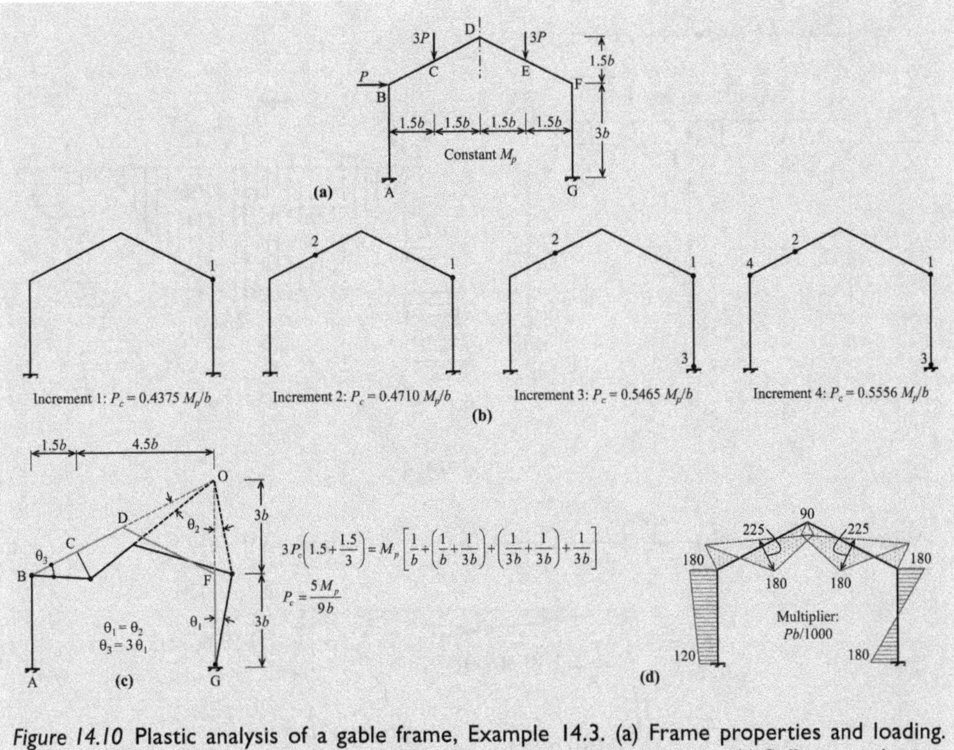

Figure 14.10 Plastic analysis of a gable frame, Example 14.3. (a) Frame properties and loading. (b) Incremental increase of load and development of plastic hinges. (c) Collapse mechanism and virtual work equation. (d) Bending moment diagram of the collapse mechanism.

Let us apply the virtual-work equation to the mechanism in Figure 14.8d, loaded as in Figure 14.11a, with the hinge in the beam assumed at mid-span. Each half of the beam is subjected to a vertical load whose resultant, $2P_{c3}$ moves through a vertical distance $l\,\theta/4$.

The internal virtual work and the external virtual work of the horizontal force are the same as before (Equation 14.11); thus:

$$M_p\left(\theta+\theta+2\theta\right)+2M_p\left(2\theta\right)=P_{c3}\left(0.6\theta l\right)+2\times 2P_{c3}\left(\frac{\theta l}{4}\right) \tag{14.12}$$

This equation gives the same value of the collapse load as Equation 14.11.

The bending moment diagram for the mechanism in Figure 14.8d with the loads in Figure 14.11a is shown in Figure 14.11b, from which it can be seen that the fully plastic moment $2M_p$ is slightly exceeded in the left-hand half of the beam.

This maximum moment occurs at a distance $x = 0.45l$ from B, and its value is $2.025M_p$. It follows that the assumed collapse mechanism is not correct; the reason for this being that the plastic hinge in the beam should not be located at mid-span. The calculated value of $P_c = 5M_p/l$ is an upper bound on the value of the collapse load. If the load P_c is given, the structure must be designed for a plastic moment M_p slightly greater than $P_c l/5$, and this value is to be considered a lower bound on the required plastic moment. It is clear that the structure will be safe if designed for:

$$M_p=\frac{2.025}{2}\left(\frac{P_c l}{5}\right)=1.0125\,\frac{P_c l}{5}$$

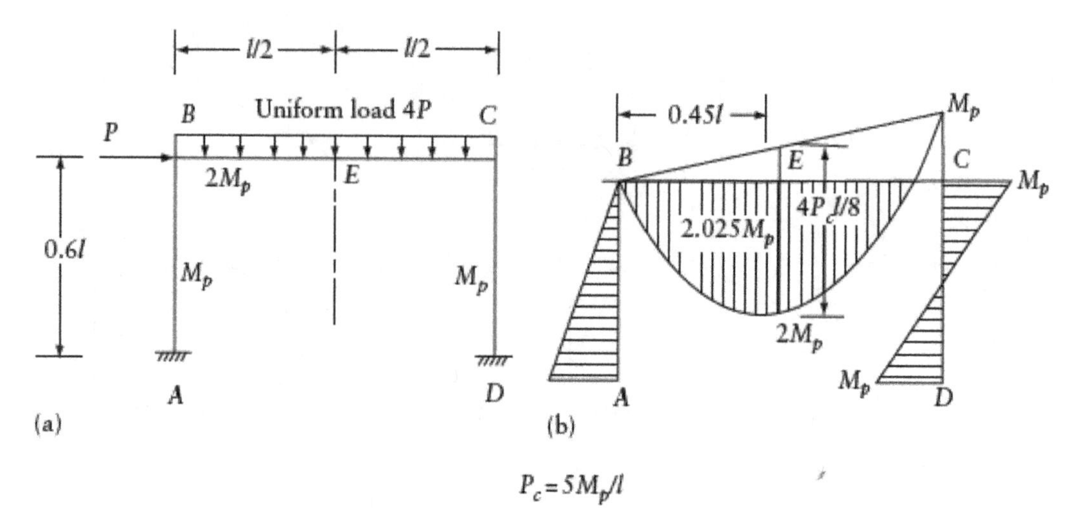

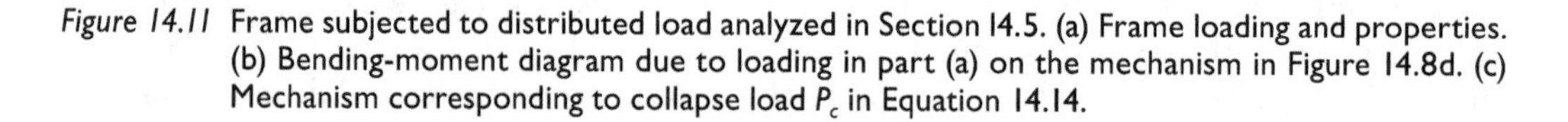

Figure 14.11 Frame subjected to distributed load analyzed in Section 14.5. (a) Frame loading and properties. (b) Bending-moment diagram due to loading in part (a) on the mechanism in Figure 14.8d. (c) Mechanism corresponding to collapse load P_c in Equation 14.14.

Therefore, the required value of M_p is:

$$\frac{P_c l}{5} < M_p < 1.0125\,\frac{P_c l}{5} \tag{14.13}$$

These limits define the value of M_p to within 1.25 percent and are accurate enough for practical purposes. However, the precise value of M_p can be calculated considering the mechanism in Figure 14.11c, with a plastic hinge F at a distance x from B. By a virtual-work equation M_p can be derived in terms of x. The value of x is then chosen so that M_p is a maximum, that is, $dM_p/dx = 0$.

The rotation at the hinges (rotation of column AB, $\theta = 1/(0.6l)$) and the translation of the loads are shown in Figure 14.11c, and the virtual-work equation is:

$$M_p\left(\theta+\theta+\frac{l}{l-x}\theta\right)+2M_p\left(\frac{l}{l-x}\theta\right)=P_c(0.6l\theta)+\frac{4P_c}{l}x\left(\frac{x\theta}{2}\right)+\frac{4P_c}{l}(l-x)\left(\frac{x\theta}{2}\right)$$

or:

$$M_p = P_c\,\frac{\left(0.6l^2+1.4lx-2x^2\right)}{(5l-2x)} \tag{14.14}$$

Putting $dM_p/dx = 0$, we obtain:

$$4x^2 - 20lx + 8.2l^2 = 0 \tag{14.15}$$

whence $x = 0.4505l$. Substituting in Equation 14.14, we obtain the maximum value of M_p:

$$M_p = 1.0061 \frac{P_c l}{5} \tag{14.16}$$

The values of x and M_p differ slightly from the approximate value $x = 0.45l$ and from the conservative value of $M_p = 1.0125\ (P_c l/5)$, obtained from the bending moment diagram in Figure 14.11b.

14.6 PLASTIC ANALYSIS BY COMPUTER

The computer program PLASTICF, companion to the present book (Chapter 20), analyzes plane frames subjected to a set of nodal forces. Without changing their magnitudes, the analysis is carried out for a unit increment of one of the nodal forces ($F_i = 1$).The corresponding member end moments are used to determine a load multiplier that causes M_p to be reached at a member end, thus developing a new plastic hinge. Change the stiffness matrix of the structure, $[S]$, accordingly, and repeat the analysis with a new load increment until $[S]$ becomes singular; its determinant becomes close to zero; or an element on the diagonal (S_{ii}) becomes zero in the process of solving the equilibrium equations or when the displacements become very large. The collapse load is the sum of the multipliers in all increments.

PLASTICF analyzes the plane frame shown in Figure 14.12a (taken from Wang, C.K., "General Computer Analysis for Limit Analysis", *Proceedings*, American Society of Civil Engineers, 89 (ST6), Dec. 1963, pp. 101–117). The main results are: Figure 14.12b shows the locations of four plastic hinges and the corresponding load increments; Figure 14.12c shows the collapse mechanism and the virtual work equation, confirming the value of P_c = 259E3 N, determined by PLASTICF; and Figure 14.12d depicts the bending moment diagram at collapse.

14.7 EFFECT OF AXIAL FORCE ON PLASTIC MOMENT RESISTANCE

In previous sections, we assumed that the fully plastic state at a hinge is induced solely by the bending moment M_p. However, in the presence of a high axial compressive or tensile force, a plastic hinge can be formed at a moment M_{pc} lower than the value M_p. We shall now derive an expression for M_{pc} for a rectangular cross section subjected to a tensile or compressive axial force; any buckling effect will be ignored.

Figures 14.13a to e show the changes in stress distribution in a rectangular cross section $b \times d$ subjected to an axial compressive force P together with a bending moment M in the vertical plane; the magnitude of M and P is increased at a constant value of M/P until the fully plastic stage is reached. The values of the axial force and the moment at this stage are:

$$P = 2\sigma_y b y_0 \tag{14.17}$$

and

$$M_{pc} = \frac{\sigma_y b}{4}\left(d^2 - 4y_0^2\right) \tag{14.18}$$

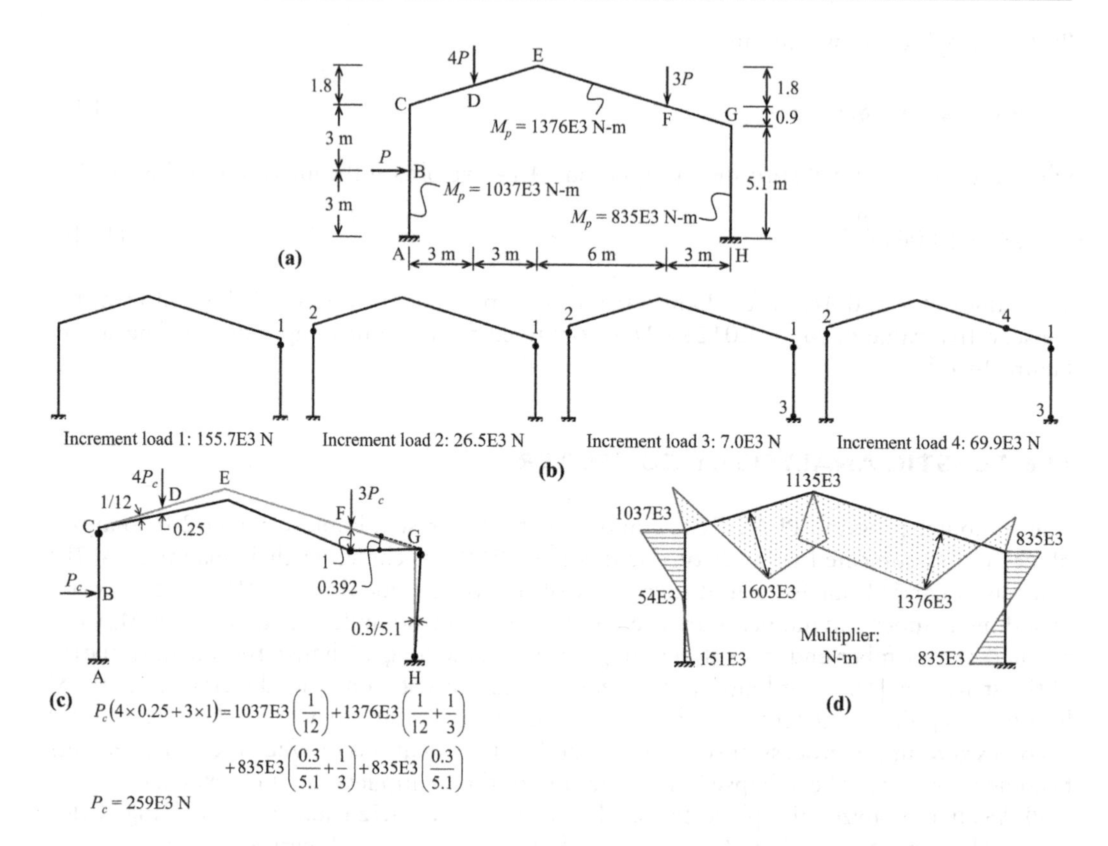

Figure 14.12 Plastic analysis of plane frame, E = 200E9 N/m² using computer program PLASTICF. (a) Frame dimensions. (b) Development of plastic hinges with load increments. (c) Collapse mechanism and virtual work equation. (d) Bending moment diagram at collapse.

where σ_y is the yield stress and y_0 is the distance from the centroid of the section to the neutral axis.

In the absence of the axial force, $y_0 = 0$, and the fully plastic moment is then:

$$M_p = \sigma_y \frac{bd^2}{4} \tag{14.19}$$

On the other hand, if the axial force alone causes a fully plastic state, its magnitude is:

$$P_y = \sigma_y bd \tag{14.20}$$

From Equations 14.17–14.20, we can obtain the interaction equation:

$$\frac{M_{pc}}{M_p} = 1 - \left(\frac{P}{P_y}\right)^2 \tag{14.21}$$

The interaction curve of M_{pc}/M_p plotted against P/P_y in Figure 14.14 can be used to determine the strength of a section under combined loading of an axial force and a bending

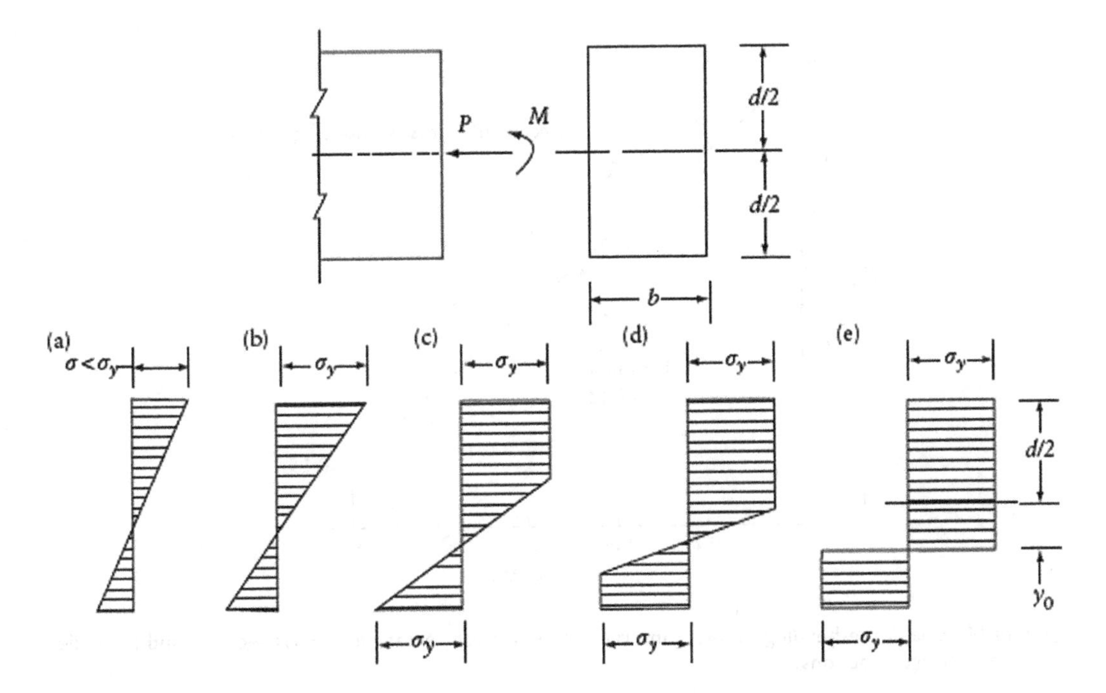

Figure 14.13 Stress distribution in a rectangular section subjected to an increasing bending moment M and axial force P at a constant value of M/P.

moment from the strengths in two simple types of loading: axial force only and bending moment only.

The shape of the interaction curve depends on the geometry of the cross section. For all wide-flange I-sections used in steel construction, the interaction curves fall within a narrow band and may be approximated by two straight lines shown in Figure 14.14. From this figure, it is apparent that when $P/P_y \leq 0.15$, the effect of the axial load is neglected, and when $P/P_y \geq 0.15$, we use the equation:

$$\frac{M_{pc}}{M_p} = 1.18\left(1 - \frac{P}{P_y}\right) \tag{14.22}$$

14.8 GENERAL

Plastic analysis presented in this chapter calculates the collapse load considering one loading case ignoring effect of axial forces and instability. Trial collapse mechanisms give an upper bound of the collapse load. The collapse load corresponding to a false mechanism is larger than the true collapse load. Through successive loading increments, the computer program PLASTICF seeks the collapse mechanism, then calculates the collapse load and the member-end moments at collapse. With PLASTICF, plane frame subjected to distributed load has to be substituted by statically equivalent nodal forces on short members.

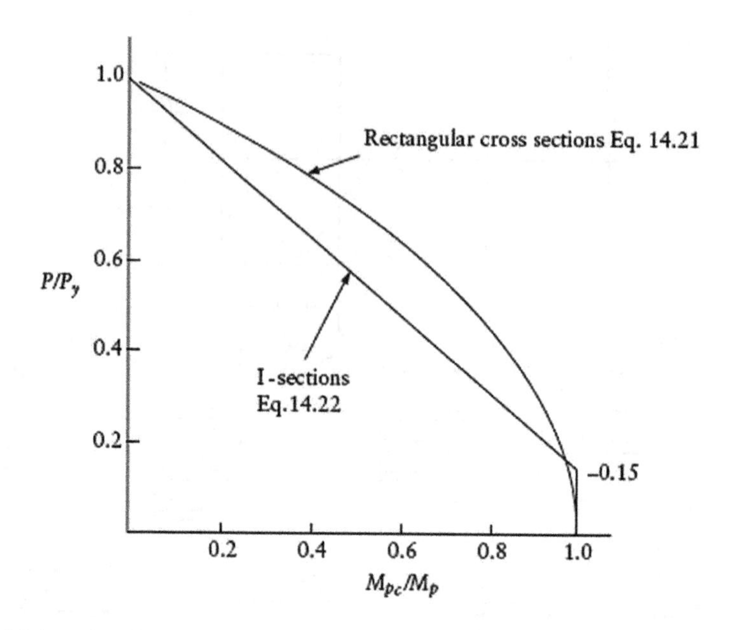

Figure 14.14 Axial force-bending moment interaction curves for rectangular cross sections and for wide-flange I-sections.

PROBLEMS

14.1 Find the required plastic moment resistance of the cross section(s) for the beam in the figure, which is to be designed to carry the given loads with a load factor of 1.7. Assume that the beam has: (a) a constant cross section, (b) two different cross sections – one from A to C and the other from C to D.

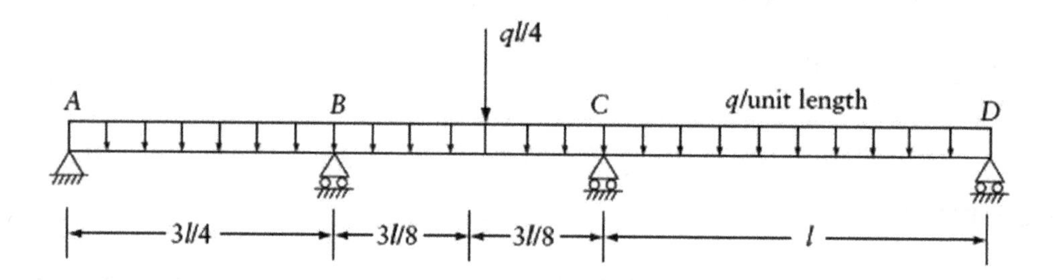

14.2 Determine the fully plastic moment for the frames shown in parts (a) to (e) with the collapse loads indicated. Ignore the effects of shear and axial forces.

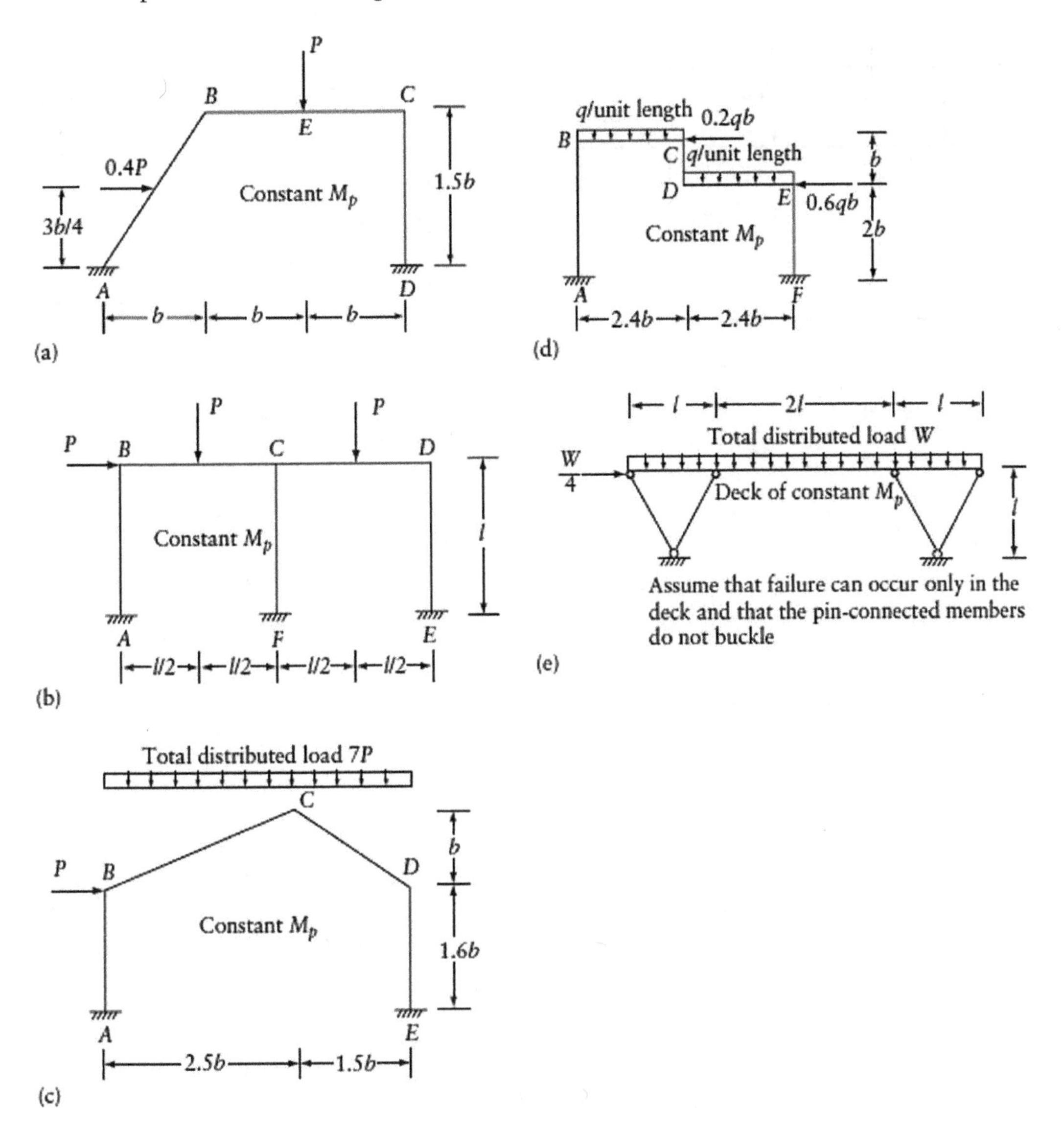

14.3 What is the value of M_p for the frame in Problem 14.2b, if the axial force effect (excluding buckling) is taken into account? Assume that the frame has a constant rectangular section $b \times d$, with $d = l/15$.

14.4 and 14.5 Use computer program PLASTIC for the shown frames.

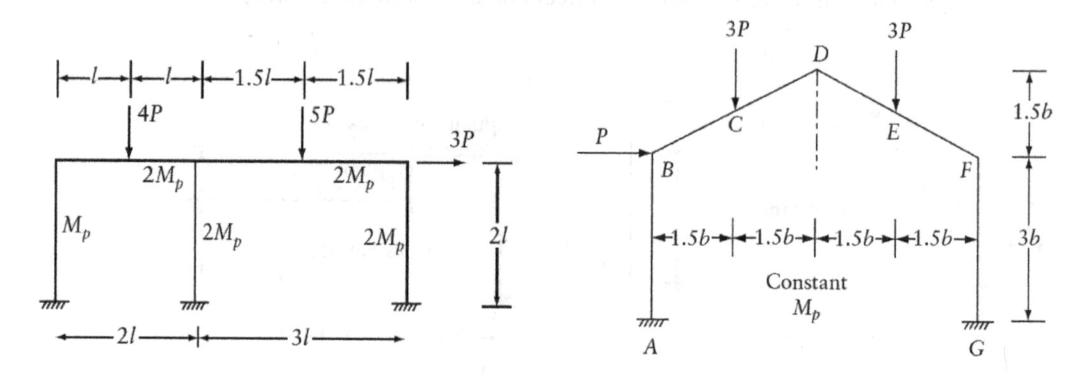

Chapter 15

Yield-line analysis of reinforced concrete slabs

15.1 INTRODUCTION

This chapter deals in a general way with plates, but many of the applications are virtually limited to reinforced concrete slabs. For this reason, the term slab will be generally used.

An elastic analysis of a reinforced concrete slab gives no indication of its ultimate load carrying capacity and further analyses have to be made for this condition. An exact solution for the ultimate flexural strength of a slab can be found only rarely, but it is possible to determine upper and lower bounds to the true collapse load.

The yield-line method of analysis gives an upper bound to the ultimate load capacity of a reinforced concrete slab by a study of assumed mechanisms of collapse. This method, developed by Johansen,[1] is a powerful tool for estimating the required bending resistance and hence the necessary reinforcement, especially for slabs of nonregular geometry or loading. Two approaches are possible in yield-line theory. The first one is an energy method in which the external work, done by the loads during a small virtual movement of the collapse mechanism, is equated to the internal work. The alternative approach is by the study of the equilibrium of the various parts of the slab into which the slab is divided by the yield lines. We may note that it is the equilibrium of slab parts that is considered and not the equilibrium of forces at all points of the yield line.

In contrast to the above, lower bound solutions to the collapse load are obtained by satisfying equilibrium at *all* points in the slab and necessitate the determination of a complete bending moment field in equilibrium with the applied loading.

15.2 FUNDAMENTALS OF YIELD-LINE THEORY

The slab is assumed to collapse at a certain ultimate load through a system of yield lines or fracture lines, called the pattern of fracture. The working load is obtained by dividing this ultimate load by the required load factor. For design, the working load is multiplied by the load factor, and the required ultimate moment of resistance is determined.

The basic fundamentals and main assumptions of the yield-line theory are as follows:

1. At fracture, the bending moment per unit length along all the fracture lines is constant and equal to the yield value corresponding to the steel reinforcement. The fracture is assumed to occur due to the yield of the steel.
2. The slab parts rotate about axes along the supported edges. In a slab supported directly on columns, the axes of rotation pass through the columns. Figure 15.1 shows some typical fracture patterns.

DOI: 10.1201/9780429286858-15

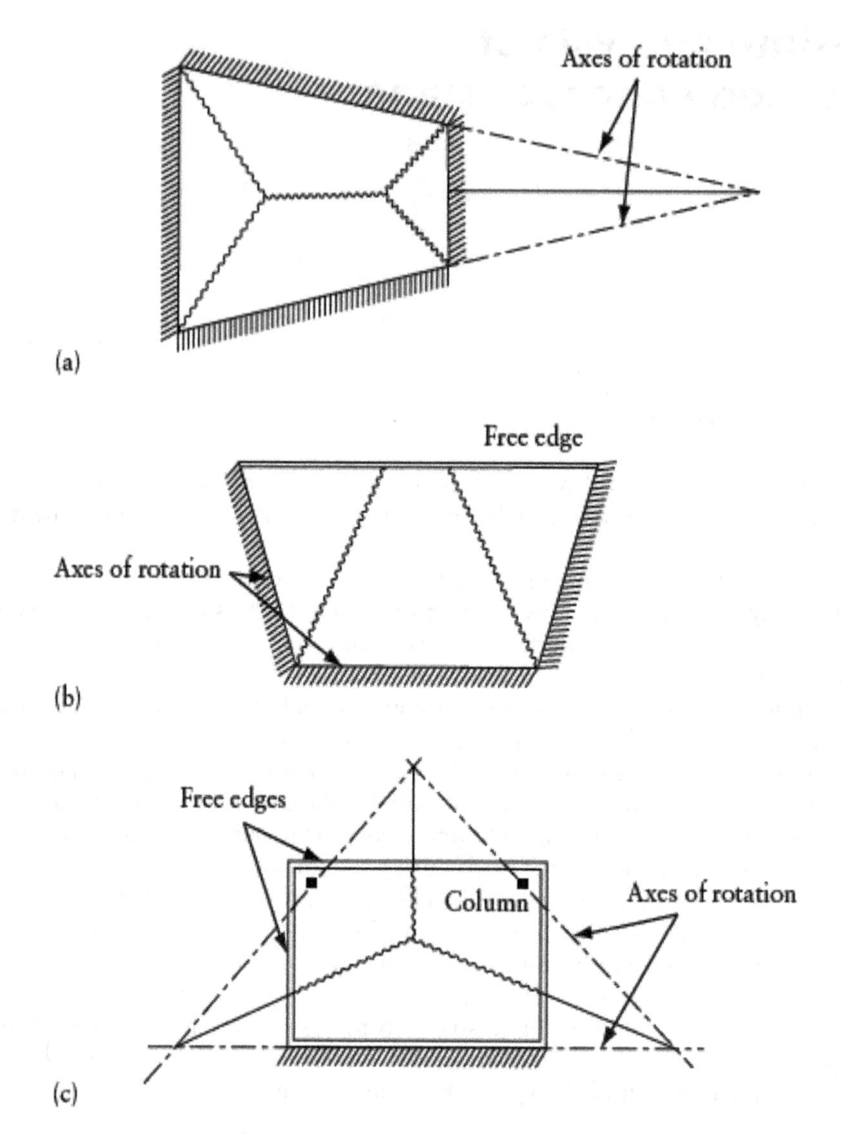

Figure 15.1 Typical fracture patterns in slabs. (a) Slab simply supported on four sides. (b) Slab simply supported on three sides and free along the fourth side. (c) Slab simply supported on two columns and on one side.

3. At fracture, elastic deformations are small compared with the plastic deformations and are therefore ignored. From this assumption and the previous one, it follows that fractured slab parts are plane, and therefore they intersect in straight lines. In other words, the yield lines are straight.
4. The lines of fracture on the sides of two adjacent slab parts pass through the point of intersection of their axes of rotation.

Figure 15.2 shows the fracture pattern of a uniformly loaded slab simply supported on three sides. It is readily seen that the pattern satisfies the requirements given above. Each of the three slab parts rotates about its axis of rotation by an angle θ_i which is related to the

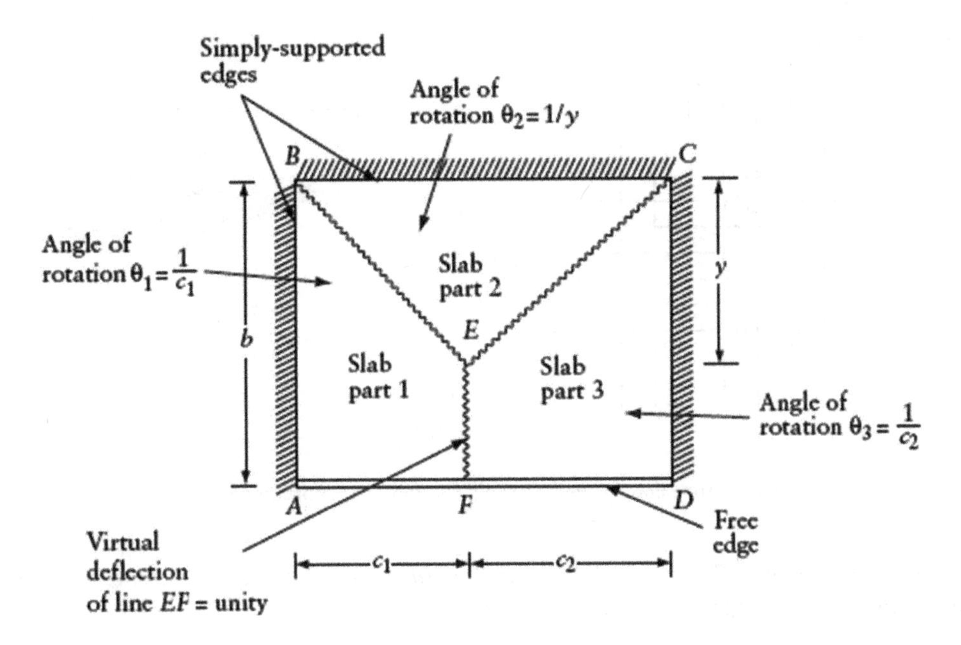

Figure 15.2 Angular rotation of slab parts.

rotation of the other parts. Let points E and F have a virtual downward displacement $w = 1$. The rotations of the slab parts are then:

$$\theta_1 = \frac{1}{c_1}; \ \theta_2 = \frac{1}{y}; \ \theta_3 = \frac{1}{c_2}$$

15.2.1 Convention of representation

The different conditions of supports will be indicated thus:

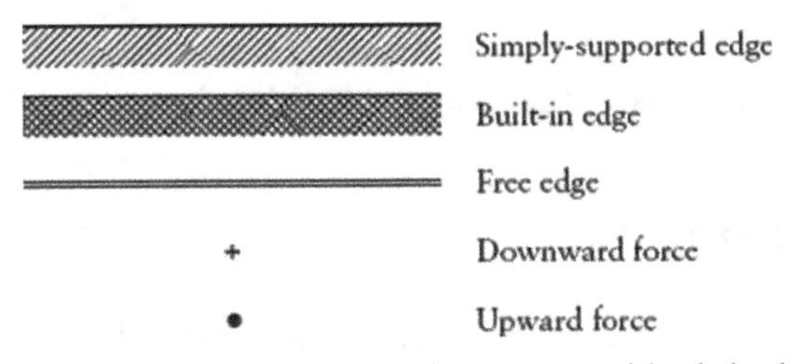

A positive ultimate moment m per unit length causes yield of the bottom reinforcement. A yield-line formed by a positive moment is referred to as a positive yield-line. Negative ultimate moment m' per unit length causes yield of the top reinforcement along a negative yield-line.*

15.2.2 Ultimate moment of a slab equally reinforced in two perpendicular directions

Consider a slab reinforced in two perpendicular directions, x and y, with different reinforcement corresponding to ultimate positive moments m_1 and m_2 (Figure 15.3a). For equilibrium

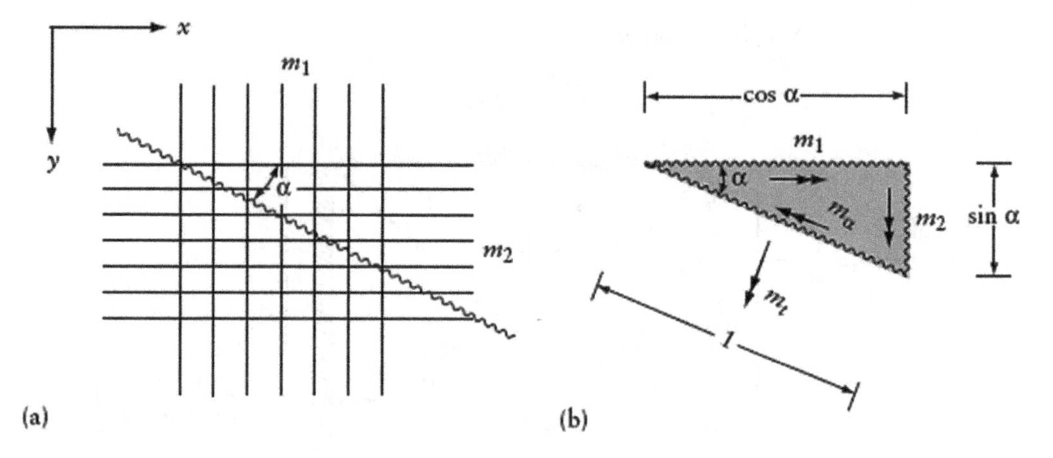

Figure 15.3 Moments at a fracture line inclined to the direction of reinforcement.

of the element shown in Figure 15.3b, the bending (m_α) and twisting (m_t) moments at a fracture line making an angle a with the x-axis are:

$$\left.\begin{aligned} m_\alpha &= m_1 \cos^2 \alpha + m_2 \sin^2 \alpha \\ m_t &= \left(m_1 - m_2\right) \sin\alpha \cos\alpha \end{aligned}\right\} \tag{15.1}$$

The moments are represented in Figure 15.3b (as usual) by double-headed arrows in the direction of the progress of a right-hand screw rotating in the same direction as the moment.

In an isotropic slab (that is, one equally reinforced in two perpendicular directions) $m_1 = m_2 = m$, the moment on any inclined fracture line is:

$$\left.\begin{aligned} m_\alpha &= m\left(\cos^2 \alpha + \sin^2 \alpha\right) = m \\ m_t &= 0 \end{aligned}\right\} \tag{15.2}$$

The values of m_α and m_t can also be determined by Mohr's circle as shown in Figure 15.4.

15.3 ENERGY METHOD

In this method, the pattern of fracture is assumed and the slab is allowed to deflect in the fractured state as a mechanism. Each slab part will rotate a small virtual angle θ about its axis of rotation. The relation between the rotation of the slab parts is defined by the choice of the fracture pattern. The internal energy dissipated on the yield lines during the virtual rotation is equated to the external virtual work done in deflecting the slab. From this equation, the value of the ultimate moment is obtained. We should note, when calculating the internal energy, that only the ultimate moments in the yield lines do work during rotation.

The virtual work equation (similar to the equation used for plastic analysis of frames, Chapter 14) gives either the correct ultimate moment or a value smaller than the correct value. In other words, if the virtual work equation is used to find the ultimate load for a slab with an assumed bending resistance, then the value obtained will be an upper bound

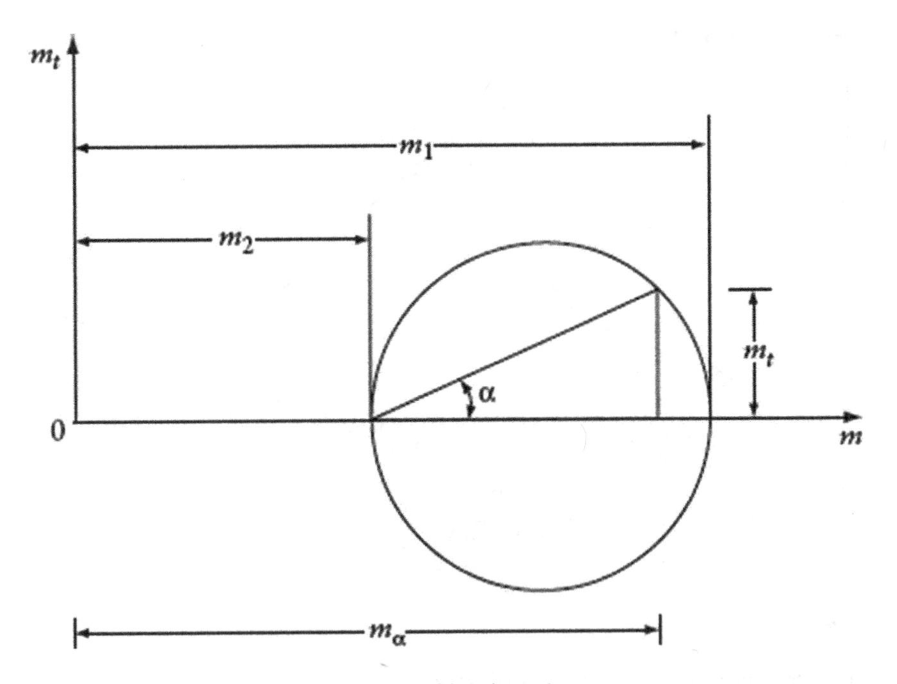

Figure 15.4 Moments on an inclined fracture line by Mohr's circle.

on the carrying capacity of the slab. This means that the solution obtained is either correct or unsafe. In practical calculations, one or two fracture patterns are assumed, and the value obtained is usually within 10 percent of the correct value. It seems to be a reasonable design procedure to increase the moment obtained by the work equation by a small percentage, depending on the number of trials and on the uncertainty of the chosen fracture pattern. The theoretical exact pattern is that for which the ultimate moment is a maximum. This can be reached, if we define the fracture pattern by certain parameters x_1, x_2, ...; the work equation will then give the value of m as a function of these parameters, i.e. $m = f(x_1, x_2, \ldots)$. The value of the parameters corresponding to the maximum moment is determined by partial differentiation: $(\partial f/\partial x_1) = 0$, $(\partial f/\partial x_2) = 0$, etc. This process can become laborious except for simple slabs in which the designer can define a reasonable pattern and proceed as suggested above.

The internal work done during a virtual rotation θ of a slab part is equal to the scalar product of a vector $\vec{M} = \overrightarrow{ml}$ and a vector $\vec{\theta}$ along the axis of rotation (Figure 15.5). The internal work for this slab part is then $\vec{M} \cdot \vec{\theta} = ml\,(\cos\alpha)\theta$, where α = angle between the two vectors. This means that, for any part of the slab, the internal work is equal to the rotation of that part multiplied by the projection of the ultimate moment upon the axis of rotation. It is sometimes convenient to consider the components of the moments in two perpendicular directions x and y in the plane of the slab. The total internal virtual work for all the slab parts is then:

$$U = \sum \vec{M} \cdot \vec{\theta} = \sum M_x \cdot \theta_x + \sum M_y \cdot \theta_y \tag{15.3}$$

where θ_x and θ_y = the x and y components of the rotation vector; M_x and M_y = the x and y components of the vector $\vec{M} = \overrightarrow{ml}$. The symbol U represents the work done by the plastic moment to produce rotation of the plastic hinge. The arrow shown on the yield-line in

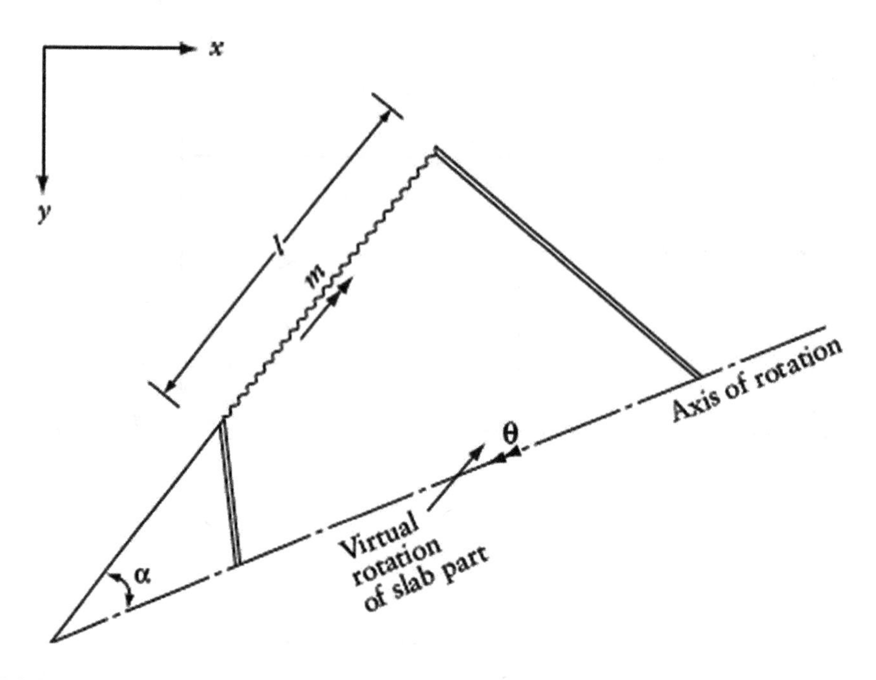

Figure 15.5 Data for calculation of internal virtual work.

Figure 15.5 represents the effect of the plastic hinge on the slab part; the moment exerted by the slab part on the plastic hinge is in the opposite direction.

If x and y are the coordinates of any point on the slab and w is the vertical displacement corresponding to the virtual rotation of the slab parts, then the total external virtual work is:

$$W = \iint qw \, dx \, dy \tag{15.4}$$

where q = the load intensity. The virtual work equation is:

$$W = U \tag{15.5}$$

whence:

$$\sum \left(M_x \cdot \theta_x + M_y \cdot \theta_y \right) = \iint qw \, dx \, dy \tag{15.6}$$

EXAMPLE 15.1: ISOTROPIC SLAB SIMPLY SUPPORTED ON THREE SIDES

Determine the ultimate moment of a square isotropic slab simply supported on three sides and subjected to a uniform load q per unit area.

Because of symmetry, the fracture pattern is fully determined by one parameter x, as shown in Figure 15.6. Let the junction of the three fracture lines have a virtual displacement, $x = 1$. The rotations of the slab parts are:

$$\theta_1 = \frac{1}{x}; \quad \theta_2 = \theta_3 = \frac{2}{l}$$

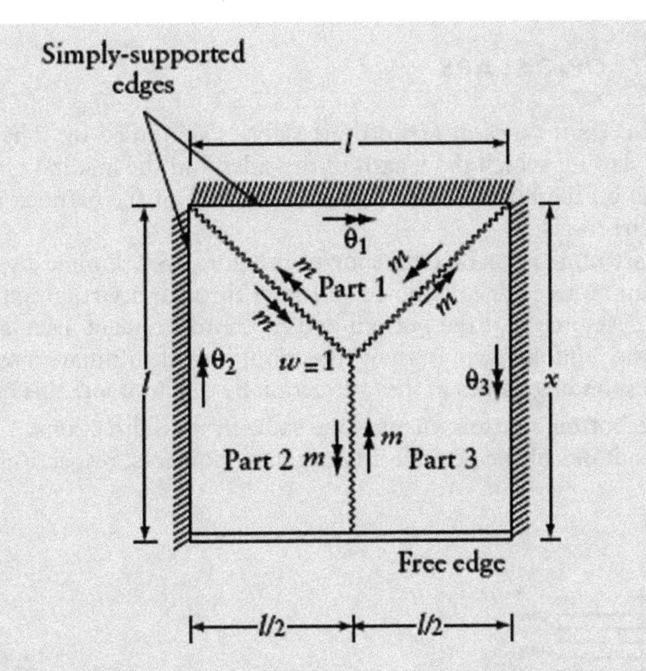

Figure 15.6 Slab considered in Example 15.1.

The internal virtual work is:

$$U = ml\frac{1}{x} + 2ml\frac{2}{l}$$

where the first term applies to Part 1 and the second to Parts 2 and 3.

The work done by the distributed load is equal to the resultant on each part multiplied by the vertical displacement of its point of application. Thus:

$$W = q\left\{l\frac{x}{2}\frac{1}{3} + 2\left[(l-x)\frac{l}{2}\frac{1}{2} + \frac{xl}{4}\frac{1}{3}\right]\right\}$$

Here again, the first term is for Part 1, and the second term is for Parts 2 and 3. Equating the internal and external work:

$$m\left(\frac{l}{x} + 4\right) = ql^2\left(\frac{1}{2} - \frac{x}{6l}\right)$$

Whence:

$$m = ql^2\left[\frac{3 - \frac{x}{l}}{6\left(\frac{l}{x} + 4\right)}\right]$$

The maximum value of m is obtained when $\mathrm{d}m/\mathrm{d}x = 0$, which gives $x/l = 0.65$. The corresponding ultimate moment is:

$$m = ql^2/14.1$$

15.4 ORTHOTROPIC SLABS

The analysis of certain types of orthotropic slabs is simplified by Johansen to that of an isotropic affine slab for which the length of the sides and the loading are altered in certain ratios depending on the ratio of the ultimate resistance of the orthotropic slab in the two perpendicular directions.

Consider a part of a slab ABCDEF shown in Figure 15.7, limited by positive and negative yield-lines and a free edge, assumed to rotate through a virtual angle θ about an axis of rotation R-R. Assume that the bottom and top reinforcement are placed in the x and y directions. Let the reinforcement in the y direction† provide ultimate moments of m and m', and let the corresponding values in the x direction be ϕm and $\phi m'$; this means that the ratio of the top to the bottom reinforcement is the same in both directions. The vectors $\vec{c}$ and $\vec{b}$ represent the resultants of the positive and negative moments, respectively.

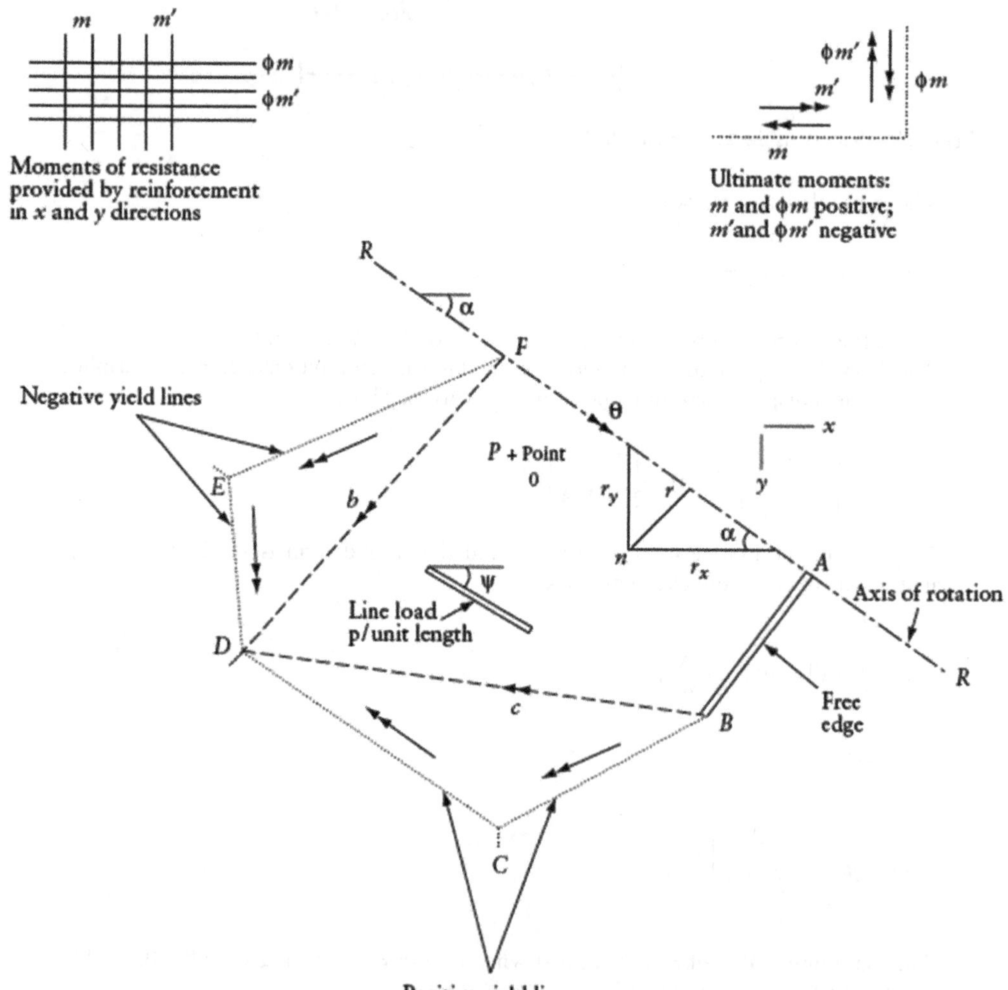

Figure 15.7 Conversion of an orthotropic slab to an isotropic affine slab.

The internal virtual work for this slab part is:

$$U = \left(m c_x + m' b_x\right) \theta_x + \phi \left(m c_y + m' b_y\right) \theta_y \tag{15.7}$$

where (c_x, c_y) and (b_x, b_y) = projections in the x and y directions of the lengths c and b; θ_x and θ_y = x and y components of the rotation vector $\vec{\theta}$.

Assuming that the virtual deflection at a point n, distance r from the axis R-R, is unity, the rotation θ and its components can be written as:

$$\theta = \frac{1}{r}; \quad \theta_x = \theta \cos\alpha = \frac{1}{r_y}; \quad \theta_y = \theta \sin\alpha = \frac{1}{r_x} \tag{15.8}$$

where α = the angle between the x axis and the axis of rotation. Substituting Equation 15.8 into Equation 15.7, we obtain:

$$U = \theta\left[\left(m c_x + m' b_x\right) \cos\alpha + \phi \left(m c_y + m' b_y\right) \sin\alpha\right] \tag{15.9}$$

Let us assume now that the loading on the slab part consists of a uniformly distributed load q per unit area, a line load p per unit length on a length l in a direction making an angle ψ with the x-axis and a concentrated load P at point O. The external virtual work of the loads on this slab part is:

$$W = \iint q w \, \mathrm{d}x \, \mathrm{d}y + \int p w_l \, \mathrm{d}l + P w_P \tag{15.10}$$

where w = deflection at any point (x, y); w_l = deflection of any point below the line load; w_P = deflection below the concentrated load. The integration is carried out for the whole area and for the loaded length. The virtual work equation is:

$$\begin{aligned} &\sum\left[\left(m c_x + m' b_x\right) \frac{1}{r_y} + \phi \left(m c_y + m' b_y\right) \frac{1}{r_x}\right] \\ &= \sum\left(\iint q w \, \mathrm{d}x \, \mathrm{d}y + \int p w_l \, \mathrm{d}l + P w_P\right) \end{aligned} \tag{15.11}$$

where the summation is for all slab parts.

Consider an affine slab equally reinforced in the x and y directions so that the ultimate positive and negative moments are m and m', respectively. Suppose that this affine slab has all its dimensions in the x direction equal to those of the actual slab multiplied by a factor λ. The pattern of fracture remains similar, and the corresponding points can still have the same vertical displacements. The internal virtual work for the part of the affine slab is:

$$U' = \left(m \lambda c_x + m' \lambda b_x\right) \frac{1}{r_y} + \left(m c_y + m' b_y\right) \frac{1}{\lambda r_x} \tag{15.12}$$

Let the loading on the affine slab be a distributed load of q' per unit area, a line load p' per unit length, and a concentrated load P'. The external work for this part of the affine slab is:

$$W' = \iint q' w \lambda \, \mathrm{d}x \, \mathrm{d}y + \int p' w_l \sqrt{(\mathrm{d}y)^2 + \lambda^2 (\mathrm{d}x)^2} + P' w_P \tag{15.13}$$

Dividing both the internal and external work by λ will not change the work equation, which then becomes:

$$\sum(mc_x + m'b_x)\frac{1}{r_y} + \frac{1}{\lambda^2}\sum(mc_y + m'b_y)\frac{1}{r_x}$$
$$= \sum\left(\iint q'w\,dx\,dy + \int p'w_l\sqrt{\frac{(dy)^2}{\lambda^2} + (dx)^2} + \frac{P'}{\lambda}w_P\right) \tag{15.14}$$

All terms of the virtual work Equations 15.11 and 15.14 are identical provided that:

$$\phi = \frac{1}{\lambda^2} \quad \text{or} \quad \lambda = \sqrt{\frac{1}{\phi}} \tag{15.15}$$

$$q' = q \tag{15.16}$$

$$p'\sqrt{\frac{(dy)^2}{\lambda^2} + (dx)^2} = p\,dl$$

or:

$$p' = \frac{p}{\sqrt{\phi\sin^2\psi + \cos^2\psi}} \tag{15.17}$$

and:

$$P' = P\sqrt{\frac{1}{\phi}} \tag{15.18}$$

It follows that an orthotropic slab with positive and negative ultimate moments m and m' in the x direction and ϕm and $\phi m'$ in the y direction, can be analyzed as an isotropic slab with moments m and m' but with the linear dimensions in the x direction multiplied by $\sqrt{1/\phi}$. The intensity of a uniformly distributed load remains the same. A linear load has to be multiplied by $(\phi\sin^2\psi + \cos^2\psi)^{1/2}$ with ψ being the angle between the load line and the x axis. A concentrated load has to be multiplied by $\sqrt{1/\phi}$.

EXAMPLE 15.2: RECTANGULAR SLAB

A rectangular orthotropically reinforced slab is shown in Figure 15.8a. Find the dimensions of an isotropic affine slab.

We have ϕ = ½, so that side AB is to be changed to $l\sqrt{1/\phi} = 1.414l$ for an isotropic affine slab with ultimate moments of $2m$ and $2m'$ (Figure 15.8b). The same orthotropic

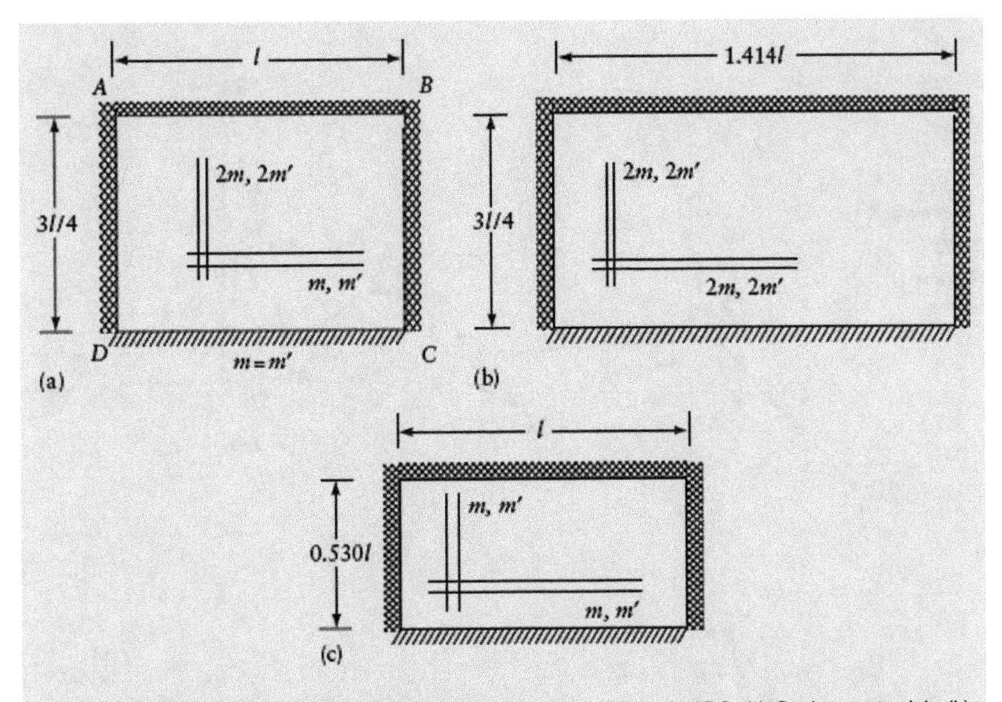

Figure 15.8 Orthotropic slab and affine slabs considered in Example 15.2. (a) Orthotropic slab. (b) Affine slab. (c) Alternative affine slab.

slab can also be analyzed as an isotropic slab for which side AB remains unchanged and side AD is changed to $(3l/4)/\sqrt{2} = 0.530l$, the ultimate moments being m and m' (Figure 15.8c).

15.5 EQUILIBRIUM OF SLAB PARTS

The energy method of the last two sections gives an upper bound value to the collapse load and can always be used for any assumed mechanism of collapse. Where the mechanism is complex and its layout is defined by several initially unknown dimensions, the algebraic manipulation necessary to obtain a solution can be long and tedious. There can, however, be a saving of work in many cases by considering the equilibrium of the slab parts.

In this approach, we abandon the virtual work equations of the energy method and consider instead the equilibrium of each slab part when acted upon by the external applied load and by the forces acting at a fracture line. In general, these are: bending moment, shearing force acting perpendicular to the slab plane, and twisting moment.

To establish the equilibrium conditions, it is not necessary to know the precise distribution of the shear and of the twisting moment; they can be replaced by two forces perpendicular to the plane of the slab, one at each end of the fracture line. These two forces are referred to as nodal forces and are denoted by V; they are considered positive when acting upwards.

15.5.1 Nodal forces

The formulas given here follow the original theory of Johansen. There are some restrictions on their use; although, in the majority of cases, they give a satisfactory solution. These restrictions have been studied by Wood and Jones.[2]

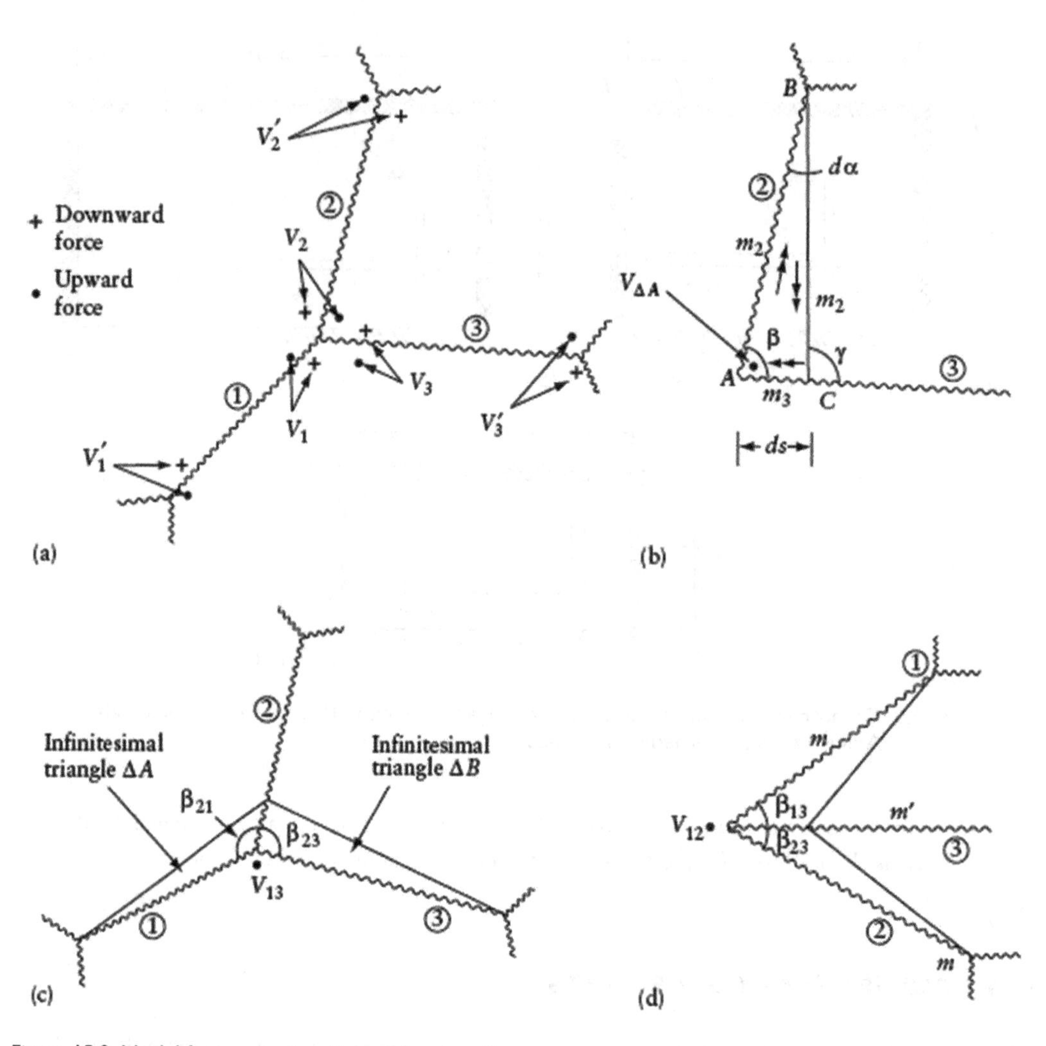

Figure 15.9 Nodal forces at a junction of fracture lines.

In Figure 15.9a the shears and twists on fracture lines (1), (2), and (3) are represented by the equivalent nodal forces V_1 and V_1', etc., on ends of lines (1), (2), and (3). The nodal forces are equal and opposite on the two sides of each fracture line. It follows that a summation of all the nodal forces at any junction of the fracture lines is zero.

Consider an elemental triangle ABC of area ΔA limited by positive fracture lines (2) and (3) (Figure 15.9b) and any adjacent line at a small angle $d\alpha$ to fracture line (2). This adjacent line is assumed to have the same bending moment m_2 as line (2) to a first order approximation. The resultant moment on the triangle $d\alpha$ is $(m_3 - m_2)$ ds directed from C to A. For equilibrium of the triangle ABC, the moments about BC vanish:

$$V_{\Delta A}\,\mathrm{d}s\sin\gamma + (m_3 - m_2)\mathrm{d}s\cos(180^\circ - \gamma) - \mathrm{d}P\frac{\mathrm{d}s\sin\gamma}{3} = 0$$

where $V_{\Delta A}$ = nodal force at A replacing the twisting moment and shearing force on the fracture line (2) and the length ds of the fracture line (3); dP = external load on the triangle

ABC, assumed to be uniformly distributed. As the triangle ΔA tends to zero, $\gamma \rightarrow \beta$ and $dP \rightarrow 0$, so that:

$$V_{\Delta A} = (m_3 - m_2)\cot\beta \tag{15.19}$$

The general form of this equation is:

$$V_{\Delta A} = (m_{ds} - m_{\text{long side}})\cot\beta \quad \text{(acting upward)} \tag{15.20}$$

where the subscripts of m indicate the bending moment per unit length on the sides ds and on the long side of the infinitesimal triangle.

Equation 15.20 can be used to find the value of the nodal force V between any two lines at the junction of fracture lines. The nodal force between lines (1) and (3) in Figure 15.9c is equal and opposite to the sum of the two nodal forces $V_{\Delta A}$ and $V_{\Delta B}$ of the infinitesimal triangles ΔA and ΔB, that is:

$$V_{13} = V_{\Delta A} - V_{\Delta B}$$

or:

$$V_{13} = (m_2 - m_1)\cot\beta_{21} - (m_2 - m_3)\cot\beta_{23} \tag{15.21}$$

It follows from the above that, when the reinforcement is equal in two orthogonal directions, the moments are $m_1 = m_2 = m_3 = m$, and all the nodal forces are zero at the junction of fracture lines of the same sign.

At a junction where two positive fracture lines (1) and (2) meet one negative fracture line (3) in an isotropic slab (Figure 15.9d) the nodal forces are:

$$V_{12} = -(-m' - m)\cot\beta_{13} - (-m' - m)\cot\beta_{23}$$

or:

$$V_{12} = (m' + m)(\cot\beta_{13} + \cot\beta_{23}) \tag{15.22}$$

where m' = absolute value of the bending moment on the negative fracture line. Similar equations can be written for the forces between other lines. When a fracture line meets a free or simply supported edge (Figure 15.10a), we have $V_{12} = m \cot \beta$. When the fracture line is positive, the nodal force V is downward in the acute angle.

If the edge rotation is restrained or otherwise subjected to a negative moment m' (Figure 15.10b), the nodal force is $V_{12} = (m + m') \cot \beta$, acting downward in the acute angle.

15.6 EQUILIBRIUM METHOD

As mentioned earlier, the slab parts are in equilibrium under the effect of: the external loading, the moments on the yield-lines, the nodal forces, and the support reactions. For each

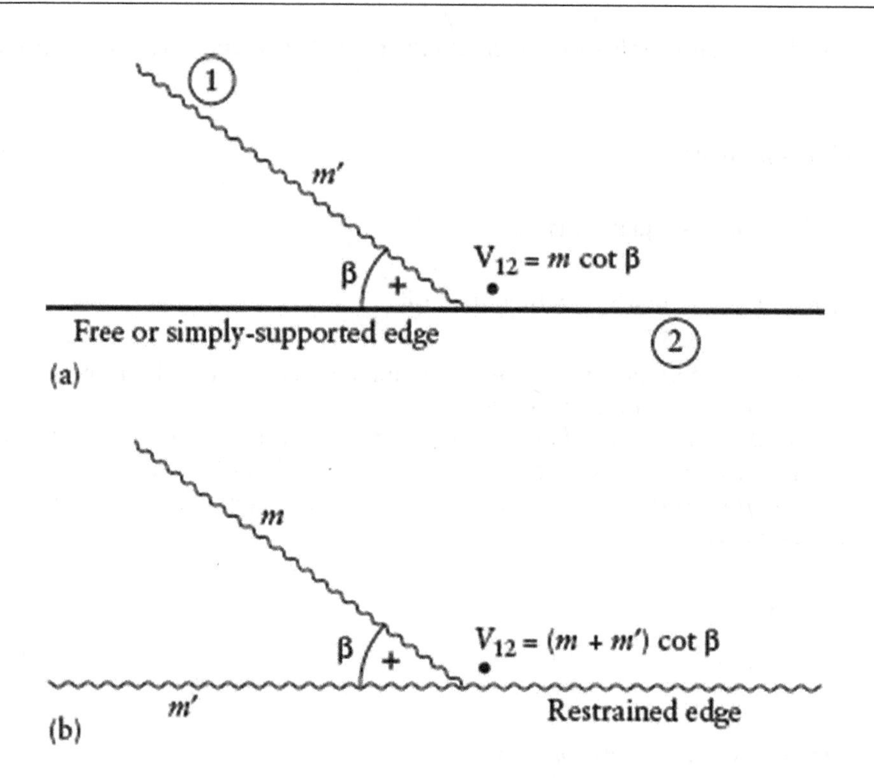

Figure 15.10 Nodal forces at a free or simply supported edge and at a restrained edge.

slab part, three equations of equilibrium can be written, viz., two moment equations about two axes in the plane of the slab and an equation for the forces perpendicular to the plane of the slab, each of which adds up to zero.

The fracture pattern for a slab is completely defined if the axes of rotation are known, together with the ratios of the rotations $\theta_1, \theta_2, \ldots, \theta_n$ of slab parts when the mechanism acquires a small virtual deflection. For n parts, we require $(n-1)$ ratios. The fracture pattern in Figure 15.11a is determined by drawing contour lines of the deflected mechanism. The contour line of deflection w is composed of n straight segments parallel to the axes of rotation and distant from them by $w/\theta_1, w/\theta_2, \ldots, w/\theta_n$ (Figure 15.11b). The intersections of the segments define points on the fracture lines.

For a part supported on one side, the position and magnitude of the reaction are unknown, thus representing two unknowns. For a part supported on a column, the axis of rotation passes through the column, but its direction is unknown and so is, of course, the magnitude of the reaction; hence, again, there are two unknowns for the part of the slab. For an unsupported part, the direction and position of the axis of rotation are unknown, so that, once again, we have two unknowns.

For n parts of a slab, the unknowns are: the value of the ultimate moment m, $(n-1)$ relations between the rotations of the parts, and two unknowns for each part. Hence, the total number of unknowns is $3n$, that is, the same as the number of equations of equilibrium (three for each part). The formulation of the equilibrium equations becomes complicated except in simple cases such as the slabs considered below.

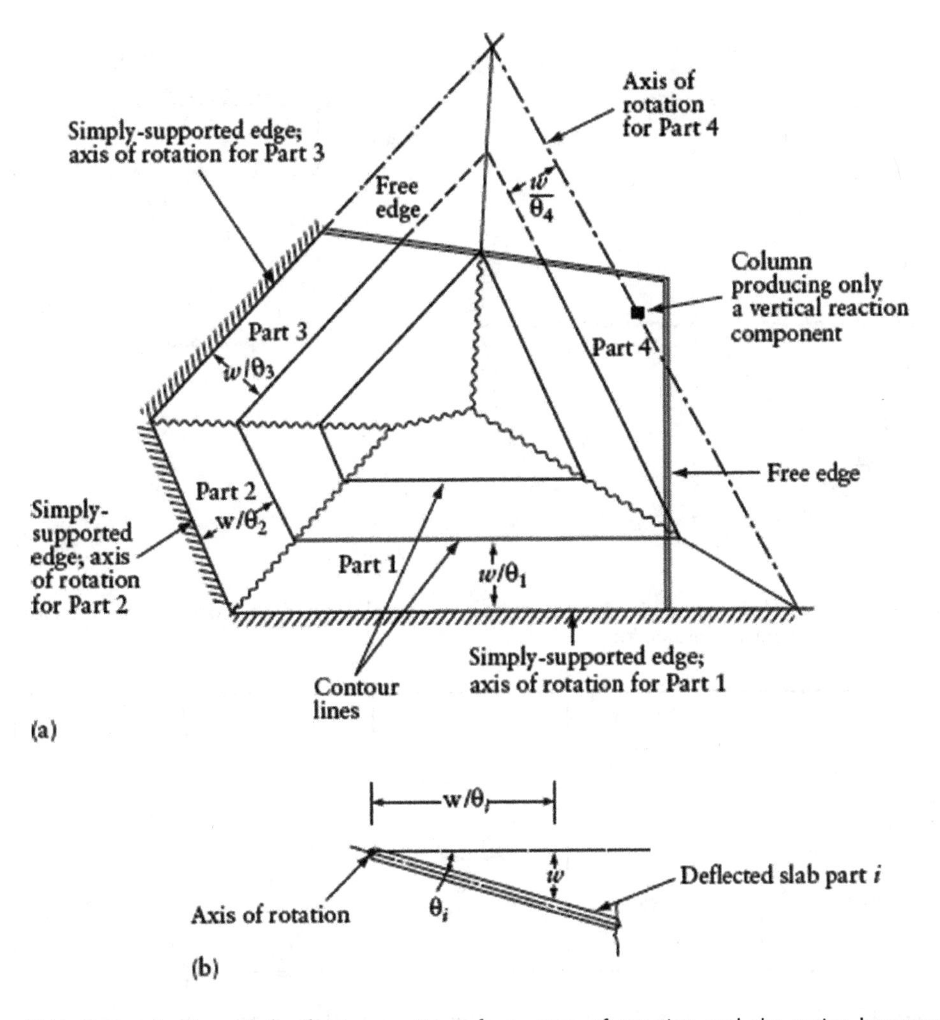

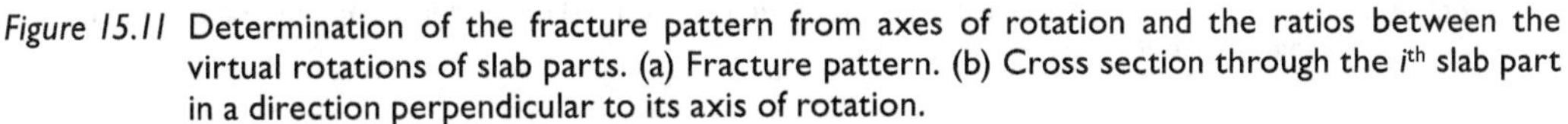

Figure 15.11 Determination of the fracture pattern from axes of rotation and the ratios between the virtual rotations of slab parts. (a) Fracture pattern. (b) Cross section through the ith slab part in a direction perpendicular to its axis of rotation.

EXAMPLE 15.3: SIMPLY SUPPORTED POLYGONAL SLAB

Find the ultimate moment, m in an isotropic polygonal slab simply supported on n equal sides, carrying a uniform load whose total magnitude is W.

The fracture pattern is shown in Figure 15.12. Taking moments about the edge for any slab part gives:

$$m = \frac{W}{6\, n \tan(\pi/n)} \tag{15.23}$$

The values of the ultimate moments for simply supported slabs in the form of regular polygons are given in Figure 15.13 (Equation 15.23). As n tends to infinity, m tends to $W/(6\pi)$ of a circular slab.

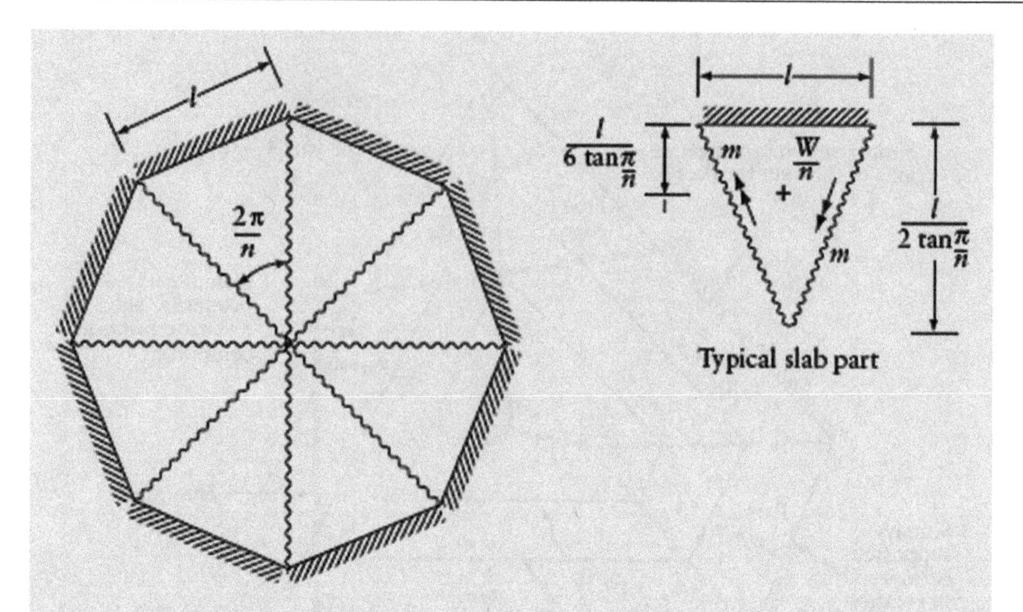

Figure 15.12 Equilibrium condition for an isotropic polygonal slab under uniformly distributed load of total magnitude W.

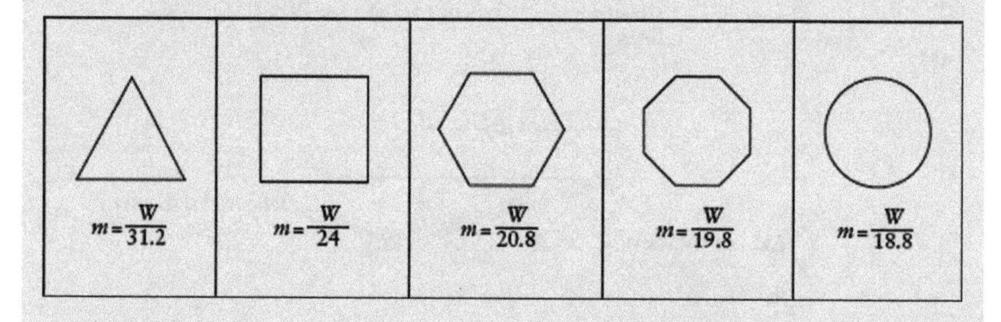

Figure 15.13 Ultimate moment for simply supported polygonal slabs under a uniform load of total magnitude W.

For isotropic square slab simply supported at corners, (Figure 15.14) gives:

$$m = W/8 \tag{15.24}$$

This is derived by taking moments for one slab part about its axis of rotation.

EXAMPLE 15.4: YIELD-LINE MECHANISM AT INTERIOR COLUMN

Figure 15.15a represents an assumed yield-line mechanism at an interior column transferring to the slab an upward force = V. The cross-sectional area of the column = c^2; the column is idealized as a circular column with radius = $c/\sqrt{\pi}$. Figure 15.15b depicts the forces and moments in one of the segments forming the mechanism.

Taking moments about the inner edge of the segment in Figure 15.15b gives:

$$m + m' = \frac{V}{0.4\,\pi\,l_c}\left(0.2\,l_c - \frac{c}{\sqrt{\pi}}\right) = \frac{V}{2\,\pi}\left(1 - 2.8\,\frac{c}{l_c}\right) \tag{15.25}$$

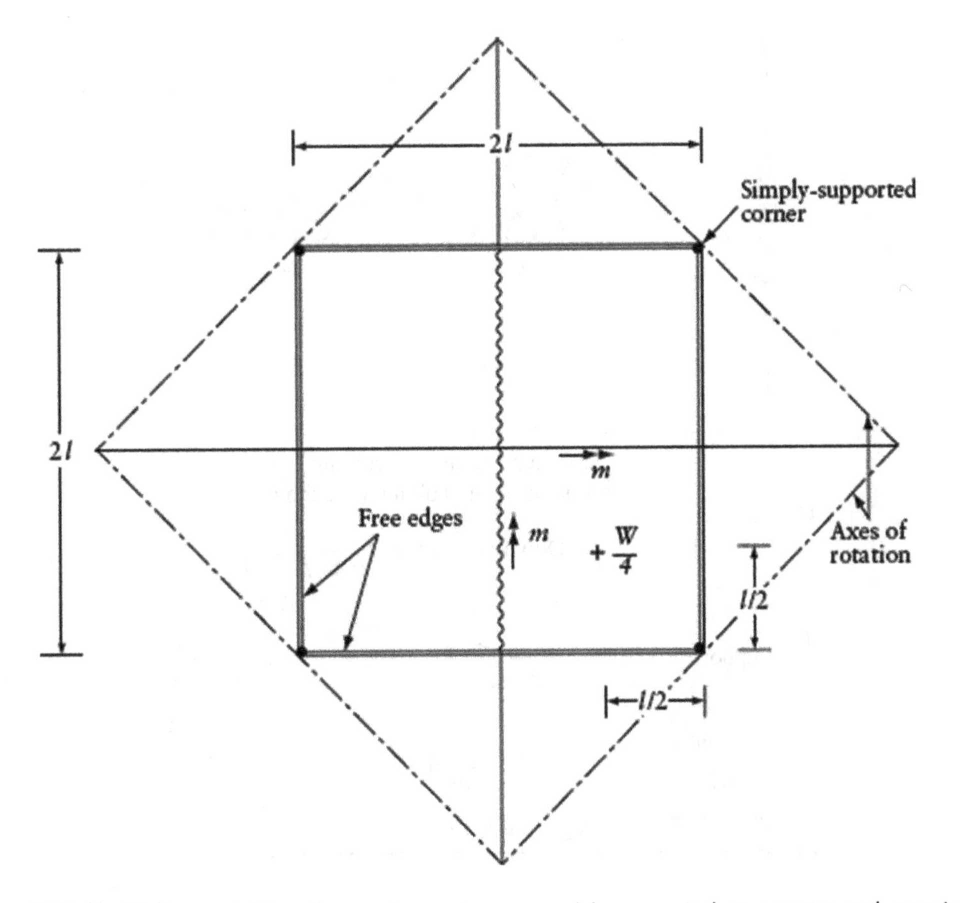

Figure 15.14 Equilibrium condition for an isotropic square slab supported at corners and carrying a uniformly distributed load of total magnitude W.

m and m' = absolute values of flexural strengths per unit width provided by bottom and top reinforcement, respectively; l_c = distance between column centers in two orthogonal directions. For a non-prestressed isotropic slab, m' (or m) can be expressed in terms of flexural reinforcement ratio ρ' $(=A_s'/(bd))$, with b = unit width; d = distance between centroid of tension reinforcement and extreme compression fiber; A_s or A_s' = cross-sectional area of bottom or top reinforcement within width = b, respectively:

$$m' = \rho' f_y d^2 \left[1 - 0.59\left(\rho' f_y / f_c'\right)\right] \tag{15.26}$$

f_c' and f_y = specified compressive strength of concrete and yield strength of flexural reinforcement, respectively.

Equations 15.25 and 15.26 can be used in design to find the reinforcement ratio ρ' or ρ for a given value of V and ratio of m/m'.

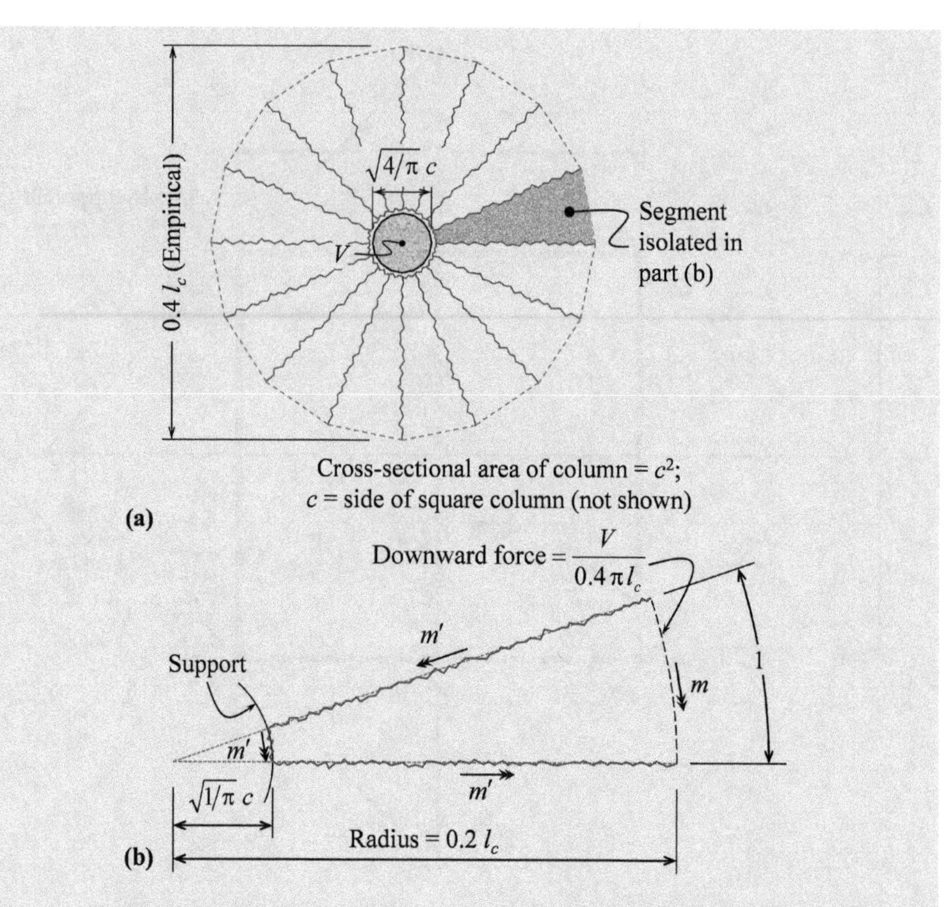

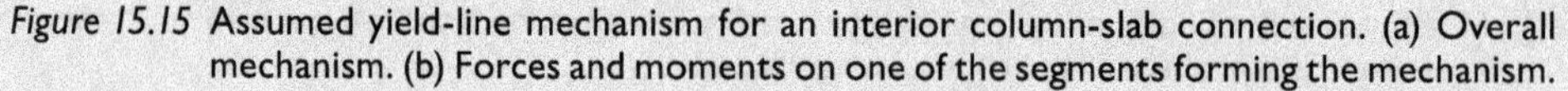

Figure 15.15 Assumed yield-line mechanism for an interior column-slab connection. (a) Overall mechanism. (b) Forces and moments on one of the segments forming the mechanism.

15.7 IRREGULAR SLABS

In the case of slabs of irregular shape or loading, the calculation of the ultimate moment is as follows.

a. A fracture pattern is assumed, and the corresponding ultimate moment is computed by considering the equilibrium of each slab part. A moment equation about the axis of rotation of each part will give different values of the ultimate moment. For the correct fracture pattern, all yield moment values must be equal.
b. In general, the ultimate moments computed for the first assumed fracture pattern are not equal and the values will indicate how the pattern should be corrected. The procedure is then repeated with a "more correct" pattern until the exact pattern is obtained. This then is an iterative method.
c. If, for an assumed pattern, the values of the ultimate moment obtained by equilibrium considerations do not differ much from one another, the application of the work equation for this fracture mechanism will give a value of the ultimate moment very close to the correct answer.

EXAMPLE 15.5: RECTANGULAR SLAB WITH OPENING

Find the ultimate moment for the isotropic slab shown in Figure 15.16a. The ultimate positive and negative moments are equal.

For the first trial, we assume the fracture pattern A of Figure 15.16b. The only nodal forces that need to be considered are those at the intersection of the inclined fracture line from the slab corner with the free edge. A moment equation about the axis of rotation of each slab part gives

Part 1:

$$2m(1.6b) = \frac{q(1.6b)b^2}{6}; \quad m = 0.0833qb^2$$

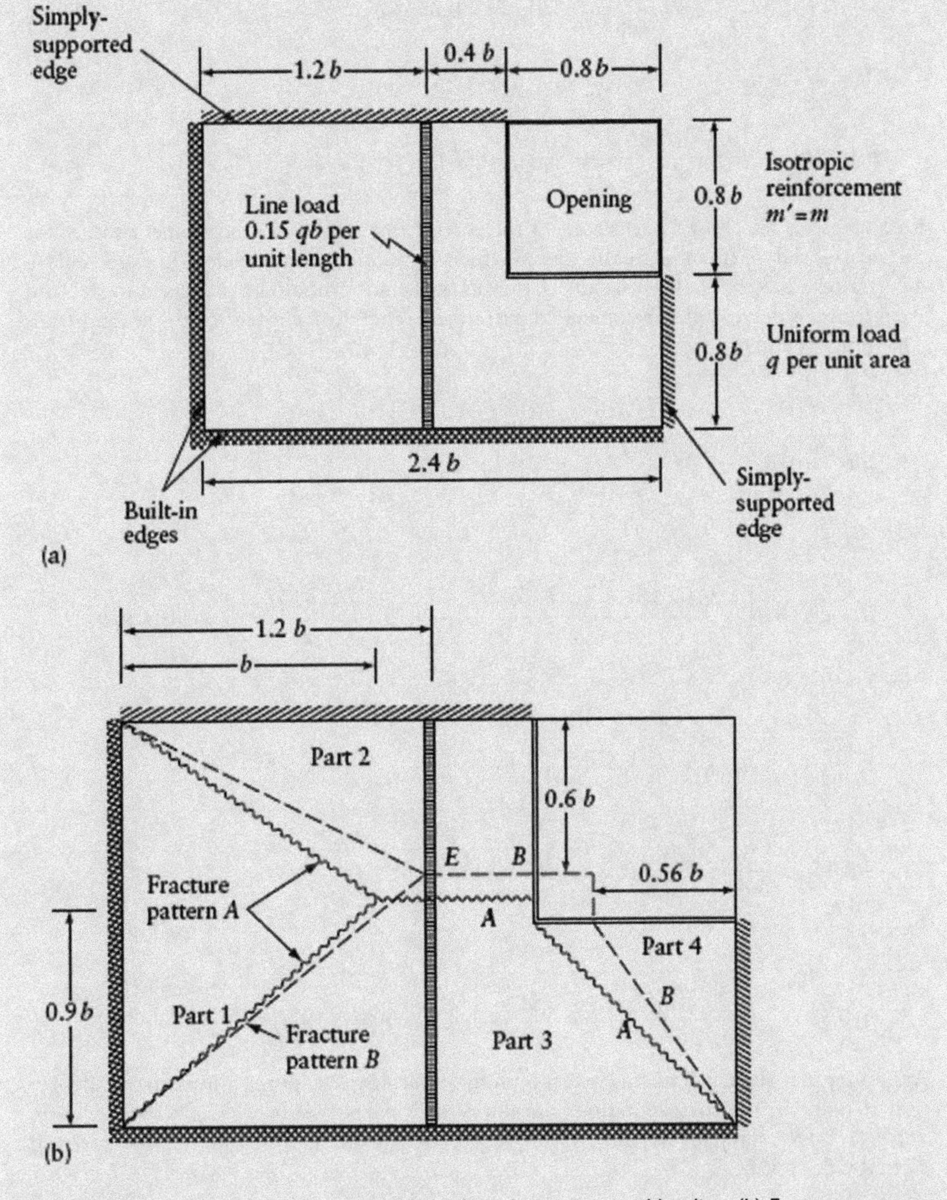

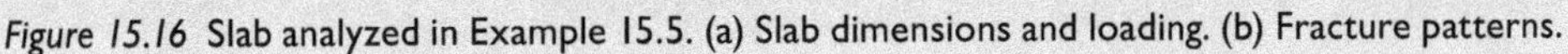
Figure 15.16 Slab analyzed in Example 15.5. (a) Slab dimensions and loading. (b) Fracture patterns.

Part 2:

$$m(1.6b)=q\left[\frac{b(0.7b)^2}{6}+\frac{0.6b(0.7b)^2}{2}+0.15qb\frac{(0.7b)^2}{2}\right];\ m=0.1659qb^2$$

Part 3:

$$2m(2.4b)=q\left[\frac{b(0.9b)^2}{6}+\frac{0.6b(0.9b)^2}{2}+\frac{0.8b(0.8b)^2}{6}\right]$$

$$+0.15qb\frac{(0.9b)^2}{2}-m\frac{0.8b}{0.8b}(0.8b);\ m=0.0936qb^2$$

Part 4:

$$m(0.8b)=q\frac{(0.8b)(0.8b)^2}{6}+m\frac{0.8b}{0.8b}(0.8b);\ m=\infty$$

It is clear that the chosen pattern is not the correct one. However, the moment values for the various slab parts indicate the way in which we should move the yield-lines to achieve a better result. Specifically, we can see that Parts 1 and 3 should be increased in size and Parts 2 and 4 decreased. The amended pattern B is therefore chosen for the second trial. The moment equation gives

Part 1:

$$2m(1.6b)=q\frac{(1.6b)(1.2b)^2}{6};\ m=0.1200qb^2$$

Part 2:

$$m(1.6b)=q\left[\frac{1.2b(0.6b)^2}{6}+\frac{0.4b(0.6b)^2}{2}\right]+0.15qb\frac{(0.6b)^2}{2};\ m=0.1069qb^2$$

Part 3:

$$2m(2.4b+2.16b)=q\left[\frac{1.2bb^2}{6}+\frac{0.4bb^2}{2}+\frac{0.24b(0.8b)^2}{2}+\frac{0.56b(0.8b)^2}{6}\right]$$

$$+0.15qb\frac{b^2}{6}-m\frac{0.56b}{0.8b}(0.8b);\ m=0.1194qb^2$$

Part 4:

$$m(0.8b)=q\frac{(0.8b)(0.5b)^2}{6}+m\frac{0.56b}{0.8b}(0.56b);\ m=0.1025qb^2$$

Assuming that the mechanism corresponding to pattern B acquires a unit virtual deflection at point E, the corresponding rotation of slab parts will be $1/(1.2b)$, $1/(0.6b)$, $1/b$, and $1/(1.7b)$ for slab Parts 1, 2, 3, and 4, respectively. Equating the internal and external virtual work gives:

$$2m(1.6b)\frac{1}{1.2b}+m(1.6b)\frac{1}{1.6b}$$

$$+(2.4b+2.6b)m\frac{1}{b}+m(0.8b)\frac{1}{0.7b}$$

$$=q\left[1.2(1.6b)\frac{1}{3}+0.4b(1.6b)\frac{1}{2}+0.24b(0.8b)\left(\frac{0.8b}{b}\right)\frac{1}{2}+0.56b(0.8b)\left(\frac{0.8b}{b}\right)\frac{1}{3}\right]$$

$$+0.15qb(1.6b)\frac{1}{2}$$

whence $m = 0.1156\ qb^2$.

15.8 GENERAL

The methods considered in this chapter are of two basic types: the energy and equilibrium methods of yield-line theory; these are analytical methods. For simple cases of slab geometry and loading, the yield-line method can be safely used as a design method, since the fracture pattern to give an upper bound close to the correct collapse load can be readily obtained. For complex cases of geometry and loading, care must be exercised in the choice of fracture patterns, particularly where concentrated loads occur, since local modes of collapse, called fan modes, can occur. A fracture pattern involving fan modes will generally give lower collapse load than that obtained from the corresponding straight line pattern.

Although the collapse loads given by the yield-line method are theoretically upper bound values, in practice the actual collapse load of a reinforced concrete slab may be above the calculated value because of the presence of various secondary effects.

It has been assumed in the analysis that Equation 15.1 can be used for calculating the moment of resistance in a yield-line at any angle to a set of orthogonal reinforcement. This method tends to underestimate the true moment of resistance which can be increased by up to 14 percent for the case of a yield-line at 45° to an isotropic orthogonal set of reinforcement. The reason for this is that the analysis takes no account of kinking of reinforcement, which places the reinforcement almost at right angles to the fracture line and therefore increases the moment of resistance.

Where the boundaries of any slab are restrained from horizontal movement (as might occur in the interior panels of a continuous slab) the formation of the collapse mechanism develops high compressive forces in the plane of the slab with a consequent increase in carrying capacity. At very large deflections of the slab, it is possible finally to develop tensile membrane action where cracks go right through the slab so that the load is supported on the net of reinforcement.

The use of the yield-line method with variable reinforcement involves the consideration of a large number of differing yield-line patterns. The yield-line method does not guarantee satisfactory deflection and cracking behavior at working loads, though tests carried out on slabs designed by this method generally exhibit satisfactory behavior. Distribution of reinforcement within the slab, which is not too far removed from that expected with an elastic distribution of bending moments, will generally ensure satisfactory working load behavior. In the context of elastic analyses, we should also remember that the bending moment distribution can change markedly for changes in the relative stiffnesses of the slab and its supporting beams and that many of the design tables based on elastic analyses are consistent only with non-deflecting supporting beams.

REFERENCES

1. Johansen, K.W., *Yield-Line Theory*, Cement and Concrete Association, London, 1962, p. 181.
2. Wood, R.H. and Jones, L.L., "Recent Developments in Yield Line Theory," *Magazine of Concrete Research*, May 1965. See also Jones, L.L. and Wood, R.H., *Yield Line Analysis of Slabs*, Thames and Hudson, Chatto and Windus, London, 1967, p. 405.

PROBLEMS

The representation of the different edge conditions of the slabs in the following problems is made according to the convention indicated in Section 15.2.1.

15.1 Using the yield-line theory, find the ultimate moment for the isotropic slabs shown in the figures under the action of uniformly distributed load. Show in a pictorial view, the forces on the fractured slab parts in Problem 15.1(a). The columns in Problems 15.1(d) and (e) are assumed to produce only a normal concentrated reaction component.

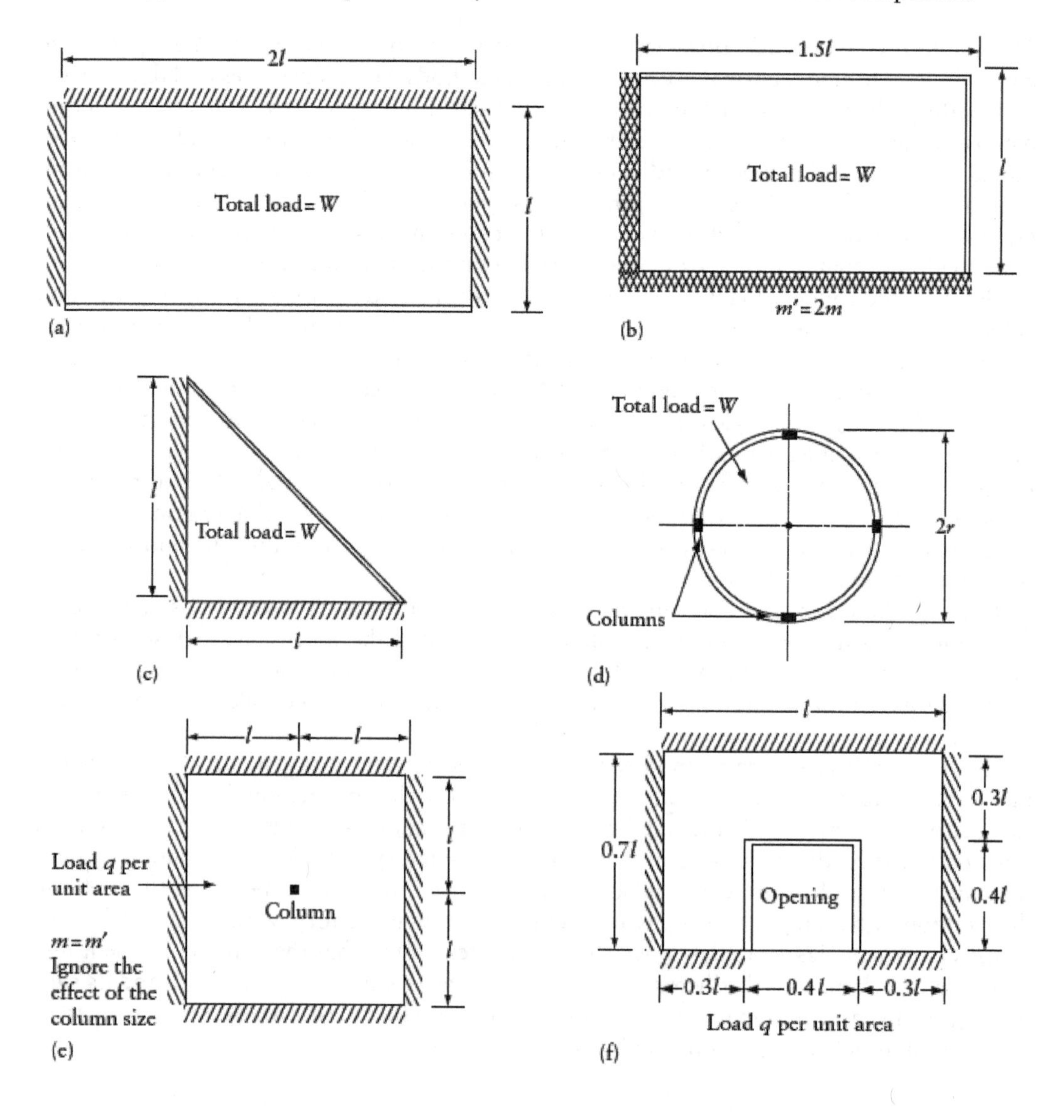

15.2 Using the yield-line theory, find the ultimate moment for the slabs shown in parts (a) to (f), assuming the given ratio of reinforcement in each case. Assume that the columns in part (b) produce only a normal concentrated reaction component.

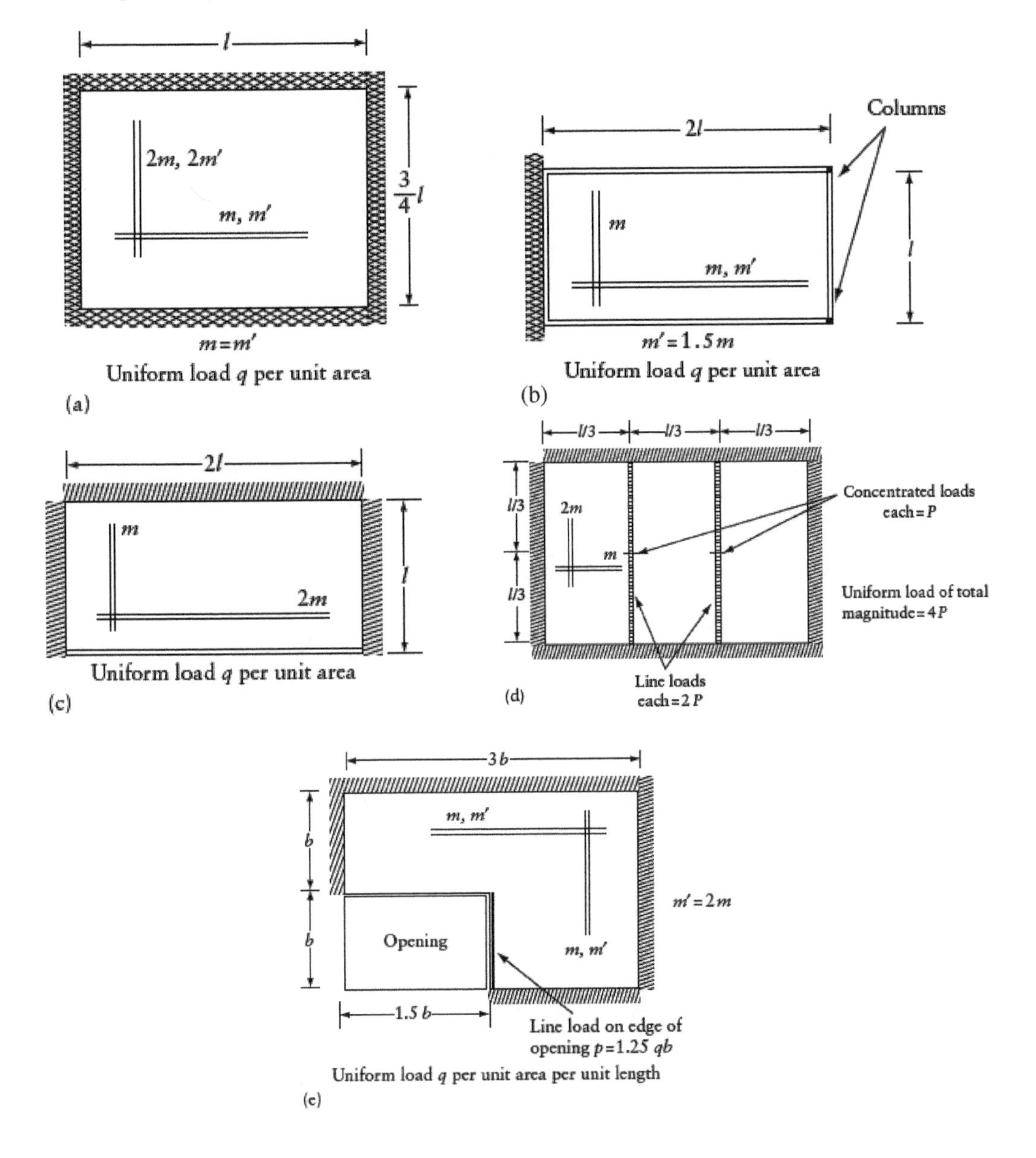

NOTES

* There are several methods of calculation of the value of the ultimate moment in terms of the depth, area of reinforcement, and the strengths of concrete and of steel. Refer, for example, to the American Concrete Institute Standard 318-19, Building Code Requirements for Structural Concrete.

† The ultimate moments in the x and y directions are indicated by vectors in the right-hand top corner of Figure 15.7.

Chapter 16

Structural dynamics

16.1 INTRODUCTION

The previous chapters dealt with structures subjected to static forces producing displacements that do not vary with time. We now consider dynamic problems, in which the forces are time-dependent and cause vibration of the structure; hence, it is necessary to take into account the forces produced by the inertia of the accelerating masses. For this purpose, we use Newton's second law of motion, which states that the product of the mass and its acceleration is equal to the force. In practice, seismic forces, non-steady wind, blast, reciprocating machinery, or impact of moving loads produce dynamic loading.

An elastic structure, disturbed from its equilibrium condition by the application and removal of forces, will oscillate about its position of static equilibrium. Thus, the displacement at any point on the structure will vary periodically between specific limits in either direction. The distance of either of these limits from the position of equilibrium is the "amplitude of the vibration." In the absence of external forces, the motion is *free vibration* and may continue with the same amplitude for an indefinitely long time. In practice, energy is lost through effects, such as friction, air resistance, imperfect elasticity, and those causing the amplitude to diminish gradually until motion ceases. These effects are termed *damping*, and this type of motion is *damped vibration*.

If external forces are applied, we have *forced motion*. This may be *damped* or *undamped*, depending on the presence or absence of these energy losses. In structural analysis, damping effects are modeled as *damping forces*, which are often assumed to be viscous (i.e. they are assumed to be proportional to the velocity). This is "viscous damping" because the resistance of a liquid or a gas to a moving mass (at a low velocity) is proportional to the velocity.

16.2 LUMPED MASS IDEALIZATION

In any structure, the mass is distributed throughout the structure and its elements, and, thus, the *inertial forces* are distributed. This can cause difficulties as seen in Figure 16.1a, where an infinite number of coordinates are required to define the displacement configuration. However, if we imagine that the beam mass is lumped into two bodies, with the force-displacement properties of the beam unchanged (Figure 16.1b), and assume that the external forces causing the motion are applied at these two masses, the deflected shape of the beam at any time can be completely defined by 12 coordinates, 6 at each mass (Figure 16.1c). We use the idealized model in Figure 16.1b, which has 12 degrees-of-freedom in dynamic analysis, as a substitute in place of the actual structure. A larger number of masses at closer intervals represents distributed mass more accurately; but this requires a greater number of coordinates.

DOI: 10.1201/9780429286858-16

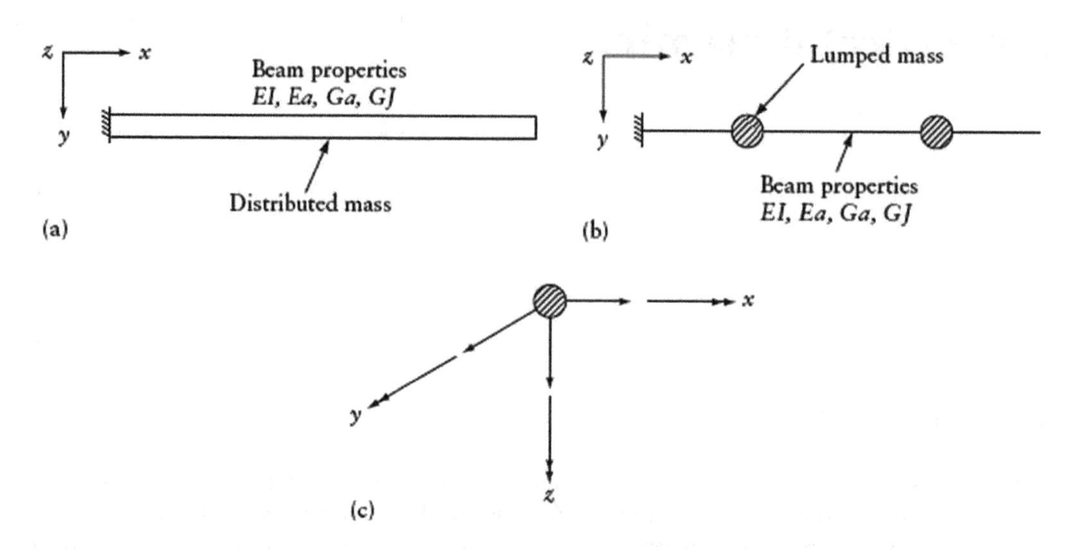

Figure 16.1 Idealization of a distributed mass by a lumped-mass system. (a) A vibrating beam has infinite degrees-of-freedom. (b) Idealization of the beam in part (a) into a lumped-mass system. (c) Typical degrees-of-freedom at a lumped mass.

To demonstrate further the use of lumped mass idealization, consider the motion of the multistory frame in its own plane (Figure 16.2a). For the idealized model, we ignore axial deformations of the members, distribute the mass of the columns to the adjacent floors, and assume that the total mass is mounted on a simple beam (Figure 16.2b) so that the mass can sidesway only. Typically, this mass will be the mass of the floor plus one-half of the mass of the columns and walls above and below the floor.

Let the horizontal displacements at floor levels at any time be $D_1, D_2, \ldots, D_i, \ldots, D_n$ and let the corresponding masses be $m_1, m_2, \ldots, m_i, \ldots, m_n$. As each degree-of-freedom of the frame is displaced, restraining forces will be developed as a function of the stiffnesses of the connecting elements (in this case, the columns), see Figure 16.2d. There may also be external forces, P_i, acting at the coordinates, which generally are functions of time. Thus, the total force acting on the ith mass is:

$$P_i - S_i\, D_i \tag{16.1}$$

Applying Newton's second law to the ith mass, we get:

$$m_i\, \ddot{D}_i = P_i - S_i\, D_i \tag{16.2}$$

where $\ddot{D}_i$, the acceleration of the ith mass, is equal to the second derivative of the displacement D_i with respect to time t, (d^2D_i/dt^2). Rearranging this equation for the ith mass gives the corresponding *equation of motion*:

$$m_i\, \ddot{D}_i + S_i\, D_i = P_i \tag{16.3}$$

For n masses, we can write this equation in matrix form:

$$[m]\{\ddot{D}\} + [S]\{D\} = \{P(t)\} \tag{16.4}$$

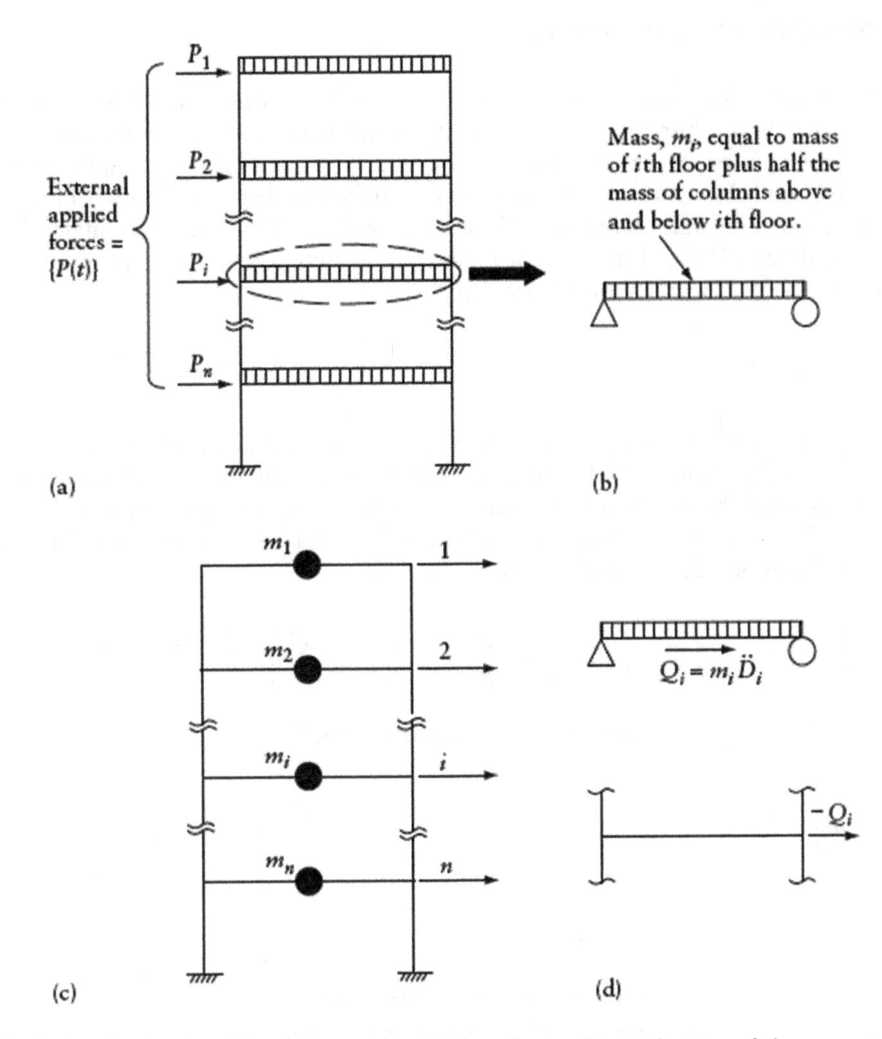

Figure 16.2 A multi-degree-of-freedom system. (a) Plane frame. (b) Idealization of the mass at a typical floor. (c) Coordinate system representing positive directions of forces and displacements. (d) Application of Newton's second law to the mass of the ith floor.

where $[m]$ is a diagonal matrix:

$$[m] = \begin{bmatrix} m_1 & & & \\ & m_2 & & \\ & & \cdots & \\ & & & m_n \end{bmatrix} \tag{16.5}$$

When the system is vibrating freely, $\{P(t)\} = \{0\}$ and the equation of motion becomes (undamped free vibration):

$$[m]\{\ddot{D}\} + [S]\{D\} = \{0\} \tag{16.6}$$

Equations 16.4 and 16.6 are termed the *undamped multi-degree-of-freedom* equations of motion for forced and free vibration, respectively. By writing the equations in scalar form, we can derive a substitute equation of a single-degree-of-freedom system.

16.3 CONSISTENT MASS MATRIX

In the preceding section we saw that, when a structure is vibrating, the mass of the various structural elements is arbitrarily lumped at nodes, resulting in a diagonal mass matrix. The motion of these lumped masses are related to the translations and rotations of the structural degrees-of-freedom. We use the *consistent* mass matrix to substitute the distributed mass. The element, m_{ij}, of the mass is the inertial force in the ith coordinate due to unit acceleration at the jth coordinate. Using displacement interpolation functions L_i (see Section 12.9) and the principle of virtual work, we determine m_{ij} by:

$$m_{ij} = \int_{vol} \gamma L_i L_j \, dv \tag{16.7}$$

where $\gamma(x, y, z)$ is the mass per unit volume, and dv is the elemental volume. It is apparent from the form of Equation 16.7, that $m_{ij} = m_{ji}$, and the consistent mass matrix is symmetric.

As an example of the development of the consistent mass, consider a prismatic beam of cross-sectional area a and the four coordinates in Figure 16.3a and use the displacement functions of Equation 12.55, which can be written as:

$$[L] = \left[\left(1 - \frac{3x^2}{l^2} + \frac{3x^3}{l^3}\right) \quad \left(x - \frac{2x^2}{l} + \frac{x^3}{l^2}\right) \quad \left(\frac{3x^2}{l^2} - \frac{2x^3}{l^3}\right) \quad \left(-\frac{2x^2}{l} + \frac{x^3}{l^2}\right) \right] \tag{16.8}$$

where l is the length of the beam. We rewrite Equation 16.7:

$$m_{ij} = \gamma a \int_0^l L_i L_j \, dx \tag{16.9}$$

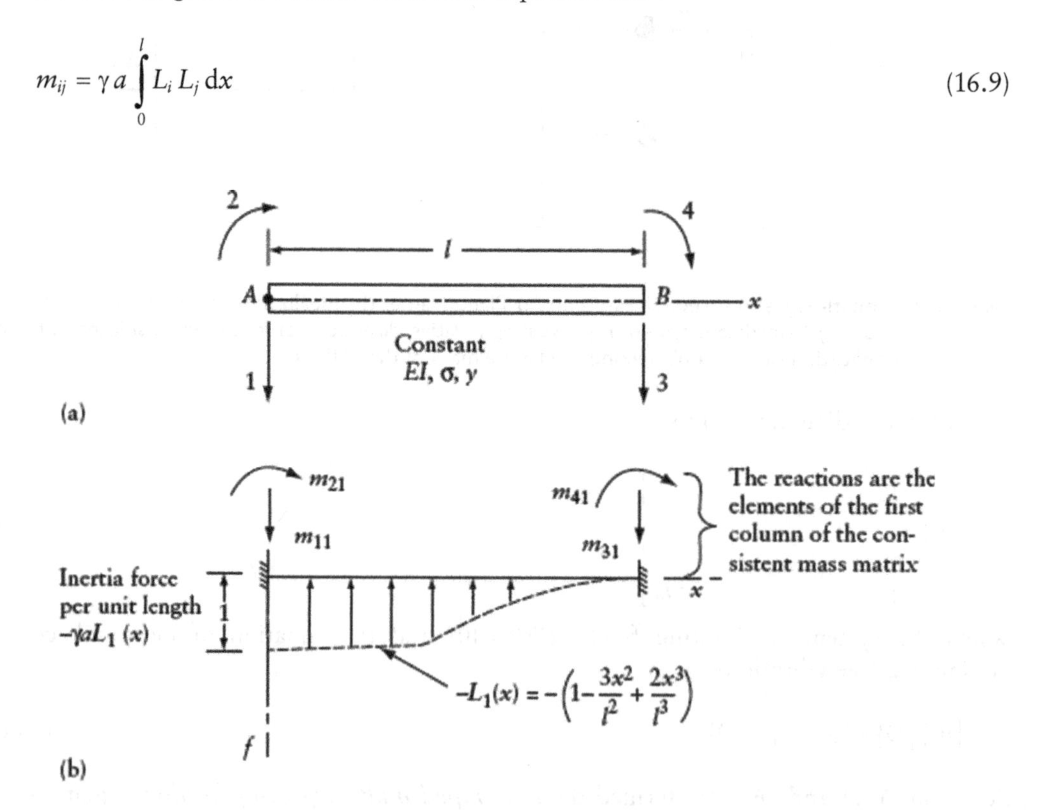

Figure 16.3 Derivation of the consistent mass matrix for a prismatic beam. (a) Prismatic beam of total mass γal. (b) Distributed inertia forces associated with acceleration $\ddot{D}_1 = 1$.

Substituting for L_i and L_j from Equation 16.8 into Equation 16.9, the following consistent mass matrix is obtained:

$$[m] = \frac{\gamma a l}{420}\begin{bmatrix} 156 & & & \text{Symm.} \\ 22l & 4l^2 & & \\ 54 & 13l & 156 & \\ -13l & -3l^2 & -22l & 4l^2 \end{bmatrix} \quad (16.10)$$

Figure 16.3b shows the inertia forces associated with $\ddot{D}_1 = 1$, with the motion prevented at the other coordinates defined in Figure 16.3a.

16.4 UNDAMPED VIBRATION: SINGLE-DEGREE-OF-FREEDOM SYSTEM

For a single-degree-of-freedom system, the undamped equation of motion, using the scalar version of Equation 16.4, is:

$$m\,\ddot{D} + S\,D = P(t) \quad (16.11)$$

Examples of such a system are the frame in Figure 1.13c with the mass concentrated at the top, and the beam in Figure 16.1 with one lumped mass. The homogeneous case of this equation, with $P(t) = 0$, defines free vibration of a single-degree-of-freedom system:

$$m\,\ddot{D} + S\,D = 0 \quad (16.12)$$

We rewrite this equation as:

$$\ddot{D} + \omega^2\,D = 0 \quad (16.13)$$

where ω is the *natural frequency* of the single-degree-of-freedom system, with units of radian/sec:

$$\omega = \sqrt{S/m} \quad (16.14)$$

The solution to the differential equation (Equation 16.13) is:

$$D = C_1 \sin\omega t + C_2 \cos\omega t \quad (16.15)$$

This equation represents periodic motion, where the *natural period of vibration*, T, is given by:

$$T = 2\pi/\omega \quad (16.16)$$

The reciprocal of the period T, $f = \omega/(2\pi)$, is called the *cyclic natural frequency* and has units of cycle/s = Hertz.

The integration constants C_1 and C_2 in Equation 16.15 can be determined from the initial conditions at time $t = 0$, of displacement D_0, and velocity $\dot{D}_0$. Substituting these values into Equation 16.15 and its first derivative with respect to time, we obtain:

$$C_1 = \dot{D}_0/\omega; \; C_2 = D_0 \tag{16.17}$$

The complete solution for the free vibration of a single-degree-of-freedom system is then:

$$D = \frac{\dot{D}_0}{\omega} \sin \omega t + D_0 \cos \omega t \tag{16.18}$$

The amplitude of vibration (Figure 16.4) is the maximum value of displacement that Equation 16.18 gives, that is:

$$D_{max} = \bar{D} = \sqrt{\left(\frac{\dot{D}_0}{\omega}\right)^2 + D_0^2} \tag{16.19}$$

We rewrite Equation 16.18 to express D in terms of the amplitude:

$$D = \bar{D} \sin(\omega t + \alpha) \tag{16.20}$$

where α is called the *phase angle*, given as:

$$\alpha = \tan^{-1}\left(D_0 \, \omega/\dot{D}_0\right) \tag{16.21}$$

Figure 16.4a shows the variation of D with t of undamped single-degree of freedom system.

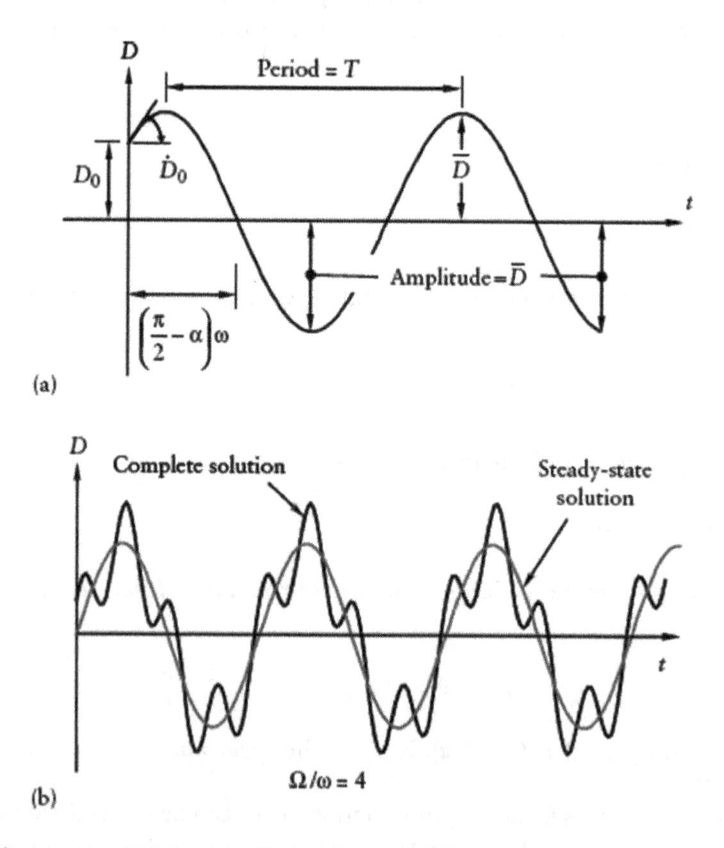

Figure 16.4 Displacement versus time of an undamped single-degree-of-freedom system. (a) Free vibration. (b) Motion forced by a force varying harmonically.

EXAMPLE 16.1: LIGHT CANTILEVER WITH HEAVY WEIGHT AT TOP

A water tower is idealized as a light prismatic cantilever fixed at the base having a lumped mass $m = W/g$ at the top, where $W = 400$ kN and g = gravity acceleration = 9.81 m/s^2. The length of the cantilever is $l = 50$ m, and its flexural rigidity is $EI = 30 \times 10^9$ N-m^2. Find the natural angular frequency ω and the natural period of vibration T. If motion is initiated by displacing the mass horizontally at distance D_0, then the system is left to vibrate, what is the displacement at $t = T/8$? What are the values of T with l reduced to 25 m or increased to 75 m?

The static force that produces unit drift at the tip of the cantilever (Appendix C):

$$S = 3EI/l^3 = 3\left(30\times 10^9\right)/\left(50\right)^3 = 720\times 10^3 \text{ N/m}$$

Equations 16.14, 16.16, and 16.18 give:

$$\omega = \sqrt{S/m} = \left[720\times 10^3/\left(400\times 10^3/9.81\right)\right]^{1/2} = 4.202 \text{ rad/s}$$

$$T = 2\pi/\omega = 2\pi/4.202 = 1.495 \text{ s};\quad D_{t=T/8} = D_0 \cos\left(\pi/4\right) = 0.707\, D_0$$

With $l = 25$, 50, and 75 m, $T = 0.529$, 1.495, and 2.747 s, respectively; we can see that T is proportional to $l^{1.5}$.

16.4.1 Forced motion of an undamped single-degree-of-freedom system: harmonic force

When an external harmonic force is applied to a single-degree-of-freedom system, the equation of motion becomes:

$$m\,\ddot{D} + S\,D = P_0 \sin\Omega t \tag{16.22}$$

where P_0 is the force amplitude and Ω is the *impressed* or *forcing frequency*. Centrifugal forces resulting from imbalance in rotating machines, induce this type of loading. Periodic loads expressed as sum of harmonic loads, using Fourier analysis. Using the definition of ω from Equation 16.14, we rewrite Equation 16.22 as:

$$\ddot{D} + \omega^2\,D = \frac{P_0}{m}\sin\Omega t \tag{16.23}$$

The solution of this equation consists of two components: the complementary solution (Equation 16.15) and a particular solution, which is of the form:

$$D = C_3 \sin\Omega t \tag{16.24}$$

Substituting this solution into Equation 16.24 and solving for C_3, and adding this solution to the homogeneous solution, the complete solution becomes:

$$D = C_1 \sin\omega t + C_2 \cos\omega t + \frac{P_0}{S}\left[\frac{1}{1-\left(\Omega/\omega\right)^2}\right]\sin\Omega t \tag{16.25}$$

The constants C_1 and C_2 are again determined from the initial conditions. The first and second terms of Equation 16.25 are the *transient response*, while the third term is the *steady-state response*. The transient response is so-called because under damped conditions, these terms rapidly decay to zero. However, for the undamped condition, for a single-degree-of-freedom system, which starts from rest at its equilibrium position, the constants are:

$$C_1 = -\frac{\Omega\, P_0}{\omega\, S}\left[\frac{1}{1-(\Omega/\omega)^2}\right] \quad ; \quad C_2 = 0 \tag{16.26}$$

The complete solution is then:

$$D = \frac{P_0}{S}\left[\frac{1}{1-(\Omega/\omega)^2}\right]\left(\sin\Omega t - \frac{\Omega}{\omega}\sin\omega t\right) \tag{16.27}$$

Figure 16.4b plots variation of D versus t; we see that the transient solution is oscillating about the steady-state response (the last term in Equation 16.25). The quantity P_0/S is the displacement of the system under a static load, P_0. The first bracket, $1/\{1-(\Omega/\omega)^2\}$, is termed the *amplification factor.* This factor is ~1.0 when the *frequency ratio* Ω/ω is close to zero but is infinity when this ratio is 1.0. This condition is termed *resonance*, in which the amplitude of vibration increases indefinitely. In practice, damping will limit the maximum amplitude to finite values, but these may still be large enough to cause damage to the system. As the frequency ratio becomes large, $\Omega/\omega >> 1$, the magnitude of the amplification factor decreases – eventually to zero.

16.4.2 Forced motion of an undamped single-degree-of-freedom system: general dynamic forces

Analysis for a general dynamic loading resulting from blasts, wind gusts, or seismic effects is accomplished by considering the loading as a sequence of *impulse* loads and integrating for the effect of these impulses to obtain the system response. When the load function is simple, we express the integral in closed form; otherwise, we evaluate the integral numerically.

Consider the time-dependent loading shown in Figure 16.5, which begins to act at $t=0$, when the mass is at rest; thus, $D_0=0$ and $\dot{D}_0=0$. At time $t=\tau$, we show an impulse: $I=P(\tau)\,d\tau$.

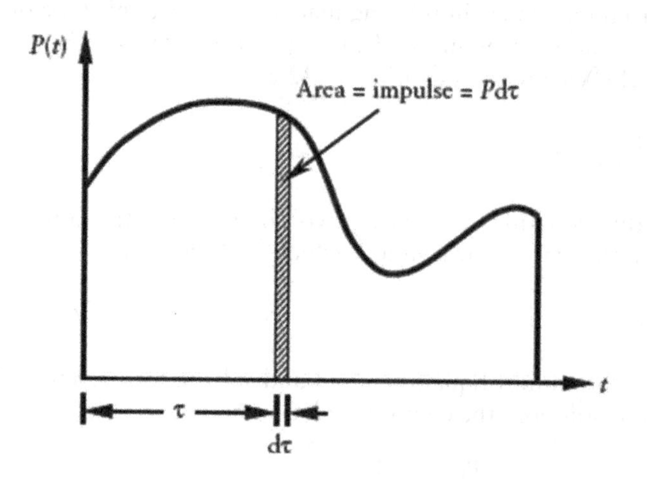

Figure 16.5 General force–time relation.

Such an impulse causes a change in momentum of the system. As the mass is constant, the change is in the velocity of the system:

$$d\dot{D} = \frac{P(\tau)\,d\tau}{m} \tag{16.28}$$

This change in velocity is initial condition at time $t=\tau$; the subsequent displacement is (Equation 16.20):

$$dD = \frac{d\dot{D}}{\omega}\sin\omega(t-\tau) \tag{16.29}$$

Substituting Equation 16.29 and integrating from time $t=0$, we obtain the displacement at time t for any arbitrary time-dependent load:

$$D = \frac{1}{m\omega}\int_0^t P(\tau)\sin\omega(t-\tau)\,d\tau \tag{16.30}$$

This is the *Duhamel integral*. Using the trigonometric identity:

$$\sin\omega(t-\tau) = \sin\omega t\cdot\cos\omega\tau - \cos\omega t\cdot\sin\omega\tau \tag{16.31}$$

we can rewrite Equation 16.30 in the form:

$$D = \frac{1}{m\omega}\left[\psi_1(t)\sin\omega t - \psi_2(t)\cos\omega t\right] \tag{16.32}$$

where $\psi_1(t) = \int_0^t P(\tau)\cos(\omega\tau)\,d\tau$; $\psi_2(t) = \int_0^t P(\tau)\sin(\omega\tau)\,d\tau$.

Depending on the nature of the function $P(\tau)$, these integrals may be evaluated in closed form or numerically, using any numerical integration system (e.g. Simpson's rule).

16.5 VISCOUSLY DAMPED VIBRATION: SINGLE-DEGREE-OF-FREEDOM SYSTEM

Generally, the most widely used form of damping is *viscous damping*, where we replace damping effects by forces proportional to the velocity. Thus, the damping force is:

$$F_c = c\,\dot{D} \tag{16.33}$$

where c is the *damping coefficient*. The equation of motion for a viscously damped single-degree-of-freedom then becomes (from Equation 16.11):

$$m\,\ddot{D} + c\,\dot{D} + S\,D = P(t);\quad \ddot{D} + 2\zeta\omega\,\dot{D} + \omega^2\,D = P(t)/m \tag{16.34}$$

$$\zeta = c/(2m\omega) \tag{16.35}$$

where ω is the natural frequency (Equation 16.14); ζ is the damping ratio = ratio between the actual damping coefficient and the critical damping coefficient (= $2m\omega$, discussed below).

16.5.1 Viscously damped free vibration

Setting $P(t)$ equal to zero in Equation 16.34, we obtain the free vibration equation of motion for a damped system:

$$\ddot{D} + 2\zeta\omega\,\dot{D} + \omega^2\,D = 0 \tag{16.36}$$

Substituting $D = e^{\lambda t}$, we obtain the characteristic equation:

$$\lambda^2 + 2\zeta\,\omega\lambda + \omega^2 = 0 \tag{16.37}$$

which has the roots:

$$\lambda_{1,2} = \omega\left(-\zeta \pm i\sqrt{1-\zeta^2}\right) \tag{16.38}$$

where $i = \sqrt{-1}$ and the solution to Equation 16.36 becomes:

$$D = e^{-\zeta\omega t}\left(C_1\,e^{i\,\omega_d\,t} + C_2\,e^{-i\,\omega_d\,t}\right) \tag{16.39}$$

$$\omega_d = \omega\sqrt{1-\zeta^2} \tag{16.40}$$

where ω_d is the *damped natural frequency*; C_1 and C_2 are constants. We rewrite Equation 16.39 using the relationships between trigonometric functions and exponential functions as:

$$D = e^{-\zeta\omega t}\left(\bar{C}_1\sin\omega_d\,t + \bar{C}_2\cos\omega_d\,t\right) \tag{16.41}$$

Using the initial conditions, D_0 and $\dot{D}_0$, Equation 16.41 is written as:

$$D = e^{-\zeta\omega t}\left[\left(\frac{\dot{D}_0 + \zeta\omega D_0}{\omega_d}\right)\sin\omega_d\,t + D_0\,\cos\omega_d\,t\right] \tag{16.42}$$

When $\zeta = 1$, the system is said to be critically damping, with critical damping coefficient = $2m\omega$. At critical damping, the initial displacement D_0 of a system, starting from rest, dies out gradually without oscillation; $D = D_o\,e^{-\omega t}$. For buildings and bridges, ζ varies between 0.01 and 0.2. Figure 16.6 illustrates this equation for a system, having $\zeta = 0.1$, that starts at rest with displacement D_0. The *natural period of damped vibration* is:

$$T_d = 2\pi/\omega_d = T/\sqrt{1-\zeta^2} \tag{16.43}$$

The ratio of the displacement at time t to that at time $t + T_d$ is constant, and the natural logarithm of this ratio is the *logarithmic decrement*, δ:

$$\delta = \ln\left(\frac{D_t}{D_{t+T_d}}\right) = \ln\left(e^{\zeta\omega T_d}\right) = \frac{2\pi\zeta}{\sqrt{1-\zeta^2}} \cong 2\pi\zeta \tag{16.44}$$

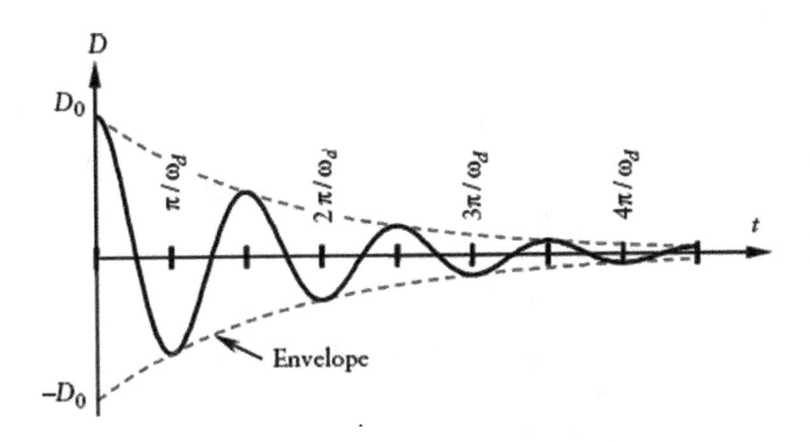

Figure 16.6 Damped-free vibration, with $\dot{D}=0$ at $t=0$.

We use Equation (16.44) to estimate the damping ratio of real structures from displacement measurements of the free vibration.

16.5.2 Viscously damped forced vibration – harmonic loading: single-degree-of-freedom system

When a damped single-degree-of-freedom system is subject to a harmonically applied load, the equation of motion becomes:

$$\ddot{D}+2\zeta\omega\,\dot{D}+\omega^2\,D=\frac{P_0}{m}\sin\Omega t \tag{16.45}$$

The complete solution of Equation 16.45 consists of two parts, the complementary solution (as given by Equation 16.41) and the particular solution, representing the response to the harmonic force, given by:

$$D_{\text{Steady state}}=\frac{P_0}{S}\left\{\frac{\left[1-(\Omega/\omega)^2\right]\sin\Omega t-(2\zeta\Omega/\omega)\cos\Omega t}{\left[1-(\Omega/\omega)^2\right]^2+(2\zeta\Omega/\omega)^2}\right\} \tag{16.46}$$

The complementary solution (Equation 16.41) is also the *transient solution* and is damped out after a few cycles. The particular solution of Equation 16.46 is called the *steady-state displacement*, and it is this solution that defines the response of the system to the external load *P*(*t*). Equation 16.46 can be written in terms of its displacement amplitude, $\overline{D}$, and phase angle, α:

$$D_{\text{Steady state}}=\overline{D}\sin(\Omega t+\alpha) \tag{16.47}$$

$$\overline{D}=\chi\frac{P_0}{S} \tag{16.48}$$

$$\alpha=\tan^{-1}\left[\frac{-2\zeta\Omega/\omega}{1-(\Omega/\omega)^2}\right] \tag{16.49}$$

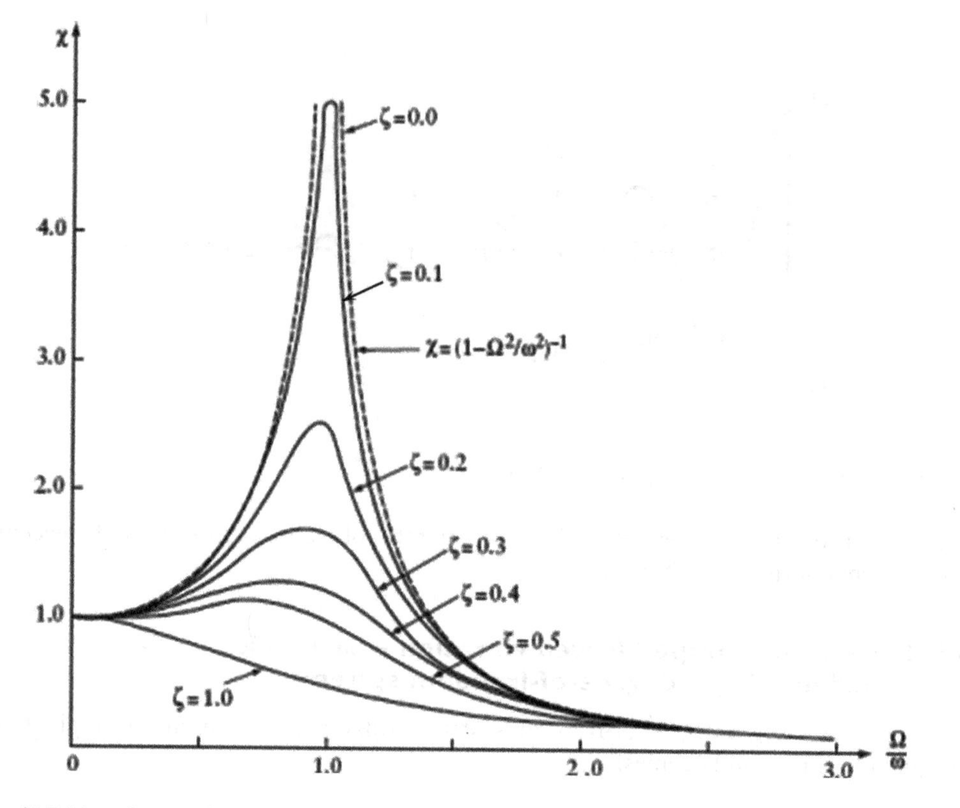

Figure 16.7 Magnification factor defined by Equation 16.50 for a damped system subjected to harmonic disturbing forces.

$$\chi = \left[\left\{1-(\Omega/\omega)^2\right\}^2 + (2\zeta\Omega/\omega)^2\right]^{-1/2} \tag{16.50}$$

The complete solution of Equation 16.45 is the sum of Equations 16.41 and 16.47:

$$D = e^{-\zeta\omega t}\left(\bar{C}_1 \sin\omega_d t + \bar{C}_2 \cos\omega_d t\right) + \bar{D}\sin\left(\Omega t + \alpha\right) \tag{16.51}$$

The amplitude $\bar{D}$ of the steady-state displacement is equal to the static displacement (P_0/S) multiplied by the amplification factor, χ (Equation 16.50). Figure 16.7 represents χ versus Ω/ω for various values of the damping ratio, ζ. At resonance, $\Omega=\omega$, the amplification factor $=1/(2\zeta)$. The dashed curve in Figure 16.7 is for the theoretical case of undamped motion (Equation 16.27). The other curves indicate that, even with a small amount of damping, finite displacements occur at resonance. Nevertheless, we avoid this condition generally.

EXAMPLE 16.2: DAMPED FREE VIBRATION OF A LIGHT CANTILEVER WITH A HEAVY WEIGHT AT TOP

For the system in Example 16.1, find the damped natural frequency, ω_d, the natural period of damped vibration, T_d, and the displacement at $t=3T_d$. Assume a damping ratio $\zeta=0.05$.

Using the results of Example 16.1 and applying Equations 16.40, 16.43, and 16.42 gives:

$$\omega_d = \omega\sqrt{1-\zeta^2} = 4.202\left[1-(0.05)^2\right]^{1/2} = 4.197 \text{ rad/s}$$

$$T_d = 2\pi/\omega_d = 1.497 \text{ s}$$

$$D_{t=3T_d} = D_0\, e^{-\zeta\omega(3T_d)} = D_0\, e^{-0.05(4.202)(3\times1.497)} = 0.389\, D_0$$

We can see that with $\zeta = 0.05$ the amplitude drops to $0.39D_0$ in three cycles; in six cycles ($t = 6T_d$), it drops to $0.15D_0$.

16.5.3 Viscously damped forced vibration – general dynamic loading: single-degree-of-freedom system

Using the procedure in Section 16.4.2, we can obtain the solution for a damped single-degree-of-freedom system subject to a randomly varying time-dependent force, $P(\tau)$. The displacement at time t caused by the impulse applied at earlier time τ (see Figure 16.5 and Equation 16.29) can be obtained from Equation 16.42 by considering as the initial velocity:

$$\mathrm{d}D = \frac{1}{m\,\omega_d}\, P(\tau)\, e^{-\zeta\omega(t-\tau)}\, \sin\omega_d(t-\tau)\,\mathrm{d}\tau \tag{16.52}$$

Integrating, we obtain the displacement:

$$D = \frac{1}{m\,\omega_d} \int_0^t P(\tau)\, e^{-\zeta\omega(t-\tau)}\, \sin\omega_d(t-\tau)\,\mathrm{d}\tau \tag{16.53}$$

Using Equation 16.42 for the solution for the initial conditions, D_0 and $\dot{D}_0$, the complete solution is given by:

$$D = e^{-\zeta\omega t}\left[\left(\frac{\dot{D}_0 + \zeta\omega D_0}{\omega_d}\right)\sin\omega_d t + D_0\cos\omega_d t\right] + \frac{1}{m\,\omega_d}\int_0^t P(\tau)\, e^{-\zeta\omega(t-\tau)}\,\sin\omega_d(t-\tau)\,\mathrm{d}\tau \tag{16.54}$$

Equation 16.54 is usually evaluated numerically, except in unusual cases where $P(\tau)$ can be described by a simple function.

16.6 UNDAMPED FREE VIBRATION OF MULTI-DEGREE-OF-FREEDOM SYSTEMS

In Section 16.2, we developed the equation of motion for the free vibration of a system with n degrees-of-freedom (Equation 16.6). This is a system of second-order differential equations; the solution is a set of equations of the form:

$$\{D\} = \{\overline{D}\}\sin(\omega t + \alpha) \tag{16.55}$$

The elements of $\{\overline{D}\}$ are the amplitudes of vibration corresponding to n degrees-of-freedom. Differentiation of Equation 16.55, with respect to time, gives the acceleration vector:

$$\{\ddot{D}\} = -\omega^2 \{D\} \tag{16.56}$$

Substituting Equation 16.56 into Equation 16.6, we obtain:

$$[S]\{D\} = \omega^2 [m]\{D\} \tag{16.57}$$

Pre-multiplying both sides by $[m]^{-1}$ gives:

$$[B]\{D\} = \omega^2 \{D\} \tag{16.58}$$

$$[B] = [m]^{-1} [S] \tag{16.59}$$

Equation 16.58 is called an *eigenvalue equation*, and its solution provides n characteristic values (*eigenvalues*) of the quantity ω^2 (computer program EIGEN, Chapter 20). The ω values are the *natural* or *modal frequencies* of the system, and the lowest is the *first mode* frequency. We can find an *eigenvector* or *mode shape* $\{\phi\}$ for each characteristic value. A value is selected (e.g. unity) for one of the elements of $\{\phi\}$ and the remaining $(n-1)$ values are multiples of the selected value. This results in a vector $\{\phi\}$ of modal amplitudes that define the relationship between the magnitudes of D of the n degrees-of-freedom.

16.6.1 Mode orthogonality

The mode shapes of a multi-degree-of-freedom system have one important property that greatly facilitates analysis of such systems. Apply Equation 16.57 to the rth mode:

$$\omega_r^2 [m]\{\phi_r\} = [S]\{\phi_r\} \tag{16.60}$$

Pre-multiplying by $\{\phi_s\}^T$, we obtain:

$$\omega_r^2 \{\phi_s\}^T [m]\{\phi_r\} = \{\phi_s\}^T [S]\{\phi_r\} \tag{16.61}$$

Now, consider s^{th} mode, $\{\phi_s\}$:

$$\omega_s^2 [m]\{\phi_s\} = [S]\{\phi_s\} \tag{16.62}$$

Pre-multiplying by $\{\phi_r\}^T$, we obtain:

$$\omega_s^2 \{\phi_r\}^T [m]\{\phi_s\} = \{\phi_r\}^T [S]\{\phi_s\} \tag{16.63}$$

Transpose this equation, and recall that $[m]$ and $[S]$ are symmetrical matrices:

$$\omega_s^2 \{\phi_s\}^T [m]\{\phi_r\} = \{\phi_s\}^T [S]\{\phi_r\} \tag{16.64}$$

Subtraction of Equation 16.64 from Equation 16.61 gives:

$$\left(\omega_r^2 - \omega_s^2\right) \{\phi_s\}^T [m]\{\phi_r\} = 0 \tag{16.65}$$

When $\omega_r \neq \omega_s$, then:

$$\{\phi_s\}^{\mathrm{T}}[m]\{\phi_r\}=0;\ \{\phi_s\}^{\mathrm{T}}[S]\{\phi_r\}=0 \tag{16.66}$$

Equation 16.66 means that the natural modes are orthogonal with respect to $[m]$ or $[S]$.

$$[\phi]^{\mathrm{T}}[m][\phi]=[M];\ [\phi]^{\mathrm{T}}[S][\phi]=[K] \tag{16.67}$$

where $[M]$ and $[K]$ are diagonal matrices; the rth elements on the diagonals are:

$$M_r=\{\phi_r\}^{\mathrm{T}}[m]\{\phi_r\};\ K_r=\{\phi_r\}^{\mathrm{T}}[S]\{\phi_r\} \tag{16.68}$$

$$K_r=\omega_r^2\ M_r \tag{16.69}$$

M_r and K_r are, respectively, *generalized mass* and *generalized stiffness* for the rth mode. Equation 16.69 is the same as the equation of free vibration of a non-damped single-degree-of-freedom system having mass, stiffness, and natural frequency m, S, and ω^2, respectively. Thus, the analysis of a multi-degree-of-freedom is analogous to that of a system vibrating in a single coordinate.

16.6.2 Normalized mode matrix

Equation 16.60 can be solved for the mode shape $\{\phi_r\}$ by setting a value for one of its elements (i.e. $\phi_{ir}=1.0$) then solving $(n-1)$ simultaneous equations to obtain the remaining elements of vector, where n = number of degrees-of-freedom of the system. The solution gives ratios between the elements, thus it defines the shape of the rth mode. The normalized mode shape of the rth mode is:

$$\{\Phi_r\}=\left(\{\phi_r\}_{n\times 1}^{\mathrm{T}}[m]_{n\times n}\{\phi_r\}_{n\times 1}\right)^{-1/2}\{\phi_r\}_{n\times 1} \tag{16.70}$$

This means that dividing the elements of $\{\phi_r\}$ by a constant whose units are that of $(\text{mass})^{1/2}$ gives the normalized vector $\{\Phi_r\}$. The orthogonality property applied to the normalized mode matrix gives:

$$[\Phi]^{\mathrm{T}}[m][\Phi]=[\mathrm{I}];\ [\Phi]^{\mathrm{T}}[S][\Phi]=\left[\omega^2\right] \tag{16.71}$$

where $[\mathrm{I}]$ and $[\omega^2]$ are diagonal matrices. Each element on the diagonal of $[\mathrm{I}]=1.0$. The elements on the diagonal of $[\omega^2]$ are: ω_1^2, ω_2^2, …, ω_n^2. (Equation 16.70, combined with Equations 16.68 and 16.69, shows that with the normalized mode matrix, $M_r=1.0$ and $K_r=\omega_r^2$.)

16.7 MODAL ANALYSIS OF DAMPED OR UNDAMPED MULTI-DEGREE-OF-FREEDOM SYSTEMS

The equation of motion of an undamped n-degree-of-freedom system is (Equation 16.4 and Figure 16.2):

$$[m]\{\ddot{D}\}+[S]\{D\}=\{P(t)\} \tag{16.72}$$

Use the normalized modal matrix to express $\{D\}$ as a sum of its modal contributions:

$$\{D\}=[\Phi]\{\eta\} \tag{16.73}$$

This equation transforms the displacements $\{D\}$ into modal coordinates $\{\eta\}$. The substitution of Equation 16.73 into Equation 16.72 and pre-multiplication by $[\Phi]^T$ gives:

$$[\Phi]^T[m][\Phi]\{\ddot{\eta}\}+[\Phi]^T[S][\Phi]\{\eta\}=[\Phi]^T\{P(t)\} \tag{16.74}$$

The substitution of Equation 16.71 in this equation gives the uncoupled set of equations of motion in modal coordinates:

$$\{\ddot{\eta}\}+\left[\omega^2\right]\{\eta\}=\{P^*(t)\} \tag{16.75}$$

$\{P^*(t)\}$ = vector of generalized forces, given by:

$$\{P^*(t)\}=[\Phi]^T\{P(t)\} \tag{16.76}$$

The typical r^{th} equation of the set in Equation 16.75 is the equation of motion of the rth mode for the undamped system:

$$\ddot{\eta}_r+\omega_r^2\,\eta_r=P_r^*(t) \tag{16.77}$$

The equation of motion of a damped n-degree-of-freedom system is:

$$[m]\{\ddot{D}\}+[C]\{\dot{D}\}+[S]\{D\}=\{P(t)\} \tag{16.78}$$

where $[C]$ is the damping matrix. We limit our discussion to classical damping, which is reasonable for many structures, with $[C]=a_0[m]+a_1[S]$, where a_0 and a_1 are constants. Pre-multiplication by $[\Phi]^T$, post-multiplication by $[\Phi]$, and substituting Equation 16.71 gives:

$$[\Phi]^T[C][\Phi]=a_0[I]+a_1\left[\omega^2\right] \tag{16.79}$$

Substituting Equations 16.71, 16.73, 16.76, and 16.79 in Equation 16.78, we can derive:

$$\{\ddot{\eta}\}+2[\zeta\omega]\{\dot{\eta}\}+\left[\omega^2\right]\{\eta\}=\{P^*(t)\} \tag{16.80}$$

where $[\zeta\omega]$ is a diagonal matrix of the products of the modal damping ratio times the modal frequency.

This is a set of uncoupled equations; a typical uncoupled equation of motion of the rth mode of the damped system is:

$$\ddot{\eta}_r+2\zeta_r\,\omega_r\,\dot{\eta}_r+\omega_r^2\,\eta_r=P_r^*(t) \tag{16.81}$$

The uncoupled equations of motion (Equation 16.77 and 16.81) make it possible to determine the response of each mode as that of a single-degree-of-freedom system. The contribution of the n-modes is then summed up by Equation 16.73 to give $\{D\}$ at any time. Comparison of Equation 16.77 or 16.81 with Equation 16.11 or 16.34, respectively, indicates that we can solve for η_r by using the equations of single-degree-of-freedom system by replacing D, S, m, and $P(t)$ by η_r, ω_r^2, 1.0, and $P^*(t)$, respectively.

In analysis of soil-structure interaction, a higher damping ratio is considered for soil (e.g. $\zeta = 0.15$ to 0.20). Equation 16.79 cannot be used to uncouple modal equations of motion (to substitute $[C]$ with a diagonal matrix).

EXAMPLE 16.3: CANTILEVER WITH THREE LUMPED MASSES

Find the natural frequencies $\{\omega\}$ and the normalized mode shapes $[\Phi]$ for the cantilever in Figure 16.8a. Verify the orthogonality of Equation 16.71.

The system has three degrees-of-freedom indicated by the three coordinates in Figure 16.8b. The corresponding stiffness matrix is (using Table D.3, Appendix D):

$$[S] = \frac{EI}{l^3}\begin{bmatrix} 1.6154 & & \text{Symm.} \\ -3.6923 & 10.1538 & \\ 2.7692 & -10.6154 & 18.4615 \end{bmatrix} \tag{a}$$

The mass matrix is:

$$[m] = \frac{W}{g}\begin{bmatrix} 4 & 0 & 0 \\ 0 & 1 & 0 \\ 0 & 0 & 1 \end{bmatrix} \tag{b}$$

where g is the acceleration due to gravity. Substituting in Equation 16.59, we obtain:

$$[B] = \frac{gEI}{Wl^3}\begin{bmatrix} 0.4039 & -0.9231 & 0.6923 \\ -3.6923 & 10.1538 & -10.6154 \\ 2.7692 & -10.6154 & 18.4615 \end{bmatrix} \tag{c}$$

Substituting this equation in Equation 16.58 and solving for the eigenvalues, ω^2 (or solving the eigenvalue problem in the form of Equation 16.57), we obtain:

$$\omega_1^2 = 0.02588\frac{gEI}{Wl^3};\quad \omega_2^2 = 3.09908\frac{gEI}{Wl^3};\quad \omega_3^2 = 25.89419\frac{gEI}{Wl^3} \tag{d}$$

or

$$\omega_1 = 0.1609\sqrt{\frac{gEI}{Wl^3}};\quad \omega_2 = 1.7604\sqrt{\frac{gEI}{Wl^3}};\quad \omega_3 = 5.0886\sqrt{\frac{gEI}{Wl^3}} \tag{e}$$

If we choose $D_1 = 1$, the modal vectors corresponding to the above angular frequencies are:

$$[\phi] = \begin{bmatrix} 1.0 & 1.0 & 1.0 \\ 0.5224 & -6.3414 & -13.1981 \\ 0.1506 & -4.5622 & 19.2222 \end{bmatrix} \tag{f}$$

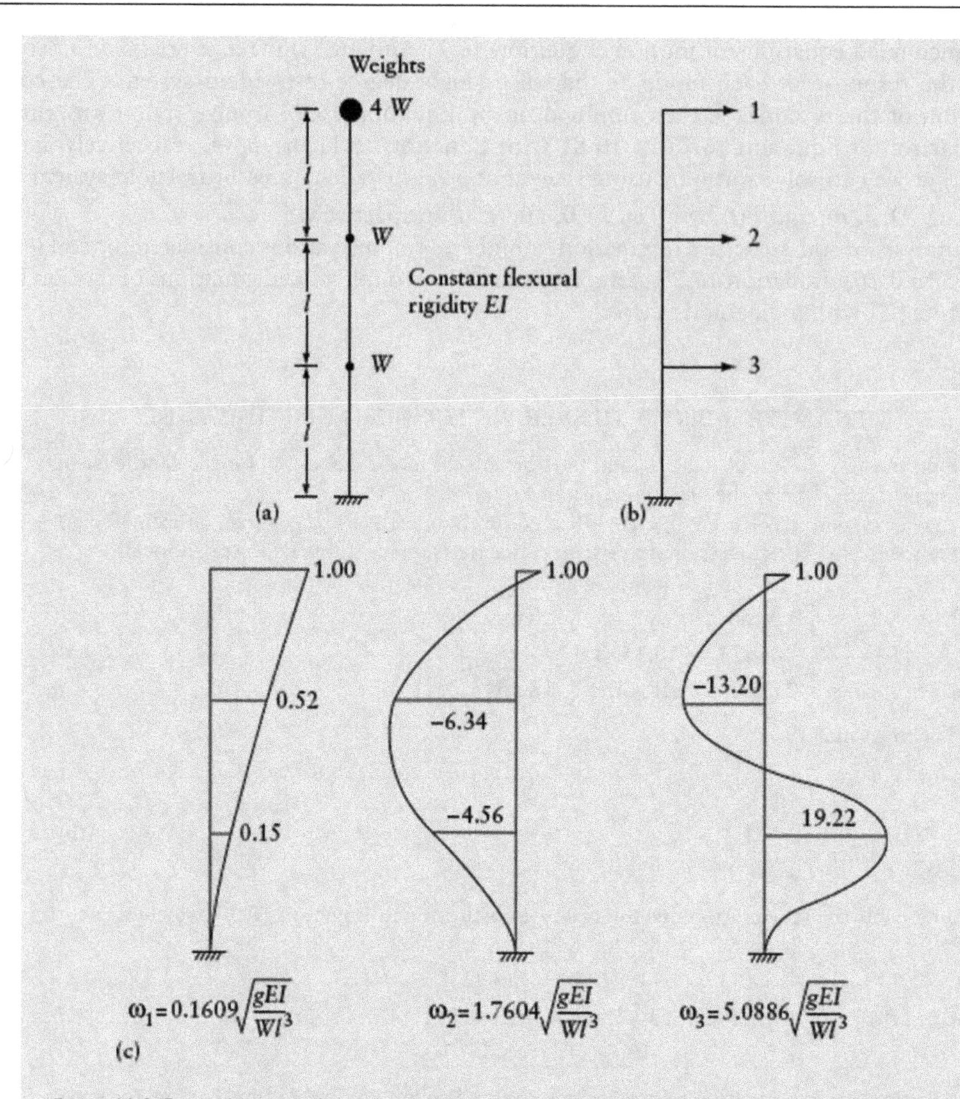

Figure 16.8 Free vibration of the beam of Example 16.3. (a) System properties. (b) Coordinate system. (c) Mode characteristic shapes.

The values of $\sqrt{M_r} = \sqrt{\{\phi_r\}^T[m]\{\phi_r\}}$, with $r = 1$, 2, and 3, respectively $= 2.073\sqrt{W/g}$, $8.064\sqrt{W/g}$, and $23.400\sqrt{W/g}$; multiply columns 1, 2, and 3 of $[\phi]$, respectively, by $(2.073)^{-1}$, $(8.064)^{-1}$, and $(23.400)^{-1}$ to obtain:

$$[\Phi] = \begin{bmatrix} 0.4825 & 0.1240 & 0.0427 \\ 0.2521 & -0.7864 & -0.5640 \\ 0.0727 & -0.5657 & 0.8214 \end{bmatrix} \left(\frac{g}{W}\right)^{1/2} \tag{g}$$

We can verify $[\Phi]$ by Equation 16.71:

$$[\Phi]^T \begin{bmatrix} 4 & 0 & 0 \\ 0 & 1 & 0 \\ 0 & 0 & 1 \end{bmatrix} [\Phi] \frac{W}{g} = \begin{bmatrix} 1 & 0 & 0 \\ 0 & 1 & 0 \\ 0 & 0 & 1 \end{bmatrix}$$

$$[\Phi]^T \begin{bmatrix} 1.6154 & & \text{Symm.} \\ -3.6923 & 10.1538 & \\ 2.7692 & -10.6154 & 18.4615 \end{bmatrix} [\Phi] \frac{EI}{l^3} = \begin{bmatrix} 0.02588 & 0 & 0 \\ 0 & 3.09908 & 0 \\ 0 & 0 & 25.89419 \end{bmatrix} \frac{gEI}{Wl^3}$$

EXAMPLE 16.4: HARMONIC FORCES ON A CANTILEVER WITH THREE LUMPED MASSES

Find the steady-state displacements of the system shown in Figure 16.8a subjected to harmonic forces $\{P(t)\} = P_0 \sin \Omega t \ \{2, 1, 1\}$ with $\Omega = \omega_1/2$, where ω_1 = frequency of the first natural mode. Use the results of Example 16.3.

The steady-state displacement of undamped single-degree-of-freedom system is (the last term of Equation 16.25):

$$D = \frac{P_0 \sin \Omega t}{S} \left(1 - \frac{\Omega^2}{\omega^2}\right)^{-1} \tag{16.82}$$

Replace D by η_r, $P_0 \sin \Omega t$ by $P_r^*(t)$, and S by ω_r^2 to obtain the rth mode of displacement in modal coordinate:

$$\eta_r = \frac{P_r^*(t)}{\omega_r^2} \left(1 - \frac{\Omega^2}{\omega_r^2}\right)^{-1} \tag{16.83}$$

In this problem, $\Omega^2 = (\omega_1)^2/4 = 6.47 \times 10^{-3}$ $[gEI/Wl^3]$. The generalized forces can be expressed as (Equation 16.76):

$$\{P^*(t)\} = [\Phi]^T \{P(t)\}$$

$$= \left(\frac{g}{W}\right)^{\frac{1}{2}} \begin{bmatrix} 0.4825 & 0.2521 & 0.0727 \\ 0.1240 & -0.7864 & -0.5657 \\ 0.0427 & -0.5640 & 0.8214 \end{bmatrix} \begin{Bmatrix} 2 \\ 1 \\ 1 \end{Bmatrix} P_0 \sin \Omega t$$

$$= \left(\frac{g}{W}\right)^{\frac{1}{2}} \begin{Bmatrix} 1.2898 \\ -1.1042 \\ 0.3428 \end{Bmatrix} P_0 \sin \Omega t$$

The steady-state displacements in natural coordinates are:

$$\eta_1 = \left(\frac{W}{g}\right)^{\frac{1}{2}} \frac{P_0 \sin \Omega t \,(1.2898)}{(EI/l^3)(0.02588)} \left[1 - \frac{6.47 \times 10^{-3}}{0.02588}\right]^{-1} = 66.44 \left(\frac{W}{g}\right)^{\frac{1}{2}} \frac{P_0 l^3}{EI} \sin \Omega t$$

$$\eta_2 = \left(\frac{W}{g}\right)^{\frac{1}{2}} \frac{P_0 \sin \Omega t \,(-1.1042)}{(EI/l^3)(3.0991)} \left[1 - \frac{6.47 \times 10^{-3}}{3.0991}\right]^{-1} = -0.3570 \left(\frac{W}{g}\right)^{\frac{1}{2}} \frac{P_0 l^3}{EI} \sin \Omega t$$

$$\eta_3 = \left(\frac{W}{g}\right)^{\frac{1}{2}} \frac{P_0 \sin \Omega t\,(0.3428)}{(EI/l^3)(25.8942)}\left[1-\frac{6.47\times10^{-3}}{25.8942}\right]^{-1} = 0.0132\left(\frac{W}{g}\right)^{\frac{1}{2}} \frac{P_0 l^3}{EI}\sin \Omega t$$

The displacements are (Equation 16.73):

$$\{D\} = \left(\frac{W}{g}\right)^{\frac{1}{2}} [\Phi]\begin{Bmatrix} 66.45 \\ -0.3570 \\ 0.0132 \end{Bmatrix} \frac{P_0 l^3}{EI}\sin \Omega t$$

$$= \left\{\begin{Bmatrix} 32.0588 \\ 16.7482 \\ 4.8282 \end{Bmatrix} + \begin{Bmatrix} -0.0443 \\ 0.2808 \\ 0.2020 \end{Bmatrix} + \begin{Bmatrix} 0.5659\times10^{-3} \\ -7.4694\times10^{-3} \\ 0.0109 \end{Bmatrix}\right\} \frac{P_0 l^3}{EI}\sin \Omega t$$

$$= \begin{Bmatrix} 32.015 \\ 17.022 \\ 5.041 \end{Bmatrix} \frac{P_0 l^3}{EI}\sin \Omega t$$

The contributions of the modes are smaller for higher modes.

16.7.1 Modal damping ratio: Rayleigh damping

Damping forces are greater with higher natural frequency, ω. Uncoupling the modes employs a single damping coefficient, c (= $\zeta/(2m\omega)$) in the equation of motion of the rth mode:

$$\ddot{\eta}_r + 2\zeta_r\,\omega_r\,\dot{\eta}_r + \omega_r^2\,\eta_r = P_r^*(t) \tag{h}$$

where η_r = modal coordinate related to displacement $\{D\}_r$; $[D] = [\Phi]\,\{\eta\}$; with $[\Phi]$ being modal matrix. The use of the same damping ratio ζ for different modes involves approximation. Rayleigh (classical) damping eliminates the errors for two modes, ω_1 and ω_2; but the damping ratio will be somewhat smaller than ζ for $\omega_1 < \omega < \omega_2$. For the rth mode, the damping ratio is assumed as (Figure 16.9):

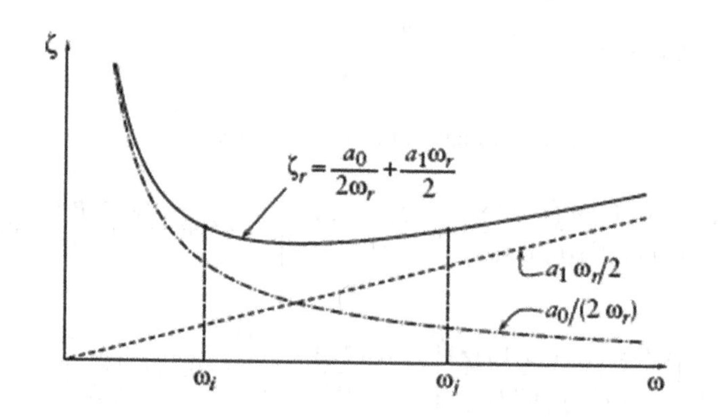

Figure 16.9 Rayleigh damping (sum of straight line and hyperbola).

$$\zeta_r = \frac{a_0}{2}\frac{1}{\omega_r} + \frac{a_1}{2}\omega_r \tag{i}$$

Experiments indicate that we can use Equation (i) to calculate ζ_r. Application of Equation (i) for two values and solving the resulting equations give:

$$a_0 = \zeta\frac{2\,\omega_i\,\omega_j}{\omega_i + \omega_j};\quad a_1 = \zeta\frac{2}{\omega_i + \omega_j} \tag{j}$$

Equations (i) and (j) are used in modal analysis of response of structures to earthquakes. If for example the first four modes are considered, ω_1 and ω_3 can be used to give a_0 and a_1; then Equation (i) can give ζ_r for the four modes; the results would overestimate ζ_2 and underestimate ζ_4. The overestimation has no effect on the accuracy of seismic analysis because the response of mode higher than 5 would be relatively small because of high damping.

EXAMPLE 16.5: RAYLEIGH DAMPING RATIO: THREE DEGREES-OF-FREEDOM SYSTEM

Given $\{\omega_1, \omega_2, \omega_3\} = \{0.2, 1.8, 5.1\}$, find Rayleigh damping ratios $\{\zeta_1, \zeta_2, \zeta_3\}$ using a single value $\zeta = 0.05$ (suitable for reinforced concrete structures with considerable cracking).

Substitution of ω_1, ω_3, and ζ in Equation (i) gives:

$$\frac{1}{2}\begin{bmatrix} 1/\omega_1 & \omega_1 \\ 1/\omega_3 & \omega_1 \end{bmatrix}\begin{Bmatrix} a_0 \\ a_1 \end{Bmatrix} = \begin{Bmatrix} \zeta \\ \zeta \end{Bmatrix} \tag{k}$$

Solution gives a_0 and a_1 (Equation (j)):

$$a_0 = 0.05\frac{2(0.2)(5.1)}{0.2+5.1} = 0.01925;\quad a_1 = 0.05\frac{2}{0.2+5.1} = 0.01887$$

Equation (i) gives:

$$\zeta_1 = \frac{0.01925}{2}\frac{1}{0.2} + \frac{0.01887}{2}(0.2) = 0.05$$

$$\zeta_2 = \frac{0.01925}{2}\frac{1}{1.8} + \frac{0.01887}{2}(1.8) = 0.027$$

$$\zeta_3 = \frac{0.01925}{2}\frac{1}{5.1} + \frac{0.01887}{2}(5.1) = 0.05$$

As expected, the use of $\zeta = 0.05$ in modal analysis would be an overestimation of ζ_2.

16.8 SINGLE- OR MULTI-DEGREE-OF-FREEDOM SYSTEMS SUBJECTED TO GROUND MOTION

The present section is an introduction to "Response of structures to earthquakes," Chapter 17. Consider the response of the single-degree-of-freedom damped system in Figure 16.10 subjected to support motion described in terms of its acceleration $\ddot{u}_g(t)$. At any instant, the displacement of the mass m relative to the ground is:

$$D = u - u_g \tag{16.84}$$

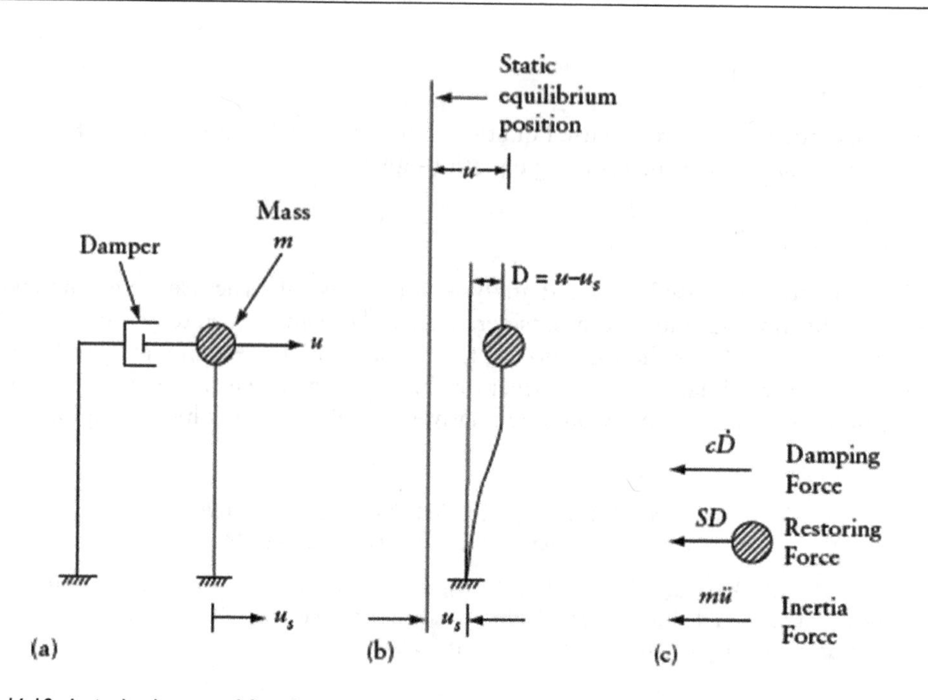

Figure 16.10 A single-degree-of-freedom system subjected to support movements. (a) Positive directions of u and u_s. (b) Deformed shape at any time t. (c) Forces acting on mass at any time t.

The forces acting on the mass are: an inertial force $m\ddot{u}$, a damping force $c\dot{D}$, and a restoring force SD; where S is the stiffness of the structure with respect to the coordinate shown. The equation of motion is:

$$m\ddot{u} + c\dot{D} + SD = 0 \tag{16.85}$$

Substituting for c and S and Equation 16.84 in 16.85, the equation of motion of a single-degree-of-freedom damped system subjected to ground motion becomes:

$$\ddot{D} + 2\zeta\omega\dot{D} + \omega^2 D = -\ddot{u}_g \tag{16.86}$$

Comparison of Equation 16.86 with Equation 16.34 shows that the effect of ground motion is the same as that of the force:

$$P(t) = -m\,\ddot{u}_g \tag{16.87}$$

The modal analysis (Section 16.7) can determine the response of a damped multi-degree-of-freedom system (Figure 16.2) subjected to ground motion, described by its acceleration, $\ddot{u}_g$. For this purpose, apply Equation 16.81:

$$\ddot{\eta}_r + 2\zeta_r\,\omega_r\,\dot{\eta}_r + \omega_r^2\,\eta_r = P_r^*(t)$$

Here, D is replaced by η and $(-\ddot{u}_g)$ by the generalized force $P_r^*(t)$; where

$$\{P_r^*(t)\} = -\ddot{u}_g\,[\Phi]^{\mathrm{T}} \begin{Bmatrix} m_1 \\ m_2 \\ \dots \\ m_n \end{Bmatrix} \tag{16.88}$$

Equation 16.81, combined with the rth element of $\{P_r^*(t)\}$, gives the uncoupled equation of the rth mode.

The effect of a particular earthquake can be analyzed by the substitution of its record of $\ddot{u}_g$ in Equation 16.86 or 16.88. Solution of these differential equations is commonly done by numerical "time-stepping" methods.

EXAMPLE 16.6: CANTILEVER SUBJECTED TO HARMONIC SUPPORT MOTION

Find the steady-state displacements of the system shown in Figure 16.8a subjected to harmonic support acceleration $\ddot{u}_g = (g/5) \sin \Omega t$, where $\Omega = \omega_1/2$ and ω_1 is the frequency of the first natural mode. Use the results of Example 16.3.

The vector of generalized forces is (Equation 16.88):

$$\{P_r^*(t)\} = -\frac{g}{5}\sin\Omega t \begin{bmatrix} 0.4825 & 0.2521 & 0.0727 \\ 0.1240 & -0.7864 & -0.5640 \\ 0.0427 & -0.5640 & 0.8214 \end{bmatrix} \begin{Bmatrix} 4 \\ 1 \\ 1 \end{Bmatrix} \left(\frac{W}{g}\right)^{\frac{1}{2}}$$

$$= -\frac{g}{5}\sin\Omega t \begin{Bmatrix} 2.2548 \\ -0.8545 \\ 0.4282 \end{Bmatrix} \left(\frac{W}{g}\right)^{\frac{1}{2}} = g\sin\Omega t \begin{Bmatrix} -0.45096 \\ 0.17122 \\ -0.08564 \end{Bmatrix} \left(\frac{W}{g}\right)^{\frac{1}{2}}$$

Given: $\Omega = \omega_1/2$; $\Omega^2 = \omega_1^2/4 = \dfrac{0.02588}{4}\dfrac{gEI}{Wl^3}$. Equation 16.80 gives the displacements of the three modes in natural coordinates:

$$\eta_1 = \left(\frac{W}{g}\right)^{\frac{1}{2}} \frac{-0.45096}{0.02588}\left[1 - \frac{0.02588}{4(0.02588)}\right]^{-1} \frac{Wl^3}{EI}\sin\Omega t = -23.23\left(\frac{W}{g}\right)^{\frac{1}{2}} \frac{Wl^3}{EI}\sin\Omega t$$

$$\eta_2 = \left(\frac{W}{g}\right)^{\frac{1}{2}} \frac{0.17090}{3.09908}\left[1 - \frac{0.02588}{4(3.09908)}\right]^{-1} \frac{Wl^3}{EI}\sin\Omega t = 0.0554\left(\frac{W}{g}\right)^{\frac{1}{2}} \frac{Wl^3}{EI}\sin\Omega t$$

$$\eta_3 = \left(\frac{W}{g}\right)^{\frac{1}{2}} \frac{-0.08564}{25.89415}\left[1 - \frac{0.02588}{4(25.89415)}\right]^{-1} \frac{Wl^3}{EI}\sin\Omega t = -0.0033\left(\frac{W}{g}\right)^{\frac{1}{2}} \frac{Wl^3}{EI}\sin\Omega t$$

The displacements are (Equation 16.71):

$$\{D\} = [\Phi]\{\eta\} = \begin{Bmatrix} -11.202 \\ -5.898 \\ -1.722 \end{Bmatrix} \frac{Wl^3}{EI}\sin\Omega t$$

16.9 SUBSTITUTE SINGLE-DEGREE-OF-FREEDOM SYSTEM

Results of the modal analysis in Section 16.7 can be obtained by the use of physical single-degree-of-freedom systems (Figure 16.11). The motion of the multi-degree-of-freedom system in Figure 16.8a due to applied forces $\{P(t)\}$ can be analyzed by n substitute systems, each having a single-degree-of-freedom. The rth substitute system gives the displacement $D_r(t)$;

Figure 16.11 Single-degree-of-freedom undamped system in free vibration.

where $D_r(t)$ = the displacement of the actual system at the rth coordinate with the system vibrating in the rth natural mode shape. The steps of analysis are as follows:

Step 1: For the actual system, find the squares of the natural frequencies $\{\omega^2\}$ and the corresponding mode shapes $[\phi]$. Choose the values of $\phi_{rr} = 1.0$ for $r = 1, 2, \ldots n$.

Step 2: Let the actual system acquire virtual displacement in the rth mode shape. The corresponding virtual work:

$$W = \{\phi_r\}^T [S] \{\phi_r\} \tag{16.89}$$

where $[S]$ = stiffness matrix of the actual system. Figure 16.11 shows a single-degree-of-freedom substitute system having displacement D_{sr} equal to the displacement D_r of the actual system when vibrating in the rth mode shape. Let the substitute system acquire a virtual displacement $D_{sr} = 1$; the corresponding virtual work $= S_{sr} (D_{sr})^2$, with S_{sr} being the stiffness of the substitute system. Equating the virtual work of the actual and substitute systems gives the stiffness of the substitute system for the rth mode:

$$S_{sr} = \bar{\phi}_{rr}^{-2} \left(\{\bar{\phi}_r\}^T [S] \{\bar{\phi}_r\} \right) \tag{16.90}$$

where $[S]$ and $\{\bar{\phi}_r\}$ are the stiffness and the rth mode's vector of the multi-degree-of-freedom system. Equation 16.90 applies, with $\bar{\phi}_{rr}$ = any value. When the eigenvector $\{\bar{\phi}_r\}$ is normalized such that $\bar{\phi}_{rr} = 1.0$, the stiffness of the substitute system becomes:

$$S_{sr} = \{\phi_r\}^T [S] \{\phi_r\} \tag{16.91}$$

The mass m_{sr} of the substitute system for the rth mode has to be equal to:

$$m_{sr} = S_{sr} / \omega_r^2 \tag{16.92}$$

ω_r^2 = square of the natural frequency of the rth mode of the multi-degree-of-freedom system; while m_{sr} and S_{sr} are properties of the substitute system. The substitute system can have a viscous damper (not shown in Figure 16.11) with damping coefficient c_{sr} and damping ratio ζ:

$$c_{sr} = 2\zeta m_{sr} \omega_r \tag{16.93}$$

c_{sr} is property of the substitute system; the damping ratio ζ is assumed the same for single as well as multi-degree-of-freedom system.

Step 3: Apply a substitute force $P_{sr}(t)$ on the substitute system. The virtual work $P_{sr}(t)\ D_{sr}$ (with $D_{sr} = 1$) of the substitute system is equal to the virtual work of the actual system with displacements $\{D_r\} = \{\phi_r\}$; thus, the force to be applied on the substitute system is:

$$P_{sr}(t) = \{\phi_r\}^{T}\{P(t)\} = \sum_{i=1}^{n}\left(\phi_{ir}\ P_i(t)\right) \tag{16.94}$$

Step 4: Find the displacement $D_{sr}(t)$ of the substitute system. This will be equal to the contribution of the rth mode to the displacement at coordinate r of the actual system. At any coordinate i, the contribution of the rth mode to the displacement in the actual system is:

$$D_{ir}(t) = \phi_{ir}\ D_{sr}(t);\quad D_i(t) = \sum_{r=1}^{n}\phi_{ir}\ D_{sr}(t) \tag{16.95}$$

Condensed stiffness matrix is commonly used to determine natural frequencies, ω^2 and modes of vibration, $\{\phi\}$ of structural systems. The purpose of the following example is to determine ω^2 and $\{\phi\}$ without stiffness matrix calculation.

EXAMPLE 16.7: SUBSTITUTE SINGLE-DEGREE-OF-FREEDOM SYSTEM APPLICATION

Solve Example 16.6 using substitute single-degree-of-freedom systems.

Step 1: Parameters determined in Example 16.3:

$$[S] = \frac{EI}{l^3}\begin{bmatrix} 1.6154 & & \text{Symm.} \\ -3.6923 & 10.1538 & \\ 2.7692 & -10.6154 & 18.4615 \end{bmatrix};\ [m] = m\begin{bmatrix} 4 & 0 & 0 \\ 0 & 1 & 0 \\ 0 & 0 & 1 \end{bmatrix};\ m = W/g$$

$$\omega_1^2 = 0.02588\frac{EI}{ml^3};\ \omega_2^2 = 3.09908\frac{EI}{ml^3};\ \omega_3^2 = 25.89419\frac{EI}{ml^3}$$

Choosing $\phi_{rr} = 1$, with $r = 1$, 2 and 3, the modal vectors are:

$$[\phi] = \left[\{\phi_1\}\quad\{\phi_2\}\quad\{\phi_3\}\right] = \begin{bmatrix} 1.0 & -0.1577 & 0.0520 \\ 0.5224 & 1.0 & -0.6866 \\ 0.1506 & 0.7194 & 1.0 \end{bmatrix}$$

Step 2: The stiffness and masses of the substitute systems in Figure 16.11 are (Equations 16.91 and 16.92):

$$S_{sr} = \{\phi_r\}^{T}[S]\{\phi_r\}\ \text{ with } r = 1, 2 \text{ and } 3.$$

$$S_{s1} = 0.1112\frac{EI}{l^3};\ m_{s1} = \frac{0.1112}{0.02588}m = 4.2956m$$

$$S_{s2} = 5.0113\frac{EI}{l^3};\ m_{s2} = \frac{5.0113}{3.09908}m = 1.6170m$$

$$S_{s3} = 38.3818\frac{EI}{l^3};\ m_{s3} = \frac{38.3818}{25.89419}m = 1.4823m$$

Step 3: The applied loads of the actual system are: $\{P(t)\} = P_0 \sin \Omega t\ \{2,1,1\}$

$$\Omega = \omega_1/2 = \frac{(0.02588)^{1/2}}{2}\left(\frac{EI}{ml^3}\right)^{1/2} = 0.08044\left(\frac{EI}{ml^3}\right)^{1/2}$$

The force to apply on each substitute system is (Equation 16.94):

$$P_{s1}(t)=\{\phi_1\}^{\mathrm{T}}\{2,1,1\}\,P_0\,\sin\Omega t=2.6730\,P_0\,\sin\Omega t$$

$$P_{s2}(t)=\{\phi_2\}^{\mathrm{T}}\{2,1,1\}\,P_0\,\sin\Omega t=1.4040\,P_0\,\sin\Omega t$$

$$P_{s3}(t)=\{\phi_3\}^{\mathrm{T}}\{2,1,1\}\,P_0\,\sin\Omega t=0.4174\,P_0\,\sin\Omega t$$

Step 4: For the rth substitute system, the steady-state displacement due to $P_{sr}(t)$ is (the last term of Equation 16.25):

$$D_{sr}=\frac{P_{sr}(t)}{S_{sr}}\left(1-\frac{\Omega^2}{\omega_r^2}\right)^{-1}\quad \text{with } r=1,2\text{ and }3.$$

$$D_{s1}=\frac{2.6730}{0.1112}\left(1-\frac{(0.08044)^2}{0.02588}\right)^{-1}\frac{P_0\,l^3}{EI}\sin\Omega t=32.0588\,\frac{P_0\,l^3}{EI}\sin\Omega t$$

$$D_{s2}=\frac{1.4040}{5.0113}\left(1-\frac{(0.08044)^2}{3.09908}\right)^{-1}\frac{P_0\,l^3}{EI}\sin\Omega t=0.2808\,\frac{P_0\,l^3}{EI}\sin\Omega t$$

$$D_{s3}=\frac{0.4174}{38.3818}\left(1-\frac{(0.08044)^2}{25.89419}\right)^{-1}\frac{P_0\,l^3}{EI}\sin\Omega t=0.0109\,\frac{P_0\,l^3}{EI}\sin\Omega t$$

The displacements of the actual system are:

$$\{D(t)\}=\{\phi_1\}\,D_{s1}+\{\phi_2\}\,D_{s2}+\{\phi_3\}\,D_{s3}$$

$$=\left\{\begin{Bmatrix}32.0588\\16.7482\\4.8282\end{Bmatrix}+\begin{Bmatrix}-0.0443\\0.2808\\0.2020\end{Bmatrix}+\begin{Bmatrix}0.5659\times10^{-3}\\-7.469\times10^{-3}\\0.0109\end{Bmatrix}\right\}\frac{P_0\,l^3}{EI}\sin\Omega t$$

$$=\begin{Bmatrix}32.015\\17.022\\5.041\end{Bmatrix}\frac{P_0\,l^3}{EI}\sin\Omega t$$

The contributions of the modes and their summation are the same as the answers obtained by the use of modal coordinates and generalized forces (Example 16.4).

EXAMPLE 16.8: DAMPED MULTI-DEGREE-OF-FREEDOM SYSTEM ANALYZED BY SUBSTITUTE SYSTEMS

Determine the displacement $D_1(t)$ (at coordinate 1, mid-span) of the simple beam in Figure 16.12a. The applied forces are: $\{P(t)\}=P_0\sin\Omega t\{1,1,1\}$, with $\Omega=0.75\,\omega_1$. Consider damping with $\zeta=0.02=$ ratio between the damping of the system and the critical damping coefficient $=2m\omega_r$; where $\omega_r=$ the natural frequency of the substitute system for the rth

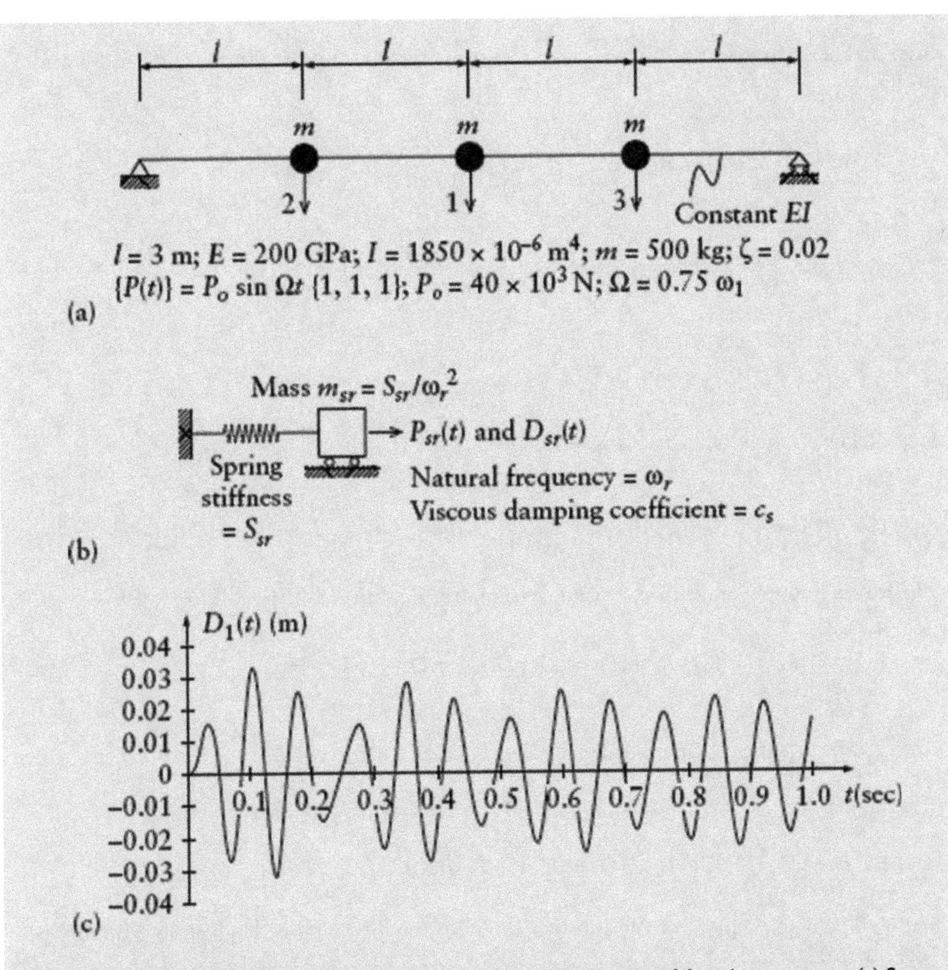

Figure 16.12 Substitute systems for modal analysis of multi-degree-of-freedom systems. (a) System analyzed in Example 16.8. (b) Substitute single-degree-of-freedom system for the rth mode. (c) Deflection at mid-span, $D_1(t)$ of the system in (a).

mode with $r = 1, 2,$ and 3. Assume that the motion starts from rest. Data: $l = 3$ m; $E = 200$ GPa; $I = 1850 \times 10^6$ mm^4; $m = 500$ kg; $P_0 = 40 \times 10^3$ N.

Step 1: The stiffness and mass matrices are (Appendix D, Table D.1):

$$[S] = \frac{EI}{l^3}\begin{bmatrix} 13.7143 & & \text{Symm.} \\ -9.4286 & 9.8571 & \\ -9.4286 & 3.8571 & 9.8571 \end{bmatrix}; \quad [m] = m\begin{bmatrix} 1 & 0 & 0 \\ 0 & 1 & 0 \\ 0 & 0 & 1 \end{bmatrix}$$

$$\omega_1^2 = 0.3803\frac{EI}{ml^3}; \quad \omega_2^2 = 6.0000\frac{EI}{ml^3}; \quad \omega_3^2 = 27.0483\frac{EI}{ml^3}$$

$$[\phi] = \begin{bmatrix} 1.0 & 0 & -1.4142 \\ 0.7071 & 1.0 & 10 \\ 0.7071 & -1.0 & 1.0 \end{bmatrix}$$

Step 2: The stiffness and the masses of the substitute systems are (Equations 16.88 and 16.89):

$$S_{s1} = \{\phi_1\}^T [S] \{\phi_1\} = 0.76038 \frac{EI}{l^3}; \; m_{s1} = \frac{0.76038}{0.3803} m = 2.00\, m$$

$$S_{s2} = \{\phi_2\}^T [S] \{\phi_2\} = 12.000 \frac{EI}{l^3}; \; m_{s2} = \frac{12.000}{6.000} m = 2.00\, m$$

$$S_{s3} = \{\phi_3\}^T [S] \{\phi_3\} = 108.19 \frac{EI}{l^3}; \; m_{s3} = \frac{108.19}{27.0483} m = 4.00\, m$$

Step 3: The applied loads are: $\{P(t)\} = P_0 \sin\Omega t \, \{1, 1, 1\}$

$$\Omega = 0.75\, \omega_1 = 0.75 \left(0.3803\right)^{1/2} \left(\frac{EI}{ml^3}\right)^{1/2} = 0.4625 \left(\frac{EI}{ml^3}\right)^{1/2}$$

The forces to apply on each substitute system are (Equation 16.91):

$$P_{s1}(t) = \{\phi_1\}^T \{1, 1, 1\} P_0 \sin\Omega t = 2.4142\, P_0 \sin\Omega t$$

$$P_{s2}(t) = \{\phi_2\}^T \{1, 1, 1\} P_0 \sin\Omega t = 0$$

$$P_{s3}(t) = \{\phi_3\}^T \{1, 1, 1\} P_0 \sin\Omega t = 0.5858\, P_0 \sin\Omega t$$

Step 4: The displacement of the substitute system for the rth mode is given by Equations 16.47 to 16.51:

$$D_{sr}(t) = e^{-\zeta \omega_r t} \left(\bar{C}_{1r} \sin \omega_{dr} t + \bar{C}_{2r} \cos \omega_{dr} t\right) + \bar{D}_r \sin\left(\Omega t + \alpha_r\right)$$

$$\alpha_r = \tan^{-1}\left[\frac{-2\zeta\Omega/\omega_r}{1-(\Omega/\omega_r)^2}\right]$$

$$\bar{D}_r = \frac{\chi_r}{S_{sr}} \left(\frac{P_{sr}(t)}{\sin\Omega t}\right)$$

$$\chi_r = \left[\left\{1-(\Omega/\omega_r)^2\right\}^2 + (2\zeta\Omega/\omega_r)^2\right]^{-1/2}$$

The constants $\bar{C}_{1r}$ and $\bar{C}_{2r}$ are given by the conditions at $t=0$, $D_{sr}=0$ and $\dot{D}_{sr} = 0$. In the present example, the contribution of the second mode to the displacement at coordinate 1 is nil. $D_{s1}(t)$ is a contributed displacement of the actual system at coordinate 1. $D_{s3}(t)$ is a contributed displacement of the actual displacement at coordinate 3. At coordinate 1 (mid-span), the total displacement of the system is (Equation 16.95):

$$D_1(t) = D_{s1}(t) + \phi_{13}\, D_{s3}(t) = D_{s1}(t) - 1.4142\, D_{s3}(t)$$

The parameters used to calculate $D_{sr}(t)$, with $r=1$ and 3 are:

$r=1$; $P_0=40$ kN; $S_{s1}=0.76038$ EI/l^3; $\overline{D}_r=21.13\times10^{-3}$; $\omega_r=0.1021\times10^3$; $\omega_{dr}=0.1021\times10^3$; $\alpha_r=-68.46\times10^{-3}$; $\overline{C}_{1r}=-15.78\times10^{-3}$; $\overline{C}_{2r}=1.446\times10^{-3}$.

$r=3$; $S_{s3}=108.19$ EI/l^3; $\overline{D}_r=15.93\times10^{-6}$; $\omega_r=0.8610\times10^3$; $\omega_{dr}=0.8608\times10^3$; $\alpha_r=-3.585\times10^{-3}$; $\overline{C}_{1r}=-1.416\times10^{-6}$; $\overline{C}_{2r}=57.11\times10^{-9}$.

Figure 16.12c shows the variation of mid-span deflection. Solution of the problem using SAP 2000 gives close answers.

16.10 SUBSTITUTE SINGLE-DEGREE-OF-FREEDOM SYSTEM FOR STRUCTURES HAVING NUMEROUS DEGREES-OF-FREEDOM

For general structural analysis, we use practice models having numerous degrees-of-freedom; let n be the number of degrees-of-freedom. We can adequately determine the dynamic response to lateral forces on the same models considering displacements at a smaller number of coordinates, k. For general analysis of a plane frame (Figure 16.13a), two translations in x- and y-axes and a rotation about z-axis are commonly considered at each node. Symmetric-plan buildings can be idealized as plane frames (Chapter 11). The response to dynamic lateral forces can be analyzed by considering a horizontal coordinate at each floor level. Then, we can use Equation 16.88 to calculate stiffness, S_{sr}, of a single-degree-of-freedom substitute system vibrating in the rth mode:

$$S_{sr} = \{\phi_r\}^{\mathrm{T}}[S]\{\phi_r\}$$

In this application, $[S]_{k\times k}$ can be the stiffness matrix corresponding to k-chosen coordinates. In Figure 16.13c, the chosen coordinates are two translations in the x-direction. $[S]$ can be determined by static condensation (Equation 5.55). Alternatively, the jth column of $[S]$ can be determined as the forces at the k-coordinates when $D_j=1$ while $D_i=0$; where $i\neq j$. Computer program PLANEF can determine these forces in separate runs, one for each column of $[S]$. To find natural frequencies and modes, a diagonal matrix $[m]$ is generated and entered with $[S]$ as input of a computer program that solves the eigenvalue problem (e.g. EIGEN, Chapter 20). The mass matrix $[m]$ consists of k-masses on the diagonal (with the remaining elements = 0); these correspond to translational coordinates only. When the eigenvector $\{\phi_r\}$ has $\overline{\phi}_{rr}\neq 1.0$ the stiffness of a single-degree-of-freedom substitute system can be expressed as (Equation 16.90):

$$S_{sr} = \left(\overline{\phi}_{rr}\right)^{-2}\left(\left\{\overline{\phi}_r\right\}^{\mathrm{T}}[S]\left\{\overline{\phi}_r\right\}\right)$$

When the stiffness matrix $[S]_{n\times n}$ is huge, with n much greater than k, a computer is commonly used to generate the stiffness and mass matrices and continues to determine ω_r^2 and $\left\{\overline{\phi}_r\right\}$ with $r=1, 2, \ldots, k$; where ω_r^2 is eigenvalue; $\left\{\overline{\phi}_r\right\}$ = eigenvector with $\overline{\phi}_{rr}\neq 1.0$. In Example 16.10, a three-dimensional finite-element model is analyzed. The model has 990 nodes.

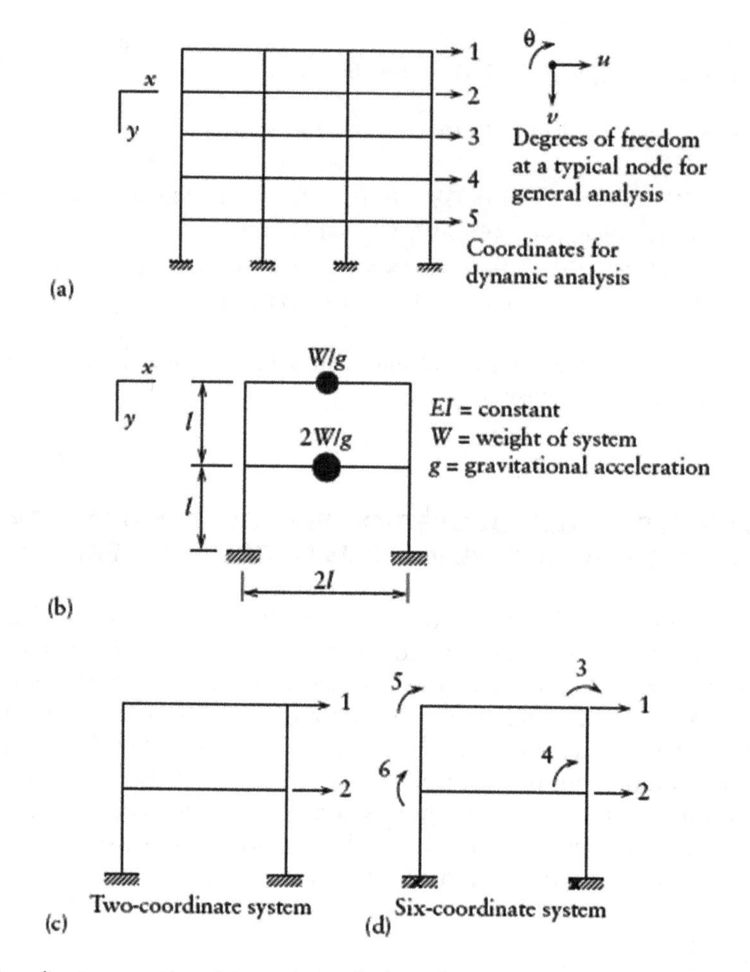

Figure 16.13 Coordinates considered in analysis of plane frames. (a) Coordinates considered for general analysis and coordinates retained for dynamic effect of lateral forces. (b) Coordinates considered for a plane frame in Example 16.9. (c) Two and six-coordinate systems.

Each node has 3 rotational coordinates and 3 translational coordinates. The mass matrix $[m]$ is diagonal, with zeros for the rotational coordinates. The product $\left([S]\{\bar{\phi}_r\}\right)$ is a vector representing inertia forces when the lumped masses are moving in the rth mode. Thus, we expect that the vector will have zero elements at the rotational degrees-of-freedom, because they have no lumped masses. We apply this concept to a plane frame in Example 16.9.

EXAMPLE 16.9: TWO-COORDINATE SYSTEMS FOR DYNAMIC ANALYSIS OF A PLANE FRAME

Consider the plane frame in Figure 17.13b. We use computer program PLANEF (Chapter 20) to find the elements on the first column of $[S]$; these are equal to the forces at coordinates 1 and 2 with prescribed displacements $D_1 = 1.0$, while $D_2 = 0$. We generate the second column in a similar way. The result is:

$$[S] = \frac{EI}{l^3}\begin{bmatrix} 10.192 & -15.123 \\ -15.123 & 37.149 \end{bmatrix} \tag{1}$$

$$[m] = \frac{W}{g}\begin{bmatrix} 1.0 & 0 \\ 0 & 2.0 \end{bmatrix} \tag{m}$$

Equation of motion in free vibration:

$$\left[[S] - [\omega^2][m]\right][\phi] = \{0\} \tag{n}$$

For non-trivial solution:

$$\left|[S] - [\omega^2][m]\right| = 0 \tag{o}$$

$$(10.192 - \omega^2)(37.149 - 2\omega^2) - (15.123)^2 = 0$$

$$2\omega^4 - 57.529\omega^2 + 149.917 = 0$$

$$\omega_1^2 = 2.898\,\frac{gEI}{Wl^3};\quad \omega_2^2 = 25.865\,\frac{gEI}{Wl^3}$$

Setting $\phi_{11} = 1.0$ and $\omega_1^2 = 2.898$ in Equation (n) gives: $\phi_{21} = 0.4824$.
Setting $\phi_{22} = 1.0$ and $\omega_2^2 = 25.865$ in Equation (n) gives: $\phi_{12} = -0.9649$.

$$\{\phi_1\} = \begin{Bmatrix} 1.0 \\ 0.4824 \end{Bmatrix};\quad \{\phi_2\} = \begin{Bmatrix} -0.9649 \\ 1.0 \end{Bmatrix} \tag{p}$$

The stiffnesses of substitute single-degree-of-freedom systems are (Equation 16.91):

$$S_{s1} = \{\phi_1\}^T[S]\{\phi_1\} = \begin{Bmatrix} 1.0 \\ 0.4824 \end{Bmatrix}^T \begin{Bmatrix} 2.8967 \\ 2.7977 \end{Bmatrix}\frac{EI}{l^3} = 4.246\,\frac{EI}{l^3}$$

$$S_{s2} = \{\phi_2\}^T[S]\{\phi_2\} = \begin{Bmatrix} -0.9649 \\ 1.0 \end{Bmatrix}^T \begin{Bmatrix} -24.957 \\ 51.741 \end{Bmatrix}\frac{EI}{l^3} = 75.820\,\frac{EI}{l^3}$$

Six-coordinate system (Figure 16.13d)
Using Appendix C, we generate the stiffness matrix:

$$[S] = \frac{EI}{l^3}\begin{bmatrix} 24 & -24 & -6l & -6l & -6l & -6l \\ -24 & 48 & 6l^2 & -6l^2 & 0 & 0 \\ -6l & 6l^2 & 6l^2 & l^2 & 2l^2 & 0 \\ -6l & -6l^2 & l^2 & 6l^2 & 0 & 2l^2 \\ -6l & 0 & 2l^2 & 0 & 10l^2 & l^2 \\ -6l & 0 & 0 & 2l^2 & l^2 & 10l^2 \end{bmatrix}$$

The mass matrix assigning zero mass for the rotational coordinates 3 to 6:

$$[m] = \frac{W}{g}\begin{bmatrix} 1 & & & & & \\ & 2 & & & & \\ & & 0 & & & \\ & & & 0 & & \\ & & & & 0 & \\ & & & & & 0 \end{bmatrix}$$

Computer program EIGEN gives (Chapter 20):

$$\omega_1^2 = 2.898; \ \omega_2^2 = 25.865$$

$$\{\bar{\phi}_1\} = \begin{Bmatrix} 0.8261 \\ 0.3985 \\ 0.2509 \\ 0.2509 \\ 0.4050 \\ 0.4050 \end{Bmatrix}; \ \{\bar{\phi}_2\} = \begin{Bmatrix} -0.5636 \\ 0.5842 \\ -0.9450 \\ -0.9450 \\ -0.1356 \\ -0.1356 \end{Bmatrix}$$

Inertia forces at the nodes corresponding to $\{\bar{\phi}_1\}$ and $\{\bar{\phi}_2\}$:

$$\{F_1\} = [S]\{\bar{\phi}_1\} = \{2.392, 2.312, 0, 0, 0, 0\}$$

$$\{F_2\} = [S]\{\bar{\phi}_2\} = \{-14.578, 30.266, 0, 0, 0, 0\}$$

As expected, F_3 to F_6 are nil, because m_3 to m_6 are equal to zero. The stiffness of the substitute system is (Equation 16.90):

$$S_{s1} = (0.8261)^{-2}\left(\{\bar{\phi}_1\}^T [S]\{\bar{\phi}_1\}\right) = 4.246\,\frac{EI}{l^3}$$

$$S_{s2} = (0.5842)^{-2}\left(\{\bar{\phi}_2\}^T [S]\{\bar{\phi}_2\}\right) = 75.816\,\frac{EI}{l^3}$$

Figure 16.14 shows the substitute systems and the corresponding mode shapes. This example shows that the dynamic characteristics of a structure can be determined when $[S]$ is large without the use of a condensed stiffness matrix. This concept is applied in the following example.

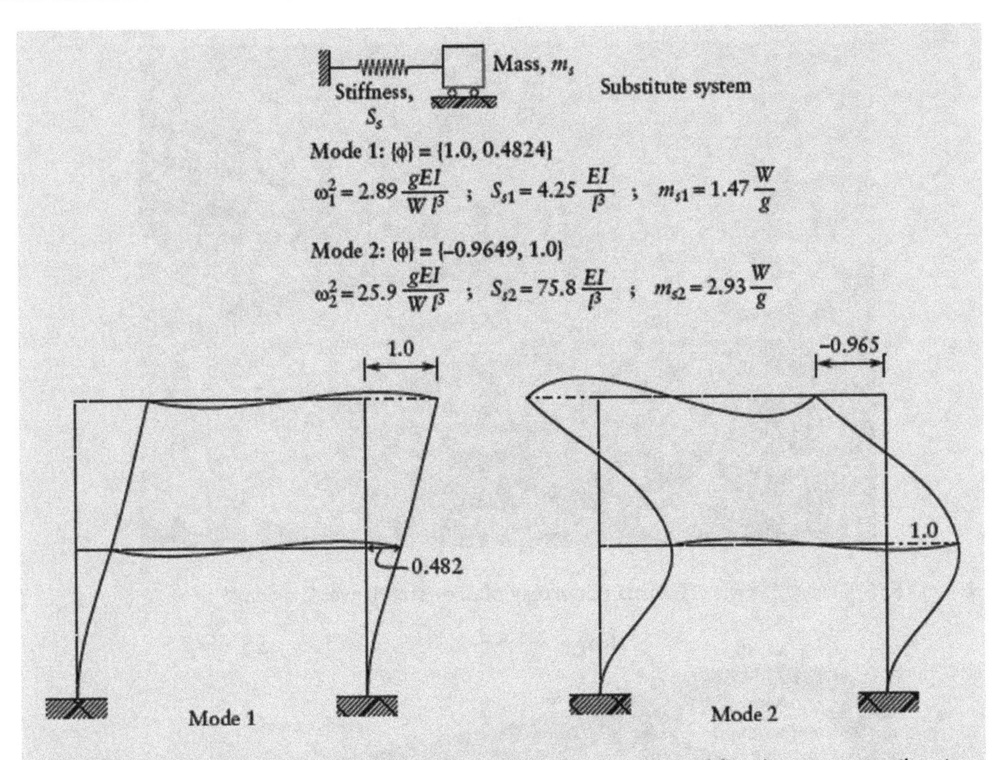

Figure 16.14 Plane frame of Figure 16.13a. Substitute single-degree-of-freedom systems vibrating in modes 1 and 2.

EXAMPLE 16.10: PEDESTRIAN-INDUCED VIBRATION: GLACIER SKYWALK

The stiffness and the mass of single-degree-of-freedom system are derived for analysis of pedestrian-induced vibration of a complex structure, Glacier Skywalk, Alberta, Canada (Figure 16.15). For a wide panorama, the structure overhangs 30 m (100 ft) above a mountain slope; it is light and flexible, thus vulnerable to vibration. The skywalk is idealized as a three-dimensional assembly of bars and plate bending elements (Figure 16.16). The finite-element model has 990 nodes, with three translational and three rotational coordinates at each node. The square of the natural frequencies, ω_r^2 and corresponding vector $\{\phi_r\}$ with $r=1, 2, \ldots s$ were determined by the software SAP2000. The largest translation is expected at the r coordinates, which are selected by preliminary study; $s=4=$number of the considered modes. A single-degree-of-freedom substitute system vibrating in the rth mode of the finite-element model has a stiffness given by (Equation 16.90):

$$S_{sr} = \left(\bar{\phi}_{rr}\right)^{-2} \left(\left\{\bar{\phi}_r\right\}^{\mathrm{T}} \left[S\right] \left\{\bar{\phi}_r\right\}\right) \tag{q}$$

$[S]$ is stiffness matrix for n-degrees-of-freedom system, with $n=6\times 990$; $\left\{\bar{\phi}_r\right\}$ is eigenvector, whose r^{th} element$= \bar{\phi}_{rr} \neq 1.0$. The product $\left(\bar{\phi}_{rr}\right)^{-2} \left\{\bar{\phi}_r\right\}$ represents a vector of n displacements of which the displacement at coordinate $r=1.0$. Steps 1 to 4, of Section 16.10 are:

Figure 16.15 Glacier Skywalk, Canada (Courtesy of Brewster Travel Canada).

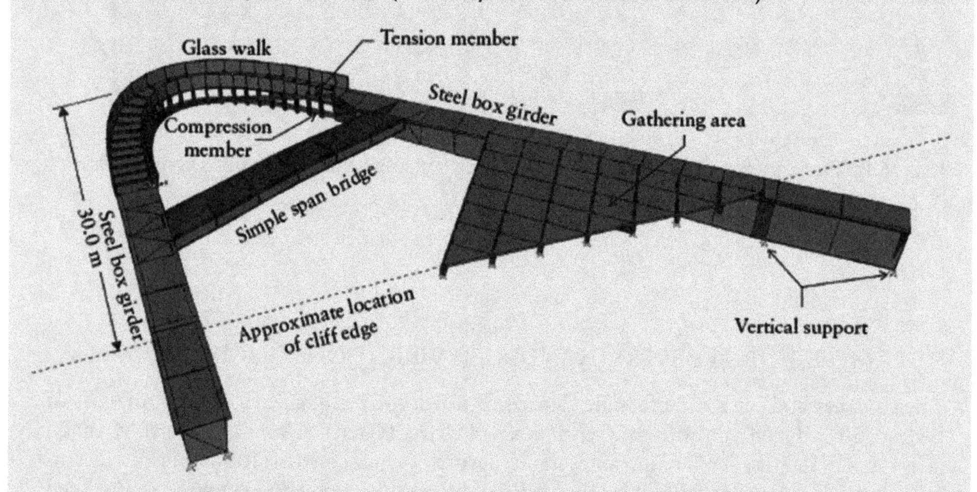

Figure 16.16 Finite-element model of the Glacier Skywalk.

Steps 1 and 2

The elements of $\{\bar{\phi}_r\}$, with $r=1$ (the fundamental mode), determined by SAP2000, are listed in columns A to F of Table 16.1. The product $[S]\{\bar{\phi}_r\}$, also generated by SAP2000, is a vector of forces at the coordinates due to virtual prescribed displacements $\{\bar{\phi}_r\}$; columns G to M of Table 16.1 list the elements of the vector. As expected, the elements corresponding to the rotational degrees-of-freedom are nil (because no lumped masses are assigned to these coordinates). The virtual work of the forces at the coordinates is calculated in columns N to T of Table 16.1. The last column of the table sums up the work of the forces at a node (Columns N to T); sum of columns N to T $= W_1 = 145.89 \times 10^3$ N-m. This is the virtual work due to prescribed displacements $\{\bar{\phi}_r\}$, with $r=1$; the largest element of this vector ($\bar{\phi}_{rr} = 0.282$) corresponds to vertical displacement at node 156. This node is close to the highest point on the outer edge of the skywalk. The virtual work for the rth mode, with unit displacement at the rth coordinate:

Table 16.1 Calculation of virtual work associated with the first mode shape

	Joint displacements						Joint reactions						Virtual work						
Column	A	B	C	D	E	F	G	H	J	K	L	M	N	P	Q	R	S	T	U
Function													A×G	B×H	C×J	D×K	E×L	F×M	Σ *Cols* N *to* T
Displacement	u	v	w	θ_x	θ_y	θ_z	F_1	F_2	F_3	F_4	F_5	F_6	W_1	W_2	W_3	W_4	W_5	W_6	W_{node}
Node		10^{-3} *m*			10^{-3} *radians*			10^3 *N*			10^{-9} *N-m*					*N-m*			
1	10.1	11.5	183	12.8	−18.6	2.05	0.01	0.01	0.19	−6.93	−22.6	−2.66	0.11	0.14	35.3	0.00	0.00	0.00	35.6
2	−5.08	0.00	10.3	−1.08	0.51	−1.60	−0.38	0.00	0.78	18.4	−7.42	−0.84	1.95	0.00	8.08	0.00	0.00	0.00	10.0
3	−4.65	1.52	10.4	−1.28	0.44	0.49	−220	−42.3	551	4.27	3.45	1.33	1025	−6.41	5750	0.00	0.00	0.00	6711
155	−38.7	2.81	215	−2.71	−23.8	0.09	−0.21	0.01	1.14	−0.01	0.32	0.00	7.98	0.04	245	0.00	0.00	0.00	253
156	−41.7	0.38	282	−1.38	−26.3	0.01	−0.31	0.00	2.06	−0.19	−0.21	0.00	12.7	0.00	582	0.00	0.00	0.00	595
157	7.18	8.65	185	13.3	−18.5	1.50	0.01	0.01	0.19	−15.0	−72.6	1.19	0.05	0.08	35.9	0.00	0.00	0.00	36.1
988	0.11	−9.42	−0.11	2.28	0.07	1.59	0.02	−1.85	−0.02	−5.59	−0.13	0.16	0.00	17.5	0.00	0.00	0.00	0.00	17.5
989	0.02	−4.04	−0.27	1.86	0.01	1.40	0.00	−0.66	−0.04	−6.13	2.00	0.70	0.00	2.68	0.01	0.00	0.00	0.00	2.69
990	0.00	−0.39	−0.64	0.27	0.04	−2.15	−895	−0.03	−0.05	0.53	−0.75	−0.04	0.00	0.01	0.03	0.00	0.00	0.00	0.04

Total virtual work (N-m) = 145890

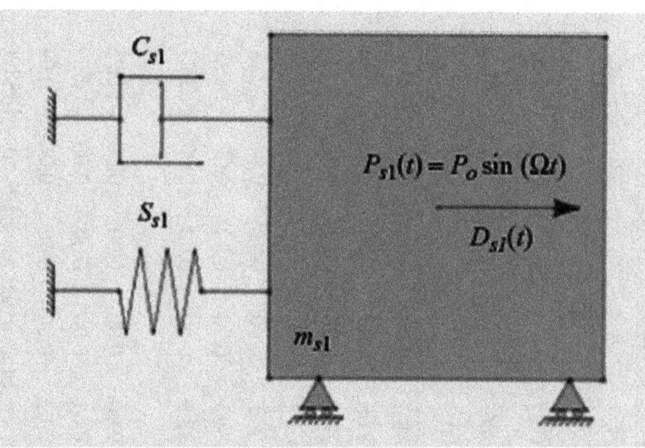

Figure 16.17 Substitute single-degree-of-freedom system for first mode.

$$W_r = \left(\bar{\phi}_{rr}\right)^{-2}\left(\{\bar{\phi}_r\}^T[S]\{\bar{\phi}_r\}\right) \tag{r}$$

$$W_r = S_{sr}\,(1.0)^2 \tag{s}$$

Equations (q), (r), and (s) give the characteristic parameters of a substitute system for mode 1 (Figure 16.17):

$$S_{s1} = (0.282)^{-2}\left(145.89\times10^3\right) = 1.836\times10^6\ \text{N/m}\left(10.48\times10^3\ \text{lb/in}\right)$$

SAP2000 gives for the first mode $\omega_1^2 = (12.08)^2\ \text{s}^{-2}$. The mass for the first substitute system is (Equation 16.92):

$$m_{sr} = S_{sr}/\omega_r^2;\ \text{with}\ r = 1$$

$$m_{s1} = 1.836\times10^6/(12.08)^2 = 12.586\times10^3\ \text{kg}$$

Assuming damping ratio $\zeta = 0.005$ for all modes, the damping coefficient for the first substitute system (Equation 16.93):

$$c_r = 2\zeta m_{sr}\,\omega_r;\ \text{with}\ r = 1$$

$$c_1 = 2(0.005)\left(12.586\times10^3\right)(12.08) = 1520\ \text{N/m}$$

Step 3
To study vibration of the skywalk, consider a driving vertical force on the substitute system for the first mode:

$$P_{s1}(t) = P_0\sin(\Omega t);\ P_0 = 205\ \text{N}\,(46.1\ \text{lb})$$

where Ω = the frequency of the driving force; the amplitude = 205 N; $\Omega = 8\pi$ (=4 Hz) is assumed an upper limit. This is close to the average of athletes in a 100 m race (4.7 Hz). For peak response, we set $\Omega = \omega_1 = 12.08\ \text{s}^{-1}$ = the fundamental natural frequency.
Step 4
The displacement $D_{sr}(t)$ of the substitute system, with $r = 1$ is calculated by Equation 16.51:

$$D = e^{-\zeta\omega t}\left(\bar{C}_1\sin\omega_d t + \bar{C}_2\cos\omega_d t\right) + \bar{D}\sin(\Omega t + \alpha) \tag{t}$$

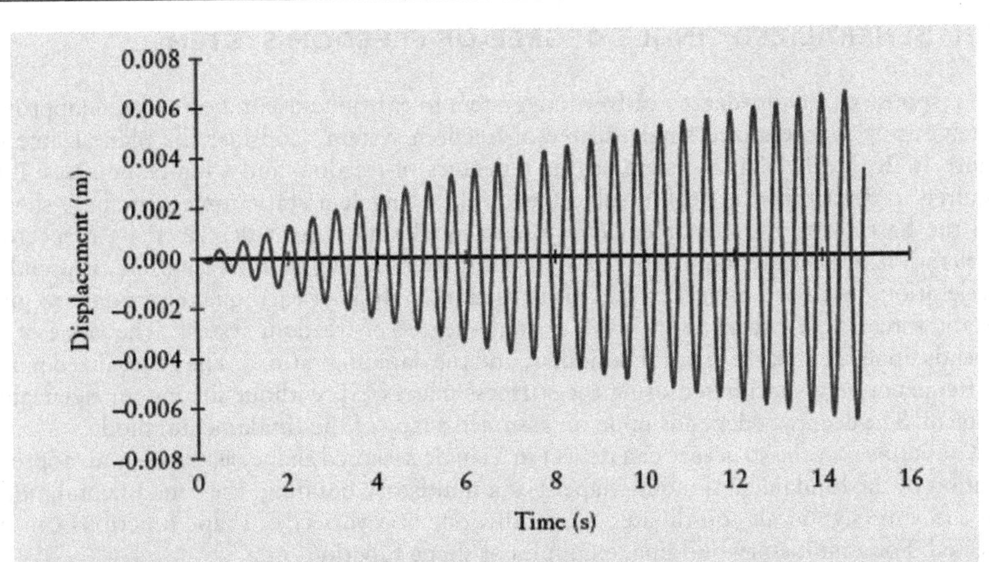

Figure 16.18 Response for first mode excitation.

where ω_d = damped natural frequency (Equation 16.40); $\bar{D}$ = amplitude of steady-state displacement (Equation 16.48); α = phase angle (Equation 16.49) = $\pi/2$; $\bar{C}_1$ and $\bar{C}_2$ = constants. The subscript r, referring to the rth mode is omitted in Equation (t) for simplicity. Assuming that the motion starts from rest, $\bar{C}_1 = -5.58\times10^{-5}$ m; $\bar{D} = 11.2\times10^{-3}$ m. Figure 16.18 is a graph of $D(t)$ versus t, from which it is seen that the peak displacement is in the steady-state: $D_{s1\text{-peak}} = \bar{D} = 11.2\times10^{-3}$ m (0.441 in.). By differentiation, the peak acceleration:

$$\ddot{D}_{s1-peak} = \bar{D}_1\,\Omega^2 = 11.2\times10^{-3}\,(12.08)^2 = 1.63\ \text{m/s}^2\ \left(64\ \text{in./s}^2\right)$$

This is 17 percent of the gravitational acceleration, g; a target limit of $0.05g$ is exceeded. Tuned mass dampers are needed to reduce the acceleration induced by pedestrians. A substitute system having two dampers and two degrees-of-freedom is shown in Figure 16.19.

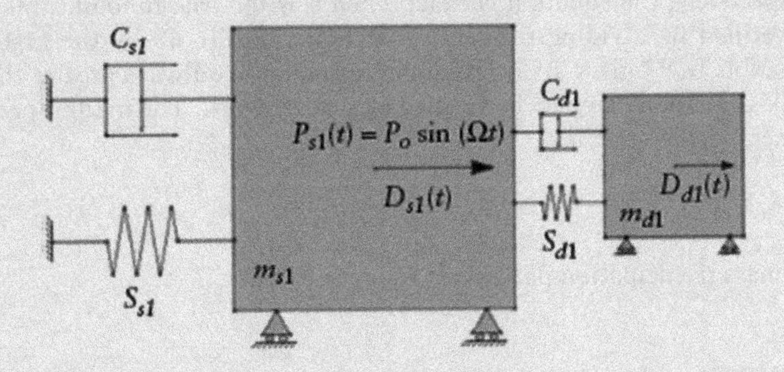

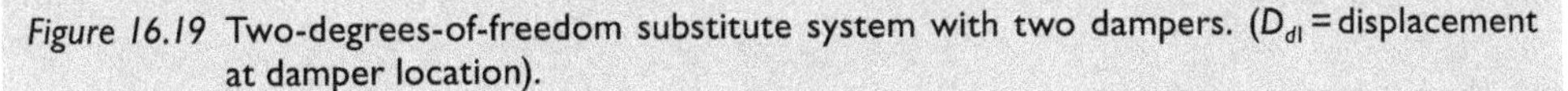

Figure 16.19 Two-degrees-of-freedom substitute system with two dampers. (D_{d1} = displacement at damper location).

16.11 GENERALIZED SINGLE-DEGREE-OF-FREEDOM SYSTEM

The response of a multi-degree-of-freedom system to earthquake can be analyzed (approximately) using a generalized single-degree-of-freedom system. Consider the plane frame in Figure 16.2c, idealized as a system having n degrees-of-freedom and n lumped masses. It is required to determine the peak displacements, the equivalent static forces, the base shear, and the base overturning moment due to earthquake movement described by a spectral acceleration, S_a, as function of ζ and T. The parameter S_a, having units length/s², is pseudo-acceleration, discussed in Section 17.7. It is used in analysis of earthquake response to give a static force $= S_a\, m$; where $m =$ mass of a single-degree-of-freedom system. The value of S_a depends upon the site, the natural period, T, and the damping ratio, ζ. The generalized natural frequency, ω_n, is estimated using the stiffness matrix $[S]$, without solving an eigenvalue problem. The accuracy depends upon an assumed shape of the fundamental mode.

We assume that the structure can deflect in a single assumed shape, ψ, which is an approximation of the fundamental mode shape. For a multistory building, the fundamental mode has the same sign at all coordinates. Generally, one of a variety of shape functions can be selected. For a multistory building, examples of shape functions are:

$$\psi = \xi; \quad \psi = \left(3\xi^2 - \xi^3\right)/2; \quad \psi = \left(6\xi^2 - 4\xi^3 + \xi^4\right)/3 \tag{16.96}$$

where $\xi = x/l$, with x and l being the distance between the base and any coordinate and between the base and the top coordinate, respectively. Each function gives $\psi = 0$ and 1 at the base and the top, respectively. The first function is simply linear; the second and the third represent the deflected shape of a prismatic cantilever subjected to a lateral force at its tip and to a uniform load, respectively (see Equations A.22 and A.29, Appendix A). The shape function must satisfy the boundary conditions that coincide with any of the n degrees-of-freedom. Sometimes, the shape function ψ is taken as the deflected shape of the frame due to static forces mg at the coordinates, where g is the gravitational acceleration. The natural frequency of the generalized system is given by:

$$\omega_n^2 = \frac{S_n}{m_n}; \quad S_n = \{\psi\}^{\mathrm{T}}[S]\{\psi\}; \quad m_n = \{\psi\}^{\mathrm{T}}[m]\{\psi\} \tag{16.97}$$

This is known as Rayleigh's equation; S_n and m_n are called *generalized stiffness* and *generalized mass*, respectively; the equation is exact when ψ is the true fundamental mode shape. This can be verified by dividing the second of Equations 16.68 by the first. When ψ is assumed, Equation 16.97 gives an approximate value for ω_n, always greater than the exact value. Rayleigh's equation can also be verified by virtual work. The natural period of vibration of the generalized system is:

$$T_n = 2\pi/\omega_n \tag{16.98}$$

Calculate the mass participation parameter L_n, defined as:

$$L_n = \sum_{i=1}^{n} \psi_i\, m_i \tag{16.99}$$

The peak displacement of the generalized system is:

$$D_{n\,\text{peak}} = L_n\, S_a/\left(m_n\, \omega_n^2\right) \tag{16.100}$$

This is the peak displacement at the point whose $\xi = 1$. The equivalent static force at any coordinate i is:

$$p_{ni} = L_n \, m_i \, \psi_i \, S_a / m_n \tag{16.101}$$

The equivalent static shearing force and overturning moment at the base and the effective height are:

$$V_{\text{base}} = \sum_{i=1}^{n} p_{ni}; \; M_{\text{base}} = \sum_{i=1}^{n} p_{ni} \, h_i; \; h_{\text{effective}} = M_{\text{base}} / V_{\text{base}} \tag{16.102}$$

where h_i is the distance from the base to coordinate i.

The generalized system is an approximation of the fundamental (first) mode. When the equations included in Figure 17.11, for modal spectral analysis, are applied for mode $r = 1$, they should give approximately the same answers as Equations 16.99 to 16.102. For this purpose $\{\psi\}$ has to be normalized:

$$\{\Phi\}_1 = \{\psi\} \left(\sum_{i=1}^{n} m_i \, \psi_i^2 \right)^{-1/2} \tag{16.103}$$

$\{\Phi\}_1$ is the normalized vector of the first mode, obtained by dividing each element of $\{\psi\}$ by the square root of $\left(\sum_{i=1}^{n} m_i \, \Psi_i^2 \right)$.

16.11.1 Cantilever idealization of a tower with variable cross section

Consider a vertical tower idealized as a cantilever totally fixed at the base (Figure 16.20a). Assume the length $= l$; the flexural rigidity $EI(x)$ and the mass/unit length $m(x)$ vary arbitrarily with x, the distance from the base. We approximate the fundamental mode shape in Figure 16.19b by Equation 16.96 or by another equation, such as:

$$\psi = 1 - \cos\left(\xi \frac{\pi}{2} \right) \tag{16.104}$$

where $\xi = x/l$; Equation 16.104 = 0 and 1 at $x = 0$ and l. The natural frequency of the generalized system is given by:

$$\bar{\omega}^2 = \frac{\bar{S}}{\bar{m}}; \; \bar{m} = \int_0^l m(x) \psi^2 \, dx; \; \bar{S} = \int_0^l EI(x) \left(\frac{d^2\psi}{dx^2} \right)^2 dx \tag{16.105}$$

$$dx = l \, d\xi; \; m(x) = m_0 \, \xi_m; \; EI(x) = EI_0 \, \xi_{EI} \tag{16.106}$$

m_0 and EI_0 are constants of unit mass and (force-length^2), respectively; ξ_m and ξ_{EI} are arbitrary dimensionless functions describing the variation of $m(x)$ and $EI(x)$ over the length l. For a simple example of the functions ξ_m and ξ_{EI}, consider a hollow circular cylindrical tower whose thickness varies with $\xi = x/l$; assume that the thickness at the top is half the thickness at the base. Considering only the own mass of the tower, $\xi_m = \xi_{EI} = 1 - (\xi/2)$; m_0 and EI_0 = mass/unit length and flexural rigidity at $\xi = 0$.

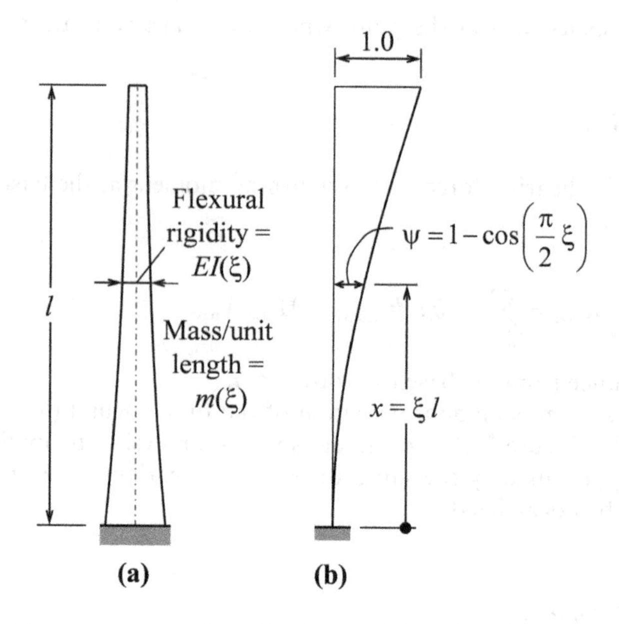

Figure 16.20 Tower subjected to earthquake. (a) Variation of flexural rigidity and mass per unit length. (b) Lateral displacement relative to ground: assumed shape function, ψ.

For convenience, we express the generalized mass and generalized stiffness in Equation 16.105 as:

$$\overline{m} = m_0\, l \int_0^1 \xi_m\, \psi^2\, \mathrm{d}\xi;\ \overline{S} = \left(EI_0/l^3\right)\int_0^1 \xi_{EI}\left(\frac{\mathrm{d}^2\psi}{\mathrm{d}\xi^2}\right)^2 \mathrm{d}\xi \tag{16.107}$$

The mass participation parameter is defined as:

$$\overline{L} = m_0\, l \int_0^1 \xi_m\, \psi\, \mathrm{d}\xi \tag{16.108}$$

The equivalent shearing force and bending moment at the base are:

$$\overline{V}_{\text{base}} = \frac{\overline{L}}{\overline{m}} S_a;\ \overline{M}_{\text{base}} = \frac{\overline{L}}{\overline{m}} m_0\, l^2\, S_a \int \xi_m\, \psi\, \xi\, \mathrm{d}\xi \tag{16.109}$$

The integrals in Equations 16.107 and 16.109 are dimensionless; in Equation 16.109, they depend upon the shape of vibration of mass (ξ_m), flexural rigidity (ξ_{EI}), and the assumed fundamental mode of vibration ψ.

The analysis in the present section uses a fictitious weightless cantilever with a lumped mass at the top as substitute of the tower; the base of the substitute cantilever is subjected to acceleration S_a; its mass and stiffness are calculated by Equations 16.107. The peak displacement at the top of the tower is given by:

$$D_{\text{peak at top}} = \frac{\overline{L}}{\overline{S}} S_a \tag{16.110}$$

Equation 16.109 can be divided by equating external and internal virtual work corresponding to virtual lateral translation $= \psi\, \delta_u$; where δ_u = arbitrary small displacement at the top of the tower.

EXAMPLE 16.11: A CANTILEVER WITH THREE LUMPED MASSES SUBJECTED TO EARTHQUAKE: USE OF A GENERALIZED SINGLE-DEGREE-OF-FREEDOM SYSTEM

The structure in Figure 16.8a is subjected to earthquake characterized by the design spectrum in Figure 17.8a ($\zeta = 0.05$). Find the shearing force and the bending moment at the base and the peak displacement at the top. Use a generalized single-degree-of-freedom system. Assume $Wl^3/(gEI) = 950 \times 10^{-6}\ \text{s}^2$; $g = 9.81\ \text{m/s}^2$.

The stiffness matrix of the three-degrees-of-freedom system in Figure 16.8b is (Table D.3, Appendix D):

$$[S] = \frac{EI}{l^3}\begin{bmatrix} 1.6154 & & \text{Symm.} \\ -3.6923 & 10.1538 & \\ 2.7692 & -10.6154 & 18.4615 \end{bmatrix}$$

The mass matrix is:

$$[m] = \frac{W}{g}\begin{bmatrix} 4 & 0 & 0 \\ 0 & 1 & 0 \\ 0 & 0 & 1 \end{bmatrix}$$

We select the shape function:

$$\psi = \left(3\xi^2 - \xi^3\right)/2 \quad \text{with } \xi = x/(3l)$$

$$\psi_1 = 1 \quad ; \quad \psi_2 = 0.5185; \quad \psi_3 = 0.1481$$

The generalized stiffness and the generalized mass are (Equation 16.97):

$$S_n = \{\psi\}^{\mathrm{T}}[S]\{\psi\} = 0.1112\ EI/l^3; \quad m_n = \{\psi\}^{\mathrm{T}}[m]\{\psi\} = 4.291\ W/g$$

$\omega_n^2 = S_n/m_n = 0.1112\, gEI/\left(4.291 W l^3\right) = 25.92 \times 10^{-3}\ gEI/\left(W l^3\right)$ (slightly greater than ω_1^2, for the fundamental mode determined in Example 16.3)

$$T_n = \frac{2\pi}{\omega_n} = \frac{2\pi}{\sqrt{25.92 \times 10^{-3}}}\left(\frac{gEI}{Wl^3}\right)^{-1/2}$$

$$= 39.03\left(\frac{gEI}{Wl^3}\right)^{-1/2} = 39.03\left(950 \times 10^{-6}\right)^{1/2} = 1.2\ \text{second}$$

From Figure 17.8a, $S_a = 0.122g = 0.122(9.81) = 1.197\ \text{m/s}^2$

The mass participation parameter (Equation 16.99):

$$L_n = \sum_{i=1}^{n} \psi_i\, m_i = \frac{W}{g}\left[1.0(4.0) + 0.5185(1.0) + 0.1481(1.0)\right] = 4.667\ \frac{W}{g}$$

The peak displacement of the generalized system is:

$$D_{n\,\text{peak}} = L_n\, S_a \Big/ \left(m_n\, \omega_n^2\right) = \frac{4.667\,(1.197)}{4.291\left[25.92\times10^{-3}/\left(950\times10^{-6}\right)\right]} = 0.0477 \text{ m}$$

The equivalent static forces at the coordinates are (Equation 16.101):

$$\{p_n\} = \frac{L_n\, S_a}{m_n}\{m_i\, \psi_i\} = \frac{4.667\,(0.122)}{4.291}\begin{Bmatrix} 4.0\,(1.0) \\ 1.0\,(0.5185) \\ 1.0\,(0.1481) \end{Bmatrix} W = \begin{Bmatrix} 0.5309 \\ 0.0688 \\ 0.0197 \end{Bmatrix} W$$

The equivalent static shearing force and overturning moment at the base and the effective height $h_{\text{effective}}$ are (Equation 16.102):

$$V_{\text{base}} = W\left(0.5309 + 0.0688 + 0.0197\right) = 0.6195\, W$$

$$M_{\text{base}} = Wl\left(0.5309\times3 + 0.0688\times2 + 0.0197\right) = 1.789\, Wl$$

$$^*h_{\text{effective}} = M_{\text{base}}/V_{\text{base}} = \left(1.789/0.6195\right)l = 2.888\, l$$

16.11.2 Cantilever with distributed mass

Consider the cantilever in Figure 16.20; mass per unit length $= m(\xi)$ and the flexural rigidity $= EI\,(\xi)$; where $\xi = x/l$. The response to earthquake, characterized by spectral acceleration, S_a (as function of ξ and T) can be analyzed approximately using a generalized single-degree-of-freedom system. The selected shape function is:

$$\psi = 1 - \cos\frac{\pi}{2}\xi \;;\; \xi = \frac{x}{l} \tag{16.111}$$

At $\xi = 0$, $\psi = 0$; $d\psi/d\xi = 0$; at $\xi = 1$, $\psi = 1$.

The square of natural frequency and period of the generalized system is:

$$\omega_n^2 = \frac{S_n}{m_n};\; T_n = \frac{2\pi}{\omega_n} \tag{16.112}$$

S_n and m_n, the generalized stiffness and mass are given by ("Virtual work principle," Section 7.6):

$$m_n = l\int_0^1 m(\xi)\psi^2\, d\xi;\; S_n = \frac{1}{l^3}\int_0^1 EI(\xi)\left(\frac{d^2\psi}{d\xi^2}\right)^2 d\xi \tag{16.113}$$

The mass participation parameter, L_n:

$$L_n = l\int_0^1 \psi\, m(\xi)\, d\xi \tag{16.114}$$

The peak displacement at the top:

$$D_{n\,\text{peak}} = L_n\, S_a \Big/ \left(m_n\, \omega^2\right) \tag{16.115}$$

The equivalent static force at any point:

$$p_n(\xi) = S_a\,(L_n/m_n)\,m(\xi)\,\psi \tag{16.116}$$

The equivalent static shearing force and overturning moment at base are:

$$V_{\text{base}} = l\int_0^1 p_n\,d\xi;\quad M_{\text{base}} = l\int_0^1 \xi\,p_n\,d\xi \tag{16.117}$$

EXAMPLE 16.12: TOWER HAVING CONSTANT EI AND UNIFORMLY DISTRIBUTED MASS: RESPONSE TO EARTHQUAKE

The tower in Figure 16.20 having uniform mass m per unit length and constant flexural rigidity, EI, is subjected to earthquakes characterized by the design spectrum in Figure 17.8a ($\zeta=0.05$). Find the peak displacement, the shearing force, and the bending moment at the base. Use the shape function of Equation 16.111. Assume $m=7500$ kg/m (420.1 lb/in.); $EI=50\times10^9$ N-m^2 (17.42×10^9 kip-in.^{2}); $l=50$ m (1970 in.); $g=9.81$ m/s^2 (386 in./s^2).

The generalized properties (Equations 16.112 to 16.114):

$$S_n = \frac{EI}{l^3}\int_0^1\left(\frac{d^2\psi}{d\xi^2}\right)^2 d\xi = \frac{EI}{l^3}\int_0^1\left(\frac{\pi^2}{4}\right)^2\cos^2\frac{\pi\xi}{2}\,d\xi = 3.044\,\frac{EI}{l^3}$$

$$m_n = l\,m\int_0^1\left(1-\cos\frac{\pi\xi}{2}\right)^2 d\xi = 0.227\,ml$$

$$L_n = l\,m\int_0^1\left(1-\cos\frac{\pi\xi}{2}\right) d\xi = 0.363\,ml$$

The equivalent static force at any point (Equation 16.115):

$$p_n(\xi) = S_a\left(\frac{0.363}{0.227}\right)m\left(1-\cos\frac{\pi\xi}{2}\right)$$

The natural frequency and period are (Equation 16.112):

$$\omega_n^2 = \frac{3.044}{0.227}\frac{EI}{ml^4} = 14.3\ \text{s}^{-2};\quad T_n = \frac{2\pi}{\sqrt{14.3}} = 1.66\ \text{s}$$

The graph in Figure 17.8a gives: $S_a=0.0785\,g=0.770$ m/s^2 (=30.3 in./s^2). The equivalent static force at any point (Equation 16.115):

$$p_n(\xi) = 0.0785g\left(\frac{0.363}{0.227}\right)m\left(1-\cos\frac{\pi\xi}{2}\right) = 9236\left(1-\cos\frac{\pi\xi}{2}\right)\ \text{N/m}$$

$$= 52.74\left(1-\cos\frac{\pi\xi}{2}\right)\ \text{lb/in.}$$

The shearing force and the bending moment at the base (Equation 16.117)

$$V_{\text{base}} = 50\,(9236)\int_0^1\left(1-\cos\frac{\pi\xi}{2}\right)d\xi = 50\,(9236)\,0.363$$

$$= 167.6\times10^3\ \text{N}\ \ (37.68\ \text{kips})$$

$$M_{\text{base}} = 50\,(9236)\int_0^1 \xi\left(1-\cos\frac{\pi\xi}{2}\right)d\xi = 50\,(9236)\,0.2687$$

$$= 124.1\times10^3\ \text{N-m}\ \ \left(10.9\times10^3\ \text{kips-in.}\right)$$

16.12 GENERAL

This chapter is only as an introduction to the subject of structural dynamics. Chapter 17 studies the response of structures to earthquakes.

REFERENCES

Chopra, A.K., *Dynamics of Structures*, Pearson Prentice-Hall, Upper Saddle River, NJ, 2007, 876 pp.

Clough, R.W. and Penzien, J., *Dynamics of Structures*, 2nd ed., McGraw-Hill, New York, 1993, 738 pp.

Humar, J., *Dynamics of Structures*, 3rd ed., CRC Press – Taylor & Francis Group, Boca Raton, FL, 2012, 1058 pp.

Paz, M., *Structural Dynamics, Theory and Computation*, 3rd ed., Van Nostrand Reinhold, New York, 1991, 626 pp.

Penelis, G.G. and Kappos, A.J., *Earthquake-Resistant Concrete Structures*, E&FN Spon, London, 1997, 572 pp.

Priestly, M.J., Seible, F. and Calvi, G.M., *Seismic Design and Retrofit of Bridges*, John Wiley & Sons, London, 1996, 686 pp.

SAP 2000, *Software*, Computers and Structures, Berkeley, CA.

PROBLEMS

Take the acceleration of gravity $g=32.2$ ft/sec^2 (386 in./sec^2) or 9.81 m/sec^2 whenever it is needed in the solution of the following problems.

16.1 Compute the natural angular frequency of vibration in sidesway for the frame in the figure, and calculate the natural period of vibration. Idealize the frame as a one-degree-of-freedom system. Neglect the axial and shear deformations and the weight of the columns. If initially the displacement is 1 in. (25 mm) and the velocity is 10 in./sec (0.25 m/sec), what is the amplitude and what is the displacement at $t=1$ sec?

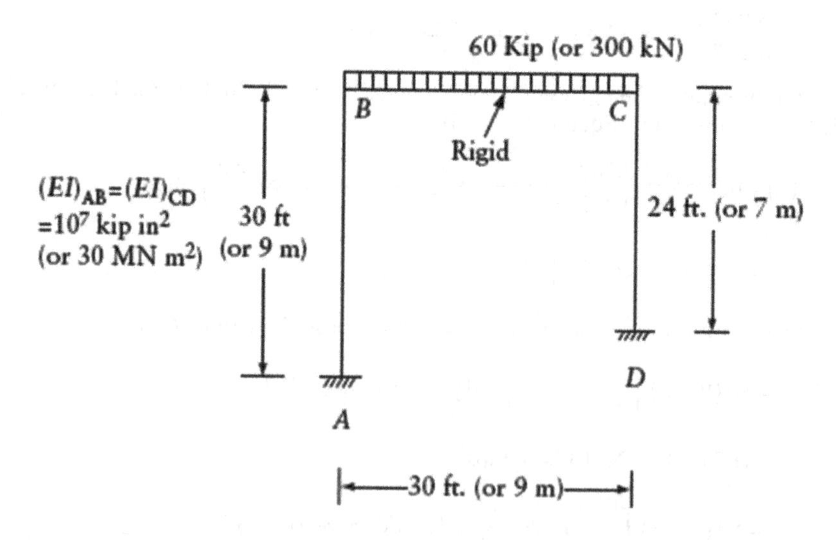

16.2 Solve Problem 16.1 assuming that BC has a flexural rigidity $(EI)_{BC}=10^7$ kip-in.^{2} (30 MN-m^2).

16.3 The prismatic cantilever AB is idealized by the two-degrees-of-freedom system shown in the figure. Using the consistent mass matrix, find the first natural angular frequency and the corresponding mode. The beam has a total mass m, length l, and flexural rigidity EI. Consider bending deformation only.

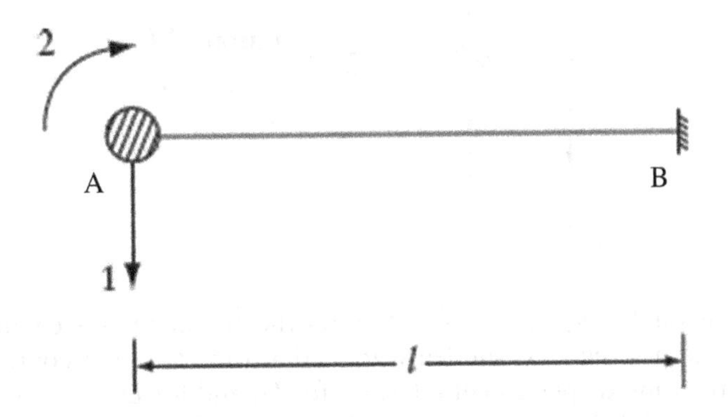

16.4 The frame in Problem 16.1 is disturbed from rest by a horizontal force of 8 kip (40 kN) at C, suddenly applied at time $\tau = 0$ and removed at time $\tau = T/2$, where T is the natural period of vibration. What is the displacement and velocity at the removal of the force? What is the displacement at time $\tau = 11T/8$?

16.5 Solve Problem 16.4 assuming that the disturbing force increases linearly from zero at $\tau = 0$ to 8 kip (40 kN) at $\tau = T/4$, then decreases linearly to zero at $\tau = T/2$, at which time the force is removed.

16.6 If the system of Problem 16.1 has a damping coefficient $\zeta = 0.1$, what are the damped natural circular frequency ω_d and the natural period of damped vibration T_d? What is the displacement at $t = 1$ sec, if $D_0 = 1$ in. (25 mm) and $\dot{D}_0 = 10$ in./sec (0.25 m/sec)?

16.7 If the amplitude of free vibration of a system with one degree-of-freedom decreases by 50 percent in 3 cycles, what is the damping coefficient?

16.8 Determine the maximum steady-state sidesway in the frame of Problem 16.1 when it is subjected to a harmonic horizontal force at the level of BC of magnitude 4 sin 14t (kip) [20 sin 14t (kN)], and (a) no damping is present, (b) the damping coefficient = 0.10.

16.9 Assume that the frame in Problem 16.1 has a damping coefficient = 0.05, and it is disturbed from rest by a horizontal force of 8 kip (40 kN) at C. The force is suddenly applied at time $\tau = 0$ and removed at time $\tau = T_d/2$, where T_d is the natural period of damped vibration. What are the displacements at the removal of the force and at time $\tau = 11T_d/8$? (Compare the answers with the undamped case, Problem 16.4.)

16.10 and 16.11

Determine the natural circular frequencies and characteristic shapes for the two-degrees-of-freedom systems shown in the figures.

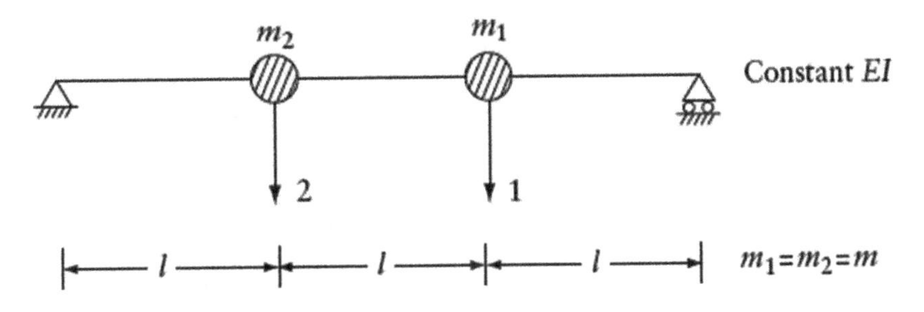

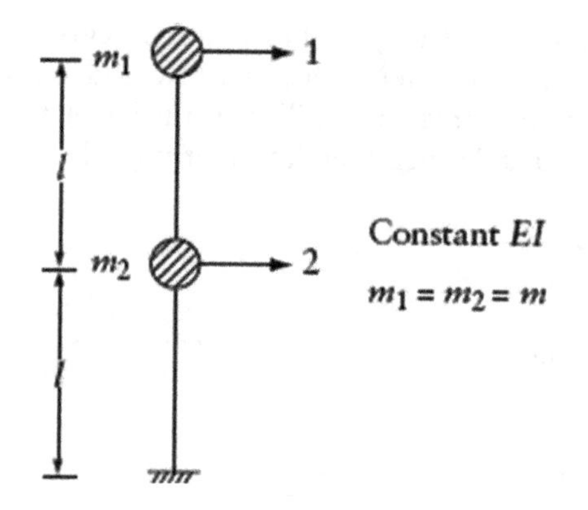

16.12 Write the uncoupled equations of motion for the undamped system in Problem 16.11, assuming that a force P_1 is suddenly applied at time $t=0$ and continues to act after this. Find the time-displacement relations for D_1 and for D_2.

16.13 If in Problem 16.12 the force $P_1 = P_0 \sin \Omega t$, what is the amplitude of the steady-state vibration of m_1?

16.14 The supports of the frame in Problem 16.1 move horizontally with an acceleration indicated in the figure. What is the maximum displacement of BC relative to the support? Neglect damping.

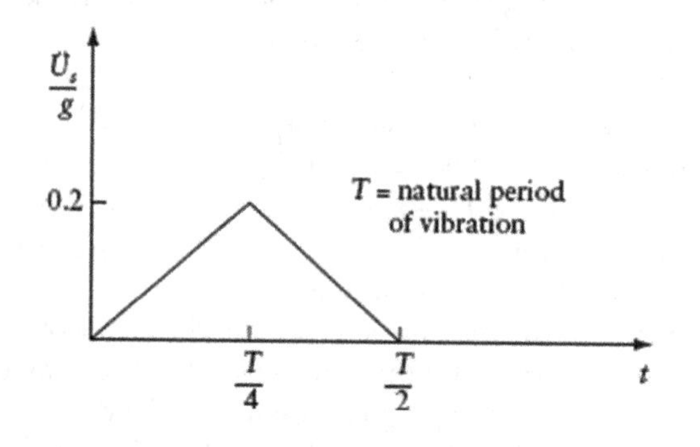

16.15 If the support in Problem 16.11 has a horizontal acceleration $\ddot{u}_s = (g/4) \sin \Omega t$ determine the maximum displacement of the mass m_1 relative to the support.

16.16 Solve Example 16.8 using a generalized single-degree-of-freedom system with the shape function ψ taken as the deflected shape of the frame due to static forces $\{m\}g$ at the coordinates.

Chapter 17

Response of structures to earthquakes

17.1 INTRODUCTION

Chapter 16 includes a section on the response of a single-degree-of-freedom system due to the shaking of its support. The current chapter presents the analysis needed in the design of seismic-resistant structures. For convenience, we review the relevant material in Chapter 16.

The shaking of the ground induces inertial forces derived from a record of acceleration versus time. We use records of earthquakes in the design of new structures. Numerical integration of the acceleration gives displacement. The spectrum of response to several recorded ground accelerations enables prediction of the response of a new structure, without numerical integration.

The objective of analyzing the response to an earthquake is to design structures that will not collapse during a specified design ground motion. Earthquake induces damage without collapse. Strong ground shaking is a rare event; it is uneconomical to design a building to remain elastic in such an event. Dynamic linear analysis gives higher forces due to ground shaking compared to the forces determined by considering deformations beyond linear limits.

The analysis gives the peak response subjecting the structure to specified ground acceleration. For a given acceleration, the peak response depends upon the natural period of vibration and the level of damping. Many records of real earthquakes give a spectrum of the pseudo-acceleration to be expected. The pseudo-acceleration is a function of the fundamental period of vibration and specified level of damping. Dynamic analysis determines the frequencies and the natural vibration modes. Then, the effect of equivalent static forces gives an estimate of the peak internal forces used in the design of members.

The equivalent static forces are acceptable by codes when a structure is uniform along its height and has a relatively small fundamental vibration period (e.g. one second). Irregular buildings of heights of less than 20 m commonly have a fundamental period of less than 0.5 s; then, the equivalent static forces are acceptable even when the building is irregular. Certain codes consider that these buildings can have periods exceeding 1.5 seconds.

Section 17.2 derives the equation motion induced by ground acceleration of single-degree-of-freedom systems. Section 17.3 discusses multi-degree-of-freedom systems. The present chapter is limited to structures idealized as a plane frame.

17.2 SINGLE-DEGREE-OF-FREEDOM SYSTEM

Consider a single-degree-of-freedom system (Figure 17.1a) subjected to support acceleration $\ddot{u}_s(t) = \ddot{u}_g(t)$. The displacement D of the mass m relative to the support is:

$$D = u - u_s = u - u_g \tag{17.1}$$

DOI: 10.1201/9780429286858-17

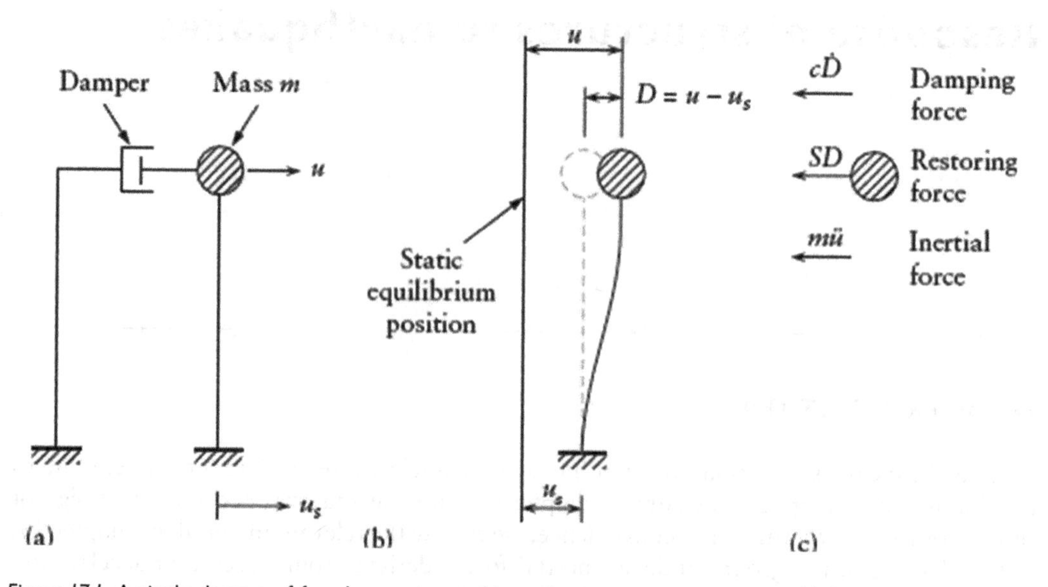

Figure 17.1 A single-degree-of-freedom system subjected to support movements. (a) Positive directions of u and u_s. (b) Deformed shape at any time t. (c) Forces acting on mass at any time t.

where u = total displacement of the mass, and $u_s = u_g$ = displacement of the ground at the support (Figure 17.1b). The forces acting on the mass (Figure 17.1c) are an inertial force, $m\ddot{u}$, a damping force, $c\dot{D}$, and a restoring force, SD, where c = viscous damping coefficient and S = the stiffness of the system with respect to the coordinate u. The equilibrium equation of motion is:

$$m\ddot{u} + c\dot{D} + SD = 0 \tag{17.2}$$

Differentiation of Equation 17.1 with respect to time gives: $\ddot{u} = \ddot{D} + \ddot{u}_g$; substitution in Equation 17.2 and dividing by m gives:

$$\ddot{D} + \frac{c}{m}\dot{D} + \frac{S}{m}D = -\ddot{u}_g \tag{17.3}$$

We defined the viscous damping coefficient c as a fraction ζ of the critical damping coefficient ($2\,m\omega$) (Equation 16.35).

$$c = 2\zeta m\omega \tag{17.4}$$

ζ is the damping ratio; ω is the natural frequency (Equation 16.14):

$$\omega = \sqrt{S/m} \tag{17.5}$$

The natural period of vibration is:

$$T = 2\pi/\omega = 2\pi\sqrt{m/S} \tag{17.6}$$

With these definitions, the equation of motion of a viscously damped single-degree-of-freedom system subjected to ground motion becomes:

$$\ddot{D} + 2\zeta\omega\,\dot{D} + \omega^2 D = -\ddot{u}_g \tag{17.7}$$

We rewrite this equation as:

$$m\ddot{D} + c\,\dot{D} + SD = P(t) \tag{17.8}$$

The solution of Equation 17.8 depends upon the initial conditions $D(0)$ and $\dot{D}(0)$. The ground shaking has the same effect as that of a force:

$$P(t) = -\ddot{u}_g\, m \tag{17.9}$$

This is the *effective force*. For a given ground acceleration $\ddot{u}_g(t)$, the deformation response, $D(t)$, depends only upon the natural frequency, ω (or natural period, T), of the system and its damping ratio, ζ. Two systems having the values of the parameters (T and ζ) will have the same deformation response, $D(t)$, when one system is more massive or stiffer than the other. (But ω and ζ affect the relationship between D and u in Equation 17.7.) The effect of T and ζ on $D(t)$ is demonstrated numerically in Section 17.4.

17.3 MULTI-DEGREE-OF-FREEDOM SYSTEM

We analyze the response of a regular multi-story structure to earthquake by considering a plane-frame having n degrees-of-freedom (Figure 17.2). For this system, all nodal displacements are translations in the same direction as the support motion. Derivation of the condensed stiffness matrix corresponding to the n-translations accounts for joint rotations. The displacement configuration is defined by n translations at floor levels; n lumped masses represent the distributed mass of the structure (Section 16.2). With this idealization, the equation of motion of the system is:

$$[S]\{D\}_r = \omega_r^2 [m]\{D\}_r \text{ with } r = 1,2,\ldots,n \tag{17.10}$$

where $[S]_{n\times n}$ = condensed stiffness matrix (Section 5.12); $[m]_{n\times n}$ = mass matrix; $\{D\}_r$ = displacement vector; ω_r = natural frequency of mode $r = 2\pi/T_r$, with T_r being the natural period of vibration. Solution of Equation 17.10 is an eigenvalue problem (computer program EIGEN1

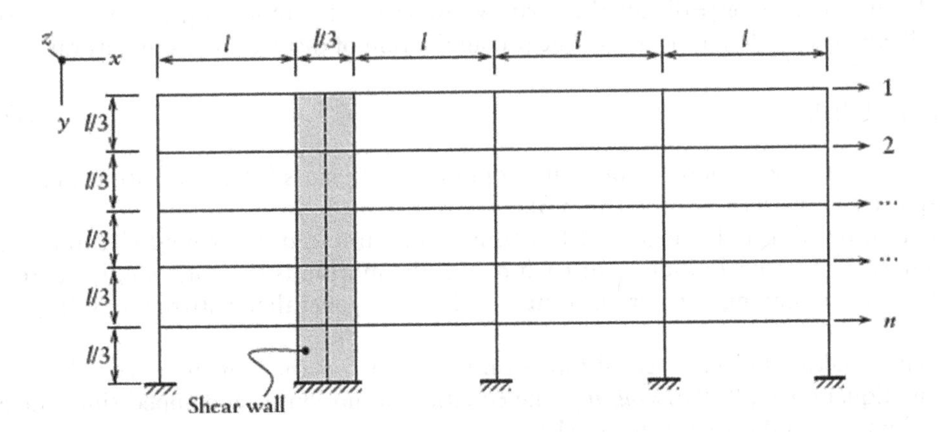

Figure 17.2 Multi-degree-of-freedom system: a plane-frame idealization of a multi-story building subjected to horizontal ground shaking in x-direction. In Problem 17.7, $n=4$; for all beams and columns, $I_{beam} = I_{column} = I_{wall}/400$; E = constant.

or EIGEN2, Chapter 20), which provides n eigenvalues $= \omega_1^2, \omega_2^2, \ldots, \omega_n^2$ (characteristic values of the system). For the rth characteristic value ω_r^2, the solution gives the eigen vector $\{\phi\}_r$, representing a mode shape of vibration; the elements ϕ_{ir} define the ratio of displacement amplitude at coordinate i to the displacements at other coordinates. Substituting $\{\phi\}_r$ for $\{D\}_r$ gives:

$$[S]\{\phi\}_r = \omega_r^2 [m]\{\phi\}_r \tag{17.11}$$

The natural modes of an undamped system are orthogonal with respect to $[m]$ or $[S]$, meaning that (Section 16.6.2):

$$\{\phi\}_r^T [m]\{\phi\}_s = \text{nonzero only with } r = s \tag{17.12}$$

$$\{\phi\}_r^T [S]\{\phi\}_s = \text{nonzero only with } r = s \tag{17.13}$$

The nonzero values in Equation 17.12 are equal to 1.0 when the modal matrix $[\phi]$ is replaced by normalized modal matrix $[\Phi]$; the rth normalized mode vector is:

$$\{\Phi\}_r = \left(\{\phi\}_r^T [m]\{\phi\}_r\right)^{-1/2} \{\phi\}_r \tag{17.14}$$

Equation 17.14 means that the rth column of $[\Phi]$ (of unit mass$^{-1/2}$) is equal to the corresponding column of $[\phi]$ divided by the square root of the product$\left(\{\phi\}_r^T [m]\{\phi\}_r\right)$. The normalization results in:

$$[\Phi]^T [m][\Phi] = [I] \tag{17.15}$$

$$[\Phi]^T [S][\Phi] = \left[\omega^2\right] \tag{17.16}$$

$[I]$ = identity matrix; $[\omega^2]$ = diagonal matrix whose elements are $\omega_1^2, \omega_2^2, \ldots, \omega_n^2$. The use of $[\Phi]$ will enable calculation of $\{D\}$ as linear combination of displacements in natural modes:

$$\{D\} = [\Phi]\{\eta\} \tag{17.17}$$

where $\{\eta\}$ is a vector of modal coordinates (of unit [length mass$^{1/2}$]), representing amplitudes of displacement patterns (Section 16.5.3).

The motion of a multi-degree-of-freedom system induced by ground shaking is as a combination of natural modes. For each mode, the motion is that of a single degree-of-freedom system having a generalized mass, M, and a generalized stiffness, K (Equation 16.69).

Ground shaking induces inertial forces on the lumped masses; at the ith coordinate, the force is (Equation 17.9): $P_i(t) = -\ddot{u}_g\, m_i$. The equation of motion of undamped single-degree-of-freedom system due to ground acceleration, $\ddot{u}_g$ is:

$$m\ddot{D} + SD = -\ddot{u}_g\, m \tag{17.18}$$

For a system having n coordinates (Figure 17.2) subjected to forces $\{P(t)\}_{n\times1}$, the equation of motion is:

$$[m]\{\ddot{D}\}+[S]\{D\}=\{P(t)\} \tag{17.19}$$

Equation 17.19 is a set of coupled equations relating forces to displacements at n-coordinates. The coupling means that any displacement D_i is the combined effect of forces at n coordinates. Substitute Equation 17.17 in Equation 17.19 and pre-multiply by $[\Phi]^T$:

$$[\Phi]^T[m][\Phi]\{\ddot{\eta}\}+[\Phi]^T[S][\Phi]\{\eta\}=[\Phi]^T\{P(t)\} \tag{17.20}$$

Equations 17.15 and 17.16 indicate that the products of the three matrices in the first and second terms in Equation 17.20 are equal to [I] and $[\omega^2]$, respectively; both matrices are diagonal. For the rth mode, we define:

$$M_r=\{\Phi\}_r^T[m]\{\Phi\}_r=\sum_{i=1}^{n}m_i\,\Phi_{ir}^2=1 \tag{17.21}$$

$$\{\Phi\}_r^T[S]\{\Phi\}_r=\omega^2 \tag{17.22}$$

Equation 17.20 is a set of uncoupled equations; for mode r, the rth equation is:

$$\ddot{\eta}_r+\omega_r^2\,\eta_r=\{\Phi\}_r^T\{P(t)\}(1/M_r) \tag{17.23}$$

The ground shaking induces, at the ith coordinate, an inertial force $P_i(t)=-m_i\,\ddot{u}_g(t)$ and:

$$\{P(t)\}=-\begin{Bmatrix}m_1\\m_2\\\cdots\end{Bmatrix}_{n\times1}\ddot{u}_g(t) \tag{17.24}$$

Define modal response factor, L_r as:

$$L_r=\{\Phi\}_r^T\begin{Bmatrix}m_1\\m_2\\\cdots\end{Bmatrix}_{n\times1}=\sum_{i=1}^{n}m_i\,\Phi_{ir} \tag{17.25}$$

Equation 17.25 treats the elements of $\{\Phi\}_r$ as the influence coefficient to be used (in Section 17.13, Equation 17.59) to obtain the inertial forces induced by an earthquake. Substitution of Equations 17.24 and 17.25 in Equation 17.23 gives:

$$\ddot{\eta}_r+\omega_r^2\,\eta_r=-\frac{L_r}{M_r}\ddot{u}_g(t) \tag{17.26}$$

For each modal response, Equation 17.17 gives:

$$\{D\}_r=\{\Phi\}_r\,\eta_r \tag{17.27}$$

$\{\Phi\}_r$ represents a mode shape (a pattern of translations) in free vibration in the rth mode; η_r is an amplitude.

For viscously damped multi-degree-of-freedom system, the equation of motion in the rth mode is:

$$\ddot{\eta}_r + 2\zeta_r\,\omega_r\,\dot{\eta}_r + \omega_r^2\,\eta_r = -\frac{L_r}{M_r}\,\ddot{u}_g(t) \qquad (17.28)$$

where ζ_r = damping ratio for the rth mode; $\ddot{u}_g\ (t)$ = the ground acceleration in direction of coordinate r. In normal practice, one damping ratio, ζ, is assumed but not necessarily for all modes. Equation 17.28 is a modal equation of motion of a single-degree-of-freedom system with natural frequency ω_r and damping ratio ζ_r subjected to ground acceleration $\ddot{u}_g$, factored by (L_r/M_r). The solution of Equation 17.28 (numerically or using Duhamel's integral) gives $\eta_r\ (t)$; the contribution of the rth mode to the displacements at the n degrees-of-freedom (Figure 17.2) can be calculated by Equation 17.27. This is *modal response history analysis*; it can give the variation of the response over a period.

17.3.1 Damping ratio

The damping ratio, $\zeta = c/(2\,m\omega)$, where $(2\,m\omega)$ is the critical damping coefficient; c = damping coefficient = damping force/velocity (Section 16.5). It is not possible to determine an accurate value for ζ to use in the seismic design of new structures. The recommended value of ζ in Table 17.1 depends upon the type and condition of the structure. Two ranges of values are given for ζ, depending upon the expected stress level. The lower range is for stress ≤ ½ the yield stress. The higher range for ζ applies when: ½ yield stress < stress level < yield stress. Rayleigh damping is for more accuracy of analysis of response of structures to earthquakes (Section 16.7.1). Damping force is greater for higher ω.

17.4 TIME-STEPPING ANALYSIS

Problem statement: Find numerically the displacement $D(t)$ of a damped single-degree-of-freedom system whose equation of motion is:

$$m\ddot{D} + 2\zeta m\omega\dot{D} + m\omega^2 D = P(t) \qquad (17.29)$$

Given: The damping ratio, ζ, the natural period of vibration, T and the values of the applied force = P_i at t_0, t_1, …, t_n, with constant time interval = $\Delta t = t_{i+1} - t_i$. The values of $\ddot{u}_{g0}$, D_0, and $\dot{D}_0$ at time t_0 are also given.

Table 17.1 Recommended damping ratio

Stress level	*Type and condition of structure*	*Damping ratio (%)*
Working stress, no more than about ½ yield point	Welded steel, prestressed concrete, well-reinforced concrete (only slight cracking)	2–3
	Reinforced concrete with considerable cracking	3–5
	Bolted and/or riveted steel, wood structures with nailed or bolted joints	5–7
At or just before yield point	Welded steel, prestressed concrete (without complete loss of prestress)	5–7
	Prestressed concrete with no prestress left	7–10
	Reinforced concrete	7–10
	Bolted and/or riveted steel, wood structures with bolted joints	10–15
	Wood structures with nailed joints	15–20

The mass, m, and the stiffness, S, relate to T and ζ as:

$$T = \frac{2\pi}{\omega};\ \omega = \sqrt{\frac{S}{m}};\ \frac{S}{m} = \left(\frac{2\pi}{T}\right)^2 \tag{17.30}$$

$$\text{Damping force} = c\dot{D}\ ;\ \frac{c}{m} = 2\zeta\omega \tag{17.31}$$

We repeat Equation 17.7, and then express its terms in finite differences:

$$\ddot{D} + 2\zeta\omega\dot{D} + \omega^2 D = -\ddot{u}_g \tag{17.32}$$

Central finite difference expressions:

$$\dot{D}_i = \frac{D_{i+1} - D_{i-1}}{2(\Delta t)};\ \ddot{D}_i = \frac{D_{i-1} - 2D_i + D_{i+1}}{(\Delta t)^2} \tag{17.33}$$

Apply Equation 17.32 at t_i in finite difference form:

$$\frac{D_{i-1} - 2D_i + D_{i+1}}{(\Delta t)^2} + (2\zeta\omega)\frac{D_{i+1} - D_{i-1}}{2(\Delta t)} + \omega^2 D_i = -\ddot{u}_{gi} \tag{17.34}$$

This equation is a recurring expression that gives D_{i+1} using values at two immediately preceding instants; the procedure is referred to as *time-stepping analysis*:

$$D_{i+1} = \left(-\ddot{u}_{gi} + C_1 D_i - C_2 D_{i-1}\right)/C_3 \tag{17.35}$$

where

$$C_1 = \frac{2}{(\Delta t)^2} - \omega^2;\ C_2 = \frac{1}{(\Delta t)^2} - \frac{\zeta\omega}{(\Delta t)};\ C_3 = \frac{1}{(\Delta t)^2} + \frac{\zeta\omega}{(\Delta t)} \tag{17.36}$$

Using Equation 17.6, we can write:

$$C_1 = \frac{2}{(\Delta t)^2} - \left(\frac{2\pi}{T}\right)^2;\ C_2 = \frac{1}{(\Delta t)^2} - \frac{2\pi\zeta}{T(\Delta t)};\ C_3 = \frac{1}{(\Delta t)^2} + \frac{2\pi\zeta}{T(\Delta t)} \tag{17.37}$$

Application of the recurring expression, Equation 17.35 at t_1 requires knowledge of D_{-1} and D_0; with D_0 and $\dot{D}_0$ given values, we express D_{-1} as:

$$D_{-1} = D_0 - (\Delta t)\dot{D}_0 + \frac{(\Delta t)^2}{2}\ddot{D}_0 \tag{17.38}$$

$$\ddot{D}_0 = -\ddot{u}_{g0} - \frac{4\pi\zeta}{T}\dot{D}_0 - \left(\frac{2\pi}{T}\right)^2 D_0 \tag{17.39}$$

The finite difference expressions (Equation 17.33) give Equation 17.38. Equation 17.39 is an application of Equation 17.32 at t_0.

EXAMPLE 17.1: TIME-STEPPING ANALYSIS: RESPONSE TO A FICTITIOUS RECORD OF GROUND ACCELERATION

Find the displacement $D(t)$ due to ground shaking defined by a fictitious acceleration presented in Figure 17.3a. Given data: damping ratio, $\zeta = 0.05$; natural vibration period, $T = 1$ s; assume $\Delta t = 0.1$ s; consider motion starts from rest:

At $t_0 = 0$, $D_0 = 0$, and $\dot{D}_0 = 0$; the subscript 0 refers to the starting time. For clarity, we assume fictitious acceleration and use a limited number of steps.

INITIAL CONDITIONS:

At t_0, the ground acceleration, the displacement and the velocity are zero. Apply Equation 17.39 at t_0 to get: $D'_0 = 0$; Equation 17.38 gives $D_{-1} = 0$.

The constants of the recurrence equation are (Equation 17.37):

$$C_1 = \frac{2}{(\Delta t)^2} - \left(\frac{2\pi}{T}\right)^2 = \frac{2}{(0.1)^2} - \left(\frac{2\pi}{1.0}\right)^2 = 160.52$$

$$C_2 = \frac{1}{(\Delta t)^2} - \frac{2\pi\zeta}{T(\Delta t)} = \frac{1}{(0.1)^2} - \frac{2\pi(0.05)}{1.0(0.1)} = 96.858$$

$$C_3 = \frac{1}{(\Delta t)^2} + \frac{2\pi\zeta}{T(\Delta t)} = \frac{1}{(0.1)^2} + \frac{2\pi(0.05)}{1.0(0.1)} = 103.14$$

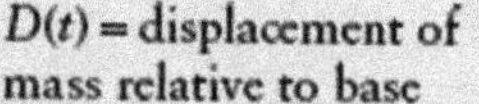

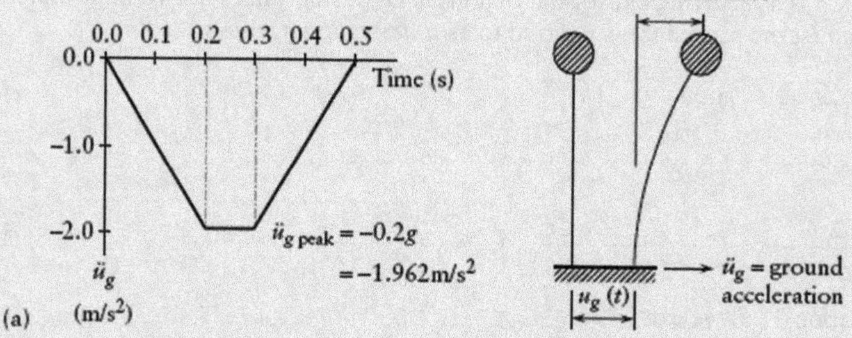

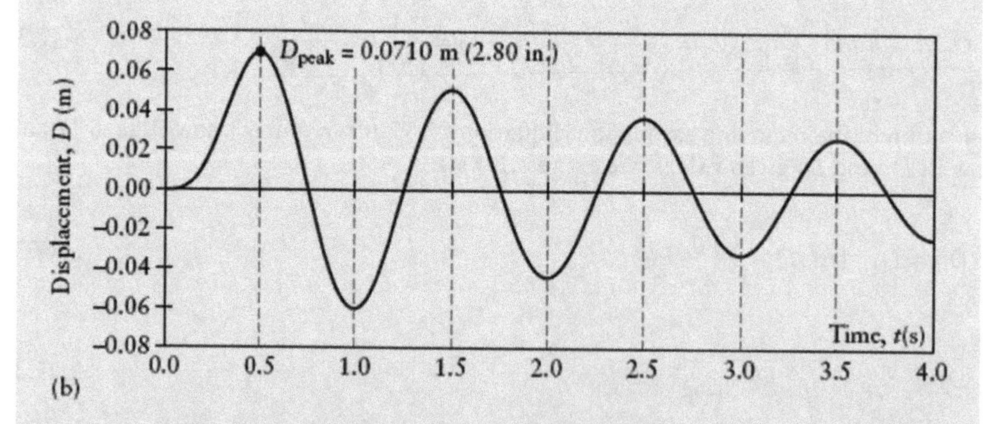

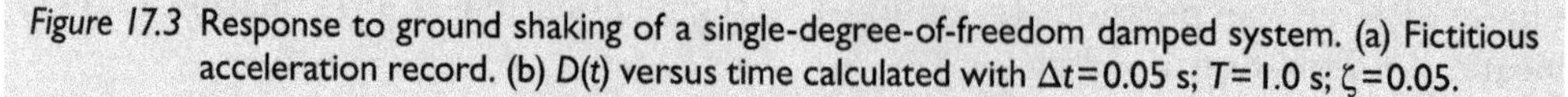
Figure 17.3 Response to ground shaking of a single-degree-of-freedom damped system. (a) Fictitious acceleration record. (b) $D(t)$ versus time calculated with $\Delta t = 0.05$ s; $T = 1.0$ s; $\zeta = 0.05$.

Table 17.2 Time-stepping calculation of response using central finite differences, Example 17.1; $\Delta t = 0.1$ s

i	t_i s	$\ddot{u}_{gi}$ m/s^2	D_{i-1} m	D_i m	D_{i+1} m
0	0	0	0.00000	0.00000	0.00000
1	0.1	−0.981	0.00000	0.00000	0.00951
2	0.2	−1.962	0.00000	0.00951	0.03382
3	0.3	−1.962	0.00951	0.03382	0.06273
4	0.4	−0.981	0.03382	0.06273	0.07538
5	0.5	0	0.06273	0.07538	0.05840
6	0.6	0	0.07538	0.05840	0.02011
7	0.7	0	0.05840	0.02011	−0.02355
8	0.8	0	0.02011	−0.02355	−0.05554
9	0.9	0	−0.02355	−0.05554	−0.06432
10	1.0	0	−0.05554	−0.06432	−0.04794
11	1.1	0	−0.06432	−0.04794	−0.01422
12	1.2	0	−0.04794	−0.01422	0.02290
13	1.3	0	−0.01422	0.02290	0.04899
14	1.4	0	0.02290	0.04899	0.05474
15	1.5	0	0.04899	0.05474	0.03918

The time-stepping calculations are done in the period $t=0$ to $t=1.5$ s in Table 17.2 using the recurrence Equation 17.35. Figure 17.3b plots $D(t)$ versus time in the first 4 s. For more accuracy, the graph is prepared with $\Delta t=0.05$ s; from the graph, we can estimate $D_{peak}=0.0710$ m, where D_{peak} is the largest absolute value of the displacement $D(t)$.

17.5 EFFECTS OF DAMPING AND NATURAL PERIOD OF VIBRATION ON RESPONSE TO GROUND SHAKING

Consider the response of a damped linear single-degree-of-freedom system to ground shaking. For a given record of the variation of ground acceleration with time, we can determine by time-stepping the response $D(t)$. We have done this calculation in Example 17.1 using for clarity a fictitious ground acceleration record (Figure 17.3a). We have seen that $D(t)$ depends upon the damping ratio, ζ, and the natural period of vibration, T. In Figure 17.4a, $\zeta=0.05$, while the natural period $T=1.0$, 2.0, or 4.0 s. D_{peak} increases substantially with the increase of T. Figure 17.4b indicates the effect of varying the damping ratio ζ on response. With $T=1.0$ s and $\zeta=0.0$, 0.05, and 0.10, the peak displacement becomes smaller with the increase of ζ. For short periods or very long periods, a damping increase does not reduce the response in the same manner. This is observed by considering real records of ground motion.

17.6 PSEUDO-ACCELERATION: STATIC EQUIVALENT LOADING

Consider a weightless cantilever having a lumped mass, m, at its tip (Figure 17.5). In response to a specific earthquake, the absolute value of the largest displacement $= D_{peak}$ (T, ζ); this

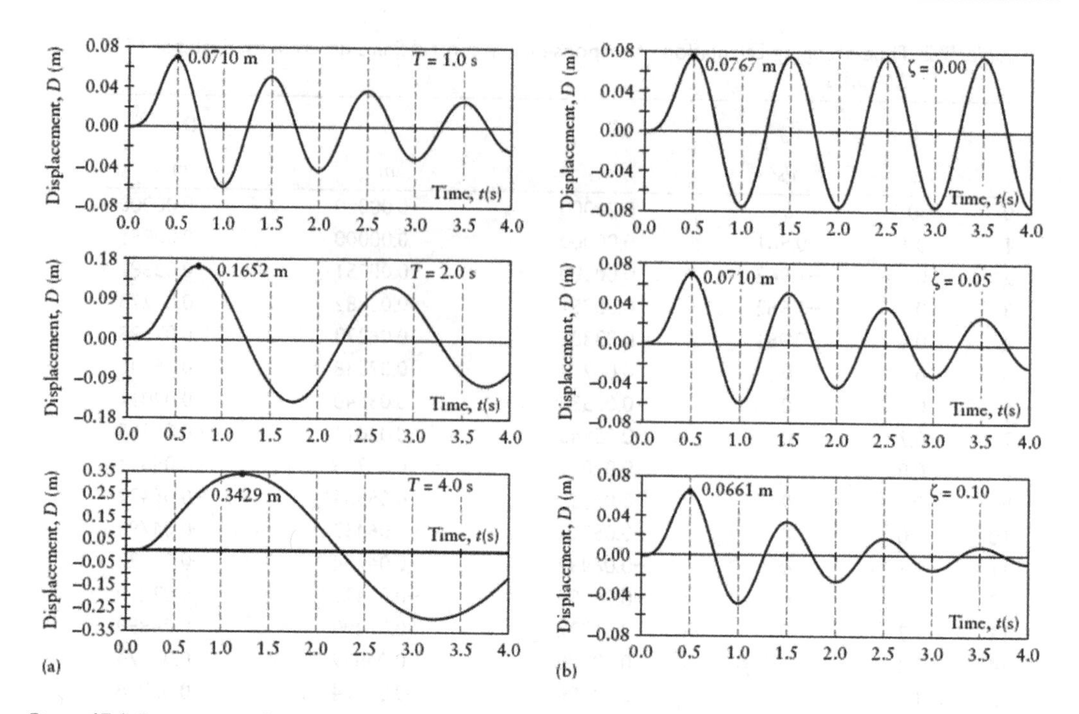

Figure 17.4 Response to fictitious ground acceleration record, Figure 17.3a. (a) Variation of response with T, while $\zeta = 0.05$. (b) Variation of response with ζ, while $T = 1.0$ s. (1 m = 39.4 in.)

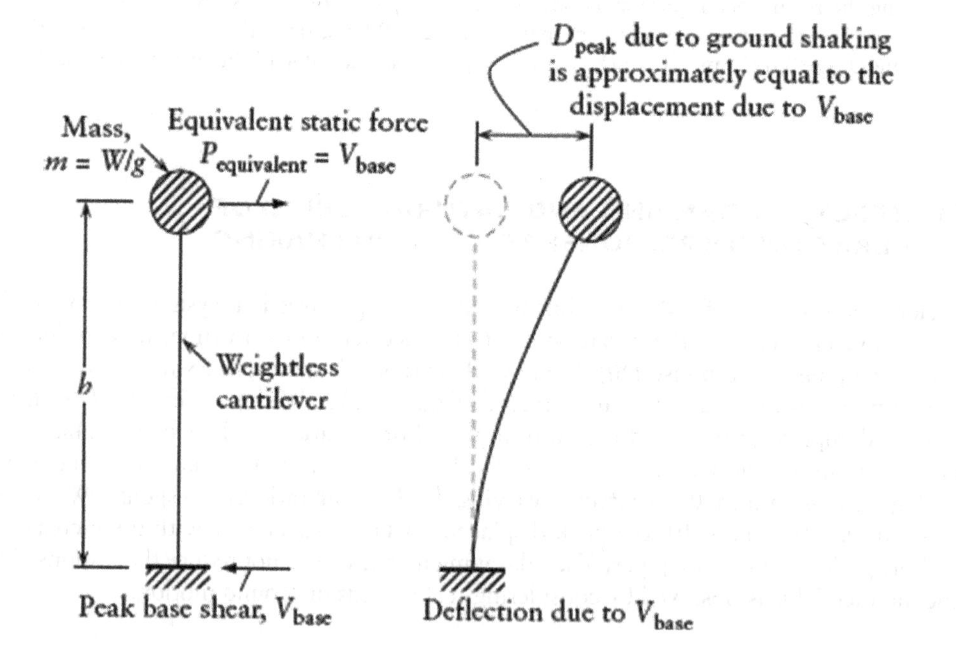

Figure 17.5 Equivalent static force producing displacement $\approx D_{peak}$; where D_{peak} is the largest absolute value of displacement induced by earthquake.

value is determined by time-stepping analysis of a single-degree-of-freedom system having a natural period of vibration $= T$ and damping ratio $= \zeta$. In design, we assume that an equivalent static force equal to $(S_a\, m)$ produces displacement $\approx D_{peak}$ and shearing force at the base equal to (Figure 17.5): $V_{base} = S_a\, m$. The *pseudo-acceleration*, S_a, is defined as:

$$S_a = \omega^2\, D_{peak} = \left(\frac{2\pi}{T}\right)^2 D_{peak} \tag{17.40}$$

ω is the natural frequency and T ($=2\pi/\omega$) is the natural period of vibration; ω and m are related (Equation 16.14): $\omega = \sqrt{S/m}$. With S being the stiffness of the system, relating the static force V_{base} to the displacement it produces, which is assumed equal to D_{peak}.

$$V_{base} = S\, D_{peak} = \omega^2\, m\, D_{peak} \tag{17.41}$$

Substitution of $m = W/g$ gives the peak base shear:

$$V_{base} = \left(\frac{S_a}{g}\right) W \tag{17.42}$$

The term (S_a/g) is a dimensionless *lateral force coefficient* used in design as a multiplier to the weight of the system, W. Time-stepping response, using real earthquake records expected to occur in the site, determines the lateral force coefficient (S_a/g) of a structure for which T and ζ are known. Seismic design codes give the lateral force coefficient. The graphs in Figures 17.6a and b show the variation of D_{peak} and S_a with T determined by time-stepping using the fictitious acceleration. In practice, D_{peak} and S_a are true displacement and pseudo-acceleration, respectively. The difference between the pseudo-acceleration and the true acceleration is small (except when T and ζ are relatively high, e.g. $T = 10$ second and $\zeta = 0.2$).

17.7 PSEUDO-VELOCITY

The parameter *pseudo-velocity*, S_v, is:

$$S_v = \omega\, D_{peak} = \left(\frac{2\pi}{T}\right) D_{peak} \tag{17.43}$$

The graph in Figure 17.6c is pseudo-velocity spectra. The graph gives S_v for a single-degree-of-freedom system having $\zeta = 0.05$ and any T; but this is valid only for the fictitious acceleration record in Figure 17.3a. S_a, S_v, and S_d for real earthquake records are available.

Any of the parameters D_{peak}, S_a, or S_v (peak displacement, pseudo-acceleration or pseudo-velocity) is sufficient to calculate the peak forces required in structural design. In summary, we repeat the relations between the three parameters:

$$S_v = \omega\, D_{peak} = \left(\frac{2\pi}{T}\right) D_{peak} \tag{17.44}$$

$$S_a = \omega^2\, D_{peak} = \left(\frac{2\pi}{T}\right)^2 D_{peak} \tag{17.45}$$

$$\frac{S_a}{S_v} = \omega = \frac{2\pi}{T} \tag{17.46}$$

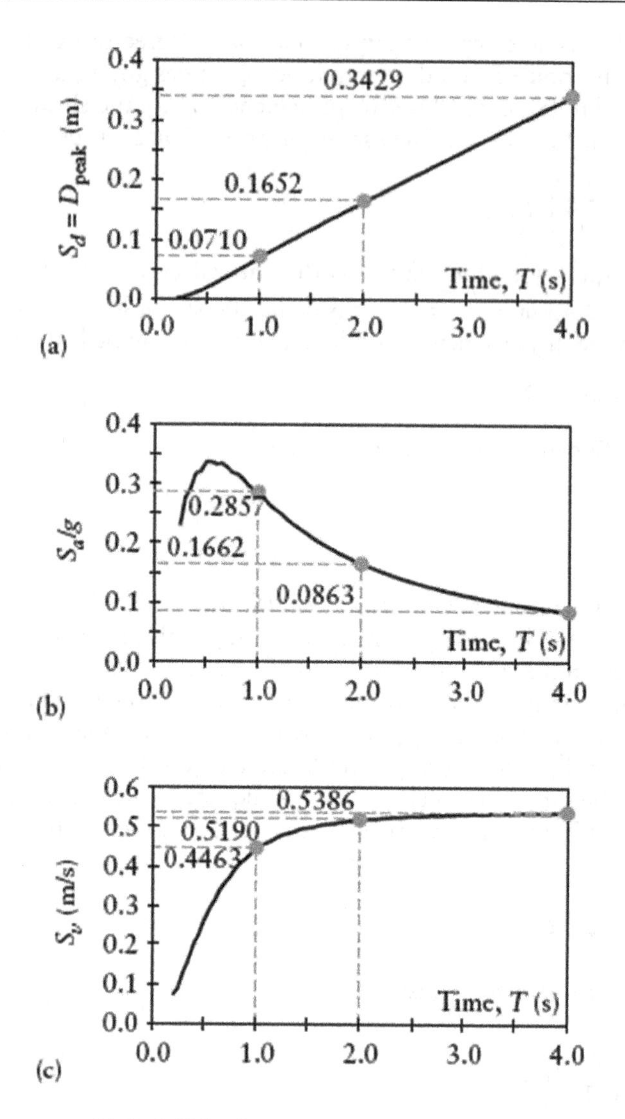

Figure 17.6 Response spectra with $\zeta = 0.05$ for fictitious acceleration record (Figure 17.3a). (a) Peak displacement. (b) Pseudo-acceleration in terms of gravitational acceleration. (c) Pseudo-velocity. The peak displacement values for $T = 1.0$, 2.0, and 4.0 are the same as in Figure 17.4a. $S_a = (T/2\pi)^2\, S_d$; $S_v = (2\pi/T)\, S_d$.

In the long periods (e.g. $T = 10$ seconds), the pseudo-velocity can be significantly smaller than the relative velocity. There is little motive to study the difference. The peak deformation and forces in a system can be adequately determined from the pseudo-acceleration. The deformation, the pseudo-velocity, and pseudo-acceleration are different ways of presenting the same information about structural response to a particular earthquake.

17.8 GRAPHS FOR PSEUDO-ACCELERATION AND PSEUDO-VELOCITY

Time-stepping analysis of acceleration records of a real earthquake gives graphs of pseudo-acceleration and pseudo-velocity versus natural period of vibration, T of single-degree-of-freedom systems. Such graphs, when done for several earthquake records expected for a site, give spectral acceleration graphs to determine the loading in seismic design of new structures. Certain codes require techniques that are more sophisticated.

17.9 EARTHQUAKE RESPONSE SPECTRA

For simplicity of presentation, we derive response spectra (Figure 17.6) using a fictitious record of ground acceleration versus time. Records of real earthquakes are available electronically (e.g. from Pacific Earthquake Engineering Center (PEER), *Ground Motion Database*). Response spectra derived from these records are used in codes specifying earthquake load and effect. The concept is reviewed in this section for easy reference.

The equation of motion of a single-degree-of-freedom system (Equation 17.3) indicates that for a given ground acceleration $\ddot{u}_g$ (t), the response is dependent only upon the natural frequency, ω and the damping ratio, ζ. Thus, systems having the same ω (or $T=2\pi/\omega$) and the same ζ have the same displacement response although they may differ in mass and stiffness. We use this principle in Example 17.2 and in the nonlinear static (pushover) analysis for seismic design of certain structures (Section 17.18).

For displacements, a structure will respond to typical ground motions primarily in fundamental mode, with a period close to its fundamental T. Structures with higher fundamental period T will experience greater peak displacement. Damping decreases the peak displacement; higher ζ corresponds to smaller peak displacement; but this is not true for systems with very long periods.

The variations of ground acceleration $\ddot{u}_g$ (t), recorded during strong earthquakes, have been used to derive *pseudo-acceleration response spectra*. The jagged line in Figure 17.7a is the spectrum derived from the record of the earthquake at El Centro, California, in 1940. The spectrum shown in Figure 17.7a is for $\zeta=0.02$. For any T-value, a numerical solution (in time steps) of the equation of motion (Equation 17.3) gives the variation of the acceleration for the whole duration of the earthquake record; the absolute peak value of the acceleration gives the ordinate S_a of the pseudo-acceleration spectrum. Repetition of the solution using several earthquake records and varying T (keeping $\zeta=0.02$) gives the spectrum shown in Figure 17.7a. Commonly, curves for three to five ζ values are plotted on the same graph (e.g. $\zeta=0$, 0.02, 0.05, 0.10, and 0.20).

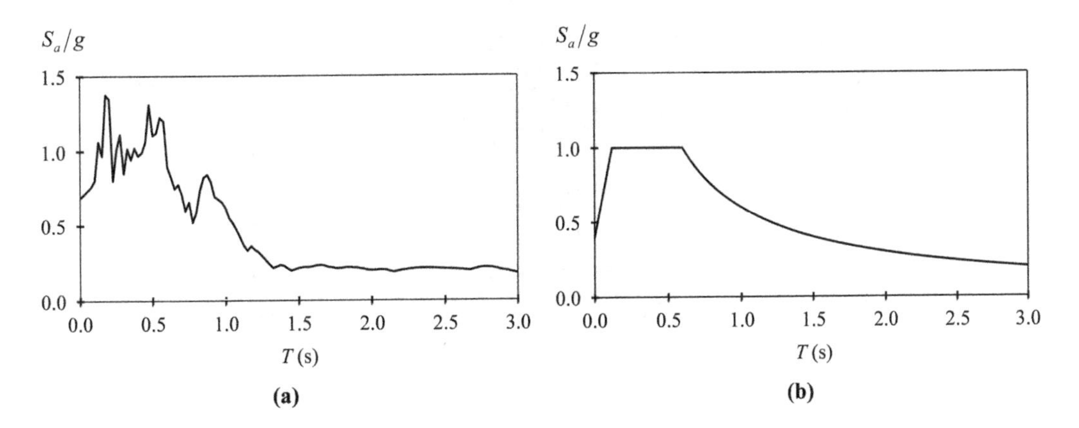

Figure 17.7 Pseudo-acceleration response spectrum. (a) Response spectrum for El Centro (1940) ground motion with $\zeta=0.02$. (b) Simplified spectrum.

From the same ground acceleration record, it is also possible to prepare *pseudo-velocity response* and *true displacement response* spectra. For specific values of ζ and T, the pseudo-velocity and -displacement spectra, S_v and S_d, are related to S_a:

$$S_v = S_a\,(T/2\pi);\; S_d = S_a\,(T/2\pi)^2 \tag{17.47}$$

For multi-degree-of-freedom system, the shearing force at the base is considered equal to the sum of the contributions of the considered modes. An equation similar to Equation 17.42 gives the contribution of the rth mode; V_{base}, S_a, and W are replaced by $V_{r\,\text{base}}$, $S_{ra}(T_r, \zeta_r)$, and W_r; with T_r and ζ_r being the natural period and the damping ratio for the rth mode. W_r is a fraction of the total weight of the building (Section 17.14).

Figure 17.7b is a simplified version of the spectrum in Figure 17.7a. Codes develop simplified design spectra giving S_a (T, ζ) that vary with the geographic location of the structure; the codes also provide soil multipliers to S_a to account for the conditions of the soil (higher multipliers for softer soil). Figure 17.8a is an example of acceleration design spectra for $\zeta = 0.05$; Figure 17.8b is for a site of high seismicity. The use of acceleration design spectra is demonstrated in Examples 17.2, 17.3, and 17.4. D_{peak}, and equivalent static force, $P_{\text{equivalent}}$, are given by:

$$D_{\text{peak}} = S_a\,(T/2\pi)^2 \tag{17.48}$$

$$P_{\text{equivalent}} = m\,S_a \tag{17.49}$$

The fundamental natural period, T, should not be overestimated. A value of T larger than the real underestimates design forces; smaller T generally corresponds to larger S_a and larger $P_{\text{equivaent}}$.

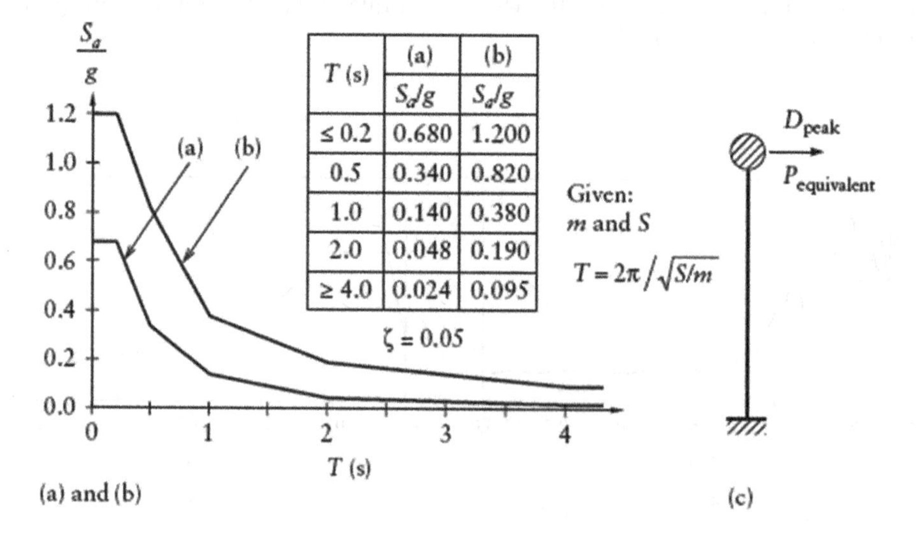

T (s)	(a) S_a/g	(b) S_a/g
≤ 0.2	0.680	1.200
0.5	0.340	0.820
1.0	0.140	0.380
2.0	0.048	0.190
≥ 4.0	0.024	0.095

Figure 17.8 Typical design specification. (a) Spectral acceleration. (b) (S_a/g) for high seismicity site. (c) Single degree-of-freedom.

EXAMPLE 17.2: A TOWER IDEALIZED AS A SDOF SYSTEM: RESPONSE TO EARTHQUAKE

A tower is idealized as a weightless vertical cantilever of length $l = 40$ m (1575 in.), carrying a weight $W = 160$ kN (36.0×10^3 lb) at its tip. Find the equivalent static force at the tip, the peak drift at the tip, and the shearing force and bending moment at the base. Given $E = 200 \times 10^9$ N/m² (29.0×10^6 lb/in.²), $I = 30 \times 10^{-3}$ m⁴ (72×10^3 in.⁴); the damping ratio, $\zeta = 0.05$; the spectral acceleration for the site is shown in Figure 17.8a. Gravitational acceleration, $g = 9.81$ m/s² (386 in./s²).

The stiffness is (see Appendix A, inverse of Equation A.19):

$$S = 3EI/l^3 = 3\left(200\times10^9\right)\left(30\times10^{-3}\right)/\left(40\right)^3 = 281\times10^3 \text{ N/m}$$

$$\omega^2 = S/m = S/\left(W/g\right) = 281\times10^3/\left[\left(160\times10^3\right)/9.81\right] = 17.23 \ \left(\text{rad/s}\right)^2$$

$$T = 2\pi/\omega = 2\pi/\sqrt{17.23} = 1.51 \text{ s}$$

Figure 17.8a gives: $S_a = 0.093\,g = 0.093\,(9.81) = 0.913$ m/s²

Peak displacement, $D_{peak} = S_a\,(T/2\pi)^2 = 0.913\,(1.51/2\pi)^2 = 0.053$ m

Equivalent static force, $P_{equivalent} = m\,S_a = \dfrac{W}{g}S_a = \dfrac{160\times10^3}{9.81}(0.913) = 14.9\times10^3 \text{ N}$

(3.35×10^3 lb)

Shearing force at base $= 14.9 \times 10^3$ N (3.35×10^3 lb)

Bending moment at base $= 14.9 \times 10^3\,(40) = 596 \times 10^3$ N-m (5.28×10^6 lb-in.)

EXAMPLE 17.3: THE TOWER OF EXAMPLE 17.2 WITH HEIGHT REDUCED

Compare the shearing force and the bending moment at the base when the tower of Example 17.2 is considered for a nearby location with a height, $l = 30$ m, instead of 40 m. Other data are the same.

The stiffness is:

$$S = 3EI/l^3 = 3\left(200\times10^9\right)\left(30\times10^{-3}\right)/\left(30\right)^3 = 666.7\times10^3 \text{ N/m}$$

$$\omega^2 = S/m = S/\left(W/g\right) = 666.7\times10^3/\left[\left(160\times10^3\right)/9.81\right] = 40.88 \ \left(\text{rad/s}\right)^2$$

$$T = 2\pi/\omega = 2\pi/\sqrt{40.88} = 0.983 \text{ s}$$

Figure 17.8a gives: $S_a = 0.147\,g = 0.147\,(9.81) = 1.44$ m/s²

$$\text{Equivalent static force} = m\,S_a = \frac{W}{g}S_a = \frac{160\times10^3}{9.81}(1.441) = 23.5\times10^3 \text{ N}$$

Shearing force at base, $V_{base} = 23.5 \times 10^3$ N

Bending moment at base $= 23.5 \times 10^3\,(30) = 705 \times 10^3$ N-m

The bending moment in the 30 m tower is 18 percent higher than that in the 40 m tower. This is because of the stiffness increase due to shorter height. Assuming that the computed bending moment is too high and we want to reduce it to the same level as that of the 40 m tower, what would be the solution? Avoiding high bending stresses by increasing the size of members can be ineffective because it results in higher stiffness, S, shorter period, T, and significant increase in spectral acceleration, S_a, and equivalent static force, V_{base}. When the structural response goes into the inelastic range, the seismic forces are smaller. Detailing the structure to ensure ductile behavior can be the practical solution (Section 17.10).

17.10 EFFECT OF DUCTILITY ON FORCES DUE TO EARTHQUAKES

The peak drift due to earthquake excitation is roughly the same for elastic or plastic structures. However, when yielding occurs, the peak forces are limited by the yield strengths. The dashed and solid lines in Figure 17.9 compare the load-displacement graphs of an elastic and a perfectly elasto-plastic pendulum subjected to the same maximum displacement $D > D_y$; where D_y is the displacement at yielding. For the elastic member, the loading and unloading paths are the same (O to A and A to O), without loss of energy. When the yield strength F_y is reached, the paths are OBC and CH for loading and unloading, respectively. The area enclosed in OBCH is the energy lost. In seismic design of structures, codes specify forces smaller than the elastic forces and require that the structures possess the ductility that ensures that plastic deformations can occur without failure. We see that the force-displacement relationships in Figure 17.9 do not apply after the system has yielded. In consecutive oscillation cycles, the system vibrates differently.

17.11 COMPARISON OF ELASTIC AND PLASTIC RESPONSES

Figure 17.10 represents the variation of force F in a perfectly elasto-plastic system subjected to imposed displacement, D. When the displacement is increased from zero to D_o, the force variation will follow the bilinear graph ACG; at G, the displacement is D_o and the force $= F_y$; where F_y = yield strength. If the system were linear, the force variation would follow a straight line ACH; at H, the hypothetical displacement and force would be D_o and F_o. If the displacement of the elasto-plastic system is increased beyond D_o, the force-displacement graph will be the extension of line CG to a failure point E; at this point the displacement $= D_{failure}$. At

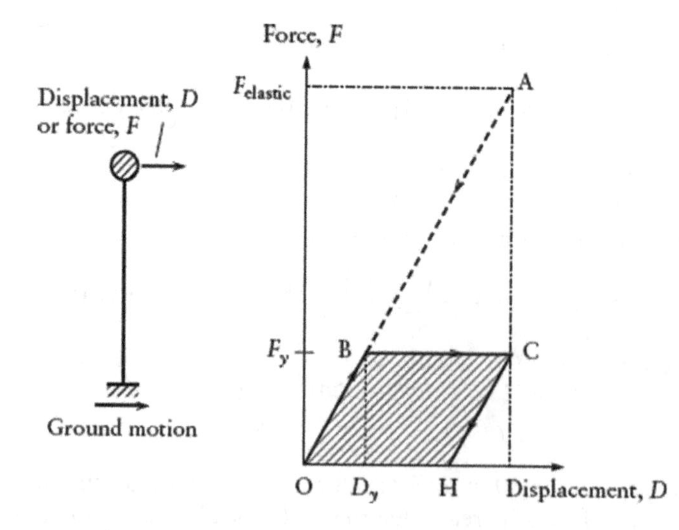

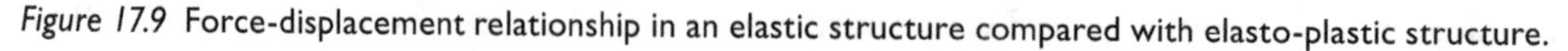
Figure 17.9 Force-displacement relationship in an elastic structure compared with elasto-plastic structure.

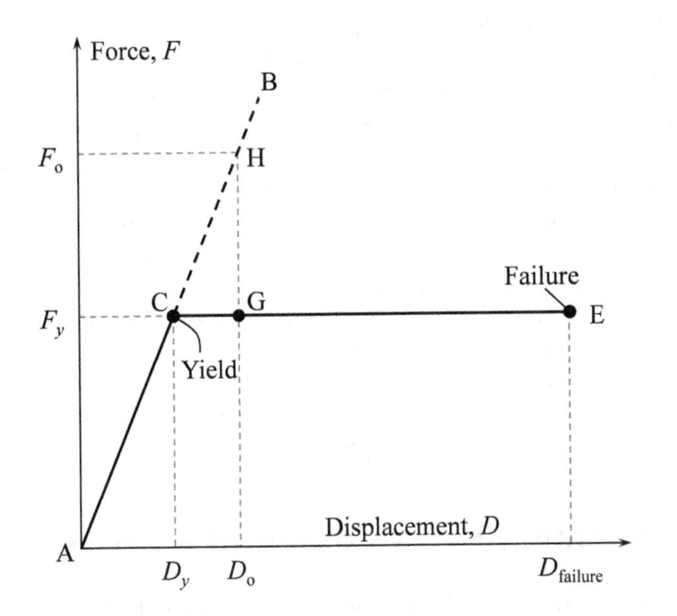

Figure 17.10 Elasto-plastic and linear systems permitted in design for same earthquake.

point C, the yield strength F_y and the yield displacement D_y are reached. The slope of AC is the stiffness before yielding; immediately after, the stiffness is zero.

Definitions: Yield strength reduction factor:

$$R = \frac{F_o}{F_y} = \frac{D_o}{D_y} \tag{17.50}$$

Ductility factor:

$$\mu = \frac{D_{failure}}{D_y} \tag{17.51}$$

The response of an elasto-plastic system can be designed for a force $= F_y$ and displacement D_o. The ductility factor μ represents the ductility demand; the design can be considered safe when the system can acquire the displacement, D_o without failure (when $D_o \leq D_{failure}$). The strength reduction factor, R, is a required ductility characteristic of the system. When $R=1$, $F_o=F_y$ and the system is elastic (failure occurs when the displacement reaches D_y); thus, $D_{failure}$ has to be considered equal to D_y. But, with a ductile system, safe design can be based on linear analysis combined with static force $= F_o$ divided by a factor greater than 1.0. Also, the design can be based on linear analysis with an increased yield force. Section 17.12 discusses the two concepts.

17.12 REDUCTION OF EQUIVALENT STATIC LOADING: DUCTILITY AND OVER-STRENGTH FACTORS

Seismic forces become smaller when stresses exceed yielding. For this reason, codes include a ductility factor in the expression used to calculate V_{base}, such as:

$$V_{base} = \frac{S_a/g}{R_d\ R_o} W \tag{17.52}$$

where R_d = ductility modification factor, reflecting capacity to dissipate energy through inelastic behavior. R_d varies from 1.0 for non-ductile structures to 5.0 for the most ductile systems. R_o = over-strength reduction factor that varies from 1.0 to 1.7 and is based on parameters governing reliable over-strength typically found in a structure (e.g. rounding of sizes or dimensions, strain hardening, ratio of actual yield to minimum specified value, and mobilization of full capacity through formation of consequent plastic hinges). The product (R_o R_d), in the denominator of the expression for V_{base} (Equation 17.52), is sometimes replaced by the symbol R (refer to applicable code). For structures with high (R_o R_d) values, codes typically specify strict detailing requirements to ensure high ductility capacity. Design of members for the required level of ductility is the subject of specialized references.

A severe earthquake is a rare event in which the actual capacity of a structure can be fully utilized. However, for loading other than earthquake, we apply in design factored load combinations and reduced strength of materials. Thus, members possess greater strength than the strength admitted in a non-seismic design.

17.13 MODAL SPECTRAL ANALYSIS OF LINEAR SYSTEMS

The response spectrum analysis reduces the response to earthquake to a series of static analyses using the natural frequencies, the natural modes, and modal damping ratios. The dynamic characteristic of the ground motion is given through spectral acceleration graph $S_a(T, \zeta)$ or its design spectrum; the graph determines the pseudo-acceleration of the rth mode, S_{ar}; the value read from the graph depends upon T_r, the period of vibration of the considered mode. Commonly, we need to consider a small number of modes. Then, S_{ar} is used to get equivalent static forces $\{p_r\}$ to calculate displacements, internal forces, or reactions of the actual multi-degree-of-freedom systems. Consider the plane frame in Figure 17.2 having n degrees-of-freedom and n lumped masses (not shown). The steps of analysis, provided below, are justified in Section 17.14 and the main equations are included in Figure 17.11.

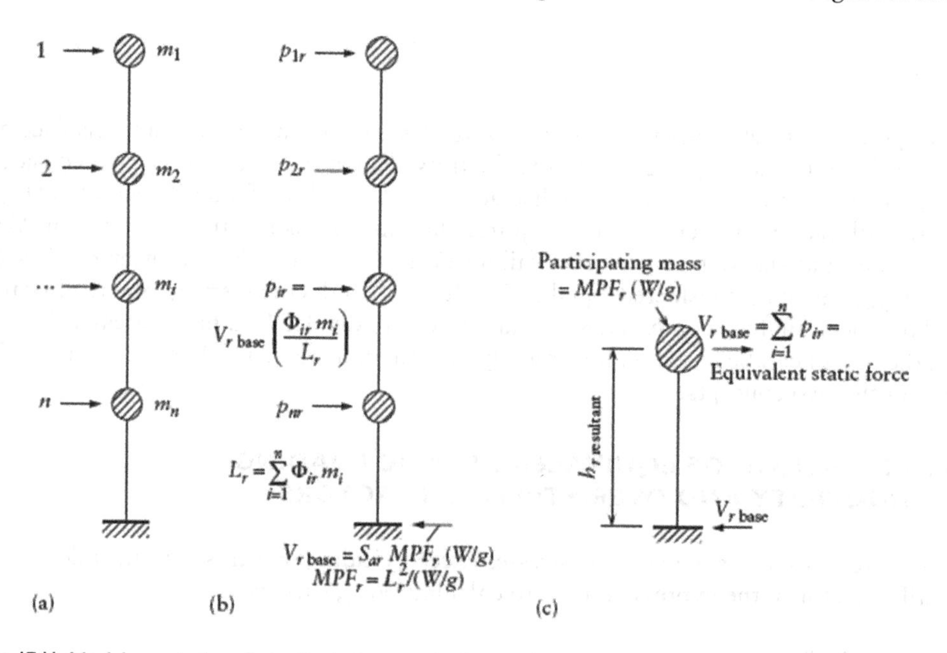

Figure 17.11 Modal spectral analysis: Equivalent static forces due to earthquake. (a) Multi-degree-of-freedom system. (b) Equivalent static forces at the coordinates for the rth mode. (c) Equivalent static resultant.

1. Generate the stiffness and mass matrices, $[S]$ and $[m]$. Solve the free vibration equation:

$$[S][\Phi] = \begin{bmatrix} \omega_1^2 & & \\ & \omega_2^2 & \\ & & \cdots \end{bmatrix}_{n \times n} [m][\Phi] \tag{17.53}$$

The solution gives the eigenvalues $\{\omega^2\}$ and the corresponding eigenvectors $[\Phi]$. The normalized matrix $[\Phi]$ has the property:

$$\{\Phi\}_r^T [m] \{\Phi\}_k = 1.0 \text{ when } k = r \tag{17.54}$$

Program EIGEN1 (or EIGEN2) (Chapter 20) can be used to provide the solution when the number of the degrees-of-freedom is small. Calculate the natural vibration periods $\{T_r\} = \{2\pi/\omega_r\}$, with $r = 1, 2, \ldots, s$; where s is the number of modes considered.

2. Calculate the *mass participation parameters* $\{L\}$ and the dimensionless *mass participation factors* $\{MPF\}$ defined as:

$$\begin{Bmatrix} L_1 \\ L_2 \\ \cdots \end{Bmatrix}_{n \times 1} = [\Phi]^T [m] \begin{Bmatrix} 1 \\ 1 \\ \cdots \end{Bmatrix}_{n \times 1} \tag{17.55}$$

$$\begin{Bmatrix} MPF_1 \\ MPF_2 \\ \cdots \end{Bmatrix}_{n \times 1} = \begin{Bmatrix} L_1^2 \\ L_2^2 \\ \cdots \end{Bmatrix}_{n \times 1} \left(\sum_{i=1}^{n} m_i \right)^{-1} \tag{17.56}$$

The sum of the mass participation factors = 1.0. Equation 17.57 verifies Steps 1 and 2:

$$\sum_{i=1}^{n} MPF_i = 1.0 \tag{17.57}$$

3. Read the spectral accelerations $\{S_a(T, \zeta)\}$ from graphs similar to Figure 17.8a or b.
4. The contribution of the modes to the equivalent static base shear is given by:

$$\begin{Bmatrix} V_{\text{base } 1} \\ V_{\text{base } 2} \\ \cdots \end{Bmatrix}_{n \times 1} = \begin{Bmatrix} (L^2 S_a)_1 \\ (L^2 S_a)_2 \\ \cdots \end{Bmatrix}_{n \times 1} = \begin{Bmatrix} (MPF \cdot S_a)_1 \\ (MPF \cdot S_a)_2 \\ \cdots \end{Bmatrix}_{n \times 1} \left(\sum_{i=1}^{n} m_i \right) \tag{17.58}$$

A typical r^{th} element $V_{r\,\text{base}}$ of this vector is equal to the sum of the equivalent static forces at the coordinates for the rth mode; to find the force at individual coordinates, partition $V_{r\,\text{base}}$ proportional to $m_i\ \Phi_{ir}$ with $i = 1, 2, \ldots, n$. The contribution of the rth mode to the equivalent static forces at the coordinates is:

$$\{p_r\}_{n\times1} = \frac{V_{r\ \text{base}}}{\sum_{i=1}^{n} m_i\ \Phi_{ir}} \begin{Bmatrix} m_1\ \Phi_{1r} \\ m_2\ \Phi_{2r} \\ \dots \end{Bmatrix}_{n\times1} = \frac{V_{r\ \text{base}}}{L_r} \begin{Bmatrix} m_1\ \Phi_{1r} \\ m_2\ \Phi_{2r} \\ \dots \end{Bmatrix}_{n\times1}$$

$$= (L\ S_a)_r \begin{Bmatrix} m_1\ \Phi_{1r} \\ m_2\ \Phi_{2r} \\ \dots \end{Bmatrix}_{n\times1} \quad (17.59)$$

The equivalent static overturning moment and *the modal height* for the rth mode are:

$$OM_r = \sum_{i=1}^{n} p_{ir}\ h_i; \quad h_r = OM_r / V_{r\ \text{base}} \quad (17.60)$$

5. Apply the equivalent static forces $\{p_r\}$ on the structure to calculate displacements, internal forces, or reactions (as required) for the rth mode.
6. Find the contribution of each mode *separately* to any required action (e.g. displacement or internal force) before combining the modal contributions, by one of the methods of Section 17.15, to obtain the peak value of the action.

17.14 MASS PARTICIPATION FACTOR

For easy reference, this section includes all the equations for modal spectral analyses of linear systems. The vibration of a system (Figure 17.11) during ground shaking induces inertial force at coordinate i having a mass m_i. The equations presented in this section assume that the n coordinates are translational and the lumped mass has the unit of mass (e.g. kg).

The vibration of the system tends to occur in the natural mode shapes defined by the normalized mode matrix $[\Phi]_{n\times n}$. The rth column of the matrix indicates the rth mode, $\{\Phi\}_r$. The first mode is a fundamental mode corresponding to the smallest natural frequency, (largest vibration period, $T=2\pi/\omega$). Successive modes have greater ω (shorter T).

The equivalent static forces are in equilibrium with the reaction at the base, which is referred to as $V_{r\ \text{base}}$, meaning the shearing force at the base for the rth mode. Thus, $V_{r\ \text{base}}$ is an estimate of the peak sum of inertial forces at the rth mode. The inertial force at the ith coordinate is proportional to the product $(\Phi_{ir}\ m_i)$; thus, Φ_{ir} represents influence ordinate of the participation of the ith mass on $V_{r\ \text{base}}$. The equivalent static force on the ith mass for the rth mode is:

$$p_{ir} = V_{r\ \text{base}} \frac{\Phi_{ir}\ m_i}{\sum_{i=1}^{n} (\Phi_{ir}\ m_i)} \quad (17.61)$$

$$p_{ir} = V_{r\ \text{base}} \frac{\Phi_{ir}\ m_i}{L_r} \quad (17.62)$$

L_r is mass participation parameter (Equation 17.55). The mass participation factor for the rth mode:

$$MPF_r = \frac{L_r^2}{\sum_{i=1}^{n} m_i} = \frac{L_r^2}{W/g} \tag{17.63}$$

where W = weight of the structure and g = gravitational acceleration. Since the first vibration mode has the highest contribution (maximum MPF), modern codes permit the use of a triangular distribution of static equivalent lateral forces (proportional to floor height from ground). Such distribution is used in Example 17.6 to simplify Equations 17.61 and 17.62 when substituting values of the vector $\{\phi_1\}$.

The normalized mode matrix $[\Phi]$ is derived by dividing the rth column of dimensionless matrix $[\phi]$ by $\left(\{\phi\}_r^{\mathrm{T}}\,[m]\,\{\phi\}_r\right)^{1/2}$; thus, $[\Phi]$ has the units of $(\text{mass})^{-1/2}$. It follows that L_r has the units $(\text{mass})^{1/2}$ and MPF_r is dimensionless.

The peak responses in the modes occur at different instances. The sum of $V_{r\,\text{base}}$ in the n modes is an upper bound of the shearing force at the base:

$$\left(V_{\text{base}}\right)_{\text{upper bound}} = \left(V_1 + V_2 + \cdots + V_n\right)_{\text{base}} \tag{17.64}$$

The use of the upper bound in design overestimates the effect of earthquake. Section 17.15 calculates a more probable value of V_{base}. $V_{r\,\text{base}}$ is the contribution of the rth mode to V_{base}; the fundamental mode ($r = 1$) is the major participant.

We assume that the pseudo-acceleration, $(S_a(T, \zeta))_r$, is known for the site (e.g. taken from a code). The contribution of the rth mode to the equivalent shearing force at the base is:

$$V_{r\,\text{base}} = S_{ar}\; MPF_r\,\left(W/g\right) = S_{ar}\; L_r^2 \tag{17.65}$$

Equation 17.61 or 17.62 gives the equivalent static forces p_{ir}, with $i = 1, 2, \ldots, n$. Static analysis of the structure subjected to $\{p_r\}$ gives the internal forces. It is not appropriate to sum up $\{p_r\}$ for the modes and use the sum to calculate the internal forces. Commonly, a small number of modes is considered. Figure 17.11a–c is a schematic representation of Equations 17.60 to 17.68. Figure 17.11c shows the resultant of the equivalent static forces and its position. The resultant can be considered equal to the product of the acceleration S_{ar} and the participating mass, $[MPF_r\,(W/g)]$.

The resultant of equivalent static forces and its height above base are:

$$V_{r\,\text{base}} = \sum_{i=1}^{n} p_{ir};\;\; h_{r\,\text{resultant}} = \frac{1}{V_{r\,\text{base}}} \sum_{i=1}^{n} \left(p_{ir}\, h_i\right) = \left(\frac{OM}{V_{\text{base}}}\right)_r \tag{17.66}$$

where OM = overturning moment; h_i = height of ith lumped mass above base.

The peak displacements at the n-coordinates in Figure 17.2 can be calculated by Equation 17.67 or by the solution of Equation 17.68:

$$\{D\}_r = \{\Phi\}_r \left(L\, S_a/\omega^2\right)_r \tag{17.67}$$

$$[S]\{D\}_r = \{p\}_r \tag{17.68}$$

$\{D\}_r$ is the rth column of $[D]$. Application of static forces $[p]$ on the structure gives displacements $[D]$ such that $[S]\;[D] = [p]$; thus, solving this equation for $[D]$ should give the same answer as Equation 17.67. This can verify ω^2-values and $[\Phi]$, determined by solving Equation 17.53 (eigenvalue problem).

17.15 MODAL COMBINATIONS

The peak values for different modes do not occur at the same instant. The sum of the absolute peak-values is an upper bound of the total response:

$$r_{\text{upperr bound}} = \sum_{i=1}^{s} |r_{Ei}| \tag{17.69}$$

where r_E refers to generalized earthquake response (displacement or force); i refers to the mode number. The use of the upper bound in design is conservative. The *square root of the sum of the squares* (SRSS) gives a more realistic estimate of the peak response:

$$(r_E)_{SRSS} = \left(\sum_{i=1}^{s} r_i^2 \right)^{1/2} \tag{17.70}$$

This equation is appropriate except when the natural frequencies are close to each other. In this case, the *complete quadratic combination* (CQC) method is more appropriate.

The contribution of the fundamental mode to the response is commonly much higher than the remaining modes. When the number of degrees-of-freedom n is high, it is commonly sufficient to use the contribution of s modes, with $s < n$.

The complete quadratic combination estimates the absolute value of the maximum response by the equation:

$$(r_E)_{CQC} = \left(\sum_{i=1}^{s} \sum_{n=1}^{s} r_{Ei}\, \rho_{in}\, r_{En} \right)^{1/2} = \left(\{r_E\}^{\mathrm{T}} [\rho] \{r_E\} \right)^{1/2} \tag{17.71a}$$

$[\rho]_{s\times s}$ = symmetrical matrix whose element ρ_{in} is:

$$\rho_{in} = \frac{8\,\zeta^2\,(1+\alpha)\,\alpha^{3/2}}{\left(1-\alpha^2\right)^2 + 4\,\zeta^2\,\alpha\,(1+\alpha)^2} \tag{17.71b}$$

s = number of modes; r_{Ei} and r_{En} = generalized response in modes i and n; ζ = damping ratio, assumed the same for all modes; $\alpha = \omega_i/\omega_n$ = natural frequency ratio in modes i and n. $\rho_{in} = 1.0$ for $i = n$; for $i \neq n$, ρ_{in} and α vary between 0 and 1.0. CQC, based on Random Vibration Theory, is closer to the "exact" answer; but it can be conservative or non-conservative.

EXAMPLE 17.4: RESPONSE OF A MULTI-STORY PLANE FRAME TO EARTHQUAKE: MODAL SPECTRAL ANALYSIS

A three-story building is idealized as elastic plane frame with masses lumped at three coordinates (Figure 17.12). Calculate the equivalent static shearing force and overturning moment at the base for the effect of an earthquake whose design acceleration response spectrum, $S_a(T, \zeta)$ is given in Figure 17.8a (with $\zeta = 0.05$). Assume rigid floors and consider only bending deformation for the columns. Gravitational acceleration, $g = 9.81\ \text{m/s}^2$; $EI = 31.3 \times 10^6\ \text{N-m}^2$ = constant. It is more appropriate to consider the flexural rigidity of floors (Example 17.5). To simplify presentation, we assume rigid floor.

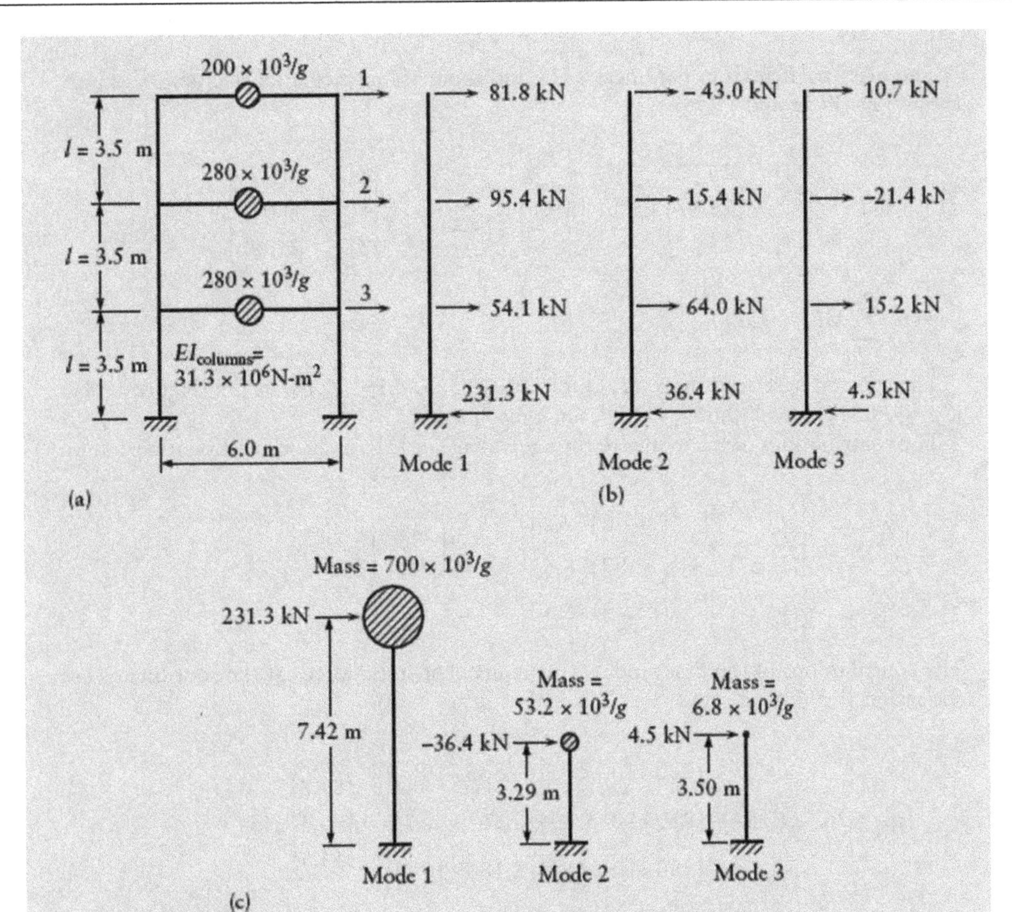

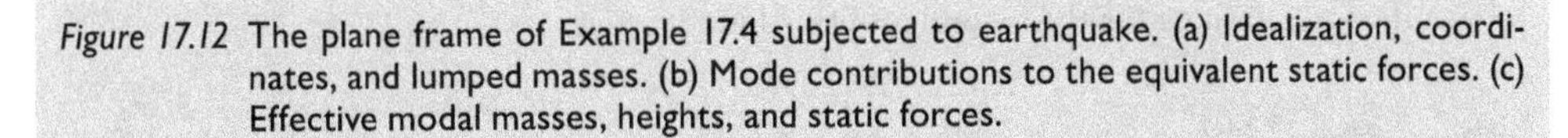

Figure 17.12 The plane frame of Example 17.4 subjected to earthquake. (a) Idealization, coordinates, and lumped masses. (b) Mode contributions to the equivalent static forces. (c) Effective modal masses, heights, and static forces.

The stiffness and the mass matrices are:

$$[S]=\frac{12EI}{l^3}\begin{bmatrix}2 & -2 & 0\\ -2 & 4 & -2\\ 0 & -2 & 4\end{bmatrix};[m]=\frac{10^3}{g}\begin{bmatrix}200 & 0 & 0\\ 0 & 280 & 0\\ 0 & 0 & 280\end{bmatrix}$$

The computer program EIGEN1 (Chapter 20) gives:

$$\{\omega_1^2,\omega_2^2,\omega_3^2\}=\{143.3, 1079, 2091\}\ \ (\text{rad/s})^2$$

The natural periods of vibration are ($T=2\pi/\omega$):

$$\{T\}=\{0.525,\ 0.191,\ 0.137\}\ \text{s}$$

The modal shapes normalized with respect to the masses are:

$$[\Phi]=\begin{bmatrix}4.6348 & 4.2858 & 3.0330\\ 3.8616 & -1.0981 & -4.3495\\ 2.1861 & -4.5508 & 3.0898\end{bmatrix}10^{-3}\ \text{kg}^{-1/2}$$

We can verify that $\{\Phi\}_r^{\text{T}}[m]\{\Phi\}_r=1.0$, with $r=1$, 2, or 3.

The mass participation parameters $\{L\}$ and the mass participation factors $\{MPF\}$ are (Equations 17.55 and 17.56):

$$\{L\}=[\Phi]^T[m]\begin{Bmatrix}1\\1\\1\end{Bmatrix}=\begin{Bmatrix}267.2\\73.87\\25.88\end{Bmatrix}\text{kg}^{1/2};\ \{MPF\}=\left(\sum_{i=1}^{n}m_i\right)^{-1}\begin{Bmatrix}L_1^2\\L_2^2\\L_3^2\end{Bmatrix}=\begin{Bmatrix}0.921\\0.070\\0.009\end{Bmatrix}$$

Check: $\sum_{i=1}^{3}MPF_i=1.000$.

The spectral accelerations $S_a\,(T,\zeta)$ for $\{T\}=\{0.525, 0.191, 0.137\}$ s are (Figure 17.8a): $\{S_a\}=g\ \{0.330, 0.680, 0.680\}=\{3.24, 6.67, 6.67\}$ m/s²

The contributions of the modes to the equivalent static base shear are (Equation 17.58):

$$\begin{Bmatrix}V_1\\V_2\\V_3\end{Bmatrix}_{\text{base}}=\begin{Bmatrix}\left(L^2\,S_a\right)_1\\\left(L^2\,S_a\right)_2\\\left(L^2\,S_a\right)_3\end{Bmatrix}=\begin{Bmatrix}(267.2)^2\ 3.24\\(73.87)^2\ 6.67\\(25.88)^2\ 6.67\end{Bmatrix}=10^3\begin{Bmatrix}231.3\\36.4\\4.5\end{Bmatrix}\text{N}$$

The contributions of the first mode to the equivalent static forces at the coordinates are (Equation 17.59):

$$\{p\}_1=267.2\,(3.24)\begin{Bmatrix}\left(200\times10^3/9.81\right)\left(4.635\times10^{-3}\right)\\\left(280\times10^3/9.81\right)\left(3.862\times10^{-3}\right)\\\left(280\times10^3/9.81\right)\left(2.186\times10^{-3}\right)\end{Bmatrix}=\begin{Bmatrix}81.8\\95.4\\54.1\end{Bmatrix}\ \text{kN}$$

Check: $(\text{Sum})_1=231.3$ kN

Similar calculations for the second and third modes give:

$$\{p\}_2=\begin{Bmatrix}-43.0\\15.4\\64.0\end{Bmatrix}\text{kN};\ \{p\}_3=\begin{Bmatrix}10.7\\-21.4\\15.2\end{Bmatrix}\text{kN}$$

Check: $(\text{Sum})_2=36.4$ kN; $(\text{Sum})_3=4.5$ kN

The effective height, the distance between the resultant of the equivalent static forces above the base, and the overturning moment for the modes are (Equation 17.60):

$h_1=[81.8\ (3\times3.5)+95.4\ (2\times3.5)+54.1\ (3.5)]/231.3=7.42$ m; $OM_1=1716$ kN-m
$h_2=[-43.0\ (3\times3.5)+15.4\ (2\times3.5)+64.0\ (3.5)]/36.4=-3.29$ m; $OM_2=-120$ kN-m
$h_3=[10.7\ (3\times3.5)-21.4\ (2\times3.5)+15.2\ (3.5)]/4.5=3.50$ m; $OM_3=16$ kN-m

Figure 17.12b shows the mode contributions to the equivalent static forces at the nodes. Figure 17.12c is a physical interpretation of the results. The effective mass for typical mode r is equal to $\left(\sum_{i=1}^{n}m_i\right)MPF_r$; this mass multiplied by S_{ar} gives the equivalent static force $\left(\sum_{i=1}^{n} p_{ir}\right)$. We use the SRSS (Equation 17.70) to calculate the required answers; the equivalent static shearing force and overturning moment at the base are:

$$V_{base} = \left[(231.3)^2 + (36.4)^2 + (4.5)^2\right]^{1/2} = 234.2 \text{ kN}$$

$$OM = \left[(1716)^2 + (120)^2 + (16)^2\right]^{1/2} = 1720 \text{ kN-m}$$

The sum of the mass participation factors of the second and third modes is 8 percent of the total mass. But, in this example, considering the two modes has increased V_{base} and OM by a negligible amount. However, considering only the fundamental first mode is not always adequate.

The resultant of the equivalent static forces above the base in the first mode is $h_1 = 7.42$ m = 0.71 (the height of the structure); this is greater than half the height of the building. A similar result can be seen for the regular building in Example 17.5.

The peak displacements in three modes, determined by Equation 17.67 are:

$$[D] = 10^{-3}\begin{bmatrix} 28.00 & 1.957 & 0.2504 \\ 23.33 & -0.5014 & -0.3591 \\ 13.202 & -2.078 & 0.2551 \end{bmatrix}$$

$[S]\,[D] = [p]$; $EI/l^3 = 730.03 \times 10^3$; Equation 17.68 gives:

$$\frac{12\,EI}{l^3}\begin{bmatrix} 2 & -2 & 0 \\ -2 & 4 & -2 \\ 0 & -2 & 4 \end{bmatrix}[D] = 10^3\begin{bmatrix} 81.8 & -43.07 & 10.7 \\ 95.4 & 15.4 & -21.4 \\ 54.1 & 64.0 & 15.2 \end{bmatrix}$$

Solution gives almost the same $[D]$ as above (sign reversal of any column of $[D]$ has no significance in response to earthquakes).

17.16 MASS LUMPING

We apply effective earthquake forces in multi-story buildings subjected to horizontal ground motion only at selected translational degrees-of-freedom. At these coordinates, the distributed mass of the structure is lumped. The floor slabs (with or without beams) are assumed to be rigid in their own plane. We ignore effective earthquake forces associated with vertical displacements. Thus, each floor moves as a rigid body; masses are lumped only at coordinates representing horizontal translations. These selected degrees-of-freedom are used in the dynamic analysis to determine the eigenvalues and eigenvectors (ω_1^2, ω_2^2, ..., and $\{\phi\}_1$, $\{\phi\}_2$, ...) and these need to be calculated for a limited number of modes. However, in static analysis, we consider all the degrees-of-freedom of the idealized structure.

EXAMPLE 17.5: REINFORCED CONCRETE FLAT PLATE BUILDING: MODAL SPECTRAL ANALYSIS

Figure 17.13a is a plan of a typical floor of a ten-story flat plate building (composed of solid slabs supported directly on columns and shear walls, without beams). The structure is idealized by the plane system in Figure 17.13b. A similar structure, having the same plan and member dimensions is analyzed for the effect of static lateral forces in Example 11.3. The properties of the members of the plane system in Figure 17.13b are the same as those in the lower ten stories of the system in Figure 11.11; calculation of the properties of the members is discussed in Example 11.3 and is not repeated here.

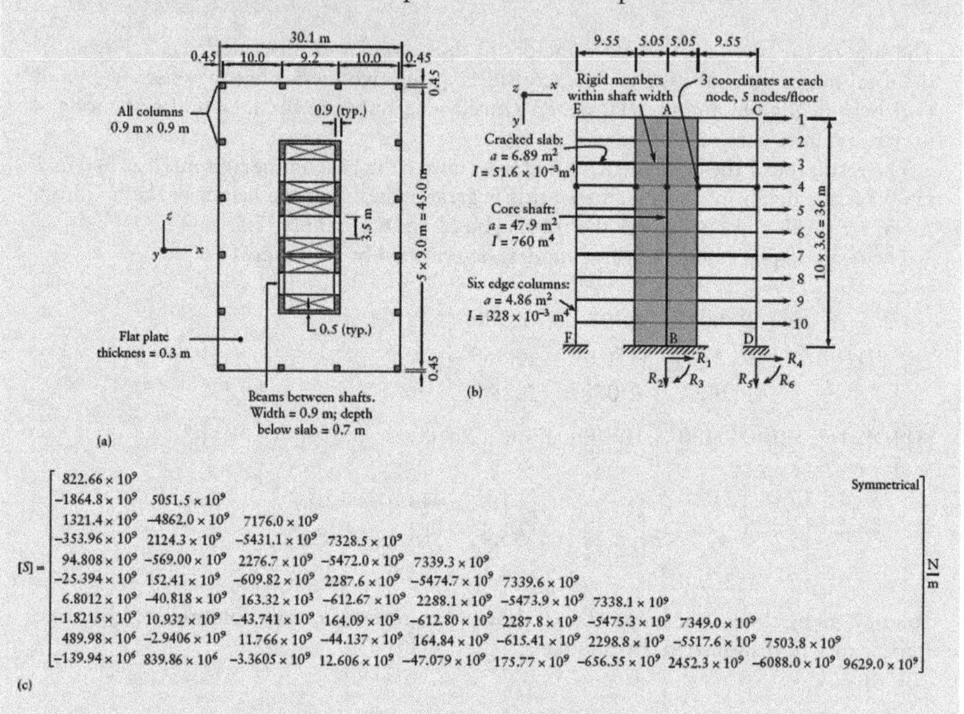

Figure 17.13 Flat plate building of Example 17.5. (a) Plan of typical floor, (b) Plane frame idealization; selected coordinates for dynamic analysis of response to ground shaking in x-direction. (c) Stiffness matrix. (1 m = 39.37 in.)

It is required to determine the maximum absolute values of the reactions at B and D (bases of shear walls and outer columns) induced by horizontal ground shaking in the x-direction. Given data: the weight of the building, $W = 14.926 \times 10^6$ N (3.3554×10^6 lb); the elasticity modulus, $E = 32.0 \times 10^9$ N/m^2 (4.64×10^6 lb/in.^{2}); gravitational acceleration, $g = 9.81$ m/s^2 (386 in./s^2). Use the spectral acceleration graph in Figure 17.8b in a linear modal spectral analysis (high seismicity site).

For dynamic analysis, the mass of the structure (W/g) is divided in masses, each = 0.1 (W/g) at coordinates 1 to 10 in Figure 17.13b; the stiffness matrix $[S]_{10 \times 10}$ is generated in separate runs of computer program PLANEF (Chapter 20). The ith column of $[S]$ is a vector of the forces at the selected coordinates with translation = 1.0 at i and = 0 at the remaining coordinates. For this analysis, the model has five nodes at each floor; two translations and one rotation can occur at each node. Axial and shear deformations are ignored.

Figure 17.13c presents the calculated stiffness matrix. The mass matrix is diagonal; each element on the diagonal = 0.1 W/g = 0.15215×10^6 kg. Computer program EIGEN1 (Chapter 20) gives the eigenvalues and the eigenvectors for the first four modes, presented in Table 17.3

Table 17.3 Reinforced concrete ten-story building, Example 17.5. Square of natural frequencies and natural modes

	r	Mode *r* 1	2	3	4
ω_r^2 ((rad/s)^2)		4152.8	0.13743×10^6	1.0802×10^6	4.1748×10^6
	1	1.4765×10^{-3}	1.2465×10^{-3}	1.0423×10^{-3}	-0.85355×10^{-3}
	2	1.2720×10^{-3}	0.53549×10^{-3}	-69.7757×10^{-6}	0.56941×10^{-3}
	3	1.0691×10^{-3}	-0.12192×10^{-3}	-0.85196×10^{-3}	1.0515×10^{-3}
	4	0.87010×10^{-3}	-0.65860×10^{-3}	-1.0348×10^{-3}	0.34035×10^{-3}
$\{\Phi\}_r$	5	0.67874×10^{-3}	-1.0114×10^{-3}	-0.58953×10^{-3}	-0.75909×10^{-3}
(kg$^{-1/2}$)	6	0.49962×10^{-3}	-1.1425×10^{-3}	0.21433×10^{-3}	-1.1000×10^{-3}
	7	0.33828×10^{-3}	-1.0538×10^{-3}	0.93751×10^{-3}	-0.32039×10^{-3}
	8	0.20088×10^{-3}	-0.79213×10^{-3}	1.2053×10^{-3}	0.80882×10^{-3}
	9	94.033×10^{-6}	-0.44598×10^{-3}	0.91930×10^{-3}	1.1912×10^{-3}
	10	24.704×10^{-6}	-0.13569×10^{-3}	0.33825×10^{-3}	0.57979×10^{-3}

Note: The mode vectors are normalized such that $\{\Phi\}_r^T\,[m]\,\{\Phi\}_r = 1.0$. Figure 17.14a is a graphical presentation of the modes.

and Figure 17.14a. For each eigenvector ω_r^2, we calculate T_r and read S_{ar} in the graph in Figure 17.8a. We also calculate the mass participation parameter L_r, the mass participation factor, MPF_r, and the shearing force at the base, $V_{r\,\text{base}}$; the results are included in Table 17.4. We use $V_{r\,\text{base}}$ to calculate the equivalent static forces $\{p_r\}$ at n coordinates. The equations used are:

$$T_r = \frac{2\,\pi}{\omega_r} \tag{a}$$

$$L_r = \sum_{i=1}^{n} (\Phi_{ir}\ m_i) \tag{b}$$

$$MPF_r = L_r^2 \Big/ \sum_{i=1}^{n} m_i \tag{c}$$

$$V_{r\,\text{base}} = (S_{ar}/g)\ MPF_r\ W \tag{d}$$

$$p_{ir} = \frac{V_{r\,\text{base}}}{L_r} (\Phi_{ir}\ m_i) \tag{e}$$

ω_r = natural frequency; T_r = natural period; L_r = mass participation parameter; MPF_r = mass participation factor; $V_{r\,\text{base}}$ = shearing force at base (reaction); p_i = equivalent static force at coordinate i. Application of Equations (a) to (e) for the first mode ($r = 1$) gives:

$$L_1 = 0.9926 \times 10^3\ \text{kg}^{1/2};\ MPF_1 = \left(0.9926 \times 10^3\right)^2 / 1.5220 \times 10^6 = 0.6478$$

$$T_1 = 2\,\pi / \sqrt{4.1528 \times 10^3} = 0.0975\ \text{s};\ S_{a1}/g\ \text{(Fig. 17.8b)} = 1.2$$

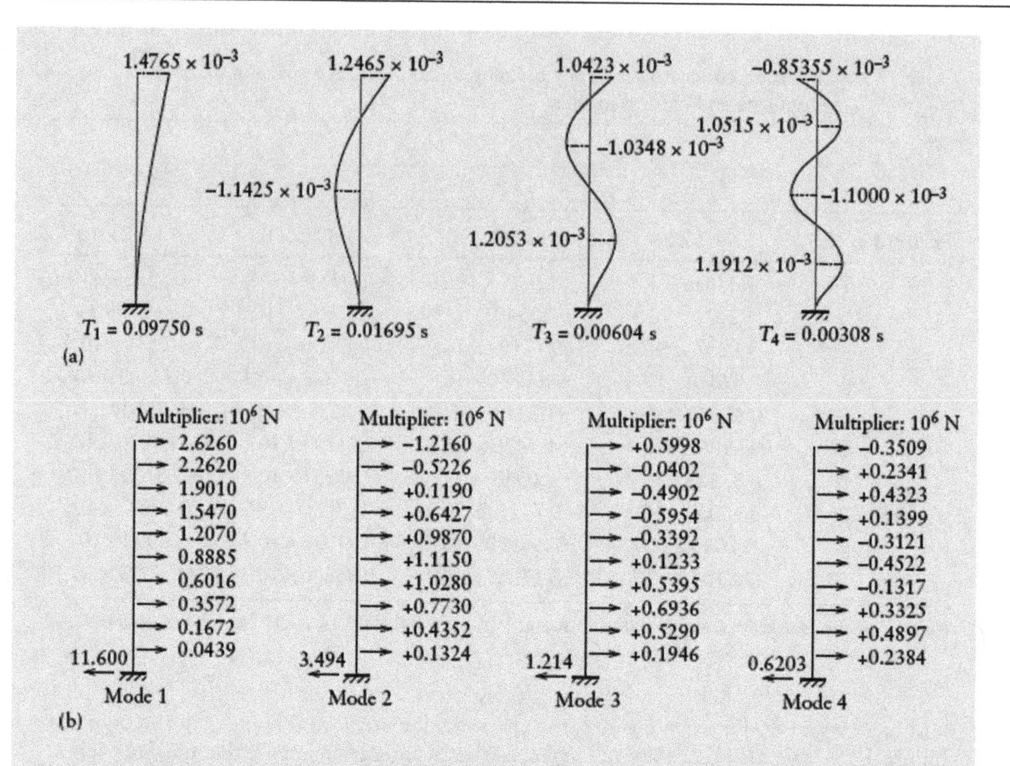

Figure 17.14 Ten-story flat plate building. (a) Mode shapes. (b) Equivalent static loading.

$$V_{1\,\text{base}} = 1.2\,(0.6478)\,14.926\times10^6 = 11.60\times10^6\ \text{N}$$

$$p_{i1} = \frac{11.60\times10^6}{0.9926\times10^3}\left(\Phi_{i1}\times0.1522\times10^6\right) = 1778.062\times10^6\ \Phi_{i1};\ \text{with } i=1,2,\ldots n.$$

Substituting values of Φ_{i1} from Table 17.3 gives the equivalent static forces for the first mode, $\{p_1\}$. Application of Equations (a) to (e), with $r=2$, 3, and 4 gives the equivalent static forces for modes 2, 3, and 4. Table 17.4 lists the calculated parameters for four modes. Figure 17.14b shows the calculated static forces.

Table 17.4 Values of ω_r^2, T_r, L_r, MPF_r, $V_{r\,\text{base}}$ with $r=1$, 2, 3, and 4 for ***modal*** spectral analysis of the building in Figure 17.13

	Mode r			
	1	*2*	*3*	*4*
ω_r^2 ((rad/s)²)	4.1528×10^3	0.13743×10^6	1.0807×10^6	4.1748×10^6
T_r (s)	0.0975	0.0170	0.0060	0.0031
S_{ar}/g	1.2	1.2	1.2	1.2
L_r (kg$^{1/2}$)	0.9926×10^3	-0.5447×10^3	0.3212×10^3	0.2295×10^3
MPF_r	0.6476	0.1950	0.0678	0.0346
$V_{r\,\text{base}}$ (N)	11.6000×10^6	3.4930×10^6	1.2140×10^6	0.6203×10^6

Table 17.5 Peak absolute values of reaction components of shear walls and exterior columns in response to earthquake, Example 17.5, Figure 17.13

	Shear walls (core of building)			*Six edge columns*		
Mode number	R_1 *(10^6 N)*	R_2 *(10^6 N)*	R_3 *(10^6 N-m)*	R_4 *(10^3 N)*	R_5 *(10^3 N)*	R_6 *(10^3 N-m)*
1	−11.57	0	−304.9	−14.57	−420.0	−145.7
2	−3.488	0	−27.17	−2.300	−5.177	−12.88
3	−1.213	0	−5.812	−0.6806	−0.2308	−2.735
4	−0.3508	0	−2.124	−0.3192	−0.0374	−0.9911
Square root of the sum of squares	12.18	0	306.2	14.76	420.0	146.3

The equivalent static loads for the four modes are applied on the system in Figure 17.13b, with three coordinates at each of its fifty nodes. This gives the reactions at the base and the peak internal forces in the members induced by earthquake for the four modes. Table 17.5 lists the reaction components R_1 to R_6; it also gives the square root of the sum of the squares for each component. One computer run of program PLANEF (Chapter 20), with four load cases, one for each mode, determines the values in Table 17.5.

This example demonstrates linear modal spectral procedure. Its results do not represent the level of stress in an earthquake. For simplicity of presentation, the same plan of a 50-story building (Chapter 11) is employed. However, for ten stories, smaller and thinner core and columns would be more representative.

17.17 DUCTILITY AND STRENGTH REQUIREMENT

In seismic design, buildings that deform beyond elasticity in strong ground shaking are subjected to smaller forces compared to buildings designed to remain elastic. Most modern buildings are designed for the smaller design forces; members should undergo plastic deformations without failure. Thus, the demand in seismic design is for greater ductility and smaller strength. Lateral drift of the roof, inter-story drift ratio (the relative drift of consecutive floors divided by story height), and plastic hinge rotation are means to evaluate ductility demands. The detailing of structures for ductile behavior is a design issue beyond the scope of structural analysis.

For multi-story buildings (modeled as plane or space frames), it is generally required to have the plastic hinges at the ends of beams, rather than in columns; this is to eliminate formation of a story-sway mechanism (see Example 17.6). While in bridges, it is neither practical nor desirable to have the plastic hinges in the superstructure; piers are commonly chosen as the site for the inelastic deformation (refer to Example 17.7). To ensure ductile inelastic flexural response, it is essential to prevent non-ductile deformation; for example, if shear strength of a bridge pier develops before its flexural strength, brittle response can result; strength and stiffness would degrade rapidly. Formation of plastic hinges in any structure is associated with reduction in its stiffness, increased damping, and longer natural period. Accounting for these effects requires elaborate computational nonlinear capabilities –

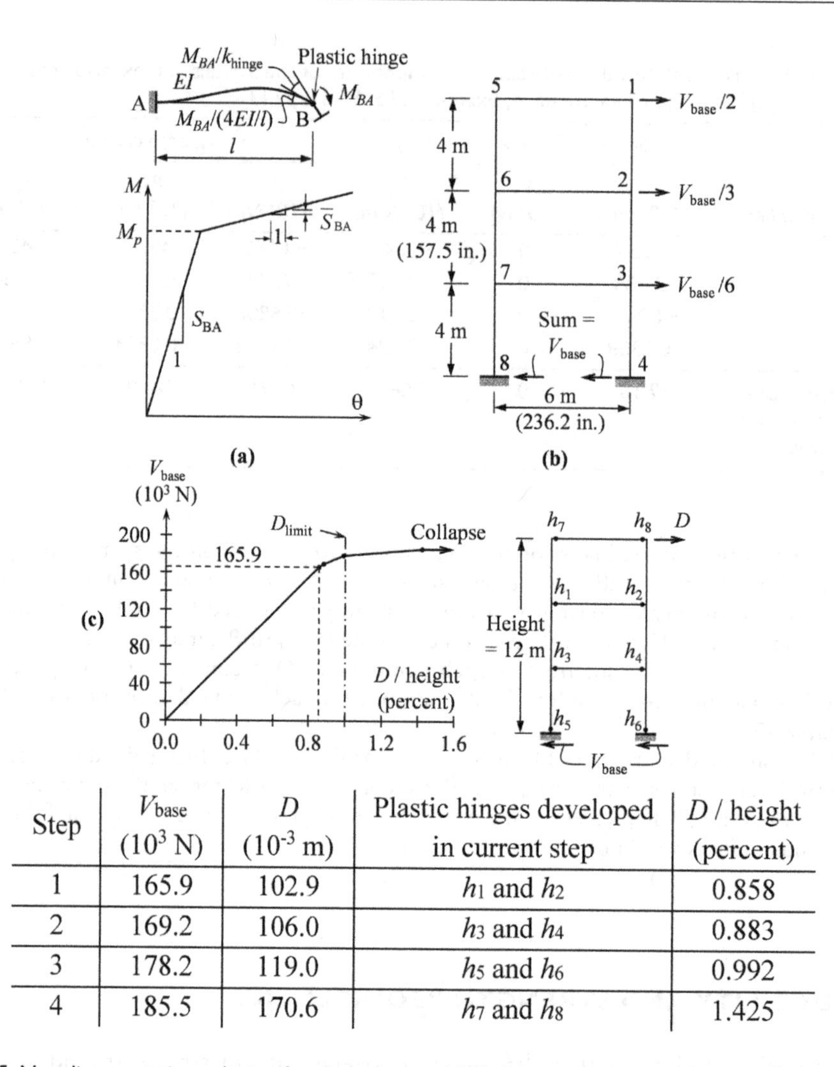

Step	V_{base} (10^3 N)	D (10^{-3} m)	Plastic hinges developed in current step	D / height (percent)
1	165.9	102.9	h_1 and h_2	0.858
2	169.2	106.0	h_3 and h_4	0.883
3	178.2	119.0	h_5 and h_6	0.992
4	185.5	170.6	h_7 and h_8	1.425

Figure 17.15 Non-linear static analysis of a plane frame. (a) End moment-rotation relationship of member AB. (b) Plane frame of Example 17.6: load distribution in terms of base shear, V_{base}. (c) Roof drift, D versus V_{base}; location of plastic hinges consequently developed in four pushover steps.

adopted in many computer programs (e.g. SAP 2000) – is, however, beyond the scope of this chapter.

Figure 17.15a presents the relation between member-end moment M_{BA} and angular rotation θ, where M_{BA} = end moment caused by an earthquake (additional to the initial service moments before the earthquake); θ = corresponding rotation of the joint at end B. The M-θ graph is a straight line with slope $k = dM/d\theta$ until M_{BA} reaches M_p, corresponding to the fully plastic state. Beyond M_p, the relationship is linear with $dM/d\theta = \alpha_p\, k$, the end rotational stiffness of the member having a plastic hinge at a section adjacent to the joint. The following section discusses analysis of plane frames having the bilinear relationship in Figure 17.15a.

When allowing plastic deformation in design, the P-delta effect of axial forces in the columns (Chapter 10) can be more significant compared to design that maintains elastic behavior.

17.18 NONLINEAR STATIC (PUSHOVER) ANALYSIS

Elastic analysis of structures – allowed by most codes – gives elastic deformations and an indication where yielding might first occur in members; however, such analysis does not account for redistribution of forces during progressive yielding that follows. A non-linear static (*pushover*) analysis can predict more accurately the structural behavior and deformations as the structure's stiffness decreases with yielding (formation of plastic hinges) of its members (Chapter 14). In the pushover analysis, we push the structure with monotonically increasing lateral force (in predetermined steps and distribution up to a certain value). For nonlinear numerical analysis, a number of members and their failure mechanisms are well-defined, such that the stiffness of the system $S(t)$ can be varied in the time steps. Thus, nonlinear and linear analyses differ in the restoring force, (SD), in the equation of motion (Equation 17.8); in nonlinear analysis, both S and D are time-dependent. In general, calculating the stiffness of the system in the time steps requires much computing.

For regular structures with relatively short natural periods, it is possible to use specified distribution of lateral static forces to investigate earthquake response. The load distribution adopted in codes emulates dynamic behavior (i.e. use of the procedure in Section 17.13 with a specified response spectrum graph, $S_a(T, \zeta)$). In the nonlinear analysis, the mass is constant and the damping characteristics are indeterminate. For simplicity, we employ constant damping and natural period of vibration. We apply load in steps; in step i, the applied forces are:

$$\{\Delta F\}_i = \alpha_{di}\{\Delta F\}_{\text{ref}} \tag{17.72}$$

α_d = ductility parameter used as a variable multiplier of reference forces $\{\Delta F\}_{\text{ref}}$. The reference forces have constant relative values. In each step, the stiffness depends on the plastic deformations at the joints.

The M-θ graph (Figure 17.15a) is for a prismatic member encastré at end A and subjected to end moment M_{BA} at B. The symbol θ = discontinuity angle at the plastic hinge. The initial slope of the bilinear graph is $S_{\text{AB}} = 4EI/l$ (see Figure 9.1). When the value of M_{BA} reaches M_p, the graph becomes a straight line with slope equal to:

$$\bar{S}_{\text{BA}} = \frac{4EI}{l}\left(1 + \frac{4EI/l}{k_{\text{hinge}}}\right)^{-1}; \quad \bar{S}_{\text{BA}} \to 0 \text{ when } k_{\text{hinge}} \to 0 \tag{17.73}$$

where k_{hinge} = value of the end moment, inducing unit angular discontinuity at the plastic hinge. Without the plastic hinge, an end moment $M = 1$ produces end rotation = $1/(4EI/l)$. An additional rotation = $1/k_{\text{hinge}}$ would occur if M were applied with an existing plastic hinge. The reduced end rotational stiffness, $\bar{S}_{\text{BA}}$ = (sum of two angles)$^{-1}$. The carryover moment, at end A = $\bar{t} = \bar{S}_{\text{BA}}/2$. $\bar{S}_{\text{BA}}$ and $\bar{t}$ can be used to generate the stiffness matrix of a member having a plastic hinge at end B. When the stress–strain relationship is idealized by a bilinear graph with zero slope beyond yielding, $k_{\text{hinge}} = 0$ and the M-θ graph in Figure 17.15a will have zero slope beyond M_p. Replacing the fixed support at A in Figure 17.15a with a hinge changes the slope of M-θ graph beyond the formation of the plastic hinge to:

$$\bar{S}_{\text{BA}} = \frac{3EI}{l}\left(1 + \frac{3EI/l}{k_{\text{hinge}}}\right)^{-1}; \quad \bar{S}_{\text{BA}} \to 0 \text{ when } k_{\text{hinge}} \to 0 \tag{17.74}$$

The computer program PLASTICF (Chapter 20) applies concentrated loads, increasing their magnitude in steps, and uses a new stiffness matrix in each step. For pushover analysis, the solution can give the displacements, the base shear, the member end forces, and the location of plastic hinges after each step. The step analysis can be stopped when a target criterion is reached; e.g. when the lateral drift at the roof exceeds a specified limit. The structure should have the strength to carry the sum of the load increments. PLASTICF applies in pushover analysis of steel plane frames having perfect plastic moment-curvature relation (zero slope beyond M_p). For this purpose, we apply monotonically increasing lateral forces, having variable resultant, V_{base} and invariant distribution.

EXAMPLE 17.6: PUSHOVER ANALYSIS OF A MULTI-STORY STRUCTURE MODELED AS A PLANE FRAME

The steel frame in Figure 17.15b is subjected to monotonically increasing lateral forces having resultant = V_{base} and invariant distribution. Determine V_{base} versus the lateral drift of the roof, D. The upper limit of D = height/100. Consider only bending deformation. Use member end moment-rotation relationship in Figure 17.15a, with $k_{hinge} = 0$. Data: Elasticity modulus, $E = 200 \times 10^9$ N/m² (29.0 × 10⁶ lb/in.²). The cross-sectional properties of columns and beams are: Columns: $I = 146.7 \times 10^6$ mm⁴ (352.4 in.⁴); $M_p = 366.0 \times 10^3$ N-m (2.974×10^6 lb-in.); Beams: $I = 95.80 \times 10^6$ mm⁴ (230.2 in.⁴); $M_p = 176.6 \times 10^3$ N-m (1.563×10^6 lb-in.).

Computer program PLASTICF (Chapter 20) gives the results presented in Figure 17.15c. The structure becomes a mechanism (unstable) in the fourth step; this would not be the case when $k_{hinge} > 0$.

EXAMPLE 17.7: PUSHOVER ANALYSIS OF A CONCRETE BRIDGE

Figure 17.16a shows a plane frame idealization of a three-span concrete bridge, with dimensions and cross-sectional properties of its members. Use pushover analysis to calculate $(V_o)_{base}$ corresponding to a limiting displacement, D_o; assume $D_o/D_y = 4$ with D_o and D_y as defined in Figure 17.10. Use spectral acceleration, $S_a(T, \zeta)$ in Figure 17.8, graph b. Assume effective mass (the superstructure plus half the piers) = 2.78×10^6 kg. Other data: $E_c = 27 \times 10^9$ N/m²; $\zeta = 0.05$. Assume that the deck remains elastic; plastic moment for piers, $(M_p)_{AB} = 36 \times 10^6$ N-m and $k_{hinge} = 0.1\ (EI/l)_{AB}$.

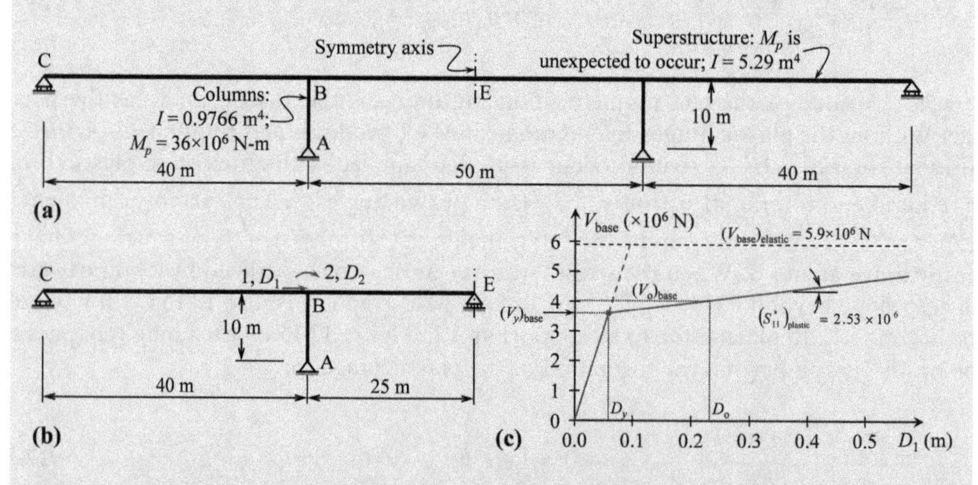

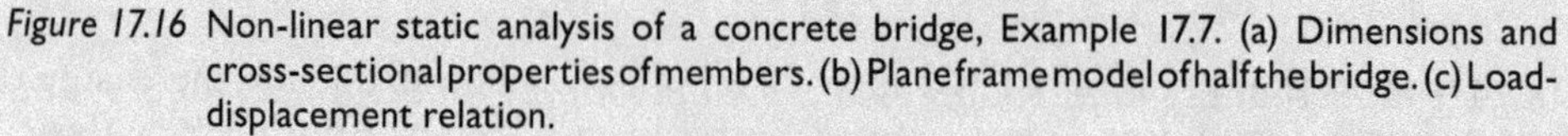
Figure 17.16 Non-linear static analysis of a concrete bridge, Example 17.7. (a) Dimensions and cross-sectional properties of members. (b) Plane frame model of half the bridge. (c) Load-displacement relation.

Consider one-half of the bridge, with the two coordinates shown in Figure 17.16b. The elastic stiffness matrix corresponding to the two coordinates in Figure 17.16b is:

$$[S] = 10^6 \begin{bmatrix} 79.1 & -791 \\ -791 & 35.762 \times 10^3 \end{bmatrix}$$

Since $F_2 = 0$, the condensed stiffness, $S_{11}^* = S_{11} - \left(S_{12}^2/S_{22}\right) = 61.6 \times 10^6$ N/m. Substituting

S_{11}^* and m in Equation 17.6 gives $T = 0.94$ sec. From graph b of Figure 17.8, the corresponding spectral acceleration, $S_a/g = 0.433$. The equivalent base shear, $(V_{base})_{elastic} = 0.433 \times 9.81 \times (0.5 \times 2.78 \times 10^6) = 5.9 \times 10^6$ N; this is a hypothetical shearing force for a structure remaining elastic during the spectral (design) earthquake.

When the bending moment in the pier at B reaches $M_p = 36 \times 10^6$ N-m, the displacement:

$$D_y = \frac{(M_p/l)_{AB}}{S_{11}^*} = \frac{\left(36 \times 10^6/10\right)}{61.6 \times 10^6} = 0.058 \text{ m}$$

With the displacement D_o exceeding the D_y-value (at the end of the elastic range of behavior), the stiffness matrix for the structure with reduced stiffness of the pier is:

$$[S] = \begin{bmatrix} \left(\frac{3EI}{l^3}\right)_{AB} (1+30)^{-1} & -\left(\frac{3EI}{l^2}\right)_{AB} (1+30)^{-1} \\ -\left(\frac{3EI}{l^2}\right)_{AB} (1+30)^{-1} & \left(\frac{3EI}{l}\right)_{AB} (1+30)^{-1} + (3EI)_{beam}\left(\frac{1}{l_{BC}} + \frac{1}{l_{BE}}\right) \end{bmatrix}$$

$$= 10^6 \begin{bmatrix} 2.55 & -25.5 \\ -25.5 & 28.1 \times 10^3 \end{bmatrix}$$

where the value $30 = \left(3EI/l\right)_{AB}/k_{hinge}$. Considering plasticity, the condensed stiffness,

$\left(S_{11}^*\right)_{plastic} = S_{11} - \left(S_{12}^2/S_{22}\right) = 2.53 \times 10^6$ N/m; this is the slope of the graph in Figure

17.16c past the elastic limit. The shearing force at the base corresponding to D_o (=4 $D_y = 0.232$ m), and accounting for ductile pier, $(V_o)_{base}$ (Figure 17.16c):

$$(V_o)_{base} = \left(\frac{M_p}{l}\right)_{AB} + \left(S_{11}^*\right)_{plastic} (D_o - D_y) = 4.04 \times 10^6 \text{ N}$$

From Figure 17.10, $(V_{limit}/V_y)_{base} = D_o/D_y = 4.0$; where $(V_{limit})_{base}$ = hypothetical elastic shearing force corresponding to D_o.

17.18.1 P-delta effect

In an earthquake, gravity load can have significant influence on laterally deformed buildings in their inelastic range. The pushover graph of base shear versus displacement displays constant strength after yielding (e.g. see Figure 17.15c). However, when P-delta effect is included, the graph would show rapid decrease of resistance shortly after yielding.

In Example 17.7, the demanded $(V_o)_{base} = 4.04 \times 10^6$ N, corresponding to $D_o = 0.232$ m, while the reaction at A due to gravity load, $(R_A)_{vertical} = (27.3 \times 10^6)$ N and pier height 10 m.

With P-delta effect included, the demanded bending moment at $B = 40.4 \times 10^6 + 27.3 \times 10^6 (0.232) = 46.7 \times 10^6$ N-m, and the shear resistance demand $= 4.67 \times 10^6$ N.

The post-yield static behavior indicates that P-delta effect is also significant in response of multi-story building to earthquake excitation (Chopra 2012).

17.18.2 Modal pushover analysis

Consider a multi-story building. Elastic response spectrum analysis gives distribution of lateral static forces, displacements, and forces at floor levels. These are determined using natural frequencies and modes. For convenience, we repeat a few equations, referring to Figure 17.11:

$$V_{r\,\text{base}} = S_{ar}\, L_r^2$$

$$L_r = \sum_{i=1}^{n} (\Phi_{ir}\; m_i)$$

$$\{p_r\} = \frac{V_{r\,\text{base}}}{L_r} \{\Phi_{1r}\, m_1, \Phi_{2r}\, m_2, \cdots, \Phi_{nr}\, m_n\}$$

$$\{D_r\} = \frac{L_r}{\omega_r^2} \{\Phi\}_r\, S_{ar}$$

The subscript r refers to the rth mode. ω_r = natural frequency. $\{\Phi_r\}$ = mode shape vector, normalized such that $\{\Phi\}_r^{\mathrm{T}} [m] \{\Phi\}_r = 1.0$. $[m]$ = diagonal matrix of lumped masses; n = number of degrees of freedom. $\{p_r\}$ = equivalent static forces at floor levels; $V_{r\,\text{base}}$ = shearing force at base. $\{D_r\}$ = lateral floor displacements relative to base. S_{ar} = pseudo-acceleration spectrum, corresponding to the natural vibration period, damping ratio, and plasticity.

The nonlinear modal pushover analysis is a set of static analyses in which the lateral force-resisting system is subjected to monotonically increasing lateral forces, having resultant $V_{r\,\text{base}}$ and invariant distribution identical to that of $\{p_r\}$. The pushover analysis is repeated for as many modes as required. The modal contribution Equation 17.70 or 17.71a gives an estimate of the total peak response. When plastic range is reached, the use of S_{ar} and distribution of lateral forces of an elastic system involves approximation. This analysis is acceptable for many structures.

17.18.3 Limitations

The static non-linear pushover analysis predicts the structural behavior and deformations of a structure designed for ductility. Stiffness of the structure decreases with the development of plastic hinges. Analysis determines the variation of displacement, D versus pushover force. In practice, the displacement corresponding to the design base shear – calculated from the site acceleration spectrum, $S_a(T, \zeta)$ (Section 17.18.2) divided by a *strength reduction factor*, R – is compared with a pre-determined ductility limit; the R-factor is specified by applicable code to account for inelastic behavior of the structural system.

Examples 17.6 and 17.7 apply the pushover procedure to two structural systems differing in their characteristics of response to earthquakes. The analysis considers the stiffness of a ductile structure, induced by the development of flexural plastic hinges only. In Example 17.6, the analysis gives the load level and location of each plastic hinge, without considering the level of damage at each hinge. While in Example 17.7, we calculated the base shear corresponding to a specified limit for the ductile displacement, D_o.

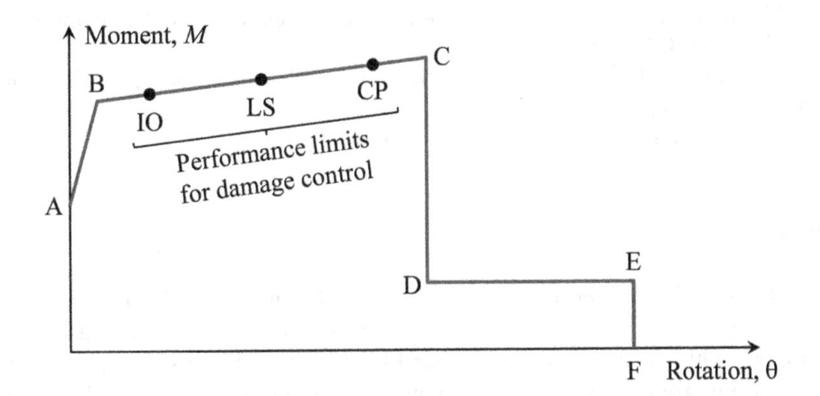

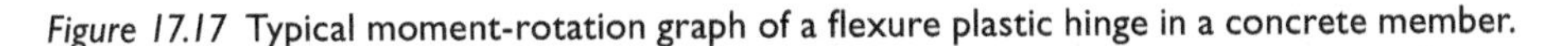

Figure 17.17 Typical moment-rotation graph of a flexure plastic hinge in a concrete member.

Certain practice considers three types of hinges: flexural, shear, and axial. During an earthquake, the flexural and shear hinges can develop at ends of beams and columns. Reversals of bending moment and shearing force typically cause cross diagonal cracks, for which flexural and shear reinforcements need proper detailing. Examples of axial hinges are masonry infill in concrete frames (modeled as compressive struts) and cable cross-braces (modeled to resist tensile forces only).

Codes provide rules for detailing members to ensure ductile performance and to prevent occurrence of non-ductile deformation prior to achieving the required ductility (D_o in Figures 17.10, 17.15c and 17.16c); e.g. the piers in Example 17.7 need reinforcement amount and detailing to ensure that the shearing force and bending at the plastic hinge do not induce collapse.

Figure 17.17 shows a typical moment-rotation relation of a flexural plastic hinge. Line BC represents linear response with reduced stiffness. Codes define three performance limits over the ductile range BC: Immediate Occupancy (IO), Life Safety (LS) and Collapse Prevention (CP); these are damage control limits, which depend upon a design philosophy of retrofitting or replacement following an earthquake event. Shear failure or concrete crushing induce sudden reduction in moment capacity, represented by line CD. Line DE represents remaining capacity due to strain hardening, followed by total loss of capacity (EF). Behavior of plastic hinges is dependent upon the amount and detailing of the reinforcement.

17.19 GENERAL

This chapter considers the response of uniform buildings, idealized as plane frames. We consider displacements of structures and ground motion in horizontal direction. Damping and plasticity reduce significantly the strength requirement; the coverage of the two subjects is limited in this book. The present chapter, combined with Chapter 16, treats the main methods needed for analyzing response of structures to earthquakes.

REFERENCES

Chopra, A.K., *Dynamics of Structures: Theory and Applications to Earthquake Engineering*, 4th ed., Prentice Hall, Pearson, 2012, 944 pp.

Ghali, A. and Neville, A.M., *Structural Analysis: A Unified Classical and Matrix Approach*, 7th ed., CRC Press, Taylor & Francis Group, London, 2017, 933 pp.

Newmark, N.M. and Hall, W.J., *Earthquake Spectra and Design*, Earthquake Engineering Research Institute, Berkeley, CA, 1982.

Priestly, M.J., Seible, F. and Calvi, G.M., *Seismic Design and Retrofit of Bridges*, John Wiley & Sons, London, 1996, 686 pp.

SAP 2000, *Software*, Computers and Structures, Berkeley, CA.

PROBLEMS

Solve the problems in SI or Imperial units. Conversion of units: 1 m = 39.710 in.; 1 N = 1 kg m/s^2 = 0.22481 lb; 1 kg = 5.7115×10^{-3} lb-s^2/in. The answers are given in SI and Imperial units. The mode matrix $[\Phi]$ is normalized such that $\{\Phi\}^T [m] \{\Phi\} = 1$; thus, $[\Phi]_{\text{Imperial}} \approx 13.2\ [\Phi]_{\text{SI}}$. In SI units, N and m are used for force and length; in Imperial units, lb and in. are adopted.

17.1 Determine the peak drift and the bending moment at the base of the single-degree-of-freedom system in Figure 17.5 subjected to horizontal ground motion. Use the spectral design graph in Figure 17.8a. Data: $m = 1000$ kg (5.710 lb s^2/in.); $h = 18$ m (709 in.); $EI = 480\ 10^6$ N-m^2 (167×10^6 kip-in.^{2}); $g = 9.81$ m/s^2 (386 in./s^2).

17.2 Determine the equivalent static forces at the coordinates of the vertical beam in Figure 16.8b in response to an earthquake. Use a generalized single-degree-of-freedom system (Section 16.11). Assume $Wl^2/(gEI) = 950 \times 10^{-6}$ s^2. Consider a fundamental mode as the translations due to nodal forces = W {4, 1, 1}. Use $[S]$ of Example 16.7 and (S_a/g) from Figure 17.8a.

17.3 Consider a hollow cylindrical concrete tower fixed at the base; from the base, its thickness varies linearly with height. Find the peak displacement at the top and the shearing force and bending moment at the base, using generalized single-degree-of-freedom system and spectral acceleration from Figure 17.8b. Assume shape function, $\Psi = 1 - \cos(\xi\pi/2)$; $\xi = x/l$, where x = distance from base, l = height of tower; its mass/unit length, $m(\xi) = m_0\,(1 - \xi/2)$; $EI(\xi) = EI_0\,(1 - \xi/2)$. Data: $EI_0 = 900 \times 10^9$ N-m^2 (313.6×10^{12} kip-in.^{2}); $l = 60.00$ m (2362 in.); $m_0 = 22620$ kg/m (3.2815 lb-s^2/in.); $g = 9.81$ m/s^2 (386 in./s^2).

17.4 Find the natural frequencies $\{\omega_1, \omega_2\}$ and the normalized mode matrix $[\Phi]$ for the plane frame shown.

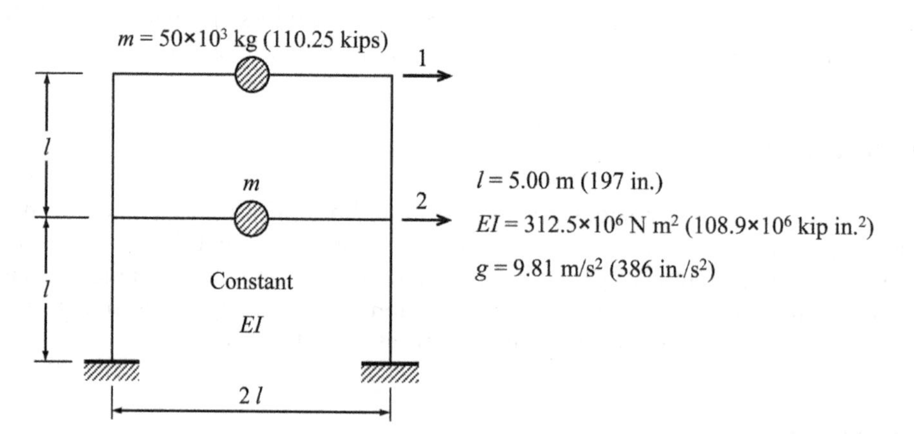

17.5 Find the shearing force at the base and the overturning moment for the plane frame of Problem 17.4 in response to the earthquake. Take (S_a/g) from Figure 17.8a. Use the answers of Problem 17.4.

17.6 The plane frame in Figure 17.2 is an idealization of the lateral force-resisting system of a reinforced concrete building, consisting of a solid slab directly supported on columns and a shear wall. Find the square of natural frequencies ω^2 and the mode matrix $[\Phi]$. Data: $n=4$; $l_1=8.000$ m (315.0 in.); $l_2=4.000$ m (157.5 in.); $I_1=0.2880$ m^4 (691.9×10^3 in.^{4}); $I_2=8.500\times10^{-3}$ m^4 (20.42×10^3 in.^{4}); $I_3=32.8$ m^4 (78.80×10^6 in.^{4}). At each slab level, assume a lumped mass $= W/g$; where $W=4.000\times10^6$ N (899.2 kips); $g=9.81$ m/s^2 (386 in./s^2); $E=30\times10^9$ N/m^2 (4.351×10^6 lb/in.^{2}). Computer program PLANEF and EIGEN (Chapter 20) can be used to generate stiffness matrix $[S]_{4\times4}$ and solve the eigenvalue problem.

17.7 For the plane frame of Figure 17.2, determine ω_n^2 using the generalized system, assuming a deflected shape, $\Psi=(3\xi^2-\xi^3)/2$, with $\xi=x/h$; $h=$height of structure $(4l/3)$, $x=$distance above base. Determine ω_1^2 the square of the natural frequency of the fundamental mode. Assume a lumped mass $=m$ (kg) at each floor level such that $ml^3/EI_{beam}=7.0$ (s^2). Find the corresponding normalized vector $\{\Phi\}_1$.

Hint: PLANEF may be used to generate stiffness matrix $[S]_{4\times4}$ and EIGEN to solve eigenvalue problem. In PLANEF, enter 1.0 for each of E, I_{beam}, and I_{column}. Enter 400.0 for I_{wall}. In EIGEN, enter 7.0 for m; the output will give $\{\Phi\}_1$ in SI units.

17.8 Considering only the first mode, find the inter-story drift ratios for the four stories of the plane frame in Figure 17.2. Use the spectral graph in Figure 17.8a for ground motion in the x-direction. The normalized mode vector $\{\Phi\}_1$ and ω_1^2 are given (see answers of Problem 17.6).

17.9 The drift at the four coordinates in Figure 17.2 in Problem 17.8 are 10^{-3} {43.6, 37.9, 27.3, 12.9} m or {1.72, 1.49, 1.07, 0.508} in. Verify that the same drift values can be determined by Equation 17.65 or by application of the nodal forces at coordinates 1 to 4. The stiffness matrix $[S]$ and the normalized mode vector $\{\Phi\}_1$ are given in the answers of Problem 17.6.

Chapter 18

Nonlinear analysis

18.1 INTRODUCTION

In linear analysis of structures, the equations of equilibrium apply to the non-deformed geometry existing before load application. This is sufficiently accurate for many practical cases. However, some structures, such as cable nets and fabrics, trusses, and frames with slender members may have large deformations, so that it is necessary to consider equilibrium in the real deformed configurations. This requires nonlinear analysis involving iteration, using Newton–Raphson technique. The equilibrium equations apply to trial displacement values. Iteration improves accuracy to satisfy the equilibrium of the nodes or the members in their displaced positions or their deformed shapes. Nonlinearity caused by large deformations is *geometric nonlinearity*. We are dealing here with large deformation, but small strain with linear stress–strain relationship. Nonlinearity can also arise when the stress–strain relationship of the material is nonlinear in the elastic or in the plastic range; this is *material nonlinearity*.

A member subjected to a large axial force combined with transverse loads, or to transverse translation or rotation at an end, represents a geometric nonlinear problem. Often, we call this a *problem*, or casually, a *P-delta problem*. In Chapter 10, we solve the beam-column by quasi-linear analysis of plane frames, including calculation of the critical buckling loads. Linear analysis is possible because of the fact that, in the presence of a constant axial force, P, the magnitude of the transverse deflection y due to any pattern of transverse loads $\alpha\{q\}$ varies linearly with the value of α (see the governing differential Equation 10.1).

We note that the P-delta analysis of Chapter 10 ignores the change in length of members associated with the transverse deflection. For example, the left-hand end of the member in Figure 10.2a acquires unit transverse translation without change in the magnitude of the axial force, in spite of the fact that the length of the curved-deflected line is longer than the initial length. Such an assumption may cause inaccuracy when the deformations are very large. This source of error is avoided in the geometric nonlinear analysis presented in this chapter for plane and space trusses, cable nets, and plane frames. The length of a deflected member of a frame is equal to the length of the straight line joining its ends (the chord). In all these cases, the geometric nonlinearity caused by axial forces, particularly prestressing of cables, can be of prime importance.

Material nonlinearity is considered in the computer plastic analysis of plane frames presented in Section 14.6. By assuming a bilinear moment-curvature relationship (Figure 18.3), the structure behaves linearly until the first hinge has developed. Under increasing load, the structure continues to behave linearly, generally with a reduced stiffness, until a second hinge is formed. The same behavior continues under increasing load until sufficient hinges have developed to form the failure mechanism.

DOI: 10.1201/9780429286858-18

In the present chapter, several techniques[1] are presented that can be used in the analysis of structures in general, considering material nonlinearity. Below, we analyze simple trusses and plane frames.

18.2 GEOMETRIC STIFFNESS MATRIX

When the load on a geometrically nonlinear structure is applied in small increments, the force-displacement relation is:

$$[S]\cdot\{\Delta D\}=\{\Delta F\} \tag{18.1}$$

where $\{\Delta F\}$ and $\{\Delta D\}$ are increments of forces and displacements at a coordinate system. The stiffness matrix $[S]$ changes with the change in geometry; moreover, in framed structures, $[S]$ depends also upon the axial forces in the members. As $\{\Delta F\}$ tends to $\{0\}$, $[S]$ tends to the *tangent stiffness matrix* $[S_t]$:

$$[S_t]=[S_e]+[S_g] \tag{18.2}$$

where $[S_e]$ is the conventional *elastic stiffness matrix*, based on the initial geometry of the structure at the start of loading increment and $[S_g]$ is the *geometric stiffness matrix*, to be generated in this chapter for prismatic members of plane and space trusses and of plane frames with rigid joints and for triangular membrane elements. $[S_g]$ depends upon the deformed geometry of the structure and the axial forces in the members. The displacements change node positions; thus, $[S_e]$ also varies during the analysis.

The initial geometric stiffness matrix is not nil when the initial axial force is nonzero, as in prestressed cables.

18.3 SIMPLE EXAMPLE OF GEOMETRIC NONLINEARITY

As a simple example, consider a wire stretched between two points with initial tension $N^{(0)}$ (Figure 18.1a). Define one coordinate at the middle as shown. Application of a force Q at the coordinate produces a displacement D and the tension in the cable becomes N. Assuming elastic material, the tension in the cable can be expressed as:

$$N=N^{(0)}+(2Ea/b)\left\{\left[(b/2)^2+D^2\right]^{1/2}-b/2\right\} \tag{18.3}$$

where b = initial length of cable; a = cross-sectional area; E = modulus of elasticity.

Considering equilibrium of node B in the deflected position, we write:

$$Q=2ND\left[(b/2)^2+D^2\right]^{-1/2} \tag{18.4}$$

For this simple structure, with a single degree-of-freedom, assumed value of D and successive use of Equations 18.3 and 18.4 gives a plot of D versus Q (Example 18.1). With multiple degrees-of-freedom, when the cable carries several concentrated forces, we use trial values of $\{D\}$, and check equilibrium by equations similar to Equation 18.4. We reach a solution when by iterative calculations, the force in various parts of the cable, satisfies elastic relation similar to Equation 18.3.

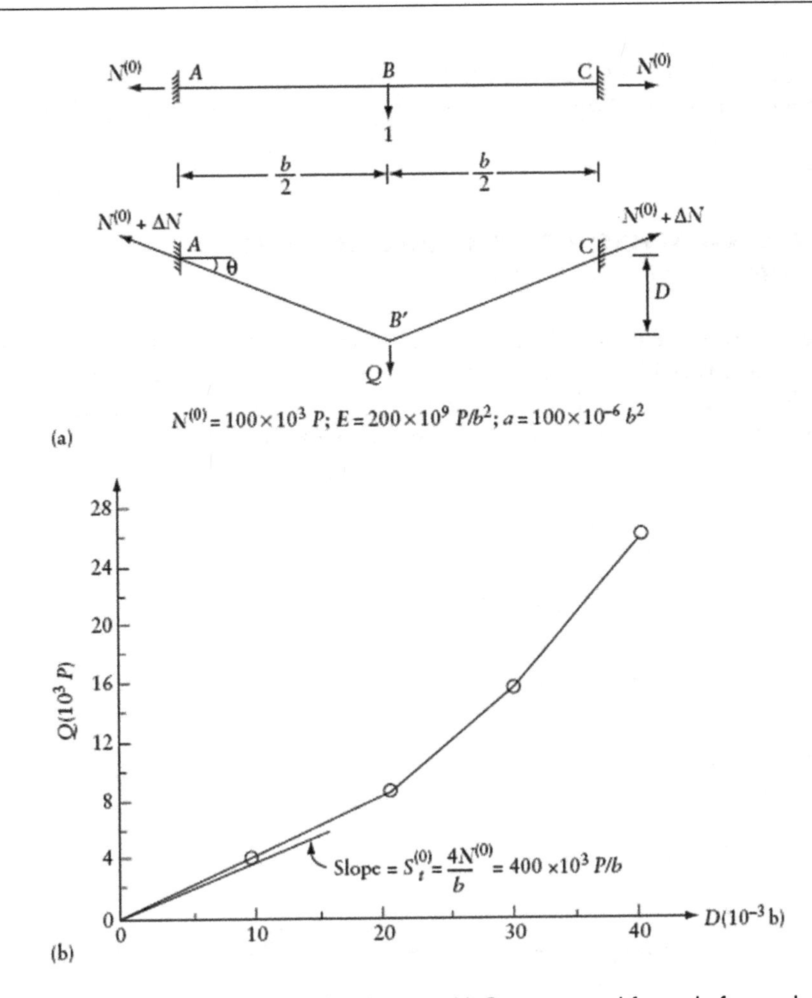

Figure 18.1 A wire stretched between two fixed points. (a) Geometry and forces before and after application of force Q. (b) Plot of force Q versus displacement D.

EXAMPLE 18.1: PRESTRESSED CABLE CARRYING CENTRAL CONCENTRATED LOAD

Plot a graph of D versus Q for the structure in Figure 18.1, assuming $N^{(0)} = 100 \times 10^3 P$, $a = 100 \times 10^{-6} b^2$, and $E = 200 \times 10^9 P/b^2$, where P and b are, respectively, force and length units. What are the initial values $S_g^{(0)}$ and $S_t^{(0)}$?

Substitution of selected values of D in Equations 18.3 and 18.4 gives the following values of Q, which are plotted in Figure 18.1b:

						Multiplier:
D	0	10	20	30	40	$10^{-3}b$
Q	0	4.159	9.272	16.287	26.140	10^3P

The tangent stiffness S_t, that is, the slope of the graph Q-D, increases with an increase in D. When $D = 0$, $N = N^{(0)}$ and the tangent stiffness is (by differentiation of Equation 18.4):

$$S_t^{(0)} = \frac{4N^{(0)}}{b} \tag{18.5}$$

In this case, $S_e^{(0)} = 0$ and $S_t^{(0)} = S_g^{(0)} = 4N^{(0)}/b = 400\times10^3\ P/b$.

Use program NLST (Chapter 20) to solve the same example as assembly of triangular membrane elements.

18.4 NEWTON–RAPHSON TECHNIQUE: SOLUTION OF NONLINEAR EQUATIONS

In the present section, we present Newton–Raphson technique in solution of nonlinear equations with a single variable or with n variables. For this purpose, consider the symmetrical plane truss in Figure 18.2a. Define a system composed of a single coordinate as shown. It is required to find the displacement D, for a given value of the force F.

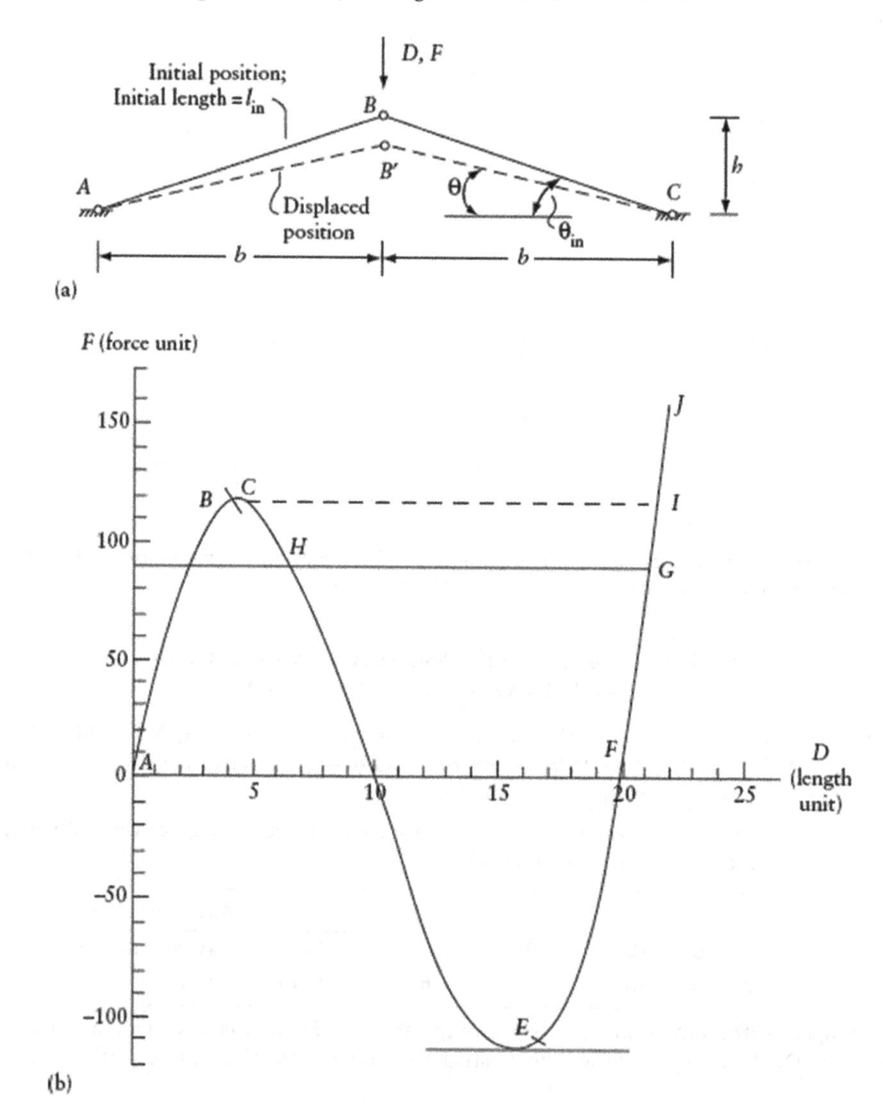

Figure 18.2 Geometric nonlinearity of a plane truss with a single degree-of-freedom. (a) Plane truss. Example values: $b=300$ and $h=10$ (length unit); $Ea=8\times10^6$ (force unit). (b) Force versus displacement at coordinate 1.

We plot the D-F diagram in Figure 18.2b, using assumed values of D and determining the value of F by equilibrium equation of the displaced node B′ (Figure 18.2a):

$$F = -2N \sin\theta \tag{18.6}$$

where

$$\sin\theta = (h - D)/l \tag{18.7}$$

$$l = \left[b^2 + (h - D)^2\right]^{1/2} \tag{18.8}$$

$$N = Ea\,(l - l_{in})/l_{in} \tag{18.9}$$

Here, N = axial force in the members; θ = angle between the horizontal and the displaced position of member (B′C); l_{in} and l = initial length (BC) and the length after deformation (B′C) of each member; E = modulus of elasticity; a = cross-sectional area. The employed stress–strain relation is linear.

In a multi-degree-of-freedom system with n unknown displacements, it is not simple to assume a set of displacements $\{D\}$ to satisfy equilibrium and geometry equations similar to Equations 18.6 and 18.9 at all nodes and for all members. We solve the problem by iteration, explained first with reference to the single-degree-of-freedom system in Figure 18.2a.

In the present section, we limit our discussion to the case when h is small compared to l_{in} (Figure 18.2a); the angle θ_{in} between the horizontal and the initial position of BC is small and the geometric nonlinearity is too important to be ignored. Crisfield[2] gives the following approximate relation between F and D:

$$F \approx \left(Ea/l_{in}^3\right)\left(2h^2 D - 3hD^2 + D^3\right) \tag{18.10}$$

We can verify that Equation 18.10 satisfies approximately Equation 18.6 for values of D in the range considered ($0 \le D \le 2.5h$), assuming as an example, $b = 300$ and $h = 10$ (length unit), respectively, and $Ea = 8 \times 10^6$ (force unit). The second moment of area of the cross section is sufficiently large to preclude buckling of individual members.

To determine D corresponding to a specified value of F by Newton–Raphson technique, we rewrite Equation 18.10:

$$g(D) = F - \left(Ea/l_{in}^3\right)\left(2h^2 D - 3hD^2 + D^3\right) = 0 \tag{18.11}$$

For the correct value of D, the function $g(D)$ should be zero. We start by an estimated answer $D^{(0)}$, for which $g(D)$ is a non-zero value representing an out-of-balance force. By recursion equations, we obtain a more accurate estimate of D:

$$D^{(1)} = D^{(0)} + \Delta D \tag{18.12}$$

$$\Delta D = \left[S_t\left(D^{(0)}\right)\right]^{-1} g(D) \tag{18.13}$$

$S_t(D^{(0)})$ = value of the tangent stiffness. When the value of the unknown is $D^{(0)}$, the tangent stiffness is (by differentiation):

$$S_t = \mathrm{d}F/\mathrm{d}D \tag{18.14}$$

Calculating g with a trial value of D and determining an improved estimate of D by Equation 18.12 represents one iteration cycle. We repeat the cycle until the out-of-balance force is sufficiently small. The procedure converges rapidly as demonstrated below.

As an example, consider the structure in Figure 18.2; it is required to find the value of D when $F=90$. The tangent stiffness is (by differentiation of Equation 18.10):

$$S_t = \mathrm{d}F/\mathrm{d}D = \left(Ea/l_{\mathrm{in}}^3\right)\left(2h^2 - 6hD + 3D^2\right) \tag{18.15}$$

This equation applies only for the simple truss in Figure 18.2, with $\theta_{\mathrm{in}} << 1.0$. in Section 18.6, we derive general equations applicable for plane or space trusses.

At $D=0$, $S_t=59.16$. We use this value to obtain the approximate answer, corresponding to linear analysis: $D^{(0)}=(59.16)^{-1}(90)=1.52$. With $D=1.52$, the out-of-balance force and the tangent stiffness are (Equations 18.11 and 18.15):

$$g\left(D^{(0)}\right) = -19.54\,(\text{force unit}); \quad S_t\left(D^{(0)}\right) = 34.23\,(\text{force/length})$$

The improved estimate of D at the end of the first cycle is (Equations 18.12 and 18.13):

$$D^{(1)} = 1.52 - (34.23)^{-1}(-19.54) = 2.09 \quad (\text{length unit})$$

The second and third iteration cycles give, respectively, $D=2.18$ and $D=2.19$. The latter value is accurate to two decimal places.

The above procedure, represented in Figure 18.3, applies to points on the curve AB before reaching the buckling load at point C in Figure 18.b. At this point the iteration fails because $S_t=0$ and its inverse cannot be used in Equation 18.13. If a model of the structure is tested experimentally, it would be possible to follow the path AB by increasing the value of F up to a value just below the buckling load; the structure will then suddenly snap through to reach the equilibrium position represented by point I. By a displacement-controlled experiment, we obtain the path CHEFI.

The same path is followed analytically by appropriate selection of the trial values of D. We can verify that the values $D=6.55$ and $D=21.27$ (length unit) are valid answers corresponding to $F=90$ (force unit) (points H and G in Figure 18.2b).

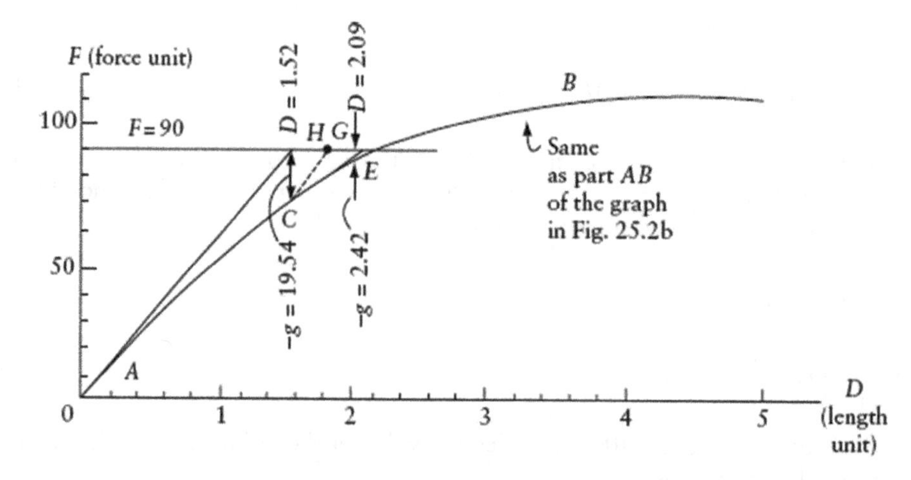

Figure 18.3 The Newton–Raphson iterative procedure applied to the structure in Figure 18.2a. Two iteration cycles are shown; the third iteration converges to the solution $D=2.19$ (point E).

Each of Equations 18.12, 18.13, and 18.14 becomes a system of n equations when we apply Newton–Raphson technique to solve a system of n equations:

$$\{D\}^{(1)} = \{D\}^{(0)} + \{\Delta D\} \tag{18.16}$$

$$\{\Delta D\} = -\left[S_t\left(\{D\}^{(0)}\right)\right]^{-1} \{g\} \tag{18.17}$$

$$[S_t] = \begin{bmatrix} \partial F_1/\partial D_1 & \partial F_1/\partial D_2 & \cdots & \partial F_1/\partial D_n \\ \partial F_2/\partial D_1 & \partial F_2/\partial D_2 & \cdots & \partial F_2/\partial D_n \\ \cdots & & & \\ \partial F_n/\partial D_1 & \partial F_n/\partial D_2 & \cdots & \partial F_n/\partial D_n \end{bmatrix} \tag{18.18}$$

18.4.1 Modified Newton–Raphson technique

In practical nonlinear analysis, we introduce the external forces in stages; Newton–Raphson technique is employed to reduce the out-of-balance forces $\{g\}$ to tolerably small values before introducing a new stage. The tangent stiffness matrix, $[S_t]$ is determined by considering the geometry and forces in the members at the start of a loading stage; the same matrix, $[S_t]$, can be used in the iteration cycles, in order to save computing a new stiffness matrix in each cycle.

For application of Equation 18.17, $[S_t]$ must be nonsingular. The matrix $[S_t]$ can become singular when we reach instability, for example, because of buckling of the system.

Figure 18.3 explains the difference between Newton–Raphson technique and the modified Newton–Raphson technique. Point E on the curve represents the value $D=2.19$ (length unit) corresponding to $F=90$ (force unit) on the structure in Figure 18.2a. The values of D obtained in three iteration cycles by Newton–Raphson technique are (Section 18.4): 1.52, 2.09, 2.18, and 2.19. The tangent at A (Figure 18.3) with slope $S_t^{(0)}=59.16$ intersects the line $F=90$ at $D=1.52$; the line CG with slope (S_t) at $D=1.52$ is equal to 34.23 an intersects the line $F=90$ at $D=2.09$; the second and third iterations cannot be represented in Figure 18.3 because $D=2.09$ is very close to the exact answer.

If the modified Newton–Raphson technique is used, line CG in the first cycle will be replaced by CH, whose slope is 59.16 and the improved value of D at the end of the first cycle will be $[1.52 - (59.16)^{-1}(-19.54)] = 1.85$. More cycles will be required to reach the correct answer. We can verify the following answers for D, determined in seven cycles: 1.52, 1.85, 2.00, 2.08, 2.13, 2.15, 2.17, and 2.18.

18.5 NEWTON–RAPHSON TECHNIQUE APPLIED TO TRUSSES

The analysis discussed below applies to plane and space trusses and cable nets. Pre-tensioned cable nets behave as trusses, when the cables remain in tension after nodal displacements. As usual, we apply the external forces only at the nodes. Thus, we apply the self-weight of a truss member or a cable as concentrated forces at the nodes. With this assumption, the cable consists of straight segments between the nodes. The tangent stiffness of a cable segment is the same as that of a truss member with the same tensile force, cross-sectional area and length, but because a cable segment cannot carry a compressive force, its tangent stiffness is null whenever, in the analysis, the force in the segment becomes compressive.

Consider a plane or a space truss with m members and n degrees-of-freedom, representing nodal displacements in directions of global axes (x, y) or (x, y, z); n is two or three times the number of nodes in a plane truss or a space truss, respectively. It is required to find the nodal displacements $\{D\}$ and the change in member forces $\{\Delta N\}$ due to external applied loads and temperature variations, considering geometric nonlinearity.

Assume that, in the initial state, the nodal displacements $\{D\}_{n\times 1} = \{0\}$, the member lengths $= \{l_{\text{in}}\}_{m\times 1}$, and the forces in the members $= \{N_{\text{in}}\}_{m\times 1}$; the initial member forces may be caused, for example, by prestressing or by initial nodal forces $\{F_{\text{in}}\}$. The analysis discussed below is not concerned with the displacements occurring during the initial prestressing or nodal forces to produce the forces $\{N_{\text{in}}\}$. Unlike linear analysis, the initial member forces are required in the input data because the tangent stiffness of individual members depends upon the geometry and material properties as well as on the magnitude of the axial forces (see Section 18.6).

As in linear analysis (Section 4.6), the forces $\{F\}$ necessary to prevent the nodal displacements are determined. For the consideration of geometric nonlinearity, the forces at the nodes must be in equilibrium in their displaced positions. This condition is expressed by:

$$-[G]^{\mathrm{T}}\{N\} = -\{F\} + \{F_{\text{in}}\} \tag{18.19}$$

$[G]_{m\times n}$ = geometry matrix composed of one row for each member. The kth row contains the components, in the directions of the global axes of the two equal and opposite forces exerted by the member, on the two nodes i and j at its ends, when $\Delta N_k = 1$. Thus, the kth row of $[G]$ is:

$$\begin{matrix} & i & & j & \\ [\cdots & [t]_k & \cdots & [-t]_k & \cdots] \end{matrix} \tag{18.20}$$

This row is composed of submatrices whose number equals the number of nodes; all submatrices except the ith and the jth are null. The submatrix $[t]_k$ is the transformation matrix given by Equations 19.17 and 19.18 and repeated here:

$$[t]_k = \begin{bmatrix}\lambda_{x^*x} & \lambda_{x^*y}\end{bmatrix}_k \quad \text{(plane truss member } k\text{)} \tag{18.21}$$

and

$$[t]_k = \begin{bmatrix}\lambda_{x^*x}, \lambda_{x^*y}, \lambda_{x^*z}\end{bmatrix}_k \quad \text{(space truss member } k\text{)} \tag{18.22}$$

where λ_{x^*x}, λ_{x^*y} and λ_{x^*z} = cosines of the angles between the global axes x, y, and z and the vector x^* joining node i to node j in their displaced positions (Figure 19.2). Thus, $[G]$ depends upon the unknown displacements $\{D\}$.

Assuming that the material is linearly elastic, the force in any member is (positive when tensile):

$$N = N_{\text{in}} + \Delta N; \quad \Delta N = N_r + \frac{Ea}{l_{\text{in}}}\left(l - l_{\text{in}}\right) \tag{18.23}$$

where E = modulus of elasticity; a = cross-sectional area; l_{in} and l = member lengths before and after the displacements of the nodes; N_r = force in the member due to temperature variation with the nodal displacements artificially restrained.

No approximation needs to be involved in calculating the change in length of members; Equation 19.3 or 19.6 can be applied to give l_{in} and l, using the x, y and z coordinates of the nodes in their initial and displaced positions, respectively. Thus, $\{N\}$ generated by Equation 18.23 is dependent upon the unknown displacements $\{D\}$.

We employ the iterative Newton–Raphson technique to determine $\{D\}$ to satisfy Equation 18.19. Each iteration cycle involves the calculations specified below.

18.5.1 Calculations in one iteration cycle

The cycle starts with trial displacements $\{D^{(0)}\}$ and the corresponding out-of-balance forces $\{g^{(0)}\}$. The cycle terminates with more accurate displacements $\{D^{(1)}\}$ and smaller out-of-balance forces $\{g^{(1)}\}$. A cycle is complete in four steps:

Step 1 Generate the structure stiffness matrix $\left[S_t^{(0)}\right]$, expressed as the sum of the elastic stiffness matrix $\left[S_e^{(0)}\right]$ and the geometric stiffness matrix $\left[S_g^{(0)}\right]$. Both matrices are dependent upon the displacements $\{D^{(0)}\}$; the latter matrix also depends on member forces $\{N^{(0)}\}$. The superscript (0) refers to values known at the start of the current cycle. The geometric stiffness matrices of individual members of plane and space trusses are derived in Section 18.6. Assemblage of the stiffness matrices of members to obtain the stiffness matrix of the structure is discussed in Section 19.9.

Step 2 Determine the displacement increments $\{\Delta D\}$ by solving:

$$\left[S_t^{(0)}\right]\{\Delta D\} = \left\{g^{(0)}\right\} \tag{18.24}$$

Step 3 Calculate the new, more accurate displacements:

$$\left\{D^{(1)}\right\} = \left\{D^{(0)}\right\} + \{\Delta D\} \tag{18.25}$$

Using the new displacements $\{D^{(1)}\}$ and Equation 18.23, generate new member forces $\{N^{(1)}\}$. For this purpose, the (x, y) or (x, y, z) coordinates of the nodes in their new displaced positions must be calculated.

Step 4 Compute the new out-of-balance forces:

$$\left\{g^{(1)}\right\} = -\{F\} + \left[G^{(1)}\right]^{\mathrm{T}}\left\{N^{(1)}\right\} + \{F_{\mathrm{in}}\} \tag{18.26}$$

Here, $[G^{(1)}]$ and $\{N^{(1)}\}$ are to be generated by Equations 18.20 and 18.23, respectively, based on the new position of the nodes (with displacements $D^{(1)}$). This terminates the iteration cycle.

If the out-of-balance forces calculated at the end of the cycle are sufficiently small, the analysis is complete. If not, a new cycle is started with $\{D^{(0)}\}$ and $\{g^{(0)}\}$ equal to the improved values determined at the end of the preceding cycle.

18.5.2 Convergence criteria

We may terminate the iteration cycles discussed above when the following criterion is satisfied:

$$\left(\{g\}^{\mathrm{T}}\{g\}\right)^{1/2} \le \beta\left(\{F\}^{\mathrm{T}}\{F\}\right)^{1/2} \tag{18.27}$$

where β = tolerance value, say, 0.01 to 0.001. This criterion ensures that the out-of-balance forces are small compared to the forces $\{F\}$.

When the analysis is for the effect of prescribed displacements, Equation 18.27 cannot be used because $\{F\}$ can be equal to $\{0\}$, while the non-zero forces are only at the coordinates where the displacements are prescribed (including the supports). In such a case, we may terminate the iteration cycles when:

$$\left(\{g\}^{\mathrm{T}}\{g\}\right)^{1/2} \leq \beta\left(\{R\}^{\mathrm{T}}\{R\}\right)^{1/2} \tag{18.28}$$

where $\{R\}$ = vector of the forces at the coordinates for which the displacements are prescribed, including the support reactions.

As mentioned in Section 18.4.1, in practical use of nonlinear analysis, the load is introduced in stages, with iterations and convergence achieved in each stage. Slow convergence or the lack of it, indicates instability, which can be monitored by looking for a negative or zero value on the diagonal element of the tangent stiffness matrix $[S_t]$ during the Gauss elimination process to solve Equation 18.24. Thus, by introducing the load in stages, the analysis can give a range of the load level in which buckling occurs.

The Newton–Raphson rapid convergence occurs when the trial values are near the correct answer. This is why convergence occurs more quickly when the load is applied in increments; in some problems, convergence may fail when the full load is applied in one stage.

As mentioned above, each iteration cycle is started with trial displacements $\{D^{(0)}\}$, which are taken as displacements determined in the preceding iteration cycle. In the first cycle of the first load stage, the starting displacements may simply be $\{D^{(0)}\} = \{0\}$.

18.6 TANGENT STIFFNESS MATRIX OF A MEMBER OF PLANE OR SPACE TRUSS

The tangent stiffness matrix required in the Newton–Raphson iteration is an assemblage of the tangent stiffness matrices of individual members. We derive below the tangent stiffness matrix $[S_t]_m$ of a member of a space truss; deletion of appropriate rows and columns from this matrix results in the tangent stiffness matrix for a plane truss member.

Figure 18.4a represents a member of a space truss and a system of coordinates in directions of global axes x, y and z. It is required to determine the tangent stiffness matrix relating small increment of forces $\{\Delta F\}$ to the corresponding displacements $\{\Delta D\}$:

$$[S_t]_m \{\Delta D\} = \{\Delta F\} \tag{18.29}$$

Before application of $\{\Delta F\}$, the member is assumed to have an axial force N, positive when tensile. We derive first the tangent stiffness matrix $[S_t]$ for the member AB with end B supported as shown in Figure 18.4b.

The following equilibrium equation applies to node A treated as a free body (Figure 18.4c):

$$[t]^{\mathrm{T}} N = -\{F\} \tag{18.30}$$

where $[t]$ is the transformation matrix (Equation 19.18):

$$[t] = \left[\lambda_{x^*x}, \lambda_{x^*y}, \lambda_{x^*z}\right] \tag{18.31}$$

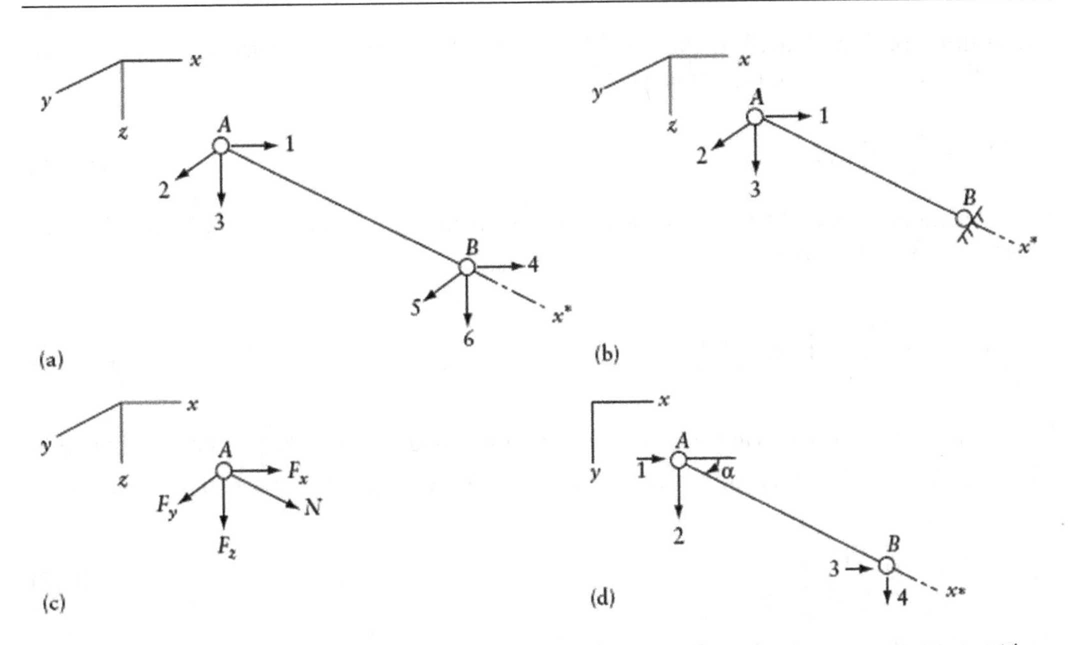

Figure 18.4 Derivation of the geometric stiffness matrices for a member of a plane or space truss with an axial force N. (a) Coordinate system for a space truss member. (b) Same as (a) but with end B supported. (c) Free-body diagram of node A. (d) Coordinate system for a plane truss member.

The terms λ are cosines of the angles between the member local axis x^* in its position before application of $\{\Delta F^*\}$ and the global axes x, y and z.

The definition of the tangent stiffness matrix is:

$$\left[\bar{S}_t\right]=\left[\frac{\partial}{\partial\{D\}}\{F\}^{\mathrm{T}}\right]^{\mathrm{T}} \tag{18.32}$$

We rewrite this equation as:

$$\left[\bar{S}_t\right]=\begin{bmatrix}\partial F_1/\partial D_1 & \partial F_1/\partial D_2 & \partial F_1/\partial D_3\\ \partial F_2/\partial D_1 & \partial F_2/\partial D_2 & \partial F_2/\partial D_3\\ \partial F_3/\partial D_1 & \partial F_3/\partial D_2 & \partial F_3/\partial D_3\end{bmatrix} \tag{18.33}$$

Substitution of Equation 18.30 into 18.32 and performing the differentiation gives $\left[\bar{S}_t\right]=\left[\bar{S}_e\right]$ when $[t]$ is considered constant. However, when $[t]$ is considered variable, the result will be $\left[\bar{S}_t\right]=\left[\bar{S}_e\right]+\left[\bar{S}_g\right]$. In linear analysis, we assume $[t]=$constant and employ the elastic stiffness:

$$\left[\bar{S}_e\right]=\left[-\frac{\partial N}{\partial\{D\}}\{t\}\right]^{\mathrm{T}} \tag{18.34}$$

where $\partial N/\partial\{D\}=\{\partial N/\partial D_1, \partial N/\partial D_2, \partial N/\partial D_3\}$. For elastic material, we have for a truss member:

$$\frac{\partial N}{\partial\{D\}}=-\frac{Ea}{l}[t]^{\mathrm{T}} \tag{18.35}$$

Substitution of Equation 18.35 into 18.34 gives the elastic stiffness matrix, used in conventional linear analysis (see Equation 19.29):

$$\left[\bar{S}_e\right] = \frac{Ea}{l}[t]^{\mathrm{T}}[t] \tag{18.36}$$

Considering both N and $[t]$ as variables and substituting Equation 18.30 in 18.32 gives the tangent stiffness matrix:

$$\left[\bar{S}_t\right] = \left[-\frac{\partial N}{\partial\{D\}}[t] - N\frac{\partial[t]}{\partial\{D\}}\right]^{\mathrm{T}} \tag{18.37}$$

The first term on the right-hand side of this equation is $\left[\bar{S}_e\right]$. Recognizing that $\left[\bar{S}_t\right] = \left[\bar{S}_e\right] + \left[\bar{S}_g\right]$ (see Equation 18.2), we conclude that the geometric stiffness matrix is:

$$\left[\bar{S}_g\right] = \left[-N\frac{\partial[t]}{\partial\{D\}}\right]^{\mathrm{T}} \tag{18.38}$$

Substitution of Equation 18.31 in this equation gives:

$$\left[\bar{S}_g\right] = -N\begin{bmatrix} \partial\lambda_{x^*x}/\partial D_1 & \partial\lambda_{x^*x}/\partial D_2 & \partial\lambda_{x^*x}/\partial D_3 \\ \partial\lambda_{x^*y}/\partial D_1 & \partial\lambda_{x^*y}/\partial D_2 & \partial\lambda_{x^*y}/\partial D_3 \\ \partial\lambda_{x^*z}/\partial D_1 & \partial\lambda_{x^*z}/\partial D_2 & \partial\lambda_{x^*z}/\partial D_3 \end{bmatrix} \tag{18.39}$$

The following geometry relations apply:

$$\lambda_{x^*x} = \frac{l_x}{l};\quad \lambda_{x^*y} = \frac{l_y}{l};\quad \lambda_{x^*z} = \frac{l_z}{l};\quad l = \sqrt{l_x^2 + l_y^2 + l_z^2} \tag{18.40}$$

$$\frac{\partial}{\partial D_1} = -\frac{\partial}{\partial l_x};\quad \frac{\partial}{\partial D_2} = -\frac{\partial}{\partial l_y};\quad \frac{\partial}{\partial D_3} = -\frac{\partial}{\partial l_z} \tag{18.41}$$

where l_x, l_y and $l_z = x$, y, and z components of a vector of magnitude l directed from node A to node B (Figure 18.4b).

Substitution of Equations 18.40 and 18.41 into 18.39 gives the geometric stiffness matrix for the space truss member in Figure 18.4b:

$$\left[\bar{S}_g\right] = \frac{N}{l}\begin{bmatrix} 1-\lambda_{x^*x}^2 & & \text{Symm.} \\ -\lambda_{x^*x}\lambda_{x^*y} & 1-\lambda_{x^*y}^2 & \\ -\lambda_{x^*x}\lambda_{x^*z} & -\lambda_{x^*y}\lambda_{x^*z} & 1-\lambda_{x^*z}^2 \end{bmatrix} \tag{18.42}$$

The forces at ends A and B are equal and opposite. We use the matrices $\left[\bar{S}_e\right]$ and $\left[\bar{S}_g\right]$ for the supported member in Figure 18.4b to generate the stiffness matrices for the space and plane truss members in Figure 18.4a and d. By deleting rows and columns associated with the coordinates in the z direction, the stiffness matrices for a space member become matrices for a member of a plane truss situated in the x–y plane. For easy reference, the member stiffness matrices required for the nonlinear analysis of space and plane trusses are:

For a member of a space or a plane truss, the tangent stiffness matrix is (Figure 18.4a or d):

$$[S_t]_m = [S_e]_m + [S_g]_m \tag{18.43}$$

$$[S_e]_m = \begin{bmatrix} [\bar{S}_e] & -[\bar{S}_e] \\ -[\bar{S}_e] & [\bar{S}_e] \end{bmatrix} \quad ; \quad [S_g]_m = \begin{bmatrix} [\bar{S}_g] & -[\bar{S}_g] \\ -[\bar{S}_g] & [\bar{S}_g] \end{bmatrix} \tag{18.44}$$

For a space truss member, the elastic and geometric stiffness submatrices are:

$$[\bar{S}_e] = \frac{Ea}{l_{\text{in}}} \begin{bmatrix} \lambda_{x^*x}^2 & & \text{Symm.} \\ \lambda_{x^*x}\lambda_{x^*y} & \lambda_{x^*y}^2 & \\ \lambda_{x^*x}\lambda_{x^*z} & \lambda_{x^*y}\lambda_{x^*z} & \lambda_{x^*z}^2 \end{bmatrix}; \quad \text{for } [\bar{S}_g], \text{ see Eq. 18.42} \tag{18.45}$$

where l_{in} is the initial length.

For a plane truss member, the elastic and geometric stiffness matrices are:

$$[\bar{S}_e] = \frac{Ea}{l_{\text{in}}} \begin{bmatrix} c^2 & cs \\ cs & s^2 \end{bmatrix}; \quad [\bar{S}_g] = \frac{N}{l} \begin{bmatrix} s^2 & -cs \\ -cs & c^2 \end{bmatrix} \tag{18.46}$$

where $c = \cos\alpha$ and $s = \sin\alpha$; angle α and its positive sign convention are defined in Figure 18.4d.

EXAMPLE 18.2: PRESTRESSED CABLE CARRYING CENTRAL CONCENTRATED LOAD: ITERATIVE ANALYSIS

Perform two Newton–Raphson iteration cycles to determine, for the prestressed cable in Figure 18.1a, the deflection D at mid-point and the tension N in the cable when the transverse force $Q = 16P$. Use the elastic and the geometric stiffness matrices in Equation 18.46. Take $E = 200 \times 10^6 P/b^2$; $a = 100 \times 10^{-6} b^2$; initial tension $= 100P$; where P and b are, respectively, force and length units.

Because of symmetry, the system has one degree-of-freedom shown in Figure 18.1a; use of Equation 18.46 gives the following stiffness for the structure:

$$S_e = \frac{4Ea}{b}\sin^2\theta; \quad S_g = 2\left(\frac{N}{l}\right)_{AB'}\cos^2\theta$$

For this structure, the out-of-balance force is (Equation 18.26)

$$g = Q - 2\, N_{AB'} \sin\theta$$

The angle θ is defined in Figure 18.1a.

We start the cycles with $D = 0$ and $g = Q$. The four calculation steps specified in Section 18.5.1 are:

ITERATION CYCLE 1

1. $D^{(0)} = 0$; $g^{(0)} = 16P$; $N_{AB'}^{(0)} = 100P$; $\theta^{(0)} = 0$; $\sin\theta^{(0)} = 0$; $\cos\theta^{(0)} = 1$; $l_{AB'} = 0.5b$; $S_e = 0$; $S_g = 400P/b$; $S_t = 0 + (400P/b) = 400P/b$.
2. $\Delta D = (400/b)^{-1}(16) = 0.04b$.
3. $D^{(1)} = 0 + 0.04b = 0.04b$.

4. Equation 18.23 gives $N_{AB'} = 163.9P$; $\sin\theta^{(1)} = 0.04[(0.04)^2 + (0.5)^2]^{-1/2} = 79.7 \times 10^{-3}$; $g^{(1)} = -10.140P$.

ITERATION CYCLE 2

1. $D^{(0)} = 0.04b$; $g^{(0)} = -10.140P$; $N^{(0)}_{AB'} = 163.9P$; $\sin\theta^{(0)} = 79.7 \times 10^{-3}$; $\cos\theta^{(0)} = 0.9968$; $l_{AB'} = 0.5016b$; $S_e = 508.7P/b$; $S_g = 649.4P/b$; $S_t = 1158.1P/b$.
2. $\Delta D = (1158.1/b)^{-1}(-10.140) = -8.76 \times 10^{-3}b$.
3. $D^{(1)} = (0.04 - 8.76 \times 10^{-3})b = 31.24 \times 10^{-3}b$.
4. Equation 18.23 gives $N_{AB'} = 139.0P$; $\sin\theta^{(0)} = 62.3 \times 10^{-3}$; $g^{(1)} = -1.330P$.

A third iteration gives: $D = 29.7 \times 10^{-3}b$; $N_{AB'} = 135.3P$. The exact answers satisfying Equations 18.3 and 18.4 are: $D = 29.6 \times 10^{-3}b$; $N = 135.2P$.

18.7 NONLINEAR BUCKLING

Consider the plane truss in Figure 18.2a and assume that its individual members have sufficient flexural rigidities such that buckling of individual members is precluded. In the analysis in Section 18.4, we saw that the system becomes unstable and the structure snaps through when point C on the *F-D* graph in Figure 18.2b is reached. This corresponds to $D = 4.23$ (length unit) and $F = 114$ (force unit). Thus, we conclude that the critical buckling value of F for this system is $F_{cr} = 114$ (force unit).

The tangent stiffness corresponding to the single coordinate shown at B in Figure 18.2a is (Equations 18.6 and 18.46):

$$S_t = S_e + S_g \tag{a}$$

$$S_t = \frac{2Ea}{l}s^2 + \left[-\frac{F}{sl}\right]c^2 \tag{b}$$

where $s = \sin\theta$ and $c = \cos\theta$, with θ being the angle between B′C (position of BC in the displaced position) and the horizontal. The buckling load F_{cr} is the value of F that makes S_t = zero; thus, we can write:

$$F_{cr} = 2Ea\frac{s^3}{c^2} \tag{c}$$

We can verify that this equation gives $F_{cr} = 114$ (force unit) for $\theta = \tan^{-1}[(h - D)/b]$, with $h = 10$, $b = 300$ and $D = 4.23$ (length unit).

The value of F_{cr} based on the displaced positions of the nodes is the *nonlinear buckling load*. For shallow plane or space trusses, the risk of buckling can be examined by nonlinear analysis that considers equilibrium of the nodes in their displaced positions.

If we use Equation (c) to calculate F_{cr} based on the initial geometry (with $\theta_{in} = \tan^{-1}(h/b)$), we would obtain the erroneous result $F_{cr} = 593$ (force unit), which is 5.2 times larger than the correct answer. Calculation of the critical buckling load based on the initial geometry gives the *linear buckling load*. For the single degree-of-freedom system considered, the buckling is calculated by setting S_t = zero and solving for the value of F. For a multi-degree-of-freedom system, the determinant $|S_t|$ is set equal to zero and the solution gives several values of F of which the lowest is the critical value.

The nonlinear buckling load for the plane truss in Figure 18.2a is[1]:

$$F_{\text{cr}} = 2Ea\left(1 - c_{\text{in}}^{2/3}\right)^{3/2} \tag{d}$$

where $c_{\text{in}} = \cos\theta_{\text{in}}$, with θ_{in} being the angle between BC (the initial position of the member) and the horizontal.

18.8 TANGENT STIFFNESS MATRIX OF A MEMBER OF PLANE FRAME

In this section, we derive the tangent stiffness matrix $[S_t]_m$ for a plane frame member AB (Figure 18.5a) with respect to coordinates at the two ends in the directions of global axes.

The tangent stiffness matrix relates small increments of forces $\{\Delta F\}$ to the corresponding displacements $\{\Delta D\}$:

$$[S_t]_m \{\Delta D\} = \{\Delta F\} \tag{18.47}$$

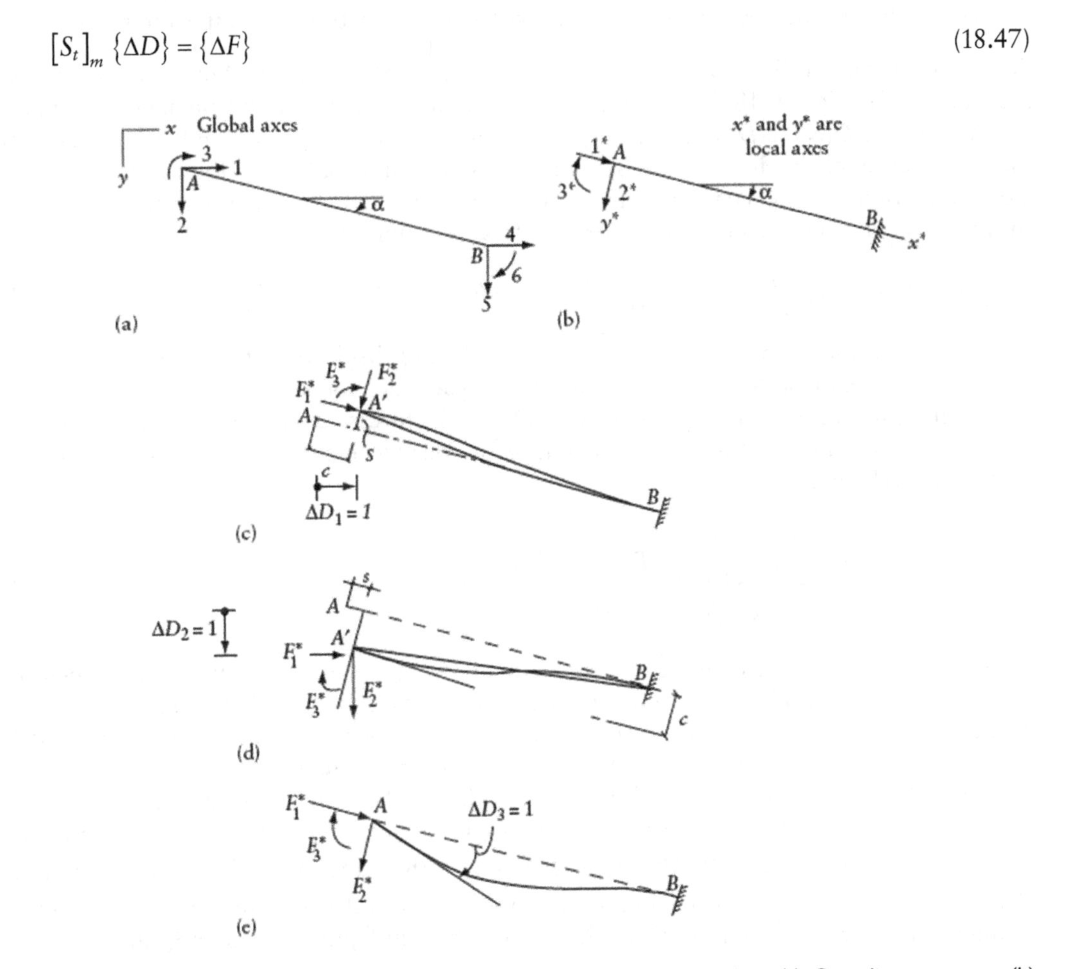

Figure 18.5 Derivation of tangent stiffness matrix for a plane frame member. (a) Coordinate system. (b) Local coordinates at end A with end B fixed. (c), (d), and (e) Introduction of unit displacements in direction of global axes at end A.

The member end forces at A existing before applications of $\{\Delta F\}$ are $\{F^*\}$ in the direction of local axes (Figure 18.5b). We derive first the tangent stiffness matrix $\left[\overline{S}_t\right]$ for the member with end B encastré, as shown in Figure 18.5b. The forces and displacements at coordinates 1, 2, 3 (Figure 18.5a) are related to the forces and displacements at coordinates 1^*, 2^*, 3^* (Figure 18.5b) by equations:

$$\{D^*\} = [t]\{D\} \quad ; \quad \{F\} = [t]^T\{F^*\} \tag{18.48}$$

where $[t]$ is transformation matrix (Equation 19.19):

$$[t] = \begin{bmatrix} c & s & 0 \\ -s & c & 0 \\ 0 & 0 & 1 \end{bmatrix} \tag{18.49}$$

with $c = \cos\alpha$ and $s = \sin\alpha$; where α is the angle between local x^* axis, before application of $\{\Delta F\}$, and the global x axis.

We derive the tangent stiffness matrix by differentiation, as was done for truss members in Section 18.6. However, an alternative approach is used here to derive the tangent stiffness matrix $\left[\overline{S}_t\right]$ for the member in Figure 18.5b (with end B encastré) with respect to three coordinates in the global directions (not shown in Figure 18.5b for clarity, but shown at end A in Figure 18.5a). Figures 18.5c, d, and e show the deformed shapes of the member with small displacements $\Delta D_1 = 1$, $\Delta D_2 = 1$, and $\Delta D_3 = 1$, respectively. The end forces $\{F^*\}$ existing before the introduction of the displacements, are shown in the figures, but the forces $\{\Delta F\}$ – in the global directions – necessary for equilibrium in the three deformed configurations are not shown. By definition, the tangent stiffness matrix Equation 18.33 is:

$$\left[\overline{S}_t\right] = \text{Lim}_{(\Delta D)\to\{0\}}\left[(\Delta D_1)^{-1}\{\Delta F\}_1 \quad (\Delta D_2)^{-1}\{\Delta F\}_2 \quad (\Delta D_3)^{-1}\{\Delta F\}_3\right] \tag{18.50}$$

where $\{\Delta F\}_i$ represents force increments corresponding to displacement change ΔD_i at coordinate i while the displacement increments are zero at other coordinates.

To maintain equilibrium at end A, when $\Delta D_1 = 1$, the forces $\{\Delta F^*\}$ – in the x^* and y^* directions – must be added to the forces existing before displacement (Figure 18.5c). The force increments $\{\Delta F^*\}$ are:

$$\{\Delta F^*\}_{\Delta D_1=1} = \begin{Bmatrix} c\,(Ea/l_{\text{in}}) \\ -2s(S+t)/l^2 \\ -s(S+t)/l \end{Bmatrix} + \frac{F_1^*}{l}\begin{Bmatrix} 0 \\ s \\ 0 \end{Bmatrix} + \frac{F_2^*}{l}\begin{Bmatrix} -s \\ c \\ 0 \end{Bmatrix} \tag{18.51}$$

where E = modulus of elasticity; a = cross-sectional area; l_{in} = initial member length before any deformation; S and t = end moments at A and B, respectively, when a unit rotation is introduced at A while B is encastré (Figure 18.5e). Equations 10.16 and 10.17 give S and t when the axial force is compressive. When the axial force is tensile, Equations 10.20 and 10.21 give S and t; l = member length before introducing ΔD_1. In verifying Equation 18.51, note that as ΔD_1 tends to zero, the cosine of the angle between the chord A′B and the original member direction tends to 1.0. Also, note that F_1^* is in the direction of the chord before and after displacement. F_2^* is perpendicular to the chord before and after displacement. Similarly, equilibrium equations for end A of the deformed beams in Figures 18.5d and e are:

$$\Delta F^* = \left[\{\Delta F^*\}_{\Delta D_1=1} \quad \{\Delta F^*\}_{\Delta D_2=1} \quad \{\Delta F^*\}_{\Delta D_3=1}\right] \tag{18.52}$$

$$\left[\Delta F^*\right]=\begin{bmatrix} c\left(Ea/l_{in}\right) & cs\left(Ea/l_{in}\right) & 0 \\ -2s\left(S+t\right)/l^2 & 2c\left(S+t\right)/l^2 & \left(S+t\right)/l \\ -s\left(S+t\right)/l & c\left(S+t\right)/l & S \end{bmatrix}$$

$$+\frac{F_1^*}{l}\begin{bmatrix} 0 & 0 & 0 \\ s & -c & 0 \\ 0 & 0 & 0 \end{bmatrix}+\frac{F_2^*}{l}\begin{bmatrix} -s & c & 0 \\ c & s & 0 \\ 0 & 0 & 0 \end{bmatrix} \tag{18.53}$$

The forces $[\Delta F^*]$ in local directions can be transformed to the global directions by Equation 18.48 to give the tangent stiffness matrix for the beam in Figure 18.5b, corresponding to coordinates at A, in global directions, shown in Figure 18.5a.

$$\left[\bar{S}_t\right]=\left[t\right]^{\mathrm{T}}\left[\Delta F^*\right] \tag{18.54}$$

Combining Equations 18.50 to 18.54 gives:

$$\left[\bar{S}_t\right]=\left[\bar{S}_e\right]+\left[\bar{S}_g\right] \tag{18.55}$$

where $\left[\bar{S}_e\right]$ is the elastic stiffness matrix:

$$\left[\bar{S}_e\right]=\begin{bmatrix} c^2Ea/l_{in}+2s^2\left(S+t\right)/l^2 & & \text{Symm.} \\ sc\left[Ea/l_{in}-2\left(S+t\right)/l^2\right] & s^2Ea/l_{in}+2c^2\left(S+t\right)/l^2 & \\ -s\left(S+t\right)/l & c\left(S+t\right)/l & S \end{bmatrix} \tag{18.56}$$

where l_{in} = initial length of the member, and $\left[\bar{S}_g\right]$ = geometric stiffness matrix:

$$\left[\bar{S}_g\right]=\frac{F_1^*}{l}\begin{bmatrix} -s^2 & sc & 0 \\ sc & -c^2 & 0 \\ 0 & 0 & 0 \end{bmatrix}+\frac{F_2^*}{l}\begin{bmatrix} -2sc & c^2-s^2 & 0 \\ c^2-s^2 & 2sc & 0 \\ 0 & 0 & 0 \end{bmatrix} \tag{18.57}$$

Considering the fact that the forces in each column of the stiffness matrices are in equilibrium, the tangent stiffness for a plane frame member (Figure 18.5a) is generated:

$$\left[S_t\right]_m=\left[S_e\right]_m+\left[S_g\right]_m \tag{18.58}$$

where $[S_e]_m$ is the elastic stiffness matrix:

$$\left[S_e\right]_m=\begin{bmatrix} \frac{c^2Ea}{l_{in}}+\frac{2s^2}{l^2}\left(S+t\right) & & & & & \text{Symm.} \\ cs\left[\frac{Ea}{l_{in}}-\frac{2}{l^2}\left(S+t\right)\right] & \frac{s^2Ea}{l_{in}}+\frac{2c^2}{l^2}\left(S+t\right) & & & & \\ -\frac{s}{l}\left(S+t\right) & \frac{c}{l}\left(S+t\right) & S & & & \\ -S_{11} & -S_{21} & -S_{31} & S_{11} & & \\ -S_{21} & -S_{22} & -S_{32} & -S_{42} & S_{22} & \\ -\frac{s}{l}\left(S+t\right) & \frac{c}{l}\left(S+t\right) & t & \frac{s}{l}\left(S+t\right) & -\frac{c}{l}\left(S+t\right) & S \end{bmatrix} \tag{18.59}$$

The parameters S and t are dependent upon the value of the axial force F_1^*. When F_1^* = zero, $S=4EI/l$, and $t=2EI/l$, Equation 18.59 becomes the same as Equation 19.31; here, I = second moment of cross-sectional area.

The geometric stiffness matrix corresponding to the coordinates in Figure 18.5a is:

$$\left[S_g\right]_m = \begin{bmatrix} \left[\bar{S}_g\right] & -\left[\bar{S}_g\right] \\ -\left[\bar{S}_g\right] & \left[\bar{S}_g\right] \end{bmatrix} \tag{18.60}$$

where $\left[\bar{S}_g\right]$ is given by Equation 18.57.

The tangent stiffness derived above differs from the stiffness matrices generated in Chapter 10 only by the presence of the last term containing F_2^* in Equation 18.57. Thus, if the angle $\alpha=0$ ($s=0$; $c=1$) and the last term in Equation 18.57 is ignored, $[S_t]$ generated by the equations derived in this section will be the same as that given by Equation 10.19 when the axial force is compressive or by Equation 10.23, when the axial force is tensile. Alternatively, if the stiffness matrix in Equation 10.19 or 10.23 is transformed to correspond to global coordinates (by Equation 19.23), the resulting matrix will be the same as the matrix obtained by the equations derived in this section if the transverse force F_2^* is assumed equal to zero.

In derivation of $[S_t]_m$ (Equation 18.58), the member local x^* axis is assumed to be the chord joining the member ends. However, in general, the member axis is curved due to earlier loading. To reduce the difference between the chord and the curved member axis, divide the member into segments. Unlike linear analysis, in nonlinear analysis of a frame, adding a node at mid-length of a member, generally gives results that are more accurate.

18.9 APPLICATION OF NEWTON–RAPHSON TECHNIQUE TO PLANE FRAMES

Consider a plane frame subjected to forces applied only at the nodes; m = number of members and n = degrees-of-freedom = three times the number of nodes. The members may also be subjected to temperature changes. We discuss below the geometrically nonlinear analysis to determine the displacements $\{D\}_{n\times1}$ and the corresponding changes in the member end forces caused by the applied forces combined with the thermal effects. First, the displacements are artificially restrained by restraining nodal forces $\{F\}_{n\times1}$; in this state, the temperature change produces member end forces $\{F^*\}_r$. The asterisk refers to the directions of the local axes x^* and y^*.

At all stages of the analysis, the three forces at the second node of a member can be determined from the three forces at the first node by the equilibrium equation (Figure 18.6a):

$$\begin{Bmatrix} F_4^* \\ F_5^* \\ F_6^* \end{Bmatrix}_{\text{mem}} = [R]_{\text{mem}} \begin{Bmatrix} F_1^* \\ F_2^* \\ F_3^* \end{Bmatrix}_{\text{mem}} \; ; \; [R]_{\text{mem}} = \begin{bmatrix} -1 & 0 & 0 \\ 0 & -1 & 0 \\ 0 & l & -1 \end{bmatrix} \tag{18.61}$$

where l = member length at the stage considered; the subscript "mem" refers to an individual member.

Define vector $\{F^*\}_{3m\times1}$ containing the required member restraining forces at end 1 of all members. The input data includes: initial geometry of the structure, the elasticity modulus E, the cross-sectional area a, the second moment of area I of members, and $\left\{F_{\text{in}}^*\right\}_{3m\times1}$. The

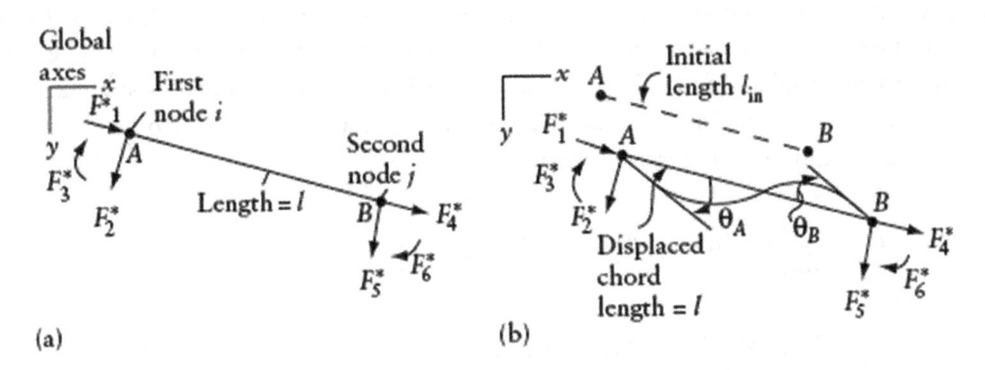

Figure 18.6 Member end forces. (a) Definition of symbols and positive directions of $\{F^*\}$. (b) A typical member in initial and displaced positions.

vector $\left\{F_{\text{in}}^*\right\}$ contains initial member end forces at the first node of all members. Assume that in the initial state, the displacements $\{D\}_{n\times 1}=\{0\}$, and the member lengths $=\{l_{\text{in}}\}_{m\times 1}$. The initial member forces may be caused by prestressing or by initial nodal forces $=\{F_{\text{in}}\}_{n\times 1}$. The analysis discussed below is to determine the displacement changes $\{\Delta D\}_{n\times 1}$. It is also required to determine the final member end forces.

Use the same procedure of analysis for trusses discussed in Sections 18.5 and 18.5.1, with the following differences. For equilibrium of nodes, replace Equation 18.19 by:

$$[G]^{\mathrm{T}}\left\{F_{\text{final}}^*\right\}_{3m\times 1}=-\{F\}_{n\times 1}+\{F_{\text{in}}\}_{n\times 1} \tag{18.62}$$

where $\left\{F_{\text{final}}^*\right\}$ represents the changes in end forces at the first node of all members from the initial stage to the final values. Note that $\left\{F^*\right\}\neq\left\{F_{\text{final}}^*\right\}-\left\{F_{\text{in}}^*\right\}$ because not all the initial and the final member end forces are in the same direction.

The geometry matrix $[G]_{3m\times n}$ can be divided into m submatrices, one for each member. The submatrix for the kth member is:

$$\begin{matrix} & i & & j & \\ [\cdots & [\bar{t}]_k & \cdots & [\bar{t}]_k & \cdots] \end{matrix} \tag{18.63}$$

In this equation the $3\times n$ submatrix is partitioned into submatrices, each 3×3, all of which are null with the exception of the ith and the jth, where i and j are, respectively, the first and second node of the kth member. Equation 18.49 gives the submatrix $[t]_k$. The submatrix $\left[\bar{t}\right]_k$ is:

$$\left[\bar{t}\right]_k=[t]_k\,[R]^{\mathrm{T}}=\begin{bmatrix}-c & -s & sl\\ s & -c & cl\\ 0 & 0 & -1\end{bmatrix} \tag{18.64}$$

where $c=\cos\alpha$ and $s=\sin\alpha$, with α being the angle between the global x axis and the local member x^* axis after deformation.

After r iteration cycles, the axial forces at member ends become:

$$F_{1r}^{*(1)}=F_{1\,\text{in}}^{*(1)}-\left(Ea/l_{\text{in}}\right)\left(l_r^{(1)}-l_{\text{in}}\right)\quad ;\quad F_{4r}^{*(1)}=-F_{1r}^{*(1)} \tag{18.65}$$

As before, the superscript (1) refers to the state at cycle end. The first subscript (e.g. 1 or 4) refers to the local coordinates (Figure 18.6a), the second subscript r refers to the cycle number. Approximate values of the end moments F_3^* and F_6^* can be determined by summation of the increments occurring in the iteration cycles. Then, employ F_3^* and F_6^* to determine the corresponding equilibrating forces F_2^* andF_5^*:

$$\left.\begin{aligned} F_{3r}^{*(1)} &= F_{3\,\text{in}}^* + \sum_{i=1}^{r}\left[S^{(0)}\left(\theta_\text{A}^{(1)} - \theta_\text{A}^{(0)}\right) + t^{(0)}\left(\theta_\text{B}^{(1)} - \theta_\text{B}^{(0)}\right)\right]_i \\ F_{6r}^{*(1)} &= F_{6\,\text{in}}^* + \sum_{i=1}^{r}\left[S^{(0)}\left(\theta_\text{B}^{(1)} - \theta_\text{B}^{(0)}\right) + t^{(0)}\left(\theta_\text{A}^{(1)} - \theta_\text{A}^{(0)}\right)\right]_i \\ F_{2r}^{*(1)} &= \left(F_{3r}^{*(1)} + F_{6r}^{*(1)}\right)\Big/ l_r^{(1)} \\ F_{5r}^{*(1)} &= -F_{2r}^{*(1)} \end{aligned}\right\} \tag{18.66}$$

where θ_A and θ_B = angles between the chord connecting ends A and B and the tangents of the deflected member at the same sections (Figure 18.6b). The subscript i refers to the iteration cycle. The magnitudes of the end-rotational stiffness $S_i^{(0)}$ and the carryover moment $t_i^{(0)}$ are dependent upon the value of the axial force at the start of the ith cycle$\left(= F_{4i}^{(0)}\right)$.

Equation 18.67 gives the out-of-balance forces at the end of each cycle (corresponding to Equation 18.26 used for trusses):

$$\left\{g^{(1)}\right\} = -\{F\} - \left[G^{(1)}\right]^\text{T}\left\{F_\text{final}^{(1)}\right\} + \{F_\text{in}\} \tag{18.67}$$

The end forces $\left\{F_3^* \text{ and } F_6^*\right\}_r^{(1)}$ and the equilibrating forces $\left\{F_2^* \text{ and } F_5^*\right\}_r^{(1)}$ may be determined more accurately by using the value of the axial force $\left(F_{4r}^{*(1)}\right)$ to calculate values of $S_r^{(1)}$ and $t_r^{(1)}$. Combined with the final values of end rotations, these can be used to determine the member end moments:

$$\left.\begin{aligned} F_{3r}^{*(1)} &= F_{3\,\text{in}}^* + S_r^{(1)}\left(\theta_{\text{A}r}^{(1)} - \theta_{\text{A in}}\right) + t_r^{(1)}\left(\theta_{\text{B}r}^{(1)} - \theta_{\text{B in}}\right) \\ F_{6r}^{*(1)} &= F_{6\,\text{in}}^* + S_r^{(1)}\left(\theta_{\text{B}r}^{(1)} - \theta_{\text{B in}}\right) + t_r^{(1)}\left(\theta_{\text{A}r}^{(1)} - \theta_{\text{A in}}\right) \\ F_{2r}^{*(1)} &= \left(F_{3r}^{*(1)} + F_{6r}^{*(1)}\right)\Big/ l_r^{(1)} \\ F_{5r}^{*(1)} &= -F_{2r}^{*(1)} \end{aligned}\right\} \tag{18.68}$$

However, employing Equations 18.68, successively, to replace Equations 18.66 may hamper convergence. This is so because the values of $S^{(0)}$ and $t^{(0)}$ employed in the generation of the stiffness matrix are not the same as $S^{(1)}$ and $t^{(1)}$ used in the calculation of the member end forces and the corresponding out-of-balance forces.

In the computer program NLPF (Nonlinear Analysis of Plane Frames) described in Chapter 20, Equations 18.66 are used during the iteration with $S = 4EI/l_\text{in}$ and $t = S/2$, where E = modulus of elasticity; I = second moment of cross-sectional area. This implies ignoring the influence of axial force on member end-rotational stiffness and carryover moment.

After convergence, the analysis gives approximate values of the axial forces. Subsequently determine more accurate values of S and t; use Equations 10.16, 10.17, 10.20, and 10.21. Equations 18.65 and 18.68 gives new member end forces. Application of Equation 18.67 commonly gives new out-of-balance forces, which are eliminated by a few iteration cycles in which S and t values are unchanged. Usually the refinement cycles produce small changes in the values of axial forces that are too small to justify further analysis. However, if this is not the case, repeat the refinement cycles using an updated set of axial forces.

EXAMPLE 18.3: FRAME WITH ONE DEGREE-OF-FREEDOM

Perform two iteration cycles to determine the vertical downward displacement D at joint B of the toggle shown in Figure 18.7a due to a load $Q=120$. What are the corresponding forces at the ends of member BC? Take $Ea=8.40\times10^6$; $EI=26.6\times10^6$. The data (with units newton and mm) represent a model in an experiment reported by Williams[3] (see also Salami and Morley[4]).

We give below the equations used in the iteration cycles. Because of symmetry, there is a single degree-of-freedom, representing the deflection D at B. The elastic and the geometric stiffnesses are (Equations 18.59 and 18.60):

$$S_e = 2\left(s^{(0)}\right)^2 \left(Ea/l_{\text{in}}\right) + \left[4\left(c^{(0)}\right)^2 \Big/ \left(l^{(0)}\right)^2\right]\left(S^{(0)} + t^{(0)}\right) \tag{e}$$

$$S_g = -2\left(c^{(0)}\right)^2 \left(F_1^{*(0)}/l^{(0)}\right) + 4\,s^{(0)}\,c^{(0)}\left(F_2^{*(0)}/l^{(0)}\right) \tag{f}$$

The superscript (0) refers to the geometry and member end forces known at the start of the cycle; $s=\sin\alpha$, $c=\cos\alpha$, α is the angle between the x direction and the chord B′C. $\{F^*\}$ are forces at end B′ (Figure 18.7b); S and t=end-rotational stiffness and carryover moment. The values of S and t are dependent upon the axial force (Equations 10.16 and 10.17)), l_{in}=initial length of BC, and $l^{(0)}$=length of the chord B′C.

Each cycle starts with trial displacement $D^{(0)}$ and ends with a more accurate value $D^{(1)}$. In a typical r^{th} iteration, the forces at end B of member BC are (Equations 18.65 and 18.66):

$$F_1^{*(1)} = -\frac{Ea}{l_{\text{in}}}\left(l^{(1)} - l_{\text{in}}\right);\quad F_3^{*(1)} = \sum_{i=1}^{r}\left[\left(S^{(0)} + t^{(0)}\right)\left(\theta^{(1)} - \theta^{(0)}\right)\right]_i;\quad F_2^{*(1)} = \frac{2F_3^{*(1)}}{l^{(1)}} \tag{g}$$

The term between the square brackets is the change in end-moment in a single iteration. The summation applies for cycles 1, 2… r.

$$\theta^{(1)} = \tan^{-1}\left[D^{(1)}\,c_{\text{in}} \Big/ \left(l_{\text{in}} - D^{(1)}\,s_{\text{in}}\right)\right];\quad l^{(1)} = \left[\left(h - D^{(1)}\right)^2 + \left(b/2\right)^2\right]^{1/2} \tag{h}$$

The subscript "in" refers to the initial geometry before any displacement; s and c are, respectively, sine and cosine of the angle between the x axis and BC. $s_{\text{in}}=h(h^2+b^2/4)^{-1/2}$; $c_{\text{in}}=0.5b(h^2+b^2/4)^{-1/2}$; $l_{\text{in}}=328.68$.

The out-of-balance force (Equation 18.67) is:

$$g^{(1)} = Q - 2\left(s^{(1)}\,F_1^{*(1)} + c^{(1)}\,F_2^{*(1)}\right) \tag{i}$$

where $s^{(1)}$ and $c^{(1)}$ are based on the updated geometry with the displacement equal to $D^{(1)}$.

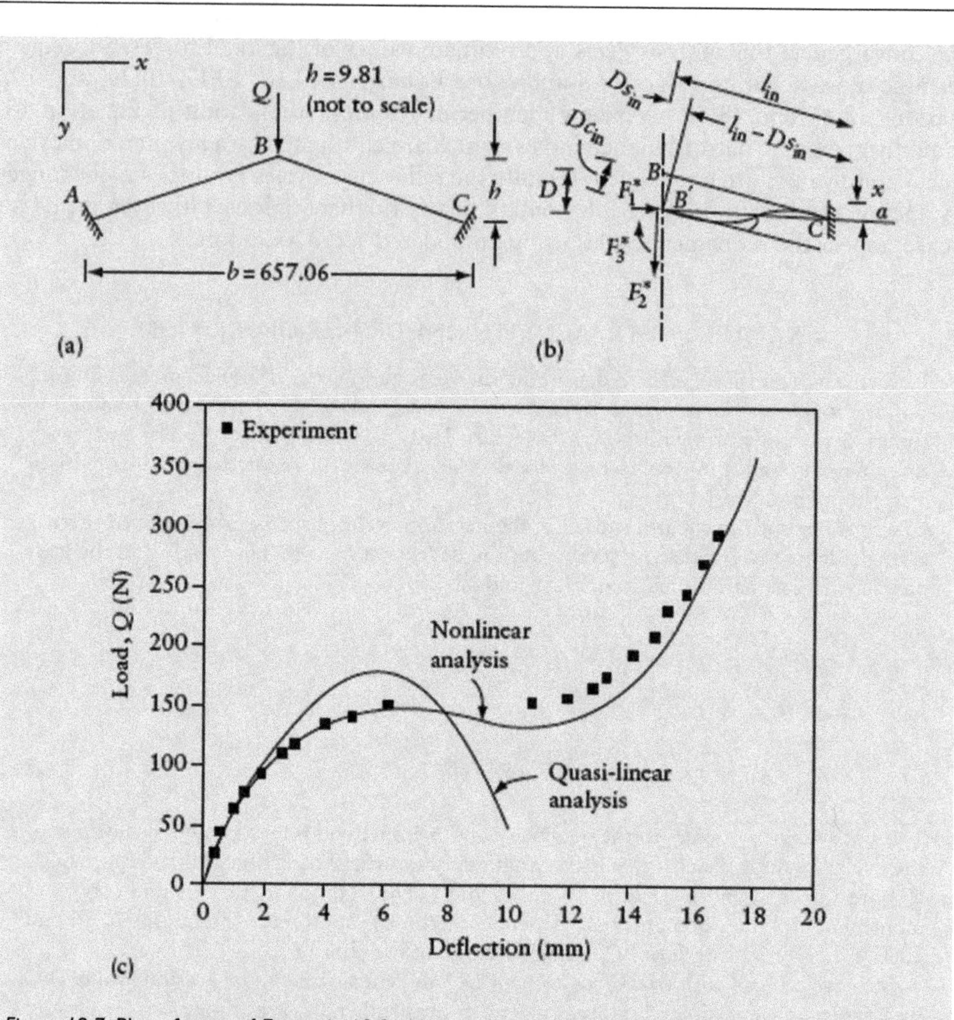

Figure 18.7 Plane frame of Example 18.3. (a) Initial shape. (b) Deformed shape of the right-hand half. (c) Comparison of analytical and experimental results.

ITERATION CYCLE 1

$s^{(0)} = s_{in} = 29.85 \times 10^{-3}$; $c^{(0)} = c_{in} = 0.9996$; $l_{in} = 328.68$.

1. $D^{(0)} = 0$; $g^{(0)} = 120$; $\{F^{*(0)}\} = \{0\}$.
2. $S^{(0)} = 4EI/l_{in} = 323.7 \times 10^3$; $t^{(0)} = 2EI/l_{in} = 161.9 \times 10^3$; $S^{(0)} + t^{(0)} = 485.6 \times 10^3$;

$$S_e = 2\left(29.85 \times 10^{-3}\right)^2 \left(\frac{8.4 \times 10^6}{328.68}\right) + \frac{4\left(0.9996\right)^2}{\left(328.68\right)^2}\left(485.6 \times 10^3\right) = 63.50$$

$$S_g = 0;\ S_t = S_e + S_g = 63.50$$

$$\Delta D = \left(63.50\right)^{-1}\left(120\right) = 1.890$$

3. $D^{(1)} = D^{(0)} + \Delta D = 1.890$; $s^{(1)} = 24.10 \times 10^{-3}$; $c^{(1)} = 0.9997$; $l^{(1)} = 328.63$.
4. The difference between the length of the chord $l^{(1)}$ and the initial member length l_{in} is -50.98×10^{-3}. From Equation (h), $\theta^{(1)} = 5.748 \times 10^{-3}$. The member end forces (Equation (g)) are:

$$\{F^*\}^{(1)} = \left\{ -\frac{8.4\times10^6}{328.68}\left(-50.98\times10^{-3}\right), \frac{2\left(5.749\times10^{-3}\right)\left(485.6\times10^3\right)}{328.63}, 5.749\times10^{-3}\left(485.6\times10^3\right) \right\}$$

whence $\{F^*\}^{(1)} = \{1303, 16.99, 2792\}$

The out-of-balance force (Equation (i)) is:

$$g = 120 - 2[24.10\times10^{-3}(1303) + 0.9997(16.99)] = 23.20$$

ITERATION CYCLE 2

$s^{(0)} = 24.1\times10^{-3}$; $c^{(0)} = 0.9997$

1. $D^{(0)} = 1.890$; $g^{(0)} = 23.2$; $\{F^{*(0)}\} = \{1303, 16.99, 2792\}$.
2. With an axial compressive force of 1303, Equations 10.16 and 10.17 give:
   $S^{(0)} = 3.2395EI/l_{in} = 262.2\times10^3$; $t^{(0)} = 2.2102EI/l_{in} = 178.9\times10^3$;
   $S^{(0)} + t^{(0)} = 441.1\times10^3$; $S_e = 46.03$; $S_g = -7.92$; $S_t = 38.11$;
   $\Delta D = (38.11)^{-1}(23.2) = 0.609$.
3. $D^{(1)} = D^{(0)} + \Delta D = 2.499$; $s^{(1)} = 22.25\times10^{-3}$; $c^{(1)} = 0.9998$; $l^{(1)} = 328.61$.
4. $l^{(1)} - l_{in} = -65.09\times10^{-3}$; $\theta^{(1)} - \theta^{(0)} = 1.853\times10^{-3}$. The updated member end forces are (Equation (g)):
   $F_1^{*(1)} = 1664$; $F_3^{*(1)} = 2792 + 441.1\times10^3\left(1.853\times10^{-3}\right) = 3609$;
   $F_2^{*(1)} = 2(3609)/328.61 = 21.96$

$$\Delta\{F^*\} = \{361, 4.97, 817\}; \quad \{F^*\} = \{1664, 21.96, 3609\}$$

The out-of-balance force (Equation (i)) is:

$$g = 120 - 2\,[22.25\times10^{-3}(1664) + 0.9998\,(21.96)] = 2.05$$

After the third iteration, the updated downward displacement at B and the end forces at end B of member BC are:

$$D = 2.566; \quad \{F^*\} = \{1701, 22.49, 3695\}$$

The corresponding out-of-balance force is calculated by Equation (i):

$$s = 22.04\times10^{-3}; \; c = 0.9998; \; g = 0.02$$

With member BC subjected to an axial force of −1701 (compressive), end-rotational stiffness and carryover moment are (Equations 10.16 and 10.17):

$$S = 2.9807\,\frac{EI}{l_{in}}; \quad t = 2.2913\,\frac{EI}{l_{in}} \quad \text{and} \quad (S+t) = 426.67\times10^3$$

Application of Equations 18.65, 18.68, and Equation (i) shows that, with the displacement configuration obtained, the more accurate values of the forces at end B of member BC and the corresponding out-of-balance force are:

$$\theta_A = \theta_B = 7.8053\times10^{-3}; \quad \{F^*\} = \{1701, 20.27, 3330\}; \quad g = 4.46$$

The computer program NLPF, discussed in the preceding section, performs refinement cycles to eliminate the new out-of-balance force and gives answers that are more accurate:

$$D = 2.729; \quad \{F^*\} = \{1792,\ 21.39,\ 3514\}$$

We may now verify the accuracy of the answers. The change in member length $= -70.11 \times 10^{-3}$, corresponding to an axial force $= -1792$ (compressive); $(S+t) = 5.2310EI/l_{in}$ (Equations 10.16 and 10.17), $\theta = 8.297 \times 10^{-3}$; and member end-moments at each of B and C $= (S+t)\theta = 3514$. Length of BC after deformations is $l^{(1)} = 328.61$, and the absolute value of the shearing force in the member is $= 2\theta\,(S+t)/l^{(1)} = 21.39$. The resultant of the forces at end B of members AB and BC is the same as the external force applied at B, indicating zero out-of-balance force.

As mentioned earlier, the accuracy of the solution can be improved by introducing the load in stages and partitioning members AB and BC. The results of analyses, with each of these two members subdivided into four segments of equal length, are compared with the experimental results in Figure 18.7c. In the analyses, prescribed D values are used to determine the corresponding Q values. When $D > h$, B is situated below A and C, and the compressive axial force in AB and BC decreases. This is accompanied by an increase in stiffness, represented by increase in the slope of the Q-D graph.

Figure 18.7c includes the results of quasi-linear analyses (P-delta, Chapter 10). These analyses, which do not rigorously consider the geometry of the structure after deformation, cannot give accurate results when the displacements are large – in this example, when D is greater than, say, 4. Instead of using Equations (h) and (i), the quasi-linear analyses consider $l^{(1)} = l_{in} - D \sin \alpha_{in}$ and $Q = 2\left(F_1^* \sin\alpha_{in} + F_2^* \cos\alpha_{in}\right)$, where α_{in} is the angle between the global x axis and the initial direction of member BC (Figure 18.7a).

EXAMPLE 18.4: LARGE DEFLECTION OF A COLUMN

Determine the horizontal and vertical displacements at the top of the column shown in Figure 18.8a. What are the reaction components at the bottom? Take $l_{in} = 75$ (length unit); $Q = P = 150$ (force unit); $Ea = 10 \times 10^6$ (force unit); $EI = 1.0 \times 10^6$ (force × length²). Assume elastic material. The forces P and Q are assumed to maintain their original directions after deformation.

We give below the equations to be used in the iteration cycles; the superscripts (0) and (1) refer to the values available at the start of a cycle and those determined at its end, respectively. In a typical cycle, the updated member end forces at A are (Figure 18.8c and Equations 18.65 and 18.66):

$$F_1^* = -\frac{Ea}{l_{in}}\left(l^{(1)} - l_{in}\right); \quad F_3^{*(1)} = S\,\theta_A^{(1)} + t\,\theta_B^{(1)}; \quad F_2^{*(1)} = \frac{(S+t)}{l^{(1)}}\left(\theta_A^{(1)} + \theta_B^{(1)}\right) \tag{j}$$

where $S = 4EI/l_{in}$ and $t = 2EI/l_{in}$ (effects of axial force on S and t ignored)

$$\left.\begin{aligned} \theta_A^{(1)} &= D_3^{(1)} + \tan^{-1}\left[D_2^{(1)}\Big/\left(l_{in} - D_1^{(1)}\right)\right] \\ \theta_B^{(1)} &= \tan^{-1}\left[D_2^{(1)}\Big/\left(l_{in} - D_1^{(1)}\right)\right] \end{aligned}\right\} \tag{k}$$

The out-of-balance forces are:

$$\left\{g^{(1)}\right\} = \begin{Bmatrix} P \\ Q \\ 0 \end{Bmatrix} - \begin{bmatrix} c^{(1)} & -s^{(1)} & 0 \\ s^{(1)} & c^{(1)} & 0 \\ 0 & 0 & 1 \end{bmatrix} \left\{F^{*(1)}\right\} \tag{l}$$

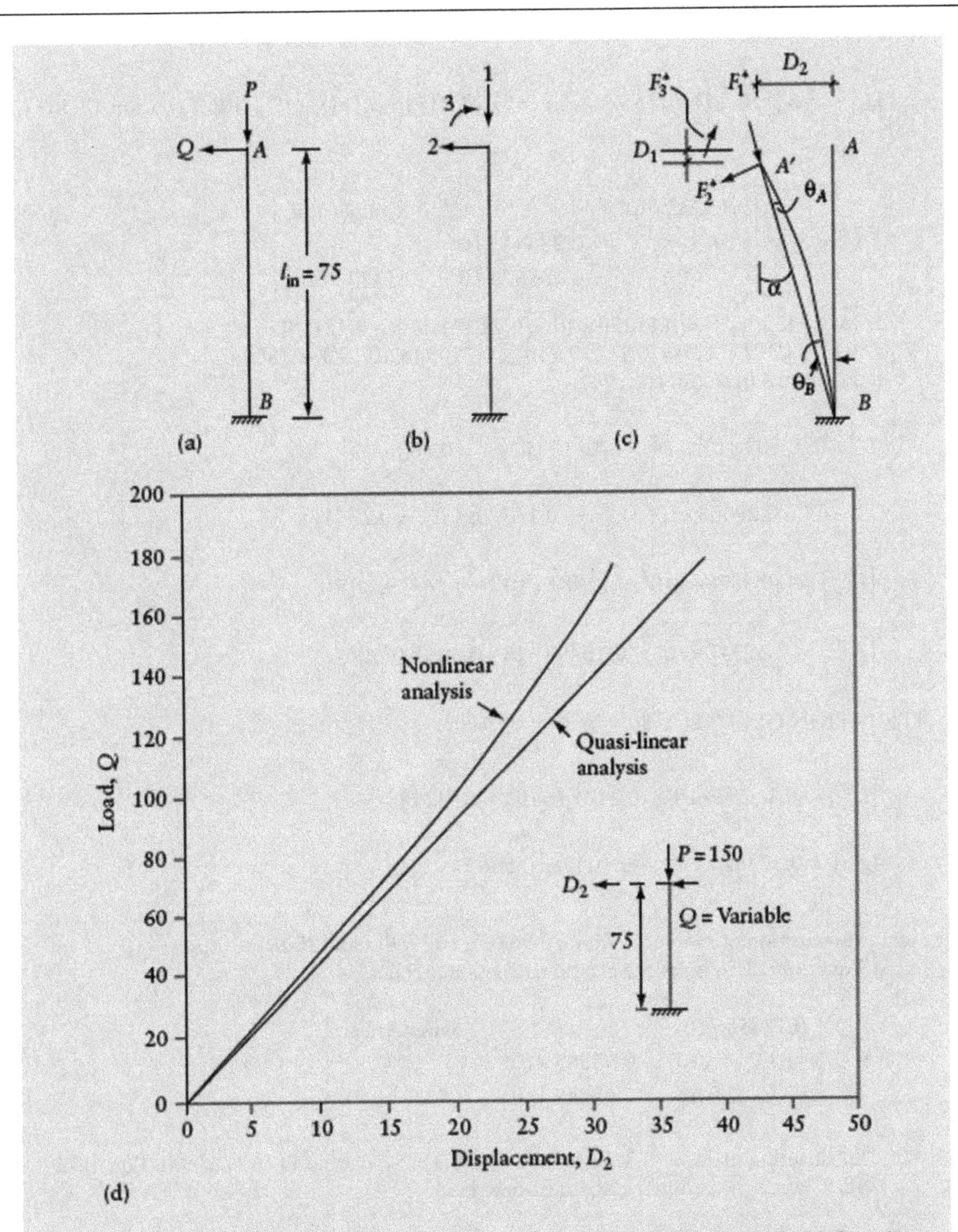

Figure 18.8 Column analyzed in Example 18.4. (a) Initial geometry and applied loads. (b) Coordinate system. (c) Deflected shape and forces at top end. (d) Q-D_2 graph with P constant.

$$s^{(1)} = \frac{D_2^{(1)}}{l^{(1)}} \quad ; \quad c^{(1)} = \frac{l_{in} - D_1^{(1)}}{l^{(1)}} \quad ; \quad l^{(1)} = \left[\left(l_{in} - D_1^{(1)}\right)^2 + \left(D_2^{(1)}\right)^2\right]^{1/2} \tag{m}$$

ITERATION CYCLE 1

$$\left\{D^{(0)}\right\} = \{0\}; \quad \{F^{*(0)}\} = \{0\}; \quad s^{(0)} = 0; \quad c^{(0)} = 1$$

1. $\{g^{(0)}\} = \{150, 150, 0\}$; $S = 4EI/l_{in}$; $t = 2EI/l_{in}$; $[S_t] = [S_e] + [S_g]$; $[S_g] = [0]$. Equation 18.56 gives:

$$[S_t] = [S_e] = \begin{bmatrix} 0.13333\times10^6 & & \text{Symmetrical} \\ 0 & 0.28444\times10^2 & \\ 0 & 0.10667\times10^4 & 0.53333\times10^5 \end{bmatrix}$$

2. $\{\Delta D\} = [S_t]^{-1}\{g^{(0)}\} = \{0.11250\times10^{-2}, 0.21094\times10^2, -0.42188\}$.
3. $\{D^{(1)}\} = \{D^{(0)}\} + \{\Delta D\} = \{0.11250\times10^{-2}, 0.21094\times10^2, -0.42188\}$.
4. Equations (j) to (m) give:

$$l^{(1)} = 0.77909\times10^2;\quad s^{(1)} = -0.27075;\quad c^{(1)} = 0.96265$$

$$l^{(1)} - l^{(0)} = 0.29088\times10;\quad \theta_A^{(1)} = -0.14770;\quad \theta_B^{(1)} = 0.27417$$

$$\{F^{*(1)}\} = \{-0.38784\times10^6, 0.12987\times10^3, -0.56629\times10^3\}$$

$$\{g^{*(1)}\} = \{0.37347\times10^6, -0.10498\times10^6, 0.56629\times10^3\}$$

ITERATION CYCLE 2

$$\{D^{(0)}\} = \{0.11250\times10^{-2},\ 0.21094\times10^2,\ -0.42188\};$$

$$\{g^{(0)}\} = \{0.37347\times10^6,\ -0.10498,\ 0.56629\times10^3\}$$

1. Substitution of $s = -0.27075$, $c = 0.96265$, $l = 77.909$ and $(Ea/l) \equiv (Ea/l_{in}) = 10^6/75$ in Equation 18.56 gives the elastic stiffness matrix:

$$[S_e] = \begin{bmatrix} 0.12356\times10^6 & & \text{Symmetrical} \\ -0.34745\times10^5 & 0.97985\times10^4 & \\ 0.27802\times10^3 & 0.98849\times10^3 & 0.53333\times10^5 \end{bmatrix}$$

Substitution of $F_1^* = -0.38784\times10^6$; $F_2^* = 0.12987\times10^3$; and $l = 77.909$ in Equation 18.57 gives the geometric stiffness matrix:

$$[S_g] = \begin{bmatrix} 0.36579\times10^3 & & \text{Symm.} \\ 0.12989\times10^4 & 0.46123\times10^4 & \\ 0 & 0 & 0 \end{bmatrix}$$

$$[S_t] = [S_e] + [S_g]$$

2. $\{\Delta D\} = [S_t]^{-1}\{g^{(0)}\} = \{0.28030\times10, -0.78016, 0.10466\times10^{-1}\}$.
3. $\{D^{(1)}\} = \{D^{(0)}\} + \{\Delta D\} = \{0.28042\times10, 0.20314\times10^2, -0.41141\}$.
4. Equations (j) to (l) give:

$l^{(1)} = 0.74999\times10^2$; $s^{(1)} = -0.27085$; $c^{(1)} = 0.96262$.

$l^{(1)} - l_{in} = 0.77790\times10^{-3}$; $\theta_A^{(1)} = -0.13713$; $\theta_B^{(1)} = 0.27428$.

The updated member forces at end A in this cycle are:

$$\{F^{*(1)}\} = \{0.10372\times10^3,\ 0.14629\times10^3,\ 0.31483\}$$

The out-of-balance forces (Equation (l)) are:

$$\{g^{(1)}\} = \{0.10538\times10^2,\ 0.37271\times10^2,\ -0.31483\}.$$

After six iteration cycles and refinement cycles in which S and t are considered functions of the axial force, the solution is:

$$\{D\} = \{0.52711\times10,\ 0.27619\times10^2, -0.57027\}$$

The member end forces are:

$$\{F^*\} = \{0.84221\times10^2, 0.19470\times10^3, 0, -0.84221\times10^2, -0.19470\times10^3, 0.14602\times10^5\}$$

Quasi-linear analysis

For comparison, we give below the results of a quasi-linear analysis, in which the stiffness matrix is determined by Equation 10.14, for a member subjected to an axial compressive force of 150. Furthermore, this analysis ignores the downward movement of the tip A associated with the lateral deflection. The displacements at A and the six member end forces obtained by the quasi-linear analysis are:

$$\{D\} = \{1.125\times10^{-3},\ 31.908,\ -0.64754\}$$

$$\{F^*\} = \{150,\ 150,\ 0,\ -150,\ -150,\ 16036\}$$

The graph in Figure 18.8d shows the results of nonlinear and quasi-linear analyses for the structure in Figure 18.8a, maintaining P = constant = 150 and varying Q. As expected, the quasi-linear analysis is represented by a straight line. The nonlinear analysis gives smaller values of the lateral deflection because, in the displaced position, the force Q is closer to the fixed end than in its initial position.

18.10 TANGENT STIFFNESS MATRIX OF TRIANGULAR MEMBRANE ELEMENT

The stiffness of membrane elements is discussed here because it is required in the analysis of shell structures and fabric structures composed of cable nets and fabric material. We generate below the tangent stiffness matrix for a triangular plane element whose corner nodes i, j, and k are defined by their (x, y, z) coordinates with respect to global orthogonal axes (Figure 18.9a). The element local orthogonal axes (x^*, y^*, z^*) have the same origin O as the global axes and the x^*-y^* plane is parallel to the plane of the element (Figure 18.9b). Furthermore, the local x^* axis is parallel to the vector connecting node i to node j; the local y^* axis is the perpendicular to x^*, pointing toward node k in direction away from line ij.

We will derive the tangent stiffness matrix with respect to coordinates $1^*, 2^*, \ldots, 9^*$ in the directions of the local axes[1] (Figure 18.9b). The tangent stiffness matrix is the sum of the

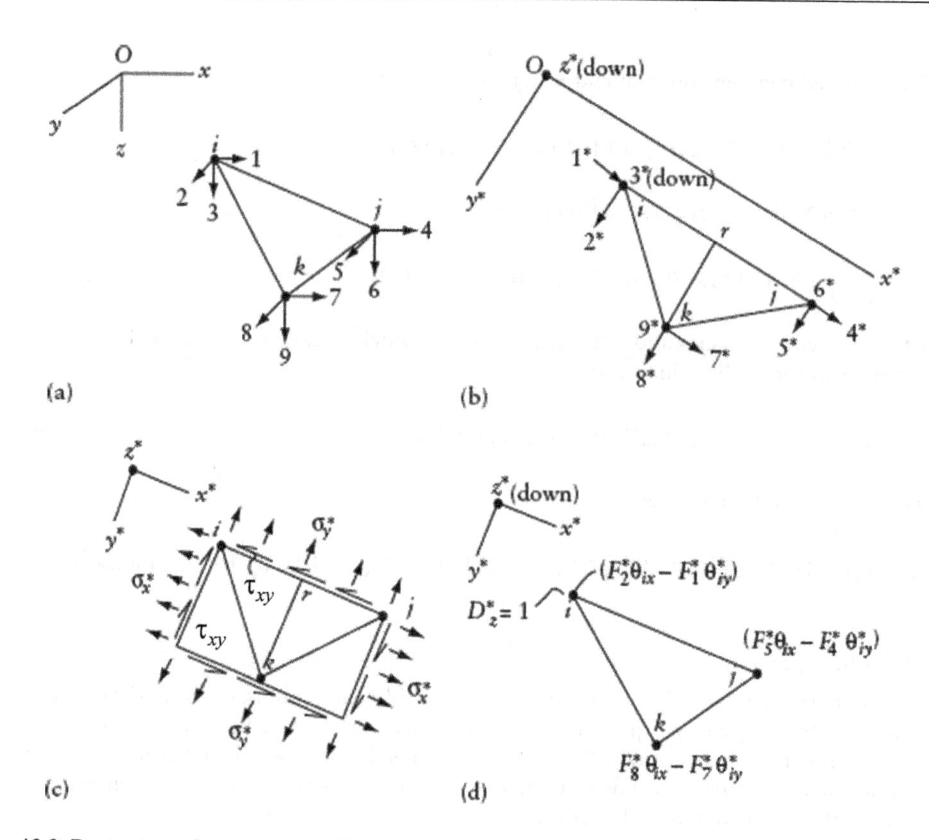

Figure 18.9 Derivation of geometric stiffness matrix of triangular membrane element. (a) Coordinate system in global directions. (b) View of the plane of the element and local coordinate system. (c) Stresses on element edges. (d) Forces necessary to maintain equilibrium when $D_z^* = 1$.

elastic stiffness matrix $\left[S_e^*\right]$ and the geometric stiffness matrix $\left[S_g^*\right]$. The stiffness matrices corresponding to the nine coordinates in the global directions can be determined by the transformation equation:

$$[S]_{e \text{ or } g} = [T]^{\text{T}} \left[S^*\right]_{e \text{ or } g} [T] \tag{18.69}$$

where

$$[T] = \begin{bmatrix} [t] & & \\ & [t] & \\ & & [t] \end{bmatrix} \quad \text{Submatrices not shown are null}$$

with $[t]$ being a 3×3 matrix consisting of cosines of the angles between the local and global axes (Equation 19.22). At any corner of the triangle, the nodal coordinates, displacements, and forces in the local and global directions are related as follows:

$$\begin{Bmatrix} x^* \\ y^* \\ z^* \end{Bmatrix} = [t] \begin{Bmatrix} x \\ y \\ z \end{Bmatrix} \tag{18.70}$$

$$\begin{Bmatrix} D_x^* \\ D_y^* \\ D_z^* \end{Bmatrix} = [t] \begin{Bmatrix} D_x \\ D_y \\ D_z \end{Bmatrix} \quad ; \quad \begin{Bmatrix} F_x \\ F_y \\ F_z \end{Bmatrix} = [t]^{\mathrm{T}} \begin{Bmatrix} F_x^* \\ F_y^* \\ F_z^* \end{Bmatrix} \tag{18.71}$$

The stresses on the edges of the element are constant and are represented in their positive directions in Figure 18.9c (see Section 13.11 and Figure 13.11c). The nodal forces in the x^* and y^* directions at a node are equal to the sum of one-half of the distributed load on each of the two sides connected to the node (see Equation 13.53). Thus:

$$\begin{Bmatrix} F_1^* \\ F_2^* \\ F_3^* \\ F_4^* \\ F_5^* \\ F_6^* \\ F_7^* \\ F_8^* \\ F_9^* \end{Bmatrix} = \frac{h}{2} \begin{bmatrix} -(y_k^* - y_i^*) & 0 & -(x_j^* - x_k^*) \\ 0 & (x_k^* - x_i^*) & -(y_k^* - y_j^*) \\ 0 & 0 & 0 \\ -(y_i^* - y_k^*) & 0 & -(x_k^* - x_i^*) \\ 0 & (x_i^* - x_k^*) & -(y_i^* - y_k^*) \\ 0 & 0 & 0 \\ 0 & 0 & -(x_i^* - x_j^*) \\ 0 & (x_j^* - x_i^*) & 0 \\ 0 & 0 & 0 \end{bmatrix} \begin{Bmatrix} \sigma_x^* \\ \sigma_y^* \\ \tau_{xy}^* \end{Bmatrix} \tag{18.72}$$

where h is the thickness of the element, or:

$$[F^*] = [G] \cdot [\sigma^*] \tag{18.73}$$

where $[G]$ is the 9×3 matrix in Equation 18.72, multiplied by $h/2$.

The elastic stiffness matrix for the constant-strain triangle was derived in Section 13.10 (Equation 13.48). The same equation can be used here, by insertion of three rows and three columns composed of zeros, corresponding to coordinates 3^*, 6^*, and 9^*. This is so because the membrane element has no out-of-plane elastic stiffness.

The geometric stiffness matrix will be derived by differentiation of Equation 18.73, holding $\{\sigma^*\}$ constant. A general element of the geometric stiffness matrix (see Equation 18.33) is:

$$S_{gnm}^* = \frac{\partial F_n^*}{\partial D_m^*} \qquad \left(\text{with } \{\sigma^*\} = \text{constant}\right) \tag{18.74}$$

Noting that differentiation with respect to D^* is the same as differentiation with respect to x^* or y^* at the appropriate node, substitution of the nth row of Equation 18.72 into Equation 18.74 gives:

$$S_{gnm}^* = \frac{\partial}{\partial D_m^*} [G_{n1} \quad G_{n2} \quad G_{n3}] \{\sigma^*\} \tag{18.75}$$

For example, for $n=4$ and $m=2$, we obtain:

$$S_{g42}^* = \frac{h}{2} \frac{\partial}{\partial y_i^*} \left[-(y_i^* - y_k^*) \quad 0 \quad -(x_k^* - x_i^*)\right] \{\sigma^*\} \tag{18.76}$$

or

$$S^*_{g42} = -\frac{h}{2}\sigma^*_x \tag{18.77}$$

Equation 18.75 can generate columns 1, 2, 4, 5, 7, and 8 of S^*_g. The remaining columns of S^*_g represent changes in the nodal forces resulting from separate out-of-plane displacement $D^*_z = 1$ at each of the three nodes. For example, to derive the third column of $[S_g]$, we introduce, $D^*_3 = D^*_{zi} = 1$ resulting in rotations of the plane of the element through angles θ^*_{ix} and θ^*_{iy} about the x^* and y^* axes. To maintain equilibrium, the following non-zero changes in the nodal forces must occur (Figure 18.9d):

$$S^*_{g33} = F^*_2\,\theta^*_{ix} - F^*_1\,\theta^*_{iy};\quad S^*_{g63} = F^*_5\,\theta^*_{ix} - F^*_4\,\theta^*_{iy};\quad S^*_{g93} = F^*_8\,\theta^*_{ix} - F^*_7\,\theta^*_{iy} \tag{18.78}$$

To verify the first of these three equations, consider the equilibrium of the membrane element in the rotated position. Forces F^*_1 and F^*_2 at node i are in a plane parallel to the initial plane of the element. Addition at node i of upward and downward forces of magnitudes $F^*_1\,\theta_{iy}$, and$F^*_2\,\theta_{ix}$, respectively, brings the resultant nodal forces to the rotated plane of the element. For the same purpose, forces are required at nodes j and k expressed by the second and the third of Equations 18.78.

The complete geometric stiffness matrix of the membrane triangular element with respect to the local coordinates in Figure 18.9b is:

$$[S_g] = \begin{array}{c} 1\\2\\3\\4\\5\\6\\7\\8\\9 \end{array}\left[\begin{array}{ccc|ccc|ccc}
0 & 0 & 0 & -h\tau^*_{xy}/2 & h\sigma^*_x/2 & 0 & h\tau^*_{xy}/2 & -h\sigma^*_x/2 & 0 \\
0 & 0 & 0 & -h\sigma^*_y/2 & h\tau^*_{xy}/2 & 0 & h\sigma^*_y/2 & -h\tau^*_{xy}/2 & 0 \\
0 & 0 & F^*_2\theta^*_{ix} - F^*_1\theta^*_{iy} & 0 & 0 & F^*_2\theta^*_{jx} - F^*_1\theta^*_{jy} & 0 & 0 & F^*_2\theta^*_{kx} \\ \hline
h\tau^*_{xy}/2 & -h\sigma^*_x/2 & 0 & 0 & 0 & 0 & -h\tau^*_{xy}/2 & h\sigma^*_x/2 & 0 \\
h\sigma^*_y/2 & -h\tau^*_{xy}/2 & 0 & 0 & 0 & 0 & -h\sigma^*_y/2 & h\tau^*_{xy}/2 & 0 \\
0 & 0 & F^*_5\theta^*_{ix} - F^*_4\theta^*_{iy} & 0 & 0 & F^*_5\theta^*_{jx} - F^*_4\theta^*_{jy} & 0 & 0 & F^*_5\theta^*_{kx} \\ \hline
-h\tau^*_{xy}/2 & -h\sigma^*_x/2 & 0 & h\tau^*_{xy}/2 & -h\sigma^*_x/2 & 0 & 0 & 0 & 0 \\
-h\sigma^*_y/2 & -h\tau^*_{xy}/2 & 0 & h\sigma^*_y/2 & -h\tau^*_{xy}/2 & 0 & 0 & 0 & 0 \\
0 & 0 & F^*_8\theta^*_{ix} - F^*_7\theta^*_{iy} & 0 & 0 & F^*_8\theta^*_{jx} - F^*_7\theta^*_{jy} & 0 & 0 & F^*_8\theta^*_{kx}
\end{array}\right] \tag{18.79}$$

where θ^*_{ix} and θ^*_{iy} = rotations of the element, with respect to axes x^* and y^*, when $D^*_{zi} = 1$; in a similar way, θ^*_{jx}, θ^*_{jy}, θ^*_{kx} and θ^*_{ky} are defined. We can verify the following geometric relations:

$$\theta^*_{ix} = -\left(\frac{x^*_j - x^*_k}{x^*_j - x^*_i}\right)\left(\frac{1}{y^*_k - y^*_j}\right);\quad \theta^*_{iy} = \frac{1}{x^*_j - x^*_i} \tag{18.80}$$

$$\theta^*_{jx} = -\left(\frac{x^*_j - x^*_i}{x^*_j - x^*_i}\right)\left(\frac{1}{y^*_k - y^*_j}\right);\quad \theta^*_{jy} = \frac{1}{x^*_j - x^*_i} \tag{18.81}$$

$$\theta^*_{kx} = 1/\left(y^*_k - y^*_j\right);\quad \theta^*_{ky} = 0 \tag{18.82}$$

We note that, in this case, $\left[S^*_g\right]$ is a non-symmetric matrix.

The iterative Newton–Raphson technique (Section 18.5.1) can be employed to determine the nodal displacements, which define the geometry of triangular element after deformation.

Let the line rk in Figure 18.9b be parallel to the y^* axis passing through node k in its initial position and in its displaced position. The element strains may be calculated from the length changes (Figure 18.9b):

$$\delta_{ij} = l_{ij} - \left(l_{ij}\right)_{\text{in}}; \quad \delta_{rk} = l_{rk} - \left(l_{rk}\right)_{\text{in}}; \quad \delta_{ir} = l_{ir} - \left(l_{ir}\right)_{\text{in}}$$

where the subscript "in" refers to initial lengths. With respect to the local axes x^*, y^*, and z^* (Figures 18.9b and c), the changes in strain from the initial state are given by:

$$\varepsilon_x^* = \frac{\delta_{ij}}{\left(l_{ij}\right)_{\text{in}}}; \quad \varepsilon_y^* = \frac{\delta_{rk}}{\left(l_{rk}\right)_{\text{in}}}; \quad \gamma_{xy}^* = \frac{1}{\left(l_{rk}\right)_{\text{in}}}\left[\delta_{ir} - \varepsilon_x^*\left(l_{ir}\right)_{\text{in}}\right] \tag{18.83}$$

In verifying the equation for γ_{xy}^*, note that δ_{ir} represents the combined effects of ε_x^* and τ_{xy}^* on the translation in the x^* direction of k with respect to i; $\varepsilon_x^*\left(l_{ir}\right)_{\text{in}}$ represents the part of this translation attributed to ε_x^*.

Figure 18.9c gives stresses after deformation:

$$\begin{Bmatrix} \sigma_x^* \\ \sigma_y^* \\ \tau_{xy}^* \end{Bmatrix} = [d] \begin{Bmatrix} \varepsilon_x^* \\ \varepsilon_y^* \\ \gamma_{xy}^* \end{Bmatrix} + \begin{Bmatrix} \left(\sigma_x^*\right)_{\text{in}}\left[\left(l_{rk}\right)_{\text{in}}/l_{rk}\right] \\ \left(\sigma_y^*\right)_{\text{in}}\left[\left(l_{ij}\right)_{\text{in}}/l_{ij}\right] \\ \left(\tau_{xy}^*\right)_{\text{in}}\left(\Delta_{\text{in}}/\Delta\right) \end{Bmatrix} \tag{18.84}$$

where $[d]$ = material elasticity matrix (Equation 7.28); $\left\{\sigma_x^*, \sigma_y^*, \tau_{xy}^*\right\}_{\text{in}}$ = stresses existing in the initial state; Δ_{in} and Δ = areas of the element in initial and deformed states, respectively.

The last term in Equation 18.84 represents adjusted stresses equivalent to the initial stresses, $\left\{\sigma_x^*, \sigma_y^*, \tau_{xy}^*\right\}_{\text{in}}$. Considering the rectangular element in Figure 18.9c, we see that the resultants of the initial and the adjusted stress normal to any of the sides are the same. Also, the initial and the adjusted shear stress on each pair of opposite sides have equal resultant couples $\left(\tau_{xy}\, l_{ij}\, l_{rk}\right)$.

18.11 ANALYSIS OF STRUCTURES MADE OF NONLINEAR MATERIAL

Figure 18.10 shows examples of the shape of the stress–strain relation for nonlinear materials in practice. In this section, we discuss the simplest analysis technique known as the *incremental method*.

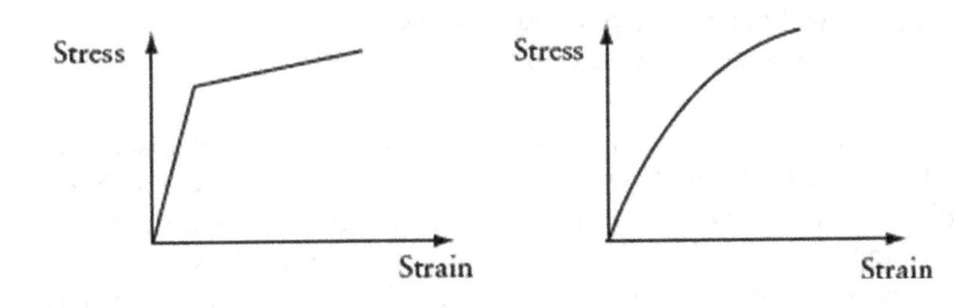

Figure 18.10 Examples of stress–strain relations.

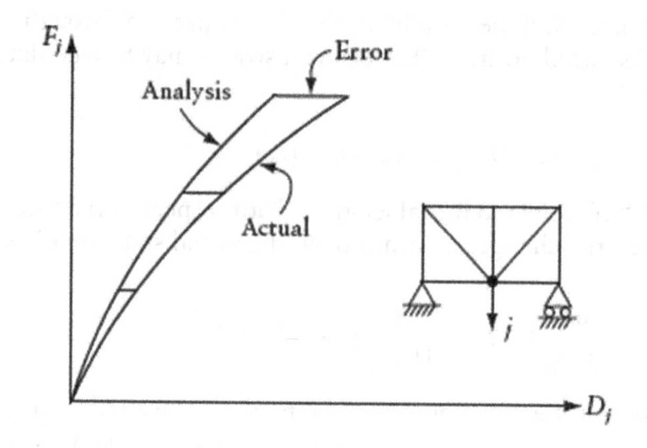

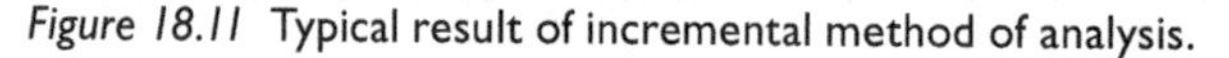
Figure 18.11 Typical result of incremental method of analysis.

The applied forces $\{F\}$ are divided into increments $\beta_i\{F\}$. The load increments are applied one at a time, and an elastic analysis is used. For the ith load increment, the following equilibrium equation is solved:

$$[S]_i \{\Delta D\}_i = -\{\Delta F\}_i \tag{18.85}$$

The stiffness matrix $[S]_i$ depends upon the stress level reached in the preceding increment; thus, for the ith increment, the modulus of elasticity is the slope of the stress–strain diagram at the stress level reached in the increment $(i - 1)$. The displacements obtained by the solution of Equation 18.85 for each load increment are summed to give the final displacements:

$$\{D\} = \sum \{\Delta D\}_i \tag{18.86}$$

Similarly, the final stresses or stress resultants are determined by the sum of the increments calculated in all the linear analyses.

A typical plot of the displacement at any coordinate and the corresponding nodal force is composed of straight segments (Figure 18.11). It can be seen that the error in the results of the incremental method increases as the load level is increased. However, the accuracy can be improved by using smaller load increments.

The advantage of the incremental method is its simplicity. It can also be used for geometrically nonlinear analysis. For this purpose, the stiffness matrix $[S]_i$ for increment i is based on the geometry of the structure and the member end forces determined in the preceding increment, $(i - 1)$.

18.12 ITERATIVE METHODS FOR ANALYSIS OF MATERIAL NONLINEARITY

Newton–Raphson and the modified Newton–Raphson techniques discussed in Sections 18.4, 18.4.1, 18.5, and 18.9 can also be used to analyze structures with material nonlinearity. Figure 18.12 represents two analysis schemes. In Figure 18.12a, the full load is introduced, and an approximate solution is obtained and corrected by a series of iterations. A linear analysis is performed in each iteration, using approximate stiffness and giving

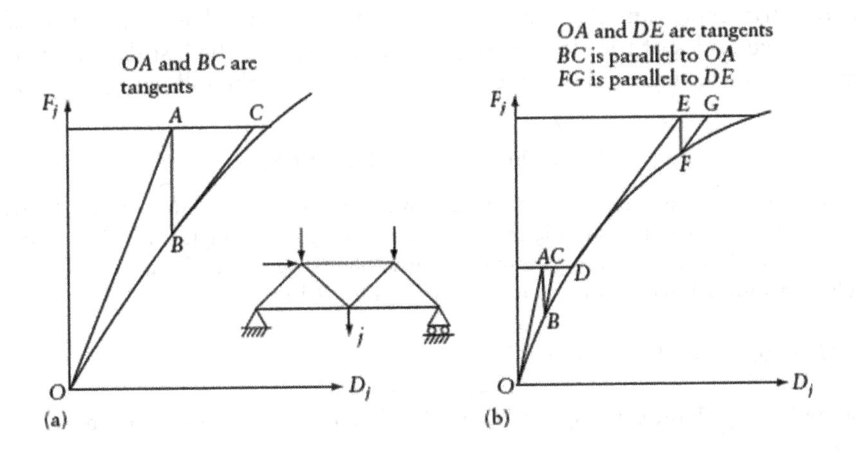

Figure 18.12 Load-displacement analysis in two consecutive iterative cycles. (a) Use of new tangent stiffness matrix in each iteration. (b) Application of the load in increments and use of new tangent stiffness matrix only in the first cycle after introduction of each new load increment.

approximate displacements and hence the strain in each member (or in each finite-element). Using the stress–strain relation, the corresponding stress is determined. From the stress, we determine the forces at member ends by Equation 18.88 (or at the local coordinates of finite-elements). These forces are assembled to determine a vector of nodal forces $\{F\}$ that maintain the equilibrium of the structure in the approximate deformed configuration determined in this iteration cycle. The difference between $\{F\}$ and the known nodal forces represents out-of-balance forces to be eliminated by iteration:

$$\{g\} = -\{F\} - \{\bar{F}\} \tag{18.87}$$

where $\{F\}$ represents the forces which artificially prevent the nodal displacements; $\{\bar{F}\}$ represents the forces balanced by the internal forces in the members in the trial deformed configuration.

In each iteration cycle, we use a new stiffness matrix based on the stress level in each member (or finite-element) at the end of the preceding cycle. This is represented in Figure 18.12a, which is a graph of the force versus displacement at a typical coordinate. The slopes of OA and BC are different, indicating different stiffnesses in consecutive iterations.

The scheme presented in Figure 18.12b employs the modified Newton–Raphson technique. The load is introduced in stages, and in each stage, iteration cycles are employed to determine the nodal displacements. To avoid generating a new stiffness matrix in each iteration cycle, the stiffness matrix determined in the first cycle in each load stage is employed in all subsequent cycles until convergence has been achieved, before proceeding to the next load stage. This is schematically represented in Figure 18.12b in which the slopes of OA and BC are the same, indicating the same stiffness in two cycles.

In both schemes described above, we need to determine the nodal forces for a member (or finite-element) for which the stress $\{\sigma\}$ is known. The unit-displacement theorem is used for this purpose (Equation 5.39). For a finite-element, the nodal forces are:

$$\{\bar{F}\}_{\text{element}} = \int_v [B]^{\mathrm{T}} \{\sigma\}\, \mathrm{d}v \tag{18.88}$$

where $[B]$ = matrix expressing the strain at any point within the element in terms of the displacements at the nodes of the element (Equation 12.61); $\{\sigma\}$ = the last determined stress at any point. Analysis for material nonlinearity is illustrated in the following example.

EXAMPLE 18.5: PLANE TRUSS

Find the displacements at A and the forces in members AB, AC, and AD of the truss shown in Figure 18.13a, assuming the stress–strain diagram in Figure 18.13b. All members have the same cross-sectional area, a. The applied force $P = 1.75\sigma_y a$.

The restraining forces at the coordinates in Figure 18.13c are:

$$\{F\} = \{0, -P\} = -1.75\,\sigma_y\, a\,\{0, 1\}$$

In accordance with the σ-ε graph in Figure 18.13b, the axial force in any member is:

$$N = \begin{cases} a\,E_1\,\varepsilon & \text{with } \varepsilon \le \varepsilon_y \\ a\left[\sigma_y + E_2\left(\varepsilon - \varepsilon_y\right)\right] & \text{with } \varepsilon > \varepsilon_y \end{cases}$$

From the geometry of the structure (Figure 18.13a), the strain in the members is expressed in terms of the nodal displacements:

$$\{\varepsilon\} = \begin{Bmatrix} \varepsilon_{AB} \\ \varepsilon_{AC} \\ \varepsilon_{AD} \end{Bmatrix} = \frac{1}{l}\begin{bmatrix} -1 & 0 \\ 0 & 1 \\ 0.707 & 0.707 \end{bmatrix}\{D\}$$

The stiffness matrix for m members connected at one node (Figure 18.13b) is:

$$\left[S_t\right] = \sum_{i=1}^{m}\left[\frac{Ea}{l}\begin{bmatrix} c^2 & sc \\ sc & s^2 \end{bmatrix}\right]_i \tag{18.89}$$

where $s_i = \sin\theta_i$, $c_i = \cos\theta_i$, and θ_i is the angle, measured in the clockwise direction, between the x axis and the member. For the same type of structure, the forces at the nodes balanced by the axial forces $\{N\}$ in the m members are:

$$\left\{\bar{F}\right\} = -\sum_{i=1}^{m}\left\{N\begin{Bmatrix} c \\ s \end{Bmatrix}\right\}_i \tag{18.90}$$

The four calculation steps outlined in Section 18.5.1 follow below.

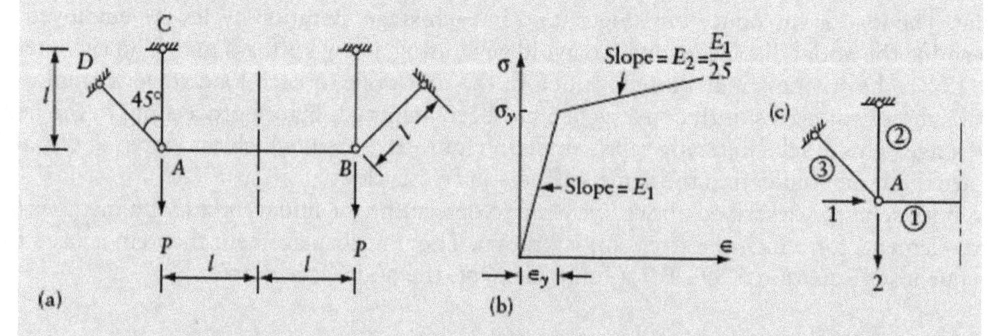

Figure 18.13 Plane truss of Example 18.5. (a) Truss geometry. (b) Stress–strain relation. (c) Coordinate system.

ITERATION CYCLE 1

1. $\{D^{(0)}\} = \{0\}$; $\{g^{(0)}\} = 1.75\ \sigma_y\ a\{0, 1\}$.
2. $\{\varepsilon^{(0)}\} = \{0\}$ and E for all members equals E_1. Equation 18.89 gives:

$$[S_t] = \frac{E_1\ a}{l}\begin{bmatrix} 1.5 & 0.5 \\ 0.5 & 1.5 \end{bmatrix}$$

$$\{\Delta D\} = [S_t]^{-1}\left\{g^{(0)}\right\} = (\sigma_y\ l/E_1)\{-0.438, 1.313\}.$$

3. $\left\{D^{(1)}\right\} = \left\{D^{(0)}\right\} + \{\Delta D\} = (\sigma_y\ l/E_1)\{-0.438, 1.313\}$.

4. $\left\{\varepsilon^{(1)}\right\} = \varepsilon_y\ \{0.438, 1.313, 0.619\}$.

$$\left\{N^{(1)}\right\} = a\,\sigma_y\ \{0.438, 1.013, 0.619\}.$$

Application of Equation 18.90 gives:

$$\left\{g^{(1)}\right\} = -\{F\} - \left\{\bar{F}^{(1)}\right\} = a\,\sigma_y\ \{0, 0.299\}$$

ITERATION CYCLE 2

1. $\{D^{(0)}\} = (\sigma_y l/E_1)\{-0.438, 1.313\}$; $\{g^{(0)}\} = a\sigma_y\ \{0, 0.299\}$; $\{\varepsilon^{(0)}\} = \varepsilon_y\ \{0.438, 1.313, 0.619\}$. At this strain level, E is equal to E_1 for members 1 and 3, and to E_2 for member 2. Use of Equation 18.89 gives:

$$[S_t] = \frac{E_1\ a}{l}\begin{bmatrix} 1.5 & 0.5 \\ 0.5 & 0.54 \end{bmatrix}$$

2. $\{\Delta D\} = [S_t]^{-1}\left\{g^{(0)}\right\} = (\sigma_y\ l/E_1)\{-0.267, 0.801\}$
3. $\left\{D^{(1)}\right\} = \left\{D^{(0)}\right\} + \{\Delta D\} = (\sigma_y\ l/E_1)\{-0.705, 2.114\}$
4. $\{\varepsilon^{(1)}\} = \varepsilon_y\{0.705,\ 2.114,\ 0.996\}$

$\{N^{(1)}\} = a\ \sigma_y\ \{0.705, 1.045, 0.996\}$; $\left\{g^{(1)}\right\} = -\{F\} - \left\{\bar{F}^{(1)}\right\} = \{0\}$.

The out-of-balance forces are zero, indicating that the exact solution has been reached.

EXAMPLE 18.6: SINGLE-BAY GABLE FRAME

Apply computer program PLASTICF to the plane frame in Figure 18.14a. Assume moment-curvature relationship as shown in Figure 18.14b. For the seven members in Figure 18.14a, assume the values of M_p in terms of Pb as: {765, 765, 1275, 1275, 1015, 1015, 616}. From the results of PLASTICF, plot the load multiplier versus the downward displacement at node 6 (Figure 18.14a).

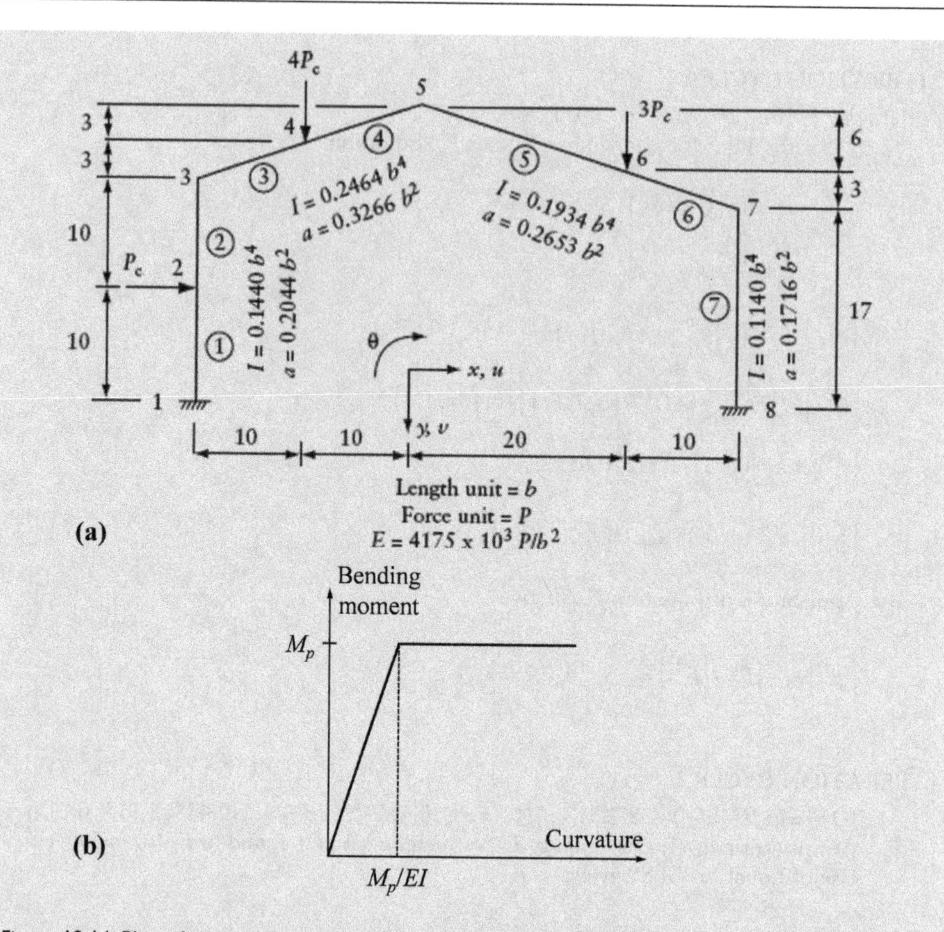

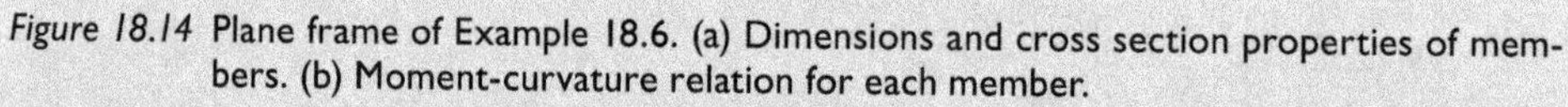

Figure 18.14 Plane frame of Example 18.6. (a) Dimensions and cross section properties of members. (b) Moment-curvature relation for each member.

The load multiplier increments that develop a new plastic hinge are determined by linear analysis. Four increments induce collapse (see Figure 18.15). The sums of increments versus the downward displacement at node 6 are:

Sum of increments in terms of P	*Displacement v at node 6 in terms of b*	*New plastic hinge developed at:*
0	0	—
34.46	0.053	Node 7 of member 7
40.33	0.079	Node 3 of member 2
41.86	0.089	Node 8 of member 7
57.34	0.210	Node 6 of member 5

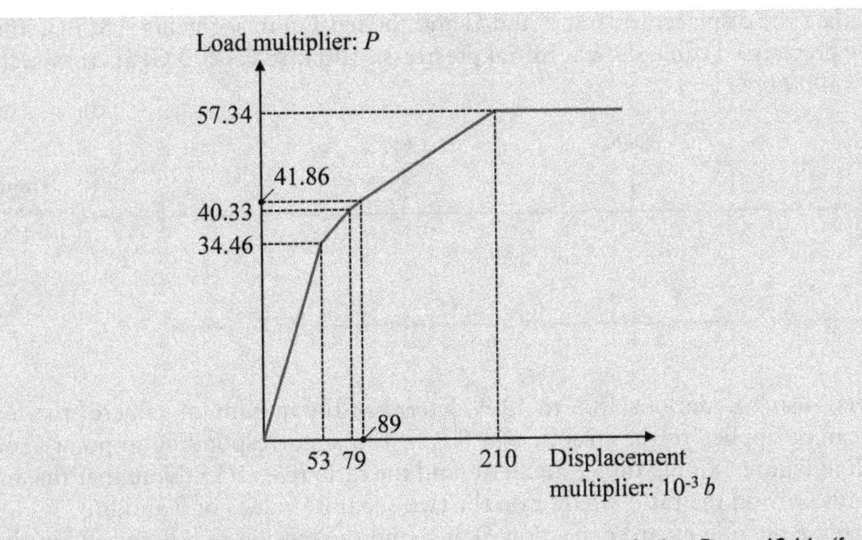

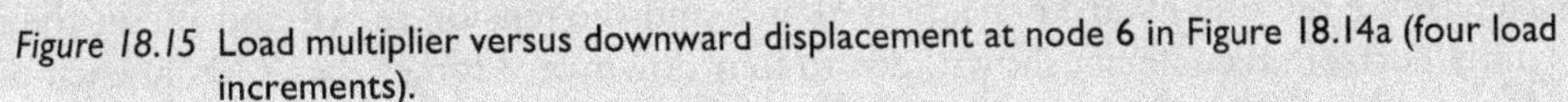
Figure 18.15 Load multiplier versus downward displacement at node 6 in Figure 18.14a (four load increments).

18.13 GENERAL

Except for simple structures, use of a computer is necessary to perform nonlinear analysis, generally requiring iterations. Understanding of nonlinear behavior can be enhanced by the use of computer programs, which permit the introduction of loads in multi-stages, allow changing the number of iteration cycles, and give the out-of-balance forces at the end of each iteration cycle. Available computer programs, described in Chapter 20, may be used for this purpose.

REFERENCES

1. Levy, R. and Spillers, W.R., *Analysis of Geometrically Nonlinear Structures*, Chapman and Hall, New York, 1995, 199 pp.
2. Crisfield, M.A., *Non-linear Finite Element Analysis of Solids and Structures*, Vol. 1, Wiley, Chichester, 1991, 345 pp.
3. Williams, F.W., "An Approach to the Nonlinear Behaviour of the Members of A Rigid-Jointed Plane Framework with Finite Deflections," *Q. J. Mech. Appl. Math.*, XVII(Pt. 4), (1964), pp. 451–469.
4. Salami, A.T. and Morley, C.T., "Finite Element Analysis of Geometric Nonlinearity in Plane Frameworks," *Struct. Eng.*, 70(15), (1992), pp. 268–271.

PROBLEMS

18.1 Perform two iteration cycles to determine, for $F=90$, the corresponding value of D and the forces in the members of the truss in Figure 18.2, using the tangent stiffness matrix derived in Section 18.6. Approximate answers are given in Figure 18.3.

18.2 Calculate the displacements at B and C and the tension in segments AB, BC, and CD of the prestressed cable shown. Initial prestress = 100 kN, $E = 200$ GPa; cross-sectional area = 100 mm^2; $b = 1$ m.

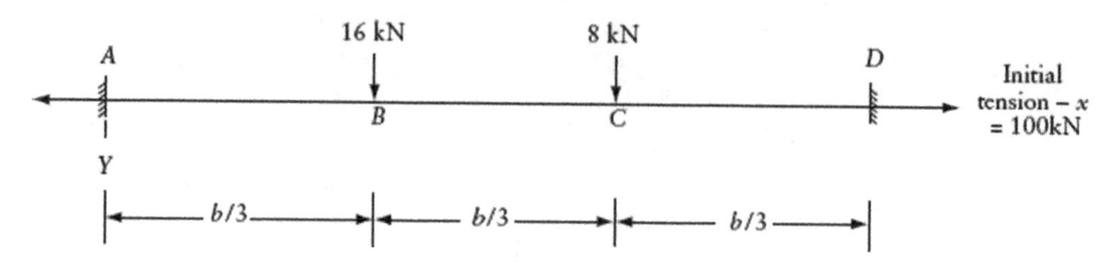

18.3 By trial, using Equations 18.6 to 18.9, determine the maximum value of the force F that can be applied to the structure in Figure 18.2a (corresponding to point C on the graph in Figure 18.2b). Take $Ea = 8 \times 10^6$ and the ratio $h/b = 1/15$. Note that the answer depends only on the ratio h/b, not on the two separate values of h and b.

18.4 Determine the downward deflection D at B and the tension in AB and BC in the initially horizontal cable net shown, due to four downward forces $P = 12$ kN at each of B, C, F, and G. Take initial tension in each cable to be 180 kN and $Ea = 60$ MN.

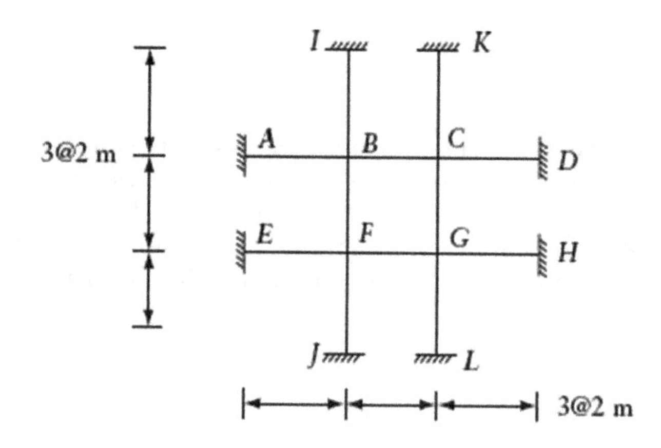

18.5 The figure shows the top view and elevation of a shallow space truss. Determine the vertical deflection of node A and the forces in the members due to a downward force $P = 100$ at the same node. Take $Ea = 40{,}000$ for all members.

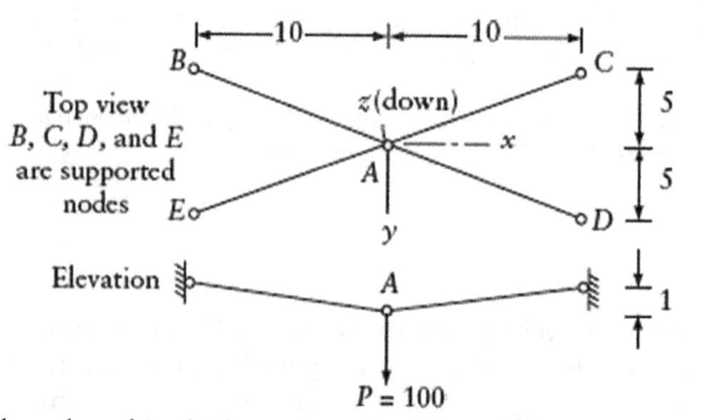

18.6 Determine the value of P which produces buckling of the space truss of Problem 18.5, with the direction of the force P reversed.

18.7 Determine the displacements at B and the member end forces for AB and BC of the structure shown. Take $b=100$; $h=2$; $Q=0.8$; $Ea=1.0\times10^6$; $EI=0.1\times10^6$, where a and I are the cross-sectional area and the second moment of area of all members.

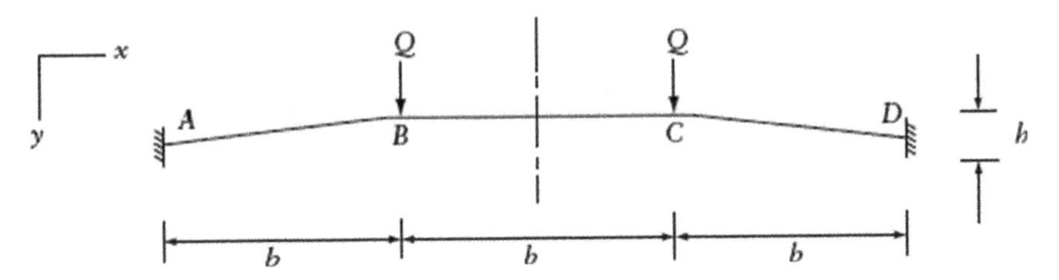

18.8 The plane frame of Problem 10.12 is subjected, in addition to the downward forces Q at B and C, to two horizontal forces at the same nodes, each equal to $Q/10$ pointing to the right-hand side. Find the nodal displacements and the member end forces. Take $l=6c$; $h=2.4c$; for all members, $a=25\times10^{-3}c^2$, $I=60\times10^{-6}c^4$, and $E=40\times10^9$ (F/c^2); and $Q=1.5\times10^6$ F, where c and F are force and length units, respectively.

18.9 Find the displacement D due to an axial force F applied at the end of a straight prismatic bar of length l, assuming the stress–strain relation $\sigma=C(\varepsilon-\varepsilon^2/4)$, where C=constant. Take $F=0.7Ca$, with a being the cross-sectional area. Apply the Newton–Raphson technique and give the answers for two iterations.

18.10 The figure shows the top view of a cable network. The z coordinates of the nodes in their initial positions are given by $z=xy/(8l)$. This equation describes a hyperbolic surface, having concave and convex curvatures along AC and BD, respectively; the nodes having x or y=constant lie on straight lines. The nodes on the external perimeter ABCD are supported, while each of the internal nodes is subjected to a downward force $F_z=Q$. Prior to application of F_z at the nodes, each cable is pre-tensioned with a force, $P=30$ Q. Determine the displacements at nodes 5, 6, 7, and 11 and the forces after loading the cable segments 5–6, 6–7, 1–5, 2–6, 6–11, and 3–7. Data: $l=10$ m; cross-sectional area of cables, $a=l^2/10^4$; modulus of elasticity of cables=200×10^9 N/m²; $Q=200\times10^3$ N. Program NLST can solve this problem; the input is given in file NLST.INM. Verify equilibrium of selected nodes using the displacement values given in the answers.

Parabolic paraboloid roof: Use NLST to analyze a saddle-shaped roof, adding two triangular membrane elements in each panel. The answers to this exercise are not provided.

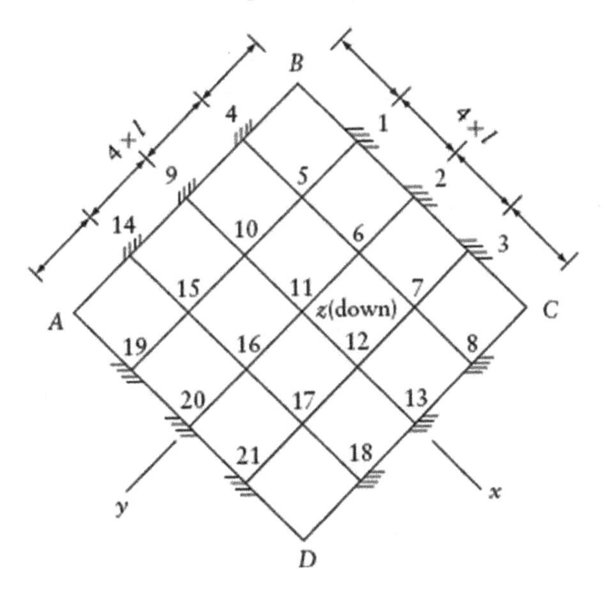

Chapter 19

Computer analysis of framed structures

19.1 INTRODUCTION

This chapter discusses the use of computers in the analysis of large structures by the displacement (stiffness) method. We consider six types of framed structures: plane and space trusses with pin joints, plane and space frames, plane grids, and eccentric grids with rigid joints. The technique that applies to frames apply also to structures idealized as finite-elements.

We number the nodes (joints), the members, and the coordinates sequentially. Figure 19.1 indicates the sequence of coordinate numbering. For example, the coordinates at the ith node of a plane truss are coordinates numbers $(2i - 1)$ and $2i$, respectively. The location of the origin and the directions of the orthogonal x, y (and z) axes are arbitrary.

The analysis gives the nodal displacements in the global directions, and the member end forces in local directions of individual members (Section 19.2). Before proceeding with this chapter, some readers may find it beneficial to go over the summary given in Section 19.13.

19.2 MEMBER LOCAL COORDINATES

We generate the stiffness matrix $[S^*]$ of a member of a framed structure with respect to local coordinates. These are in the directions of the centroidal axis and the principal axes of the member cross-section; then, we transform $[S^*]$ to coordinates parallel to the global directions. The stiffness matrix of the structure = the algebraic sum of stiffness matrices of individual members. This is possible when all matrices refer to coordinates in global directions.

Solution of a set of simultaneous equations gives the nodal displacements in global directions. For each member we transform the nodal coordinates to local directions to determine the member end forces.

Figures 19.1 and 19.2 show nodal coordinates in global directions and member coordinates in local directions. We specify the first arbitrarily by the order in which the two nodes are listed in the member information in the input data (see Section 19.4). Each set of global and local axes is an orthogonal right-hand triad.

19.2.1 Plane and eccentric grids

Analysis of a plane grid, in which the longitudinal axis of each member is on a horizontal plane, can be done considering three degrees-of-freedom per node: vertical deflection and two rotations about horizontal orthogonal axes (only $\{v, \theta_x, \theta_z\}$ of Figure 19.1). The results would be adequate for horizontal concrete flat plates of constant thickness, subjected to gravity loads and no prestressing. Analysis of a grid of short beam elements having five DOFs per node[1] (Figure 19.3) applies to prestressed floors subjected to gravity and

DOI: 10.1201/9780429286858-19

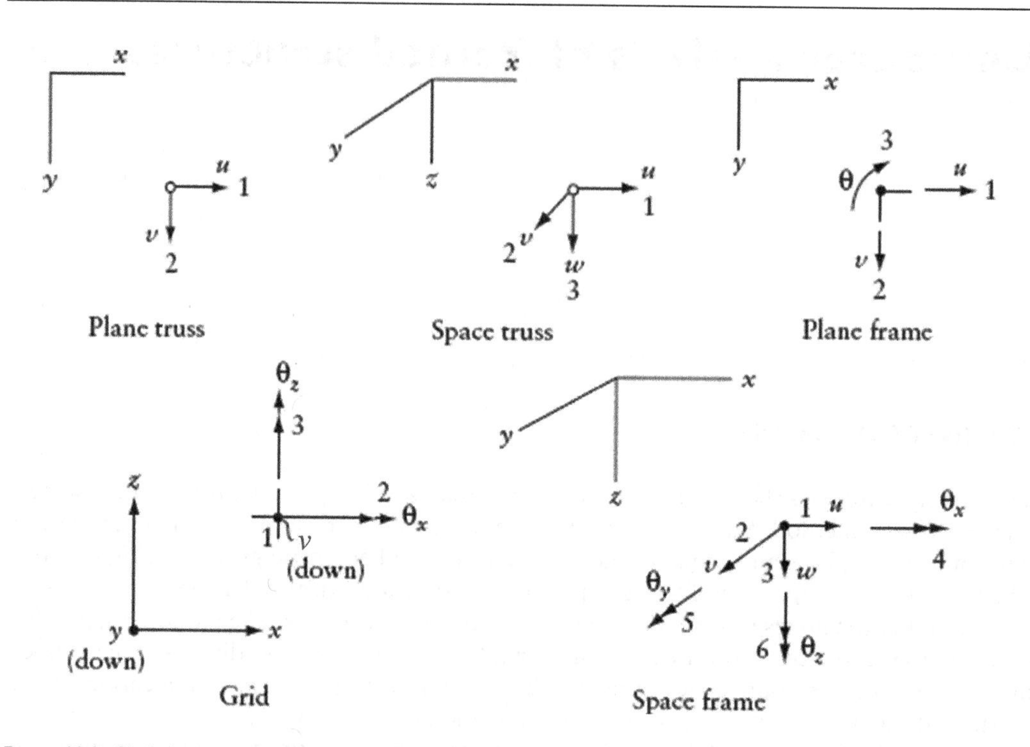

Figure 19.1 Global axes, degrees of freedom, and order of numbering of the coordinates at typical nodes of framed structures.

prestressing forces. The nodes are in a horizontal reference plane; the centroidal axis of each member, x^* runs horizontally at a distance y_c below the reference plane. The displacements in local directions at member ends, $\{D^*\} = \{v^*, \theta_x^*, \theta_z^*, u^*, w^*\}$; where v^*, u^*, and w^* = translations in local y^*, x^*, and z^* directions, respectively; θ_x^* and θ_z^* = rotations about x^* and z^* axes, respectively.

19.3 BAND WIDTH

Generally, the non-zero elements of [S] are limited to a band adjacent to the diagonal (Figure 19.4). This property of [S], combined with the fact that the matrix is symmetrical, is used to conserve computer storage space and to reduce the number of computations. We generate only the diagonal elements of [S] and the part of the band above the diagonal. We store the elements in a rectangular matrix of n rows and n_b columns (Figure 19.4), where n is the number of degrees of freedom; n_b is the *half-band- width* = the number of columns:

$$n_b = s\left(|k - j|_{\text{largest}} + 1 \right) \tag{19.1}$$

where s = number of degrees-of-freedom per node. A plane truss has 2, space truss and plane frame have 3, eccentric grid has 5, and a space frame has 6 degrees-of-freedom per node. The term $|k - j|$ = absolute value of the difference between the node numbers at the two ends of a member. The member having largest $|k - j|$ determines the bandwidth.

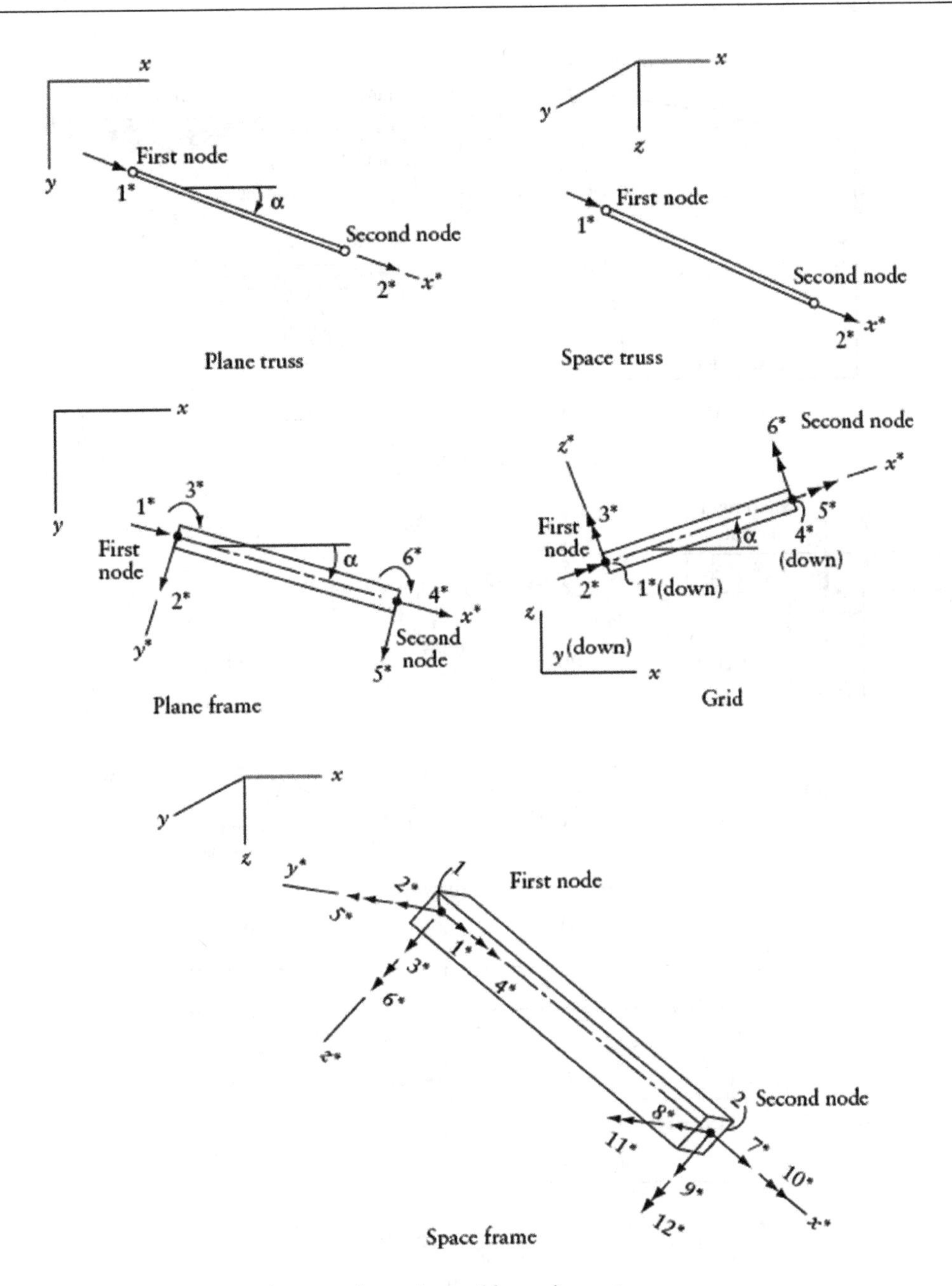

Figure 19.2 Local coordinates for typical members of framed structures.

The equilibrium equations are:

$$[S][D] = -[F]$$

The solution of the equilibrium equations gives the unknown displacements at the coordinates for a number of load cases in one computer run. Each of matrices $[D]$ and $[F]$ has n rows; where n = number of nodes multiplied by the number of degrees-of-freedom per nodes. The number of columns in each of $[D]$ and $[F]$ = number of load cases. The respective columns of $[D]$ and $[F]$ correspond to the same loading case. The number of arithmetic

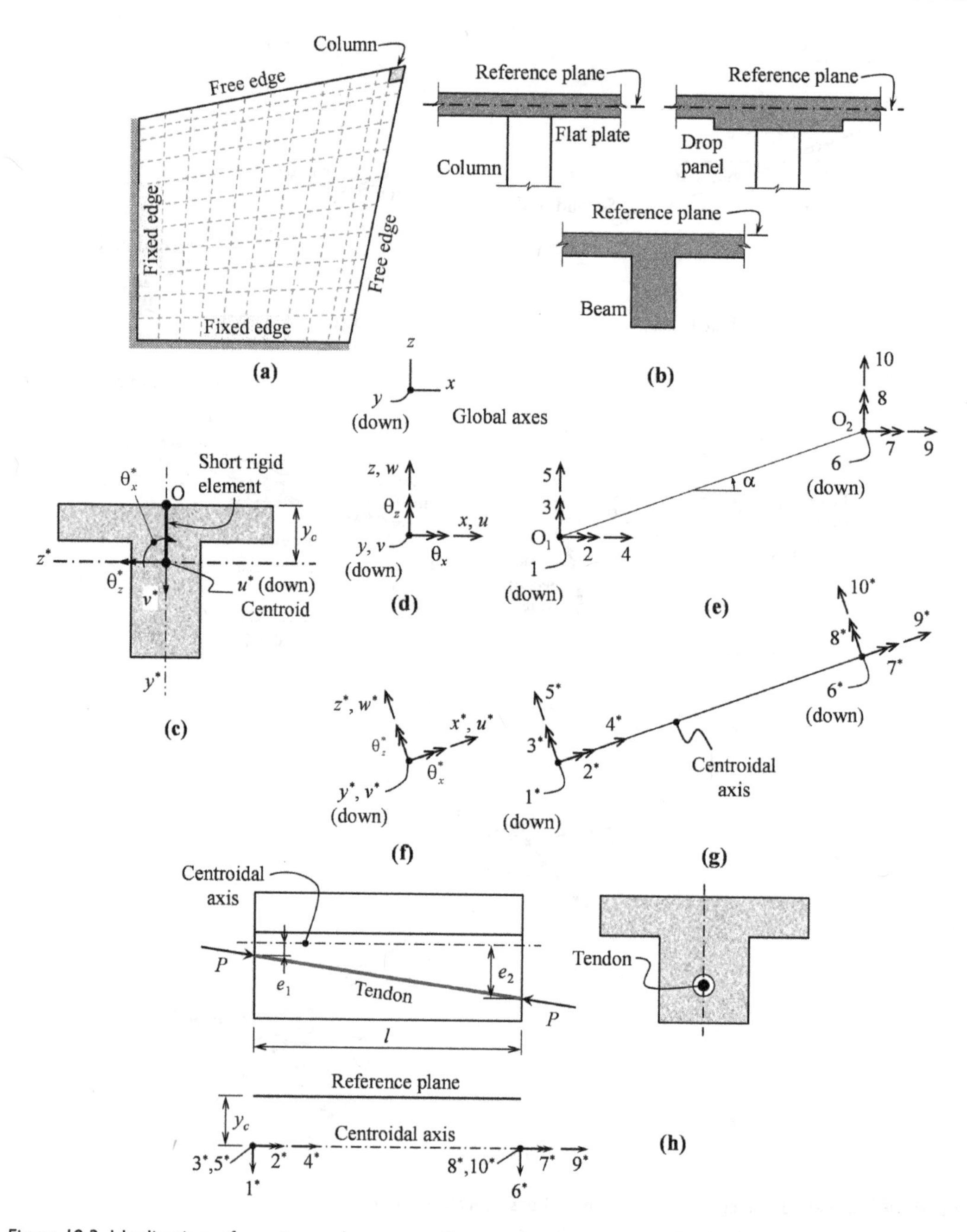

Figure 19.3 Idealization of prestressed concrete floors using eccentric grids. (a) Top view of an irregular concrete floor. (b) Arbitrary level of reference plane. (c) Fictitious rigid elements connecting member ends to nodes. (d) Global directions. (e) Global directions at reference plane. (f) Local coordinates at centroidal axis. (g) Local coordinates at centroidal axis level; positive directions of $\{D^*\}$ and $\{F^*\}$. (h) Forces to be applied for prestressing.

operations in the solution of the equilibrium equations increases linearly with n and with n_b^2. Thus, we have an interest in reducing n_b.

For a given structure, numbering the nodes sequentially across the side that has a smaller number of nodes makes a narrower bandwidth. Figure 19.4 shows suggested node numbering of a plane frame; the corresponding $[S]$ is stored in a rectangular matrix having

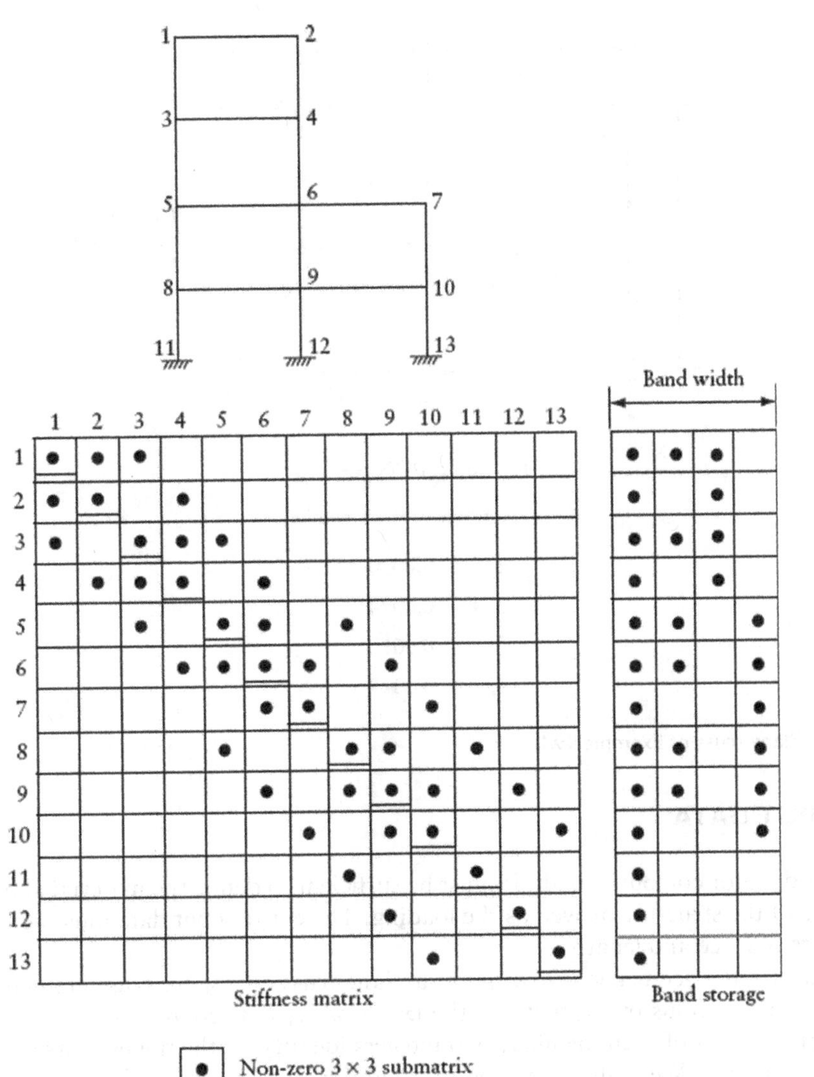

Figure 19.4 Node numbering and band width for a plane frame.

$n_b = 12$ columns. For the same structure, $n_b = 18$ (instead of 12) when the nodes are numbered sequentially down the columns. Note that [S] generally has zeros within the band (as in Figure 19.4).

There exist several techniques aimed at efficiency and accuracy in equation solvers by automatic renumbering of the nodes to minimize n_b, by generating and operating only on the nonzero entries of [S] and by estimating and controlling round-off errors.

The stiffness matrix discussed above (Figure 19.4) is for a free (unsupported) structure. Elimination of rows and columns corresponding to the coordinates where the displacements are zero, results in a condensed stiffness matrix of the supported structure (see Section 5.12).

It is, of course, possible to avoid generating the non-required rows and columns of [S]. However, for simplicity in computer programming, at the expense of more computing, it is possible to work with the stiffness matrix of the free structure by adjusting [S]. The adjustment, discussed in Section 19.10, causes the displacement to be equal to zero or equal to a prescribed value at any specified coordinates.

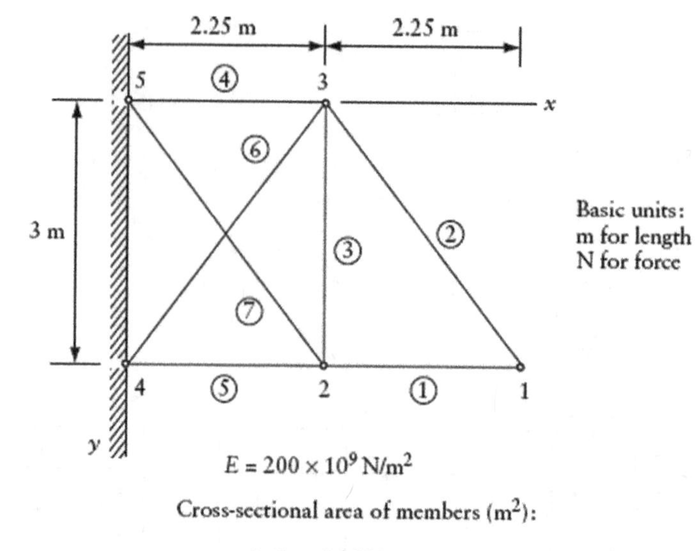

Figure 19.5 Plane truss of Example 19.1.

19.4 INPUT DATA

The input data for computer analysis must be sufficient to define the material and geometric properties of the structure as well as the loading. The entire input data must be in the same set of units of force and length.

The data for the geometric properties of a plane truss or a space truss are defined by the following: the modulus of elasticity E; the (x, y) or (x, y, z) coordinates of the nodes; the cross-sectional area of each member; two integers identifying the nodes at its ends; and the support conditions. We define the support conditions by restrain indicators: integer 1 for free (unrestrained) displacement; integer 0 for prescribed displacement. When the restraint indicator is zero, the magnitude of the prescribed displacement, we give a real value (0.0 for a support with no settlement). Table 19.1 gives the input data for the plane truss in Figure 19.5.

When preparing the data for support conditions, it is important to ensure that the structure cannot translate or rotate freely as a rigid body. This requirement for stability should be verified before the computer analysis of any structure (particularly, spatial structures).

Ideally, a truss should be loaded only at the nodes so that there are only axial forces in the members. The input data for the load at a node are the node number and the components (F_x, F_y) or (F_x, F_y, F_z).

When the analysis is for a temperature rise in a member, the data are the member number and the two end forces $\{A_r\}$ (at the local coordinates) which occur with prevented nodal displacements. This produces an axial compressive force of magnitude αTEa, where α = the coefficient of thermal expansion, T = the temperature rise, E = modulus of elasticity, and a = cross-sectional area.

When the analysis is for a prescribed displacement at a support, we may adjust the stiffness matrix (Section 19.3); however, we can do this only when the analysis is for a single loading case. To avoid this restriction, we calculate the member end forces for the members

Table 19.1 Input data for a plane truss (Figure 19.5)

$E=200\times10^9$				
Nodal coordinates				
Node	x	y		
1	4.5	3.0		
2	2.25	3.0		
3	2.25	0.0		
4	0.0	3.0		
5	0.0	0.0		
Element information				
Element	First node	Second node	a	
1	2	1	0.0030	
2	3	1	0.0024	
3	2	3	0.0012	
4	5	3	0.0024	
5	4	2	0.0030	
6	4	3	0.0048	
7	5	2	0.0048	
Support conditions				
Node	Restraint indicator		Prescribed displacement	
	u	v	u	v
4	0	0	0.0	0.0
5	0	0	0.0	0.0
Loading case 1				
Forces applied at nodes				
Node	F_x	F_y		
1	0.0	100×10^3		
Loading case 2				
Member end forces with displacements restrained				
Member	A_{r1}	A_{r2}		
4	96×10^3	-96×10^3		
Loading case 3				
Member end forces with displacements restrained				
Member	A_{r1}	A_{r2}		
6	-410×10^3	410×10^3		

connected to the displaced node and include them in the input data in the same way as the data for temperature.

EXAMPLE 19.1: PLANE TRUSS

Prepare the input data for the analysis of the plane truss in Figure 19.5 with three load cases. (1) A single downward force of 100×10^3 N at node 1; (2) a temperature rise of 20 degrees of member 4, the coefficient of thermal expansion being 10^{-5} per degree; and (3) a downward movement of 0.002 m of node 4. Table 19.1 gives the required input data.

For a plane frame, the input data differ from those of a plane truss. The second moment of area of cross-section of members are needed. We need three restrain indicators and three prescribed displacements u, v, and θ for a supported node. The forces at a node have three components (Figure 19.2). For loaded members, we need member end forces $\{A_r\}$, with the nodal displacements. Example 19.2 gives the data for a plane frame.

The input data for a grid (or eccentric grid) are similar to those for a plane frame, except that the cross-sectional torsion constant J (Appendix F) is required for the members instead of their area; the modulus of elasticity in shear G is also required. Figures 19.1 gives the displacement components at a typical node, and the local coordinates of a typical member of a grid are shown in 19.2.

The data defining the geometric properties of a space frame are: the (x, y, z) coordinates of the nodes, two integers identifying the nodes at its ends, the area a of the cross-section, its torsion constant J, and its second moments of area I_y^* and I_z^* about the centroidal principal axes parallel to the local y^* and z^* axes (Figure 19.2). While the direction of the local x^* axis is defined by the (x, y, z) coordinates of the nodes, the y^* direction needs further definition. For this reason, we include in the data for each member the direction cosines of y^*: $\lambda_{y^*x}, \lambda_{y^*y}, \lambda_{y^*z}$. Here, the λ values are the cosines of the angles between the local y^* axis and the global directions. The third direction z^* is orthogonal to the other two local axes and, thus, does not require additional data.

In most practical cases, we select the global axes in horizontal and vertical directions and the member cross sections have one of their principal axes horizontal or vertical. The y^* axis can be considered in such a direction and its direction cosines will be easy to find (Equation 19.9a).

The data for loading of a space frame are similar to those for a plane frame, noting that the forces at the nodes have six components in the global directions and each member end has six forces defined in Figures 19.1 and 19.2.

EXAMPLE 19.2: PLANE FRAME

Prepare the input data for the analysis of the plane frame in Figure 19.6 for the following loadings: (1) the loads shown; (2) displacements $u=0.2b$ and $v=0.5b$ of the support at node 1.

Table 19.2 gives the required input data. We may calculate the member end forces included in the data using Appendices B and C.

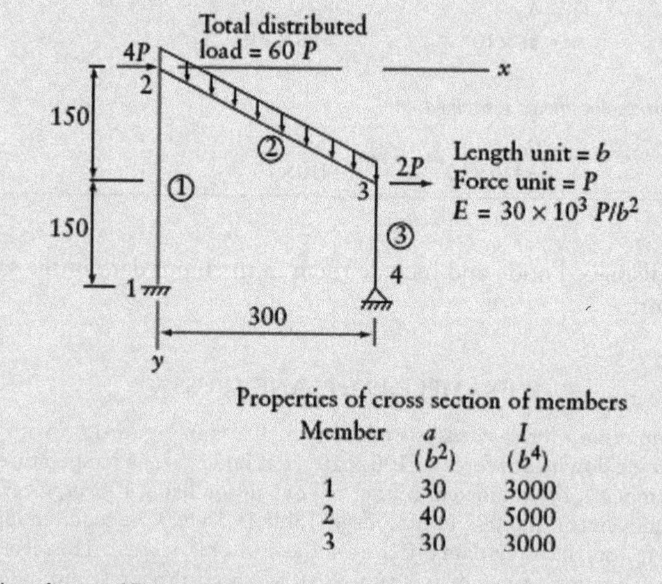

Properties of cross section of members

Member	a (b^2)	I (b^4)
1	30	3000
2	40	5000
3	30	3000

Figure 19.6 Plane frame of Example 19.2.

Table 19.2 Input data for a plane frame (Figure 19.6)

E = 30,000.0

Nodal coordinates

Node	*x*	*y*
1	0.0	300.0
2	0.0	0.0
3	300.0	150.0
4	300.0	300.0

Element information

Element	First node	Second node	*a*	*I*
1	1	2	30.0	3000.0
2	2	3	40.0	5000.0
3	3	4	30.0	3000.0

Support conditions

Node	Restraint indicator			Prescribed displacement		
	u	*v*	θ	*u*	*v*	θ
1	0	0	0	0.0	0.0	0.0
4	0	0	1	0.0	0.0	0.0

Loading case 1 *Forces applied at nodes*

Node	F_x	F_y
2	4.0	0.0
3	2.0	0.0

Member end forces with displacements restrained

Member	A_{r1}	A_{r2}	A_{r3}	A_{r4}	A_{r5}	A_{r6}
2	−13.42	−26.83	−1500.0	−13.42	−26.83	1500.0

Loading case 2

Member end forces with displacements restrained

Member	A_{r1}	A_{r2}	A_{r3}	A_{r4}	A_{r5}	A_{r6}
1	−1500.0	8.0	1200.0	1500.0	−8.0	1200.0

19.5 DIRECTION COSINES OF ELEMENT LOCAL AXES

The direction cosines λ of the member local x^*, y^*, z^* axes will be used in the transformation matrices. The coordinates (x, y) defining the position of the nodes give the λ-values for the local x^* axis. For a member of a plane truss or a plane frame:

$$\lambda_{x^*x} = \frac{x_2 - x_1}{l_{12}}; \quad \lambda_{x^*y} = \frac{y_2 - y_1}{l_{12}} \tag{19.2}$$

where

$$l_{12} = \sqrt{(x_2 - x_1)^2 + (y_2 - y_1)^2} \tag{19.3}$$

The subscripts 1 and 2 refer to the first and second nodes of the member.

In a plane frame, the y^* axis for any member is perpendicular to x^* (Figure 19.2), so that the λ values for the y^* axis are:

$$\lambda_{y^*x} = -\lambda_{x^*y}; \quad \lambda_{y^*y} = \lambda_{x^*x} \tag{19.4}$$

For a member of a grid or an eccentric grid in the x-z plane, the direction cosines required for the local x^* and z^* axes are λ_{x^*x}, λ_{x^*z}, λ_{z^*x}, and λ_{z^*z}. Use Equations 19.2–19.4, replacing y with z and y^* with z^*.

The direction cosines of the local x^* axis of a member of a space truss or a space frame are:

$$\lambda_{x^*x} = \frac{x_2 - x_1}{l_{12}}; \quad \lambda_{x^*y} = \frac{y_2 - y_1}{l_{12}}; \quad \lambda_{x^*z} = \frac{z_2 - z_1}{l_{12}} \tag{19.5}$$

where

$$l_{12} = \sqrt{(x_2 - x_1)^2 + (y_2 - y_1)^2 + (z_2 - z_1)^2} \tag{19.6}$$

As mentioned in the preceding section, the λ values for the local y^* axis of each member are included in the input data. The direction cosines of the local z^* axis are obtained from the cross product of two unit vectors in the x^* and y^* directions:

$$\lambda_{z^*x} = \lambda_{x^*y}\,\lambda_{y^*z} - \lambda_{x^*z}\,\lambda_{y^*y} \tag{19.7}$$

$$\lambda_{z^*y} = \lambda_{x^*z}\,\lambda_{y^*x} - \lambda_{x^*x}\,\lambda_{y^*z} \tag{19.8}$$

$$\lambda_{z^*z} = \lambda_{x^*x}\,\lambda_{y^*y} - \lambda_{x^*y}\,\lambda_{y^*x} \tag{19.9}$$

In practice, the global x and y axes for space frames are selected in horizontal directions and the local axis y^* for most members can be considered horizontal. To simplify the input data, the computer program SPACEF (available from a web site given in Chapter 20) requires that the direction cosines for axis y^* be given only for a special member when the assumption that its y^* axis is parallel to the global x-y plane is not acceptable. Program SPACEF (Chapter 20) does not require the direction cosines of local y^*-axis when the member runs parallel to plane x-y of global axes; the program calculates the direction cosines by:

$$\lambda_{y^*x} = \frac{y_2 - y_1}{\sqrt{(x_2 - x_1)^2 + (y_2 - y_1)^2}}; \quad \lambda_{y^*y} = \frac{x_2 - x_1}{\sqrt{(x_2 - x_1)^2 + (y_2 - y_1)^2}}; \quad \lambda_{y^*z} = 0 \tag{19.9a}$$

19.6 ELEMENT STIFFNESS MATRICES

The stiffness matrices of individual members of a framed structure with respect to local coordinates (Figure 19.2) can be generated from Appendix C.

For a member of a *plane truss* or a *space truss* (Figure 19.2), the stiffness matrix is:

$$\left[S^*\right]=\frac{Ea}{l}\begin{bmatrix}1 & -1\\ -1 & 1\end{bmatrix} \tag{19.10}$$

The stiffness matrix for a member of a *plane frame* (Figure 19.2), given in Equation 5.6 is:

$$\left[S^*\right]=\begin{bmatrix} Ea/l & & & & & \\ & 12EI/l^3 & & & & \\ & 6EI/l^2 & 4EI/l & & & \\ -Ea/l & & & Ea/l & & \\ & -12EI/l^3 & -6EI/l^2 & & 12EI/l^3 & \\ & 6EI/l^2 & 2EI/l & & -6EI/l^2 & 4EI/l \end{bmatrix} \quad \text{Symmetrical; elements not shown are zero} \tag{19.11}$$

In the above equation, each member has its centroidal axis in the x-y plane, and $I \equiv I_z^*$ with z^* perpendicular to the same plane.

The stiffness matrix for a member of a *grid* is (Figure 19.2):

$$\left[S^*\right]=\begin{bmatrix} 12EI/l^3 & & & & & \\ & GJ/l & & & & \\ 6EI/l^2 & & 4EI/l & & & \\ -12EI/l^3 & & -6EI/l^2 & 12EI/l^3 & & \\ & -GJ/l & & & GJ/l & \\ 6EI/l^2 & & 2EI/l & -6EI/l^2 & & 4EI/l \end{bmatrix} \quad \text{Symmetrical; elements not shown are zero} \tag{19.12}$$

For a member if an *eccentric grid* is (Figure 19.3):

$$\left[S^*\right]=\begin{bmatrix}\left[S^*\right]_{11} & \left[S^*\right]_{12}\\ \left[S^*\right]_{21} & \left[S^*\right]_{22}\end{bmatrix} \tag{19.13}$$

$$\left[S^*\right]_{11}=\left[S^*\right]_{22}=\begin{bmatrix} 12EI/l^3 & 0 & 6EI/l^2 & 0 & 0\\ 0 & GJ/l & 0 & 0 & 0\\ 6EI/l^2 & 0 & 4EI/l & 0 & 0\\ 0 & 0 & 0 & Ea/l & 0\\ 0 & 0 & 0 & 0 & 0 \end{bmatrix} \tag{19.14}$$

$$\left[S^*\right]_{21}=\left[S^*\right]_{12}^{T}=\begin{bmatrix} -12EI/l^3 & 0 & -6EI/l^2 & 0 & 0 \\ 0 & -GJ/l & 0 & 0 & 0 \\ 6EI/l^2 & 0 & 2EI/l & 0 & 0 \\ 0 & 0 & 0 & -Ea/l & 0 \\ 0 & 0 & 0 & 0 & 0 \end{bmatrix} \tag{19.15}$$

In Equations 19.12, 19.14, and 19.15, the centroidal axes of the members are in the x-z plane and $I \equiv I_z^*$, with z^* situated in the same plane.

Equation 5.18 gives the stiffness matrix for a member of a *space frame* (Figure 19.2). In the above matrices, we ignore shear deformations. To consider shear deformations, we adjust the elements of $[S^*]$ associated with shearing force or bending moment (Section 11.2). Equation 11.1 gives the stiffness matrix of a typical member of a plane frame, accounting for deformations due to axial force, bending, and shear.

19.7 TRANSFORMATION MATRICES

The displacements $\{D^*\}$ in local directions at the end of a member of a framed structure of any type (Figure 19.2) are related to the displacements at the same node in global directions (Figure 19.1) by the geometrical relation:

$$\{D^*\} = [t]\,\{D\} \tag{19.16}$$

where $[t]$ is a transformation matrix which is generated in terms of the direction cosines of the local axes x^*, y^*, z^* with respect to the global directions. For the different types of framed structures, the $[t]$ matrix is:

$$\text{Plane truss } [t] = \begin{bmatrix} \lambda_{x^*x} & \lambda_{x^*y} \end{bmatrix} \tag{19.17}$$

$$\text{Space truss } [t] = \begin{bmatrix} \lambda_{x^*x} & \lambda_{x^*y} & \lambda_{x^*z} \end{bmatrix} \tag{19.18}$$

$$\text{Plane frame } [t] = \begin{bmatrix} \lambda_{x^*x} & \lambda_{x^*y} & 0 \\ \lambda_{y^*x} & \lambda_{y^*y} & 0 \\ 0 & 0 & 1 \end{bmatrix} \tag{19.19}$$

$$\text{Plane grid } [t] = \begin{bmatrix} 1 & 0 & 0 \\ 0 & \lambda_{x^*x} & \lambda_{x^*z} \\ 0 & \lambda_{z^*x} & \lambda_{z^*z} \end{bmatrix} \tag{19.20}$$

$$\text{Eccentric grid } [t] = \begin{bmatrix} 1 & 0 & 0 & 0 & 0 \\ 0 & \lambda_{x^*x} & \lambda_{x^*z} & 0 & 0 \\ 0 & \lambda_{z^*x} & \lambda_{x^*x} & 0 & 0 \\ 0 & y_c\lambda_{x^*z} & -y_c\lambda_{x^*x} & \lambda_{x^*x} & \lambda_{x^*z} \\ 0 & y_c\lambda_{x^*x} & y_c\lambda_{x^*z} & \lambda_{z^*x} & \lambda_{x^*x} \end{bmatrix} \tag{19.21}$$

$$\text{Space frame } [t] = \begin{bmatrix} \lambda_{x^*x} & \lambda_{x^*y} & \lambda_{x^*z} \\ \lambda_{y^*x} & \lambda_{y^*y} & \lambda_{y^*z} \\ \lambda_{z^*x} & \lambda_{z^*y} & \lambda_{z^*z} \end{bmatrix} \tag{19.22}$$

The transformation matrix given in the last equation is used to relate the three translations $\{u^*, v^*, w^*\}$ to $\{u, v, w\}$. The same matrix can be used to relate $\{\theta^*_x, \theta^*_y, \theta^*_z\}$ to $\{\theta_x, \theta_y, \theta_z\}$.

A member stiffness matrix $[S^*]$, with respect to local coordinates, can be transformed into a stiffness matrix $[S_m]$ corresponding to coordinates in the global directions (see Equation 8.47) by:

$$[S_m] = [T]^T [S^*][T] \tag{19.23}$$

where:

$$[T] = \begin{bmatrix} [t] & [0] \\ [0] & [t] \end{bmatrix} \tag{19.24}$$

The above equation may be used for members of a plane truss, a space truss, a plane frame, or a grid substituting for $[t]$ from one of the Equations 19.17 to 19.21. However, for a member of a space frame:

$$[T] = \begin{bmatrix} [t] & & & \\ & [t] & & \\ & & [t] & \\ & & & [t] \end{bmatrix} \quad \text{Submatrices not shown are null} \tag{19.25}$$

where $[t]$ is given by Equation 19.22.

The $[T]$ matrices transform the actions in local directions at the ends of a member into equivalent forces along nodal coordinates in the global directions (see Equation 8.46b):

$$\{F\}_m = [T]^T \{A_r\}_m \tag{19.26}$$

where the subscript m refers to element number.

The elements of the matrices $[S_m]$ and $\{F\}_m$ are forces in the global directions; they represent contributions of one member. Summing the matrices for all members gives the stiffness matrix and load vector of the assembled structure, as shown in Section 19.9.

When the x-axis is parallel to x^*-axis in Fig. 19.3f, $\lambda_{x^*x} = 1$ and the remaining λ-values $= 0$ in Equation 19.21. When, in addition, the coordinates 4^* and 5^* are omitted, Equation 19.23 becomes the same as Equation 19.20.

19.8 MEMBER STIFFNESS MATRICES WITH RESPECT TO GLOBAL COORDINATES

Equation 19.23 expresses the stiffness matrix $[S_m]$ for a member of a framed structure with respect to coordinates at its two ends in the direction of the global axes. Instead of the matrix product in Equation 19.23, we give the stiffness matrix $[S_m]$ in explicit form for a member of framed structures of various types.

Plane truss:

$$[S_m] = \frac{Ea}{l}\left[\begin{array}{cc|cc} c^2 & cs & -c^2 & -cs \\ cs & s^2 & -cs & -s^2 \\ \hline -c^2 & -cs & c^2 & cs \\ -cs & -s^2 & cs & s^2 \end{array}\right] \tag{19.27}$$

where $c = \cos\alpha = \lambda_{x^*x}$; $s = \sin\alpha = \lambda_{x^*y}$; and α is the angle defined in Figure 19.2 with its positive sign convention.

Space truss:

$$[S_m] = \left[\begin{array}{c|c} [S_{11}] & [S_{12}] \\ \hline [S_{21}] & [S_{22}] \end{array}\right]_m \tag{19.28}$$

where:

$$[S_{11}]_m = \frac{Ea}{l}\begin{bmatrix} \lambda_{x^*x}^2 & & \text{Symm.} \\ \lambda_{x^*y}\lambda_{x^*x} & \lambda_{x^*y}^2 & \\ \lambda_{x^*z}\lambda_{x^*x} & \lambda_{x^*y}\lambda_{x^*z} & \lambda_{x^*z}^2 \end{bmatrix} \tag{19.29}$$

$$[S_{22}]_m = [S_{11}]_m;\quad [S_{21}]_m = -[S_{11}]_m;\quad [S_{12}]_m = [S_{21}]_m^T \tag{19.30}$$

Plane frame:

$$[S_m] = \left[\begin{array}{ccc|ccc} \frac{Eac^2}{l}+\frac{12EIs^2}{l^3} & & & & & \\ \left(\frac{Ea}{l}-\frac{12EI}{l^3}\right)cs & \frac{Eas^2}{l}+\frac{12EIc^2}{l^3} & & & \text{Symmetrical} & \\ -\frac{6EI}{l^2}s & \frac{6EI}{l^2}c & \frac{4EI}{l} & & & \\ \hline -\frac{Eac^2}{l}-\frac{12EIs^2}{l^3} & -\left(\frac{Ea}{l}-\frac{12EI}{l^3}\right)cs & \frac{6EI}{l^2}s & \frac{Eac^2}{l}+\frac{12EIs^2}{l^3} & & \\ -\left(\frac{Ea}{l}-\frac{12EI}{l^3}\right)cs & -\frac{Eas^2}{l}-\frac{12EIc^2}{l^3} & -\frac{6EI}{l^2}c & \left(\frac{Ea}{l}-\frac{12EI}{l^3}\right)cs & \frac{Eas^2}{l}+\frac{12EIc^2}{l^3} & \\ -\frac{6EI}{l^2}s & \frac{6EI}{l^2}c & \frac{2EI}{l} & \frac{6EI}{l^2}s & -\frac{6EI}{l^2}c & \frac{4EI}{l} \end{array}\right] \tag{19.31}$$

where $c = \cos\alpha = \lambda_{x^*x} = \lambda_{y^*y}$; $s = \sin\alpha = \lambda_{x^*y} = -\lambda_{y^*x}$; and α is the angle defined in Figure 19.2 with its positive sign convention.

Grid:

$$[S_m] = \left[\begin{array}{ccc:ccc} \frac{12EI}{l^3} & & & & & \\ -\frac{6EI}{l^2}s & \frac{GJc^2+4EIs^2}{l} & & & \text{Symmetrical} & \\ -\frac{6EI}{l^2}c & \left(\frac{GJ-4EI}{l}\right)cs & \frac{GJs^2+4EIc^2}{l} & & & \\ \hdashline -\frac{12EI}{l^3} & \frac{6EI}{l^2}s & -\frac{6EI}{l^2}c & \frac{12EI}{l^3} & & \\ -\frac{6EI}{l^2}s & \frac{GJc^2+2EIs^2}{l} & -\left(\frac{GJ+2EI}{l}\right)cs & \frac{6EI}{l^2}s & \frac{GJc^2+4EIs^2}{l} & \\ \frac{6EI}{l^2}c & -\left(\frac{GJ+2EI}{l}\right)cs & \frac{-GJs^2+2EIc^2}{l} & -\frac{6EI}{l^2}c & \left(\frac{GJ-4EI}{l}\right)cs & \frac{GJs^2+4EIc^2}{l} \end{array}\right] \tag{19.32}$$

where $c=\cos\alpha=\lambda_{x^*x}=\lambda_{z^*z}$; $s=\sin\alpha=\lambda_{x^*z}=-\lambda_{z^*x}$; and α is the angle defined in Figure 19.2 with its positive sign convention.

The stiffness matrices $[S_m]$ for a member of a *space frame* and *an eccentric grid* with respect to global coordinates are too long to be included here.

19.9 ASSEMBLAGE OF STIFFNESS MATRICES AND LOAD VECTORS

Consider a typical member of a framed structure whose first and second nodes are numbered, j and k respectively. Let $[S_m]$ be the stiffness matrix of the member with respect to coordinates in the global directions at the two nodes. Partition $[S_m]$:

$$[S_m] = \begin{bmatrix} S_{11} & S_{12} \\ S_{21} & S_{22} \end{bmatrix}_m \tag{19.33}$$

We make the same partitioning by the dashed lines in Equations 19.27, 19.28, 19.31, and 19.32. The elements above the horizontal dashed line represent the forces at the first node (the jth node), and the elements below the same line represent forces at the second node (the kth node). The elements to the left of the vertical dashed line correspond to unit displacements at node j, and the elements to the right of the same line correspond to unit displacements at node k. Each of the submatrices in Equation 19.33 is of size $s \times s$, where s = number of degrees of freedom per node.

We partition stiffness matrix of the structure, $[S]$ into $n_j \times n_j$ submatrices, each of size $s \times s$, where n_j is the number of nodes. Arrange the submatrices of the element stiffness matrix in Equation 19.33 in the same way as $[S]$:

$$[\bar{S}_m] = \begin{array}{c} \\ j \\ \\ k \\ \\ \end{array} \begin{bmatrix} & j & & k & \\ & \cdots & & & \\ \cdots & [S_{11}] & \cdots & [S_{12}] & \cdots \\ & \cdots & & & \\ & [S_{21}] & \cdots & [S_{22}] & \cdots \\ & \cdots & & \cdots & \end{bmatrix}_m \tag{19.34}$$

In this equation only the submatrices associated with nodes j and k are shown while all other submatrices are null. The above matrix represents the contribution of the mth member to the structure stiffness. The sum of the matrices for all members gives stiffness matrix of the structure:

$$[S] = \sum_{m=1}^{n_m} \left[\bar{S}_m\right] \tag{19.35}$$

where n_m is the number of members.

We should note that each column of the stiffness matrix $\left[\bar{S}_m\right]$ for an element represents a set of forces in equilibrium, and an assemblage of such matrices gives a structure stiffness matrix $[S]$ which has the same property.

The matrix $[S]$ generated as described above represents the stiffness of a free unsupported structure. We cannot use a singular matrix to solve for the unknown displacements, before adjusting it to account for the support conditions (Section 19.10).

When the analysis is for one case of loading, we need to solve the equilibrium equations $[S]\,\{D\} = -\,\{F\}$. We generate $\{F\}$ as sum of two vectors:

$$\{F\} = \{F_a\} + \{F_b\} \tag{19.36}$$

Here, $\{F_a\}$ is for the external forces at the nodes, and $\{F_b\}$ for other forces between the nodes, for temperature variation, and for support movements. $\{F_a\}$ = the nodal forces given in the input data. The vector $\{F_b\}$ is generated from the restraining forces $\{A_r\}_m$ given in the input data for individual members.

We transform $\{A_r\}_m$ into equivalent nodal forces $\{F\}_m$ in global directions at member ends using Equation 19.26. The vector $\{F\}_m$, which has $2s$ elements (s being the number of degrees of freedom per node), may be partitioned in two submatrices, $\{\{F_1\}$ and $\{F_2\}\}_m$, each having s elements representing forces at one end. Rearrange the two submatrices into a vector of the same size as $\{F\}$):

$$\left[\bar{F}\right]_m = \begin{Bmatrix} \cdots \\ j \quad \{F_1\} \\ \cdots \\ k \quad \{F_2\} \\ \cdots \end{Bmatrix}_m \tag{19.37}$$

This vector has n_j submatrices, all of which are null, with the exception of the jth and the kth, where j and k are the node numbers at the two ends of the mth member. The vector $\left\{\bar{F}\right\}_m$ represents the contribution of the mth member to the load vector $\{F_b\}$: thus, summing for all members (number of members $= n_m$):

$$\{F_b\} = \sum_{m=1}^{n_m} \left\{\bar{F}\right\}_m \tag{19.38}$$

Enlarging the element stiffness and the force matrices to the size of the structure stiffness matrix and of its load vector (Equations 19.34 and 19.37) requires formidable computer storage space. We start with $[S] = [0]$, then for each member, generate $[S_m]$ and add its elements to the appropriate element of $[S]$ by an algorithm. In the same way, we start with $\{F\} = \{0\}$, then for each member, generate $\left\{\bar{F}\right\}_m$ and add its elements to the appropriate element of $\{F\}$. Furthermore, we take advantage of the symmetry of the structure stiffness matrix and of its banded nature; thus, for $[S]$, only the elements on and above the diagonal are generated, as discussed in Section 19.3.

EXAMPLE 19.3: STIFFNESS MATRIX AND LOAD VECTORS FOR A PLANE TRUSS

Generate the stiffness matrix and the load vectors for the three loading cases in Example 19.1 (Figure 19.5). Table 19.1 gives the input data.

To save space, we give below the member stiffness matrices, $[S_m]$ and $\left[\bar{S}_m\right]$, for two members only. The submatrices not shown are null.

We start with the structure stiffness matrix $[S]=[0]$. For member 1, having its first node $j=2$ and second node $k=1$ (see Table 19.1), Equations 19.27 and 19.34 give:

$$\left[S_1\right]=\frac{200\times10^9(0.003)}{2.25}\left[\begin{array}{cc|cc} 1 & 0 & -1 & 0 \\ 0 & 0 & 0 & 0 \\ \hline -1 & 0 & 1 & 0 \\ 0 & 0 & 0 & 0 \end{array}\right]$$

and

$$\left[\bar{S}_1\right]=\begin{array}{c} \\ 1 \\ \\ 2 \\ \\ 3 \\ 4 \\ 5 \end{array}\begin{array}{c} \begin{array}{cccccccccc} 1 & & 2 & & 3 & & 4 & & 5 & \end{array} \\ \left[\begin{array}{cccc|c|c|c} 267 & 0 & -267 & 0 & & & \\ 0 & 0 & 0 & 0 & & & \\ \hline -267 & 0 & 267 & 0 & & & \\ 0 & 0 & 0 & 0 & & & \\ \hline & & & & & & \\ \hline & & & & & & \\ \hline & & & & & & \end{array}\right] \end{array}\times10^6$$

Adding $\left[\bar{S}_1\right]$ to the existing structure stiffness matrix $[S]$ gives the new matrix $[S]$, which, in this case, is the same as $\left[\bar{S}_1\right]$ and therefore not repeated here.

A repetition of the above calculations for member 2, which has $j=3$ and $k=1$, gives:

$$\left[S_2\right]=\frac{200\times10^9(0.0024)}{3.75}\left[\begin{array}{cc|cc} 0.36 & 0.48 & -0.36 & -0.48 \\ 0.48 & 0.64 & -0.48 & -0.64 \\ \hline -0.36 & -0.48 & 0.36 & 0.48 \\ -0.48 & -0.64 & 0.48 & 0.64 \end{array}\right]$$

and

$$\left[\bar{S}_2\right]=\begin{array}{c} \\ 1 \\ \\ 2 \\ \\ 3 \\ 4 \\ 5 \end{array}\begin{array}{c} \begin{array}{cccccccccc} 1 & & 2 & & 3 & & 4 & & 5 & \end{array} \\ \left[\begin{array}{cc|c|cc|c|c} 46 & 61 & & -46 & -61 & & \\ 64 & 82 & & -61 & -82 & & \\ \hline & & & & & & \\ \hline -46 & -61 & & 46 & 61 & & \\ -61 & -82 & & 61 & 82 & & \\ \hline & & & & & & \\ \hline & & & & & & \end{array}\right] \end{array}\times10^6$$

Adding this matrix to the existing matrix [S], we obtain the new matrix [S]:

$$
[S]=\begin{array}{c} \\ 1 \\ \\ 2 \\ \\ 3 \\ \\ 4 \\ 5 \end{array}
\left[\begin{array}{cc|cc|cc|cc|cc}
\multicolumn{2}{c}{1} & \multicolumn{2}{c}{2} & \multicolumn{2}{c}{3} & \multicolumn{2}{c}{4} & \multicolumn{2}{c}{5} \\
313 & 61 & -267 & 0 & -46 & -61 & & & & \\
61 & 82 & 0 & 0 & -61 & -82 & & & & \\
\hline
-267 & 0 & 267 & 0 & & & & & & \\
0 & 0 & 0 & 0 & & & & & & \\
\hline
-46 & -61 & & & 46 & 61 & & & & \\
-61 & -82 & & & 61 & 82 & & & & \\
\hline
 & & & & & & & & & \\
\hline
 & & & & & & & & &
\end{array}\right]\times 10^{6}
$$

Repetition of the above computations for members 3 to 7 leads to the structure stiffness matrix:

$$
[S]=\begin{array}{c} \\ 1 \\ \\ 2 \\ \\ 3 \\ \\ 4 \\ \\ 5 \\ \end{array}
\left[\begin{array}{cc|cc|cc|cc|cc}
\multicolumn{2}{c}{1} & \multicolumn{2}{c}{2} & \multicolumn{2}{c}{3} & \multicolumn{2}{c}{4} & \multicolumn{2}{c}{5} \\
313 & 61 & -267 & 0 & -46 & -61 & & & & \\
61 & 82 & 0 & 0 & -61 & -82 & & & & \\
\hline
-267 & 0 & 625 & 123 & 0 & 0 & -267 & 0 & -92 & -123 \\
0 & 0 & 123 & 244 & 0 & -80 & 0 & 0 & -123 & -164 \\
\hline
-46 & -61 & 0 & 0 & 352 & -62 & -92 & 123 & -213 & 0 \\
-61 & -82 & 0 & -80 & -62 & 326 & 123 & -164 & 0 & 0 \\
\hline
 & & -267 & 0 & -92 & 123 & 359 & -123 & & \\
 & & 0 & 0 & 123 & -164 & -123 & 164 & & \\
\hline
 & & -92 & -123 & -213 & 0 & & & & \\
 & & -123 & -164 & 0 & 0 & & & &
\end{array}\right]\times 10^{6}
$$

The load vector for case 1 is:

$$\{F\}=\{0 \quad -100\times 10^{3} \;\vdots\; 0 \quad 0 \;\vdots\; 0 \quad 0 \;\vdots\; 0 \quad 0 \;\vdots\; 0 \quad 0\}$$

For cases 2 and 3, we need a transformation matrix for member 4, which has its first node 5 and the second node 3. Equations 19.2, 19.17, and 19.24 give:

$$[t]=[0 \quad 1] \quad \text{and} \quad [T]=\begin{bmatrix} 1 & 0 & 0 & 0 \\ 0 & 0 & 1 & 0 \end{bmatrix}$$

For the same member, Equations 19.27 and 19.37 give for loading case 2:

$$\{F\}_{4}=\begin{bmatrix} 1 & 0 \\ 0 & 0 \\ 0 & 1 \\ 0 & 0 \end{bmatrix}\begin{Bmatrix} 96\times 10^{3} \\ -96\times 10^{3} \end{Bmatrix}=\begin{Bmatrix} 96 \\ 0 \\ -96 \\ 0 \end{Bmatrix}\times 10^{3}$$

$$\{\overline{F}\}_{4}=\{0 \quad 0 \;\vdots\; 0 \quad 0 \;\vdots\; -96 \quad 0 \;\vdots\; 0 \quad 0 \;\vdots\; 96 \quad 0\}\times 10^{3}$$

Because member 4 is the only member, which has restraining forces listed in the input data (Table 19.1), this vector is the global vector {F} for load case 2. The load vectors for the three cases are combined into one matrix:

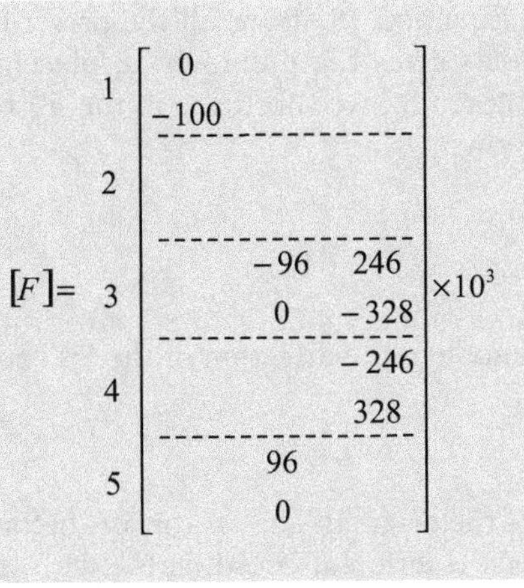

19.10 DISPLACEMENT SUPPORT CONDITIONS AND SUPPORT REACTIONS

To find the unknown displacements we need to solve the equilibrium equation:

$$[S]\{D\} = -\{F\} \tag{19.39}$$

where $[S]$ is the structure stiffness matrix; $\{D\}$ is a vector of nodal displacements; $\{F\}$ is a vector of artificial restraining forces which prevent the nodal displacements.

In general, some elements of $\{D\}$ are zero or have prescribed values at the supports, while the corresponding elements of $\{F\}$ are unknown. The total number of unknowns is n, which is equal to the number of equations.

The stiffness matrix $[S]$, generated as described in the preceding section, is the stiffness of a free (unsupported) structure. It is theoretically possible to rearrange the rows and columns in the matrices in Equation 19.39 so that the known elements of $\{D\}$ appear first. We can then divide the equation into two sets (Sections 5.9 to 5.11). We solve one set for the unknown displacements, then substitute the result in the second set to give the reactions. However, rearranging the equations increases the bandwidth.

Without disturbing the arrangement, it is possible to adjust Equation 19.39 so that its solution ensures that, at any coordinate k, the displacement $D_k = c_k$, where c_k can be zero in the case of a support without movement or can be a known value of support movement. In one method, we modify the kth row and column of the matrices in Equation 19.39 as follows:

$$\begin{matrix} & 1 & & k & & n \\ 1 & & & 0 & & \\ & & & \cdots & & \\ k & 0 & \cdots & 1 & \cdots & 0 \\ & & & \cdots & & \\ n & & & 0 & & \end{matrix} \quad \begin{Bmatrix} D_1 \\ \cdots \\ D_k \\ \cdots \\ D_n \end{Bmatrix} = \begin{bmatrix} -F_1 - S_{1k}\,c_k \\ \cdots \\ c_k \\ \cdots \\ -F_n - S_{nk}\,c_k \end{bmatrix} \tag{19.40}$$

The kth equations are now replaced by $D_k = c_k$; the terms $S_{1k}\,c_k$, $S_{2k}\,c_k$, ... represent the forces at nodes 1, 2, ... when $D_k = c_k$ and the other displacements are prevented. We repeat the modifications shown in Equation 19.40 for all the prescribed displacements. The solution of the modified equations gives the unknown displacements, but not the unknown reactions. To determine the force F_k, we substitute in the kth row of Equation 19.39 the D values, which are now all known:

$$F_k = \sum_{j=1}^{n} S_{kj} D_j \tag{19.41}$$

The reaction at a supported node in the direction of the kth coordinate is:

$$R_k = F_k + F_k' \tag{19.42}$$

where F_k' is the kth element of the original load vector, i.e. before Equation 19.39 was modified. This element represents a reaction at coordinate k with $\{D\}=\{0\}$. The value F_k will be nonzero only when an external force is applied at the supported node or over the members meeting at the node, or when the members meeting at the node are subjected to temperature change.

As mentioned in Section 19.3, computer storage space is commonly saved by storing $[S]$ in a compact form that takes advantage of its symmetry and its banded nature. The same space is used to store the reduced matrix needed for the solution; thus, the original $[S]$ is lost when $\{D\}$ is determined. To be able to use Equation 19.41 at this stage, the necessary rows of $[S]$ must be stored. Also, the element F_k' of the original load vector must be retained for use in Equation 19.42.

Another method of prescribing a displacement $D_k = c_k$ that requires fewer operations is to modify the kth row of the original equations as follows:

$$\begin{bmatrix} \cdots & & & & \\ S_{k1} & \cdots & S_{kk}\times 10^6 & \cdots & S_{kn} \\ \cdots & & & & \end{bmatrix} \begin{Bmatrix} D_1 \\ \cdots \\ D_k \\ \cdots \\ D_n \end{Bmatrix} = \begin{bmatrix} -F_1 \\ \cdots \\ S_{kk}\times 10^6 \; c_k \\ \cdots \\ -F_n \end{bmatrix} \tag{19.43}$$

where the 10^6 is an arbitrarily chosen large number. The change in Equation 19.43 is equivalent to providing a fixed support connected to the structure by a very strong spring of stiffness $(10^6-1)S_{kk}$ at the kth coordinate and applying a large force of $S_{kk}\times 10^6 c_k$. For all practical purposes, Equation 19.43 satisfies with sufficient accuracy the condition that D_k equals c_k. This is so because the division of both sides of the kth row in Equation 19.43 by $10^6 S_{kk}$ reduces the term $S_{kk}\times 10^6$ to 1; the other S-terms become negligible, and the kth row reduces to $D_k \simeq c_k$.

After modification of the original equations for each prescribed displacement, a solution will give a vector $\{D\}$ of the displacements at all nodes. Now, Equation 19.41 gives the force F_k. This requires storing the kth row of the original $[S]$ matrix. Alternatively, we obtain F_k by combining Equation 19.41 and the kth row of Equation 19.43:

$$F_k = S_{kk}\times 10^6 \left(c_k - D_k\right) + S_{kk} D_k \tag{19.44}$$

The difference $c_k - D_k$ needs to be determined with sufficient number of significant figures (three or more). After calculating F_k by the above equation, substitution in Equation 19.42 gives the support reaction.

The condition that D_k equals c_k can also be satisfied by replacing S_{kk} on both sides of Equation 19.44 by $(S_{kk}+1)$. In this case, $(S_{kk}+1)$ gives F_k using Equation 19.44 and replaces S_{kk} in the first term on the right-hand side. We use this option when S_{kk} can be zero (e.g. in trusses).

When a structure is analyzed for a number of load cases, the vectors $\{D\}$ and $\{F\}$ in Equation 19.39 are replaced by rectangular matrices $[D]$ and $[F]$, with the respective columns of the two matrices representing one load case; however, the same stiffness matrix $[S]$ is used for all cases. Modification of $[S]$, as suggested by Equation 19.40 or 19.43, indicates that a loading case representing the effect of support movements cannot be solved simultaneously with other loading cases unless the prescribed displacements are valid for all cases. This will be the condition when all the prescribed displacements are zero, thus representing supports without movement.

In order to analyze for a non-zero prescribed displacement $D_k=c_k$ at the same time as for other loading cases, we may prepare the support conditions in the input data as if c_k were zero. However, the effect of $D_k=c_k$ must then be presented in the forces $\{A_r\}$ included in the data; $\{A_r\}$ represents the end forces for the members connected to the coordinate k when $D_k=c_k$ when the displacements at the other coordinates are restrained (see Example 19.1).

The techniques presented in this section require the solution of a larger number of equations than the number of unknown displacements. The extra computing is justified by simpler coding and by the fact that the solution gives the unknown displacements as well as the unknown forces.

As mentioned in Section 19.3, it is possible to generate only the equilibrium equations corresponding to the unknown displacements. After the solution of these equations, the displacements, which become known at all coordinates, are used to determine the forces $\{A\}$ at member ends in the directions of local axes (see Section 19.12). Then, we obtain the reactions at a supported node (with or without support movement) by summing the forces at the ends of the members meeting at the node:

$$\{R\} = \sum\left([t]^T\{A\}\right)_i - \{F_s\} \tag{19.45}$$

Here, $\{R\}$ represents s reaction components in global directions; s = number of degrees of freedom per node; the $[t]$ matrix is included to transform the member end forces from local to global directions (Equation 19.26); the subscript i refers to a member number, and the summation is performed for the members meeting at the supported node; and $\{F_s\}$ represents forces applied direct at the supported node, producing no displacements and no member end forces.

19.11 SOLUTION OF BANDED EQUATIONS

The solution of the equilibrium Equation 19.39 was expressed in earlier chapters in the succinct form $\{D\}=-[S]^{-1}\{F\}$. However, a solution by matrix inversion involves more operations than the Gauss elimination or the Cholesky method, discussed later in this section.

Consider a system of n linear equations.

$$[a]\{x\} = \{c\} \tag{19.46}$$

The elements of $[a]$ and $\{c\}$ are known, and it is required to determine the vector of unknowns $\{x\}$.

The standard Gauss elimination reduces the original equations to:

$$[u]\{x\} = \{d\} \tag{19.47}$$

where $[u]$ is a unit upper triangular matrix in which all the diagonal elements are equal to 1, and the elements below the diagonal are zero. Back substitution, starting from the last row, gives successively the unknowns $x_n, x_{n-1}, \ldots, x_1$.

In Crout's procedure, the Gauss elimination can be achieved by generating auxiliary matrices $[b]$ and $\{d\}$ of the same size as $[a]$ and $\{c\}$. The elements above the diagonal in $[b]$ are the same as the corresponding elements of $[u]$; thus:

$$u_{ij} \equiv b_{ij} \quad \text{with} \quad i < j \tag{19.48}$$

For structural analysis, we are interested in the case when $[a]$ is symmetrical and banded. In such a case, $[b]$ is also banded and has the same bandwidth. In addition, any element b_{ij} above the diagonal is equal to b_{ji}/b_{ii}. Thus, when only the half-band of $[a]$ is stored in a rectangular matrix (Figure 19.4), $[b]$ can be stored in a matrix of the same size with no need to store the elements below the diagonal.

Examination of the equations that give the elements of $[b]$ will indicate that, starting from the top row, the elements of $[b]$ can be calculated one by one. Each element b_{ij} can be stored in the same space occupied by a_{ij} because the latter term is no longer needed. In other words, the $[b]$ matrix can be generated in a band form and can replace $[a]$ using the same storage space. The vector $\{d\}$ is generated in a similar way and is stored in the same space originally occupied by $\{c\}$.

The Cholesky method offers some advantages when $[a]$ is symmetrical and banded. In this method, the symmetrical matrix $[a]$ is decomposed into the product of three matrices:

$$[a] = [u]^T [e][u] \tag{19.49}$$

Here again, $[u]$ is a unit upper triangular matrix and therefore its transpose $[u]^T$ is a unit lower triangular matrix; $[e]$ is a diagonal matrix. Let:

$$[e][u] = [h] \tag{19.50}$$

where $[h]$ is an upper triangular matrix. Also, let:

$$[h]\{x\} = \{g\} \tag{19.51}$$

Substitution of Equations 19.49–19.51 into Equation 19.46 gives:

$$[u]^T \{g\} = \{c\} \tag{19.52}$$

Forward substitution in this equation gives $\{g\}$, and using the result with backward substitution in Equation 19.51 gives the required vector $\{x\}$.

The operations involved in the matrix product in Equation 19.49 and in the forward and backward substitutions are relatively small, compared with the operations involved in the decomposition to generate $[u]$ and $[e]$.

Performing the matrix product in Equation 19.49 gives:

$$a_{ij} = e_{ij}\, u_{ij} + \sum_{k=1}^{i-1} e_{kk}\, u_{kj}\, u_{ki} \quad \text{with} \quad i < j \tag{19.53}$$

and

$$a_{jj} = e_{jj} + \sum_{k=1}^{j-1} e_{kk}\, u_{kj}^2 \tag{19.54}$$

To avoid the triple products in the above summations and thereby reduce the number of operations, let us introduce the symbol:

$$\bar{u}_{ij} = e_{ii}\, u_{ij} \tag{19.55}$$

Substitution of Equation 19.55 into Equations 19.53 and 19.54 and rearrangement of terms give:

$$\bar{u}_{ij} = a_{ij} - \sum_{k=1}^{i-1} \bar{u}_{kj}\, u_{ki} \quad \text{with} \quad i < j \tag{19.56}$$

and:

$$e_{jj} = a_{jj} - \sum_{k=1}^{j-1} \bar{u}_{kj}\, u_{kj} \tag{19.57}$$

Here again, when $[a]$ is symmetrically banded, only the elements within the band, on and above the diagonal, are stored in the computer. The same storage space is reused; the elements e_{jj} replace a_{jj}, and $\bar{u}_{ij}$ replace a_{ij} (with $i<j$).

Working down the jth column, the non-zero elements $\bar{u}_{ij}$ are calculated by Equation 19.56 and temporarily stored in place of a_{ij}. Then, Equation 19.55 calculates the first non-zero u_{ij} in the column from $\bar{u}_{ij}$; the product $\bar{u}_{ij}\, u_{ij}$ is deducted from a_{jj}, and u_{ij} replaces $\bar{u}_{ij}$. Repetition of this step for the remaining u_{ij} in the jth column results in the replacement of a_{ii} by e_{ii} (see Equation 19.57).

Unlike the Gauss eliminations, with the Cholesky method, only the matrix $[a]$ is reduced, while $\{c\}$ is used in the forward substitution without change (Equation 19.50). In the following example, the same equations are solved here by the Cholesky method.

EXAMPLE 19.4: FOUR EQUATIONS SOLVED BY THE CHOLESKY METHOD

Use the Cholesky method to solve the equations $[a]\{x\}=\{c\}$. The vector $\{c\}$ is $\{1, -4, 11, -5\}$. The matrix $[a]$ is:

$$[a] = \begin{bmatrix} 5 & -4 & 1 & 0 \\ & 6 & -4 & 1 \\ & & 6 & -4 \\ \text{Symm.} & & & 7 \end{bmatrix}$$

We use Equations 19.53–19.57 to generate:

$$\begin{bmatrix} e_{11} & u_{12} & u_{13} & 0 \\ - & e_{22} & u_{23} & u_{24} \\ - & - & e_{33} & u_{34} \\ - & - & - & e_{44} \end{bmatrix} = \begin{bmatrix} 5 & -4/5 & 1/5 & 0 \\ - & 14/5 & -8/7 & 5/14 \\ - & - & 15/7 & -4/3 \\ - & - & - & 17/6 \end{bmatrix}$$

Equation 19.52 and forward substitution give:

$$\begin{bmatrix} 1 & 0 & 0 & 0 \\ -4/5 & 1 & 0 & 0 \\ 1/5 & -8/7 & 1 & 0 \\ 0 & 5/14 & -4/3 & 1 \end{bmatrix} \begin{Bmatrix} g_1 \\ g_2 \\ g_3 \\ g_4 \end{Bmatrix} = \begin{Bmatrix} 1 \\ -4 \\ 11 \\ -5 \end{Bmatrix}$$

whence:

$$\{g\} = \{1, \quad -16/5, \quad 50/7, \quad 17/3\}$$

Equation 19.50 and 19.51 and backward substitution give:

$$\begin{bmatrix} 5 & -4 & 1 & 0 \\ & 14/5 & -16/5 & 1 \\ & & 15/7 & -20/7 \\ & & & 17/6 \end{bmatrix} \begin{Bmatrix} x_1 \\ x_2 \\ x_3 \\ x_4 \end{Bmatrix} = \begin{Bmatrix} 1 \\ -16/5 \\ 50/7 \\ 17/3 \end{Bmatrix}$$

whence:

$$\{x\} = \{3, \quad 5, \quad 6, \quad 2\}$$

19.12 MEMBER END FORCES

The last step in the displacement method (see Section 5.4) is the superposition to determine the required actions:

$$\{A\} = \{A_r\} + [A_u]\{D\} \tag{19.58}$$

where $\{A_r\}$ are the values of the actions with $\{D\} = \{0\}$, and $[A_u]$ are the values of the actions due to unit displacements introduced separately at each coordinate.

In a framed structure, the required actions are commonly the forces at the member ends in the local directions for individual members. We do the superposition separately for individual members.

After solution of the equilibrium equations, we have a vector $\{D\}$ of the nodal displacements in the global directions at all the nodes. Let $\{D\}$ be partitioned into submatrices, each having s values of the displacements at a node, where s is the number of degrees of freedom per node. Equation 19.16 gives the displacements in local coordinates at the two ends of the mth member; j and k, are its first and second node, respectively:

$$\{D^*\}_m = \begin{Bmatrix} \{D^*\}_1 \\ \hline \{D^*\}_2 \end{Bmatrix}_m = \begin{bmatrix} [t] & [0] \\ [0] & [t] \end{bmatrix}_m \begin{Bmatrix} \{D\}_j \\ \hline \{D\}_k \end{Bmatrix} \tag{19.59}$$

where $\{D\}_j$ and $\{D\}_k$ are submatrices of $\{D\}$, corresponding to nodes j and k, and $[t]_m$ is a transformation matrix for the mth member given by one of Equations 19.17–19.22.

To apply the superposition Equation 19.58 in order to find member end forces, we have $\{A_r\}$ given in the input data. The values of the member end forces due to unit displacements introduced separately at each local coordinate (Figure 19.2) are the elements of the member stiffness matrix $[S^*]$, given by one of Equations 19.10–19.13. Thus, in this application, the product $[S^*]\ \{D^*\}$ stands for $[A_u]\ \{D\}$. Therefore, the end forces for any member are:

$$\{A\} = \{A_r\} + \left[S^*\right]\{D^*\} \tag{19.60}$$

We need to recalculate the stiffness matrices of individual members in this step, unless they have been stored.

EXAMPLE 19.5: REACTIONS AND MEMBER END FORCES IN A PLANE TRUSS

Find the reactions and the end forces in member 6 in the truss of Figure 19.5 for each of the three cases in Example 19.1.

The stiffness matrix and the load vectors for the three cases have been determined in Example 19.3. Zero displacement is prescribed in the x and y directions at nodes 4 and 5; therefore, multiply the diagonal coefficient in each of rows 7, 8, 9, and 10 by 10^6 and set zeros in the same rows of the load vectors (see Equation 19.43). Because we accounted for the support movement in the member end forces by $\{A_r\}$ used in the input data, we put here $c_k = 0$. The adjusted equilibrium equations are:

$$10^6 \times \begin{bmatrix} 313 & 61 & -267 & 0 & -46 & -61 & & & & \\ & 82 & 0 & 0 & -61 & -82 & & & & \\ \hline & & 625 & 123 & 0 & 0 & -267 & 0 & -92 & -123 \\ & & & 244 & 0 & -80 & 0 & 0 & -123 & -164 \\ \hline & & & & 352 & -62 & -92 & 123 & -213 & 0 \\ & & & & & 326 & 123 & -164 & 0 & 0 \\ \hline & & & & & & 359\times10^6 & -123 & & \\ & & & & & & & 164\times10^6 & & \\ \hline & & & & & & & & 305\times10^6 & 123 \\ & & & & & & & & & 164\times10^6 \end{bmatrix} [D]$$

Symmetrical; Submatrices not shown are null

$$= \begin{bmatrix} 0 & 0 & 0 \\ 100 & 0 & 0 \\ \hline 0 & 0 & 0 \\ 0 & 0 & 0 \\ \hline 0 & 96 & -46 \\ 0 & 0 & 328 \\ \hline 0 & 0 & 0 \\ 0 & 0 & 0 \\ \hline 0 & 0 & 0 \\ 0 & 0 & 0 \end{bmatrix} \times 10^3$$

The solution is:

$$[D]=\begin{bmatrix} -0.6510\times10^{-3} & -0.3223\times10^{-5} & -0.1912\times10^{-3} \\ 0.3016\times10^{-3} & 0.5688\times10^{-3} & 0.1372\times10^{-2} \\ \hdashline -0.3698\times10^{-3} & -0.3223\times10^{-5} & -0.1912\times10^{-3} \\ 0.4694\times10^{-3} & 0.9411\times10^{-4} & 0.5582\times10^{-3} \\ \hdashline 0.5925\times10^{-3} & 0.4097\times10^{-3} & -0.2390\times10^{-3} \\ 0.8627\times10^{-3} & 0.2374\times10^{-3} & 0.1408\times10^{-2} \\ \hdashline -0.4180\times10^{-9} & -0.1013\times10^{-15} & -0.6856\times10^{-9} \\ 0.4183\times10^{-9} & -0.6994\times10^{-10} & 0.1587\times10^{-8} \\ \hdashline 0.4910\times10^{-9} & 0.3142\times10^{-9} & 0.3782\times10^{-15} \\ 0.1921\times10^{-9} & 0.6994\times10^{-10} & 0.4149\times10^{-9} \end{bmatrix}\begin{matrix} 1 \\ \\ 2 \\ \\ 3 \\ \\ 4 \\ \\ 5 \\ \end{matrix}$$

The forces at the supported nodes 4 and 5 (Equation 19.44) are:

$$\begin{Bmatrix} F_1 \\ F_2 \\ F_3 \\ F_4 \end{Bmatrix} = 10^3 \times \begin{bmatrix} 150 & 0 & 246 \\ -69 & 11 & -260 \\ \hdashline -150 & -96 & 0 \\ -31 & -11 & -68 \end{bmatrix}$$

These forces are calculated by the multiplication of the nodal displacement with reversed sign by the diagonal elements of the adjusted stiffness matrix. (The last term in Equation 19.44 is negligible and $c_k = 0$.) Addition of the above forces to the corresponding rows of the original load vector (generated in Example 19.3) gives the reactions (Equation 19.42):

$$\begin{Bmatrix} R_7 \\ R_8 \\ R_9 \\ R_{10} \end{Bmatrix} = 10^3 \times \begin{bmatrix} 150 & 0 & 0 \\ -69 & 11 & 68 \\ \hdashline -150 & 0 & 0 \\ -31 & -11 & -68 \end{bmatrix}$$

The first and second nodes of member 6 are $j = 4$ and $k = 3$ (see input data, Table 19.1). The transformation matrix is (Equation 19.17):

$$[t]=[0.6 \quad -0.8]$$

Let us transform the nodal displacements for the three cases of nodes 4 and 3 from global to local directions (Equation 19.59):

$$[D^*]=\begin{bmatrix} 0.6 & -0.8 & 0 & 0 \\ 0 & 0 & 0.6 & -0.8 \end{bmatrix}$$

$$\times\begin{bmatrix} -0.4180\times10^{-9} & -0.1013\times10^{-15} & -0.6856\times10^{-9} \\ 0.4183\times10^{-9} & -0.6994\times10^{-10} & 0.1587\times10^{-8} \\ 0.5925\times10^{-3} & 0.4097\times10^{-3} & -0.2390\times10^{-3} \\ 0.8687\times10^{-3} & 0.2374\times10^{-3} & 0.1408\times10^{-2} \end{bmatrix}$$

whence:

$$[D^*]=\begin{bmatrix} -0.5854\times10^{-9} & 0.5595\times10^{-10} & -0.1681\times10^{-8} \\ -0.3347\times10^{-3} & 0.5590\times10^{-4} & -0.1270\times10^{-2} \end{bmatrix}$$

The end forces for member 6 in the three cases are (Equation 19.58):

$$\left[A\right]=10^3\times\begin{bmatrix}0 & 0 & -410\\0 & 0 & 410\end{bmatrix}+256\times10^6\begin{bmatrix}1 & -1\\-1 & 1\end{bmatrix}\left[D^*\right]$$

whence:

$$\left[A\right]=10^3\times\begin{bmatrix}86 & -14 & -85\\-86 & 14 & 85\end{bmatrix}$$

19.13 GENERAL

The analysis of framed structures by computer to give the nodal displacements and the member end forces is summarized below. The nodal displacements are in the directions of arbitrarily chosen global axes x, y, (and z) (Figure 19.1), while the member end forces are in local coordinates pertaining to individual members (Figure 19.2).

The input data are composed of the following: the material properties E (and G); the x, y (and z) coordinates of the nodes, the two nodes at the ends of each member, the cross-sectional properties a, I, J, etc. of each member, and the support conditions. The input data include the external forces applied at the nodes, and the end forces $\{A_r\}$ for the members are subjected to loads away from the nodes. The elements of $\{A_r\}$ are forces in local coordinates at the member ends when the nodal displacements are restrained.

The five steps of the displacement method summarized in Section 5.4 are executed as follows:

Step 1 The nodes and the members are numbered sequentially by the analyst, from 1 to n_j and from 1 to n_m respectively. This will automatically define the degrees of freedom and the required end actions according to the systems specified in Figures 19.1 and 19.2.

Step 2 We transform $\{A_r\}$ included in the input data into global directions. We then assemble them to give a vector $\{F_b\}$ of restraining nodal forces (Equations 19.26 and 19.38). We generate also vector $\{F_a\}$, simply by listing from the input data. The sum $\{F_a\}$ plus $\{F_b\}$ gives a vector of the restraining forces $\{F\}$ that are necessary to prevent the nodal displacements (Equation 19.36).

Step 3 To generate the stiffness matrix of the structure, start with $[S]=[0]$ and partition this matrix into $n_j \times n_j$ submatrices each of size $s \times s$, where s is the number of degrees of freedom per node. We generate the stiffness matrix of the first member by one of Equations 19.27, 19.28, 19.31, 19.32, or 5.18 and partition it into 2×2 submatrices, each of size $s \times s$. The four submatrices are added to the appropriate submatrices of $[S]$ according to Equations 19.34 and 19.35. By repeating this procedure for all the members, the stiffness matrix of a free unsupported structure is generated.

When one of Equations 19.27, 19.28, 19.31, or 19.2 is used, the stiffness matrix of a member is obtained with respect to coordinates in global directions. However, Equation 5.18 gives the stiffness matrix of a member of a space frame with respect to local coordinates. Transformation is necessary, according to Equation 19.23, before the assemblage of the stiffness matrices by Equation 19.35.

The structure stiffness matrix $[S]$ is symmetrical and generally has the non-zero elements limited to a band adjacent to the diagonal. To save computer space, only the diagonal element of $[S]$ and the elements above the diagonal within the band are stored in a rectangular matrix as shown in Figure 19.4. The node numbering selected by the analyst in step 1 affects the bandwidth and hence the width of the rectangular matrix used to store $[S]$. A narrower bandwidth is generally obtained by numbering the nodes sequentially across that side of the frame which has a smaller number of nodes (as an example, see Figure 19.4).

Step 4 Before the equilibrium equations $[S]\{D\}=-\{F\}$ can be solved, $[S]$ and $\{F\}$ must be adjusted according to the displacements prescribed in the input data (Equations 19.40 or 19.43).

Section 19.11 gives two methods of solution of the equilibrium equations. The solution gives the unknown displacements $\{D\}$; then, we use the displacements to calculate the reactions by Equation 19.42 and Equation 19.41 or 19.44.

Step 5 The end forces for each member are obtained by the superposition Equation 19.60, which sums $\{A_r\}$ given in the input data and the product $([S^*]\{D^*\})$, with $[S^*]$ and $\{D^*\}$ being, respectively, the member stiffness matrix and the displacements at the member ends in local coordinates. For this reason, the displacements obtained in step 4 have to be transformed from global directions to local directions by Equation 19.59 before the superposition can proceed.

REFERENCE

1. Ghali, A. and Gayed, R.B., "Serviceability and Strength of Concrete Floors and Bridge Decks: Grid Analogy," *J. Struct. Eng. ASCE*, 145(6), (2019), ISSN 0733-9445, 10 pp.

Chapter 20

Computer programs

20.1 INTRODUCTION

A series of computer programs is available as a companion to this book. These programs can assist in solving examples and problems by removing the burden of repetitive calculations. They are tools for teaching and learning. The textbook, combined with the comments in the source code, teaches the steps of analysis. Modification of the programs to perform additional tasks can prove to be an efficient means of fully understanding structural analysis.

20.2 AVAILABILITY OF THE PROGRAMS

This book and its companion computer programs are available through the publisher. The availability of these programs depends upon fast-developing techniques. At time of release of the current edition, the programs are provided at: www.routledge.com/9780367252618

The computer programs are also available on an electronic device that can store the files produced by the user (see Advertisement at the end of the book).

20.3 PROGRAM COMPONENTS

The programs are presented in groups: (A) Linear programs, (B) Nonlinear programs, (C) Matrix algebra, (D) Programs EIGEN1 and EIGEN2, and (E) Time-dependent analysis programs.

For each program, four files are provided:

1. The source code is in FORTRAN language: Name.FOR
2. Executable file: Name.EXE
3. Input file: Name.IN
4. Manual: Name.INM

For example, in program PLANEF, which analyzes plane frames, the four files are: PLANEF.FOR, PLANEF.EXE, PLANEF.IN, and PLANEF.INM.

The source code can be used to modify the program to perform additional tasks or to thoroughly learn the art of structural analysis. After modification, a FORTRAN compiler produces an executable file for the revised program. The input file, completed with problem data, is read by the executable file; the solution is provided in file identified as Name.OUT.

DOI: 10.1201/9780429286858-20

```
Plane frame of Example 19.2, Figure 19.6
4   3  2  2                                        NJ, NM, NSJ, NLC
30.0e3     0.0                                     Elasticity modulus, Poisson's ratio
1       0.0  300.0                                 Node no., x and y coordinates
2       0.0     0.0
3     300.0  150.0
4     300.0  300.0
1    1  2    30.0  3000.0  1.0e6                   Member no., JS, JE, cross-sec. area, I, ar
2    2  3    40.0  5000.0  1.0e6
3    3  4    30.0  3000.0  1.0e6
1    0  0    0        0.0  0.0  0.0                Node no., restr. indics., prsc. displs.
4    0  0    1        0.0  0.0  0.0
1    2  4.0  0.0      0.0                          Ld. case no., node, Fx, Fy, Mz
1    3  2.0  0.0  0.0
10   0  0.0  0.0  0.0                              Dummy, end of data of loaded nodes
1    2  -13.42  -26.83  -1500.0  -13.42  -26.83  1500.0  Case, member, {Ar}
2    1  -1500.0  8.0  1200.0  1500.0  -8.0  1200.0
10   0  0.0  0.0  0.0  0.0  0.0  0.0 Dummy, end of data of member loads
```

Figure 20.1 Image of "PLANEF.IN" for a plane frame of Example 19.2 (Figure 19.6).

The file Name.INM contains input data for example problems, and few lines serve as a manual; it specifies the function of the program and its input file.

When using the "C-command" in Microsoft, typing the name of the program will produce file Name.OUT, giving the solution of the problem whose data is at the top of the file Name.IN. The words "Program terminated" will appear on the screen; alternatively, it might indicate why the task could not be completed. The reason could be a deficient input file or unstable structure; the uncompleted output file may help to indicate the error in the input.

The input file provided contains the data for an example or a problem selected from the book. A typical line in the input file contains numbers on the left-hand side; on the right-hand side, symbols or words indicate the parameters given on the line. Figure 20.1 is an image of PLANEF.IN for a plane frame (Example 19.2). To analyze a new structure, move the data of the example downwards, and replace it with a copy to be edited with the data of the new structure. When running the program, only the data of the new structure will be read. The lower part of the input file, containing the data for Example 19.2 and definition of the symbols, may be retained for quick reference.

20.4 DESCRIPTION OF PROGRAMS

The names of the available programs and short descriptions are presented in the current section.

20.4.1 Group A. Linear analysis programs (basis: Chapter 19)

- PLANEF (plane frame)
- SPACEF (space frame)
- PLANET (plane truss)

- SPACET (space truss)
- PLANEG (plane grid)
- EGRID (eccentric grid)

These programs are for linear analysis of plane frames, space frames, plane trusses, space trusses, plane grids, and eccentric grids. The trusses consist of prismatic members pin-connected to the joints; trusses with rigid connections can be treated as frames. A grid has its members in the xy plane, and applied forces act in the z direction perpendicular to the xy plane.

Input files for analysis of quadrilateral slabs of constant thickness with EGRID can be prepared by the computer program QMESH (provided with EGRID). Simple input – inserted in a file "QMESH.IN" and followed by running QMESH6.EXE – generates an output file "QMESH.OUT." Rename the file "QMESH.OUT" to "EGRID.IN" and run the EGRID for analysis.

20.4.2 Group B. Nonlinear analysis programs

- PDELTA (basis: Chapter 10)
- PLASTICF (basis: Section 14.6)
- NLST (basis: Chapter 18)
- NLPF (basis: Chapter 18)

The program PDELTA performs quasi-linear analysis of plane frames, considering the effect of axial forces on the stiffness matrix of individual members (Equations 10.19 and 10.23). Analysis of the collapse loads of plane frames due to the development of plastic hinges (Section 14.6) can be performed with the program PLASTICF. The program increases the values of the loads gradually and indicates the load levels at which the plastic hinges are developed, the corresponding nodal displacements, and member-end forces. PLASTICF is used in the analysis of plane frames subjected to earthquakes (Chapter 17).

The program NLST performs nonlinear analysis of space trusses, employing Newton-Raphson's technique. Cable nets can be analyzed by the same program. Cable nets with membrane triangular elements can also be analyzed.

Nonlinear analysis of plane frames can be performed by the program NLPF. Similar to the program PDELTA, NLPF calculates the stiffness of the members, accounting for the magnitude of the axial forces. In addition, NLPF considers the equilibrium of the nodes in their displaced positions. The change in members' lengths is not ignored in determining the geometry of the deformed structure. NLPF can be employed to indicate the buckling loads.

20.4.3 Group C. Matrix algebra

To enhance the study of this book, folder MATALGBR consists of simple programs that perform frequently needed matrix operations:

- ADD
- MULTIPLY
- INVERT
- SOLVE
- DETERM
- MULTIPLY5

The program ADD calculates the sum of two given matrices: [C] = $\alpha[A] + \gamma[B]$, where $[A]$ and $[B]$ are the given matrices, and α and γ are the given multipliers. The program MULTIPLY determines the product of two given matrices. The program INVERT calculates the inverse of a nonsingular matrix. The program MULTIPLY5 calculates the product of two to five matrices; the computer transposes one of the matrices before multiplication (if required).

The program SOLVE gives the matrix $[X]$ in the simultaneous linear equations $[A]$ $[X]$ = $[B]$, when $[A]$ and $[B]$ are given. The program DETERM calculates the determinant of a given square matrix.

20.4.4 Group D. Programs EIGEN1 and EIGEN2

For easier access, the four files of program EIGEN1 (or EIGEN2) are in a folder with the same name, instead of the folder MATALGBR. Program EIGEN1 (or EIGEN2) solves the characteristic equation $[S]\{D\} = \lambda[m]\{D\}$, where $[S]$ and $[m]$ are square matrices. The solution gives eigenvalues, λ and the corresponding eigenvectors. In dynamics of structures (Chapters 16 and 17), λ stands for ω^2, the square of natural frequency, and eigenvectors represent mode shapes. The input of EIGEN1 includes guess eigenvalues, which can be difficult when $[S]$ and $[m]$ are large. Program EIGEN1 (or EIGEN2) is adequate for solving the characteristic equations in the examples and problems in this book. For larger matrices, specialized programs may be needed. EIGEN1 (or EIGEN2) is frequently used in the examples and problems on response of structures to earthquakes (Chapter 17). EIGEN2 does not require guess eigenvalues; it can also be used for solving dynamic analysis problems in which $[m]$ has no zero elements on the diagonal.

20.4.5 Group E. Time-dependent analysis programs

Four programs are included in this group (Chapter 6):

- CREEP
- SCS (Stresses in cracked sections)
- TDA (Time-dependent analysis)
- CGS (Cracked general sections)

The program CREEP calculates the creep and aging coefficients, the relaxation function, the autogenous shrinkage and the drying shrinkage of concrete in accordance with CEB-FIP Model Code 2010. The program SCS calculates stresses and strains in a reinforced concrete section subjected to a bending moment, M, with or without a normal force, N. The section can be composed of any number of trapezoidal concrete layers and any number of reinforcement layers; the layers can have different elasticity moduli, E_c and E_s. Prestressed and non-prestressed reinforcements are treated in the same way. First, the stresses are calculated for uncracked section. If stress in concrete at an extreme fiber exceeds the tensile strength, f_{ct}, the analysis is redone ignoring concrete in tension.

The program TDA analyzes a section that is composed of any number of trapezoidal layers and any number of prestressed or non-prestressed reinforcement layers; results include strain and stress immediately after prestressing, after occurrence of time-dependent effects, and after application of live load (introduced later after prestressing). All concrete layers have the same elasticity modulus, E_c. The section may have a single prestressed reinforcement layer, which can be pretensioned or post-tensioned. The prestressing is introduced simultaneously with a normal force, N, at the top fiber and moment, M. After a period, during which creep and shrinkage of concrete and relaxation of prestressed steel occur, additional normal force and moment are introduced, representing effect of live load.

Computer program CGS analyzes a section of any shape composed of concrete of one type, non-prestressed steel and pretensioned and/or post-tensioned tendons. At time t_0, prestressing is introduced simultaneously with a normal force, N, at an arbitrary reference point, O, combined with bending moments, M_x and M_y about orthogonal x and y axes through O. This is referred to as loading Stage (a). At time t, after occurrence of time-dependent effects, loading Stage (b) is introduced; it is composed of a normal force and bending moments, $\{N, M_x, M_y\}_b$. The program calculates the strain and stress in concrete and in reinforcements at time t_0, immediately after loading (a) and at time t before and after loading (b). It is assumed that the section is not cracked when loading (b) is introduced. When loading (b) produces cracking, the analysis ignores concrete in tension. Geometry of the section is defined by (x, y) coordinates of nodes connected by straight lines and numbered in a counterclockwise order along the section periphery. The CGS program performs the analysis that can be done by SCS and TDA and requires more input data.

20.4.6 Group F. Programs for analysis of axially symmetric loaded structures

Two computer programs are included in this group (Chapter 12):

- CTW (Cylindrical Tank Walls)
- SOR (Shells of Revolution)

The CTW program performs elastic analysis of circular-cylindrical walls of variable thickness, subjected to axially symmetrical external loads or to temperature variations. Application examples are vertical walls of circular storage tanks and silos and beams on elastic foundations. The analysis gives the variation of bending moments, M and M_ϕ in the vertical and circumferential direction (force-length/length), hoop force, N (force/length) and reactions of the edges (Figure 12.14). The top and bottom edges of the wall can rotate freely, or the rotation can be fully restrained. The translation of the edges in the radial direction can be either free or elastically restrained; the latter represents the case when the wall slides on an elastomeric pad. The wall thickness is to be specified at a minimum of three points; the program assumes parabolic variation between the three points. By giving the thickness values at additional points, practically any variation of thickness can be analyzed. A variety of loads is available that would cover all potential loading cases. The analysis is done by finite differences, using 101 equally spaced nodes.

CTW can be used for analysis of a beam on elastic foundation only when its cross-sectional moment of inertia, I and foundation modulus, K (force/lenght^2) are constants. This is possible because the beam has the same deflection and bending moment as for a strip of unit width in an analogous circular cylinder; the strip runs in the direction of the cylinder axis. To use CTW in such analysis, enter Poisson's ratio = 0, and the length of the cylinder = the length of the beam. No thermal loading can be analyzed by CTW for beams on elastic foundation. The bending moment obtained by the analysis is the same as the beam's bending moment, but the circumferential bending moment calculated for cylinders has no meaning for beams on elastic foundations. The hoop force, N_ϕ, can be used to calculate the beam's deflection or the elastic foundation reaction.

Computer program SOR analyzes shells of revolution by the finite-element method (Chapter 13). The element used has the shape of a frustum of a cone of constant thickness. The idealized structure is defined by (x, r) coordinates of nodes situated on a radial half section passing through the axis of revolution. The x-axis is vertical, pointing upward; it coincides with the axis of revolution. The analysis gives the nodal displacements, the reactions, and four stress resultants at mid-height of each element. The stress resultants are: N_s, N_ϕ, M_s, M_ϕ; where N and M = normal force and bending moment per unit length, respectively; the subscripts s and ϕ refer, respectively, to the meridian and the circumferential (hoop) directions.

20.5 INPUT AND OUTPUT OF PROGRAM PLANEF

The file "Name.INM" provided for each program contains an example input file and instructions on its preparation. Figures 20.1 and 20.2 are images of "PLANEF.IN" and "PLANEF.OUT" for Example 19.2 of a plane frame shown in Figure 16.6. The data in the input and output files of all programs are of similar format. At the top of each input file, the title of the problem is given. On subsequent lines, the program reads only the numbers on the left-hand side of the input file. The remaining parts of the lines contain words and symbols indicating required data. Thus, the input data file is a form to be completed by the user. For reference, the definition of symbols is given at the bottom of the input file. (This part is not shown in Figure 20.1.)

The symbols NJ, NM, NSJ, and NLC respectively refer to the number of joints, members, supported joints (or joints for which one or more displacement components are prescribed),

Plane frame of Example 19.2 , Figure 19.6
Number of joints = 4
Number of members = 3
Number of joints with prescribed displacement(s) = 2
Number of load cases = 2
Elasticity modulus = .30000E+05; Poissons ratio = .0000

Nodal coordinates

Node	x	y
1	.000	300.000
2	.000	.000
3	300.000	150.000
4	300.000	300.000

Element information

Element	1st node	2nd node	a	I	ar
1	1	2	.30000E+02	.30000E+04	.10000E+07
2	2	3	.40000E+02	.50000E+04	.10000E+07
3	3	4	.30000E+02	.30000E+04	.10000E+07

Support conditions

Node	Restraint indicators			Prescribed displacements		
	u	v	theta	u	v	theta
1	0	0	0	.00000E+00	.00000E+00	.00000E+00
4	0	0	1	.00000E+00	.00000E+00	.00000E+00

Figure 20.2 Image of "PLANEF.OUT" for a plane frame of Example 19.2 (Figure 19.6).

Forces applied at the nodes

Load case	Node	Fx	Fy	Mz
1	2	.40000E+01	.00000E+00	.00000E+00
1	3	.20000E+01	.00000E+00	.00000E+00
10	0	.00000E+00	.00000E+00	.00000E+00

Member end forces with nodal displacement restrained

Ld. case	Member	Ar1	Ar2	Ar3	Ar4	Ar5	Ar6
1	2	-.1342E+02	-.2683E+02	-.1500E+04	-.1342E+02	-.2683E+02	.1500E+04
2	1	-.1500E+04	.8000E+01	.1200E+04	.1500E+04	-.8000E+01	.1200E+04
10	0	.0000E+00	.0000E+00	.0000E+00	.0000E+00	.0000E+00	.0000E+00

Analysis results

Load case no. 1

Nodal displacements

Node	u	v	theta
1	-.68154E-07	.81189E-08	-.16082E-09
2	.39816E-01	.81189E-02	.71980E-03
3	.38236E-01	.59405E-02	.47227E-03
4	.27269E-07	.59405E-08	.61850E-03

Forces at the supported nodes

NODE	Fx	Fy	Mz
1	.27261E+01	-.24357E+02	.19298E+03
4	-.87261E+01	-.35643E+02	.00000E+00

Member end forces

Member	F1	F2	F3	F4	F5	F6
1	.24357E+02	.27261E+01	.19298E+03	-.24357E+02	-.27261E+01	.62486E+03
2	-.48766E+01	-.24793E+02	-.62486E+03	-.21956E+02	-.28872E+02	.13089E+04
3	.35643E+02	-.87261E+01	-.13089E+04	-.35643E+02	.87261E+01	-.22737E-12

Figure 20.2 Continued

and load cases. The node-restraint indicators for this example are the integers 0 0 0, referring to nodal displacements u, v, and θ (Figure 19.1). The integer 0 means a prescribed displacement; the alternative is the integer 1 to mean a free displacement. The three real numbers 0.0 0.0 0.0 are the prescribed displacement values. In this example, nodes 1 and 4 are totally fixed. The dummy integer 10 in this example can be any integer > NLC.

20.6 GENERAL

Each of the programs presented in this chapter performs a specific task. These programs assist in linear or nonlinear analysis of plane frames, space frames, plane trusses, space trusses, plane grids, and eccentric grids, or cable nets; four programs outlined in Section 20.4.5 perform time-dependent sectional analyses to calculated stress and strain parameters of prestressed or non-prestressed concrete sections. Easy use is essential for quick learning; it takes a few minutes to become familiar with and employ any of the programs. These programs enhance the teaching and learning of structural analysis, but no computer program can eliminate the need for structural designers to learn the subject.

Appendix A

Displacements of prismatic members

The following table gives the displacements in beams of constant flexural rigidity EI and constant torsional rigidity GJ, subjected to the loading shown on each beam. The positive directions of the displacements are downward for translation, clockwise for rotation. The deformations due to shearing forces are neglected.

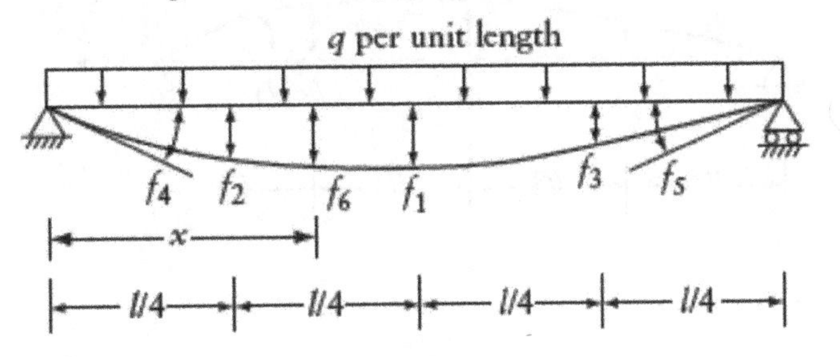

$$f_1 = \frac{5}{384}\frac{ql^4}{EI} \tag{A.1}$$

$$f_2 = f_3 = \frac{19}{2048}\frac{ql^4}{EI} \tag{A.2}$$

$$f_4 = -f_5 = \frac{ql^3}{24EI} \tag{A.3}$$

$$f_6 = \frac{qx}{24EI}\left(l^3 - 2lx^2 + x^3\right) \tag{A.4}$$

l
b
x
P
f2
f1
f3

DOI: 10.1201/9780429286858-21

$$f_1 = \frac{P(l-b)x}{6lEI}\left(2lb - b^2 - x^2\right) \text{ when } x \le b \tag{A.5}$$

$$f_1 = \frac{Pb(l-x)}{6lEI}\left(2lx - x^2 - b^2\right) \text{ when } x \ge b \tag{A.6}$$

$$f_2 = \frac{Pb(l-b)}{6lEI}(2l-b); \quad f_3 = -\frac{Pb}{6lEI}\left(l^2 - b^2\right) \tag{A.7}$$

When $b = l/2$,

$$f_2 = -f_3 = \frac{Pl^2}{16EI}; \quad f_1 = \frac{Pl^3}{48EI} \text{ at } x = l/2 \tag{A.8}$$

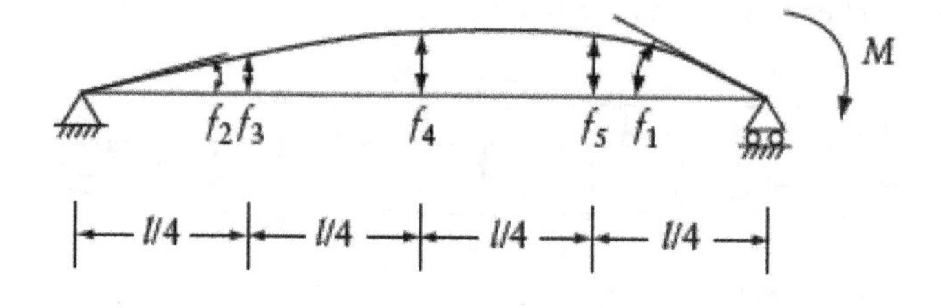

$$f_1 = \frac{Ml}{3EI} \tag{A.9}$$

$$f_2 = -\frac{Ml}{6EI} \tag{A.10}$$

$$f_3 = -\frac{15Ml^2}{634EI} \tag{A.11}$$

$$f_4 = \frac{Ml^2}{16EI} \tag{A.12}$$

$$f_5 = -\frac{21Ml^2}{384EI} \tag{A.13}$$

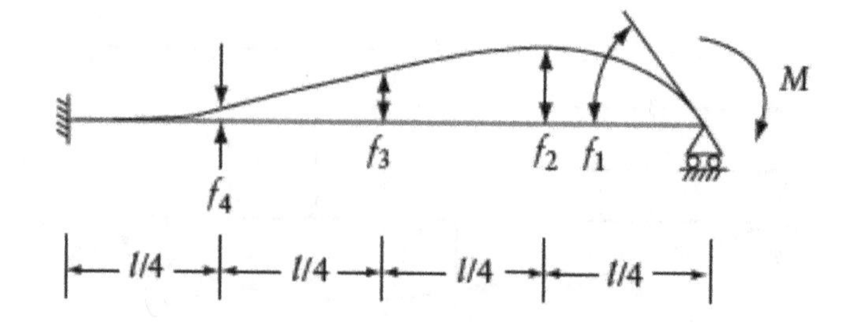

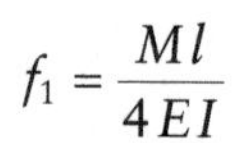

$$f_1 = \frac{Ml}{4EI} \tag{A.14}$$

$$f_2 = -\frac{9Ml^2}{256EI} \tag{A.15}$$

$$f_3 = -\frac{Ml^2}{32EI} \tag{A.16}$$

$$f_4 = -\frac{3Ml^2}{256EI} \tag{A.17}$$

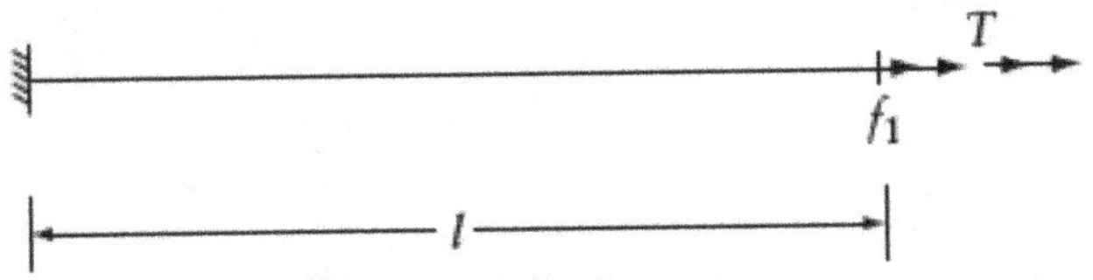

T

f_1

l

$$f_1 = \frac{Tl}{GJ} \quad \text{(Effect of warping is ignored)} \tag{A.18}$$

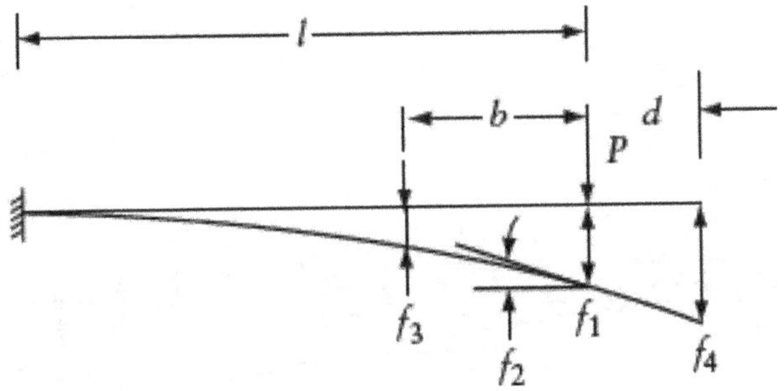

l

b d

P

f_3 f_2 f_1 f_4

$$f_1 = \frac{Pl^3}{3EI} \tag{A.19}$$

$$f_2 = \frac{Pl^2}{2EI} \tag{A.20}$$

$$f_4 = f_1 + d\,f_2 \tag{A.21}$$

$$f_3 = \frac{Pl^3}{3EI}\left(1 - \frac{3b}{2l} + \frac{b^3}{2l^3}\right) \quad \text{for} \quad 0 \le b \le l \tag{A.22}$$

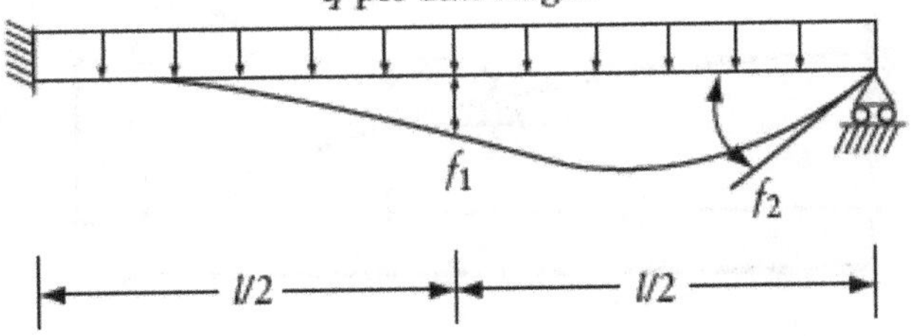

q per unit length

f_1 f_2

$l/2$ $l/2$

$$f_1 = \frac{ql^4}{192EI} \tag{A.23}$$

$$f_2 = -\frac{ql^3}{48EI} \tag{A.24}$$

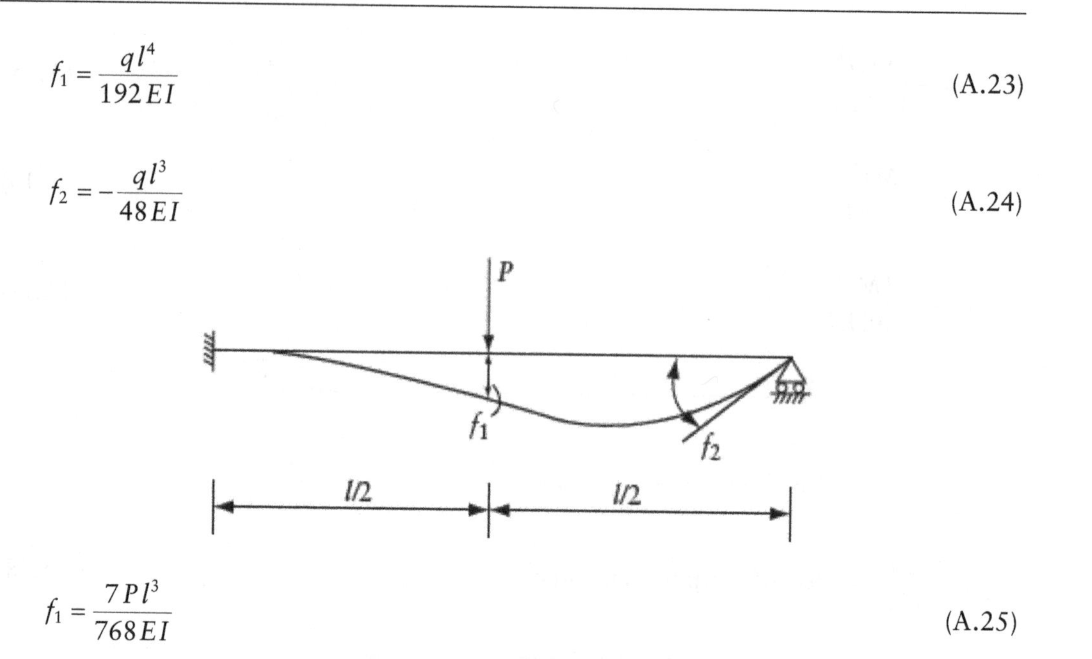

$$f_1 = \frac{7Pl^3}{768EI} \tag{A.25}$$

$$f_2 = \frac{Pl^2}{32EI} \tag{A.26}$$

q per unit length

$$f_1 = \frac{ql^4}{8EI} \tag{A.27}$$

$$f_2 = \frac{ql^3}{6EI}\left(3\xi - 3\xi^2 + \xi^3\right) \tag{A.28}$$

$$f_3 = \frac{ql^4}{24EI}\left(6\xi^2 - 4\xi^3 + \xi^4\right) \tag{A.29}$$

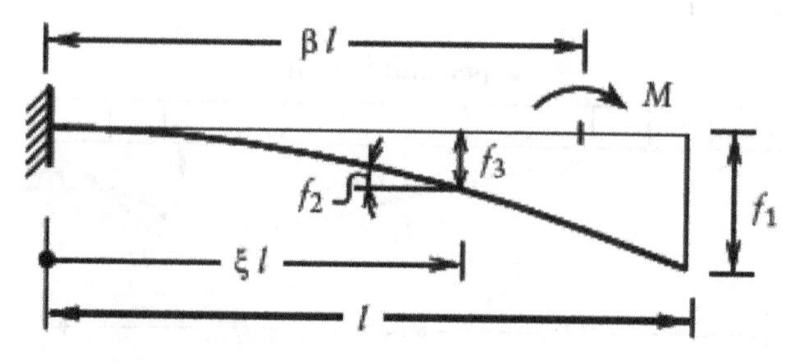

$$f_1 = \frac{Ml^2}{EI}\beta(1-0.5\beta) \tag{A.30}$$

$$f_2 = \begin{cases} Ml\xi/(EI) & \text{with } \xi \le \beta \\ Ml\beta/(EI) & \text{with } \beta \le \xi \le 1 \end{cases} \tag{A.31}$$

$$f_3 = \begin{cases} M(l\xi)^2/(2EI) & \text{with } \xi \le \beta \\ Ml^2\beta(\xi-0.5\beta)/(EI) & \text{with } \beta \le \xi \le 1 \end{cases} \tag{A.32}$$

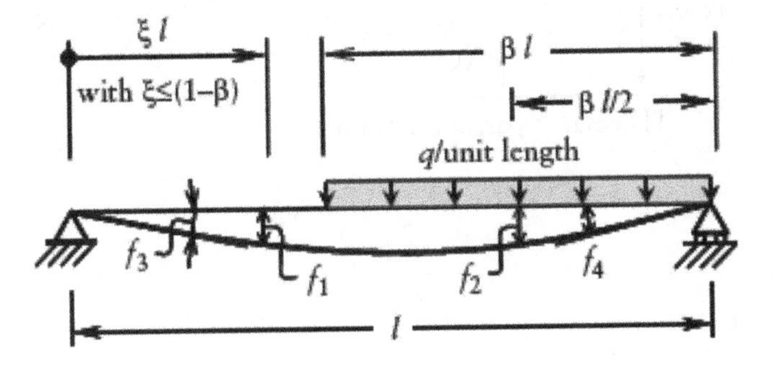

$$f_1 = \frac{ql^4}{24EI}\beta^2\xi\left(2-\beta^2-2\xi^2\right) \tag{A.33}$$

$$f_2 = \frac{ql^4}{384EI}\beta^3\left(32-39\beta+12\beta^2\right) \tag{A.34}$$

$$f_3 = \frac{ql^3}{24EI}\beta^2\left(2-\beta^2\right) \tag{A.35}$$

$$f_4 = -\frac{ql^3}{24EI}\beta^2\left(4-4\beta+\beta^2\right) \tag{A.36}$$

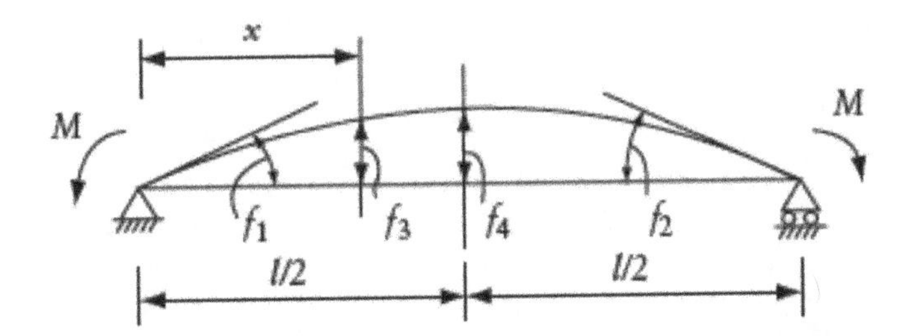

$$f_1 = -f_2 = -\frac{Ml}{2EI} \tag{A.37}$$

$$f_3 = -\frac{Mx(1-x)}{2EI}; \quad f_4 = -\frac{Ml^2}{8EI} \tag{A.38}$$

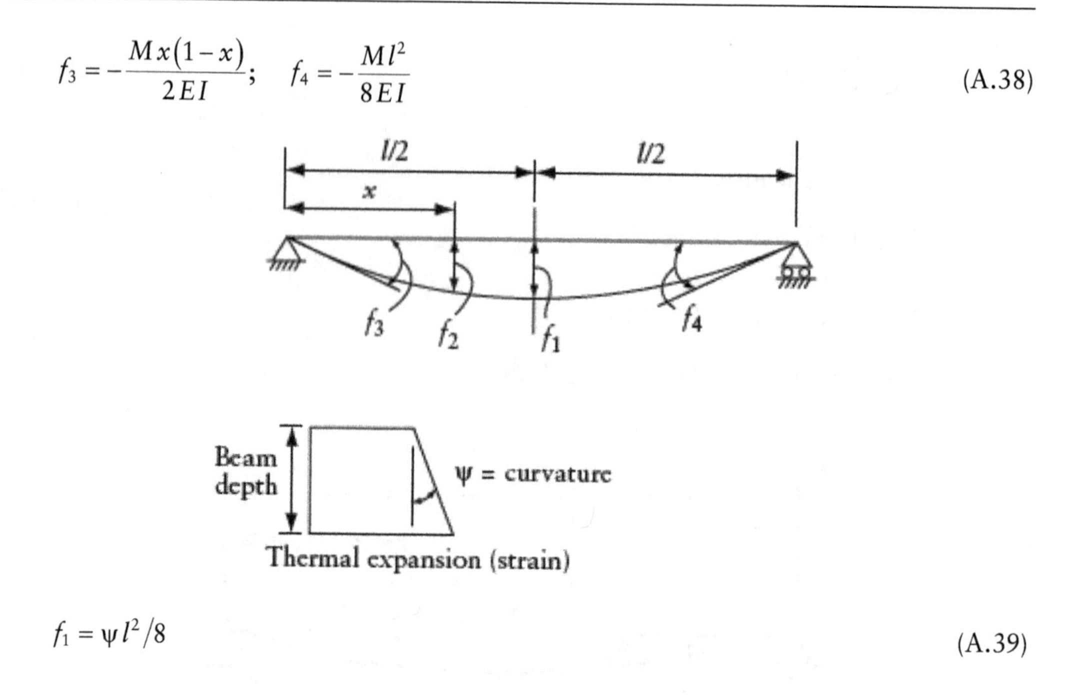

$$f_1 = \psi l^2/8 \tag{A.39}$$

$$f_2 = \frac{\psi x(l-x)}{2} \tag{A.40}$$

$$f_3 = -f_4 = \frac{\psi l}{2} \tag{A.41}$$

Appendix B

Fixed-end forces of prismatic members

The following table gives the fixed-end forces in beams of constant flexural rigidity and constant torsional rigidity due to applied loads. The forces are considered positive if upward or in the clockwise direction. A twisting couple is positive if it acts in the direction of rotation of a right-hand screw progressing to the right. When the end-forces are used in the displacement method, appropriate signs have to be assigned according to the chosen coordinate system.

DOI: 10.1201/9780429286858-22

	Fixed-end force	*Equation*
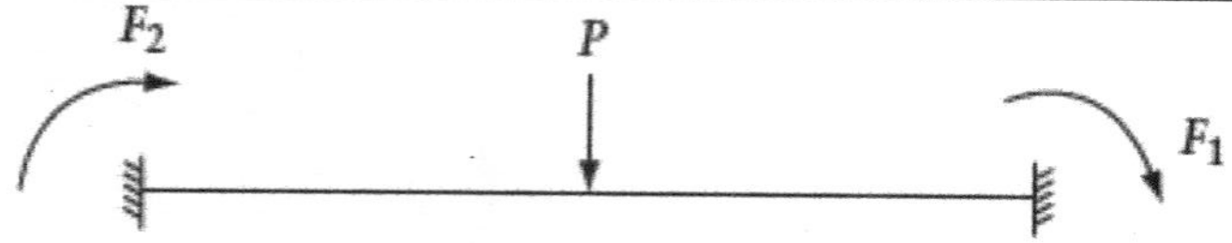	$F_1 = -F_2 = \frac{P\,l}{8}$	(B.1)
	$F_3 = F_4 = \frac{P}{2}$	(B.2)
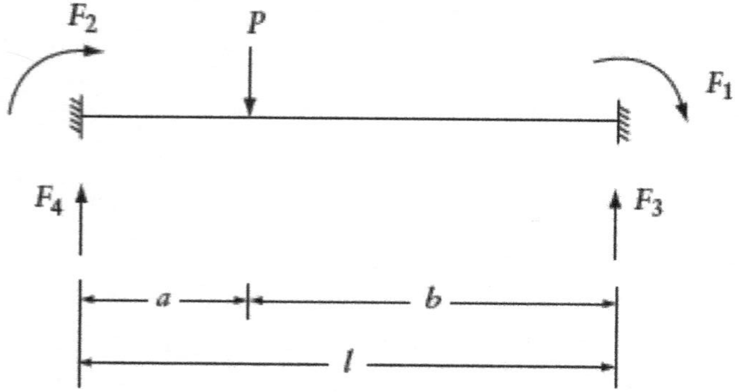	$F_1 = \frac{P\,a^2\,b}{l^2}$	(B.3)
	$F_2 = -\frac{P\,a\,b^2}{l^2}$	(B.4)
	$F_3 = P\left(\frac{a}{l} + \frac{a^2\,b - a\,b^2}{l^3}\right)$	(B.5)
	$F_3 = P\left(\frac{b}{l} + \frac{a\,b^2 - a^2\,b}{l^3}\right)$	(B.6)
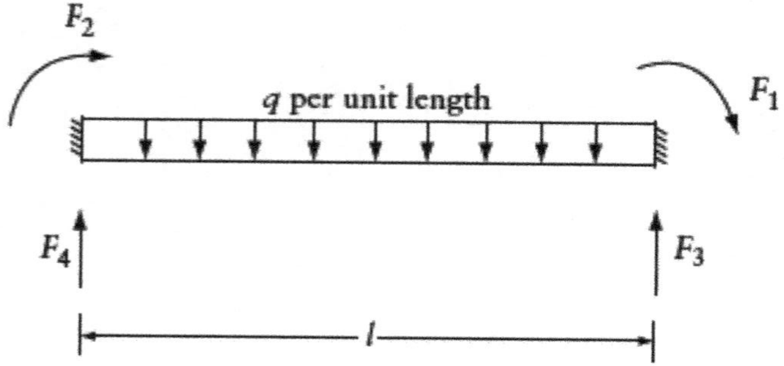 q per unit length	$F_1 = -F_2 = \frac{q\,l^2}{12}$	(B.7)
	$F_3 = F_4 = \frac{q\,l}{2}$	(B.8)

(Continued)

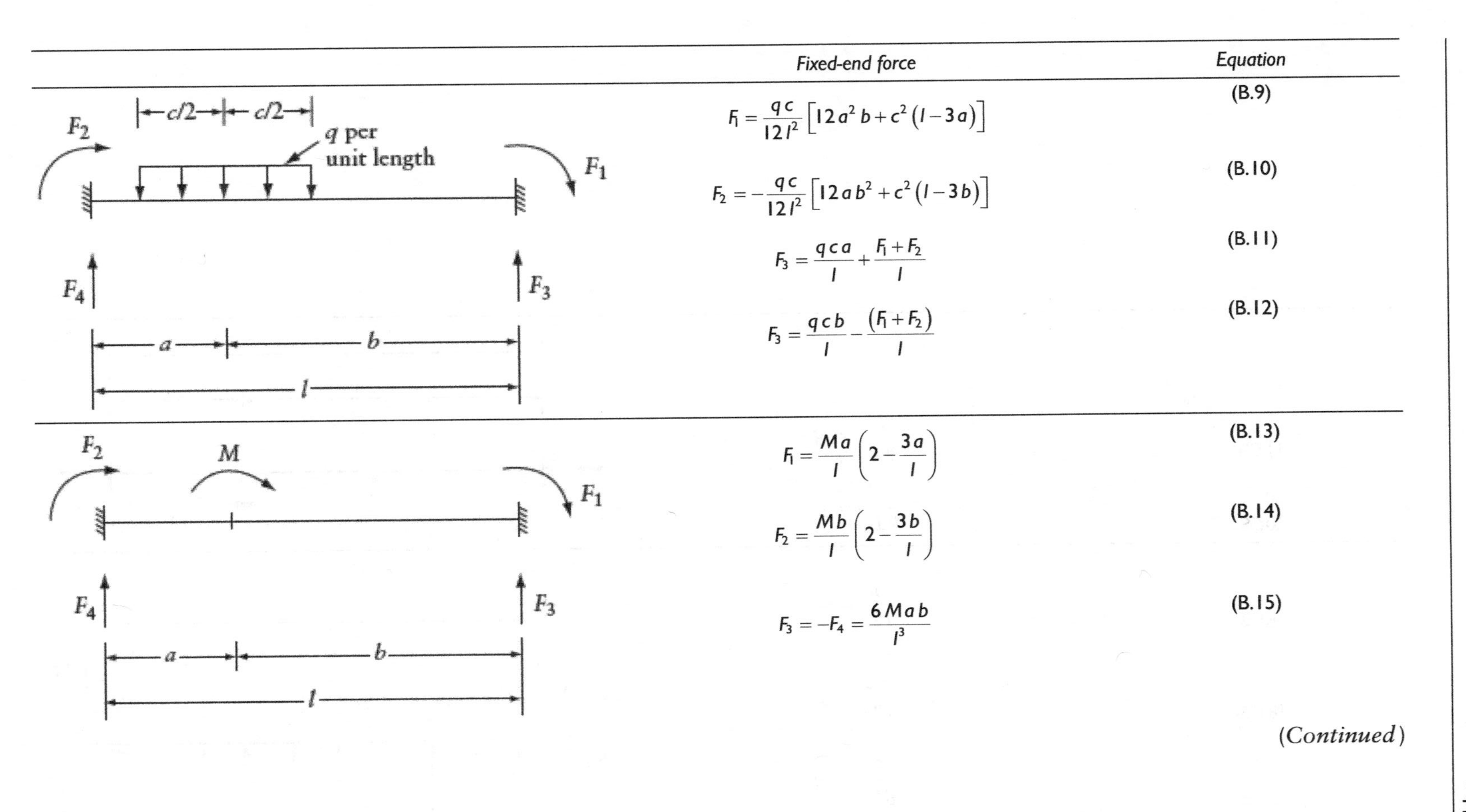

	Fixed-end force	*Equation*
	$F_1 = \frac{qc}{12l^2}\left[12a^2b + c^2(l - 3a)\right]$	(B.9)
	$F_2 = -\frac{qc}{12l^2}\left[12ab^2 + c^2(l - 3b)\right]$	(B.10)
	$F_3 = \frac{qca}{l} + \frac{F_1 + F_2}{l}$	(B.11)
	$F_3 = \frac{qcb}{l} - \frac{(F_1 + F_2)}{l}$	(B.12)
	$F_1 = \frac{Ma}{l}\left(2 - \frac{3a}{l}\right)$	(B.13)
	$F_2 = \frac{Mb}{l}\left(2 - \frac{3b}{l}\right)$	(B.14)
	$F_3 = -F_4 = \frac{6Mab}{l^3}$	(B.15)

(Continued)

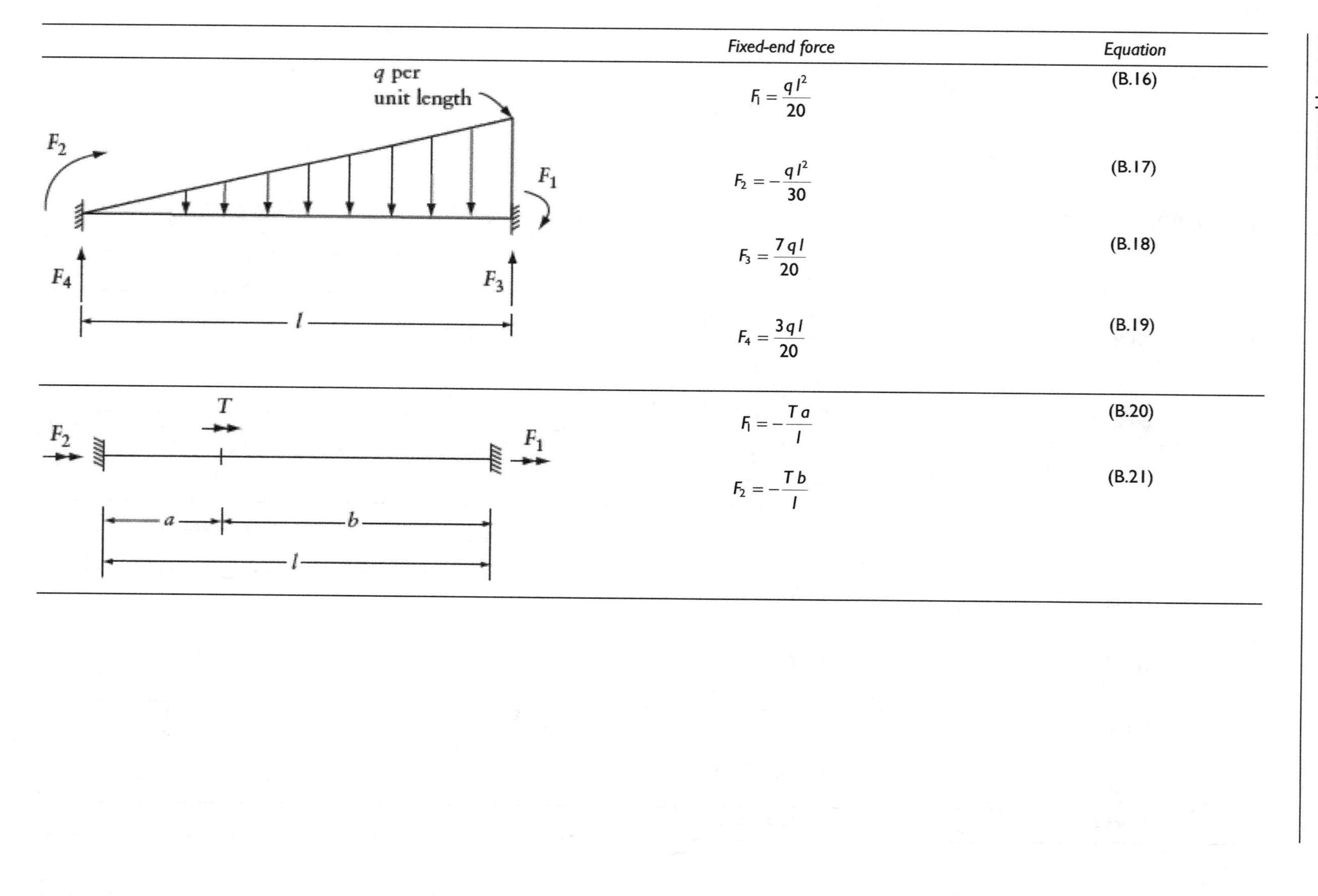

	Fixed-end force	Equation
	$F_1 = \frac{q l^2}{20}$	(B.16)
	$F_2 = -\frac{q l^2}{30}$	(B.17)
	$F_3 = \frac{7 q l}{20}$	(B.18)
	$F_4 = \frac{3 q l}{20}$	(B.19)
	$F_1 = -\frac{T a}{l}$	(B.20)
	$F_2 = -\frac{T b}{l}$	(B.21)

If the totally fixed support in any of the above cases, except the last, is changed to a hinge or a roller, the fixed-end moment at the other end can be calculated using the equations of this appendix and Equation 9.17. Examples are as follows:

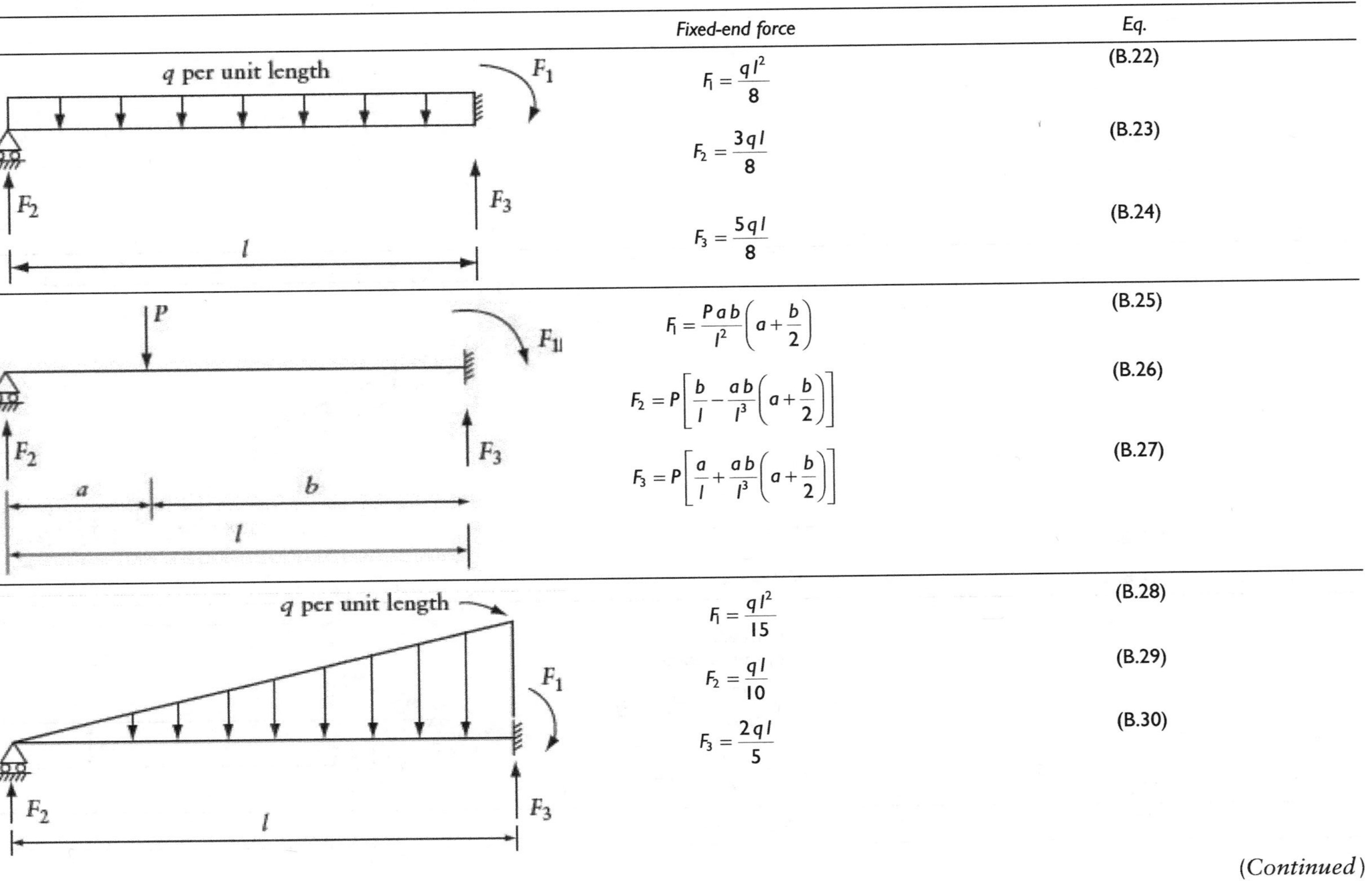

Fixed-end force	Eq.
$F_1 = \frac{ql^2}{8}$	(B.22)
$F_2 = \frac{3ql}{8}$	(B.23)
$F_3 = \frac{5ql}{8}$	(B.24)
$F_1 = \frac{Pab}{l^2}\left(a + \frac{b}{2}\right)$	(B.25)
$F_2 = P\left[\frac{b}{l} - \frac{ab}{l^3}\left(a + \frac{b}{2}\right)\right]$	(B.26)
$F_3 = P\left[\frac{a}{l} + \frac{ab}{l^3}\left(a + \frac{b}{2}\right)\right]$	(B.27)
$F_1 = \frac{ql^2}{15}$	(B.28)
$F_2 = \frac{ql}{10}$	(B.29)
$F_3 = \frac{2ql}{5}$	(B.30)

(*Continued*)

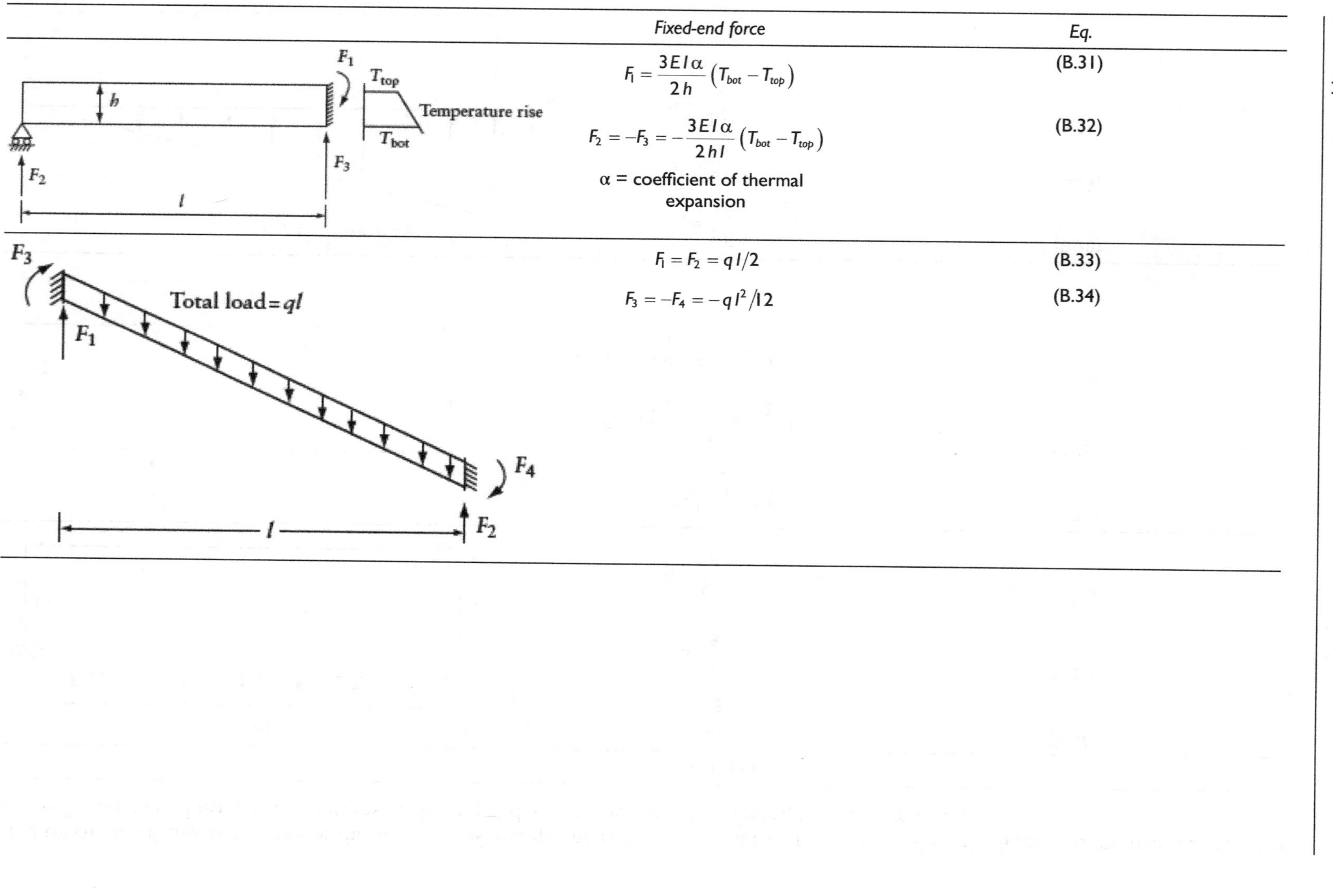

	Fixed-end force	Eq.
	$F_1 = \frac{3EI\alpha}{2h}\left(T_{bot} - T_{top}\right)$	(B.31)
	$F_2 = -F_3 = -\frac{3EI\alpha}{2hl}\left(T_{bot} - T_{top}\right)$	(B.32)
	α = coefficient of thermal expansion	
	$F_1 = F_2 = ql/2$	(B.33)
	$F_3 = -F_4 = -ql^2/12$	(B.34)

Appendix C

End-forces caused by end displacements of prismatic members

The following table gives the forces at the ends of beams due to a unit translation or unit rotation of one end. The positive directions for the forces are upward and clockwise. The effect of the deformation caused by the shearing forces is neglected; this topic is considered in Section 11.2. Moreover, the equations do not account for the bending moment due to axial forces; if a member is subjected to a large axial force, its effect may be included using Table 10.2 instead of this appendix. The beams have a constant flexural rigidity EI and a constant torsional rigidity GJ.

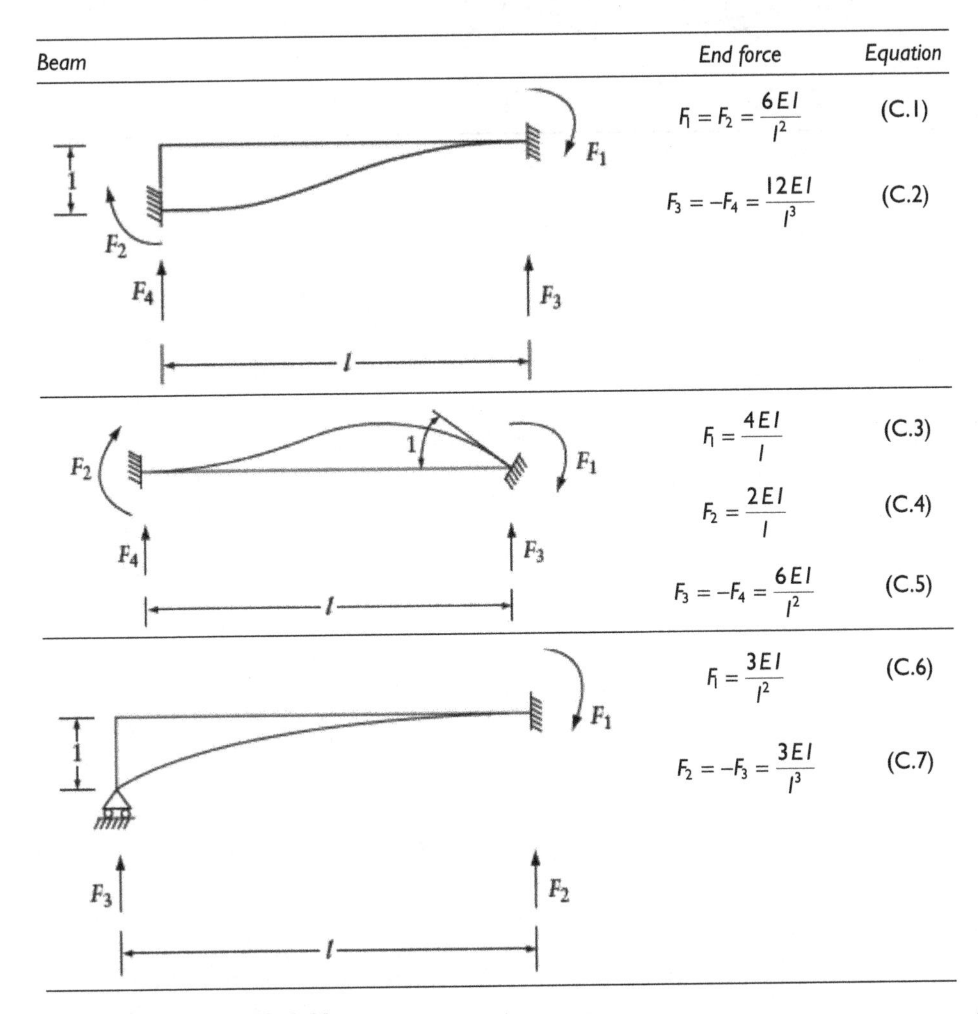

Beam	End force	Equation
(unit translation at one end, both ends fixed)	$F_1 = F_2 = \frac{6EI}{l^2}$	(C.1)
	$F_3 = -F_4 = \frac{12EI}{l^3}$	(C.2)
(unit rotation at one end, both ends fixed)	$F_1 = \frac{4EI}{l}$	(C.3)
	$F_2 = \frac{2EI}{l}$	(C.4)
	$F_3 = -F_4 = \frac{6EI}{l^2}$	(C.5)
(unit translation at hinged end, other end fixed)	$F_1 = \frac{3EI}{l^2}$	(C.6)
	$F_2 = -F_3 = \frac{3EI}{l^3}$	(C.7)

DOI: 10.1201/9780429286858-23

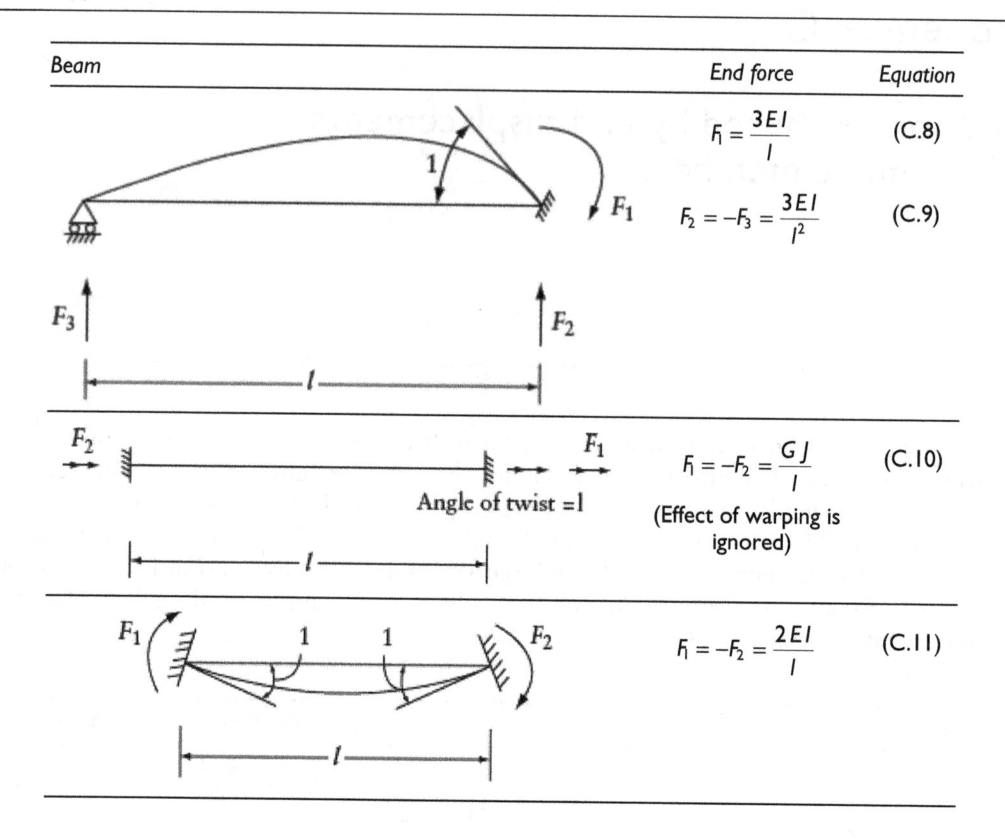

Beam	*End force*	*Equation*
	$F_1 = \frac{3EI}{l}$	(C.8)
	$F_2 = -F_3 = \frac{3EI}{l^2}$	(C.9)
	$F_1 = -F_2 = \frac{GJ}{l}$ (Effect of warping is ignored)	(C.10)
	$F_1 = -F_2 = \frac{2EI}{l}$	(C.11)

Appendix D

Reactions and bending moments at supports of continuous beams due to unit displacement of supports

The following tables give the reactions and the bending moments at the supports of a continuous beam due to a unit downward translation of each of the supports separately. All spans are of equal length l and have a constant flexural rigidity EI. The number of spans is 2 (or 1) to 5. The end supports are hinged (Table D.1), fixed (Table D.2), and fixed at the left-hand end and hinged at the right-hand end (Table D.3). The bending moment at the hinged end is known to be zero and is not listed in the tables.

The values given in each row are the bending moments or the reactions at consecutive supports starting from the left-hand end. The first row after the heading gives the effect of the settlement of the first support from the left, the second row gives the effect of the settlement of the second support from the left, and so on.

Figure D.1 shows an example of the use of the tables: the number of spans is 3, the second support from left settles a unit distance and the values are taken from the second row of the appropriate table.

In these tables, the reaction is considered positive when acting upward, and the bending moment is positive if it causes tension in the bottom fiber of the beam. When the reactions are used to generate a stiffness matrix, appropriate signs should be given according to the chosen coordinate system. The effect of deformation caused by the shearing forces is neglected.

DOI: 10.1201/9780429286858-24

Table D.1 Effect of a unit downward displacement of one support of continuous beams (end supports hinged, ***EI*** constant, all spans are of equal length ***l***)

Number of spans = 2				
Support moments in terms of EI/l^2				
−1.50000				
3.00000				
−1.50000				
Reactions in terms of EI/l^3				
−1.50000	3.00000	−1.50000		
3.00000	−6.00000	3.00000		
−1.50000	3.00000	−1.50000		
Number of spans = 3				
Support moments in terms of EI/l^2				
−1.60000	0.40000			
3.60000	−2.40000			
−2.40000	3.60000			
0.40000	−1.60000			
Reactions in terms of EI/l^3				
−1.60000	3.60000	−2.40000	0.40000	
3.60000	−9.60000	8.40000	−2.40000	
−2.40000	8.40000	−9.60000	3.60000	
0.40000	−2.40000	3.60000	−1.60000	
Number of spans = 4				
Support moments in terms of EI/l^2				
−1.60714	0.42857	−0.10714		
3.64286	−2.57143	0.64286		
−2.57143	4.28571	−2.57143		
0.64286	−2.57143	3.64286		
−0.10714	0.42857	−1.60714		
Reactions in terms of EI/l^3				
−1.60714	3.64286	−2.57143	0.64286	−0.10714
3.64286	−9.85714	9.42857	−3.85714	0.64286

(*Continued*)

Table D.1 (Continued) Effect of a unit downward displacement of one support of continuous beams (end supports hinged, ***EI*** constant, all spans are of equal length ***l***)

−2.57143	9.42857	−13.71428	9.42857	−2.57143	
0.64286	−3.85714	9.42857	−9.85714	3.64286	
−0.10714	0.64286	−2.57143	3.64286	−1.60714	
Number of spans = 5					
Support moments in terms of EI/l^2					
−1.60765	0.43062	−0.11483	0.02871		
3.64593	−2.58373	0.68900	−0.17225		
−2.58373	4.33493	−2.75598	0.68900		
0.68900	−2.75598	4.33493	−2.58373		
−0.17225	0.68900	−2.58373	3.64593		
0.02871	−0.11483	0.43062	−1.60765		
Reactions in terms of EI/l^3					
−1.60765	3.64593	−2.58373	0.68900	−0.17225	0.02871
3.64593	−9.87560	9.50239	−4.13397	1.03349	−0.17225
−2.58373	9.50239	−14.00956	10.53588	−4.13397	0.68900
0.68900	−4.13397	10.53588	−14.00957	9.50239	−2.58373
−0.17225	1.03349	−4.13397	9.50239	−9.87560	3.64593
0.02871	−0.17225	0.68900	−2.58373	3.64593	−1.60765

Table D.2 Effect of a unit downward displacement of one support of continuous beams (two end supports fixed, ***EI*** constant, all spans of equal length ***l***)

Number of spans = 1			
Support moments in terms of EI/l^2			
6.00000	−6.00000		
−6.00000	6.00000		
Reactions in terms of EI/l^3			
−12.00000	12.00000		
12.00000	−12.00000		
Number of spans = 2			
Support moments in terms of EI/l^2			
4.50000	−3.00000	1.50000	
−6.00000	6.00000	−6.00000	
1.50000	−3.00000	4.50000	
Reactions in terms of EI/l^3			
−7.50000	12.00000	−4.50000	
12.00000	−24.00000	12.00000	
−4.50000	12.00000	−7.50000	
Number of spans = 3			
Support moments in terms of EI/l^2			
4.40000	−2.80000	0.80000	−0.40000
−5.60000	5.20000	−3.20000	1.60000
1.60000	−3.20000	5.20000	−5.60000
−0.40000	0.80000	−2.80000	4.40000
Reactions in terms of EI/l^3			
−7.20000	10.80000	−4.80000	1.20000
10.80000	−19.20000	13.20000	−4.80000
−4.80000	13.20000	−19.20000	10.80000
1.20000	−4.80000	10.80000	−7.20000

(*Continued*)

Table D.2 (Continued) Effect of a unit downward displacement of one support of continuous beams (two end supports fixed, ***EI*** constant, all spans of equal length ***l***)

Number of spans = 4					
Support moments in terms of EI/l^2					
4.39286	−2.78571	0.75000	−0.21429	0.10714	
−5.57143	5.14286	−3.00000	0.85714	−0.42857	
1.50000	−3.00000	4.50000	−3.00000	1.50000	
−0.42857	0.85714	−3.00000	5.14286	−5.57143	
0.10714	−0.21429	0.75000	−2.78571	4.39286	
Reactions in terms of EI/l^3					
−7.17857	10.71428	−4.50000	1.28571	−0.32143	
10.71428	−18.85713	12.00000	−5.14285	1.28571	
−4.50000	12.00000	−15.00000	12.00000	−4.50000	
1.28571	−5.14285	12.00000	−18.85713	10.71428	
−0.32143	1.28571	−4.50000	10.71428	−7.17857	
Number of spans = 5					
Support moments in terms of EI/l^2					
4.39235	−2.78469	0.74641	−0.20096	0.05742	−0.02871
−5.56938	5.13875	−2.98564	0.80383	−0.22966	0.11483
1.49282	−2.98564	4.44976	−2.81340	0.80383	−0.40191
−0.40191	0.80383	−2.81339	4.44976	−2.98564	1.49282
0.11483	−0.22966	0.80383	−2.98564	5.13875	−5.56938
−0.02871	0.05742	−0.20096	0.74641	−2.78469	4.39235
Reactions in terms of EI/l^3					
−7.17703	10.70813	−4.47847	1.20574	−0.34450	0.08612
10.70813	−18.83252	11.91387	−4.82296	1.37799	−0.34450
−4.47847	11.91387	−14.69856	10.88038	−4.82296	1.20574
1.20574	−4.82296	10.88038	−14.69856	11.91387	−4.47847
−0.34450	1.37799	−4.82296	11.91387	−18.83252	10.70813
0.08612	−0.34450	1.20574	−4.47847	10.70813	−7.17703

Table D.3 Effect of a unit downward displacement of one support of continuous beams (support at left end fixed, hinged at right end, ***EI*** constant, all spans of equal length ***l***)

Number of spans = 1			
Support moments in terms of EI/l^2			
3.00000			
−3.00000			
Reactions in terms of EI/l^3			
−3.00000	3.00000		
3.00000	−3.00000		
Number of spans = 2			
Support moments in terms of EI/l^2			
4.28571	−2.57143		
−5.14286	4.28571		
0.85714	−1.71428		
Reactions in terms of EI/l^3			
−6.85714	9.42857	−2.57143	
9.42857	−13.71428	4.28571	
−2.57143	4.28571	−1.71428	
Number of spans = 3			
Support moments in terms of EI/l^2			
4.38462	−2.76923	0.69231	
−5.53846	5.07692	−2.76923	
1.38461	−2.76923	3.69231	
−0.23077	0.46154	−1.61538	
Reactions in terms of EI/l^3			
−7.15385	10.61538	−4.15384	0.69231
10.61539	−18.46153	10.61538	−2.76923
−4.15384	10.61538	−10.15384	3.69231
0.69231	−2.76923	3.69230	−1.61538

(*Continued*)

Table D.3 (Continued) Effect of a unit downward displacement of one support of continuous beams (support at left end fixed, hinged at right end, EI constant, all spans of equal length ***l***)

Number of spans = 4					
Support moments in terms of EI/l^2					
4.39175	−2.78350	0.74227	−0.18557		
−5.56701	5.13402	−2.96907	0.74227		
1.48454	−2.96907	4.39175	−2.59794		
−0.37113	0.74227	−2.59794	3.64948		
0.06186	−0.12371	0.43299	−1.60825		
Reactions in terms of EI/l^3					
−7.17526	10.70103	−4.45361	1.11340	−0.18557	
10.70103	−18.80411	11.81443	−4.45361	0.74227	
−4.45361	11.81443	−14.35051	9.58763	−2.59794	
1.11340	−4.45361	9.58763	−9.89690	3.64948	
−0.18557	0.74227	−2.59794	3.64948	−1.60825	
Number of spans = 5					
Support moments in terms of EI/l^2					
4.39227	−2.78453	0.74586	−0.19889	0.04972	
−5.56906	5.13812	−2.98342	0.79558	−0.19889	
1.49171	−2.98342	4.44199	−2.78453	0.69613	
−0.39779	0.79558	−2.78453	4.34254	−2.58563	
0.09945	−0.19889	0.69613	−2.58563	3.64641	
−0.01657	0.03315	−0.11602	0.43094	−1.60773	
Reactions in terms of EI/l^3					
−7.17680	10.70718	−4.47513	1.19337	−0.29834	0.04972
10.70718	−18.82872	11.90055	−4.77348	1.19337	−0.19889
−4.47513	11.90055	−14.65193	10.70718	−4.17679	0.69613
1.19337	−4.77348	10.70718	−14.05524	9.51381	−2.58563
−0.29834	1.19337	−4.17679	9.51381	−9.87845	3.64641
0.04972	−0.19889	0.69613	−2.58563	3.64641	−1.60773

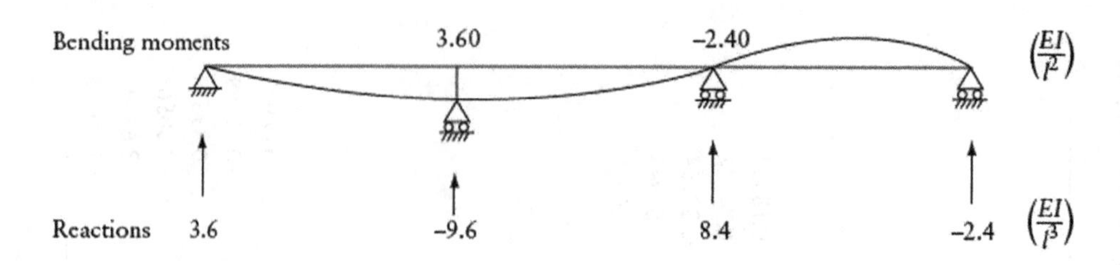

Figure D.1 Illustration of the use of tables of Appendix D.

Appendix E

Properties of geometrical figures

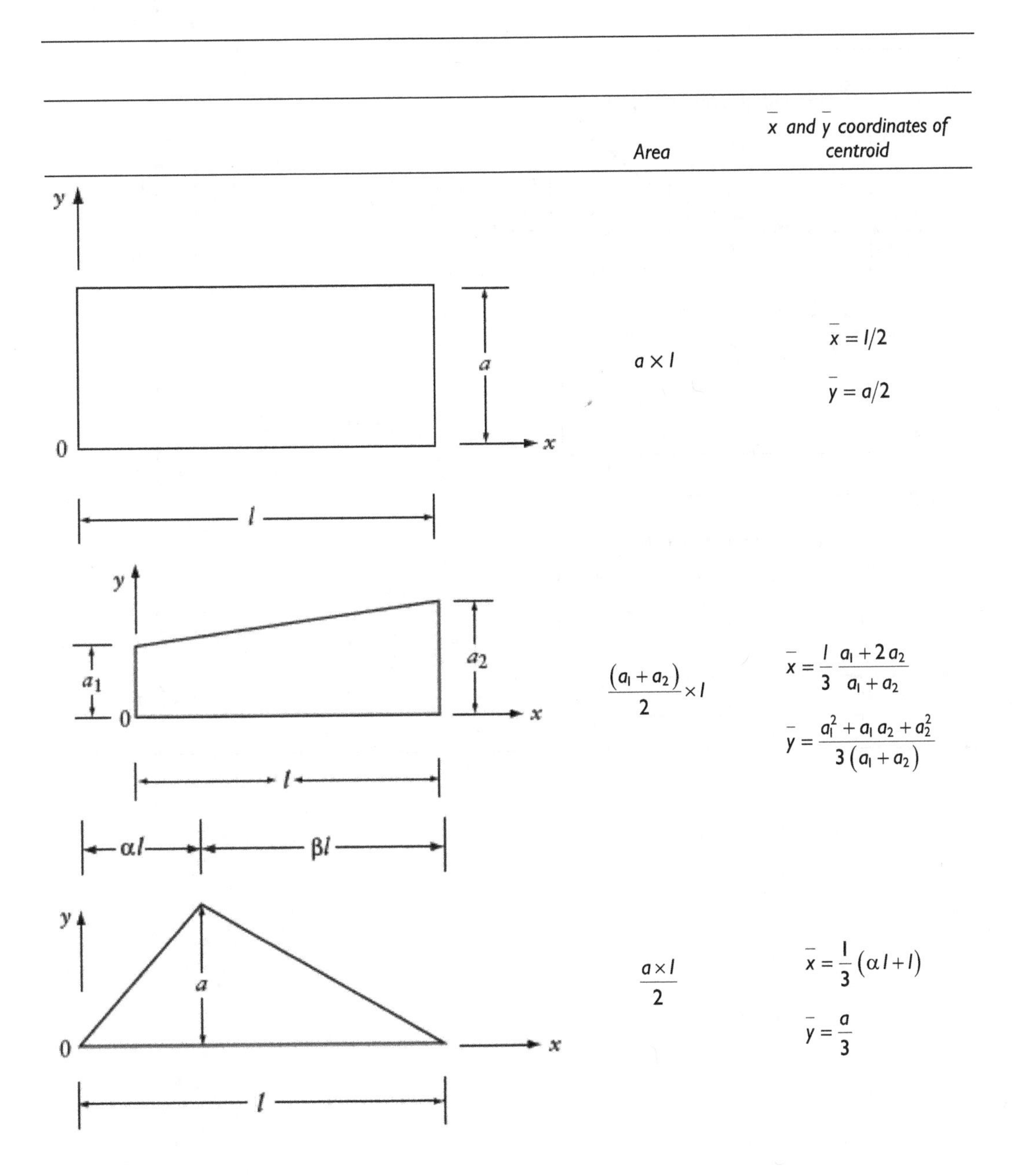

	Area	$\bar{x}$ and $\bar{y}$ coordinates of centroid
Rectangle (sides l and a)	$a \times l$	$\bar{x} = l/2$, $\bar{y} = a/2$
Trapezium (heights a_1, a_2, length l)	$\dfrac{(a_1 + a_2)}{2} \times l$	$\bar{x} = \dfrac{l}{3}\dfrac{a_1 + 2a_2}{a_1 + a_2}$, $\bar{y} = \dfrac{a_1^2 + a_1 a_2 + a_2^2}{3(a_1 + a_2)}$
Triangle (base $l = \alpha l + \beta l$, height a)	$\dfrac{a \times l}{2}$	$\bar{x} = \dfrac{1}{3}(\alpha l + l)$, $\bar{y} = \dfrac{a}{3}$

DOI: 10.1201/9780429286858-25

	Area	*$\bar{x}$ and $\bar{y}$ coordinates of centroid*
Second-degree parabola 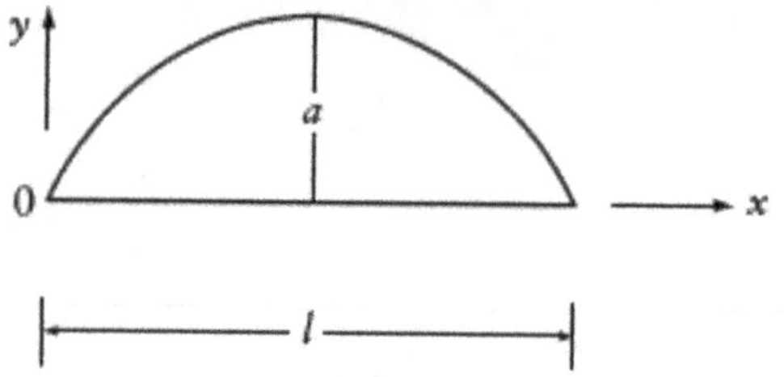	$\frac{2}{3} a \times l$	$\bar{x} = \frac{l}{2}$ $\bar{y} = \frac{2}{5} a$
Second-degree parabola 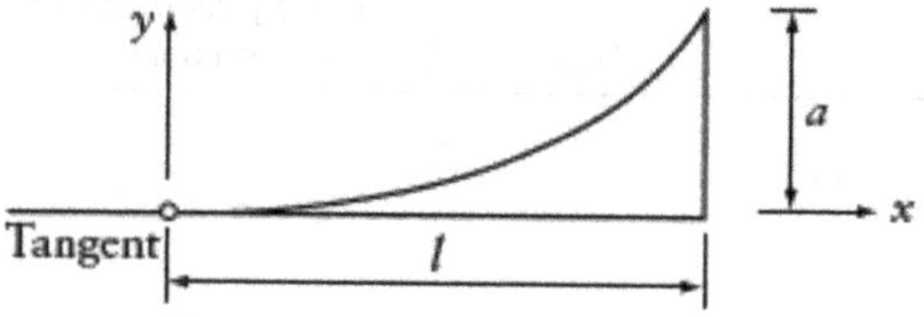	$\frac{1}{3} a \times l$	$\bar{x} = \frac{3}{4} l$ $\bar{y} = \frac{3}{10} a$
Second-degree parabola 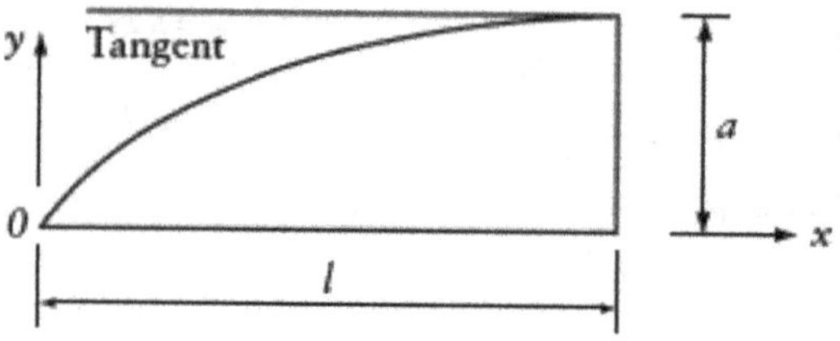	$\frac{2}{3} a \times l$	$\bar{x} = \frac{5}{8} l$ $\bar{y} = \frac{2}{5} a$
Third-degree parabola 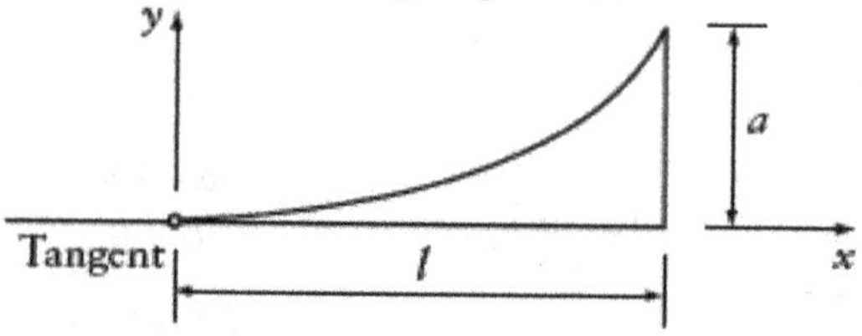	$\frac{1}{4} a \times l$	$\bar{x} = \frac{4}{5} l$ $\bar{y} = \frac{2}{7} a$

Construction of tangents of a second-degree parabola:

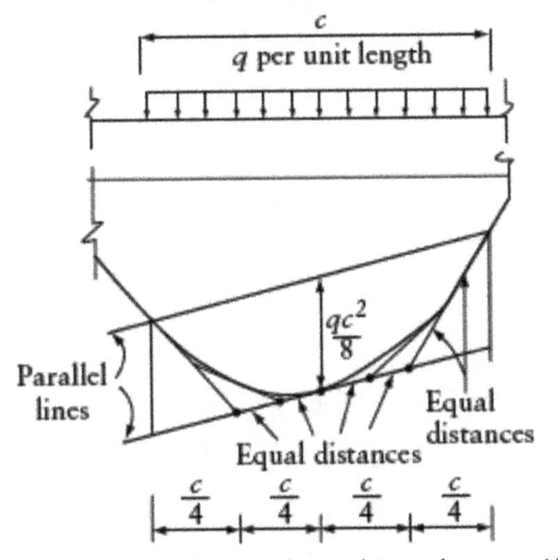

Bending moment diagram for a part of a member subjected to a uniform load.

Appendix F

Torsional constant J

If a circular bar of constant cross section and of length l is subjected to a constant torque T, the angle of twist between the two bar ends is:

$$\theta = \frac{T\,l}{G\,J}$$

where G = shear modulus; J = polar moment of inertia.

When the cross section of the bar is noncircular, plane cross sections do not remain plane after deformation, and warping will occur caused by longitudinal displacements of points in the cross section. Nevertheless, the above equation can be used with good accuracy for noncircular cross sections, but J should be taken as the appropriate torsion constant. The torsional constants for several shapes of cross sections are listed below.

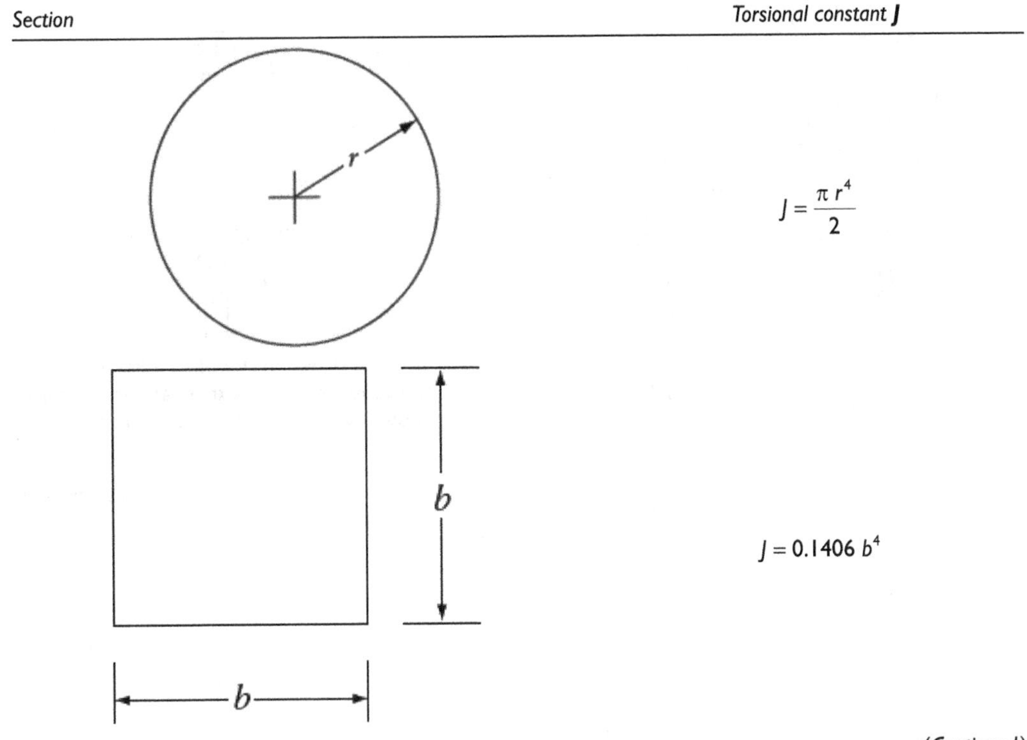

(Continued)

DOI: 10.1201/9780429286858-26

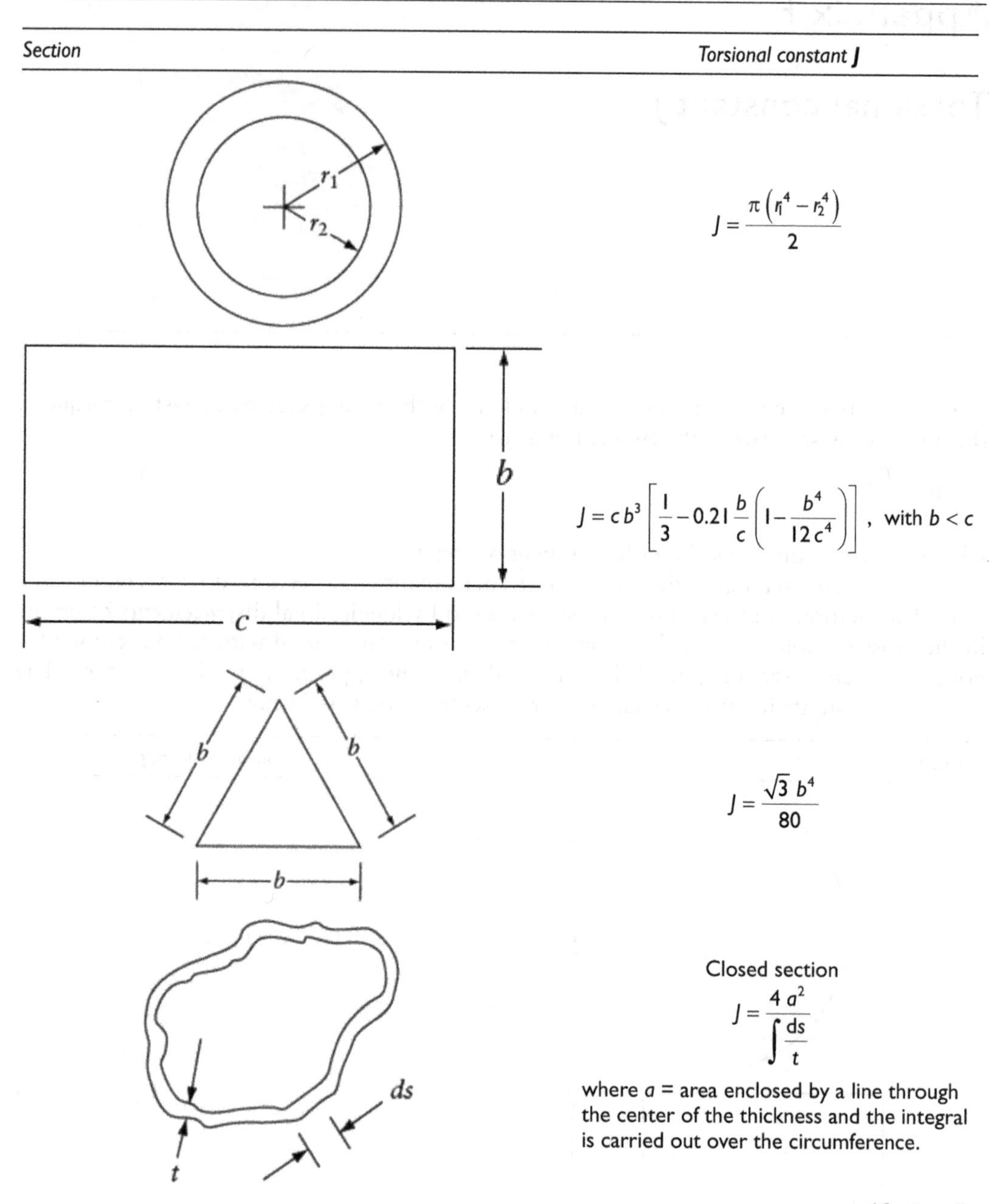

Section	*Torsional constant **J***
r_1, r_2	$J = \dfrac{\pi\left(r_1^4 - r_2^4\right)}{2}$
b, c	$J = c\,b^3\left[\dfrac{1}{3} - 0.21\,\dfrac{b}{c}\left(1 - \dfrac{b^4}{12\,c^4}\right)\right]$, with $b < c$
b, b, b	$J = \dfrac{\sqrt{3}\,b^4}{80}$
ds, t	Closed section $J = \dfrac{4\,a^2}{\int \frac{ds}{t}}$ where a = area enclosed by a line through the center of the thickness and the integral is carried out over the circumference.

(Continued)

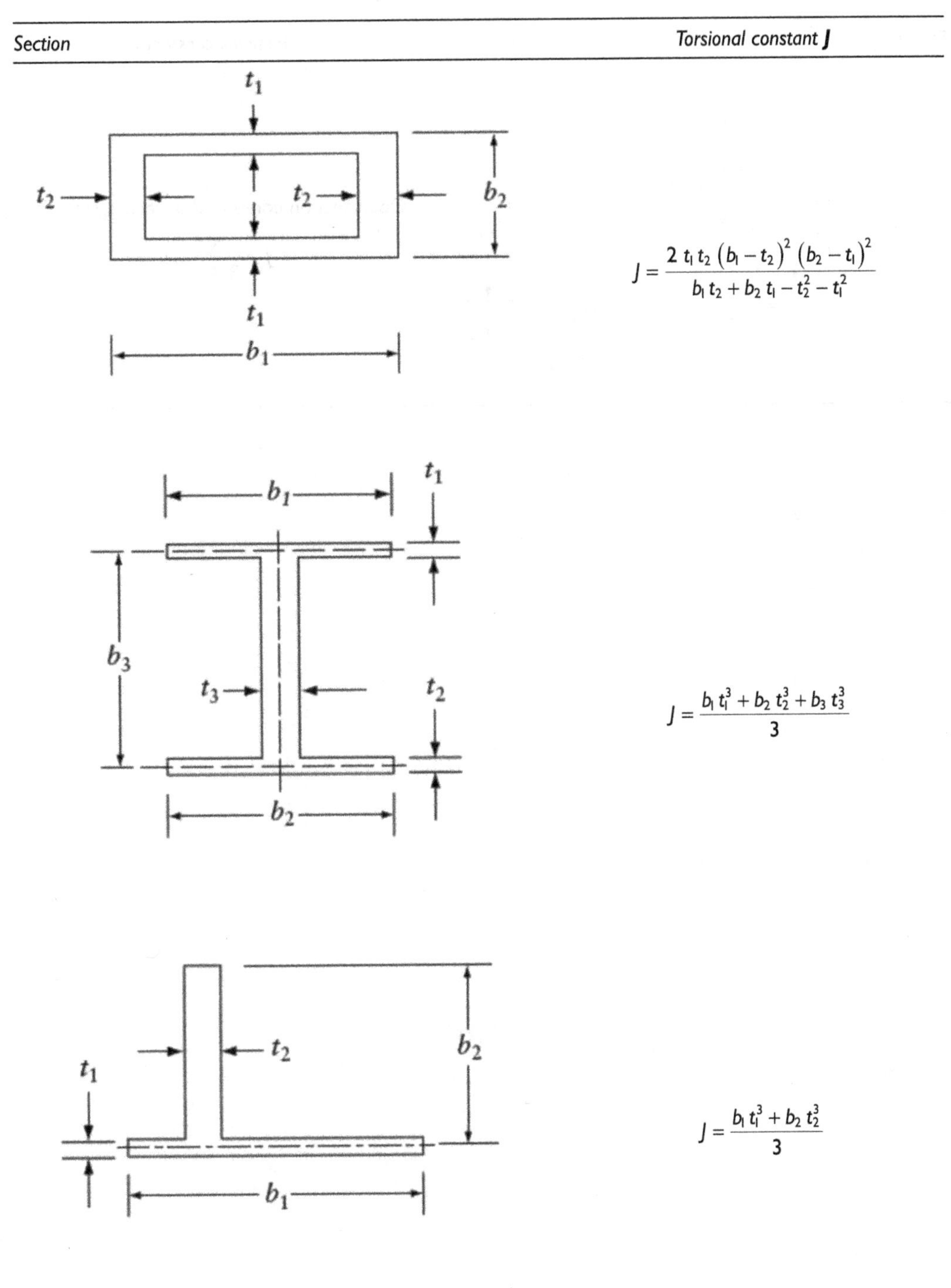

Section	*Torsional constant **J***
(hollow rectangle)	$J = \dfrac{2\, t_1 t_2 \left(b_1 - t_2\right)^2 \left(b_2 - t_1\right)^2}{b_1 t_2 + b_2 t_1 - t_2^2 - t_1^2}$
(I-section)	$J = \dfrac{b_1 t_1^3 + b_2 t_2^3 + b_3 t_3^3}{3}$
(T-section)	$J = \dfrac{b_1 t_1^3 + b_2 t_2^3}{3}$

(Continued)

Section	*Torsional constant **J***
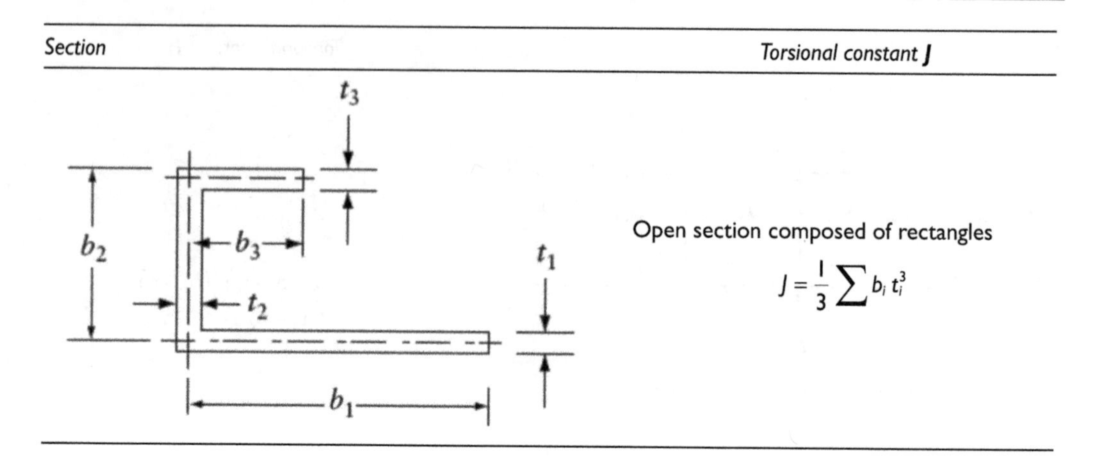	Open section composed of rectangles $J = \frac{1}{3}\sum b_i t_i^3$

Appendix G

Values of the integral $\int_l M_u M \, dl$

The following table gives the values of the integral, $\int_l M_u M \, dl$, which is needed in the calculation of displacement of framed structures by virtual work (Equation 8.4). The same table can be used for the evaluation of the integrals $\int_l N_u N \, dl$, $\int_l V_u V \, dl$, $\int_l T_u T \, dl$, or for the integral over a length l of the product of any two functions which vary in the manner indicated in the diagrams at the top and at the left-hand edge of the table.

DOI: 10.1201/9780429286858-27

M \ M_u	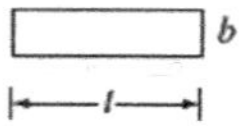	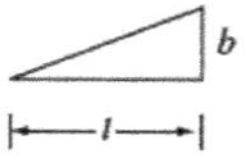	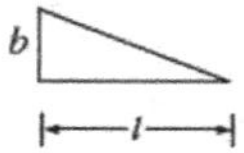	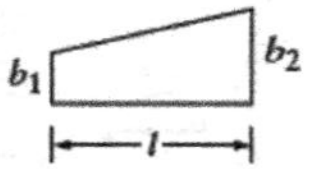	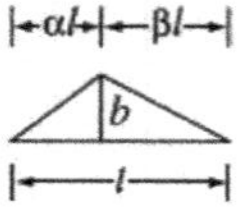
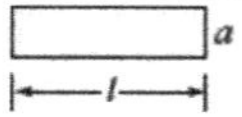	$a\,b\,l$	½ $(a\,b\,l)$	½ $(a\,b\,l)$	$\frac{a\,l}{2}(b_1+b_2)$	½ $(a\,b\,l)$
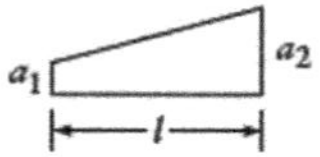	$\frac{b\,l}{2}(a_1+a_2)$	$\frac{b\,l}{6}(a_1+2\,a_2)$	$\frac{b\,l}{6}(2\,a_1+a_2)$	$\frac{l}{6}\begin{pmatrix}2\,a_1\,b_1+a_1\,b_2+\\ a_2\,b_1+2\,a_2\,b_2\end{pmatrix}$	$\frac{b\,l}{6}\begin{bmatrix}(1+\beta)\,a_1+\\ (1+\alpha)\,a_2\end{bmatrix}$
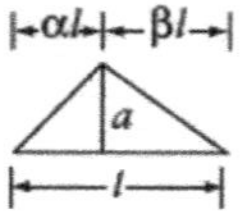	½ $(a\,b\,l)$	$\frac{a\,b\,l}{6}(1+\alpha)$	$\frac{a\,b\,l}{6}(1+\beta)$	$\frac{a\,l}{6}\begin{bmatrix}(1+\beta)\,b_1+\\ (1+\alpha)\,b_2\end{bmatrix}$	⅓ $(a\,b\,l)$
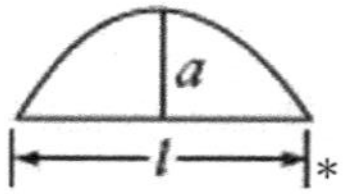	⅔ $(a\,b\,l)$	⅓ $(a\,b\,l)$	⅓ $(a\,b\,l)$	$\frac{a\,l}{3}(b_1+b_2)$	$\frac{a\,b\,l}{3}(1+\alpha\,\beta)$
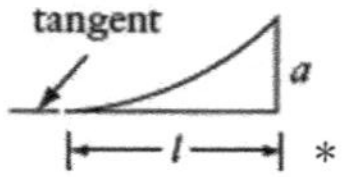	⅓ $(a\,b\,l)$	¼ $(a\,b\,l)$	$\frac{1}{12}(a\,b\,l)$	$\frac{a\,l}{12}(b_1+3\,b_2)$	$\frac{a\,b\,l}{12}(1+\alpha+\alpha^2)$
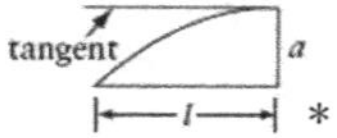	⅔ $(a\,b\,l)$	$\frac{5}{12}\,a\,b\,l$	¼ $(a\,b\,l)$	$\frac{a\,l}{12}(3\,b_1+5\,b_2)$	$\frac{a\,b\,l}{12}(5-\beta+\beta^2)$

* Second-degree parabola.

Appendix H

Forces due to prestressing of concrete members

The figures below show sets of forces in equilibrium representing the forces that prestressed tendons exert on the concrete. The tendon profiles shown are composed of one or more straight lines or of second-degree parabolas. The symbol P represents the absolute value of the prestressing force which is assumed constant, and y represents the vertical distance between the tendon and the centroidal axis of the member. θ represents the angle between a tangent to the tendon and the horizontal; θ is assumed small so that $\theta \simeq \tan\theta \simeq \sin\theta$ and $\cos\theta \simeq 1$.

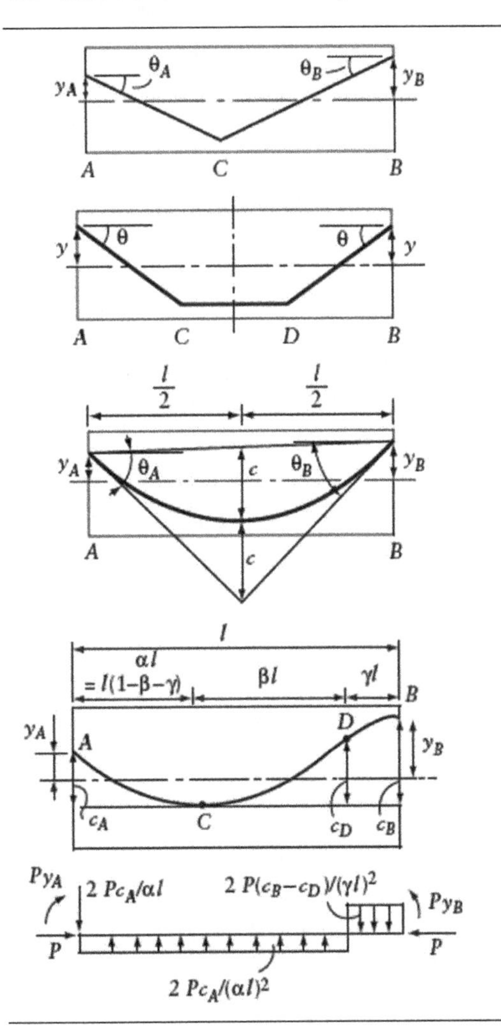

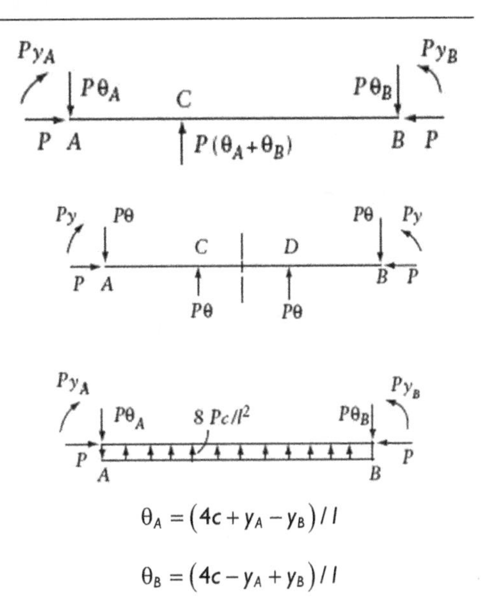

Two parabolas with a common tangent at D and horizontal tangents at B and C. The values c_A, c_B, and y may be chosen arbitrarily, but β and c_D must satisfy $\beta = \gamma\, c_D/(c_B - c_D)$ and $c_D = c_A\, \beta^2/\alpha^2$.

DOI: 10.1201/9780429286858-28

Answers to problems

CHAPTER 1

1.1 (a) $\{\theta_A, \theta_B, \theta_C, \theta_D\} = \{13.5, -9.4, 2.3, -1.2\}\ 10^{-3} ql^3/EI$. Bending moments at supports (positive moment produces tension at bottom face): $\{M_B, M_C\} = \{-32.8, 9.4\}\ 10^{-3} ql^2$.

(b) $\{\theta_A, \theta_B, \theta_C, \theta_D\} = \{-7.0, 13.9, -13.9, 7.0\}\ 10^{-3} ql^3/EI$. Bending moments at supports (positive moment produces tension at bottom face): $\{M_B, M_C\} = \{-55.6, -55.6\}\ 10^{-3} ql^2$.

(c) $\{\theta_B, \theta_C\} = \{2.4, 2.4\}\ 10^{-3} Pl^2/EI$. Member end moments: $\{M_{AB}, M_{BA}, M_{CB}, M_{DC}\} = \{-0.11, -0.09, 0.07, -0.07\}\ Pl$.

(d) $\{\theta_A, \theta_B, \theta_C, \theta_D\} = \{66.7, -8.3, 91.7, 16.7\}\ 10^{-3}\ Pl^2/EI$. $\{M_{BA}, M_{BC}, M_{CB}, M_{CD}\} = \{-0.15, 0.15, 0.35, 0.15\}\ Pl$.

(e) $\{\theta_A, \theta_B, \theta_C, \theta_D\} = \{225.0, 29.2, 54.2, 191.7\}\ 10^{-3}\ Pl^2/EI$. $\{M_{BA}, M_{BC}, M_{CB}, M_{CD}\} = \{-0.22, 0.22, 0.27, -0.27\}\ Pl$.

(f) $\{\theta_B, \theta_C\} = \{28.0, 14.8\}\ 10^{-3}\ Pl^2/EI$. $\{M_{AB}, M_{BA}, M_{BC}, M_{BE}, M_{CB}, M_{CD}\} = \{-0.24, -0.13, -0.04, 0.17, -0.09, 0.09\}\ Pl$.

1.2 $\{\theta_B, \theta_C\} = \{-0.86, -0.86\}\ \delta/l$; $\{u_B, v_B, u_C, v_C\} = \{0.429, 0, 0.429, 1\}\ \delta$; $\{M_{AB}, M_{BA}, M_{BC}, M_{CD}, M_{DC}\} = \{-0.86, 0.86, -0.86, -0.86, 0.86, -0.86\}\ EI\ \delta/l^2$.

1.3 $\{\sigma_{top}, \sigma_{bottom}\} = [N/(bh)]\ \{4, -2\}$.

1.5 $h_1 = 8l/27$; $h_2 = 10l/27$.

1.6 Horizontal component in all four segments $= Pl/(2h_C)$; vertical components in AB and BC $= 1.5P$ and $0.5P$, respectively.

1.7

	Horizontal translation at B	M_{AB}	M_{BA}
Without bracing	$27.3 \times 10^{-3}\ Pl^3/(EI)_{AB}$	$-0.223\ Pl$	$-0.154\ Pl$
With bracing as shown in part (b)	$0.968 \times 10^{-3}\ Pl^3/(EI)_{AB}$	$-7.47 \times 10^{-3}\ Pl$	$-4.51 \times 10^{-3}\ Pl$
With bracing as shown in part (c)	$1.36 \times 10^{-3}\ Pl^3/(EI)_{AB}$	$-11.2 \times 10^{-3}\ Pl$	$-7.96 \times 10^{-3}\ Pl$

1.8 Using computer program PLANET, horizontal translation at $B = 0.994 \times 10^{-3} Pl^3/(EI)_{AB}$; $\{N_{AB}, N_{BC}, N_{CD}, N_{AC}, N_{BD}\} = P\ \{0.414, -0.448, 0.336, 0.560, -0.690\}$.

1.9

Member	AG	BH	CI	AB	BC	CD	AH
Force in terms of *P*	−0.500	1.500	0.500	2.500	4.000	4.500	−3.536
Member	BI	CJ	HN	NO	IO	NI	OJ
Force in terms of *P*	−2.121	−0.707	−1.398	−1.250	−0.839	0.625	0.750

CHAPTER 2

2.1 (a) Reactions: $R_{Dx} = -0.2P$; $R_{Dy} = -0.8P$; $R_{Cy} = -1.2P$. Member forces: {AB, BC, CD, DA, AC} = P {-0.200, -1.020, 0.360, -0.816, -0.256}.

(b) Reactions: $R_{Dy} = -2P/3$; $R_{Cx} = -P/2$; $R_{Cy} = -4P/3$. Member forces: {AB, BC, CE, ED, DA, AE, EB} = P {-0.25, -1.33, -0.50, 0, -0.67, -0.42, 0.42}.

(c) Reactions: $R_{Hy} = -P$; $R_{Ky} = -4P$; $R_{Ny} = -P$. Selected member forces: {BC, IJ, BJ, BI} = P {0, 0.5, -0.707, -0.5}.

(d) Forces in members: {AB, AC, AD} = P {1.166, -6.319, -0.574}. Reactions: $\{R_x, R_y, R_z\}_B = P$ {0, -0.6, 1.0}; $\{R_x, R_y, R_z\}_C = P$ {-2.2, -2.2, -5.5}; $\{R_x, R_y, R_z\}_D = P${0.2, -0.2, -0.5}.

2.3 $\{R_1, R_3, R_4, R_5, R_6, R_7\} = P$ {-1.833, -2.75, -2.917, -3.0, -0.625, -1.375}.

2.4 (a) $R_A = 0.68ql$; $R_C = 1.42ql$; $M_B = 0.272\ ql^2$; $M_C = -0.1\ ql^2$.

(b) $R_A = 1.207P$ down and $1.207P$ to the left; $R_D = 0.5P$ up and $0.5P$ to the left; $M_B = 1.207Pl$, tension side inside frame; $M_A = M_C = M_D = 0$.

(c) $R_A = 0.57ql$; $R_B = 1.23ql$; $R_C = 1.08ql$; $R_D = 0.42ql$; $M_G = 0.16ql^2$; $M_B = M_C = -0.08ql^2$.

(d) $R_A = 0.661ql$ up; $R_C = 0.489ql$ up. Member end moments: $M_{BD} = 0.03125ql^2$; $M_{BA} = -0.06610ql^2$; $M_{BC} = 0.03485ql^2$.

(e) $R_F = 0.75ql$ up and $0.0857ql$ to the right. Member end moments: $M_{BF} = 0.0429ql^2$; $M_{BA} = 0.05ql^2$; $M_{BC} = -0.0929ql^2$.

(f) $R_A = 0.6ql$ up and $0.2ql$ to the right; $R_B = ql/\sqrt{5}$; $V_A = -V_B = ql/\sqrt{5}$; $M = 0$ at two ends and at center $M = ql^2/8$.

(g) $R_A = 2.375P$; $R_G = 1.625P$ up and $2P$ to the right. Nonzero member end moments: $M_{FG} = -M_{FE} = 1.5Pl$; $M_{EF} = -M_{ED} = 2.625Pl$; $M_{DE} = -M_{DC} = 2.75Pl$; $M_{CD} = -M_{CB} = 1.875Pl$. $V_{Br} = 1.68P$; $V_{Dl} = 0.78P$.

(h) Considering M positive producing tension at bottom fiber, $M_{Ar} = -2.18ql^2$; $M_{Bl} = -0.56ql^2$; $M_{Br} = -0.625ql^2$. Torsion in AB $= 0.28ql^2$.

(i) $R_A = 0.4ql$; $R_B = 0.98ql$; $R_C = 0.42ql$; $M_E = 0.16ql^2$; $M_B = -0.08ql^2$.

(j) $R_A = 0.4ql$; $R_C = 0.745ql$; $R_E = 0.755ql$; $R_H = 0.4ql$; M at middle of AB $= 0.08ql^2$; $M_C = -0.1ql^2$; $M_E = -0.08ql^2$; $M_G = 0.16ql^2$.

(k) $R_B = 0.62ql$; $R_C = 1.18ql$; $R_E = 0.4ql$; $M_B = -0.02ql^2$; $M_C = -0.1ql^2$; M at middle of DE $= 0.08\ ql^2$.

(l) $R_1 = 0.125ql$; $R_2 = 0.8125ql$; $R_3 = 0.375ql$; $R_4 = 1.6875ql$; $M_B = 0.125ql^2$, tension side outside the frame; end moments $\{M_{DC}, M_{DF}, M_{DE}\} = ql^2$ {0.6875, -0.1250, -0.5625}.

(m) $R_1 = -0.375ql$; $R_2 = 0.25ql$; $R_3 = 0.625ql$; $R_4 = 0.375ql^2$; $M_B = 0.375ql^2$, tension side inside the frame; $M_D = 0.625\ ql^2$, tension side outside the frame.

(n) $R_1 = 0.158ql$; $R_2 = 0.421ql$; $R_3 = 0.158ql$; $R_4 = 0.579ql$; $M_B = 0.0395ql^2$; $M_D = 0.0789ql^2$, tension side outside frame at B and D.

2.5 The force in each member running in the x, y, or z direction is $0.5T/l$ (tension). The force in each diagonal member is $-0.7071T/l$ (compression). The reaction components are nil, except: $(R_y)_I = -(R_y)_K = 0.5T/l$; $(R_z)_J = -(R_z)_L = 0.5T/l$.

2.6 Axial forces in members: $\{N_{OA}, N_{OB}, N_{OC}, N_{OD}\} = P$ {-2.405, -3.436, -2.405, -1.374}.

2.7 $\{N\} = P$ {-0.577, -0.577, 0, 0, -0.577, -0.577, 0.817, 0.817, 0, 0}

2.8 $\{N_{AB}, N_{BC}, N_{CD}, N_{DA}\} = P$ {0, 0.707, 0, 0.707}
$\{N_{AE}, N_{BF}, N_{CG}, N_{DH}\} = P$ {-0.707, 0.707, -0.707, 0.707}
$\{N_{AF}, N_{BG}, N_{CH}, N_{DE}\} = P$ {1.0, -1.0, 1.0, -1.0}

2.9 $\{R_x, R_y, R_z\}_F = P$ {0.000, -0.309, -1.051}

$\{R_x, R_y, R_z\}_G = P\ \{-0.487, -0.487, -2.949\}$
$\{N_{EA}, N_{EB}, N_{ED}\} = P\ \{0.237, 0.455, 0.455\}$

CHAPTER 3

3.1 $i=1$. Introduce a hinge in the beam at B.
3.2 $i=1$. Delete members CF or ED.
3.3 $i=3$. Make A a free end.
3.4 $i=2$. Introduce hinges in the beam at B and D.
3.5 $i=4$. Change support B to a roller and cut member DE at any section.
3.6 $i=5$. Remove support B and delete members AE, IF, FJ, and GC.
3.7 (a) $i=15$. (b) $i=7$.
3.8 $i=2$. Delete members IJ and EG.
3.9 5 degrees and 3 degrees when axial deformation is ignored.
3.10 14 degrees and 8 degrees when axial deformation is ignored.
3.11 Cut member BC at any section. The member end moments in the released structure are: $M_{AD} = -M_{AB} = Pb$, all other end moments are zero. A positive end-moment acts in a clockwise direction on the member end.
3.12 (a) Cut EF and CD; each results in three releases.
(b) Member end moments: $M_{AC} = M_{CA} = M_{BD} = M_{DB} = -Pl/2$; $M_{CE} = M_{EC} = M_{DF} = M_{FD} = -Pl/4$; $M_{EF} = M_{FE} = Pl/4$; $M_{CD} = M_{DC} = 3Pl/4$.

CHAPTER 4

4.1 (a) $[f] = \dfrac{l}{6EI}\begin{bmatrix} 4 & 1 \\ 1 & 2 \end{bmatrix}$

(b) $[f] = \dfrac{l^3}{6EI}\begin{bmatrix} 16 & 5 \\ 5 & 2 \end{bmatrix}$

4.2 (a) $\{F\} = -ql^2\left\{\dfrac{3}{28}, \quad \dfrac{1}{14}\right\}$

(b) $\{F\} = ql\left\{\dfrac{11}{28}, \quad \dfrac{8}{7}\right\}$

4.3 $M_B = 0.0106ql^2$, and $M_C = -0.0385ql^2$.
4.4 $M_B = 0.00380\,\dfrac{EI}{l}$ and $M_C = -0.00245\,\dfrac{EI}{l}$
4.5 $M_B = -0.0377ql^2 = M_C$
4.6 Forces in the springs:
(a) $\{F_A, F_B, F_C\} = -P\ \{0.209, 0.139, 0.039\}$
Reactions:
$\{R_D, R_E, R_F, R_G\} = P\ \{0.51, 0.10, 0.24, 0.15\}$
(b) $\{F_A, F_B, F_C\} = P\ \{0.5, 0, 0\}$; $R_D = R_F = 0.375P$; $R_E = R_G = 0.125P$.
4.7 Forces in the cables: $\{F_C, F_D\} = \{2.12, 23.29\}$ kN. $M_D = 1.96$ kN-m.

4.8 $M_A=-0.0556ql^2$; $M_B=-0.1032ql^2$; $M_C=-0.1151ql^2$. Upward reactions at A, B, and C are: $0.785ql$, $1.036ql$, and $0.512ql$.

4.9 At end of stage 1: $M_A=M_D=0$; $M_B=-20\times10^{-3}\ ql^2$; $R_A=0.480ql$; $R_B=0.720ql$. At end of stage 2: $M_A=M_C=0$; $M_B=-74.4\times10^{-3}\ ql^2$; $M_D=20.5\times10^{-3}\ ql^2$. $R_A=0.426ql$; $R_B=1.148ql$; $R_C=0.426ql$.

4.10 Maximum negative bending moment at each of the two interior supports is $-0.2176ql^2$. Maximum positive bending moment at the center of each exterior span is $0.175ql^2$, and at the center of the interior span is $0.100ql^2$. Maximum reaction at an interior support is $2.300ql$. Absolute maximum shear is $1.2167ql$ at the section which is at a distance l from the ends, just before reaching the interior supports.

4.12 $M_B=-0.096ql^2$; $M_C=-0.066ql^2$; $R_B=0.926ql$; deflection at center of BC $=2.88\times10^{-3}\ ql^4/(EI)$.

4.13 $\{M_{AB}, M_{BA}, M_{BC}, M_{CB}\}=\{0, 0.0899, -0.0899, 0\}\ ql^2$. Force in tie $=0.7304ql$, tension.

4.14 Maximum $R_B=2.1057W$; maximum $M_E=0.35Wl$.

4.15 For R_B, the coordinates at nodes $\{2, 4, 6\}=\{0.6826, 0.6019, -0.1174\}$. For M_E, the coordinates at nodes $\{2, 4, 6\}=l\ \{-0.0722, 0.0815, -0.0722\}$. For V_E, the coordinates at nodes $\{2, 4_l, 4_r, 6\}=\{0.0923, -0.5, 0.5, -0.0923\}$.

CHAPTER 5

5.1 Forces in members AB, AC, AD, and AE are: $\{A\}=P\ \{-0.33, 0.15, 0.58, 0.911\}$.

5.2 $A_i = PH\left(x_i \sin\theta_i \Big/ \sum_{i=1}^{n} x_i^2 \sin^2\theta_i\right)$

5.3 $$[S]=\frac{Ea}{b}\begin{bmatrix} 1.707 & & & \text{Symm.} \\ 0 & 2.707 & & \\ -1.000 & 0 & 1.707 & \\ 0 & 0 & 0 & 2.707 \end{bmatrix}$$

5.4 Force in the vertical member $=0.369P$. Force in each of the inclined members meeting at the point of application of $P=0.446P$. Force at each of the other two inclined members $=-0.261P$.

5.5 Bending moment at B $=0.298Pl$. Changes in the axial forces in AC and AD are, respectively, $0.496P$ and $-0.496P$.

5.6 Bending moment at B $=0.202Pl$. Changes in the axial forces in AE, AG, AF, and AH are, respectively, $0.299P$, $0.299P$, $-0.299P$, and $-0.299P$.

5.7 $\{M_{BC}, M_{CF}\}=Pl\ \{-0.0393, 0.00711\}$.

5.8 $\{M_{BC}, M_{CF}\} = \left(EI\Delta/l^2\right)\left\{\frac{186}{35}, \frac{-72}{35}\right\}$.

5.9 The end moments for the members are (in terms of Pl): $M_{AB}=-0.123$; $M_{BA}=0.133$; $M_{BC}=-0.133$; $M_{CB}=0.090$; $M_{CD}=-0.090$; $M_{DC}=-0.051$.

5.10 $\{D\} = \frac{Pl^2}{EI}\{0.038l, 0.022, -0.0481\}$. D_1 is a downward deflection; D_2 and D_3 are rotations represented by vectors in the positive x and z directions.

5.11 The first column of $[S]$ is $\left\{\frac{48EI}{l^3}, 0, 0, -\frac{12EI}{l^2}, 0, \frac{6EI}{l^2}, -\frac{12EI}{l^2}, \frac{6EI}{l^2}, 0, 0, 0, 0\right\}$ and the second column is $\left\{0, \left(\frac{8EI}{l}+\frac{2GJ}{l}\right), 0, 0, -\frac{GJ}{l}, 0, -\frac{6EI}{l^2}, \frac{2EI}{l}, 0, 0, 0, 0\right\}$.

5.12 Girder BE: $\{M_B, M_I, M_J, M_E\} = Pl\,\{-0.184, 0.142, -0.032, -0.035\}$. Girder AF: $\{M_A, M_K, M_L, M_F\} = Pl\,\{-0.077, 0.051, 0.006, -0.038\}$.

5.13 (a) $$[S]=\begin{bmatrix} \frac{24EI}{l^3} & & & \text{Symm.} \\ 0 & \frac{8EI}{l} & & \\ -\frac{12EI}{l^3} & -\frac{6EI}{l^2} & \frac{24EI}{l^3} & \\ \frac{6EI}{l^2} & \frac{2EI}{l} & 0 & \frac{8EI}{l} \end{bmatrix}$$

(b) $$[S]=\begin{bmatrix} \frac{12EI}{l^3} & & \text{Symm.} \\ 0 & \frac{8EI}{l} & \\ \frac{6EI}{l^2} & \frac{2EI}{l} & \frac{8EI}{l} \end{bmatrix}$$

(c) $$[S]=\frac{EI}{l^3}\begin{bmatrix} 19.2 & -13.2 \\ -13.2 & 19.2 \end{bmatrix}$$

5.14 (a) $$[S]=\begin{bmatrix} \frac{24EI}{l^3} & & & \text{Symm.} \\ 0 & \frac{8EI}{l} & & \\ -\frac{12EI}{l^3} & -\frac{6EI}{l^2} & \frac{24EI}{l^3} & \\ \frac{6EI}{l^2} & \frac{2EI}{l} & -\frac{6EI}{l^2} & \frac{4EI}{l} \end{bmatrix}$$

(b) $$[S]=\begin{bmatrix} \frac{24EI}{l^3} & & \text{Symm.} \\ 0 & \frac{8EI}{l} & \\ \frac{6EI}{l^2} & \frac{2EI}{l} & \frac{4EI}{l} \end{bmatrix}$$

(c) $$[S]=\frac{EI}{l^3}\begin{bmatrix} 13.71 & -4.29 \\ -4.29 & 1.71 \end{bmatrix}$$

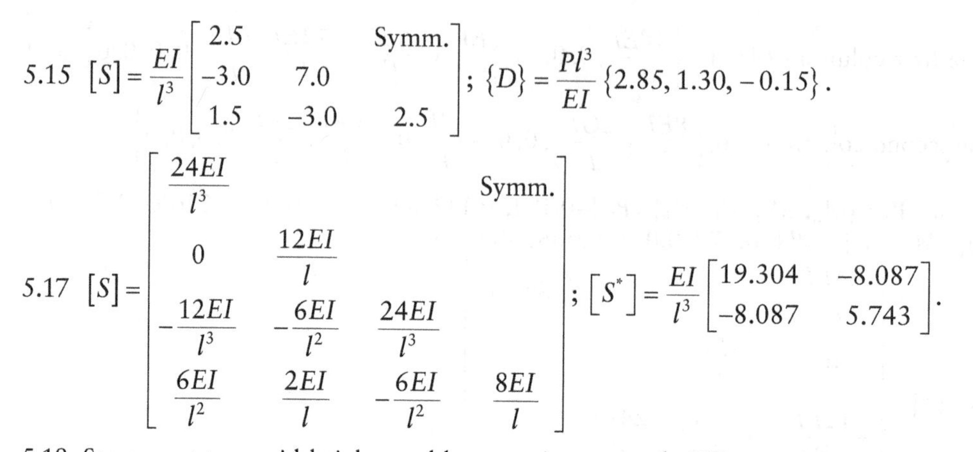

5.15 $[S] = \frac{EI}{l^3}\begin{bmatrix} 2.5 & & \text{Symm.} \\ -3.0 & 7.0 & \\ 1.5 & -3.0 & 2.5 \end{bmatrix}; \{D\} = \frac{Pl^3}{EI}\{2.85, 1.30, -0.15\}.$

5.17 $[S] = \begin{bmatrix} \frac{24EI}{l^3} & & & \text{Symm.} \\ 0 & \frac{12EI}{l} & & \\ -\frac{12EI}{l^3} & -\frac{6EI}{l^2} & \frac{24EI}{l^3} & \\ \frac{6EI}{l^2} & \frac{2EI}{l} & -\frac{6EI}{l^2} & \frac{8EI}{l} \end{bmatrix}; [S^*] = \frac{EI}{l^3}\begin{bmatrix} 19.304 & -8.087 \\ -8.087 & 5.743 \end{bmatrix}.$

5.18 Stresses at top, mid-height, and bottom, in terms of αET, are −0.25, 0.25, and −0.25 respectively. The variation between these values is linear. The change in length of centroidal axis is $0.25\alpha Tl$. The deflection at mid-span is $\alpha Tl^2/8d$ upward.

5.19 At the central support, $M = 1.5EI\alpha T/d$. The stresses at top, mid-height, and bottom are, in terms of αET: −1.00, 0.25 and 0.50, respectively. Reactions at outer supports are $1.5EI\alpha T/dl$ up and, at interior support, $3EI\alpha T/dl$ down.

5.20 Answers in terms of $|P|$: $\{M_A, M_E, M_B, M_F\} = h$ {0.15, −0.239, 0.522, −0.278}; $R_A = 0.0174h/b$ up $= -R_B$; deflection at center of AB $= 0.877hb^2/EI$ up; $V_{Ar} = -0.276h/b$; $V_{Bl} = 0.382h/b$; $V_{Br} = -0.356h/b$; $V_{Cl} = 0.356h/b$; $V_{Cr} = -0.382h/b$; $V_{Dl} = 0.276h/b$.

5.21 Answers in terms of $|P|$: $\{M_A, M_E, M_B, M_F\} = h$ {0, −0.3043, 0.3915, −0.4085}; $R_A = 0.0012$ down $= -R_B$; deflection at E $= 0.843hb^2/(EI)$ up; $V_{Ar} = -0.0869h/b$; $V_{Bl} = 0.1988h/b$; $V_{Br} = -0.1778h/b$; $V_{Cl} = 0.1778h/b$; $V_{Cr} = -0.1988h/b$; $V_{Dl} = 0.0869h/b$.

5.22 $\sigma = E\alpha T_{top}\,[0.167 + 0.357\,\mu - (0.5 - \mu)^5]$; extreme tension at bottom fiber = 3.1 MPa (or 0.45 ksi).

5.23 Forces in members OA, OB, OC, and OD are, respectively, in terms of P: 0.919, −0.306, −2.756, and −1.531. $\{R_x, R_y, R_z\}_A = P$ {−0.375, −0.375, 0.750}; $\{R_x, R_y, R_z\}_B = P$ {−0.125, 0.125, −0.250}; $\{R_x, R_y, R_z\}_C = P$ {−1.125, −1.125, −2.250}; $\{R_x, R_y, R_z\}_D = P$ {0.625, −0.625, −1.250}.

CHAPTER 6

6.1 Tip deflection, $D_1(t_0) = 0.575 \times 10^{-3}$ m; change in tip deflection in the period $(t_2 - t_1)$, $D_1(t_2, t_1) = 3.851 \times 10^{-3}$ m. Change in the cable tensile force in the period $(t_2 - t_1) = 17.3$ kN.

6.2 Mid-span deflection at time $t_0 = 14.1$ mm, and at time $t = 42.3$ mm. Change in length at time $t_0 = -4.7$ mm, and at time $t = -19.6$ mm. Effect of shrinkage cannot be done by this method and, thus, has no effect on deflection.

6.3 At time t, mid-span deflection = 56.3 mm; end moment of AB = $\{50.6 \times 10^3, 108.6 \times 10^3\}$ N-m; end moments of BC = $\{72.0 \times 10^3, -439.2\}$ N-m.

6.4 Curvature at time t, $\psi(t) = 942 \times 10^{-6}$ m^{-1}. Mid-span deflection at time $t = \psi l^2/8 = 7.54$ mm.

6.5 Curvature at time t, $\psi(t) = -697$ m^{-1}. Mid-span deflection at time $t = -5.58$ mm.

6.6 At time t_0, the concrete stress anywhere on the section, $\sigma_c(t_0) = E_c(t_0)\varepsilon(t_0) = -17.1\text{ MPa}$, and the steel stress, $\sigma_{ns}(t_0) = E_{ns}\,\varepsilon(t_0) = -114\text{ MPa}$. At time t, $\sigma_c(t) = -9.4\text{ MPa}$ and $\sigma_{ns}(t) = -362\text{ MPa}$.

6.7 Bending moment over the middle support B at age 60 days $= -69.7 \times 10^{-3}\ ql^2$; change in bending moment between $t = 60$ days and $\infty = -41.8 \times 10^{-3}\ ql^2$.

CHAPTER 8

8.1 $D_1 = 0.248$, $D_2 = 0.228$ (inch or cm).

8.2 $D = 3.73\ bP/(Ea)$

8.3 Downward deflection at E in cases (a) and (b) $= b\left\{9.84\dfrac{P}{Ea}, -0.22 \times 10^{-3}\right\}$. Forces in members in case (c) are (in terms of P): AE = 1.78, EF = 1.46, FB = 1.56, AC = -2.22, CD = -1.88, DB = -1.94, CE = 1.26, DF = 0.93, DE = 0.40, and CF = 0.12.

8.4 Vertical deflection at C in cases (a) and (b) = {3.3, −8.3} mm. Forces in members in case (c) are (kN): AB = -94.5, BC = -75.0, DE = 130.5, EC = 125.0, EA = -92.5, EB = -26.0 and BD = 32.5.

8.5 Forces in members are (in terms of P): AB = -1.95, BC = -0.38, CD = -0.38, DE = -0.62, EF = 2.06, BE = -0.33, CE = 0.54, DB = -0.88, BF = 1.34, and EA = -1.49.

8.6 $\{N\} = P\ \{-0.180, 0.320, -0.180, 0.254, -0.453, 0.641\}$.

8.7 $M_{AB} = -0.056ql^2$, $M_{BA} = 0.043ql^2$.

8.8 $M_{AB} = -0.057ql^2$, $M_{BA} = 0.130ql^2$.

8.9 Force in tie = 2.34P. Bending moment values (in terms of Pb): $M_A = 0$, $M_B = M_D = 0.099$, and $M_C = 0.132$.

8.10 Downward deflection at A $= 0.104l^3/(EI)$ and $0.0774l^3/(EI)$, respectively.

8.11 Downward deflection at D $= 0.015Pl^3/(EI)$, and rotation at A $= 0.047Pl^2/(EI)$ (clockwise).

8.12 $\{D_1, D_2, D_3\} = \{2.331, 2.992, 1.278\}\ Pl^3/(1000EI)$, in case (a); $\{D_1, D_2, D_3\} = \{353.3, -25.47, 102.7\}\ Pl^3/(1000EI)$, in case (b).

8.13 $M_A = -0.167ql^2$; $M_B = -0.080ql^2$; $M_D = 0.022ql^2$; $R_A = 0.0673ql$ (up).

8.14 $M_{BC} = 0.3915ql^2$; $R_B = 0.8115ql$ (up); deflection at A $= 0.1663ql^4/(EI)$.

8.15 $M_{center} = 39.83 \times 10^{-3}qr^2$ (producing tension at bottom fiber). The fixed-end forces are: $\{F^*\} = qr\ \{-0.5, -1.48 \times 10^{-3}r, -87.46 \times 10^{-3}r, -0.5, -1.48 \times 10^{-3}r, 87.46 \times 10^{-3}r\}$.

8.16 Chosen redundant: $R_A = 0.3471ql$ (up); $N_{AD} = 0.5362ql$; $N_{ED} = 0.5134ql$. Member end moments: $\{M_{BA}, M_{BE}, M_{EB}, M_{EC}, M_{CE}, M_{BD}\} = (ql^2/1000)\ \{8.44, -13.00, 4.58, -4.58, -9.31, 4.56\}$.

8.17 The nonzero reaction components at A are: $F_x = 0.189P$ and $M_y = 0.405Pl$. Displacement at C in the x direction is $0.2095Pl^3/(EI)$.

8.18 Forces in members in terms of $10^{-3}\ P$: AB = 2042; CD = -2591; DE = -2020; AH = -2152; HI = 439; CA = -1319; DH = -819; EI = 278; CH = 2458; DI = -542. Use symmetry for the remaining members.

8.19 Axial forces in AC and BD $= 0.5512P$ and $-0.5512P$, respectively; horizontal displacement at C $= 0.05512\ Pl^3/EI$.

8.20 $F_1 = \dfrac{\pi^4 EI}{8l^3} D_1$

CHAPTER 9

9.1 The member end moments are (in terms of Pb): $M_{AB} = -M_{AF} = -0.461$; $M_{BA} = -M_{BC} = M_{CB} = -M_{CD} = -0.149$. Use symmetry to find the moments in the other half of the frame.

9.2 The member end moments in terms of $(Pb/24)$ are: $M_{BC}=5$; $M_{CB}=-M_{CD}=1$; $M_{DC}=7$. Use symmetry about vertical and horizontal axes to find the remaining moments.

9.3 The forces at end A are: $1.6qb$ (upward), $1.07qb$ (to the right) and $0.213qb^2$ (anti-clockwise). Use the symmetry of the frame to find the forces at end B.

9.4 The end moments on the members are (in terms of $ql^2/1000$): $M_{AB}=25.3$; $M_{BA}=-M_{BC}=50.7$; $M_{CB}=72.5$; $M_{CD}=-36.3$; $M_{DC}=-18.1$; $M_{CE}=-36.3$. Reaction components at A: $0.479ql$ upward, $0.077ql$ to the right, and $0.025ql^2$ clockwise.

9.5 The values of the bending moment at support B for the four cases of loading are: ql^2 {−0.2422, −0.3672, −0.2422, −0.1016}.

CHAPTER 10

10.1 $M_{BC}=-M_{BA}=246.0$ kip-ft; $M_{AB}=-165.6$ kip-ft.

10.2 $M_{BC}=-M_{BA}=329.6$ kN-m; $M_{AB}=-226.1$ kN-m.

10.3 $\{M_{BA}, M_{BC}, M_{BD}, M_{DB}\}=\{28.1, -25.8, -2.27, -11.7\}$ kip-ft.

10.4 $\{M_{BA}, M_{BC}, M_{BD}, M_{DB}\}=\{38.6, -34.6, -3.99, -16.5\}$ kN-m.

10.5 $\{M_{AB}, M_{BA}, M_{BC}, M_{CB}\}=\{-0.589, -0.726, 0.726, 0.985\}\ EI/(10^3 l)$.

10.8 $\{M_{AB}, M_{BA}, M_{BC}, M_{CB}\}=ql^2\{-0.665, -0.366, 0.366, 0\}$.

10.9 Buckling occurs when $Q=1.82EI/b^2$.

10.10 $Q=93.2EI_{BC}/l^2$.

10.11 $Q=6.02EI/l^2$.

10.14 $Q=20.8EI/l^2$.

10.15 $Q=3.275EI/l^2$.

CHAPTER 11

11.2 The sums of moments in walls are (refer to Figure 11.7)

$$\{M_A, M_B, M_C, M_D, M_E\} = \frac{Ph}{10}\{0, 4.41, 18.80, 43.27, 77.82\}$$

The sums of moments in columns are (in terms of $Ph/10$):
$M_{FG}=-0.318$, $M_{GF}=-0.271$, $M_{GH}=-0.287$, $M_{HG}=-0.316$. $M_{HI}=-0.260$, $M_{IH}=-0.275$, $M_{IJ}=-0.196$, $M_{JI}=-0.262$.

11.3 Bending moment at Section A-A is 6.07 Ph. The moment on each end of the lower beam (at its intersection with the wall) is 0.295 Ph.

11.4 Bending moment at Section A-A is 21.247 P. The moment on each end of the lower beam (at its intersection with the wall) is 1.0086 P.

CHAPTER 12

12.1 $$\left[S^*\right] = \frac{EI_0}{l^3}\begin{bmatrix} 18 & & & \text{Symm.} \\ 8l & 5l^2 & & \\ -18 & -8l & 18 & \\ 10l & 3l^2 & -10l & 7l^2 \end{bmatrix}$$

$$\left[S^*\right]=\frac{EI_0}{l^3}\begin{bmatrix}17.45 & & & \text{Symm.}\\ 7.72l & 4.86l^2 & & \\ -17.45 & -7.72l & 17.45 & \\ 9.72l & 2.86l^2 & -9.72l & 6.86l^2\end{bmatrix}$$

12.2 Deflection by finite-element method is $0.1923Pl^3/EI_0$. Exact answer is $0.1928Pl^3/EI_0$.

12.3 $\left[S\right]=\frac{Ea}{3l}\begin{bmatrix}7 & 1 & 8\\ 1 & 7 & -8\\ -8 & -8 & 16\end{bmatrix}$; $\left[S^*\right]=\frac{Ea}{3l}\begin{bmatrix}1 & -1\\ -1 & 1\end{bmatrix}$

12.4 $D_2=0.375\alpha T_0\, l$; $D_3=0.0625\alpha T_0\, l$; $\sigma=-0.125E\alpha T_0$.

12.5 $S^*_{11}=Eh\left(0.347\frac{c}{b}+0.139\frac{b}{c}\right)$; $S^*_{21}=0.156\,Eh$; $S^*_{31}=Eh\left(-0.347\frac{c}{b}+0.609\frac{b}{c}\right)$

$S^*_{41}=-0.52\,Eh$; $S^*_{51}=Eh\left(-0.174\frac{c}{b}-0.069\frac{b}{c}\right)$

12.6
$$\left[S^*\right]=\frac{Eh}{1000}\begin{bmatrix}486 & & & & & & & \text{Symm.}\\ 156 & 486 & & & & & & \\ -278 & 52 & 486 & & & & & \\ -52 & 35 & -156 & 486 & & & & \\ -243 & -156 & 35 & 52 & 486 & & & \\ -156 & -243 & -52 & -278 & 156 & 486 & & \\ 35 & -52 & -243 & 156 & -278 & 52 & 486 & \\ 52 & -278 & 156 & -243 & -52 & 35 & -156 & 486\end{bmatrix}$$

12.7 Deflection by finite-element method is $1.202Pb/Ehc$; deflection by beam theory is $(P/Eh)\,[0.5(b/c)^3+1.44(b/c)]$.

12.8 $\{\sigma\}=\{0, 0, -0.5P/hc\}$.

12.9 $\left[S^*_{11}\right]=\frac{Eh^3}{\left(1-\nu^2\right)cb}\left[\frac{1}{3}\left(\frac{b^2}{c^2}+\frac{c^2}{b^2}\right)-\frac{\nu}{15}+\frac{7}{30}\right]$. Deflection at center of rectangular clamped plate $D^*_1=\left(S^*_{11}\right)^{-1}(P/4)$. M_x at node 2 is $-\left[Eh^3/2\left(1-\nu^2\right)b^2\right]D^*_1$.

12.10 The nonzero nodal displacements are:

$$\{v_1, v_2, v_3, u_4, v_4, u_5, v_5, u_6, v_6\}=\frac{P}{Eh}\{2.68,\ 2.21, 2.57,$$

$$-0.44, 1.16, 0.03, 1.19, 0.39, 1.28\}$$

Stresses in element 2.6.3 are: $\{\sigma_x, \sigma_y, \tau_{xy}\}=\frac{P}{bh}\{0.392, 0.358, -0.642\}$.

12.11
$$\left[\{B\}_1 \quad \{B\}_2 \quad \{B\}_3\right]=\begin{bmatrix}-(1-\eta)/b & 0 & 0\\ 0 & -\left(1-3\xi^2+2\xi^3\right)/c & \xi(\xi-1)^2\,b/c\\ -(1-\xi)/c & \left(-6\xi+6\xi^2\right)(1-\eta)/b & \left(3\xi^2-4\xi+1\right)(1-\eta)\end{bmatrix}$$

12.12

$$[S^*] = (Eh/1000)\times$$

$$\begin{bmatrix}
486 & & & & & & & & & & & \text{Symm.} \\
156 & 552 & & & & & & & & & & \\
-8.7b & 68b & 28b^2 & & & & & & & & & \\
-278 & 52 & 8.7b & 486 & & & & & & & & \\
-52 & -31 & 19b & -156 & 552 & & & & & & & \\
8.68b & -19b & -12b^2 & -8.7b & -68b & 28b^2 & & & & & & \\
-243 & -156 & -26b & 35 & 52 & 26b & 486 & & & & & \\
-156 & -219 & -40b & -52 & -302 & 47b & 156 & 552 & & & & \\
26b & 40b & 5.5b^2 & -26b & 47b & -0.3b^2 & 8.7b & -68b & 28b^2 & & & \\
35 & -52 & 26b & -243 & 156 & -26b & -278 & 52 & -8.7b & 486 & & \\
52 & -302 & -47b & 156 & -219 & 40b & -52 & -31 & -19b & 156 & 552 & \\
-26b & -47b & -0.3b^2 & 26b & -40b & 5.5b^2 & -8.7b & 19b & -12b^2 & 8.7b & 68b & 28b^2
\end{bmatrix}$$

CHAPTER 13

13.2 $L_1 = \frac{1}{32}(1-\eta)(3\xi+1)(3\xi-1)(\xi-1)$; $L_2 = -\frac{9}{32}\eta(1-\eta)(\xi+1)(3\xi-1)(\xi-1)$;

$L_{10} = -\frac{1}{16}(1-\eta^2)(3\xi+1)(3\xi-1)(\xi-1)$; $L_{11} = \frac{9}{16}(1-\eta^2)(\xi+1)(3\xi-1)(\xi-1)$.

Consistent restraining forces in the y direction at the nodes 1 to 12 are:

$$-q_y\, bch\left\{\frac{1}{48}, \frac{1}{16}, \frac{1}{16}, \frac{1}{48}, \frac{1}{12}, \frac{1}{48}, \frac{1}{16}, \frac{1}{16}, \frac{1}{48}, \frac{1}{12}, \frac{1}{4}, \frac{1}{4}\right\}.$$

13.3 $$\begin{Bmatrix} \{\varepsilon\}_1 \\ \{\varepsilon\}_2 \\ \{\varepsilon\}_3 \end{Bmatrix} = [G]\begin{Bmatrix} D_1^* \\ D_2^* \end{Bmatrix}$$

$$[G]^{\mathrm{T}} = \frac{1}{2\Delta}\begin{bmatrix} 3b_1 & 0 & 3c_1 & -b_1 & 0 & -c_1 & -b_1 & 0 & -c_1 \\ 0 & 3c_1 & 3b_1 & 0 & -c_1 & -b_1 & 0 & -c_1 & -b_1 \end{bmatrix}$$

13.4 $$\left[\{B\}_1 \quad \{B\}_2\right] = \frac{1}{2\Delta}\begin{bmatrix} 3b_1\alpha_1 & -b_1\alpha_2 & -b_1\alpha_3 & 0 \\ 0 & 0 & 0 & 3c_1\alpha_1 - c_1\alpha_2 - c_1\alpha_3 \\ 3c_1\alpha_1 & -c_1\alpha_2 & -c_1\alpha_3 & 3b_1\alpha_1 - b_1\alpha_2 - b_1\alpha_3 \end{bmatrix}$$

13.5 $S_{11}^* = \frac{h}{4\Delta}\left(d_{11}d_1^2 + d_{33}c_1^2\right)$; $S_{12}^* = \left(c_1b_1h/4\Delta\right)\left(d_{12} + d_{33}\right)$.

13.6 $$[B] = \frac{1}{l}\begin{bmatrix} -1/l & 2/l & -1/l & -1.732 & -1.732 & 0 \\ 1/l & -2/l & 1/l & -0.577 & -0.577 & 2.309 \\ -3.464/l & 0 & 3.464/l & 2 & -2 & 0 \end{bmatrix}$$

$$[S^*] = \frac{\Delta E h^3}{1000 l^4} \begin{bmatrix} 513 & & & & & \text{Symm.} \\ -256 & 513 & & & & \\ -256 & -256 & 513 & & & \\ -148l & -148l & 295l & 488l^2 & & \\ 296l & -148l & -148l & 232l^2 & 488l^2 & \\ 148l & -296l & 148l & -232l^2 & -232l^2 & 488l^2 \end{bmatrix}$$

13.7 Deflection at center = 1.154 (Pl^2/Eh^3); $\{M_x, M_y, M_{xy}\}_A = (P/1000)\,\{192, 0, 0\}$.
13.8 No deflection. $\{M_x, M_y, M_{xy}\}_A = \alpha E h^2 T\,(0.119, 0.119, 0)$.

CHAPTER 14

14.1 (a) $M_p = 0.1458ql^2$. (b) For AC, $M_p = 0.0996ql^2$. For CD, $M_p = 0.1656ql^2$.
14.2 (a) $M_p = 0.29Pb$. (b) $M_p = 0.1875Pl$
(c) $M_p = 1.2Pb$. (d) $M_p = 1.59qb^2$
(e) $M_p = 0.11Wl$.
14.3 $M_p = 0.1884Pl$
14.4 $M_p = 0.7634Pl$
14.5 $M_p = 1.6875Pb$

CHAPTER 15

15.1 (a) $m = 0.090W$. (b) $m = W/18$. (c) $m = W/6$. (d) $m = W/14.14$. (e) $m = 0.0637ql^2$. (f) $m = 0.0262ql^2$.
15.2 (a) $m = 0.00961ql^2$. (b) $m = 0.299ql^2$. (c) $m = 0.113ql^2$. (d) $m = 25P/72$. (e) $m = 0.1902qb^2$.

CHAPTER 16

16.1 $\omega = 7.0$ rad/sec (7.103 rad/sec), $T = 0.898$ sec (0.885 sec). The amplitude = 1.745 in. (43.2 mm), and $D_{t=1} = 1.692$ in. (42.8 mm).
16.2 $\omega = 5.714$ rad/sec (5.777 rad/sec), $T = 1.09$ sec (1.088 sec). The amplitude = 2.02 in. (50 mm), and $D_{t=1} = -0.101$ in. (0.9 mm).
16.3 $\omega_1 = 3.53\sqrt{\frac{EI}{ml^3}}$; $\{D^{(1)}\} = \left\{1.00, -\frac{1.38}{l}\right\}$
16.4 $D_{\tau=T/2} = 2.107$ in. (51.9 mm), $D_{\tau=T/2} = 0$, and $D_{\tau=11T/8} = 1.488$ in. (36.7 mm).
16.5 $D_{\tau=t/2} = 1.342$ in. (33.0 mm), $D_{\tau=T/2} = 0$, and $D_{\tau=11T/8} = 0.95$ in. (23.3 mm).
16.6 $\omega_d = 6.964$ rad/sec (7.067 rad/sec), $T_d = 0.902$ sec (0.889 sec), $D_{t=1} = 0.866$ in. (21.9 mm).
16.7 $\zeta = 0.0368$.
16.8 (a) $D_{max} = 0.176$ in. (4.49 mm). (b) $D_{max} = 0.174$ in. (4.45 mm).
16.9 $D_{\tau=T_d/2} = 1.953$ in. (48.1 mm), and $D_{\tau=11T_d/8} = 0.995$ in. (24.5 mm).

16.10 $\omega_1^2 = 1.2\frac{EI}{ml^3}$; $\{D^{(1)}\} = \{1, 1\}$; $\omega_2^2 = 18.0\frac{EI}{ml^3}$; $\{D^{(2)}\} = \{1, -1\}$

16.11 $\omega_1^2 = 0.341\frac{EI}{ml^3}$; $\{D^{(1)}\} = \{1, 0.32\}$; $\omega_2^2 = 15.088\frac{EI}{ml^3}$; $\{D^{(2)}\} = \{1, -3.12\}$

16.12 Equations of motion in the {η} coordinates:

$$\ddot{\eta} + 0.341 \frac{E\,I}{m\,l^3}\eta = 0.906\,P_1/m; \quad \ddot{\eta} + 15.09 \frac{E\,I}{m\,l^3}\eta = 0.103\,P_1/m$$

$$\eta_1 = \frac{P_1}{0.3758}\frac{l^3}{EI}\left[1 - \cos\left(\tau\sqrt{0.341\frac{E\,I}{m\,l^3}}\right)\right]$$

$$\eta_2 = \frac{P_1}{131.78}\frac{l^3}{EI}\left[1 - \cos\left(\tau\sqrt{15.09\frac{E\,I}{m\,l^3}}\right)\right]$$

$$D_1 = \eta_1 + \eta_2; \quad D_2 = 0.32\,\eta_1 - 3.12\,\eta_2$$

16.13 Amplitude of $m_1 = \dfrac{P_0}{0.3758 \dfrac{E\,I}{l^3} - 1.1024\, m\,\Omega^2}$

16.14 Maximum displacement of BC relative to the support = 2.01 in. (49.5 mm).

16.15 Maximum displacement of mass m_1 relative to the support $= \dfrac{(g/4)}{1.2\,E\,I/(m\,l^3) - \Omega^2}$

16.16 $V_{base} = 236.4$ kN; $M_{base} = 1728$ kN-m.

CHAPTER 17

17.1 $T = 1.26$ sec; $S_a/g = 0.116$; peak drift $= 45.9 \times 10^{-3}$ m (1.81 in.); $M_{base} = 204 \times 10^3$ N-m (1810×10^3 lb-in.).

17.2 Equivalent static forces at coordinates, $\{p\} = \{0.528, 0.070, 0.020\}\ W$; $V_{base} = 0.619W$.

17.3 $\overline{\omega}^2 = 59.48$ (rad/sec)^2; $T = 0.8147$ sec; $S_a = 5.328$ m/sec^2; $\overline{V}_{base} = 2.836 \times 10^6$ N (637×10^3 lb); $\overline{M}_{base} = 120.6 \times 10^6$ N-m (1.067×10^9 lb-in.); $D_{\text{peak at top}} = 0.1534$ m (6.039 in.).

17.4 $\{\omega^2\} = 10^3\,\{0.043, 0.517\}$ (rad/sec)^2;

$$[\Phi]_{SI} = 10^{-3}\begin{bmatrix} 1.280 & 0.595 \\ 0.595 & -1.280 \end{bmatrix}; \quad [\Phi]_{Imperial} = 10^{-3}\begin{bmatrix} 17 & 7.85 \\ 7.85 & -17 \end{bmatrix}$$

17.5 $V_{base} = 3.45 \times 10^6$ N (775×10^3 lb); overturning moment $= 23.2 \times 10^6$ N-m (205×10^6 lb-in.). Drift ratios = 0.015 and 0.013 at top and lower stories, respectively.

17.6 $\{\omega^2\} = 10^3\,\{0.03927, 0.34075, 0.92018, 1.57280\}$ (rad/sec)^2.

$$[S]_{SI} = 10^9\begin{bmatrix} 0.14570 & & & \\ -0.16313 & 0.33138 & & \\ 0.02038 & -0.18838 & 0.33554 & \\ -0.00196 & 0.02311 & -0.19121 & 0.35886 \end{bmatrix}$$

$$[\Phi]_{SI} = 10^{-3}\begin{bmatrix} 1.0472 & 0.9130 & 0.6458 & -0.3247 \\ 0.9107 & -0.0730 & -0.9474 & 0.8485 \\ 0.6559 & -0.9664 & -0.2105 & -1.0222 \\ 0.3104 & -0.8244 & 1.0458 & 0.7630 \end{bmatrix}; \quad [\Phi]_{Imperial} \cong 13.2\,[\Phi]_{SI}$$

17.7 $\omega_1^2 = 61.35$ (rad/sec)^2; $\{\Phi\}_{1,\,SI} = \{0.3029,\ 0.1991,\ 0.1029,\ 0.0297\}$; $\{\Phi\}_{1,\,Imperial} \approx 13.2$ $\{\Phi\}_{1,\,SI}$. Generalized system: $\bar{S}_n = 660.63$; $\bar{m}_n = 10.53$; $\omega_n^2 = \bar{S}_n/\bar{m}_n = 62.69$.

17.8 Inter-story drift ratios (highest to lowest story) = 10^{-3} {1.42, 2.65, 3.60, 3.23}.

CHAPTER 18

18.2 $\{D_x,\ D_y\}_B = 10^{-3}\{-0.604, 30.4\}$ m; $\{D_x,\ D_y\}_C = 10^{-3}b\{0.110,\ 24.3\}$ m. $\{N_{AB},\ N_{BC},\ N_{CD}\} = \{146.8,\ 146.2,\ 146.5\}$ kN.

18.3 The value of F at which snap through occurs is 908.3.

18.4 $D_B = 0.0605$ m. Tensions in AB and BC are 198.4 and 198.2 kN, respectively.

18.5 $D_A = 0.483$; force in any member = 190.1.

18.6 $P_{cr} = 21.86$ (upward), corresponding to upward displacement of node A, $D_{A\ cr} = -0.42$.

18.7 $\{D_x,\ D_y,\ \theta\}_B = \{1.94 \times 10^{-3},\ 0.317,\ 3.724 \times 10^{-3}\}$; forces in AB = {38.9, -0.146, -9.73, -38.9, 0.146, -4.86}; forces in BC = {38.9, 0, 4.86, -38.9, 0, -4.86}.

18.8 Displacements and member end-forces are (units: c, F, or cF):
$\{D_x,\ D_y,\ \theta\}_B = \{0.3645,\ 0.0311,\ 0.1451\}$; $\{D_x,\ D_y,\ \theta\}_C = \{0.3647,\ 0.0317,\ 0.1459\}$.
$\{F^*\}_{AB} = 10^3$ {1341.0, -379.2, -558.3, -1341.0, 379.2, -350.5}.
$\{F^*\}_{BC} = 10^3$ {-20.80, 116.9, 350.5, 20.80, -116.9, 351.2}.
$\{F^*\}_{CD} = 10^3$ {1578.4, -373.8, -351.2, -1578.4, 373.8, -544.5}.

18.9 Iteration 1: $D = 0.7l$; out-of-balance force = $0.123Ca$. Iteration 2: $D = 0.889l$; out-of-balance force = $0.009Ca$. The exact answer: $D = 0.905l$.

18.10 The displacements in x, y, and z directions are: $\{\{D_5\},\ \{D_6\},\ \{D_7\},\ \{D_{11}\}\} = 10^{-3}l$ {{2.50, 2.50, 21.17}, {3.30, -0.16, 26.84}, {2.68, -2.68, 20.93}, {0, 0, 34.65}}. Forces in the cables are $\{F_{5-6},\ F_{6-7},\ F_{1-5},\ F_{2-6},\ F_{6-11},\ F_{3-7}\} = \{31.1,\ 31.3,\ 30.9,\ 32.0,\ 32.0,\ 31.6\}\ Q$.

Advertisement

Computer Programs for Analysis of Structures, based on R.B. Gayed and A. Ghali, *Structural Analysis Fundamentals* (CRC Press, Taylor & Francis Group, 2021).

Linear analysis programs	*Nonlinear analysis programs*	*Matrix algebra programs*	*EIGEN programs*	*Time-dependent analysis programs*	*Programs for analysis of axially symmetric loaded structures*
PLANEF (plane frame)	PDELTA	ADD	EIGEN1	CREEP	CTW (Cylindrical Tank Walls)
SPACEF (space frame)	PLASTICF	MULTIPLY	EIGEN2	SCS	SOR (Shells of Revolution)
PLANET (plane truss)	NLST	INVERT		TDA	
SPACET (space truss)	NLPF	SOLVE		CGS	
PLANEG (plane grid)		DETERM			
EGRID (eccentric grid)		MULTIPLY5			

The above computer programs are described in Chapter 20 of this book. Up-to-date versions are sold together as a set. They are available on a USB memory device, containing the following files for each program: (1) The source code in FORTRAN. (2) An executable file, for use on IBM personal computers or compatibles. (3) An input-file example. The comments, which are given at the beginning of the source code for each program, include simple instructions for input data preparation. No printed manual is needed; however, for each program, a file containing a short manual is provided.

DISCLAIMER OF WARRANTIES AND ORDER FORM

The above computer programs are sold without warranty of any kind, either expressed or implied, including but not limited to implied warranties of merchantability and fitness of any particular purpose. The seller and the authors are not responsible for any damage whatsoever which may be incurred by the purchaser or any other person by reason of any defect in the software. The author reserves the right to revise the software without obligation to notify any person of revisions or changes. I hereby accept the conditions of the disclaimer.

Please send me sets of the software and invoice me for the sum of $200 Canadian per set, plus a small mailing and handling charge.

Name:

Address:

Signature: Date:

Index